Electrons and Holes in Semiconductors

WITH APPLICATIONS TO TRANSISTOR ELECTRONICS

PROBABILITY AND ITS ENGINEERING USES. *By* THORNTON C. FRY.
ELEMENTARY DIFFERENTIAL EQUATIONS. *By* THORNTON C. FRY.
TRANSMISSION NETWORKS AND WAVE FILTERS. *By* T. E. SHEA.
ECONOMIC CONTROL OF QUALITY OF MANUFACTURED PRODUCT. *By* W. A. SHEWHART.
ELECTROMECHANICAL TRANSDUCERS AND WAVE FILTERS. *By* WARREN P. MASON. Second Edition.
RHOMBIC ANTENNA DESIGN. *By* A. E. HARPER.
POISSON'S EXPONENTIAL BINOMIAL LIMIT. *By* E. C. MOLINA.
ELECTROMAGNETIC WAVES. *By* S. A. SCHELKUNOFF.
NETWORK ANALYSIS AND FEEDBACK AMPLIFIER DESIGN. *By* HENDRICK W. BODE.
CAPACITORS—THEIR USE IN ELECTRONIC CIRCUITS. *By* M. BROTHERTON.
FOURIER INTEGRALS FOR PRACTICAL APPLICATIONS. *By* GEORGE A. CAMPBELL and RONALD M. FOSTER.
APPLIED MATHEMATICS FOR ENGINEERS AND SCIENTISTS. *By* S. A. SCHELKUNOFF.
EARTH CONDUCTION EFFECTS IN TRANSMISSION SYSTEMS. *By* ERLING D. SUNDE.
THEORY AND DESIGN OF ELECTRON BEAMS. *By* J. R. PIERCE. Second Edition.
PIEZOELECTRIC CRYSTALS AND THEIR APPLICATION TO ULTRASONICS. *By* WARREN P. MASON.
MICROWAVE ELECTRONICS. *By* JOHN C. SLATER.
PRINCIPLES AND APPLICATIONS OF WAVEGUIDE TRANSMISSION. *By* GEORGE C. SOUTHWORTH.
TRAVELING-WAVE TUBES. *By* J. R. PIERCE.
ELECTRONS AND HOLES IN SEMICONDUCTORS. *By* WILLIAM SHOCKLEY.
FERROMAGNETISM. *By* RICHARD M. BOZORTH.
THE DESIGN OF SWITCHING CIRCUITS. *By* WILLIAM KEISTER, ALASTAIR E. RITCHIE, and SETH H. WASHBURN.
SPEECH AND HEARING IN COMMUNICATION. *By* HARVEY FLETCHER. Second Edition.
MODULATION THEORY. *By* HAROLD S. BLACK.
SWITCHING RELAY DESIGN. *By* R. L. PEEK, JR., and H. N. WAGAR.
TRANSISTOR TECHNOLOGY, Volume I. *Edited by* H. E. BRIDGERS, J. H. SCAFF, and J. N. SHIVE.
TRANSISTOR TECHNOLOGY, Volume II. *Edited by* F. J. BIONDI.
TRANSISTOR TECHNOLOGY, Volume III. *Edited by* F. J. BIONDI.
PHYSICAL ACOUSTICS AND THE PROPERTIES OF SOLIDS. *By* WARREN P. MASON.
THE PROPERTIES, PHYSICS AND DESIGN OF SEMICONDUCTOR DEVICES. *By* J. N. SHIVE.

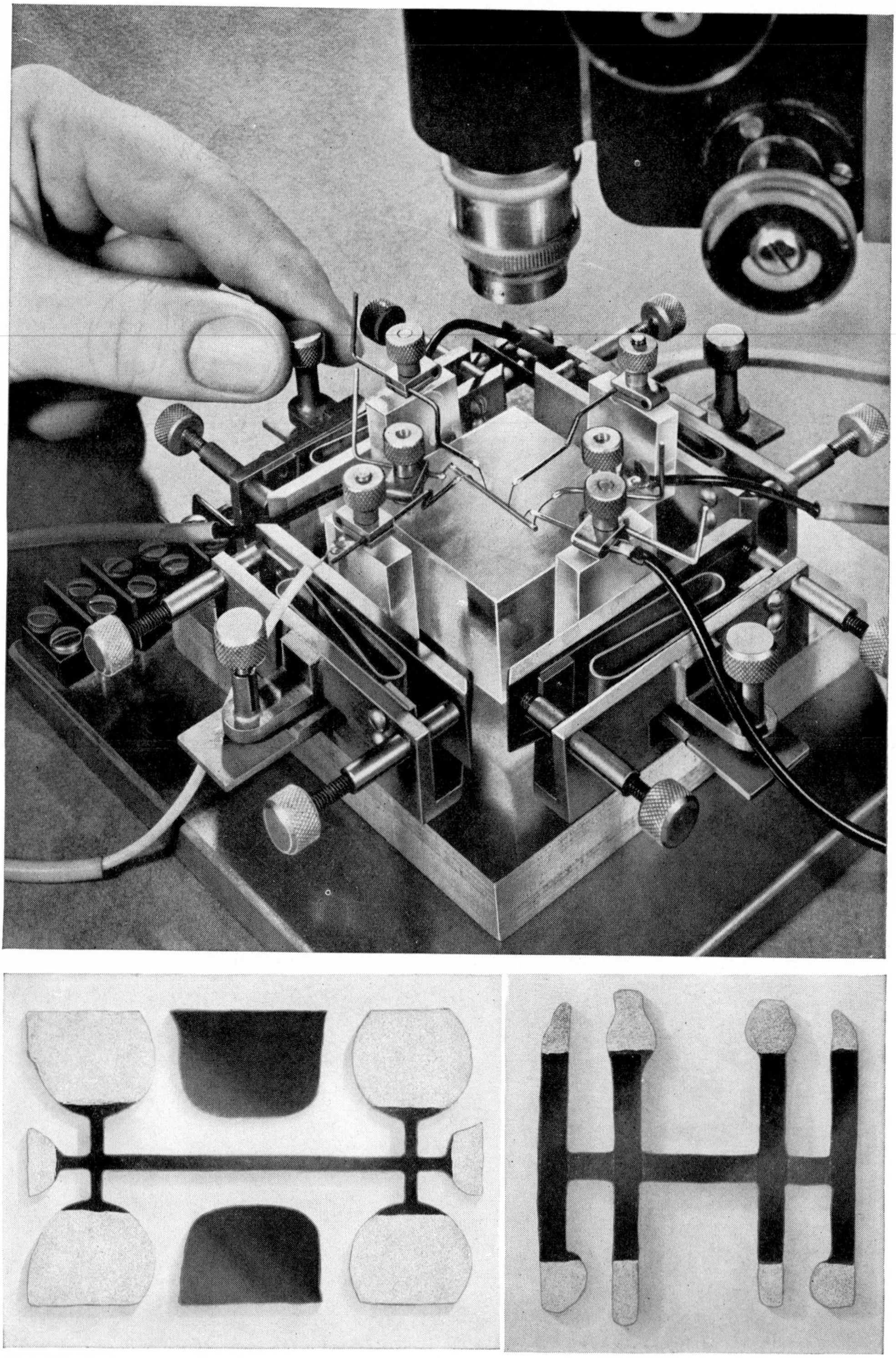

EXPERIMENTAL TECHNIQUES OF TRANSISTOR ELECTRONICS

Micromanipulator designed by W. L. Bond and used by J. R. Haynes to investigate the behavior of holes and electrons in germanium. Specimens of complex shape, consisting of thin filaments and large contact areas, are prepared for special purposes. That on the left is 7.5 mm in length, the other is 3 mm in its largest dimension.

Electrons and Holes in Semiconductors

WITH APPLICATIONS TO TRANSISTOR ELECTRONICS

By

WILLIAM SHOCKLEY

Director of Shockley Transistor, Unit of Clevite Transistor, Division of Clevite Corporation

NINTH PRINTING

D. VAN NOSTRAND COMPANY, INC.

PRINCETON, NEW JERSEY

TORONTO

LONDON

NEW YORK

D. VAN NOSTRAND COMPANY, INC.
120 Alexander St., Princeton, New Jersey (*Principal office*)
24 West 40 Street, New York 18, New York

D. VAN NOSTRAND COMPANY, LTD.
358, Kensington High Street, London, W.14, England

D. VAN NOSTRAND COMPANY (Canada), LTD.
25 Hollinger Road, Toronto 16, Canada

Published simultaneously in Canada by
D. VAN NOSTRAND COMPANY (Canada), LTD.

First Published November 1950
Reprinted November 1951, January 1953
August 1954, October 1955
August 1956, September 1959
February 1963, September 1966

PRINTED IN THE UNITED STATES OF AMERICA

To

M. B. S.

FOREWORD

If there be any lingering doubts as to the wisdom of doing deeply fundamental research in an industrial laboratory, this book should dissipate them. Dr. Shockley's purpose has been to set down an account of the current understanding of semiconductors, an understanding which incidentally is comprised in no mean degree of his own personal contributions. But he has done more than this. He has furnished us with a documented object lesson. For in its scope and detail this work is obviously a product of the power and resourcefulness of the collaborative industrial group of talented physicists, chemists, metallurgists and engineers with whom he is associated. And it is an almost trite example of how research directed at basic understanding of materials and their behavior, "pure" research if you will, sooner or later brings to the view of inventive minds engaged therein opportunities for producing valuable practical devices.

The program of work which Dr. Shockley leads was aimed at understanding a kind of materials, the semiconductors, which had already received considerable application in the communications business in the form of rectifiers, regulators and modulators. Not only were improvements in such devices hoped for but the possibility of creating an amplifier was envisioned. In the course of three years of intensive effort the amplifier has been realized by the invention of the device named the transistor.

It would be unfair to imply that any and every fundamental research program may be expected to yield commercially valuable results in so short a time as has this work in the telephone laboratories. To achieve such results, careful choice of a ripe and promising field is prudent and a clear recognition of objectives certainly helps; but there should be no illusions about the necessity of a large measure of good luck.

Solid state electronics, or transistor electronics as Dr. Shockley calls it, preceded and in one respect has always excelled vacuum electronics. This is true in communications engineering at least, for the crystal wireless detector preceded the audion and it is still the best detector when the going gets tough, as in microwaves. Nevertheless in the past forty years the vacuum tube is the tool which has shaped the whole electrical intelligence transmission art. It is an art traditionally based upon highly stable amplifiers in which distortion is rigorously suppressed and upon modulators in which distortion is precisely tailored to a useful purpose. Evidence is already strong that transistor devices will be developed having characteristics suitable for such exacting uses.

The newer methods of transmission by quantizing and time splitting, together with related fields such as electric computing, require large numbers of low power amplifiers and gating and flip-flop-circuits to handle pulses and stepwise current changes. Transistor devices are being found to have unique advantages in this type of circuitry. They are tiny, fast and efficient. Here the science of transistor electronics which Dr. Shockley and his colleagues have so effectively launched promises to lead into new areas of technology. The present volume should be a valuable guidebook.

RALPH BOWN, *Director of Research*
Bell Telephone Laboratories

PREFACE

The hole, or deficit produced by removing an electron from the valence-bond structure of a crystal, is the chief reason for existence of this book. Although the hole and its negative counterpart, the excess electron, have been prominent in the theory of solids since the work of A. H. Wilson in 1931, the announcement of the transistor in 1948 has given holes and electrons new technological significance. From the theoretical viewpoint, the hole is an abstraction from a much more complex situation and the achieving of this abstraction in a logical way appears inevitably to involve rather detailed quantum-mechanical considerations. From the experimental viewpoint, in contrast, the existence of holes and electrons as positive and negative carriers of current can be inferred directly by the experimental techniques of transistor electronics so that holes and electrons have acquired an *operational* reality in Bridgman's sense of the word. Furthermore, the new experiments have established the quantitative aspects of the behaviors of holes and electrons with sufficient accuracy for many of the purposes of transistor electronics. Thus, in the level of abstraction, there is a great difference between theory and experiment; this difference is reflected in the organization of the book.

In Part I, only the simplest theoretical concepts are introduced and the main emphasis is laid upon interpretation in terms of experimental results. This material is intended to be accessible to electrical engineers or undergraduate physicists with no knowledge of quantum theory or wave mechanics. It should serve as a basis for understanding the operation of transistor devices and for elementary design considerations.

Part III, at the other extreme, is intended to show how fundamental quantum theory leads to the abstractions of holes and electrons. In order to make the spirit, if not the details, of these mathematical investigations accessible to readers without extensive training in theoretical physics, an introductory discussion has been prepared. This material, presented in Chapters 13 and 14, is intended to answer questions that frequently arise about wave functions and their interpretation. Chapter 14 makes considerable use of electrical engineering analogues and covers most of the basic quantum-mechanical principles needed for the subsequent treatment of holes and electrons. Part III also contains an introduction to statistical mechanics and other topics applicable to the theory of electronic conduction in crystals.

Part II attempts to bridge the gap between Parts I and III by presenting the reasoning and results of Part III in pictorial and descriptive terms. The closing chapter of Part II reexpresses some of the theory in analytic form useful for quantitative studies of transistor phenomena. This chapter has also been used as a repository for some late developments in transistor electronics.

Problems follow many of the chapters. Some of these are simply numerical examples intended to give the reader a feeling for the orders of magnitude involved, others are intended to supplement or extend the mathematics in the text, and still others contain results which might well be discussed at length in the text but which have been stated as problems in the interests of brevity.

The endeavor to probe deeply into the logical consequences of the fundamental theory, to reduce these consequences to pictorial terms and to find experimental counterparts to the theoretical concepts is in keeping with the philosophy of research at Bell Telephone Laboratories. The invention of the transistor occurred in connection with a research program based on this philosophy and the development and content of this book reflect the same philosophy. The *modus operandi* of research programs like that associated with the invention of the transistor is to seek primarily for a fundamental understanding of the phenomena being investigated while at the same time remaining alert for possible applications.

I have frequently found it helpful in my own thinking to consider that fundamental understanding for many solid state problems is achieved when four questions can be answered:

(1) What are the atoms involved and how are they arranged?
(2) How did this arrangement come into being?
(3) How does this arrangement lead to certain mechanisms of electronic and atomic motion?
(4) How do these mechanisms give rise to the observed properties?

In terms of these questions, the phenomena of conduction are understood and so is carrier injection by *p-n* junctions. Many other phenomena discussed in this book are understood in so far as question (4) and part of question (3) are concerned; a completion of the understanding of some of these phenomena is currently a major research aim.

This book had its origin in a series of lectures given at Bell Telephone Laboratories in connection with the growth of the transistor program. It thus owes its existence basically to the invention of the transistor by J. Bardeen and W. H. Brattain. Its content and organization have been influenced by expositional needs discovered during the lectures. The emphasis is accordingly on those materials and phenomena that are most prominent in transistor electronics and numerous other semiconductors and effects are omitted entirely.

The preparation of this book has required support in a variety of ways. The encouragement and constructive criticism of R. Bown, J. A. Morton, R. M. Ryder, W. G. Pfann, L. A. Meacham and J. J. Markham have been of great value and have resulted in particular in the arrangement of Part I, the existence of Chapter 14 and various improvements in Chapters 15 and 17. For other assistance, I am indebted in particular to R. D. Heidenreich and also to J. Bardeen, G. C. Danielson, W. G. Dow, J. B. Johnson, C. Kittel, K. G. McKay, P. H. Miller, Miss D. J. Oxman, F. Seitz, Mrs. G. V. Smith, L. Tisza and many others. W. van Roosbroeck has commented on nearly the entire manuscript, has carried out the calculations for many of the figures in addition to undertaking responsibility for the index. The figures themselves have benefited greatly from the efforts of Sidney Lund and the late B. A. Clarke. The very tedious and difficult task of typing the several drafts has been admirably handled by our transcription department. The problem of organizing the work and keeping track of the many details has been solved for me by Mrs. E. M. Sparks. The encouragement and cooperation of Jean B. Shockley have been essential to the work.

October 21, 1950 WILLIAM SHOCKLEY

CONTENTS

CHAPTER 4

ON THE PHYSICAL THEORY OF TRANSISTORS 77

PART II DESCRIPTIVE THEORY OF SEMICONDUCTORS

CHAPTER 5

QUANTUM STATES, ENERGY BANDS, AND BRILLOUIN ZONES 119

CHAPTER 6

VELOCITIES AND CURRENTS FOR ELECTRONS IN CRYSTALS 156

CHAPTER 7

ELECTRONS AND HOLES IN ELECTRIC AND MAGNETIC FIELDS 167

CHAPTER 8

INTRODUCTORY THEORY OF CONDUCTIVITY AND HALL EFFECT 187

CHAPTER 9

DISTRIBUTIONS OF QUANTUM STATES IN ENERGY 220

PART III QUANTUM-MECHANICAL FOUNDATIONS

Chapter 17

THE THEORY OF TRANSITION PROBABILITIES FOR HOLES AND ELECTRONS 481

LIST OF SYMBOLS

Most of the symbols used fall into four broad classes distinguished by different type faces:

Lightfaced Italic. Ordinary magnitudes, components of vectors, constants of various sorts: a, B, x, E_x, P_x, p, τ_p, μ_p.

Boldfaced Italic. Vectors: $\boldsymbol{E}$, $\boldsymbol{H}$, $\boldsymbol{R}$, $\boldsymbol{P}$, $\boldsymbol{p}$.

Boldfaced Roman. Quantities whose functional dependence on other quantities is to be emphasized such as wave functions $\boldsymbol{\psi}$, $\boldsymbol{\Phi}$, $\mathbf{A}$, $\boldsymbol{\alpha}$, $\boldsymbol{\beta}$ and the Fermi-Dirac function $\mathbf{f}$.

Script. Quantities and operators on an electronic scale such as $\mathcal{H}$ the Hamiltonian operator, $\mathfrak{E}$ for energy, $\mathcal{R}$ and $\mathfrak{S}$ for operators. Also the thermodynamic functions $\mathcal{F}$ for free energy and $\mathfrak{S}$ for entropy.

SUBSCRIPTS'

The following subscripts occur frequently and are not repeated in all cases in the main list:

a, c, d, v, G, F: acceptor, conduction band, donor, valence-bond band, energy gap, Fermi level.

b, c, ϵ, J: base, collector, emitter, and junction point of Figure 4.1.

n, p: apply to electrons (negative) and holes (positive) also to n-region and p-region.

i: intrinsic, especially in Section 12.4.

s: a particular quantum state.

$\boldsymbol{P}$: Crystal Momentum.

L, M: practical and M.K.S. units, Section 8.8.

i, j, k, α, β, γ, a, b, c: sometimes stand for integers distinguishing individual members of sets of similar quantities.

MAIN LIST

In the following list chapter numbers are integers, i.e. **5**, section numbers are decimals, i.e. **15.3**, and equation numbers are in parentheses followed by the section numbers in which they are found, i.e. (3) **14.8**. Where references are given they refer to definitions or important applications of the symbols concerned.

$\boldsymbol{a}$, $\boldsymbol{a}_1$, $\boldsymbol{a}_2$, $\boldsymbol{a}_3$: lattice constant, unit vectors, or periods of other periodic structures; **5**, **3**, (3) **14.8**, **15.3**.
$\boldsymbol{a}$: a decay constant; **12.6**.
A, A_x, A_y, A_z, A_1, etc: extent of a periodic structure; **5.4**, **5.5**, **14.8**.
A: antisymmetric wave function; (6) **15.7**.
$\boldsymbol{A}$: vector potential; **15.6**.
b: symbol for base of transistor; **2.2**.
b: ratio of electron mobility to hole mobility; (4) **3.1**, **12.9**.
b: separation constant in filament; (9) **12.6**.
$\boldsymbol{B}$, $\boldsymbol{B}_M$, B: magnetic flux density.
$B = \boldsymbol{P}/h$: (11) **14.8**.
B: half width of filament; **12.6b**.
c: generally speed of light.
c: speed of sound; **11.3** and **17**.
c: separation constant in filament; (9) **12.6**.
c_{ij}: elastic constants; **17.6**.
c_{ll}: average longitudinal elastic constant; **17.6b**. Second para. p. 528.
C: capacitance; **14.2**, **15.3**.
C: half width filament; **12.6b**.
D, D_p, D_n: diffusion constants; **12.3**.
e: absolute value of the electronic charge; ($e = q$ in **12**.)
e: base of Naperian logarithms.
$\exp(x) = \mathrm{e}^x$.
$\boldsymbol{E}$, E: electric field.
$\mathcal{E}$, $\mathcal{E}(\boldsymbol{P})$, $\mathcal{E}_a$, etc.: energies of quantum states, energy differences; **5**, **9**. Thermodynamic energy of system; **16**.
f: frequency.
f, f_p: Fermi-Dirac distribution functions for electrons and holes; **10.1**, **16.1**.
$\boldsymbol{F}$: force.
$\mathcal{F}$: free energy; (44) **16.1**.
g: hole-electron pairs generated per unit volume per unit time; **12.2**.
$\boldsymbol{g}$, $\boldsymbol{G}$: reciprocal vectors; **14.9**.
$\mathrm{g}(x)$: even function; (3a) **14.8**.
$\boldsymbol{G}_{k\alpha}$: unit vector; (5) **17.3**.
h: Planck's constant.
$\hbar = h/2\pi$: Dirac's h.

$\boldsymbol{H}$, H_x, H: magnetic field.
$\mathcal{H}$: Hamiltonian operator or function; **6.4**, (26) **14.3**.
$i = \sqrt{-1}$.
i_b, i_c, etc.: a-c components of current; **2.2c**.
$\boldsymbol{i}_x$, $\boldsymbol{i}_y$, $\boldsymbol{i}_z$: unit vectors.
$\boldsymbol{I}$, $\boldsymbol{I}_p$, I_b, I_p: current densities and components, d-c total currents.
k: Boltzmann's constant; **10.1**.
$\boldsymbol{k} = \boldsymbol{P}/\hbar$: wave vector; (7) **17.3**.
K, K': constants in (1) (7) **7.5**.
$K.E.$: kinetic energy either classical or quantum; (23) **14.3**.
l_p, l_n: mean free path; **8.8**, (29) **11.3**, **11**, **17**.
ln: logarithm to the base **e**.
log: logarithm to the base 10.
L: length. Inductance in **14.2**.
$\mathfrak{L}$: linear operator; **17.4**.
m: mass of electron (except in **17**).
m_e: see **17.1**.
m_n, m_p: effective masses; (5) **7.5**, (4) **7.6**. (Also **12**.)
n: density of electrons in conduction band; (1) **1.3**, **10**, **12**.
N: a number of countable entities or a density of them.
$N(\mathcal{E})$: (1) **9.1**.
N_c, N_v: effective densities of states in bands; (12)(14) **10.3**, **16.2**.
N_a, N_d, N_I: densities of acceptors and donors; **9.2**. Ions; (32) **11.4**.
N_s: number of atoms in crystal; **5.5**, **17.3**.
N_x, N_y, N_z, N_1, N_2, N_3: number of unit cells along edge of periodic structure; (6) **14.8**.
p: density of holes in valence-bond band; (1) **1.3**, **10**, **12**.
$\mathfrak{p}$, p: momentum of electron, quantum and classical; **6.4**, (18) **14.3**.
$\mathfrak{p}_A$: (5) **15.6**.
$\boldsymbol{P}$, P, $\boldsymbol{P}'$, P_x: crystal momentum; **5.5**, **7.5**, **14.8**.
q: absolute value of electronic charge in **12**.
q, q_i: coordinate; **6.4**, **17.3**.
$\boldsymbol{q}$: position plus spin of electron; **15.7**.
Q: number of systems; **16.1**.
$\mathfrak{Q}$: Permutation operator; **15.7**.
r: resistance for small a-c signals; **2.2**.
r: hole-electron pairs recombining per unit volume per unit time; **12.2**.
$\boldsymbol{r}$, r: position vector and radius.
R: resistance.
R_H, R_L, R_M: Hall constant; **8.8**.
$\boldsymbol{R}_s$, $\delta\boldsymbol{R}$, $\boldsymbol{R}_T$: vectors describing nuclear positions, points in a crystal etc.; **14.8**, **17**.
s: surface recombination velocity; (4) **3.2**, **12.6**.

s: spin quantum number; (1) **15.7.**
s, S: number of quantum states; **11, 17.**
$\mathfrak{S}$: entropy; **16.**
t: time.
T: absolute temperature.
$\mathbf{u}(x)$: odd function; (3b) **14.6.**
$\mathbf{u}_{P}(\mathbf{r})$: periodic factor of Bloch function; (5) **5.5, 14.7, 14.8.**
$\mathfrak{U}(\mathbf{r})$: Potential energy of electron; (6) **5.2.**
$\mathbf{v}$, v: velocity; **6.**
$\mathbf{v}_g$, $\mathbf{v}_p$: group and phase velocity; **6, 15.1, 15.2.**
v_b, v_c, v_ϵ: a-c voltages; **2.2.**
V_b, V_c, V_ϵ: d-c voltages; **2.2, 12.5.**
V: volume of crystal; **6.**
dV: with various subscripts for elements of volume in various spaces.
V_P: volume in Brillouin zone.
$\mathfrak{V}$: potential energy of nuclei; **11.3.**
$\mathbf{w}$: wave packets in **15.**
W: number of ways of achieving a distribution; **16.1.**
W_i, W_{ij}: transition probabilities; **11.2, 17.2.**
$\mathfrak{W}$: see Figures 14.3 and 14.5.
x, X, y, Y, z, Z: position coordinates.
Z: nuclear charge; **9.2.** Partition function; problems **16.**
α: "alpha" for current multiplication in transistor; (3) **2.2.**
α_e, α_i: equivalent circuit alpha; Figure 2.7, (10) **4.1.** Intrinsic alpha; **3.1c, 4.5.**
$\boldsymbol{\alpha}$: spin wave function; **15.7.**
β: fraction of injected current reaching collector; **4.1, 4.4.**
$\boldsymbol{\beta}$: spin wave function; **15.7.**
γ: fraction of emitter current carried by injected carriers; (3) **3.1.**
δ_{rs}: Kronecker delta function, $=1$ for $r = s$ and $= 0$ for $r \neq s$.
Δ: dilatation; (17) **17.6.**
∇: the vector operator, subscripts indicate coordinates involved.
ϵ: symbol for emitter.
ε_0: M.K.S. permittivity.
$\varepsilon(\mathfrak{Q})$: sign of permutation; (6) **15.7.**
ζ: (13) **12.6.**
η: (13) **12.6; 15.6.**
θ_n, θ_p: Hall angles; (2) **8.6,** (16) **11.4.**
$\kappa = \kappa_e$: dielectric constant.
λ: wave length.
μ_0: M.K.S. permittivity of free space.
μ, μ_D, μ_H: mobility, drift mobility, Hall mobility; **1, 3.1,** (8) (9) **8.7.**
ν: frequency.

ν: decay constant; (5) **12.6.**

ρ: resistivity.

ρ: charge density; (12) **12.4.**

ρ_i: density of quantum states; (6) **11.2.**

σ: conductivity; (1) **1.3.**

τ_p, τ_n, τ_t: lifetimes of holes and of electrons and transit time; **4.1,** (14) **12.2.**

$\boldsymbol{\tau}, \boldsymbol{\tau}(\boldsymbol{v}), \boldsymbol{\tau}_c$: mean free time, **8**; relaxation time, **11**; mean free time between collision, (12) **11.2.**

$\boldsymbol{\tau}_H, \boldsymbol{\tau}_\mu$: Hall mean free time; mean free time for mobility; (1) (2) (3) **8.9,** (16b) (23) **11.4.**

$\varphi, \varphi_n, \varphi_p$: Fermi and quasi-Fermi levels; **12.3.**

φ: scalar potential; **15.6.**

ϕ: angle in cylindrical and spherical coordinates.

$\boldsymbol{\phi}, \boldsymbol{\phi}_i, \boldsymbol{\phi}$ (n's): wave functions; **17.1.**

$\boldsymbol{\Phi}$: wave function of crystal and one excess electron; **17.1.**

$\boldsymbol{\chi}$: wave function; (1) **15.7, 12.1.**

ψ: electrostatic potential; **12.3.**

$\boldsymbol{\psi}_P$: Bloch function; (5) **5.4,** (1) **14.8.**

$\omega = 2\pi\nu = 2\pi f$: circular frequency.

Ω: energy surface; (5) **11.2.**

$\langle\ \rangle$: $\langle x \rangle$ means average value of x.

$\dot{}$: $\dot{E} = \partial E/\partial t$.

*: $\mathcal{U}_{ij}^* =$ complex conjugate of $\mathcal{U}_{ij}$.

PART I INTRODUCTION TO TRANSISTOR ELECTRONICS

PART I

Selected Articles Closely Related to Transistor Work

(Most of the references quoted extensively in the text are omitted from this list.)

1948

J. Bardeen and W. H. Brattain, "Transistor, a Semiconductor Triode," *Phys. Rev* **74**, 230–231 (1948).

W. H. Brattain and J. Bardeen, "Nature of the Forward Current in Germanium Point Contacts," *Phys. Rev.*, **74**, 231–232 (1948).

W. Shockley and G. L. Pearson, "Modulation of Conductance of Thin Films of Semiconductors by Surface Charges," *Phys. Rev.*, **74**, 232–233 (1948).

Review articles appeared in Electronics Sept. 1948, Scientific American Sept. 1948 and a series giving some experimental results in Audio Engg. in July, Aug., Sept., Oct. and Dec. 1948.

1949

E. Aisberg, "Transistron = Transistor Plus?," *Toute la Radio*, **137**, 218–220 (1949).

S. J. Angello, "Semiconductor Rectifiers," *Elec. Engg.*, **68**, 865–872 (1949).

J. Bardeen and W. H. Brattain, "Physical Principles Involved in Transistor Action," *Phys. Rev.*, **75**, 1208–1225 (1949).

J. A. Becker, "Photoeffects in Semiconductors," *Elec. Engg.*, **68**, 937–942 (1949).

J. A. Becker and J. N. Shive, "Transistor—a New Semiconductor Amplifier," *Elec. Engg.*, **68**, 215–221 (1949). Refere

M. Becker and H. Y. Fan, "Photovoltaic Effect of *p-n* Barriers Produced in Germanium by Alpha and Deuteron Bombardment," *Phys. Rev.*, **75**, 1631 (1949).
S. Benzer, "High Inverse Voltage Germanium Rectifier," *J. App. Phys.*, **20**, 804–815 (1949).
W. C. Dunlap, "Germanium, Important New Semiconductor," *G. E. Rev.*, Feb., **52**, 15–17 (1949).
W. E. Johnson and K. Lark-Horovitz, "Neutron Irradiated Semiconductors," *Phys. Rev.*, **76**, 442 (1949).
K. Lark-Horovitz, "Conductivity in Semiconductors," *Elec. Engg.*, **68**, 1047–1056 (1949).
F. W. Lehans, "Transistor Oscillator for Telemetering," *Electronics*, Aug., **22**, 90–91 (1949).
K. Lehovec, "Testing Transistors," *Electronics*, June, **22**, 88–89 (1949).
H. J. Reich and R. L. Ungvary, "Transistor Trigger Circuit," *Rev. Sci. Instruments*, **20**, 586–588 (1949).
J. N. Shive, "New Germanium Photo-Resistance Cell," *Phys. Rev.*, **76**, 575 (1949).
W. M. Webster, E. Eberhard and L. E. Barton, "Some Novel Circuits for the Three-Terminal Semiconductors Amplifier," *R.C.A. Rev.*, **10**, 5–16 (1949).

1950

P. Aigrain et C. Dugas, "Caractéristiques des Transistors," *Comptes Rendus*, **230**, 377–378 (1950).
J. A. Becker, "Transistors," *Elec. Engg.*, **69**, 58–64 (1950).
G. Busch, "Electronenleitung in Nichtmettalen," *Zeits. f. Ang. Math. und Phys.*, **1**, 3–31 and 81–110 (1950). Bibliography.
H. Fricke, "Halbleiter-Trioden und-Tetroden als Verstarker-und Mischstufen," *E.T.Z.*, **71**, 133–137 (1950).
H. A. Gebbie, P. C. Banbury and C. A. Hogarth, "Crystal Diode and Triode Action in Lead Sulphide," *Proc. Phys. Soc. Lond.*, B, **63**, 371 (1950).
H. Heins, "Germanium Crystal Triode," *Sylvania Tech.*, Jan., **3**, 13–18 (1950).
L. Hunter, "Physical Interpretation of Type A Transistor Characteristics," *Phys. Rev.*, **77**, 558–559 (1950).
L. A. Meacham and S. E. Michaels, "Observations of the Rapid Withdrawal of Stored Holes from Germanium Transistors and Varistors," *Phys. Rev.*, **78**, 175–176 (1950).
R. P. Turner, "Crystal Receiver with Transistor Amplifier," *Radio & Telev. News*, Jan., **43**, 38–39, 153–155 (1950).

1952

In 1952 the November issue of the Proceedings of the Institute of Radio Engineers was devoted to Transistors. The fifty-one articles contain many additional references.

Books and Review Articles on Semiconductors

(For additional references see Part II.)

A. H. Wilson, "Semiconductors and Metals; an Introduction to the Electron Theory of Metals," *Cambridge University Press*, London, 1939. 120p.
N. F. Mott and R. W. Gurney, "Electronic Processes in Ionic Crystals," *Oxford, Clarendon Press*, 1940. 275p.
H. C. Torrey and C. A. Whitmer, "Crystal Rectifiers," *Massachusetts Institute of Technology, Radiation Laboratory Series, McGraw-Hill*, 1948. 443p.
H. K. Henisch, "Metal Rectifiers," *Oxford, Clarendon Press*, 1949. 155p.
F. Seitz, "The Basic Principles of Semiconductors," *J. App. Phys.*, **16**, 553–563 (1945).
R. J. Maurer, "The Electrical Properties of Semiconductors," *J. App. Phys.*, **16**, 563–570 (1945).

CHAPTER 1

THE BULK PROPERTIES OF SEMICONDUCTORS

Between 1940 and 1950, the understanding of semiconductors was very greatly increased. During that decade a substantially complete picture was developed of the energy level schemes and motions of electrons in silicon and germanium. This understanding is an outgrowth of the research and development program on crystal rectifiers undertaken in connection with the radar program during the war and continued in several laboratories thereafter. Some of the wartime work was carried out in the Radiation Laboratory of M.I.T., which operated under the supervision of the National Defense Research Committee. The Radiation Laboratories Series volume *Crystal Rectifiers*[1] by H. C. Torrey and C. A. Whitmer reports this program and mentions in particular as chief contributors to crystal research and development in England: the General Electric Company, British Thompson-Houston Ltd., Telecommunications Research Establishment and Oxford University; and in the United States: the Bell Telephone Laboratories, Westinghouse Research Laboratory, General Electric Company, Sylvania Electric Products, Inc., and the E. I. du Pont de Nemours and Company. It is also pointed out that the crystal groups at the University of Pennsylvania and Purdue University, who operated under N.D.R.C. contracts, were responsible for much fundamental work.

With the advent of the transistor, the role of semiconductors in electronic technology has assumed much greater prominence. In addition, techniques developed in connection with the transistor program have led to new experimental information concerning the basic processes of electronic conduction in semiconductors. Occupying a central position in these researches are germanium filaments and micromanipulators, like those shown in the frontispiece. These filaments play the same role as do the tubes in vacuum and gas discharge electronics. In fact a close analogy can be drawn between the science and application of conventional electronics and those of the new electronics of transistors. We shall accordingly use the phrase *transistor electronics* to designate the field in semiconductor physics that covers fundamental processes, the analysis and design of devices and the theory of their circuit applications. All of these subjects are treated to some degree in the following chapters.

The semiconductors of interest in transistor electronics are electronic

[1] McGraw-Hill Book Company, 1948.

rather than ionic conductors. In ionic conductors a substantial fraction of the current is transported with an accompanying motion of the ions. Since the positions of the ions and atoms determine the physical structure of the material and its chemical composition, ionic conduction produces radical changes in the sample. In electronic conductors, however, the atoms stay in the same positions. They may lose or gain electrons during the conduction process, but the structure and chemical composition are unaffected.

1.1 EXCESS ELECTRONS AND HOLES AS CURRENT CARRIERS

Basic to the theory of semiconductors is the idea that electrons can carry current in two distinguishable and distinctly different ways: one being called by "excess conduction", "conduction by excess electrons", or simply "conduction by electrons" and the other being called "defect conduction" or "conduction by holes". The possibility that these two processes may be simultaneously and separably active in a semiconductor affords a basis for explaining transistor action.

The quantum mechanical explanation of these processes in terms of "energy bands" and "Brillouin zones" is well developed and is one of the chief topics dealt with in the following chapters. In this introductory chapter, an elementary survey of the behavior of electrons in silicon and germanium will be presented. The general features discussed here are then explained on a more advanced level by the theory in the later chapters.

Silicon and germanium form what are called "covalent crystals", the atoms being held together by "electron-pair bonds" formed by the valence electrons. Preparatory to considering the electronic structure of the crystals, we shall first describe the covalent bond in the hydrogen molecule, which has the simplest electron pair bond. Figure 1.1 represents two hydrogen atoms and a hydrogen molecule. Each atom consists of a proton and one electron. The proton weighs approximately 2000 times as much as the electron and is a relatively immobile particle about which the electron moves in its orbit or quantum mechanical wave function. (We shall discuss the interpretation of wave functions further in Chapters 5 and 14.) In an isolated atom, this wave function has spherical symmetry, and the electronic charge is distributed on the average as a diffuse sphere centered about the proton. When the two atoms are brought close together, interaction between the wave functions of the two electrons takes place, and the electronic clouds become modified as suggested in the bottom part of Figure 1.1. The result is to produce an extra accumulation of charge between the two protons which acts to bind them together. According to quantum mechanical laws associated with the "Pauli exclusion principle", the bond is especially stable when it contains precisely two electrons. It is weakened considerably by removal of one electron and is not greatly strengthened by the addition of a third electron. This special stability of the electron-pair bond or covalent bond is a fundamental fact of chemistry which is now quite well understood on the basis of wave mechanics.[1]

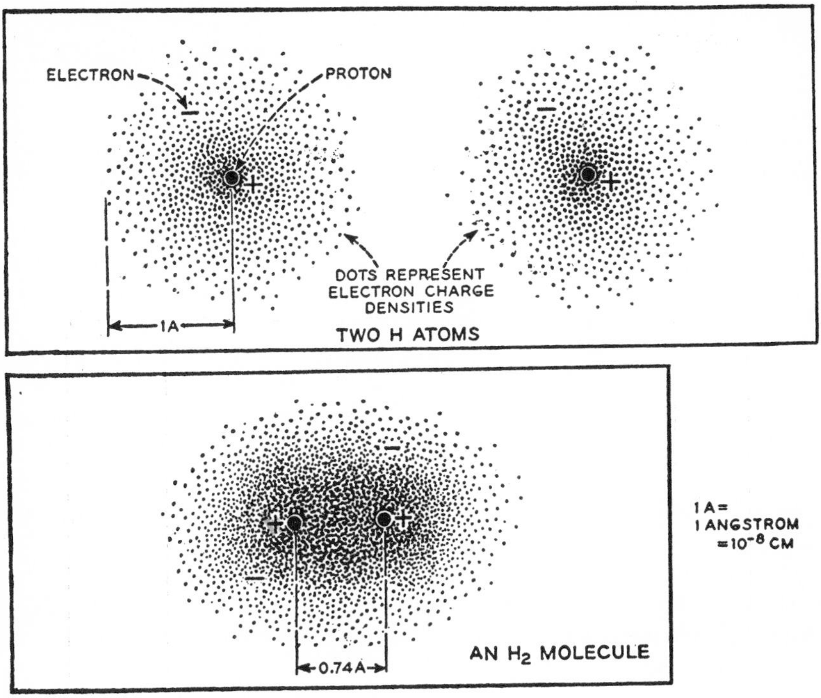

FIG. 1-1—Electron-Pair Bond in the Hydrogen Molecule.

In the periodic table of the elements, as shown in an appendix, covalent bonding is especially important in the elements of *Class* III which, according to Hume-Rothery's classification, form *valence crystals*.[1a] The elements carbon, silicon and germanium which come in *group* IV of the table have atomic numbers, usually denoted as Z, of 6, 14 and 32. These numbers are equal to the charge on the atomic nucleus measured in units of the electronic charge. A neutral atom thus has a nuclear charge of $+Ze$ surrounded by Z electrons each of charge $-e$.

The elements carbon, silicon, and germanium have the common feature

[1] See L. Pauling, *The Nature of the Chemical Bond*, Cornell University Press, 1939.

[1a] W. Hume-Rothery, "The Structure of Metals and Alloy," Institute of Metals, London (1936) revised 1950.

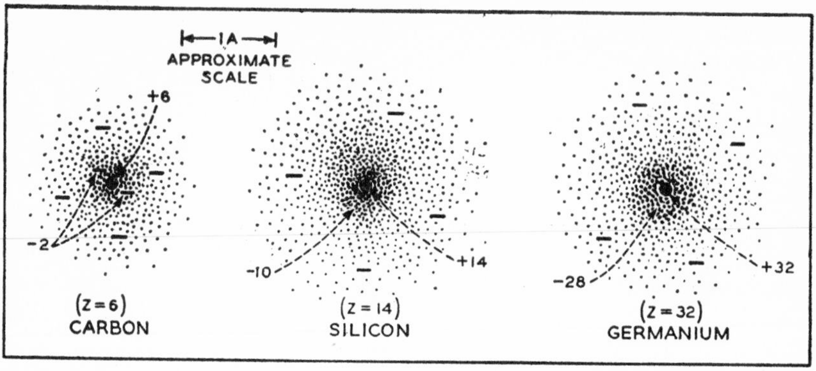

FIG. 1-2—The Electronic Structures of Carbon, Silicon, and Germanium Atoms.

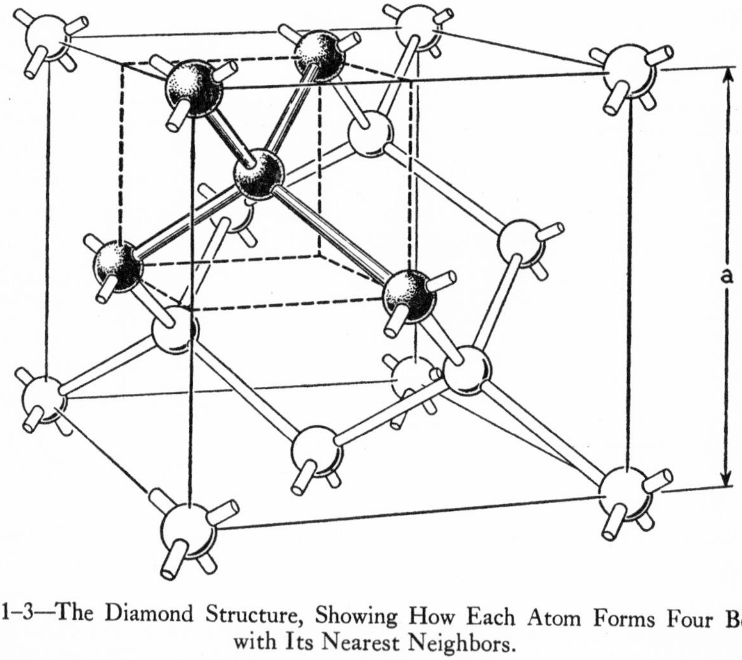

FIG. 1-3—The Diamond Structure, Showing How Each Atom Forms Four Bonds with Its Nearest Neighbors.

(The lattice constant, denoted by a, is the cube edge in this figure. For diamond, silicon, germanium, and gray tin, its value is respectively 3.56, 5.42, 5.62, and 6.46A where 1A = 1 angstrom = 10^{-8} cm. The distance between nearest neighbors is $\sqrt{3/16}\,a$. The figure shows 18 atoms, but only 8 really belong to the volume a^3; since the 8 corner atoms are each shared by 8 cubes, they contribute only 1; the 6 face atoms are each shared by 2 cubes and contribute 3; and there are 4 atoms wholly inside the cube. The number of atoms per cm^3 is thus $8/a^3$ = 17.7, 5.00, 4.52 and 2.82×10^{22} respectively.)

indicated in Figure 1.2 of being tetravalent. Although they possess respectively 6, 14, and 32 electrons all together, in each case only four of these are able to enter into chemical reactions. The remaining electrons are closely bound to the nucleus producing a stable "ionic core" having a net charge of +4 units. This core can be regarded as completely inactive so far as electronic processes in chemical reactions and in semiconductors are concerned.

Each of these atoms tends to form covalent or electron pair bonds with four other atoms. This tendency is completely satisfied in the diamond structure which is the crystalline form of all three elements. This structure, shown in Figure 1.3, is a cubic arrangement and may be regarded as made up of eight interpenetrating, simple cubic lattices like the one formed by the atoms on the eight corners of the large cube shown. Although interesting features in the theory of conduction in the diamond lattice arise from the detailed nature of its crystal structure, from the point of view of this chapter we are interested only in the feature represented in the upper left-hand corner of Figure 1.3. This part of the figure shows that each typical atom is surrounded by four neighbors regularly placed about it, with which it forms four covalent bonds. These neighbors are arranged on the corners of a regular tetrahedron in conformity with the known chemical behavior of the tetrahedral carbon atom.[2] For purposes of discussion of the conductivity in these crystals, we shall represent the three-dimensional array in two dimensions as is shown in Figure 1.4, indicating that each carbon atom forms an electron-pair bond with four neighbors. The crystal is, of course, electrically neutral as may be seen by considering one ionic core and its share of the charge in the four electron-pair bonds which surround it. Each such unit is electrically neutral as shown in (b). When impurity atoms are present, as we shall discuss later, units like (b) are not always electrically neutral.

On the basis of this valence-bond structure we can intuitively see why diamond should be an insulator. Although it contains a large number of electrons, as does a metal, the covalent bond is an entirely different structure from the metallic bond. The metallic bond is frequently described by saying that the electrons behave substantially like a gas of free electrons producing a uniform cloud of negative charge in which the positive ions of the metal float. In the case of diamond, however, the electrons do not behave like a gas permeating the crystal but instead more closely resemble structural elements which hold the crystal together. In an ideally perfect crystal, each valence bond would contain its two electrons; therefore,

[2] Long before the arrangement of atoms in the diamond crystal was established by X rays, the organic chemists had concluded that carbon formed four bonds at the tetrahedral angles—a truly remarkable inference from observations of the optical properties of solutions of organic compounds.

every electron would be tightly bound and thus unable to enter into the conduction process. The situation may be crudely represented by the analog of an automobile storage garage in which one floor is completely filled with automobiles. In this case, no flow of traffic would be possible.[3]

Conductivity can be produced in diamond, however, in a number of ways, all of which involve destroying the perfection of the valence bond

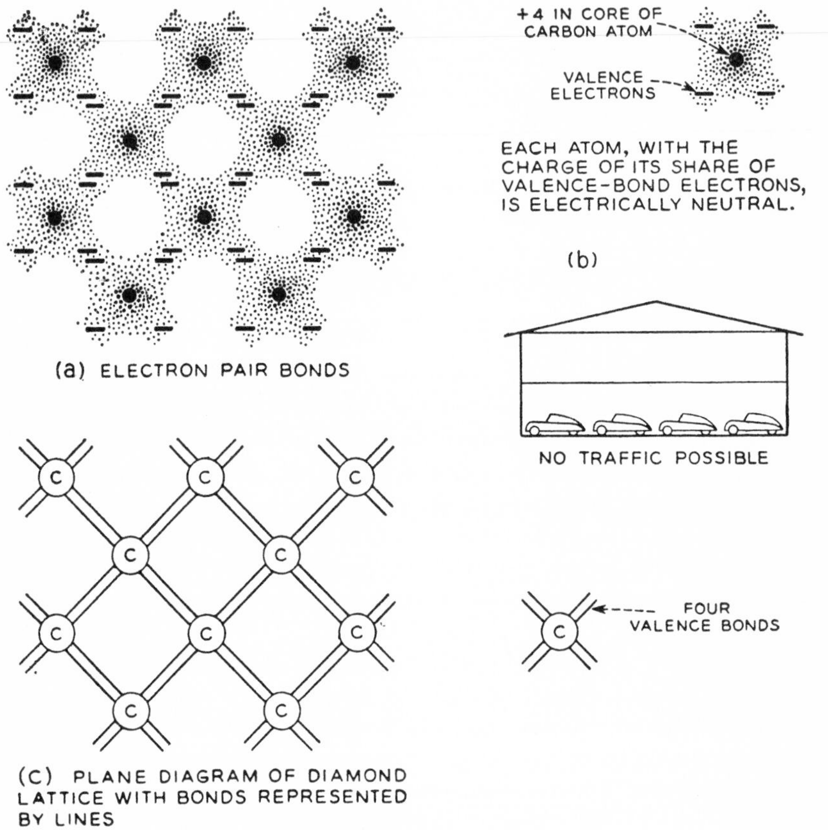

FIG. 1-4—Electrons in the Diamond Structure.

structure. Thus if high-energy particles or quanta of radiation fall upon the crystal, they can break the bonds. Conductivity in diamond induced by bombardment in this way has recently received considerable prominence in connection with "crystal counters" which have been used to detect nuclear particles and in experiments on electron-bombardment induced

[3] The theory of energy bands, discussed in Chapter 5, especially Section 5.6, shows how there may be a continuous transition from valence-bond structures to metallic structures. On this basis the metallic properties of tin and lead do not contradict the theory presented here.

conductivity.[4] In Figure 1.5(a) we represent a photon delivering its energy to an electron which is ejected from one of the bonds.[5] This ejected electron constitutes a localized negative charge in the crystal as shown in (c), since before it arrived in that part of the crystal the electron-pair bond structure was electrically neutral. Such an electron, which represents an excess over and above that required to complete the bond structure in its

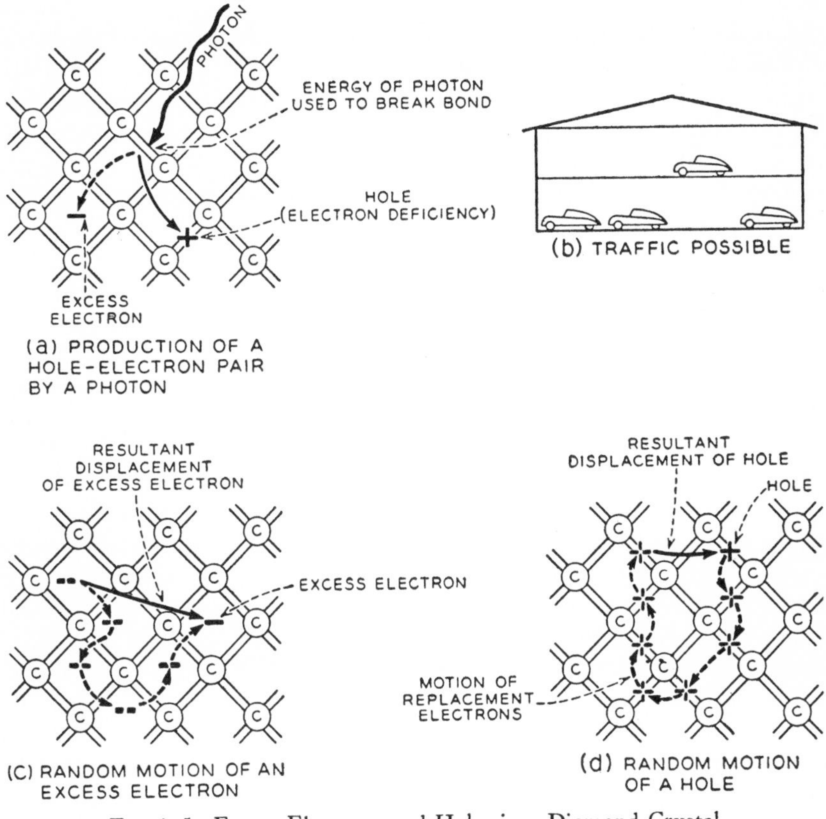

Fig. 1-5—Excess Electrons and Holes in a Diamond Crystal.

neighborhood, is called an "excess electron". Since it cannot enter any of the completed bonds in (c), it migrates about in a random manner in the crystal under the influence of thermal agitation. If an electric field is applied, it tends to drift in the direction of the applied force and to carry a current. Its behavior is represented by the vehicle on the second floor

[4] D. E. Wooldridge, A. J. Ahearn and J. A. Burton, *Phys. Rev.* **71**, 913 (1947). P. J. Van Heerden, Thesis (Utrecht, 1945). K. G. McKay, *Phys. Rev.* **77**, 816–825 (1950), references.

[5] F. S. Goucher, *Bulletin of the 298th Meeting of the American Physical Society*, has reported evidence that every photon in the wave-length range of 1.0 to 1.8μ absorbed in germanium produces a hole-electron pair.

of (b), which is now free to move. Like the vehicle, the electron also has been lifted to a state of higher energy by being removed from the bond. As mentioned earlier, this process of conduction by excess electrons is referred to simply as conduction by "electrons". This procedure serves to distinguish it from the other process whereby electrons also conduct, called conduction by "holes", discussed below.

The vacant space in (b) now permits traffic to flow on the first floor, in the analogy. A similar process takes place in the crystal through the motion of the hole left in the bond when the ejected electron moved away. As is shown in (d), this hole constitutes a net, localized, positive charge in the crystal, since before it was introduced that part of the crystal was electrically neutral. Its motion takes place, as shown in (d), by a reciprocal motion of electrons in the valence bonds (just as the vacant parking place in (b) can move owing to the successive motions of vehicles into it). Under the influence of an electric field, the random motion of the hole acquires a systematic drift, and it can also contribute to the current.[6]

Current flow in an illuminated diamond crystal is represented in Figure 1.6. The electrons and holes, produced in pairs by the photons, drift in opposite directions in the field; the electron, being negative, drifts in the opposite direction from the applied field, but the electric current it produces is, of course, in the direction of the field. In the case of the hole, the reciprocal electron motions are also opposite to the direction of the field (on the average). As a consequence, the net result is that the hole drifts to the right, and since its charge is positive it produces a current to the right. (The result obtained could be represented by raising the right side of the garage of Figure 1.5(b). The vehicle on the top floor would run to the left and the vacant place on the first floor would shift to the right.) If the source of illumination is removed, the photoconductivity dies away and the crystal then returns to its normal state. This can occur by the recombination of holes and electrons as shown in Figure 1.6(b). If the electron drops into the hole, both the hole and electron disappear and the bond structure becomes complete, the excess energy being given up to the atoms in the form of thermal vibrations as suggested in Figure 1.6(b).[7]

On the basis of quantum-mechanical theory, as discussed in Chapter 7,

[6] In this chapter we present an oversimplified picture of the conduction process. Neither the electron nor hole may be considered so localized as shown in Figure 1.5; they are even less localized than in Figure 1.14. The replacement process for the hole is not so simple either. So far as the idea of two kinds of conduction is concerned, the picture presented here is correct. Chapters 5 to 8 are required to give the same picture in more precise terms. The basic theory of semiconductors, which gives analytic form to the ideas discussed here, is due to A. H. Wilson, *Proc. Roy. Soc.* **133A**, 458 (1931).

[7] The process of recombination may actually be much more complicated and may involve intermediate stages in which the hole or the electron is trapped. For further discussion see K. G. McKay, *Phys. Rev.* **74**, 1606–1621 (1948); and Section 3.1d and Chapter 12.

it is found that a very high degree of symmetry exists between the behavior of electrons and the behavior of holes. This symmetry is not suggested by the analog of the parking garage. In a garage, it would obviously be very much easier to move the vehicle on the second floor than to move the hole on the first floor. According to the quantum mechanics of electrons in crystals, there is only a very slight difference in these two processes, and

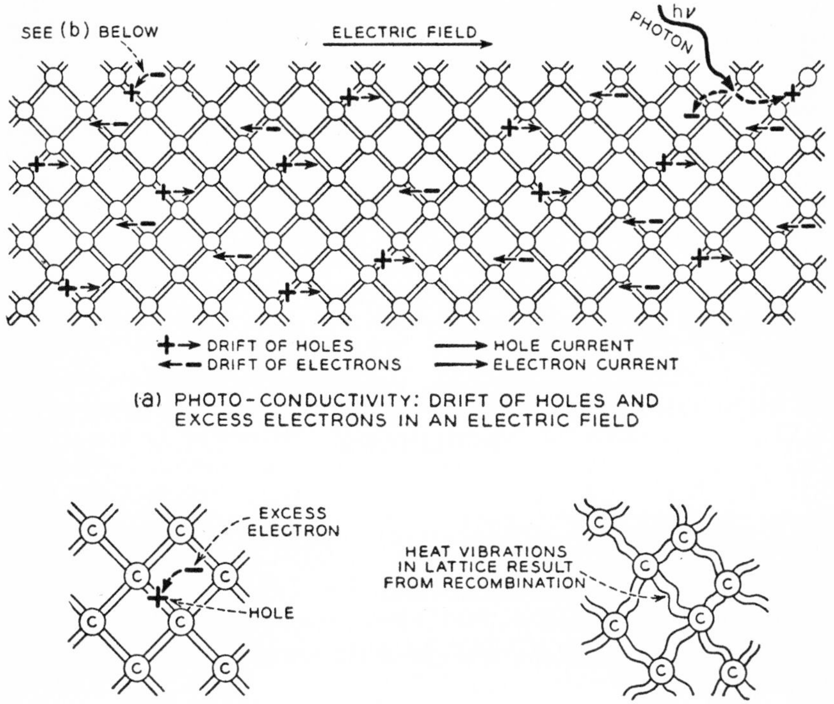

(b) RECOMBINATION OF A HOLE AND AN ELECTRON RESTORES THE NORMAL VALENCE BOND

FIG. 1–6—Photo-conductivity in Diamond.

one may think of the hole as moving through the crystal as a positively charged particle with much the same attributes as a free electron except for the sign of its charge.

It is evident that an important distinction exists between behavior of electrons which have been excited out of the valence bond structure and those which remain in it. The concepts and terminology of the theory of *Brillouin zones* and *energy bands* have been developed to describe these distinctions. According to this theory, which is discussed in Chapters 5 to 7, electrons in the valence bonds occupy a set of energy levels covering a certain band of allowed energies. All of these energy levels are occupied in the ideal crystal, and the band of energy levels is referred to as the *filled band, valence-bond band,* or simply *valence band.* It corresponds to the first floor of the garage. It is impossible for an electron to have a higher energy than the highest state in the filled band unless it is given a sufficiently large increment of energy so that it may become a free or excess electron of the sort discussed previously. The electrons in these free states may also have various energies of motion and thus give rise to the energy level scheme of the *empty band* or *conduction band* (the second floor). The filled band is also a band which permits conduction but is not referred to as a "conduction band". The filled and empty bands are separated by a region of forbidden energies for which there are no energy levels for the electrons in the crystal. In this introduction the results of the energy band theory will be used without reference to their theoretical basis.

If the temperature is sufficiently elevated, spontaneous breaking of the covalent bonds by thermal agitation will occur, producing electrons and holes in equal numbers. This effect would occur at such high temperatures in diamond that it has not been observed. However, it plays a major role in silicon and germanium at temperatures well within the range of investigation in the laboratory.

1.2 IMPURITY SEMICONDUCTORS; DONORS AND ACCEPTORS

If the only cases of conductivity open to investigation were like those already discussed, for which electrons and holes are present in equal numbers, the problem of interpreting the data would be very difficult. Fortunately, in the semiconductors silicon and germanium, there are cases in which conductivity is due to excess electrons only or to holes only. We shall discuss some specific examples for silicon[1] and indicate later how these are related to germanium.

If the conductivity of the sample is due to excess electrons it is called *n-type,* since the current carriers act like *negative charges;* if due to holes, it is called *p-type,* since the carriers act like *positive charges.* Figure 1.7 shows an example of *n*-type silicon. The conductivity arises from the presence of arsenic atoms which are termed "impurities", even though added deliberately in the otherwise pure silicon. The arsenic atom, as indicated, has five valence electrons surrounding a core having a charge of +5 units. On the basis of evidence which we shall discuss, it is believed that each arsenic atom displaces one of the silicon atoms from its regular site and forms four covalent bonds with the neighboring silicon atoms as shown in Figure 1.7, thus using four of its five valence electrons. The extra electron

[1] A systematic analysis of the behavior of silicon with impurities of the sorts discussed here was first carried out by J. H. Scaff, H. C. Theuerer, and E. E. Schumacher, *J. of Metals* **185**, 383–388 (1949). Their work was stimulated by the development of silicon detectors for microwave use by R. S. Ohl, also of Bell Telephone Laboratories, in the pre-war years.

cannot fit into these four bonds and is free to move about the crystal. As discussed previously, this excess electron constitutes a mobile, localized negative charge. The arsenic atom, on the other hand, is an *immobile*, localized positive charge, since its core, with a charge of +5 units, is not neutralized by its share (−4) of the charge in the valence bonds. Its net charge, therefore, just balances that of the excess electron it contributes to the crystal. Thus arsenic impurity atoms add excess electrons but do not disturb the over-all electrical neutrality of the crystal. The negative

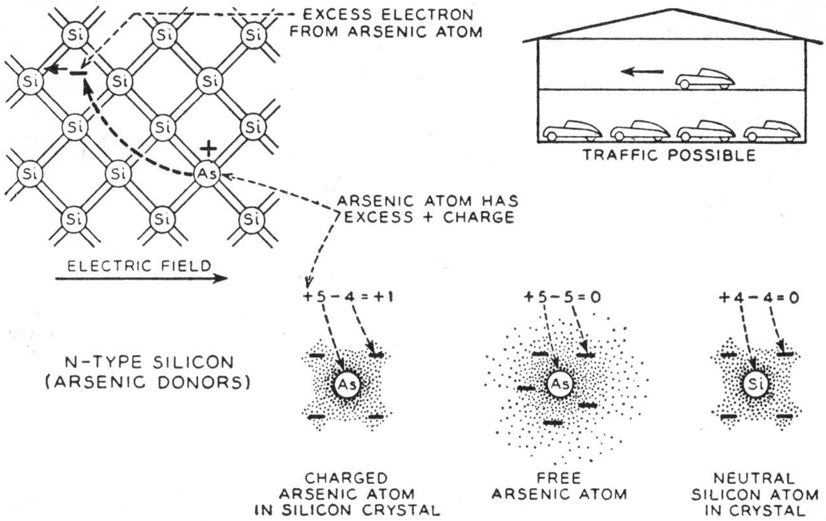

FIG. 1–7—Excess or Electron Conduction in Silicon Containing Arsenic.

electrons are attracted to the positive arsenic atoms and at low temperatures become bound to them. However, at room temperature, thermal agitation shakes them off and the electrons are free as shown in Figure 1.7. (Measurements of conductivity over wide temperature ranges yield valuable information, as we shall discuss later.)

A *p*-type semiconductor is shown in Figure 1.8. In this case the added impurity, boron, has a valence of three and therefore cannot complete the valence bond structure surrounding it. The hole in one of the bonds to the boron atom can be filled by an electron from an adjacent bond, and the hole can thus migrate away, as described in Figure 1.5(d). The boron thus becomes an *immobile*, localized negative charge. Because of the symmetry between the behavior of holes and electrons, we can describe the situation shown in Figure 1.8 by saying that the negative boron atom attracts the positively charged hole but that thermal agitation shakes the latter off at room temperature so that it is free to wander about and contribute to the conductivity.

Impurities with a valence of five are called "donor impurities" because they donate an excess electron to the crystal; those with a valence of three are called "acceptor impurities", since they accept an electron from some-

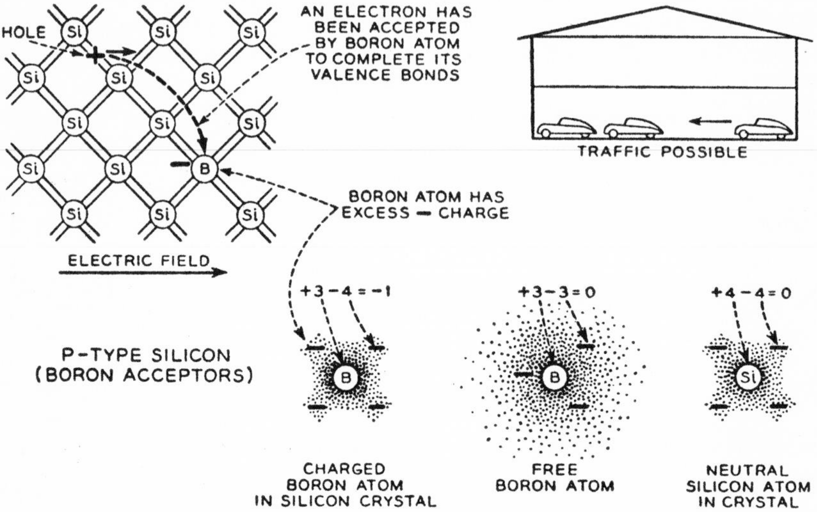

Fig. 1–8—Defect or Hole Conduction in Silicon Containing Boron.

where else in the crystal to complete the structure of the valence bonds with their neighbors, thus leaving a hole to conduct. These terms, together with the other features already discussed, are summarized in Table 1.1.

Table 1.1 Summary of Features of Impurity Conduction in Silicon and Germanium

Conductivity Type	*n*-Type or Excess	*p*-Type or Defect
Conduction by	(excess) electrons	holes
Energy band in which carrier moves	empty or conduction	full or valence-bond or valence
Sign of carrier	negative	positive
Valence of impurity atom	5	3
Name for impurity atom	donor	acceptor
Typical impurities	Elements of Group IV: Phosphorus, P Arsenic, As Antimony, Sb	Elements of Group III: Boron, B Aluminum, Al Gallium, Ga Indium, In

In Figure 1.9 we illustrate how the number of carriers depends on temperature for a sample of silicon with five donors and two acceptors. At absolute zero, the three extra electrons furnished by these seven atoms are all bound to the donors. As the temperature is raised to −125°C, or 148°K (or absolute temperature on the Kelvin scale), thermal agitation is sufficient to shake off the electrons. However, they recombine so rapidly that on the average only one electron out of three is free to conduct. At room temperature, two out of three can conduct; however, thermal agita-

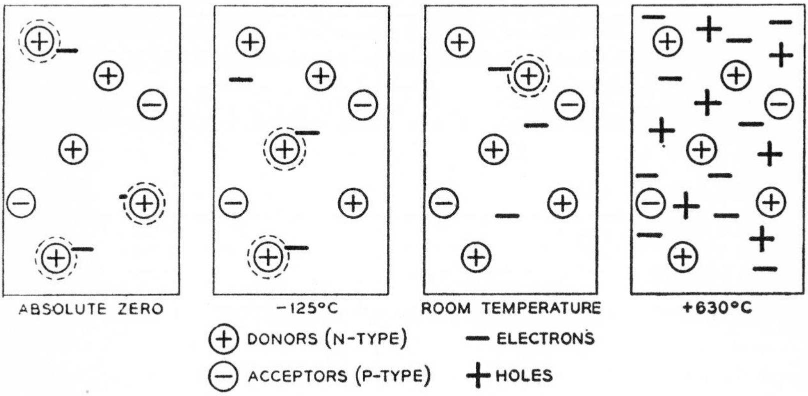

FIG. 1–9—Dependence upon Temperature of Hole and Electron Densities in *n*-Type Silicon.

tion is insufficient to break the valence bonds, and the number of holes is about 10^{-12} times the number of electrons.[2] (The situation at room temperature in germanium is quite different, and the hole concentration may be about 1 percent of the electron concentration in high resistivity samples.[3]) At higher temperatures an appreciable fraction of the valence bonds is broken thermally, and holes and electrons are created in pairs. In Figure 1.9 at 630°C, the number of carriers produced thermally outweighs those due to impurities by 6 to 1, and the conductivity is essentially the same as if there were no impurities present at all. Under these conditions the sample is said to be in the *intrinsic* range, since the conductivity exhibited is an intrinsic property of silicon itself and independent of the impurity content.

The data shown in Figure 1.9 have been chosen to fit, approximately, an actual silicon sample (see Figure 1.12) studied by Pearson and Bardeen.

[2] This value is estimated by extrapolating Figure 1.12 discussed later with the aid of Figure 16.9 in Chapter 16.

[3] The situation for germanium at room temperature is discussed in Section 10.4 and shown in Figure 10.7.

1.3 INTERPRETATION OF DATA ON CONDUCTIVITY AND HALL EFFECT

Pearson and Bardeen[1] have investigated a number of silicon samples prepared by J. H. Scaff and his colleagues containing varying amounts of added phosphorus and boron. We shall quote a number of results from their paper, to which the reader is referred for additional technical details. Before discussing the nature of these results, however, we should indicate the general outlines of the experimental procedure used for obtaining the basic date on the numbers and behaviors of the electrons and holes.

The conductivity of a semiconductor containing holes and electrons depends upon the number present and the ease with which they are moved by an applied electric field. (These subjects are discussed at length in Chapters 8, 10, and 11.) This latter property is called the "mobility" and is expressed as the drift velocity of the particle in centimeters per second in an electric field of one volt per centimeter. It has, therefore, the dimensions of cm^2/volt sec. The conductivity (in $ohm^{-1}\ cm^{-1}$) due to excess electrons is equal to the total charge per unit volume (in coulombs/cm^3) of the electrons which are free to move times the mobility. The total charge is, of course, equal to the density of electrons (in cm^{-3}) times e the electronic charge (1.60×10^{-19} coulombs). Combining this with the conductivity due to holes leads to the equation for conductivity

$$\sigma = e(n\mu_n + p\mu_p) \tag{1}$$

where n and p (for negative carriers and positive carriers) are the densities of holes and electrons, respectively, and μ_n and μ_p their respective mobilities. (This formula is derived in connection with Figure 8.6.)

A measurement of conductivity alone, therefore, gives only one item of information about the four unknown quantities occurring in this equation. Fortunately, another relationship between these same four quantities can be obtained by measuring the Hall effect.

The Hall effect, described in Chapter 8 in more detail, occurs when a magnetic field is applied at right angles to the direction of current flow in the specimen. It is found that this produces a transverse voltage, which can be measured. When the Hall effect has been measured, it gives another equation relating the four unknown quantities. However, if an n-type sample is dealt with in the impurity conduction temperature range, p may be taken as equal to zero. Under these conditions measurements of conductivity and Hall effect permit a separate determination of n and μ_n. Similarly, p and μ_p can be determined in p-type samples. Thus, by combining information on various samples it is possible to find the

[1] G. L. Pearson and J. Bardeen, *Phys. Rev.* **75**, 865–883 (1949).

behavior of μ_n and μ_p as a function of temperature. With μ_n and μ_p regarded as known, it is then possible to obtain the values of n and p even in the case where both holes and electrons are simultaneously present. The mobilities for several cases of interest are quoted in Table 1.2. The method of measurement is discussed in Section 3.1 and Chapters 8, 11, and 12.

TABLE 1.2 MOBILITIES IN CM/SEC PER VOLT/CM

	Electrons μ_n	*Holes* μ_p
Carbon (diamond)	>400[a]	>200[a]
	900[b]	—
	156[c]	—
Silicon	300[d]	100[d]
Germanium		
(Hall mobility)	2600[e]	1700[e]
(Drift mobility)	3600[f]	1700[f]

[a] K. G. McKay, *Phys. Rev.* **77**, 816–825 (1950).
[b] C. C. Klick and R. J. Maurer, *Phys. Rev.* **76**, 179 (1949).
[c] F. Seitz, *Phys. Rev.* **73**, 549–564 (1948).
[d] G. L. Pearson and J. Bardeen, *Phys. Rev.* **75**, 865–883 (1949).
[e] G. L. Pearson, *Phys. Rev.* **76**, 179 (1949).
[f] J. R. Haynes, unpublished results presented to the American Physical Society at Chicago, November 26, 1949. (See Section 3.1 and Chapter 12 for further details.)

We shall next consider the data analyzed by Pearson and Bardeen for a series of samples containing various amounts of added phosphorus. Phosphorus, like arsenic, has a valence of five and produces n-type material (it was not used in the example of Figure 1.7 in order to reduce the confusion between the symbol P for phosphorus and p-type, which is not produced by phosphorus). Figure 1.10 shows the resistivities for four samples with varying phosphorous content plotted as a function of temperature.[2]

Data from the Hall effect are plotted in Figure 1.11. The results have been expressed in terms of the density of excess electrons per cubic centimeter. Since the density of silicon atoms is 5.00×10^{22} per cubic centimeter, it is seen that sample D corresponds to one electron per 250 atoms in the crystal. For sample A at room temperature, on the other hand, there is only one conduction electron for approximately 500,000 atoms in

[2] The phosphorous content on the curves refers to that added to the molten silicon. Some of this phosphorus probably evaporates before being incorporated in the solidified silicon; furthermore, the phosphorus tends to remain in the molten silicon and be segregated in the last portions to solidify. The net result is to reduce the phosphorous content in the samples measured for Hall effect below the proportion added to the melt. Experimental evidence that one excess electron is contributed by each donor in the solid is furnished by data obtained with radioactive antimony in germanium discussed at the end of this chapter.

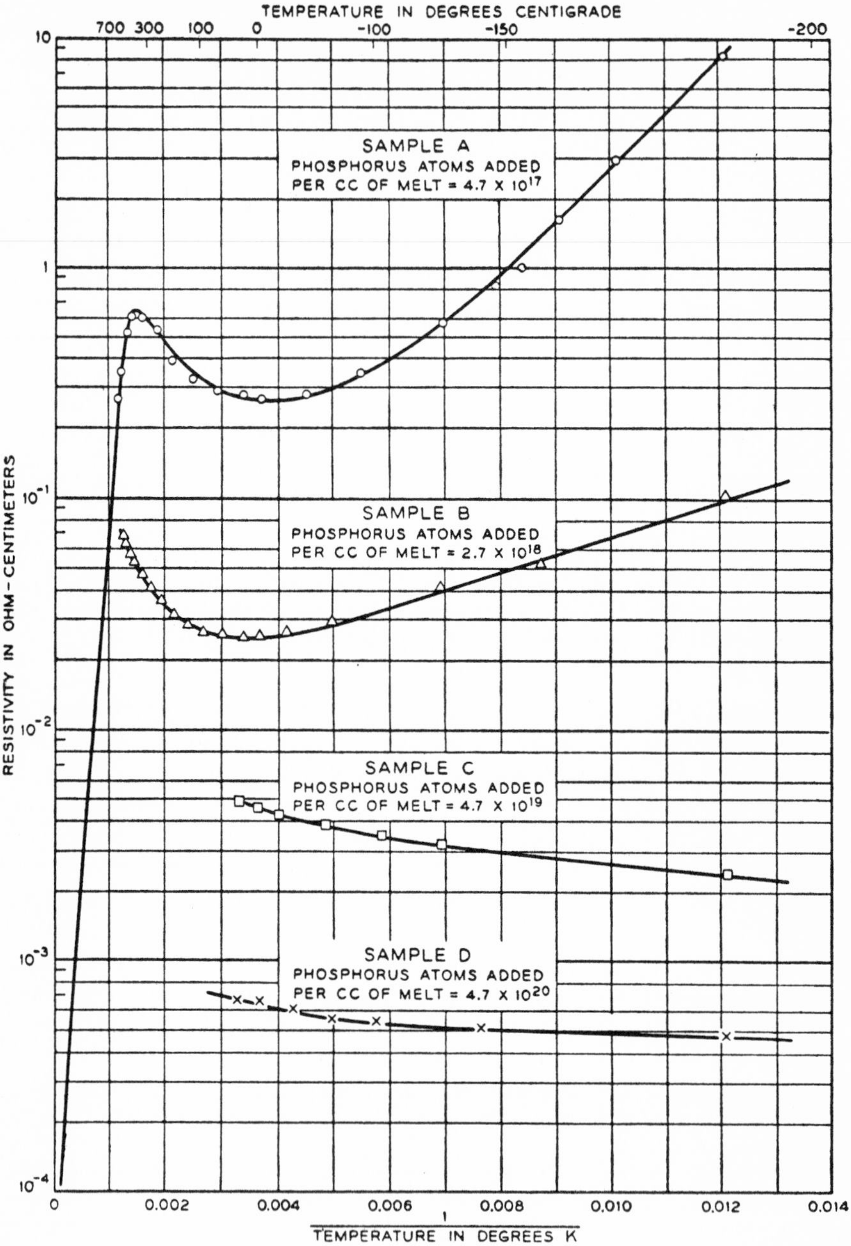

FIG. 1–10—Resistivity of Silicon-Phosphorous Alloys Versus Temperature.

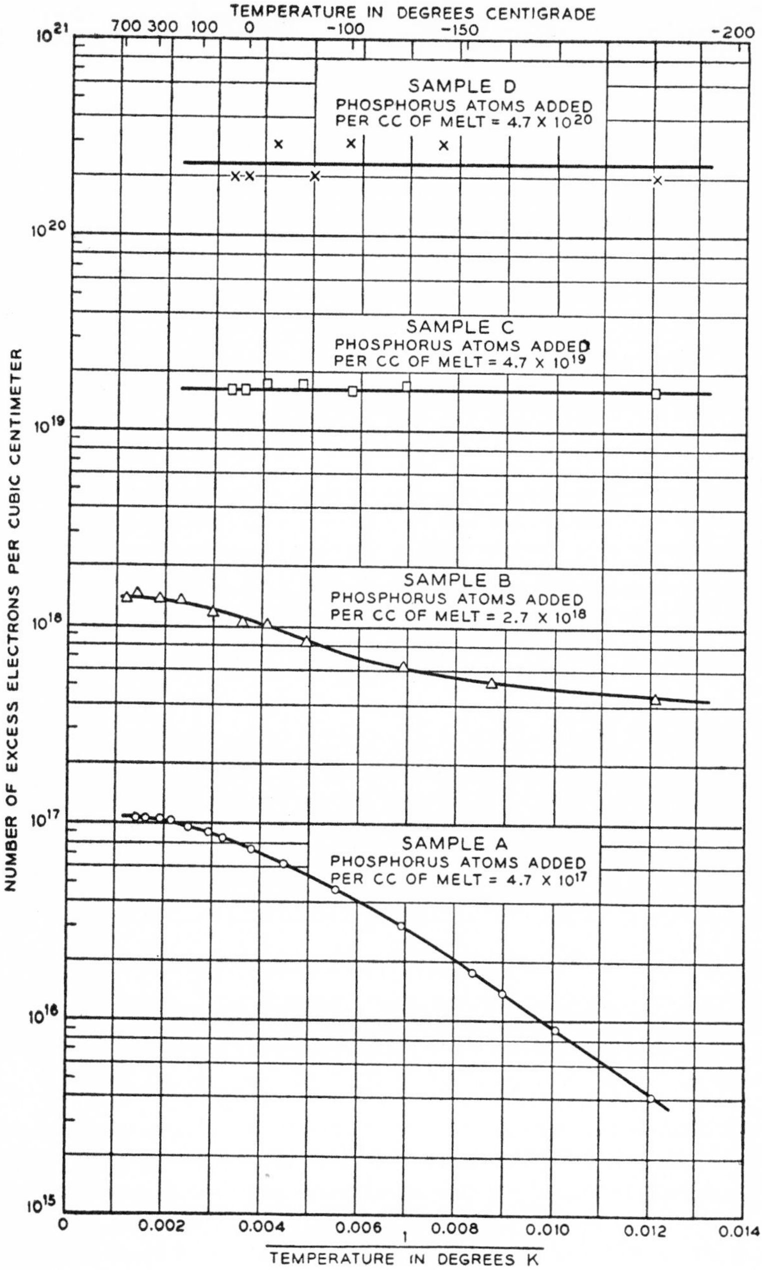

Fig. 1-11—Density of Conducting Electrons in Silicon-Phosphorous Alloys Versus Temperature.

the lattice. Figure 1.12 shows a more detailed analysis of sample A. This diagram shows both experimental points (the circles and dots) and theoretical curves (the solid lines) obtained on the basis of considerations of the energy level diagram for the semiconductor and the Fermi-Dirac

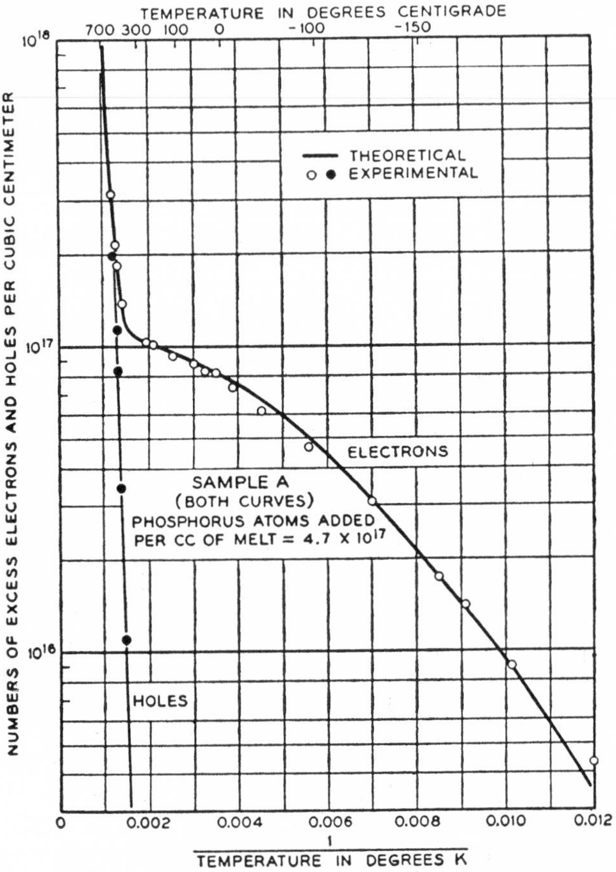

FIG. 1–12—Comparison between Predicted and Observed Electron and Hole Densities Versus Temperature.

statistics, subjects discussed in later chapters. It it seen that the theoretical interpretation of the data leaves little to be desired. The physical picture behind the theory is the one already described in connection with Figure 1.9. The concentrations of acceptors and donors and for sample A were 1.5×10^{16} and 12×10^{16} cm^{-3} (a ratio of 1 to 8 whereas in Figure 1.9 we have shown a ratio of 2 to 5).

The mobility of the electrons in the various samples is shown in Figure 1.13. It is to be noted that at high temperatures there is an "intrinsic mobility", at least for samples A and B (C and D we will discuss later), which arises from the interference to the electrons' motion due to thermal vibrations of the atoms of the crystal. For small impurity concentrations this term is independent of the impurities. For lower temperatures, the mobility is reduced by the impurities; in this case the process which limits

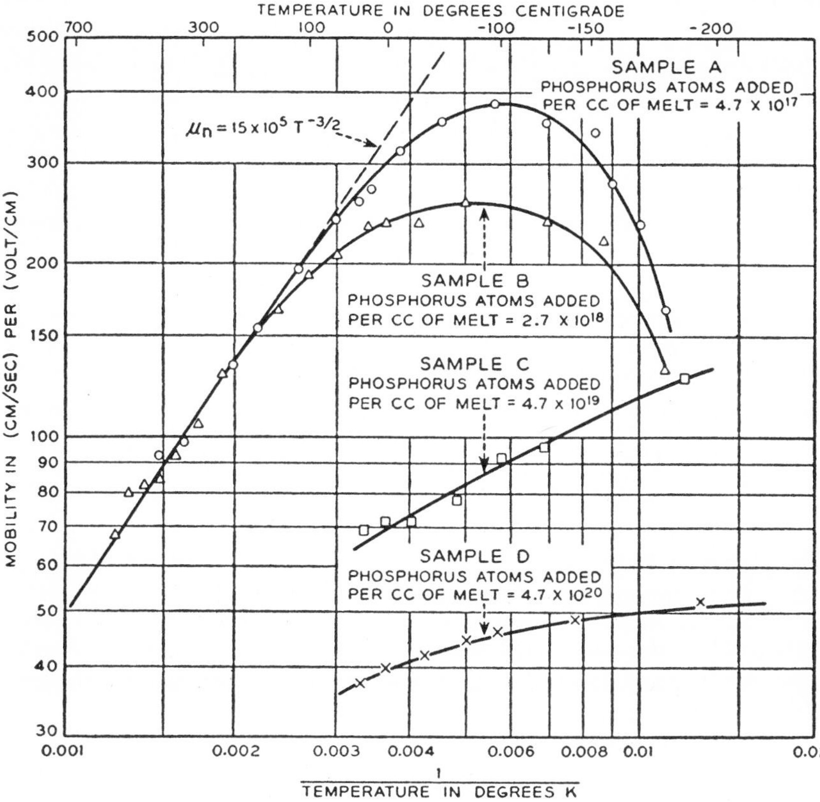

FIG. 1–13—Mobility of Electrons in Silicon-Phosphorous Alloys Versus Temperature.

the speed of drift in an applied electric field is the deflection of the drifting electrons by the electrostatic field of the charged impurity atoms. These subjects are discussed in Chapter 11.

The reader experienced with plots like Figures 1.11 and 1.12 will be aware that from them activation energies can be obtained; just as they can be obtained from the Richardson plot for thermionic emission or the plot of the temperature dependence of a chemical reaction constant. In all of these cases one encounters terms of the form

$$y = e^{-\mathcal{E}/kT} \tag{2}$$

where y is the quantity studied, $\mathcal{E}$ is the "activation energy" involved, and kT is thermal energy. If $\ln y$ is plotted against $1/T$, a straight line

$$\ln y = -(\mathcal{E}/k) \times (1/T) \tag{3}$$

is obtained whose slope is $(\mathcal{E}/k)$ from which $\mathcal{E}$ can be found since k (Boltzmann's constant) is known.

There are three energies of particular importance that can be obtained from studies of the slopes of the $\ln n$ and $\ln p$ versus $1/T$ plots. (However, the analysis involves much more than simply finding the slope of the plots.) These energies are (1) the energy required to break a valence bond and create a hole and electron pair and (2) the energy required to remove an electron from a donor or (3) a hole from an acceptor. The energy required to break a bond can be obtained almost at once from the slope of the line of Figure 1.12 in the intrinsic range. Theory shows that this slope should give $\mathcal{E}_1/2k$ where $\mathcal{E}_1$ is the energy required to break the bond. The analysis for $\mathcal{E}_2$ in n-type samples or $\mathcal{E}_3$ in p-type samples is much more involved. At very low temperatures the slopes are expected to be $\mathcal{E}_2/k$ and not $\mathcal{E}_2/2k$ (see Chapter 16, Figures 16.7 and 16.8 for details). However, values for these energies may be obtained by making a proper analysis of the data. The values obtained by Pearson and Bardeen for silicon are:

Energy to break bond	1.11 electron volts
Energy to remove electron from donor	0.054 electron volt
Energy to remove hole from acceptor	0.08 electron volt

(An equally thorough analysis for germanium has not been published; we have chosen representative values of 0.72, 0.04, and 0.04 for purposes of illustration. For diamond, the value to break the bond is 6 or 7 electron volts from ultraviolet absorption data; the energies estimated as described below to separate electrons from donors or holes from acceptors are probably about 0.35 electron volt.[3])

The small energies of binding of electrons to the donors and of holes to the acceptors can be explained on the basis of a model proposed by H. A. Bethe. The electron is supposed to move in the field of the donor in much the same way as an electron moves about the proton in a hydrogen atom.[4] There is one important difference: Because the net positive charge of the phosphorous ion is embedded in a dielectric medium, the force of attraction is reduced by the dielectric constant—about 13 for silicon. As a result of this the electron in the bound state about the

[3] Ultraviolet data from A. J. Ahearn, *Phys. Rev.* **73**, 1113 (1948). Ahearn also finds that 10 ev of α-particle energy are required for each bombardment-produced electron, and K. G. McKay, *Phys. Rev.* **77**, 816–825 (1950), finds 10 ev for 14-kv electrons. The dielectric constant of diamond is estimated as about 6.2 from its index of refraction. McKay finds trap energies of about 0.3 electron volt for electrons.

[4] This type of approximation was first published by G. Wannier, *Phys. Rev.* **52**, 191–197 (1937).

phosphorous atom moves in a very large orbit whose radius is 13 times that of an electron in a hydrogen atom, giving a binding energy of about 0.08 election volt (discussed in Chapter 9). The wave function overlaps a large number of silicon atoms as is indicated in Figure 1.14 which is

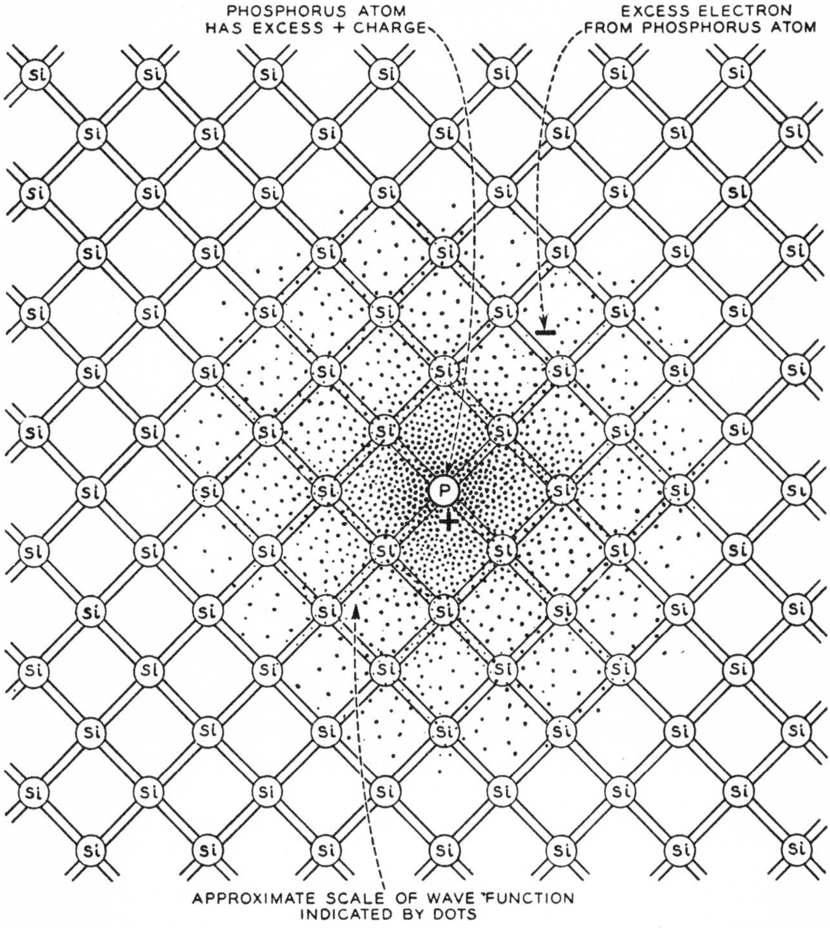

Fig. 1–14—Wave Function of Electron Bound to Phosphorous Atom in Silicon.

approximately to scale. The fact that the wave function extends over so many atoms suggests that its detailed behavior around the impurity atom itself is unimportant and that the binding energy produced by the impurity is due almost entirely to its charge. This explains the observed fact that all donor and acceptor impurities produce about the same binding energy for electrons and holes. The difference in binding energy between electrons and holes is explained by Pearson and Bardeen on the basis of a difference in effective mass, a concept discussed in Chapter 7. Furthermore, they find the activation energies quoted above only for their purest specimens; for others, the activation energy is less, a trend which they explain by considering the effects of other impurity atoms on an electron trying to escape from one impurity atom.

When relatively large concentrations of impurities are present, new effects occur; these are illustrated in Figure 1.11 by samples C and D in which the number of electrons is substantially independent of temperature. The explanation of this behavior is that the impurities have moved so close together that appreciable overlapping of the impurity wave functions, such as those shown in Figure 1.14, occurs. Under these conditions the extra electrons and the impurity atoms play much the same role as the electrons and ions in a metal, and the excess electrons form a *degenerate electron* gas which moves through the irregularly distributed array of positive ions. The theory of electrical conductivity for these semiconductors is very similar to that of metallic alloys, as is discussed in some detail in Chapter 11. The conductivity of the semiconductors is much smaller than that of the metallic alloys, however, because the density of carriers is much smaller; aside from this difference, the behavior of C and D are substantially like metallic alloys for low temperatures. As the intrinsic range is approached, however, the number of carriers and the conductivity increase rapidly with increasing temperature.

The results described for silicon for the case of electrons in connection with Figures 1.9 to 1.14 are obtained in substantially the same form for holes in silicon[5] and for electrons and holes in germanium. An important and characteristic feature of germanium is that the energies required to remove holes from acceptors and electrons from donors are so small that *a negligible fraction of the excess electrons or holes are bound to the impurities at room temperature.* In other words, the donors and acceptors are effectively *fully ionized.* This fact simplifies considerably the theory of transistor electronics for germanium.

Germanium samples show the interesting property that heat treatment at temperatures above 500°C appears to produce *p*-type impurities. Prolonged heat treatment at lower temperatures causes these impurities to disappear. One theory is that these effects are not due to chemical impurities at all but instead to disorder in the germanium lattice. The total number of disordered atoms predicted from theory is in approximate agreement with the concentrations observed.[6]

[5] G. L. Pearson and J. Bardeen, *Phys. Rev.* **75**, 865 (1949).

[6] The fraction of atoms disordered at temperature T should be approximately exp $(-\mathcal{E}/kT)$ where $\mathcal{E}$ is the energy required to disorder an atom. From data on self-diffusion in metals [J. Steigman, W. Shockley, and F. C. Nix, *Phys. Rev.* **56**, 13–21 (1939)], we may estimate that $\mathcal{E}$ is between $10kT_m$ and $20kT_m$, where T_m is the melting point. This leads to values of 10^{-7} to 10^{-10} for the exponential. The observed values correspond to about 10^{15} acceptor centers or a fraction of about 10^{-8} of the atoms.

Evidence that disorder in the lattice produces *p*-type conductivity in germanium has been obtained by K. Lark-Horovitz and his colleagues.[7] When high resistivity *n*-type germanium is bombarded with alpha particles or deuterons, it is converted to *p*-type, whose conductivity increases with increasing exposure to bombardment. It has been established that this effect is not due to transmutation of the germanium to another element but instead to the creation of disorder in the crystal, that is, germanium atoms knocked into interstitial positions leaving vacant lattice points behind. These "lattice defects" act as acceptors. There appears to be no reason why the wave functions around these defects should be like those shown in Figure 1.14, and, consequently, one might expect such samples to show different activation energies at low temperatures. The data on this subject are reviewed in Chapter 9.

Before closing this introduction, some additional experimental evidence, which supports the general picture that conductivity is produced by substitutional impurities of the sort described, should be presented. There are three types of evidence which indicate that the impurity atoms enter the lattice substitutionally in lattice sites normally occupied by atoms of the pure element rather than interstitially by squeezing into vacant spaces in the lattice:

I. The conductivity produced by the impurities is of just the sort which would be expected on the basis of their ability to complete or not to complete the valence bond structure which would surround them if they occupied normal sites. (See Table 1.1.) On the other hand, if they were to enter the lattice interstitially, they would not fit into the valence-bond scheme at all, and there would be no reason for supposing that consistent *p*- and *n*-type behavior would occur for elements with valences of three and five.

II. The activation energies required to remove electrons from the impurities, as quoted from the work of Pearson and Bardeen, show good agreement with the value to be expected theoretically on the basis of the wave functions of Figure 1.14.

III. There is evidence obtained by X-ray studies of the lattice constant for the more concentrated alloys of silicon with phosphorus and boron. The lattice constant of these alloys has been studied as a function of concentration and is found to decrease in essentially the way that would be expected if the phosphorus and boron atoms entered the lattice substitutionally, these atoms being in both cases smaller than the silicon atom. However, if they were to enter the lattice interstitially, it would be expected that they would tend to expand the lattice and pro-

[7] K. Lark-Horovitz, E. Bleuler, R. Davis, and D. Tendam, *Phys. Rev.* **73**, 1256A (1948). R. E. Davis, V. A. Johnson, K. Lark-Horovitz, and S. Siegel, *Phys. Rev.* **74**, 1255A (1948). V. A. Johnson and K. Lark-Horovitz, *Phys. Rev.* **76**, 442–443 (1949).

duce an increase in lattice constant in contradiction with the experimental facts.

In the case of germanium it has been verified that even in some relatively pure samples, in which the impurity content is much too small to determine by conventional chemical analysis, the conductivity may be controlled by minute amounts of added impurities. This investigation has been carried out with the aid of radioactive antimony which has been added to melts of germanium. It has proved possible to measure the antimony content by radioactive means and to establish that there was good agreement between the antimony content and the number of electrons produced.[8]

Problems

1. An ingot of germanium is formed by melting together 100 gm of germanium and 3.22×10^{-6} gm of antimony. Assuming that the antimony is uniformly distributed, show that the density of antimony atoms is 8.70×10^{14} cm^{-3}. [The density of germanium is 5.46 gm/cm^3 and its atomic weight is 72.60; the atomic weight of antimony is 121.76. One gm atom, that is 72.6 gm of germanium, contains Avogadro's number ($= 6.02 \times 10^{23}$) atoms. There is a slight discrepancy between the density of germanium atoms deduced from X-ray measurements, Figure 1.1, and from density and atomic weight.]

2. Assuming that the excess electrons due to the antimony atoms of problem 1 are fully excited at room temperature and that their mobility is 3600 cm^2/volt sec, show that the conductivity of the ingot of problem 1 is 0.50 ohm^{-1} cm^{-1} or 50 ohm^{-1} meter^{-1} in M.K.S. units. Hence show that a specimen 1 mm square in cross-section and 2 cm long would have a resistance of 400 ohms.

3. Assume that the ingot contains 0.78×10^{-6} gm of gallium, atomic weight 69.72, in place of antimony. Show that the density of gallium atoms will be 3.68×10^{-14} cm^{-3}. Taking μ_p as 1700 cm^2/volt sec show that the conductivity will be 0.10 ohm^{-1} cm^{-1}.

4. Assume that both the gallium and antimony discussed above are added to the same 100 gm ingot. Why will the conductivity be 0.29 ohm^{-1} cm^{-1} and not 0.6 ohm^{-1} cm^{-1}? Will the material be n-type or p-type?

5. Assume that a flash of light falls uniformly on the specimen of problem 2 and that a total number, 1.74×10^{13}, of photons are absorbed and that each produces one hole-electron pair. Show that the resistance will drop to 162 ohms.

6. Show that if 10 volts are applied from end to end of the specimen of problem 2, an electron drifts with a speed of 18,000 cm/sec and requires 1.11×10^{-4} sec to traverse the length of the specimen.

[8] G. L. Pearson, J. D. Struthers, and H. C. Theuerer, *Phys. Rev.* **75**, 344 (1949), **77**, 809–813 (1950).

CHAPTER 2

THE TRANSISTOR AS A CIRCUIT ELEMENT

2.1 IDEAS ABOUT AMPLIFICATION PRIOR TO THE TRANSISTOR

2.1a. Introduction. In Chapter 1, we have presented a description of the behavior of holes and electrons appropriate to circumstances in which their distributions in space and time are substantially uniform. In order to deal with transistor action, however, and with related phenomena in semiconductors, we must describe what happens when the concentrations of both holes and electrons vary with position and time. Studies of such phenomena have been greatly facilitated from the experimental point of view by a series of new techniques and experimental procedures developed in connection with transistor research. The experiments permit the isolation and individual analysis of a number of the separate processes which go on in connection with transistor action. For this reason we shall discuss them in Chapter 3 before presenting in Chapter 4 the theory for the conventional or type-A transistor[1] invented by J. Bardeen and W. H. Brattain.

In this chapter we shall review briefly the history of semiconductor development and research, especially in connection with amplifying devices, and show how this was culminated by the invention of the transistor. We shall then give a description of the transistor itself from a phenomenological point of view and describe its equivalent circuit. The exposition in later sections is directed in considerable measure towards giving a physical meaning to the purely formal equivalent circuit.

The general level of presentation aimed at in Part I is qualitative so far as the theory of holes and electrons is concerned, and an attempt has been made to present the necessary semiconductor theory in descriptive terms covering the thermal generation of holes and electrons, their recom-

[1] The first authoritative treatment of the type-A transistor is that of J. Bardeen and W. H. Brattain, "Physical Principles Involved in Transistor Action", *Phys. Rev.* **75**, 1208–1225 (1949). Other publications are as follows: J. Bardeen and W. H. Brattain, "The Transistor: A Semi-Conductor Triode", "Nature of the Forward Current in Germanium Point-Contact Rectifiers", *Phys. Rev.* **74**, 230, 231 (1948); R. M. Ryder, "The Type-A Transistor", *Bell Laboratories Record* **27**, 89–93 (March, 1949); J. N. Shive, "The Double Surface Transistor", *Phys. Rev.* **75**, 689–690 (1949); J. A. Becker and J. N. Shive, "The Transistor; A New Semi-Conductor Amplifier", and R. L. Wallace and W. E. Kock, "The Coaxial Transistor", both in *Electrical Engineering* **68**, 222–223 (1949); R. M. Ryder and R. J. Kircher, "Some Circuit Aspects of the Transistor", *Bell Syst. Tech. J.* **28**, 367–400 (1949).

bination, and their motions in electric and magnetic fields. The energy band theory described in Parts II and III has been developed especially for such situations; however, it is not necessary to invoke the general theory in order to get a good mental picture of the important processes in transistor action. In some cases we have given quantitative results and formulae; however, in order to make this material available at an early date in the transistor art, it has not been feasible to prepare a general

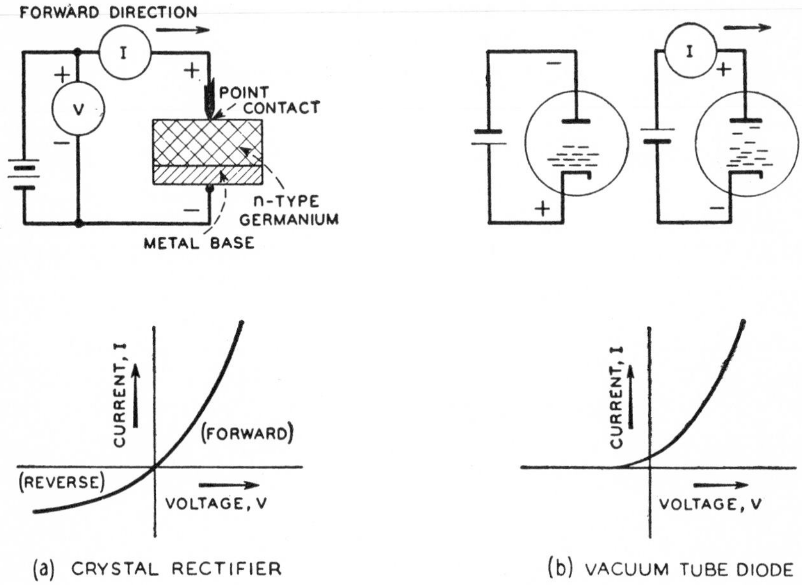

FIG. 2-1—Rectifiers and *V–I* Plots.

exposition of the quantitative theory of transistor action to the extent that has been attempted for the underlying theory of holes and electrons.

The basic idea that semiconductors may be made to amplify electrical signals is relatively old and, as we shall discuss in the next paragraph, amplification by negative resistance thermistors had actually been achieved prior to the invention of the transistor. The similarity of the current-voltage curves for vacuum-tube diodes and for crystal rectifiers, Figure 2.1, suggested to many people that it might be possible to make an amplifier from a crystal diode by somehow adding a grid. The desired aim has now been accomplished by the transistor, but, as we shall discuss, the amplifying action does not employ a grid. Another striking difference is that, in a vacuum-tube diode, it is the forward current that is controlled by the insertion of the grid. In the transistor, on the contrary, the output contact is biased in the low-current, high-resistance direction, and the current is enhanced and controlled by the input contact.

Amplifying action by use of the negative-resistance characteristics of thermistors had been achieved and analyzed relatively early in the thermistor development work. The negative resistance of a thermistor is made possible by the negative temperature coefficient of resistance of semiconductors such as, for example, that shown for the intrinsic range in Figure 1.10. As the current through a thermistor is increased, the voltage at first rises in accordance with Ohm's law. For large currents, however, appreciable heating of the thermistor results, thereby causing a decrease in resistance, and the current-voltage characteristic becomes nonlinear. A sufficiently high current increases the temperature so much that the resulting drop in resistance may actually produce a voltage decrease and, consequently, a negative differential resistance occurs. This negative-resistance characteristic can be used to make oscillators or networks with gain rather than attenuation, provided the frequencies are low enough so that the temperature of the thermistor can follow the variations of current. This requirement sets an upper frequency limit which depends on the thermal time constants under the operating conditions. By making the physical dimensions small and the thermal conductivities high, stable oscillations have been produced at frequencies as high as 100,000 cycles per second.[2] However, there is an additional mechanism involved which will limit the frequency no matter how small the structure is made: Time is required for the numbers of electrons and holes to come to equilibrium after the temperature is changed. Not much is known about this time for most semiconductors. We shall discuss later in Section 3.1d some data bearing on this time for germanium at room temperature.

2.1b. Modulation of Conductivity by Surface Charges. In the Solid State Research Group at Bell Telephone Laboratories, the hope of discovery of a purely electronic, rather than thermal, semiconductor amplifier was bolstered by the discovery of what might be called an *existence theorem*, to use a mathematical term, that such an amplifier was possible, or at least not contradictory to the theories of semiconductors and statistical mechanics. The particular device envisaged is shown in Figure 2.2. It consists of a very thin layer of semiconductor placed on an insulating support. This layer of semiconductor constitutes one plate of a parallel-plate capacitor, the other being a metal plate in close proximity to it. If this capacitor is charged, as shown in part (c), with the metal plate positive, then the additional charge on the semiconductor will be represented by an increased number of electrons. At room temperature, in germanium, a negligible number of these electrons will be bound to the donors (see Section 10.4 for details). Consequently, the added electrons should be free to move and should contribute to the conductivity of the semiconductor. In this way, the conductivity in the semiconductor can

[2] Personal communication from J. A. Becker.

be modulated, electronically, by a voltage put on the capacitor plate. Since this input signal requires no power if the dielectric is perfect, power gain will result.

The effects to be expected from such an arrangement are very appreciable, a conclusion which we shall illustrate for its own sake, and also for the purpose of showing how concentrations of electrons and holes may be

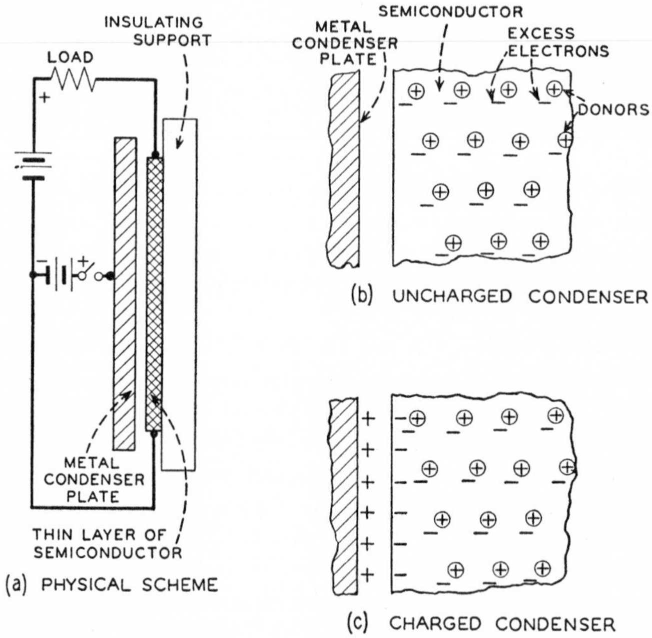

FIG. 2–2—Modulation of Conductance by Surface Charges.

(The extra electrons induced on the surface in (c) carry an added current.)

used in calculations. Suppose the layer consists of n-type germanium with a resistivity of 2.4 ohm cm.[3] The concentration of electrons is obtained from the formula for conductivity in $\text{ohm}^{-1}\ \text{cm}^{-1}$ (or coulombs/volt sec cm)

$$\sigma = \frac{1}{\rho} = ne\mu = n\ \text{cm}^{-3} \times 1.6 \times 10^{-19}\ \text{coulombs} \times 2600\ \text{cm}^2/\text{volt sec}$$

$$= n \times 4.2 \times 10^{-16}\ \text{coulombs/volt sec cm} \quad (1)$$

[3] In M.K.S. units, this is a layer 10^{-7} m $= T$ thick with conductivity $\sigma = 10^2/2.4 = 42\ \text{ohm}^{-1}\ \text{m}^{-1}$ with electrons having mobility of 0.26 m^2/volt sec. The number of electrons per unit volume is given by $\sigma = ne\mu$ so that $n = 42/0.26 \times 1.6 \times 10^{-19}\ \text{m}^{-3}$ giving 1.0×10^{14} per square meter of surface. A field of 30,000 volts/cm or 3×10^6 volts/m produces a surface charge for a medium of dielectric constant 2 of $2 \times \epsilon_0 \times 3 \times 10^6 = 5.3 \times 10^{-5}$ coulombs m^{-2} or 3.3×10^{14} electrons m^{-2}.

giving

$$n = 1/2.4 \times 4.2 \times 10^{-16} = 1.0 \times 10^{15} \text{ electrons per cm}^3. \quad (2)$$

In a layer $T = 1000A = 10^{-5}$ cm thick there will thus be $nT = 1.0 \times 10^{10}$ electrons per cm^2. We shall next calculate how many electrons will be induced by a field in the dielectric of 30,000 volts/cm or 100 electrostatic volts/cm. This field will produce a displacement of $\kappa E/4\pi$ electrostatic coulombs. Taking $\kappa = 2$, for a typical dielectric between the capacitor plates, we calculate a charge of $2 \times 100/4\pi = 16$ stat. coulombs per cm^2, corresponding to $16/4.8 \times 10^{-10} = 3.3 \times 10^{10}$ electrons per cm^2, or three times the number normally present. Such a field should thus quadruple the conductivity of the germanium layer.

Experiments have been carried out with layers of various semiconductors, and effects of the sort discussed have been observed. However, the degree of modulation has been somewhat less than that calculated above. For layers of germanium about 5000A thick, in which the mobility was only about 40 cm^2/volt sec, it was apparent upon analysis of the data that only about 10 per cent of the induced charges (holes in this instance since the evaporated germanium was *p*-type) were effective in changing the conductance.[4] This reduced effectiveness can be accounted for on the basis of a theory dealing with the behavior of the current carriers at a semiconductor surface.

2.1c. Bardeen's Theory of Surface States. J. Bardeen has proposed that the ineffective portion of the induced charge is lodged in states localized at the surface, employing for this purpose his theory of *surface states*, which he has so fruitfully applied to the explanation of a number of otherwise disconnected facts about semiconductor surfaces. A thorough discussion of his surface state theory would require the mathematics of energy bands and Fermi-Dirac statistics to an extent not in keeping with the level of exposition of Part I. We shall effect a compromise by quoting conclusions based on his theory as it applies to the modulation of conductivity by surface charges without elaborating upon the underlying reasoning. We shall adopt the same procedure later in discussing the nature of transistor action with metal point contacts.

The conclusion reached by Bardeen is that electrons which move in the body of the semiconductor can become tightly bound in *surface states* on the surface of the semiconductor and thus become immobilized. The Surface States Diagram, Figure 2.3, represents the free surface of an *n*-type semiconductor and shows, in part (a), eight electrons trapped in the surface states. These electrons repel other electrons in the conduction band and thus produce a layer of depleted conductivity below the surface. In this layer there are eight unneutralized donors, so that the semiconductor

[4] W. Shockley and G. L. Pearson, *Phys. Rev.* **74**, 232–233 (1948).

surface as a whole is neutral. (In the diagram the donor ions are arranged in a regular scheme so as to facilitate counting them; in a sample of germanium, such as we have just discussed, the donors will be distributed at

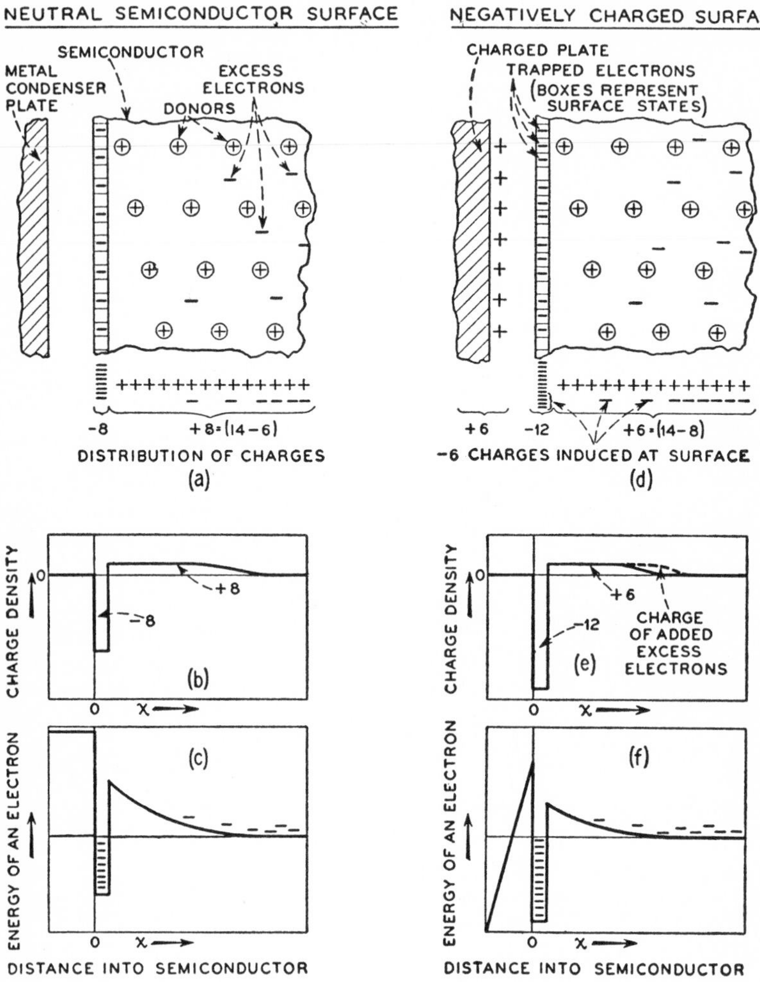

FIG. 2-3—Surface States Diagram, Showing Bardeen's Theory of the Role of Surface States in Immobilizing Induced Charge.

random and will be on the average about 500 times as far apart as normal atoms in the lattice.) In part (b) the net charge density is plotted as a function of distance below the surface. This charge density gives a variation of electrostatic potential from point to point in accordance with the energy curve for an electron shown at (c). The effect of the repulsion of the electrons in the surface states and the attraction of the unneutralized donors are represented by the approximately parabolic curve of energy versus distance. The surface states bind the electrons tightly in energies even lower than they would have in the conduction band, deep inside the crystal. Semiconductors, like metals, have work functions; the energy required to remove an electron from inside the semiconductor to a point in free space in front of it is represented by the height of the energy curve above zero on the left edge of part (c).

It may be helpful at this point to make a few remarks on the relationship between *energy of an electron* shown in Figure 2.3 and the *electrostatic potential.* One obvious difference is that, because of the negative charge of the electron, the energy diagram and the electrostatic potential diagrams are inverted in respect to each other and electrons tend to fall uphill in an electrostatic potential diagram. A more important difference, especially when an electron moves near a polarizable surface such as that of a metal, is that the electron induces charges by its presence and thus changes the electrostatic potential.[5] The energy diagram includes these interactions so that the energies shown represent the way in which the energy of the entire system depends on the position of the electron. Not all of the energy difference between states arise from differences in electrostatic interactions, however. The change in energy between an electron in the conduction band and one in a deep lying surface state is in part electrostatic and in part kinetic. In the hydrogen atom of Figure 1.1, for example, the electron has a negative potential energy which has twice the magnitude of the kinetic energy. In general, differences in energy between different electronic states involve mixtures of kinetic and potential energies and it is usually not practical or useful to try to sort them out.

When the metal capacitor plate is charged positively, a compensating negative charge flows through the charging circuit and appears as six extra electrons on the surface of the semiconductor, in the Surface States Diagram, at (d). Four of these electrons go into the surface states, and two remain as excess electrons in the conduction band. The resulting modifications of the space charge distribution (e) and energy for an electron (f) follow directly from part (d). The electric field between the capacitor plates is represented in (f) by the slope of the energy curve outside the semiconductor. The example shown in the Surface States Diagram is, of course, qualitative, and it is reasonable to suppose that 90 per cent of the charge goes into the surface states in accordance with the results measured by Pearson.[6]

[5] See, for example, W. G. Dow, *Fundamentals of Electronic Engineering,* John Wiley & Sons, New York, 1937, Section 94.

[6] W. Shockley and G. L. Pearson, *Phys. Rev.* **74**, 232–233 (1948).

A number of experiments have been performed which give information on various features of the surface state theory just presented.[7]

2.2 THE TYPE-A TRANSISTOR

2.2a. Its Discovery and External Aspect. Experimental and theoretical work on the properties of surface states was given emphasis in the research

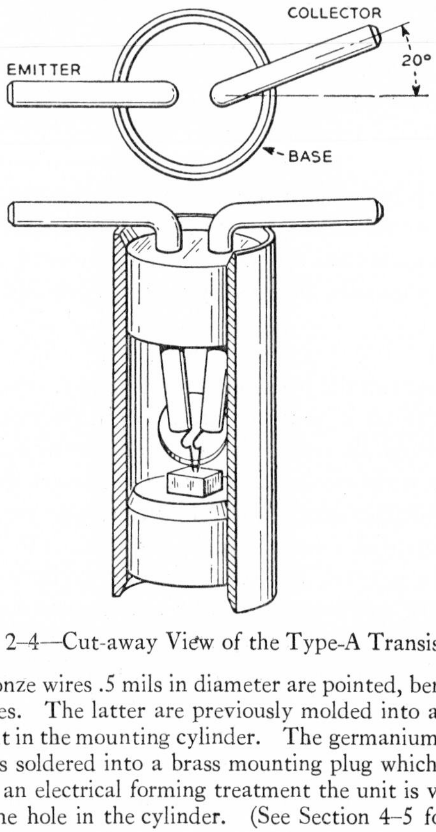

FIG. 2-4—Cut-away View of the Type-A Transistor.

Two phosphor bronze wires .5 mils in diameter are pointed, bent and welded to the nickel mounting wires. The latter are previously molded into a plastic plug, which makes a light press fit in the mounting cylinder. The germanium wafer 20 mils thick and 50 mils square is soldered into a brass mounting plug which is also pressed into the cylinder. After an electrical forming treatment the unit is vacuum impregnated with wax through the hole in the cylinder. (See Section 4–5 for some information on forming.)

program at Bell Telephone Laboratories. Stimulated by Bardeen's theory, the group undertook a number of experiments to measure characteristics of the surface states. In many instances the effects sought were below the threshold of sensitivity of the methods employed. While exploring certain aspects of this problem, John Bardeen and Walter Brattain encountered

[7] For a summary of several experiments, see the original paper by J. Bardeen, *Phys. Rev.* **71**, 717–727 (1947). For information on contact potential see W. H. Brattain and W. Shockley, *Phys. Rev.* **72**, 345 (1947). For photoelectric emission see L. Apker, E. Taft, and J. Dickey, *Phys. Rev.* **74**, 1462–1474 (1948), which contains many references.

some new effects, and, branching off into a new area of theory and experiment, they invented the transistor.[1]

We shall next discuss the transistor from a phenomenological viewpoint and will present, thereafter, the theory of its operation.

Figure 2.4 shows a cut-away view of a transistor. The schematic diagram of Figure 2.5 shows the bias battery supplies which activate the transistor so that it can amplify an input signal. The two point contacts of

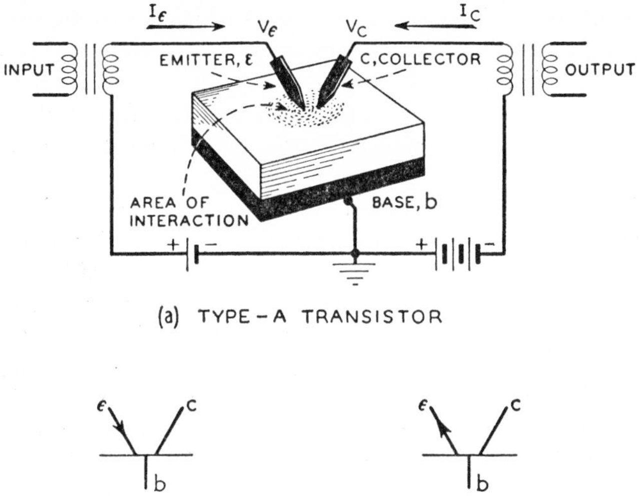

(a) TYPE-A TRANSISTOR

(b) SYMBOL FOR n-TYPE (c) SYMBOL FOR p-TYPE

FIG. 2-5—Schematic Diagram of Transistor.

The area of interaction which surrounds the emitter point, the conventional symbols and the sign conventions for current flow and voltage are shown. (Typical values for a type-A transistor are $I_\epsilon = 0.6$ ma, $V_\epsilon = 0.7$ volt, $I_c = -2.0$ ma, $V_c = -40$ volts.)

the transistor are, individually, rectifying points of the general type discussed with Figure 2.1. Under operating conditions, the input or emitter point is operated in the plus (forward or low-resistance) direction and the output or collector is operated in the minus (reverse or high-resistance) direction. An important discovery of Bardeen and Brattain was that, when the input point is biased for forward current flow, it becomes surrounded by an area of interaction; if the output point is placed within this area, the input current controls the output current in such a way that power gain results. The region of interaction is not sharply defined, and the power

[1] For additional details of this history together with a discussion of many theoretical and experimental aspects of transistors, see J. Bardeen and W. H. Brattain, "Physical Principles Involved in Transistor Action," *Phys. Rev.* **75**, 1208–1225 (1949).

gain in one conventional circuit arrangement varies from about 100-fold (or 20 db) at 0.005-cm spacing to unity or (0 db or unit power transmission) at about 0.025 cm.

The theory of transistor action, which follows later in this chapter, is that the emitter point emits holes into the *n*-type material, which otherwise would contain only electrons. (A few holes are normally present, see Chapter 10, but they make a negligible contribution to the conductivity.) These holes are drawn to the negative collector point and add to its current. In Section 3.1 we shall describe experiments which are consistent with this picture and give detailed information about the numbers and behaviors of the holes. In this section we shall describe the transistor characteristics and show how they lead to power gain, postponing a further description of the physical theory of the type-A transistor itself until the simpler cases have been studied.

2.2b. The Static Characteristics. The transistor, being a three-terminal device, requires the specification of two voltages and two currents to describe its operating condition. The conventions selected for this purpose are those standard for three-terminal networks. As shown in Figure 2.5, the voltages are considered as positive when positive in respect to the base, and the currents are considered as positive when they flow from metal point to semiconductor (that is, when electrons flow from semiconductor to metal). This situation is more complicated than for a conventional vacuum-tube triode because for the latter the grid current is usually taken as zero so that it suffices for its static characteristics to specify the plate current as a function of plate and grid voltages. In the transistor, values for both voltages and both currents are significant. There is another important difference: For a given set of voltages, there may be two or three possible sets of currents in the transistor. That is, the currents are multi-valued functions of the voltages. (The reverse is true in a vacuum tube suffering from grid emission, and there may be two grid voltages for a given grid current.) However, for the transistor, there is only one set of voltages for a given pair of currents in the normal operating range. It is thus appropriate in the case of the transistor to express the voltages as functions of the currents.

The symbols chosen for emitter, base and collector are ϵ, b and c. The choice of ϵ rather than e reduces the likelihood of confusion with c when subscripts are used, as the reader may verify by considering V_ϵ, V_e and V_c.

Figure 2.6 shows a plot for a typical transistor, for d-c conditions. The two upper plots show V_ϵ as a function of the currents, two conventions of presentation being employed, and the lower plots similarly show V_c. These particular plots have been chosen for exhibiting transistor characteristics because of the convenient relationship of their slopes to parameters in the small signal theory of transistor operation.

One extra line is shown in the upper left plot, which corresponds to a fixed collector voltage of −40 volts. This line has been extended beyond the range of the operating characteristics. It shows that, at fixed collector bias, the emitter voltage first increases with emitter current and then, somewhat beyond the operating range, it decreases. In other words, there are two sets of values of emitter current and collector current which correspond to $V_c = -40$ volts and $V_\epsilon = -0.1$ volt. In a presentation of the data in the form of currents as functions of the voltages, this multivalued feature would cause considerable inconvenience; so far as the operation of the transistor is concerned, however, the situation is even more serious. If the transistor were operated at these two voltages, that is, from low-impedance external sources, it would be conditionally stable at the first point and unstable at the second. If a transient impulse should happen to take it out of the stable range, it would tend to run away to currents still higher than the upper point and a burned-out unit would result. To avoid such damage, transistor circuits are designed so as to eliminate the instability in amplifiers and limit the currents in flip-flop circuits and thus to overcome or to control the instability. In Section 4.1(b) we shall show that the negative resistance in the input point is associated with a current amplification between input and output circuits and that the multivalued aspect of the plots is related to the decrease in the resistance of the rectifiying emitter contacts as the emitter current increases.

The static characteristic lines are of importance in determining the nonlinear behavior of transistors and in designing switching and counting circuits using them. Some problems illustrating the principles involved are given at the end of this chapter.

2.2c. The Equivalent Circuit.[2] If we write the functional relationship of Figure 2.6 in the form

$$V_\epsilon = f_1(I_\epsilon, I_c) \tag{1a}$$

$$V_c = f_2(I_\epsilon, I_c), \tag{1b}$$

then the small a-c voltages v_ϵ and v_c produced by small a-c currents i_ϵ and i_c are[3]

$$v_\epsilon = \frac{\partial f_1}{\partial I_\epsilon} i_\epsilon + \frac{\partial f_1}{\partial I_c} i_c = r_{11} i_\epsilon + r_{12} i_c \tag{2a}$$

$$v_c = \frac{\partial f_2}{\partial I_\epsilon} i_\epsilon + \frac{\partial f_2}{\partial I_c} i_c = r_{21} i_\epsilon + r_{22} i_c \tag{2b}$$

[2] General reviews of the theory of networks will be found for passive circuits in E. A. Guillemin, *Communication Networks: Vol. II*, John Wiley & Sons, New York, 1935 and for active circuits in L. C. Peterson, *Bell Syst. Tech. J.* **27**, 593–622 (1948).

[3] In this section we shall assume that the frequency is so low that all circuit parameters are resistive.

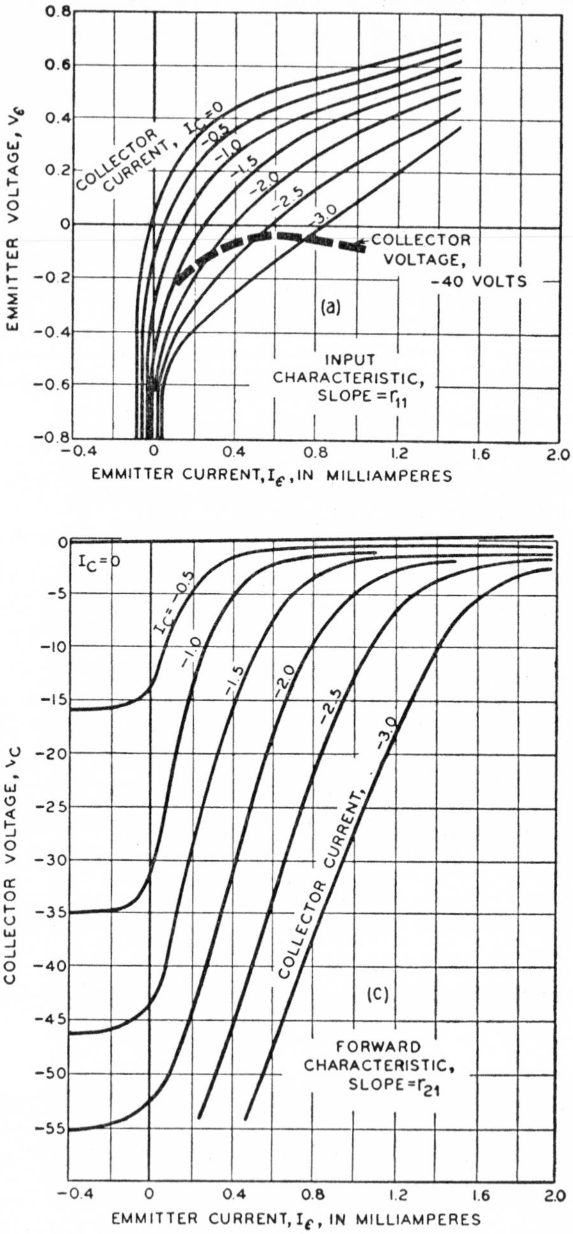

FIG. 2-6—Four Sets of Static Characteristics for the Type-A Transistor.

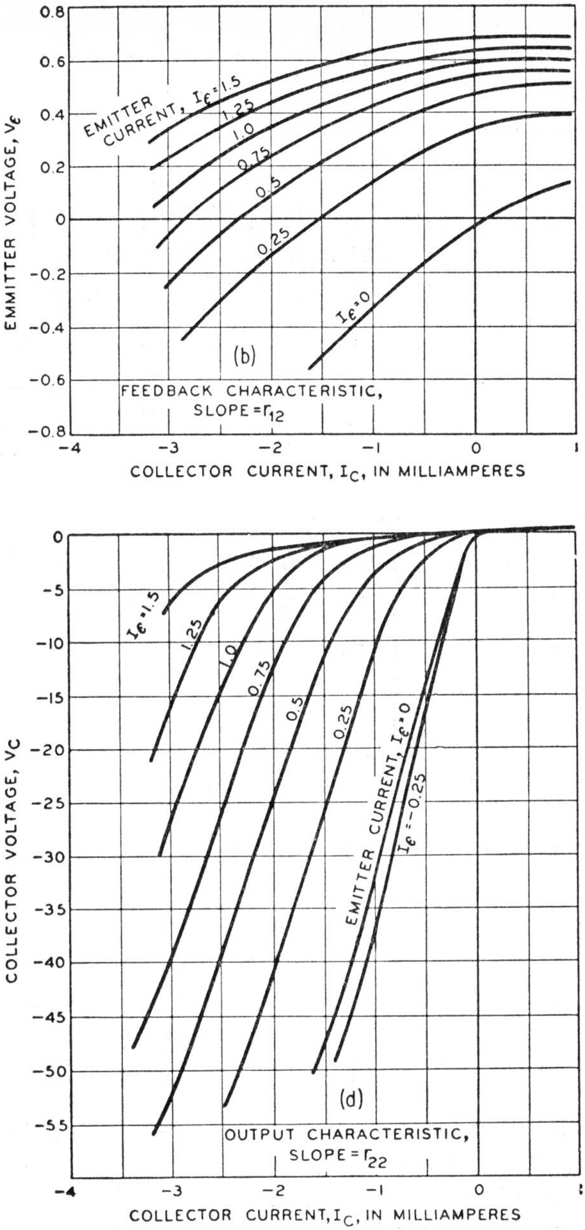

Fig. 2-6—*Continued.*

where the r's, which are abbreviations for the partial derivatives, are simply the slopes of the curves of Figure 2.6. A system described by the equations just given can be represented as a passive network of resistances if $r_{12} = r_{21}$ and $r_{11} > r_{12} < r_{22}$; otherwise negative resistances or voltage

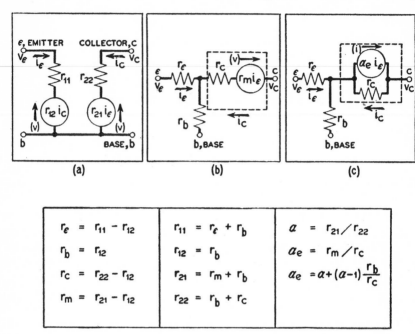

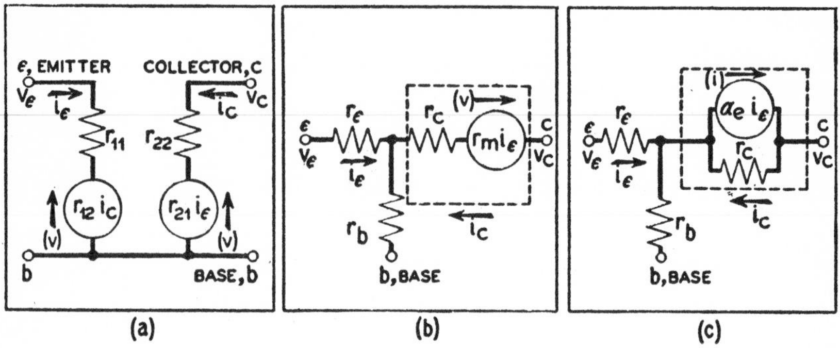

$r_\epsilon = r_{11} - r_{12}$	$r_{11} = r_\epsilon + r_b$	$\alpha = r_{21}/r_{22}$
$r_b = r_{12}$	$r_{12} = r_b$	$\alpha_e = r_m/r_c$
$r_c = r_{22} - r_{12}$	$r_{21} = r_m + r_b$	$\alpha_e = \alpha + (\alpha - 1)\frac{r_b}{r_c}$
$r_m = r_{21} - r_{12}$	$r_{22} = r_b + r_c$	

FIG. 2-7—Equivalent Circuits for the Transistor and Relationships Among the Parameters.

(Generators inside dashed rectangles are equivalent if $r_m = \alpha_e r_c$.)

or current generators must be introduced. Figure 2.7 shows three equivalent circuits together with the relationships between the various parameters involved.

The quantity α

$$\alpha = -\left[\frac{\partial I_c}{\partial I_\epsilon}\right]_{V_c = \text{const}} = \frac{r_{21}}{r_{22}} \tag{3}$$

is a measure of the effect which the hole current from the emitter has upon the collector. We shall discuss its physical meaning in later sections. The quantity r_{12}, which represents a reaction at the emitter produced by the collector, is called the base resistance; it represents a positive feedback in the grounded base amplifier circuit and, as such, it increases the gain and decreases the stability of the circuit.

In Figure 2.7(b), the resistance r_m in the voltage generator $v_m i_\epsilon$ is the *active* mutual resistance. As we shall see at the end of this section, the power gain is approximately proportional to r_m^2.

From the static characteristics shown in Figure 2.6, values of the quantities r_{11}, etc., can be obtained for any operating point. The following table gives typical values together with other information on operating conditions.

TABLE 2.1. PRELIMINARY DATA FOR TYPE-A TRANSISTOR

Typical Operating Conditions		Average Equivalent Circuit Parameters (ohms)	
Emitter Current....	0.6 ma	$r_{11} =$ 530	Emitter Resistance $r_e =$ 240
Emitter Voltage.....	0.7 volt	$r_{12} =$ 290	Base Resistance $r_b =$ 290
Collector Current....	−2 ma	$r_{21} =$ 34,000	Collector Resistance $r_c =$ 19,000
Collector Voltage....	−40 volts	$r_{22} =$ 19,000	Mutual Resistance $r_m =$ 34,000
		$\alpha_e \doteq \alpha \equiv r_{21}/r_{22} = 1.8$	

Maximum Ratings: Not to be exceeded in continuous operation. Voltages relation to base.

Collector Voltage.............................. −70 volts
Collector Dissipation.......................... 0.2 watt

Grounded Base Operation: Class A, working from a 500-ohm generator into a 20,000-ohm load.

Operating Power Gain............................ ~17 db
Power Output.................................... ~5 mw

From this data it is evident that the input impedance is of the order of hundreds of ohms. From the values of α it is seen that the output current, in equivalent circuit (c) for example, is even larger than the input current and is available from an impedance of the order of 20,000 ohms. From this it is at once evident that there is the possibility of power gain.

The source of the power gain resides in the current or voltage generators of the equivalent circuits of parts (b) and (c) of Figure 2.7. (Power gain may occur even if α_e is less than unity.) In later sections we shall discuss the internal mechanism of the transistor and relate the physical picture to the equivalent circuit so that the mechanism of power gain will be given a physical interpretation. In this section, however, we shall deal with the equivalent circuit *per se* making no use of the underlying physics peculiar to transistors. The treatment given, however, will reflect an interest in the later interpretation, and partly for this reason no use will be made of the specialized techniques of circuit theory. Although much of the discussion of this section can be compactly expressed in terms of results of the theory of feedback amplifiers, little use of these ideas and terminology is introduced. On the one hand, the reader familiar with these ideas will have little difficulty in recasting the discussion in terms of general circuit theory and, on the other hand, it would be inappropriate to incorporate a sufficient discussion of circuit theory to make it of benefit to the reader primarily concerned with the fundamental mechanisms of semiconductors. To the latter the circuit characteristics of the transistor are of interest chiefly as manifestations of more basic processes.

As an example of the use of the equivalent circuit and as a further exhibition of the behavior of the type-A transistor, we shall carry out calculations for the transistor as a grounded base amplifier. This is, of course, only one of a number of the possible circuits in which the transistor may be used. The grounded base amplifier is shown in Figure 2.8. The

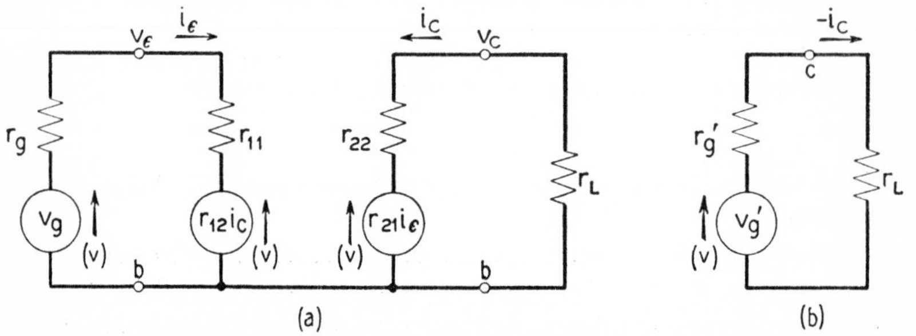

FIG. 2–8—The Transistor as a Grounded Base Amplifier.

(a) Working from a signal generator, v_g, r_g into a load r_L.
(b) Equivalent output circuit.

emitter is connected to an input generator v_g, r_g represented by a voltage generator v_g in series with a resistance r_g. The output load resistance is r_L. The bias power supplies are not shown. The generator has a certain power capacity which is defined as its *available power.* This power is defined as the maximum power which the generator can deliver and, as is well known from circuit theory, this power is a maximum when the generator looks into a matched load. For this case it has the value

$$v_g^2/4r_g = \text{available power from input generator.} \tag{4}$$

All power gains which we shall discuss will be defined in terms of the ratio of the output power to the power available from the input generator.

The power gain in the grounded base circuit is obtained by a straightforward analysis of the circuit of Figure 2.8. In order to describe the output from the transistor in terms of an equivalent generator v_g', r_g' as shown in part (b), we shall carry out the analysis by writing down expressions for the voltages at terminals ϵ and c rather than by applying Kirchhoff's law to the loops. We thus obtain for terminal ϵ:

$$v_\epsilon = v_g - r_g i_\epsilon \qquad \text{(from generator)} \tag{5a}$$

$$v_\epsilon = r_{11} i_\epsilon + r_{12} i_c \qquad \text{(from transistor).} \tag{5b}$$

Similarly we obtain for terminal c:

$$v_c = -r_L i_c \qquad \text{(from load)} \tag{6a}$$

$$v_c = r_{21} i_\epsilon + r_{22} i_c \qquad \text{(from transistor).} \tag{6b}$$

We shall next solve the last equation by expressing i_ϵ in terms of v_g. Having done this, we shall find that the equation for v_c becomes very similar to the equation for the generator working into ϵ so that we can interpret the transistor as converting the generator v_g, r_g into a new generator v_g', r_g'. Solving the v_ϵ equations for i_ϵ gives

$$i_\epsilon = (v_g - r_{12} i_c)/(r_{11} + r_g). \tag{7}$$

Inserting this in the last equation gives

$$v_c = \left(r_{22} - \frac{r_{12} r_{21}}{r_{11} + r_g}\right) i_c + \frac{r_{21}}{r_{11} + r_g} v_g$$

$$= v_g' + r_g' i_c, \tag{8}$$

the latter form corresponding to replacing the transistor by v_g', r_g', the change in sign in the $r_g' i_c$ term as compared to $r_g i_\epsilon$ in equation (5a) being due to the current convention (i_ϵ is current out of v_g, r_g and i_c is current into r_g', i_c'). Hence we conclude that the equations

$$v_g' = r_{21} v_g/(r_{11} + r_g) \tag{9a}$$

$$r_g' = r_{22} - r_{12} r_{21}/(r_{11} + r_g) \tag{9b}$$

describe the transistor as viewed from the load.

The power delivered by the generator into the load is the output current $i_c = v_g'/(r_g' + r_L)$ operating into r_L:

$$r_L i_c^2 = r_L[v_g'/(r_g' + r_L)]^2. \tag{10}$$

The ratio of this power to the available power of v_g, r_g is the *operating gain:*

$$\text{Operating gain} = G(r_g, r_L) = \frac{r_L v_g'^2}{(r_g' + r_L)^2} \frac{4r_g}{v_g^2}$$

$$= \frac{4 r_L r_g}{(r_g' + r_L)^2} \left[\frac{r_{21}}{r_{11} + r_g}\right]^2 = 4 r_g r_L \left[\frac{r_{21}}{\Delta}\right]^2 \tag{11}$$

where

$$\Delta = (r_{11} + r_g)(r_{22} + r_L) - r_{12} r_{21} \tag{12}$$

is the *circuit determinant.* As the notation $G(r_g, r_L)$ implies, the operating gain depends on the external input and output circuits as well as on the constants of the transistor. The problem of maximizing $G(r_g, r_L)$ by proper choices of r_g and r_L will be discussed below. The circuit determinant is an index of the stability of the circuit. If $\Delta < 0$, the circuit will spontaneously build up an oscillation even though $v_g = 0$. If $\Delta > 0$, the circuit may be stable; however, the value of Δ given in (12) is the low-frequency circuit determinant, and its behavior at high frequency is involved in determining the stability of the circuit; we shall return to this point briefly below.

Let us next consider the constant of the output generator in terms of the values of Table 2.1 The ratio of the voltage generators is, approximately,

$$\frac{v_g'}{v_g} = \frac{r_{21}}{r_{11} + r_g} = \frac{34{,}000}{530 + r_g} \doteq 30, \tag{13}$$

the approximation corresponding to $r_g = 500$ ohms, a representative value for many applications. The ratio v_g'/v_g would be the voltage gain into an infinite impedance load and would be analogous to the μ of a tube. The value of r_g' is influenced by the feedback due to r_{12} and is

$$\begin{aligned} r_g' &= r_{22}\left[1 - \frac{\alpha r_{12}}{r_{11} + r_g}\right] = r_{22}\left[1 - \frac{1.8 \times 290}{530 + r_g}\right] \\ &\doteq r_{22}[1 - 0.45] = 10{,}500 \text{ ohms}, \end{aligned} \tag{14}$$

the last two terms corresponding to $r_g = 500$ ohms. It is evident that the effect of positive feedback through the term involving αr_{12} acts to reduce the output generator impedance and thus to increase the power available from the generator v_g', r_g'. In principle the generator impedance r_g' can be reduced to zero or made negative by adding an external feedback resistance to the base. This reduction is equivalent to increasing all of the r's in the equivalent circuit by the added resistance, the contributions being important chiefly for r_{11} and r_{12}. As soon as r_g' passes through zero and becomes negative, the impedance between terminals c and b, being $r_g'r_L/(r_g' + r_L)$, will also become negative, and oscillations associated with stray capacitances and inductances may occur. This is an example of instability arising before the low-frequency Δ vanishes; Δ does not vanish until $r_g' + r_L = 0$. The problem of finding the optimum value of the feedback resistance is similar to other problems of feedback amplifier design.[4]

Using the value 10,500 for r_g' found in (14) and the circuit constants for *Grounded Base Operation* of Table 2.1 we find from (11) that

$$\text{Operating gain} = 47 \text{ or } 16.7 \text{ db.} \tag{14a}$$

Two other definitions of gain which are frequently referred to are the *available gain* and the *maximum available gain*. The available gain is found by adjusting r_L so as to maximize $G(r_g, r_L)$; this adjustment ob-

[4] H. W. Bode, *Network Analysis and Feedback Amplifier Design*, D. Van Nostrand Co., Inc., New York, 1945.

viously corresponds to setting $r_L = r_g'$ and gives

$$G(r_g, r_g') = \frac{r_g}{r_g'}\left[\frac{r_{21}}{r_{11} + r_g}\right]^2 = \text{available gain.} \tag{15}$$

For the values of circuit parameters previously used, its value is

$$\frac{500}{10{,}500}\left[\frac{34{,}000}{1030}\right]^2 = 52 \text{ or } 17.2 \text{ db.} \tag{16}$$

For the 20,000-ohm load resistor of the Table (which gives greater stability than the value 10,500 which maximizes the operating gain) the value of the operating gain is less by a factor equal to (11) ÷ (16):

$$\frac{4r_L r_g'}{(r_g' + r_L)^2} = \frac{4 \times 20{,}000 \times 10{,}500}{(30{,}500)^2} = 0.9 \tag{17}$$

a loss of only 0.5 db.

The available gain is still a function of the external circuit through r_g. The *maximum available gain* is obtained by maximizing this available gain. When this is done, it is found that r_g, as well as r_L, is matched to the transistor, and r_g and r_L have values

$$r_g = [(r_{11}r_{22} - r_{12}r_{21})r_{11}/r_{22}]^{1/2} \tag{18a}$$

$$r_L = [(r_{11}r_{22} - r_{12}r_{21})r_{22}/r_{11}]^{1/2}. \tag{18b}$$

If the positive feedback is too great, the foregoing expressions become imaginary. The critical condition is

$$r_{11}r_{22} - r_{12}r_{21} = r_{22}(r_{11} - \alpha r_{12}) = 0. \tag{19}$$

In terms of the matched generator and load the maximum available gain is given in terms of r_g and r_L of (18a) and (18b) as follows:

$$\text{maximum available gain} = \frac{r_{21}^2}{(r_{11} + r_g)(r_{22} + r_L)}. \tag{20}$$

Inserting the values from Table 1.3, we obtain

$$r_g = 76 \text{ ohms}, \quad r_L = 2750 \text{ ohms} \tag{21}$$

and a maximum available gain of 88 or 19 db. The extra 2 db of gain as compared to the situation of $r_g = 500$ ohms, $r_L = 20{,}000$ has been obtained by reducing the external resistances to such low values that the margin of stability has been reduced to a dangerous level. It is better, therefore, to sacrifice some gain for greater stability.

For purposes of discussion in later sections we shall consider the maximum gain neglecting the feedback term r_{12}. Under these conditions, the values for r_g and r_L for the maximum available gain reduce to r_{11} and r_{22}

respectively, and the formula for maximum available gain becomes

$$\text{maximum available gain} = \frac{r_{21}^2}{4r_{11}r_{22}} = \frac{\alpha^2 r_{22}}{4r_{11}} \tag{22}$$

since $r_{21} = \alpha r_{22}$. This formula indicates that large current amplification, represented by large α, and a large ratio of collector to emitter impedance both favor gain.

It should be pointed out in closing these remarks on the equivalent circuit that we have discussed only the "grounded base" application of the transistor. This mode of operation is in a general way analogous to the grounded grid operation of a vacuum tube; in fact for $\alpha = 1$, the circuit theories become very similar, at least at low frequencies. Two other arrangements are the grounded emitter, corresponding to grounded cathode, and the grounded collector, corresponding to the grounded plate or cathode-follower circuit. The equivalent circuit of the transistor is, of course, equally applicable to all of these cases. For further circuit details the reader is referred to the literature.[5]

2.2d. Noise. The type-A transistor exhibits noise having the frequency spectrum associated with *contact* or current noise. The noise power per cycle varies inversely with frequency so that the noise per octave is the same for all octaves, at least over the frequency range for which the equivalent circuit parameters are independent of frequency.

A detailed discussion of noise theory is beyond the scope of the present treatment.[6] We shall, however, discuss how the equivalent circuit may be modified to include the effect of noise. This can be done by adding two noise voltage generators of voltage $v_{n\epsilon}$ and v_{nc} into the emitter and collector branches of the equivalent circuit as shown in Figure 2.9(a).[7] Typical values for the noise generators are as follows: the root mean square noise voltages in a 1 cycle per second band at 1000 cycles per second are

$$v_{n\epsilon} = 1 \text{ microvolt} \tag{23a}$$

$$v_{nc} = 100 \text{ microvolts.} \tag{23b}$$

Since the noise power varies as $1/f$, the values of $v_{n\epsilon}$ and v_{nc} at another frequency, f, are $(1000/f)^{1/2}$ times their values at 1000 cycles.

In the grounded base amplifier circuit of Figure 2.8, the effect of $v_{n\epsilon}$ will be the same as v_g; and v_{nc} will simply combine additively with v_g'. Hence the equivalent circuit looking into the transistor from the load will be as

[5] R. M. Ryder and R. J. Kircher, *Bell Syst. Tech. J.* 28, 367–400 (1949).

[6] For a discussion of the theory of noise in crystal rectifiers, see H. C. Torrey and C. A. Whitmer, *Crystal Rectifiers*, McGraw-Hill Book Co., New York, 1948, Chapter 6.

[7] L. C. Peterson, "Space-Charge and Transit-Time Effects on Signal and Noise in Microwave Tetrodes", *Proc. IRE*, November 1947, 1264–1274.

shown in Figure 2.9(b) with v_{ng} given by the equation

$$v_{ng} = \frac{r_{21}v_{n\epsilon}}{r_{11} + r_g} + v_{nc}$$
$$= 33v_{n\epsilon} + v_{nc} \tag{24}$$

where we have used the values in Table 2.1 for the circuit constants and 500 ohms for r_g. There is evidence that the noise generators $v_{n\epsilon}$ and v_{nc} are "correlated" to a certain degree, indicating that for both of them the noise arises in part from a common source. For our example the contributions of v_{nc} and $v_{n\epsilon}$ to v_{ng} are respectively 100 and 33 microvolts. If

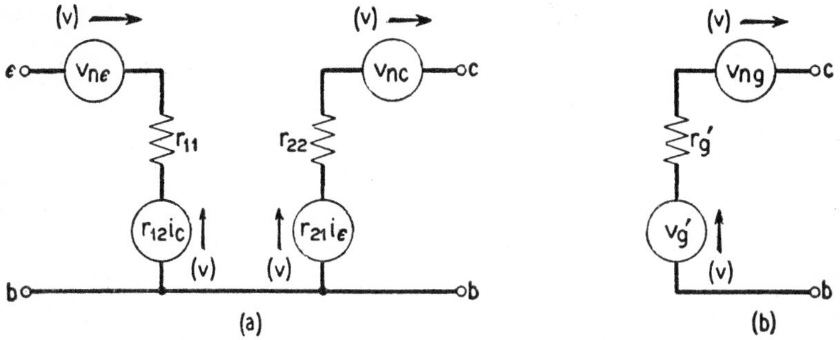

Fig. 2-9—Noise in Transistors.

(a) Noise generators in the equivalent circuit.
(b) Equivalent output circuit.

they were perfectly correlated constructively, the contribution would be 133 as compared to 67 for perfect negative correlation, a difference of 6 db in v_{ng}. The correlation coefficient C of two sources is defined for our example by the equation

$$v_{ng}^2 = (33)^2 v_{n\epsilon}^2 + v_{nc}^2 + 2 \times 33C\sqrt{v_{n\epsilon}^2 v_{nc}^2}. \tag{25}$$

Correlation coefficients ranging from -0.8 to $+0.4$ have been observed for type-A transistors.

If correlation effects are neglected, the mean square noise voltage in frequency range from f_1 to f_2 will be[8]

$$v_{ng}^2(f_1, f_2) = (1100v_{n\epsilon}^2 + v_{nc}^2)1000 \ln_e f_2/f_1. \tag{26}$$

Over the range of 1000 to 10,000 cycles this formula gives

$$v_{ng}^2(10) = (0.25 + 2.3) \times 10^{-5} \text{ volts}^2 = 2.5 \times 10^{-5} \text{ volts}^2 \tag{27}$$

[8] The contribution of v_{nc} to v_{ng}^2 can be obtained by summing the contributions from each frequency band df, thus obtaining

$$\int_{f_1}^{f_2} v_{nc}^2\,(1000/f)df = v_{nc}^2 1000 \ln_e (f_2/f_1).$$

corresponding to an r.m.s. value of 5 millivolts. The quantity (10) in $v_{ng}^2(10)$ indicates that a frequency band one decade wide is involved. It is a characteristic of the $1/f$ power spectrum that the noise power depends only on the ratio f_2/f_1 and not on the location of either frequency alone.

It is instructive to compare this with the value of v_g' when the maximum of 5 mw of undistorted power is being delivered to the r_L of 20,000 ohms. This comparison leads to a determination of v_g' as follows:

$$5 \times 10^{-3} = \frac{v_g'^2 r_L}{(r_L + r_g')^2} = \frac{v_g'^2 20{,}000}{(30{,}500)^2}$$

giving

$$v_g'^2 = 230 \text{ volts}^2 \tag{28}$$

corresponding to an r.m.s. voltage of 15 volts or about $\frac{1}{3}$ the bias voltage on the collector. The ratio of this (voltage)^2 to the (noise voltage)^2 is 9.2×10^6, or 70 db. This figure is the most significant figure of merit for repeater applications, and we shall refer to it as the *gain range decade figure*. It may also be defined as the "signal to noise ratio at maximum output for a frequency band one decade wide". For our example, output signals greater than 15 r.m.s. volts will produce distortion, and output signals less than 5 r.m.s. millivolts will be no larger than noise. Usable output signals will, therefore, lie in a range about 70 db wide, the exact value being set by the tolerance limits on signal-to-noise ratio and distortion.

The value of 70 db corresponds to a band one decade wide. For a wider band, there will be more noise power, and the *gain range figure* for the wider band will be smaller. For example, the band from 100 to 10,000 cycles per second is two decades wide and has 3 db more noise so that the gain range figure is 67 db. The gain range figure for any band may be obtained by computing 10 times the logarithm to the base 10 of the number of decades in the band and subtracting this from the *gain range decade figure*. As another example consider the band from 5 to 6 mc. This contains $\log \frac{6}{5} = \log 1.2 = 0.080$ decades. The gain range for this band would thus be $70 - 10 \log 0.080 = 70 - 10(-1.1) = 81$ db.

The *gain range figure* is obviously closely related to the permissible attenuation between repeaters. If the attenuation exceeds the gain range figure, then either the output of one repeater is overloaded or the first stage of the next repeater will have a signal output smaller than noise, it being assumed that the problems of impedance matching have been solved. In an actual design problem the considerations would include impedance matching, tolerances for distortion and signal-to-noise ratio, and the total number of repeaters in tandem.

Another quantity frequently used to describe noise is the *noise figure.* This is defined in terms of an amplifying circuit, like that of Figure 2.8, for example, as follows: If the only source of noise were the thermal or Johnson noise in the external generator resistance, which we represent by r_g, v_{nT} (T for thermal), then a certain apparent noise generator v_{nT}' would appear in the output. The ratio of $(v_{ng})^2$ actually observed in the output to $(v_{nT}')^2$ is the *noise figure.* The value of the noise figure depends on the particular circuit constants used in the generator and has a minimum value for a certain choice. We shall not give the theory of making this choice but will consider the general expression for the noise figure and then compute it for the previously discussed case in which $r_g = 500$ ohms. The formula for the thermal noise voltage at room temperature ($kT/e \doteq 0.025$ volt) is as follows:

$$\begin{aligned} v_{nT}^2 &= 4r_gkT\Delta f = r_g(4kT/e)(e\Delta f) \\ &= r_g(0.1 \text{ volt})(1.6 \times 10^{-19}\Delta f \text{ amp}) \\ &= 1.6 \times 10^{-20} r_g \text{ (ohms) } \Delta f \text{ (sec}^{-1}\text{) volts}^2. \end{aligned} \tag{29}$$

Combining this with the expression for voltage gain and with the formula for the output generator due to $v_{n\epsilon}$ and v_{nc} leads to the expression for the total output noise generator:

$$\begin{aligned} v_{ng}^2(f_1, f_2) &= \left[\frac{r_{21}}{r_{11} + r_g}\right]^2 4r_gkT(f_2 - f_1) \\ &\quad + \left[\left[\frac{r_{21}}{r_{11} + r_g}\right]^2 v_{n\epsilon}^2 \oplus v_{nc}^2\right] 1000 \ln_e (f_2/f_1) \end{aligned} \tag{30}$$

and to a noise figure

$$F = 1 + \frac{1}{4r_gkT(f_2 - f_1)}\left[v_{n\epsilon}^2 \oplus \left[\frac{r_{11} + r_g}{r_{21}}\right]^2 v_{nc}^2\right] 1000 \ln_e (f_2/f_1). \tag{31}$$

where the $\oplus$ sign indicates that correlation is to be taken into account in combining the terms. Rather than substituting directly in the formula, we shall deal with the case of $f_1 = 1000$ and $f_2 = 10{,}000$ cycles/sec discussed earlier using $r_g = 500$ ohms. For this band, v_{nT}^2 is $1.6 \times 10^{-20} \times 500 \times 9{,}000$ volts^2 and contributes 33^2 times this amount to v_{ng}^2 or 8×10^{-11} volts^2. The value 2.5×10^{-5} volts^2 due to the transistor is larger than this by a factor of 3×10^5 or 55 db and (because of the large ratio) this is also the noise figure.

A noise figure of 55 db for the 1000 to 10,000 cycle band is much worse than that of a good electron tube, which can come close to 0 db. In view of the frequency dependence which brings the transistor noise figure down to 30 db at a megacycle, the comparison at video frequencies is less unfavorable, particularly if some developmental improvement can be made.

The noise figure depends upon the circuit used as well as on all the circuit parameters. For further information, the reader is referred to the literature.[9]

The noise figure is evidently an index of the amount by which the transistor falls short of being a noiseless amplifier. It is evident that no appreciable advantage will be achieved by lowering the noise added by the transistor lower than say 3 db below that contributed by the generator. Since the noise generator in the equivalent circuit is much larger than thermal noise in the corresponding resistive components of the equivalent circuit (and also much larger than shot noise), there appears to be no basic physical principle which will prevent lowering the noise in transistors greatly.

Problems

1. Consider the circuit in which the emitter of the transistor is grounded and the input signal is applied to the base from an input generator r_g, v_g and the output is delivered to a load r_L. Show that the characteristics of this amplifier are as follows:

Circuit determinant:

$$\Delta = (r_g + r_b + r_\epsilon)(r_L + r_\epsilon + r_c - r_m) + r_\epsilon(r_m - r_\epsilon)$$
$$> 0 \text{ for stability}$$

$$\text{Input impedance} = r_{\text{in}} = r_b + r_\epsilon + \frac{r_\epsilon(r_m - r_\epsilon)}{r_L + r_\epsilon + r_c - r_m}$$

$$\text{Output impedance} = r_{\text{out}} = r_c + r_\epsilon - r_m + \frac{r_\epsilon(r_m - r_\epsilon)}{r_g + r_b + r_\epsilon}$$

$$\text{Operating gain} = G_F = 4r_g r_L[(r_m - r_\epsilon)/\Delta]^2$$
$$\text{Backward operating gain} = G_R = 4r_g r_L(r_\epsilon/\Delta)^2$$

Show that for the transistor of Table 2.1, $r_g = 500$ ohms and $r_L = 20{,}000$ ohms that $r_{\text{in}} = 2100$ ohms, $r_{\text{out}} = -6900$ ohms, $G_F = 24$ db, $G_R = -19$ db.

2. Consider the grounded collector amplifier as for problem 1 and show that the equations determining its behavior are as follows:

Circuit determinant:

$$\Delta = (r_g + r_b + r_c)(r_L + r_\epsilon + r_c - r_m) + r_c(r_m - r_c)$$
$$> 0 \text{ for stability}$$

$$\text{Input impedance} = r_{\text{in}} = r_b + r_c + \frac{r_c(r_m - r_c)}{r_L + r_\epsilon + r_c - r_m}$$

$$\text{Output impedance} = r_{\text{out}} = r_\epsilon + r_c - r_m + \frac{r_c(r_m - r_c)}{r_g + r_b + r_c}$$

$$\text{Operating gain (forward)} = G_F = 4r_g r_L(r_c/\Delta)^2$$
$$\text{Backward operating gain} = G_R = 4r_g r_L[(r_m - r_c)/\Delta]^2 = (1 - \alpha)^2 G_F$$

[9] See R. M. Ryder and R. J. Kircher, *Bell Syst. Tech. J.* **28**, 367–400 (1949). Formulae for several cases are quoted.

Show that for the transistor of Table 2.1, $r_g = 20{,}000$ ohms, and $r_L = 10{,}000$ ohms that $r_{\text{in}} = -41{,}000$ ohms, $r_{\text{out}} = -7500$ ohms, $G_F = 15$ db, $G_R = 13$ db.

Remark: The following problems are intended to give the reader some familiarity with the magnitudes of voltages and currents and with the non-linear behavior of the transistor of Figure 2.6.

3. Construct a plot of emitter voltage versus emitter current for $V_c = -10$, -20 and -30 volts with the base grounded. (*Hint:* For fixed values of V_c read pairs of values of I_ϵ and I_c from Figure 2.6 and plot in the r_{11} subplot.) Verify that near $I_\epsilon = +1.0$ ma the input resistance is about 80 ohms.

4. Construct a plot, similar to problem 3, of V_ϵ versus I_ϵ but assuming that V_c is biased from a -60 volt battery through 20,000 ohms. (*Hint:* Draw a load line on the r_{22} subplot and proceed as for problem 3.) Verify that for I_ϵ in the range 0.5 to 1.5 ma the input impedance is about 300 ohms. (This problem shows that adding resistance in the collector circuit increases the input impedance of the emitter.)

5. This problem is intended to show how adding resistance between base and ground can lead to a reduction in input impedance from positive to negative values. For this purpose consider the $V_c = -20$ volt curve of problem 3. Verify that as I_ϵ increases from 0.5 to 1.0 ma, I_c changes from -1.8 to -2.7 ma so that $I_b = -(I_\epsilon + I_c)$ changes from $+1.3$ to $+1.7$ ma. Next suppose that a resistance is inserted between the base and ground and let V_b' equal the voltage between base and ground and $V_\epsilon' = V_\epsilon + V_b'$ and $V_c' = V_c + V_b'$ be corresponding emitter and collector voltages above ground. Conclude that if the inserted resistance is 500 ohms, the increase of I_ϵ from 0.5 to 1.0 ma will produce a negative change in V_b' given by $500 \times (1.7 - 1.3) \times 10^{-3} = 0.2$ volts and hence that V_ϵ' will decrease so that the input impedance is negative and equal to about -300 ohms. (The small correction due to the effect of V_b' on V_c is negligible in this example.)

6. Consider the V_ϵ versus I_ϵ plot for a circuit in which the collector is biased at -40 volts and the base is connected to ground through 10,000 ohms. For $I_\epsilon = 0$, show that $I_c \doteq -1.0$ ma and $V_b' \doteq -10$ volts and $V_\epsilon' = V_b' + V_\epsilon \doteq -10 + (-0.3) = -10.3$ volts. For negative emitter currents, large negative emitter voltages will result. The other operating points are found by the following graphical construction on the r_{21} plot: For each value of I_ϵ, the equation

$$-40 \text{ volts} = 10{,}000(I_\epsilon + I_c) + V_c(I_\epsilon, I_c),$$

must hold; this equation states that the battery voltage must equal the sum of the drop across the resistance plus the drop between base and collector. The equation may be rewritten as

$$V_c\,(I_\epsilon,I_c) = [-40 \text{ volts} - 10{,}000I_c] - 10{,}000I_\epsilon.$$

For a fixed value of I_c, the left side is simply one of the parametric curves of the r_{21} plot. For the same value of I_c, the right side is simply a straight line of slope $-10{,}000$ ohms on the r_{21} plot. Thus for each value of I_c, the solution is given by the point of intersection of the straight line with the parametric curve. Verify that the following points will be obtained.

I_c	I_ϵ	I_b	V_c	V_b'	V_ϵ'
-1	0	1 ma	-30	-10	-10.3 volts
-2	0.5	1.5 ma	-25	-15	-14.9 volts
-3	$+1.2$	1.8 ma	-22	-18	-17.8 volts

The small difference between V_b' and V_ϵ' is due to V_ϵ. Estimate that the negative resistance of V_ϵ' versus I_ϵ will continue with a slope of about -6000 ohms until V_b' approaches -35 volts, corresponding to the bunched curves near the top of the r_{21} plot extrapolated to about $I_\epsilon = 3.5$ ma and $I_c \doteq -7$ ma. For higher values of I_ϵ, I_c will increase only slowly, V_c will change little and a positive resistance will be produced at the input. This shows that the V_ϵ', I_ϵ plot is an N-shaped curve so that for a range of fixed emitter biases there will be two stable solutions and one unstable solution. This feature can be used to make flip flop circuits by biasing the emitter to a negative voltage through a suitable resistor. (This treatment is due to B. G. Farley.)

7. (a) A grounded base transistor amplifier is coupled through a transformer to a generator v_g of internal resistance r_g. Show that if emitter noise is neglected, the greatest signal to noise ratio in the load is obtained when the transformer impedance ratio is $r_g : r_{11}$.

(b) Suppose that the generator is a typical moving coil microphone for which $r_g = 22$ ohms and that when the microphone is picking up speech from a distance of 3 feet, $v = 5 \times 10^{-5}$ volt. Show that when proper matching is used the signal to noise ratio in the load is approximately 0.6 db. Assume a band extending from 50 to 10,000 cps.

(c) Show by a qualitative argument that the final signal to noise ratio is not much different when the amplifier contains several stages provided the gain per stage is of the order of 20 db.

8. In view of the results of problem 7, how much gain can be built into a transistor amplifier designed for the frequency range 50–10,000 cps if it is required that the noise at the output must be 40 db below the maximum signal output? Assume that max. signal output is 10 milliwatts.

Ans. Approximately 46 db.

9. Consider a transistor operated as a grounded base amplifier with a load of 10,000 ohms in series with a 40 volt battery.

(a) Using the curves of Figure 2.6, find the maximum peak-to-peak output current which is comparatively free from distortion.

(b) Assuming that the input (emitter) current is sinusoidal, what is the d.c. emitter bias?

(c) If the a.c. component of the input is removed, what is the power dissipation in the collector? What is the power drawn from the battery?

(d) Let the a.c. component of the input be restored but in such a way that, unlike (b), the *output* current is *sinusoidal* with the maximum peak to peak amplitude. What is the a.c. power delivered to the load?

(e) Under the conditions in (d), what are the collector dissipation and the power drawn from the battery?

Ans. (a), 2.2 ma.; (b), .75 ma.; (c), 39.6 mw.; 88 mw.; (d), 6.1 mw.; (e), 33.9 mw., 80 mw.

10. Show by procedures similar to those of problem 7 that the signal to noise ratio for optimum matching is not affected by connecting several similar transistors in parallel; in other words, the noise figure of a matched transistor amplifier is unchanged by paralleling similar transistors.

CHAPTER 3

QUANTITATIVE STUDIES OF INJECTION OF HOLES AND ELECTRONS

3.1 CARRIER INJECTION IN GERMANIUM*

3.1a. Introduction. The invention of the transistor[1,2,3] has given great stimulus to research on the interaction of holes and electrons in semiconductors. The techniques discussed in this section for investigating the behavior of holes in *n*-type germanium were devised in part to aid in analyzing the emitter current in transistors. The early experiments of Bardeen and Brattain suggested that the hole flow from the emitter to the collector took place in a surface layer.[1,2] The possibility that transistors could also be produced by hole flow directly through *n*-type material was proposed in connection with the *p-n-p* transistor.[4] Quite independently, J. N. Shive[5] obtained evidence for hole flow through the body of *n*-type germanium by making a transistor with points on opposite sides of a thin germanium specimen. Such hole flow is also involved in the coaxial transistor of W. E. Kock and R. L. Wallace.[6] Further evidence for hole injection into the body of *n*-type germanium under conditions of high fields was obtained by E. J. Ryder.[7]

In keeping with these facts it is concluded[3] that with two points close together on a plane surface, as in the type-A transistor,[8] holes may flow either in a surface layer or through the body of the germanium. For surface flow to be large, special surface treatments appear to be necessary; such treatments were not employed in the experiments described in this section and the results are consistent with the interpretation that the hole current from the emitter point flows in the interior.

* This section appeared in slightly modified form as a part of a paper by W. Shockley, G. L. Pearson, and J. R. Haynes in the *Bell Syst. Tech. J.*, **28**, 344–365 (1949).

[1] J. Bardeen and W. H. Brattain, *Phys. Rev.* **74**, 230–231 (1948).

[2] W. H. Brattain and J. Bardeen, *Phys. Rev.* **74**, 231–232 (1948).

[3] J. Bardeen and W. H. Brattain, *Phys. Rev.* **75**, 1208–1225 (1949).

[4] W. Shockley, *Bell Syst. Tech. J.*, **28**, 453–489 (1949).

[5] J. N. Shive, *Phys. Rev.* **75**, 689–690 (1949).

[6] W. E. Kock and R. L. Wallace, *Electrical Engineering*, **68**, 222–223 (1949).

[7] E. J. Ryder and W. Shockley, *Phys. Rev.* **75**, 310 (1949).

[8] For reviews of the type-A transistor see Bardeen and Brattain, *Phys. Rev.* **75**, 1208–1225 (1949); R. M. Ryder, *Bell Laboratories Record*, **27**, 89–93 (March, 1949); J. A. Becker and J. N. Shive, *Electrical Engineering*, **68**, 215–221 (1949); and R. M. Ryder, *Bell Syst. Tech. J.*, **28**, 367–400 (1949).

The experiments described in this section, in addition to any practical implications, serve to put the action of emitter points on a quantitative basis and to open up a new area of research on conduction processes in semiconductors. It is worth while at the outset to contrast some of the new aspects of these experiments with the earlier experimental status of the bulk properties of semiconductors. Prior to the invention of the transistor, inferences about the behaviors of holes and electrons were made from measurements of conductivity and Hall effect. For both of these effects, under essentially steady state conditions, measurements were made of such quantities as lengths, currents, voltages, and magnetic fields. The measurement of time was not involved, except indirectly in the calibration of the instruments. Nevertheless, on the basis of these data, definite mental pictures were formed of the motions of holes and electrons describing in particular their drift velocity in electric fields and the transverse thrust exerted upon them by magnetic fields.[9] The new experiments show that something actually does drift in the semiconductor with the predicted drift velocity and does behave as though it had a plus or minus charge, just as expected for holes and electrons. In addition, experiments described in Section 3.2 show that the effect of sidewise thrust by a magnetic field actually is observed in terms of the concentration of holes and electrons near one side of a filament of germanium.

We shall discuss here evidence that holes are actually introduced into n-type germanium by the forward current of an emitter point and show how the numbers and lifetimes of the holes can be inferred from the data. We shall refer to this important process as *hole injection*. Discussions of the reasons why an emitter should inject holes have been given for metal-semiconductor point contacts by Bardeen and Brattain and for p-n junctions by Shockley. These theories are discussed in subsequent sections. There are other possible ways in which semiconductor amplifiers can be made without the use of hole injection into n-type material or electron injection into p-type material.[10] In this chapter, however, our remarks will be restricted to semiconductors which have only one type of carrier present in appreciable proportions under conditions of thermal equilibrium; for such cases the theoretical considerations are simplified and are apparently in good agreement with the experiments.

3.1b. Measurement of Density and Current of Injected Carriers. The experiment in its semiquantitative form is relatively simple and is shown

[9] In the sense of Chapter 14, these pictures amount to additional physical concepts which enter the theory but drop out again before Hall effect and conductivity predictions are made. The new experiments bear on other aspects of the physical concepts and give an operational reality to the pictures associated with drift velocity and Hall effect.

[10] For example see J. Bardeen and W. H. Brattain, *Phys. Rev.* **74**, 230–231 (1948), and W. Shockley and G. L. Pearson, *Phys. Rev.* **74**, 232–233 (1948).

in Figure 3.1.[11] A rod of n-type germanium is subjected to a longitudinal electric field E applied by a battery B_1. Collector and emitter point contacts are made to the germanium with the aid of a micromanipulator. The collector point is biased like a collector in a type-A transistor by the battery B_2, and the signal obtained across the load resistor R is applied to

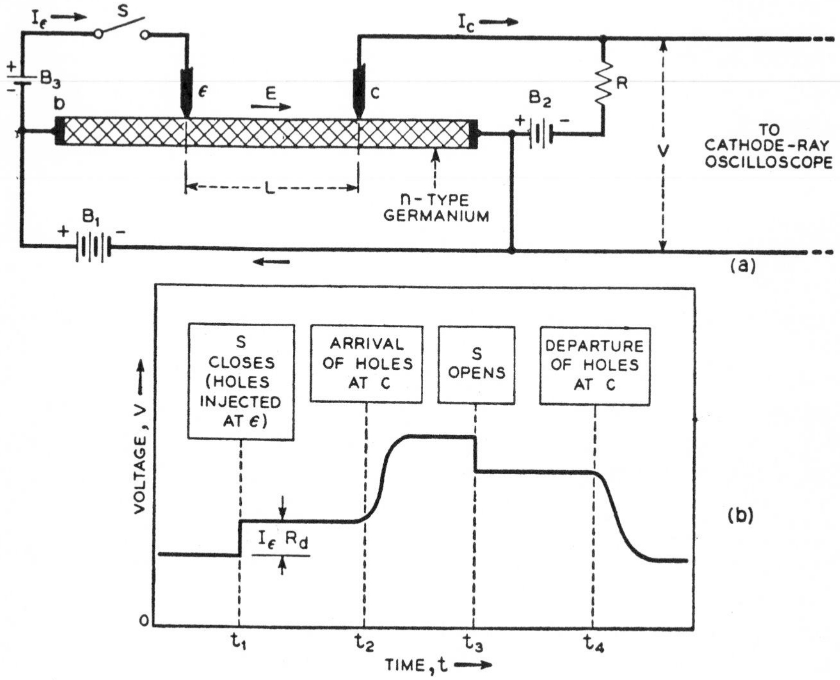

Fig. 3-1—Experiment to Investigate the Behavior of Holes Injected into n-Type Germanium.

(a) Experimental arrangement.
(b) Signal on oscilloscope showing delay in hole arrival at t_2 in respect to closing S at t_1 and delay in hole departure at t_4 in respect to opening S at t_3.

the input of an oscilloscope. At time t_1 the switch in the emitter circuit is closed so that a forward current, produced by the battery B_3, flows through the emitter point. At t_2 the switch is opened. The voltage wave at the collector, as observed on the oscilloscope, has the wave form shown in part (b) of the figure.

These data are interpreted as follows: When the emitter circuit is closed, the electrons in the emitter wire start to flow away from the germanium (that is, positive current flows into the germanium). These electrons are furnished by an electron flow in the germanium towards the point of con-

[11] Experiments of this sort were first reported by J. R. Haynes and W. Shockley, *Phys. Rev.* **75**, 691 (1949).

tact. The flow in the germanium may be either by the excess electron process or by the hole process. In Figure 3.2 we illustrate these two possibilities. At first glance it might appear that the difference between these two processes is unimportant since the net result in both cases is a transfer of electrons from the germanium to the emitter point. There is, however, an important difference, one which makes several forms of transistor action possible. In the case of the hole process an electron is transferred from the

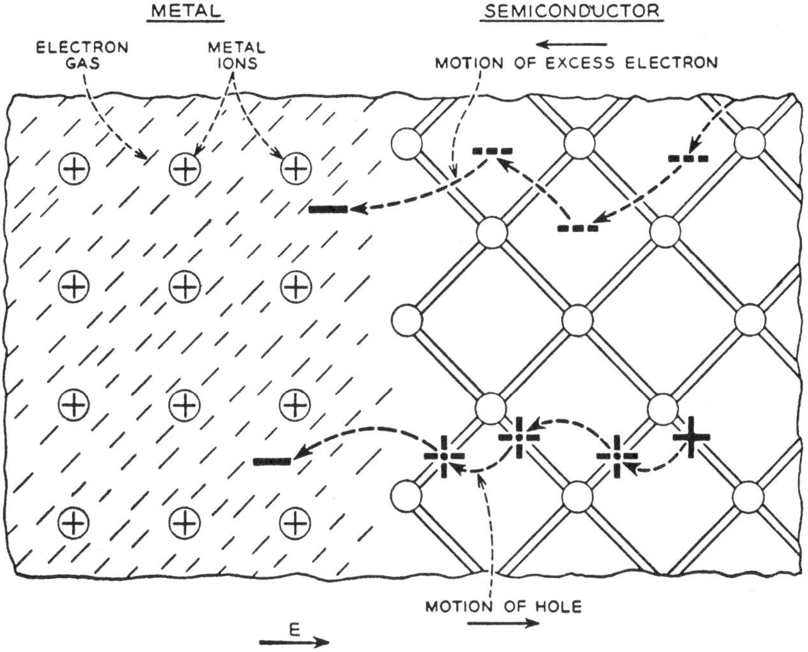

FIG. 3-2—Comparison between Electron Flow to and Hole Flow from a Point Contact.

valence band structure to the metal. After this the hole moves deeper into the germanium. As a result the electronic structure of the germanium is modified in the neighborhood of the emitter point by the presence of the injected holes.

Under the influence of the electric field E, the injected holes drift toward the collector point with velocity $\mu_p E$, where μ_p is the mobility of a hole, and thus traverse the distance L to the collector point in a time $L/\mu_p E$. When they arrive at the collector point, they increase its reverse current and produce the signal shown at t_2.

There are two important differences between the signal produced at t_2 and that produced at t_1. The signal at t_1, which is in a sense a pickup signal, would be produced even if no hole injection occurred. We shall illustrate this by considering the case of a piece of ohmic material substituted for the germanium. Conventional circuit theory applies to such a case; however, in order to contrast this purely ohmic case with that of hole injection, we shall also give a description of the conventional theory of signal transmission in terms of the motion of the carriers. According to conventional circuit theory, the addition of the current I_ϵ would simply produce an added IR drop due to current flow in the segment of the specimen to the right of the collector. This voltage drop is denoted as $I_\epsilon R_d$ in part (b), R_d representing the proper combination of resistances to take into account the way in which I_ϵ divides in the two branches. This signal will be transmitted from the emitter to the collector with practically the speed of light—the ordinary theory of signal transmission along a conductor being applicable to it. This high speed of transmission does not, of course, imply a correspondingly high velocity of motion of the current carriers. In fact, the rapidity of signal transmission has nothing to do with the speed of the carriers and comes about as follows: If the ohmic material is an electronic conductor, then the withdrawal of a few electrons by the emitter current produces a local positive charge. This positive charge produces an electric field which progresses with the speed of light and exerts a force on adjacent electrons so that they move in to neutralize the space charge. The net result is that electrons in all parts of the specimen start to drift practically instantaneously. They flow into the specimen from the end terminals to replace the electrons flowing out at the emitter point and no appreciable change in density of electrons occurs anywhere within the specimen.[12]

The distinction between the process just described and that occurring when holes are injected into germanium is of great importance in understanding many effects connected with transistor action. One way of summarizing the situation is as follows: In a sample having carriers of one type only, electrons for example, it is impossible to alter the density of carriers by trying to inject or extract carriers of the same type. The reason is, as described above (or proved in the footnote), that such changes would be accompanied by an unbalanced space charge in the sample and such an unbalance is self-annihilating and does not occur.[13]

When holes are injected into n-type germanium, they also tend to set up a space charge. Once more this space charge is quickly neutralized by an electron flow. In this case, however, the neutralized state is not the normal thermal equilibrium state. Instead the number of current carriers present

[12] This is a description in words of the result ordinarily expressed in terms of the dielectric relaxation time obtained as follows: $\nabla \cdot I = -\dot{\rho}$, $I = \sigma E = -\sigma \nabla \Psi$, $\nabla^2 \Psi = -4\pi\rho/\kappa = \dot{\rho}/\sigma$ so that $\rho = \rho_0 \exp[-(4\pi\sigma/\kappa)t]$, where I = current density, ρ = charge density, σ = conductivity, E = electric field, Ψ = electrostatic potential, κ = dielectric constant. The quantity $\kappa/4\pi\sigma$ is the dielectric relaxation time.

[13] In the case of modulation of conductivity by surface charges discussed in Section 2.1, a net charge is produced by the field from the capacitor plate. The changed charge density extends slightly into the specimen but should not be confused with the true volume effect of hole injection. Such space charge layers are discussed briefly in Section 4.3.

has been increased by the injected holes and by an equal number of electrons drawn in to neutralize the holes. The total number of electrons in the specimen will thus be increased, the extra electrons coming in from the metal terminals which complete the circuit with the emitter point. The presence of the holes and the neutralizing electrons near the emitter point modify the conductivity. As we shall show below, this modification of conductivity may be so great that it can be used to measure hole densities and also to give power gain in modified forms of the transistor. We shall summarize this situation as follows: *In a semiconductor containing substantially only one type of current carrier, it is impossible to increase the total carrier concentration by injecting carriers of the same type; however, such increases can be produced by injecting the opposite type since the space charge of the latter can be neutralized by an increased concentration of the type normally present.*

Thus we conclude that *the existence of two processes of electronic conduction in semiconductors, corresponding respectively to positive and negative mobile charges, is a major feature in several forms of transistor action.*

In a vacuum-tube triode there are also two ways in which electrons carry current: (1) ordinary metallic conduction in the leads and (2) space flow produced by thermionic emission. In Section 4.2 we shall show, by comparing a *p-n-p* transistor with a vacuum-tube triode, that there is a close analogy in the two cases between the ways in which the interaction of one form of current flow with the other produces amplification.

In terms of the description given, the experiment of Figure 3.1 is readily interpreted. The instantaneous rise at t_1 is simply the ohmic contribution due to the changing total currents in the right branch when the emitter current starts to flow. After this, there is a time lag until the holes injected into the germanium drift down the specimen and arrive at the collector. When the current is turned off at t_3, a similar sequence of events occurs.

The measured values of the time lag of $t_1 - t_2$, the field E, and the distance L can be used to determine the mobility of the holes. The fact that holes, rather than electrons, are involved is at once evident from the polarity of the effect; the disturbance produced by the emitter point flows in the direction of E, as if it were due to positive charges; if the electric field is reversed, the signal produced at t_3 is entirely lacking. The values obtained by this means are found to be in good agreement with those predicted from the Hall effect and conductivity data. The Hall mobility values obtained on single crystal filaments of n- and p-type germanium[14] are

$$\mu_p = 1700 \text{ cm/sec per volt/cm} \tag{1}$$

$$\mu_n = 2600 \text{ cm/sec per volt/cm.} \tag{2}$$

The agreement between Hall effect mobility and drift mobility, as was pointed out at the beginning of this section, is a very gratifying confirmation

[14] G. L. Pearson, *Phys. Rev.* **76**, 179 (1949).

of the general theoretical picture of holes drifting in the direction of the electric field.[15]

We shall next consider a more quantitative embodiment of the experiment just considered. In Figure 3.3, we show the experimental arrangement. In this case it is essential in order to obtain large effects that the cross section of the germanium filament be small. A thin piece of germanium is cemented to a glass backing plate and is then ground to the desired thickness. After this the undesired portions are removed by sandblasting while the desired portions are protected by suitable jigs consisting of wires, Scotch tape, metal plates, etc. After the sandblasting, the surface of the germanium is etched. In this way specimens smaller than 0.01 × 0.01 cm in cross section have been produced. The ends of the filament are usually made very wide so as to simplify the problem of making contacts.

Under experimental conditions, a battery like B_1, of Figure 3.1, applies a "sweeping" field in the filament so that any holes injected by the emitter current are swept along the filament from left to right. In the small filaments used for these experiments, the resulting concentration of holes is so high that large changes in conductivity are produced to the right of the emitter point and, as we shall describe below, these changes can be measured and the results used to determine the hole current at the emitter point. In order to treat this situation quantitatively, we introduce a quantity γ defined as follows:

$$\gamma = \textit{the fraction of the emitter current carried by holes.} \qquad (3)$$

Accordingly, a current γI_ϵ of holes flows to the right from ϵ and produces a hole density, denoted by p, which is neutralized by an equal added electron density. A fraction $(1 - \gamma)I_\epsilon$ of electrons flows to the left; these electrons do not, however, produce any increased electron density to the left of the emitter since they are of the sign normally present in the n-type material. The presence of the holes to the right in the filament increases the conductivity σ [as shown in Figure 3.3(c)] both because of their own presence and the presence of the added electrons drawn in to neutralize the space charge of the holes. The mobility of electrons is greater than the mobility of holes, the ratio being[16]

$$b = \mu_n/\mu_p = 1.5 \text{ for germanium} \qquad (4)$$

and the electrons are always more numerous than the holes[17]

$$n = n_0 + p, \qquad (5)^{18}$$

[15] See Chapter 12 for a review of the most recent data.

[16] G. L. Pearson, *Phys. Rev.* **76**, 179 (1949). See Chapter 12 for the most recent data.

[17] The notation used in the equations is as follows: n, p, n_0 = respectively density of electrons, of holes, of electrons when no holes are injected. N_d and N_a are the densities of donors and acceptors, assumed ionized so that $n_0 = N_d - N_a$. I_ϵ, I_b, I_c are as shown on Figure 3.3. (I_c used for the probe collector in Figures 3.1 and 3.9 does not enter the equations.)

[18] See Problems at the end of this chapter.

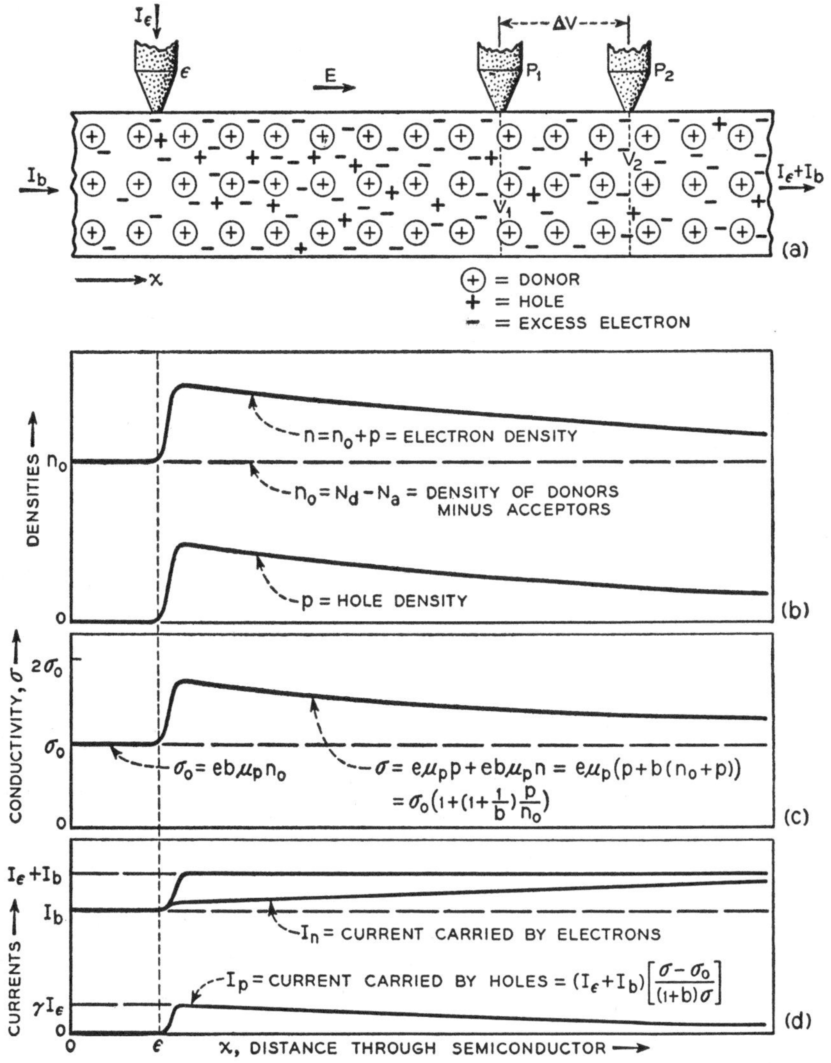

FIG. 3-3—Method of Measuring Hole Densities and Hole Currents.

(a) Distribution of holes, electrons, and donors. Acceptors, which may be present, are omitted for simplicity, the excess of donor density N_d over acceptor density N_a being n_0.

(b) To the right of the emitter the added hole density p is compensated by an equal increase in electron density.

(c) The conductivity is the sum of hole and electron conductivities.

(d) The total current $I_b + I_\epsilon$ to the right of the emitter is carried by I_p and I_n in the ratio of the hole to the electron conductivity.

where n_0 is the concentration of electrons which would be present to neutralize the donors if p were equal to zero; consequently, the current carried by electrons is greater than the current carried by holes. The concentration of holes diminishes to the right due to the fact that holes may recombine with electrons as they flow along the filament.

From this experiment the value of γ and the lifetime of a hole in the filament can be determined. The measurements are made with the aid of the two probe points P_1 and P_2, which draw no current and serve simply to measure the potential[19] on the two equi-potential surfaces V_1 and V_2. The conductance of the segment of filament between these equi-potentials is obtained by measuring the voltage difference and dividing it into the current $I_b + I_\epsilon$. The conductance of the filament between these points is obtained by measuring the voltage difference ΔV and dividing it into the current $I_b + I_\epsilon$. The necessary formulae for calculating hole density and hole current, shown on the figure, are derived as follows:

$$\text{Normal conductivity } \sigma_0 = e\mu_n n_0, \tag{6}$$

$$\begin{aligned}\text{conductivity with holes present } \sigma &= e\mu_n n + e\mu_p p \\ &= e\mu_n(n_0 + p) + e\mu_p p = \sigma_0[1 + (1 + b^{-1})(p/n_0)].\end{aligned} \tag{7}$$

This relationship is plotted in part (a) of Figure 3.4 as a function of p/n_0. The conductance,

$$G = (I_\epsilon + I_b)/\Delta V, \tag{8}$$

between P_1 and P_2 is proportional to the local conductivity, and hence to

$$1 + (1 + b^{-1})(p/n_0), \tag{9}$$

so that a measurement of the conductance gives a measurement of p/n_0. Letting G and G_0 be the conductances between the points with and without hole injection, we have

$$\frac{G}{G_0} = \frac{\sigma}{\sigma_0} = 1 + (1 + b^{-1})(p/n_0) \tag{10}$$

or

$$\frac{p}{n_0} = \frac{\sigma - \sigma_0}{\sigma_0(1 + b^{-1})} = \frac{(G/G_0) - 1}{1 + b^{-1}}. \tag{11}$$

The ratio of the hole current to the total current is, of course, simply the ratio of the hole conductance to total conductance. This ratio, $I_p/(I_n + I_p)$,

[19] When the hole and electron densities in a semiconductor do not have their equilibrium values, the potential measured by a point contact will depend on the nature of the contact, the situation being remotely analogous to that corresponding to thermoelectric potentials. These internal contact potential differences are usually so small that they can be neglected for experiments of the sort described here. For a further discussion of the theory see W. Shockley, *Bell Syst. Tech. J.* **28**, 435–489 (1949).

is plotted in part (b) of Figure 3.4. The value of I_p may then be determined graphically as follows: From a measurement of G/G_0, p/n_0 is determined from part (a); from p/n_0, the ratio $I_p/(I_n + I_p)$ is then found from (b). Since $I_n + I_p$ is known, this determines I_p. The algebraic relationship corresponding to the procedure just outlined is as follows:

$$\frac{I_p}{I_n + I_p} = \frac{e\mu_p p}{e\mu_n n + e\mu_p p} = \frac{p}{bn_0 + (1 + b)p}$$

$$= \frac{p/n_0}{b[1 + (1 + b^{-1})(p/n_0)]} = \frac{1 - (G_0/G)}{1 + b}. \qquad (12)$$

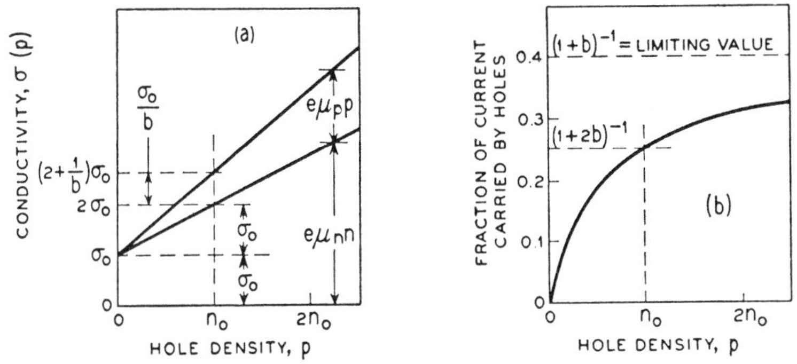

FIG. 3–4—Dependence of Conductivity and Fraction of Current Carried by Holes Upon Hole Concentration. (Drawn for $b = 1.5$.)

Hence from the measured values of G, it is possible to obtain the fraction of the current carried by holes. The hole current flowing past the probe points can be obtained by multiplying this expression by $I_\epsilon + I_b = I_n + I_p$; this leads to the formula

$$I_p = \frac{(I_\epsilon + I_b)[1 - (G_0/G)]}{1 + b} = \frac{I_\epsilon + I_b - G_0\Delta V}{1 + b} \qquad (13)$$

where ΔV is given by (8). This last expression can be given the following simple interpretation. $G_0\Delta V$ is the current carried by the normal electron density n_0. The numerator of (13) is thus the extra current carried by injected holes and added electrons, and of this extra current a fraction $1/(1 + b)$ is carried by holes so that the hole current is given by (13).[20]

If there were no decay, the current past the probe points would be γI_ϵ and, since I_ϵ is known, γ could be easily determined. Actually, however,

[20] In these calculations the formula $n = p + n_0$, corresponding to completely ionized donors and acceptors, has been used. In germanium this is a good approximation. For silicon, however, modifications will be necessary.

there may be quite an appreciable decay. However, if the current I_b is increased, the holes will be swept more rapidly from the emitter to the probes and less decay will result. Thus by increasing I_b, the effect of

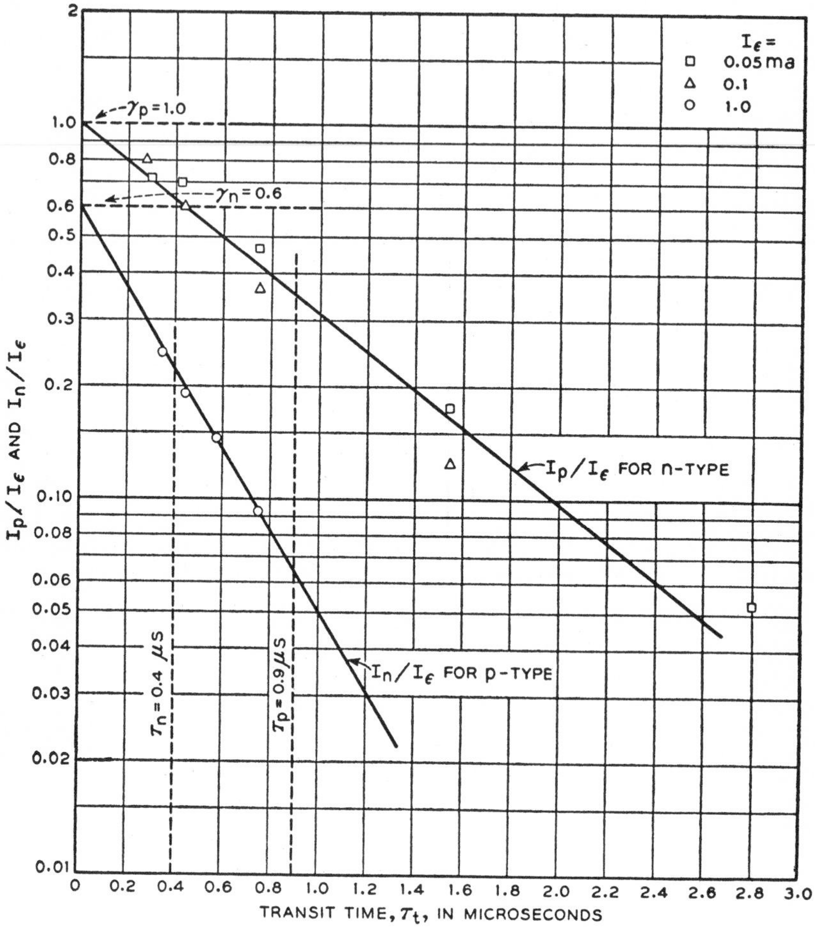

Fig. 3–5—Extrapolation of Measured Hole and Electron Currents to Zero Transit Time in Order to Determine γ.

recombination can be minimized and the value of hole current can be extrapolated to the value it would have in the absence of decay. This value is, of course, γI_ϵ.

In Figure 3.5 we show some plots of this sort. The ordinate is I_p/I_ϵ which should approach γ as the value of I_b becomes larger. The theory indicates that a logarithmic plot should be used and that the abscissa should be made proportional to transit time so that the case of no decay or zero transit time comes at the left edge.[21] The conclusion reached from this plot is that, for the case of the n-type sample, the value of γ is substantially unity, all the emitter current is holes. For the opposite case in which electrons are injected into p-type material,[22] the corresponding value of I_n/I_ϵ extrapolates to 0.6 indicating that for this case 60 per cent of the current is carried by electrons and 40 per cent by holes. For these particular specimens the lifetimes are found to be 0.9 and 0.41 microsecond respectively. There is a body of evidence, some of which we discuss below, that holes combine with electrons chiefly on the surface of the filament.

3.1c. Influence of Hole Density on Point Contacts. The presence of holes near a collector point causes an increase in its reverse current; in fact the amplification in a type-A transistor is due to the modulation of the collector current by the holes in the emitter current. The influence of hole density upon collector current has been studied in connection with experiments similar to those of Figure 3.3. After the hole current and the hole density are measured, a reverse bias of 20 to 40 volts is applied. The reverse current is found to be a linear function of the hole density. Figure 3.6 shows typical plots of such data. Different collector points, as shown, have quite different resistances. However, once data like that of Figure 3.6 have been obtained for a given point, the currents can then be used as a measure of hole density. This experimental procedure for determining hole density is simpler than that involved in using the two points and much better adapted to studies of transient phenomena. It is necessary in employing this technique to keep the current drawn by the collector point somewhat smaller than $I_b + I_\epsilon$; otherwise the disturbance in the current flow due to the collector current is too great and the sample of the hole current is not representative. Experiments have shown, however, that this condition is readily achieved and that the collector current may be satisfactorily used as a measure of hole density.

The hole density also affects the resistance of a point at low voltage. Studies of this effect have also been made in connection with the experiment of Figure 3.3. After the hole density has been determined from measurements of ΔV and $I_b + I_\epsilon$, a small additional voltage (0.015 volts) was applied between P_1 and P_2 and the current flowing externally between P_1 and P_2 was measured. From these data a differential conductance, for small currents, is obtained for the two points P_1 and P_2 in series. As is

[21] If the lifetime of a hole is τ, then the hole current at the points is $I_p = \gamma I_\epsilon \exp(-t/\tau)$ where t is the transit time to a point midway between the points, say a distance L from the emitter. If the electric field is $E = \Delta V/\Delta L$, then the transit time $t = L\Delta L/\mu_p \Delta V$. Hence if $\ln I_p$, as determined from the ratio of conductivities, is plotted against $t = L\Delta L/\mu_p\Delta V$ a straight line with intercept $\ln \gamma I_\epsilon$ and slope $-1/\tau$ should be obtained. See Chapter 12.

[22] Transistors using p-type germanium have been described by W. G. Pfann and J. H. Scaff, *Phys. Rev.* **76**, 459 (1949). Electron injection in p-type germanium has also been observed by R. Bray, *Phys. Rev.* **76**, 152, 458 (1949).

shown in Figure 3.7, this conductance is seen to be a linear function of the hole concentration. The conductance of a point contact arises in part from electron flow and in part from hole flow. From experiments using magnetic fields,[23] it has been estimated that under equilibrium conditions the two contributions to the conductance may be comparable. In connection with Figure 3.7 it should be noted that the hole concentration on the abscissa is the average hole concentration throughout the entire cross section; the hole concentration may be much less near the surface due to recombination on the surface.

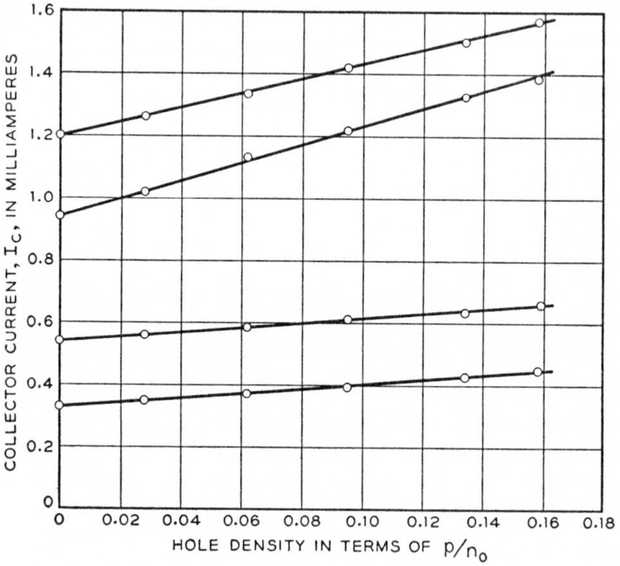

FIG. 3-6—Dependence of Collector I_c Upon Average Hole Density Being Swept by Collector Point. Collector Biased 20 Volts Reverse.

Techniques of the sort described previously can be used to measure the properties of collector points. If a collector point is placed between the emitter and P_1 in Figure 3.3, then the hole current extracted by the collector can be determined in terms of the hole current past P_1 and P_2. By these means an "intrinsic α" for the collector point can be determined. The intrinsic α is defined as follows:

α_i = intrinsic α = the ratio of change in collector current per unit change in hole current actually arriving at the collector.

3.1d. Studies of Transient Phenomena. The technique of using a collector point to measure hole concentrations has been employed in a

[23] H. Suhl and W. Shockley, *Phys. Rev.* **75**, 1617 (1949); **76**, 180 (1949).

number of experiments similar to those described in connection with Figure 3.1. These experiments give information concerning hole lifetimes, hole mobilities, diffusion and conductivity modulation.

One of the methods employed to measure hole lifetime involves the measurement of the increase in collector current, produced by the arrival of the

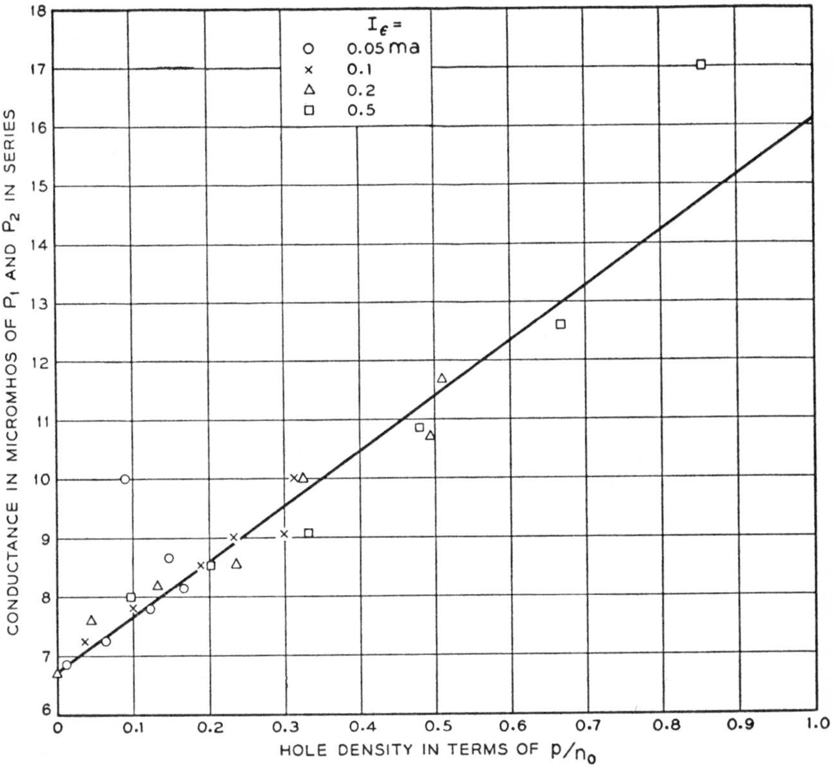

FIG. 3-7—Conductance of P_1 and P_2 of Figure 3-3 in Series as Function of p/n_0 Showing that Conductance Depends on Hole Concentration but Not on Currents in Filament. For Each Value of I_ϵ the Hole Density Was Varied by Varying $I_b + I_\epsilon$ from .038 to 0.78 ma.

leading edge of the hole pulse, as a function of the transit time of the holes from emitter to collector. This time is varied by varying the distance between the emitter and the collector points.

In Figure 3.8 we show a plot, obtained in this way, from a sample of germanium having dimensions 1.0 × .05 × .08 cm. It is seen that the increase in collector current due to hole arrival decays exponentially with a time constant of 18 microseconds. This time constant increases as the dimensions of the germanium sample are increased so that a time constant of 140 microseconds was measured, using a sample having dimensions 2.5 × .35 × .30 cm. Since the holes injected into the interior of this sample can diffuse to the surface and recombine in about 100 microseconds, the process may still be largely one of surface recombination. In any event, it may be concluded that the lifetime in the bulk material used must be at least 140 microseconds. Making use of the electron density deter-

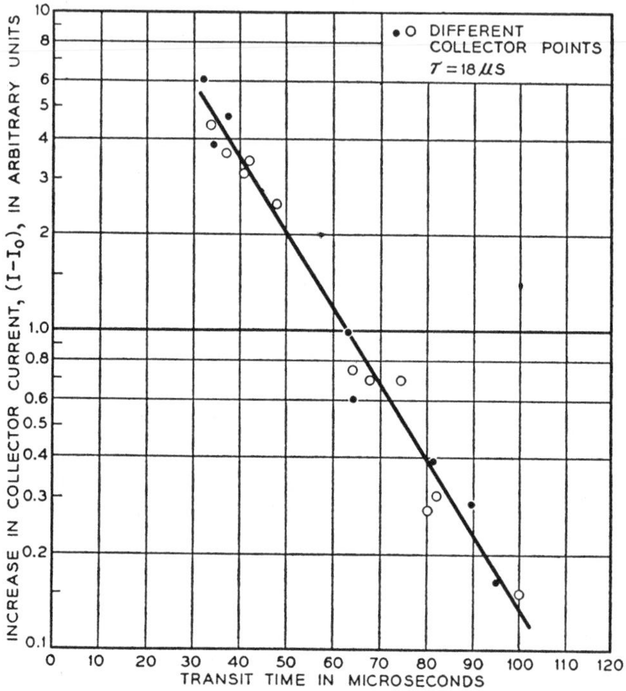

FIG. 3-8—The Decay of Injected Holes in a Sample of *n*-Type Germanium.

mined from other measurements, we conclude that the recombination cross section must be less than 10^{-18} cm^2. This cross section, which is less than 1/400 the area of a germanium atom, may be so small because a hole-electron pair has difficulty in satisfying in the crystal the conditions somewhat analogous to conservation of energy and momentum which hinder recombination of electrons and positive ions in a gas discharge. Thus it has been pointed out that a hole-electron pair will have a lowest energy state in which the two current carriers behave something like the proton and electron of a hydrogen atom.[24] Such a bound pair are called an *exciton* and the energy given up by their recombination is the "exciton energy." In order to recombine they must radiate this energy in the form of a light

[24] G. H. Wannier, *Phys. Rev.* **52**, 191–197 (1937).

quanta (photon) or a quantum of thermal vibration of the crystal lattice (phonon). The recombination time for the photon recombination process can be estimated from the optical constants for germanium and the theory of radiation density using the principle of detailed balancing, which states that under equilibrium conditions the production of hole-electron pairs by photon absorption equals the rate of recombination with photon emission; the lifetime obtained in this way is about 1 second at room temperature indicating that the photon process is unimportant.[25] As has been pointed out by A. W. Lawson,[26] the highest energy phonon will have insufficient energy to carry away the "exciton energy" of a hole-electron pair and, therefore, the release of energy will require the cooperation of several phonons with a correspondingly small transition probability.

When a square pulse of holes is injected in an experiment like that of Figure 3.1, the leading and trailing edges of the current at the collector point are deformed for several reasons. Owing to the high local fields at the emitter point, some of the holes actually start their paths in the wrong direction—i.e. away from the collector; these lines of flow later bend forward so that those holes also pass by the collector point but with a longer transit time than holes which initially started towards the collector. A spread in transit times of this sort is probably largely responsible for the loss of gain at high frequencies in transistors. For the experiments described below, however, this effect is negligible compared to two others which we shall now describe.

On top of the systematic drift of holes in the electric field, there is superimposed a random spreading as a result of their thermal motion. This would cause a sharp pulse of holes to become spread so that after drifting for a time t_d the hole concentration would extend over a distance proportional to $\sqrt{Dt_d}$ where $D =$ the diffusion constant for holes $= kT\mu_p/e \doteq 45\ \text{cm}^2/\text{sec}$. As a result of this effect, the leading and trailing edges of the square wave of emission current become spread out when they arrive at the collector. This is shown in Figure 3.9, curve A for the leading edge and B for the trailing edge. The points are 10 microsecond marker intervals traced from an oscilloscope, the time being measured from the instant at which the emitter current starts. For A and B the emitter current was so small compared to the current I_b that the holes produced a negligible modulation of conductivity and each hole moved in essentially the same electric field. It is to be observed that the wave shapes are nearly symmetrical in time about the half rise point and that the A and B waves are identical except for sign.

[25] Optical constants for germanium have been published by W. H. Brattain and H. B. Briggs, *Phys. Rev.* **75**, 1705–1710 (1940) and H. B. Briggs, *Phys. Rev.* **77**, 287 (1950). The integration over the radiation distribution was carried out by W. van Roosbroeck.

[26] Personal communication; a somewhat similar case is treated by B. Goodman, A. W. Lawson, and L. I. Schiff, *Phys. Rev.* **71**, 191–194 (1947).

This is just the result to be expected from diffusion. Furthermore, analysis shows that the spread in arrival time is in good quantitative agreement with the theoretical wave shape using the diffusion constant appropriate for holes. For this case the mid-point of the rise, corresponding to the crossing point of the curves, gives the average arrival time and has been used to obtain an accurate measure of the mobility.

Curves C and D correspond to conditions in which the emitter current was relatively large—two thirds of the base current. High impedance

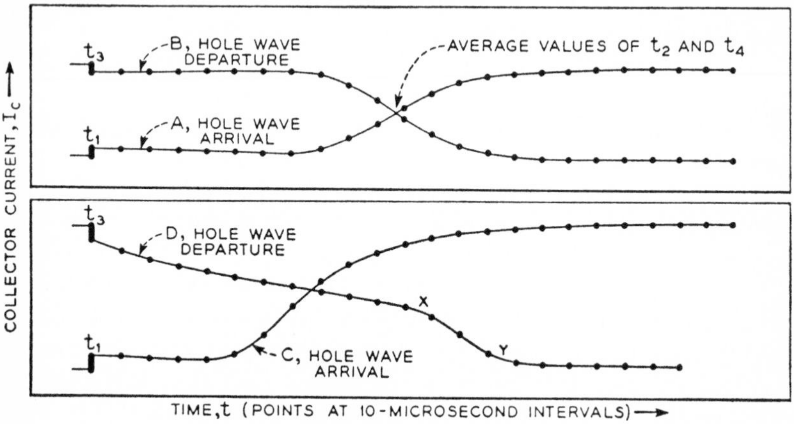

A & B EMITTER CURRENT SMALL, ABOUT 4% OF I_b, SO THAT ALL HOLES MOVE IN THE SAME FIELD.

C LEADING EDGE OF PULSE FOR $I_\epsilon = 2/3\ I_b$.

D TRAILING EDGE OF HOLE PULSE FOR $I_\epsilon = 2/3\ I_b$, SHOWING SHARPENING FROM X TO Y DUE TO TENDENCY OF LAGGING HOLES TO CATCH UP.

Fig. 3–9—Collector Current Characteristics for the Circuit Shown in Figure 3–1.

sources are used so that I_b is constant and I_ϵ is a good flat topped wave. For the currents used in this experiment, the conductivity is appreciably modulated by the presence of holes. This accounts for the shape of curve C, corresponding to the arrival of holes at the collector. It is seen that this curve is not symmetrical but is much more gradual towards later times. The reason for this is that the first holes to arrive are those which have diffused somewhat ahead of the rest and move in material of low conductivity. The later holes travel in an environment of relatively high conductivity and, consequently, in a lower electric field. (Since the current is the same at all points between emitter and collector, the field is inversely proportional to the conductivity.) The transit time for the later holes is, therefore, longer and the hole density builds up more slowly for the latter part of the incoming pulse of holes. The wave form obtained from the trailing edge of the emitter pulse, curve D, is in striking contrast with the leading edge. The first gradual decay, up to point X, is due to recombination of holes and electrons; at t_3 the emitter current becomes zero; consequently, the electric field is reduced and the holes arriving at X have taken a longer transit time than the holes arriving at t_3 and a larger fraction of them have recombined with electrons. The true trailing edge, running from X to Y, is appreciably sharper than the leading edge. The reason for this is that holes lagging behind the main body of holes are in a region of relatively low conductivity and high electric field and tend to catch up with the main body. Thus the same effect which lengthens wave C acts to shorten wave D.

C. Herring has been able to obtain mathematical solutions for the appropriate equations bearing on the matters just discussed.[27]

The delay feature discussed in connection with Figures 3.1 and 3.9 indicates interesting possibilities of using germanium filaments as delay or storage elements.

3.2 MAGNETIC CONCENTRATION OF HOLES AND ELECTRONS, THE SUHL EFFECT

Because of its great importance in affording a means of investigating nole lifetimes and related matters, we shall discuss the magnetic concentration effect observed by Suhl and Shockley.[1] This experiment represents an extension of the Hall effect beyond its usual range of application and gives direct experimental evidence that holes and electrons moving in magnetic fields are subjected to a sidewise thrust.

The experimental arrangement may consist of an n-type germanium filament which is placed in a transverse magnetic field, Figure 3.10(a). If an electric field E_x causes an electron flow parallel to the filament, then a Hall voltage will appear across the specimen and, for the polarities shown, the top surface will be charged negatively.[2] The effect is produced, according to theory, by transverse force tending to thrust electrons to one side as they drift down the filament. For the case shown, the top surface of the filament will become negatively charged, and the total electric field, having components E_x and E_y, will form angle θ_n with the lines of current flow. If a few holes are injected into this field by the current I_ϵ, they will be deflected by the magnetic field towards the same side as the electrons. (As discussed in Section 8.7, the sidewise thrust is proportional both to velocity and to charge of the particle, each of which is reversed for a hole compared to an electron so that the magnetic thrust is in the same direction.) In addition, the transverse component, E_y, of electric field which

[27] C. Herring, *Bell Syst. Tech. J.* **28**, 401–427 (1949); see also Chapter 12.

[1] H. Suhl and W. Shockley, *Phys. Rev.* **75**, 1617–1618 (1949).

[2] An elementary description of the Hall effect is given in Sections 8.6 to 8.8. Figures 8.7 and 8.9 show important features.

holds electrons away from the top surface, tends to concentrate holes there. The net result is that holes are drawn towards the top surface.

Experimental evidence for this predicted behavior of the holes is obtained with the aid of the probe points. As Figure 3.7 shows, the conductance of a probe point is a linear function of hole density. Consequently, the

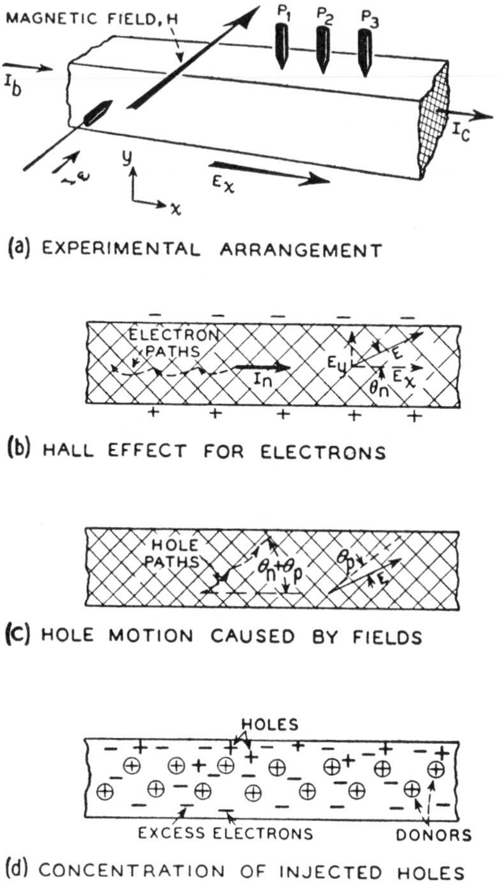

FIG. 3-10—Magnetic Concentration of Holes and Electrons.

conductance of the points can be used as a measure of hole density. The observed result can be visualized with the aid of Figure 3.11, which shows schematically the sort of effects expected. For no magnetic field, the hole current spreads out across the specimen by diffusion; anticipating the result to be established later, we have indicated that some of the lines of flow terminate on the surface where a major portion of the recombination of holes and electrons occurs. For a magnetic field H_1, the holes are strongly deflected towards the top side of the filament. For $H_2 > H_1$, the effect is still more pronounced and a second effect is indicated: the hole current dies away more rapidly to the right. The reason for this is that the holes combine with electrons on the surface of the filament and the increased concentration near the surface for H_2 compared to H_1 means an increased opportunity for recombination and a decreased lifetime.

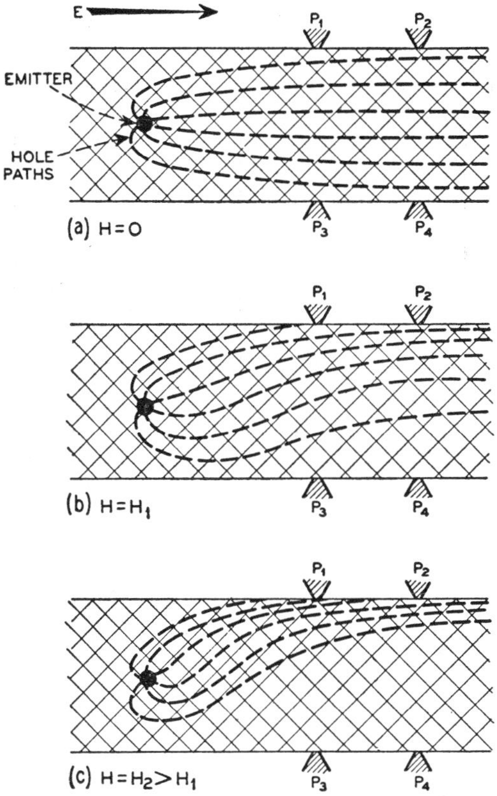

FIG. 3-11—Paths of Holes Injected into *n*-Type Filament Showing Concentration by a Magnetic Field ***H*** directed into the Plane of the Figure.

For a suitably designed experiment the effects may be quite large. The angles θ_n and θ_p between currents and electric field have values, as given in Chapter 8, of

$$\begin{aligned}\theta_n &= 10^{-8}\mu_n H = 10^{-8} \times 2600 \times 10^4 = 0.26\\ &= 15° \text{ for } 10{,}000 \text{ gauss} \qquad (1)\end{aligned}$$

$$\theta_p = 10^{-8} \times 1700 \times 10^4 = 10° \text{ for } 10{,}000 \text{ gauss.} \qquad (2)$$

For a longitudinal electric field E_x of 20 volts cm, the total thrust transverse to the filament is approximately equivalent to a field of

$$E_x \tan(\theta_p + \theta_n) = 8.5 \text{ volts/cm}, \tag{3}$$

giving an effective voltage difference of $0.025 \times 8.5 = 0.21$ volt across a filament 0.025 cm in width. This is about $8.5kT$, so that pronounced concentration will result. The total distance a hole must travel from one side of the filament to the other is $0.025/\sin 25° = 0.06$ cm. Since the drift velocity is about $1700 \times 20 = 3.4 \times 10^4$ cm/sec, the transit time will be less than 2×10^{-6} sec. This time is probably at least one order of magnitude less than the time for recombination in the interior of a good germanium specimen so that the concentration should be accomplished well before holes can get into equilibrium in the interior.

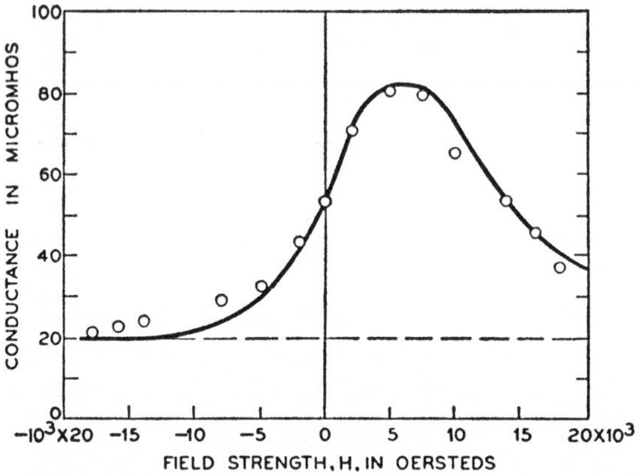

Fig. 3–12—Dependence of Conductance of Probe Point Upon Magnetic Field.

A typical set of experimental data is shown in Figure 3.12. The conductance of a point, such as P_1, is shown as a function of H. Positive H deflects holes towards P_1, and negative H away from it. With the first increase of H above zero, the conductance rises, indicating a higher hole concentration produced in the neighborhood of the probe point. Still further increase of H produces a higher degree of concentration of the holes, as shown in Figure 3.11; however, at the same time, the holes decay more rapidly so that the signal at a point like P_2 is decreased. As a result of these opposing tendencies, the curve rises to a maximum and then decreases. For negative values of H, the holes are deflected away from the point, the lifetime is decreased as for positive H, and a monotonic decrease of the response results.

The curve shown in Figure 3.12 is theoretical and has been fitted to the data by using as the law of recombination:

(Rate of recombination of holes per unit area of surface)
= (a constant s) $\times$ (hole density just next to the surface). (4)

The value of s, which has the dimensions of velocity, chosen to fit the data is 1500 cm/sec. Other methods of analysis have given comparable values for similar specimens.

The failure of the theory to fit the points for large negative values of H is significant. Practically the same points are obtained if no hole injection is used at all. The interpretation is that this part of the curve arises from holes being spontaneously generated on the surfaces. These holes are swept away from the point for negative H; however, some holes still reach the point and affect its conductance. For large negative H, the limiting value of 20 micromhos is thought to be that corresponding to a nearly complete absence of holes.

By a suitable proportioning of the specimen and the fields, it is possible to separate the effects of surface generation and volume generation. The mathematical details are not appropriate to Part I, however, and are postponed to Chapter 12. The practical use of the concentration of injected holes by magnetic fields has also been considered, particularly by R. L. Wallace.

The purpose of this section on magnetic concentration of holes and electrons has been, on the one hand, to show one more item of experimental evidence that holes and electrons really behave as the theory of the later chapters predicts and, on the other, to introduce the reader to a new and powerful experimental technique which the author predicts will play a major role in the analysis of transistor materials.

Problems

(*Remark:* Most of the problems bearing on the material of this chapter involve solving the continuity equation. A number of such problems are given at the end of Chapter 12.)

1. This problem is intended to show why deviations from

$$n = n_0 + p$$

given in equation (5) of Section 3.1 do not ordinarily occur. Suppose that in the interior of a large block of n-type germanium with a dielectric constant of $\kappa_e = 16$ and conductivity $\sigma = 0.4$ ohm^{-1} cm^{-1} the value of $n - p$ differs by 1% of n_0 throughout a sphere 10^{-3} cm in radius. Calculate the voltage and field at the surface of the sphere.

Ans. 0.36 volt, 36 volts/cm.

2. Assume that Figure 3.3 applies to the specimen of problem 2 of Chapter 1. Suppose that the current I_b is 2.9 ma and I_ϵ is 1.8 ma and $\gamma = 0.5$ and that recombination is negligible. Using $\mu_p = 1700$ and $\mu_n = 3600$ cm^2/volt sec and $b = 2.1$ (these values are in agreement with the problems of Chapter 1), show that the conductivity is increased from 0.5 ohm^{-1}cm^{-1} to 1.23 ohm^{-1}cm^{-1}. Show that the electric field is reduced from the value 0.94 volts/cm it would have if $\gamma = 0$ to 0.38 volts/cm due to the presence of injected holes and that the resistance is reduced from 400 ohms to 162 ohms. Show that the electron concentration is increased from 8.70×10^{14} cm^{-3} to 17.4×10^{14} cm^{-3}. (Compare these results with problem 5 of Chapter 1.)

3. Consider an n-type germanium filament of length L, cross-sectional area A and conductivity $\sigma_0 = en_0u_n$. Suppose that an added density of holes $p(x)$ is present so that there are $p(x)Adx$ holes in element of length dx. Show that if $p(x)$ is much smaller than n_0, the increase in conductance of the filament is

$$e(\mu_n + \mu_p)L^{-2}\int_0^L Ap(x)dx = e(\mu_n + \mu_p)L^{-2}\,p_{\text{tot.}}$$

where $p_{\text{tot.}}$ is the total number of added holes. (This result may be obtained by writing the resistance of the filament in the form

$$R = \int_0^L dx/Ae\,\{\mu_n[n_0 + p(x)] + \mu_p p(x)\}$$

and expanding the fraction in powers of $p(x)/n_0$, which is assumed to be a small quantity.) This problem proves that for a uniform filament and for small hole densities, the change in conductance depends only on the total number of added holes present and not on their distribution.

CHAPTER 4

ON THE PHYSICAL THEORY OF TRANSISTORS

In this chapter, the theories of hole and electron injection discussed in previous chapters are used to explain the internal workings of transistors. The discussion starts with the filamentary transistor. This is the simplest type from the point of view of exposition, since its ability to amplify depends chiefly on the conductivity modulation discussed in Section 3.1. The discussion then proceeds to other types of transistors which employ rectifying junctions in their output circuits. In Section 4.2 the discussion of rectifying junctions is begun with a treatment of one of the simplest types, the *p-n* junction. It is then shown how a transistor may be made by combining such junctions. In Section 4.3, metal point-contact rectifiers are described and compared with *p-n* junctions, and in Section 4.4 the theory of these rectifying junctions is applied to relate the physical picture to the equivalent circuit for the type-A transistor. In Section 4.5 a description of the electrical *forming* of the contacts in a transistor is presented together with a discussion of structures leading to high values for α. Section 4.6 discusses briefly phototransistors and counters.

4.1 THE THEORY OF FILAMENTARY TRANSISTORS

4.1a. The Equivalent Circuit. In Figure 4.1 we show a transistor with a filamentary structure.[1] Modulation is achieved in this case by injecting holes at the emitter point which flow to the right and modulate the resistance in the output branch between emitter and collector. Structures of this sort can be produced by the sand-blasting technique discussed in Section 3.1. The enlarged ends, which give the unit a dumbbell appearance, decrease the problem of making contact to the unit. The large area at the left side serves the additional purpose of reducing unwanted hole emission from the metal electrode and affords an opportunity for any emitted holes to recombine before they enter the narrow part of the unit.

The theory of this transistor is relatively simple and most of the features we shall discuss in connection with it have counterparts in the theory of the type-A transistor. We shall discuss the case for which the injected current is a small fraction of the total current in the filament. Under these con-

[1] Transistors of this type, employing *p-n* junctions as well as point contacts as emitters, have been discussed by W. Shockley, G. L. Pearson, M. Sparks, and W. H. Brattain, *Phys. Rev.* **76**, 459 (1949). This section follows W. Shockley, G. L. Pearson, and J. R. Haynes, *Bell Syst. Tech. J.* **28**, 344–366 (1949).

ditions we can use a simple linear theory. We shall show that the behavior of the transistor can be given for small a-c signals by the equivalent circuit in Figure 4.1, which shows the current and voltage relationships in a form equivalent to those used in connection with the type-A transistor. We shall carry out the analysis assuming that $V_b = 0$, the *grounded base* condition. This procedure simplifies the algebra involved in deriving the equivalent circuit by eliminating one variable; the equivalent circuit itself is, of course, applicable to cases in which the base is not grounded.

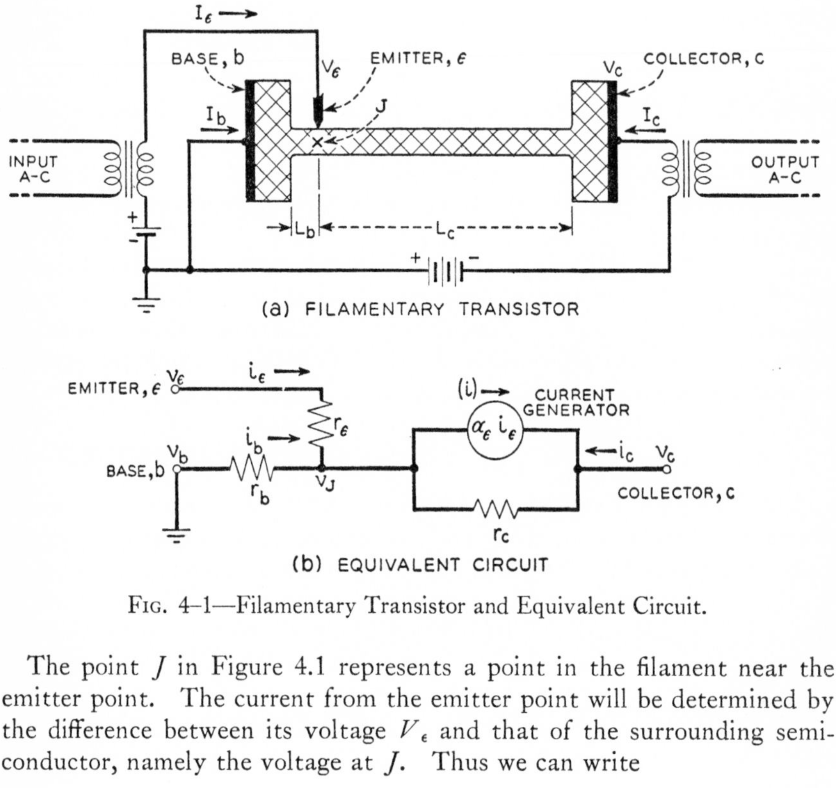

FIG. 4-1—Filamentary Transistor and Equivalent Circuit.

The point J in Figure 4.1 represents a point in the filament near the emitter point. The current from the emitter point will be determined by the difference between its voltage V_ϵ and that of the surrounding semiconductor, namely the voltage at J. Thus we can write

$$I_\epsilon = f_\epsilon(V_\epsilon - V_J). \qquad (1)$$

For small a-c variations, i_ϵ, v_ϵ and v_J, this equation leads to the relationship

$$i_\epsilon = (v_\epsilon - v_J)f_\epsilon', \qquad (2)$$

where f_ϵ' is the derivative of f_ϵ in respect to its argument. Letting $f_\epsilon' = 1/r_\epsilon$, we may rewrite this equation as

$$v_\epsilon - v_J = r_\epsilon i_\epsilon. \qquad (3)$$

This relationship is correctly represented by the r_ϵ branch of the equivalent circuit. The voltage at J, under the assumed operating conditions with I_ϵ positive and much less that I_c, will be $-I_b r_b$ where r_b is the resistance from the base to an imaginary equipotential surface passing through J. Since $v_b = 0$, corresponding to *grounded base* operation, this leads to

$$v_J = -r_b i_b = +r_b i_\epsilon + r_b i_c, \tag{4}$$

since $i_b + i_\epsilon + i_c = 0$. This relationship is obviously satisfied by the r_b branch of the equivalent circuit.

We now come to the collector branch which we have represented as a resistance r_c and a parallel current generator[2] $\alpha_e i_\epsilon$; *the value of r_c corresponds to the case of no hole injection.* (This circuit is equivalent to another in which the parallel current generator is replaced by a series voltage generator $\alpha_e r_c i_\epsilon$.) We must show that this part of the equivalent circuit represents correctly the effect of injecting holes into the right arm of the filament. We shall suppose that there is negligible recombination so that the hole current injected at the emitter point flows through the entire filament. (We consider recombination in the next section.) The current I_c in the collector branch thus contains a component $-\gamma I_\epsilon = I_p$ of hole current [minus because of the algebraic convention that positive $I_c(= -I_b - I_\epsilon)$ flows to the left]. The added hole and electron concentrations lower the resistance and r_c changes to $r_c + \delta r_c$, where δr_c is negative. The current voltage relationship for this branch of the filament then becomes

$$V_c - V_J = (r_c + \delta r_c)I_c. \tag{5}$$

Our problem is to reexpress this relationship in terms of the small a-c components and show that it reduces to the relationship

$$v_c - v_J = r_c(i_c + \alpha_e i_\epsilon) \tag{6}$$

corresponding to the equivalent circuit. For small emitter current the analysis is carried out conveniently as follows: The ratio of hole current to the total current is $-\gamma I_\epsilon/I_c$. The ratio $(r_c + \delta r_c)/r_c$ corresponds to G_0/G discussed in connection with Figure 3.3. The ratio of hole current to total current is given in (12) of Section 3.1 in terms of G_0/G and may be rewritten as

$$-\frac{\gamma I_\epsilon}{I_c} = \frac{1 - (G_0/G)}{1 + b} = \frac{-\delta r_c}{(1 + b)r_c}, \tag{7}$$

giving

$$\delta r_c = r_c(1 + b)\gamma I_\epsilon/I_c. \tag{8}$$

[2] The subscript e in α_e implies equivalent circuit. α_e differs from $\alpha = -(\partial I_c/\partial I_\epsilon)_{v_c}$ by the relationship $\alpha_e = \alpha + (\alpha - 1)(r_b/r_c)$, equivalent to equation (24).

(Since I_c is negative and I_ϵ is positive this equation shows that δr_c is negative, that is, the conductivity has been increased by the hole current.) Putting this value of $r_c + \delta r_c$ into the equation for $V_c - V_J$ gives

$$\begin{aligned} V_c - V_J &= (r_c + \delta r_c)I_c \\ &= r_c[I_c + (1 + b)\gamma I_\epsilon]. \end{aligned} \quad (9)$$

If we consider small a-c variations in the currents and voltages, this reduces to the equation given by the equivalent circuit with

$$\alpha_e = (1 + b)\gamma. \quad (10)$$

The data[3] of Section 3.1 indicate that for holes injected into n-type germanium $\gamma = 1$, and for $b = 1.5$ we obtain $\alpha_e = 2.5$.

Using $v_J = r_b(i_\epsilon + i_c)$ we eliminate v_J from (3) obtaining an equation between v_ϵ and the currents. Similarly the small signal form of (9) gives an equation for v_c:

$$v_\epsilon = (r_\epsilon + r_b)i_\epsilon + r_b i_c \quad (11a)$$

$$v_c = (r_b + \alpha_e r_c)i_\epsilon + (r_c + r_b)i_c. \quad (11b)$$

These equations are formally identical with those for the equivalent circuits of the type-A transistor and, therefore, lead to the equivalent circuit of Figure 4.1, which is identical with that of Figure 2.7.

It should be emphasized that although hole injection into n-type germanium plays a role in both the type-A and the particular form of filamentary transistor shown in Figure 4.1, there are differences in the principles of operation. One important feature of the type-A is the high impedance of the rectifying collector contact which, however, does not impede hole flow and another important feature is the current amplification occurring at the collector contact. Neither of these features is present in the filamentary type shown. Instead, the high impedance at the collector terminal arises from the small cross-section of the filament. The modulation of the output current takes place through the change in body conductivity due to the presence of the added holes, a change which appears to be unimportant in the type-A transistor. In the filamentary type, current amplification is produced by the extra electrons whose presence is required to neutralize the space charge of the holes. Current amplification in the type-A transistor is, probably, also produced by the space charge of the holes[4] but the details of the mechanism are not as easily understood. We shall discuss theories of forming and of high α's in Section 4.5.

4.1b. The Origin of the Positive Feedback and Instability. The filamentary transistor exhibits the same positive feedback as the Type-A and

[3] See Chapter 12 for the best estimates.

[4] J. Bardeen and W. H. Brattain, *Phys. Rev.* **75**, 1208–1225 (1949).

for much the same reason. Because of its simple structure, however, the description is somewhat more straightforward for the filamentary type.

We shall describe in words some of the phenomena associated with the positive feedback, next attempt a physical description of the processes involved, and finally derive the results in terms of the equations.

Suppose the collector is maintained at a fixed d-c bias V_c, say -20 volts. Next suppose that the emitter current is gradually increased from zero to a larger value. It is found, as discussed in connection with the static characteristics of Figure 2.6, that the emitter voltage rises to a maximum and then falls. In other words, the emitter connection exhibits a nonlinear response and a negative resistance; although the emitter voltage is uniquely determined by the emitter current, the reverse is not true and, for a given value of emitter voltage, there is more than one possible value for the emitter current. This gives rise to possible instability if the transistor input is operated from a low-impedance voltage source, and a transient may cause a transition from a low-current condition to a high-current condition. With external circuit parameters properly chosen to limit the currents, both the high- and low-current conditions may be within the allowable dissipations. Such circuit arrangements may be employed in "flip-flop" and counting circuits. However, if the external resistors were chosen as discussed in Section 2.2 so as to give maximum available gain, then, although the circuit might be stable at the operating point, this stability might be conditional and a large transient could cause a transition to the high-current condition with possible destruction of the unit.

In order to explain the particular behavior of V_ϵ as a function of I_ϵ shown in the static characteristics of Figure 2.6, we must show why the characteristic is nonlinear with a resistance which drops from positive to negative values. The explanation thus involves two parts, one having to do with the negative resistance and one with the nonlinearity. The negative resistance arises from the presence of positive feedback in the equivalent circuit coupled with values of α_e greater than unity, and the nonlinearity arises chiefly from the varying resistance of the rectifying emitter contact. We shall discuss the negative resistance first, basing the description on the physical picture of current flow previously described.

The negative resistance feature may be understood by considering the value of V_J as I_ϵ is increased. (For purposes of this exposition, which is intended to give the physical picture involved without the burden of mathematical refinements, we shall assume that $\gamma = 1$ so that the hole current is I_ϵ and shall neglect recombination so that α_e equals $b + 1$ and is independent of the currents.) If V_c remained constant so that the current through r_c in the equivalent circuit of Figure 4.1 also remained constant, then the effect of I_ϵ would be to increase I_c by $-\alpha_e I_\epsilon$. Thus, in addition to the injected hole current I_ϵ, which flows in the right branch in part (a) of Figure 4.1, there will be an added electron current $(\alpha_e - 1)I_\epsilon$ due to electrons moving to the left. These added electrons must flow out through r_b increasing the current from its initial value with $I_\epsilon = 0$ by an amount $(\alpha_e - 1)I_\epsilon$. This requires an added voltage across r_b of $r_b(\alpha_e - 1)I_\epsilon$, the sign being such as to make V_J change in the negative direction. Thus the injection of a positive current into the germanium at ϵ makes the adjacent point go not more positive, as it would for the case of a positive resistance, but negative instead. If the drop across r_ϵ were negligible, this would mean that ϵ would exhibit a negative resistance.

However, r_ϵ is the differential resistance, $d(V_\epsilon - V_J)/dI_\epsilon$, of a rectifying point operated in the forward direction. Thus it initially exhibits a large positive resistance, which decreases as I_ϵ is increased. For small I_ϵ, the positive resistance r_ϵ is much larger than the negative resistance $(\alpha_e - 1)r_b$; however, for large values of I_ϵ, r_ϵ may be less than $(\alpha_e - 1)r_b$, and the input resistance at ϵ will shift from positive to negative, leading to the multivalued behavior discussed previously. In this treatment, the effect of changing V_J upon the current through r_c was neglected. If $r_b \ll r_c$, this, in effect, introduces an unimportant correction as we shall next show by dealing algebraically with the equations of the equivalent circuit.

We shall now assume that all the parameters in the equivalent circuit are constant except r_ϵ and shall calculate the differential resistance v_ϵ/i_ϵ when V_c is maintained at constant bias. The constancy of V_c requires that $v_c = 0$ and hence that

$$v_c = (r_b + \alpha_e r_c)i_\epsilon + (r_c + r_b)i_c = 0 \tag{12}$$

and, hence, that

$$i_c = -\frac{\alpha_e r_c + r_b}{r_c + r_b}\, i_\epsilon. \tag{13}$$

Using this value of i_c in equation (11a) for v_ϵ gives

$$v_\epsilon = (r_\epsilon + r_b)i_\epsilon + r_b\left(-\frac{\alpha_e r_c + r_b}{r_c + r_b}\, i_\epsilon\right). \tag{14}$$

This leads to the differential resistance

$$\frac{v_\epsilon}{i_\epsilon} = r_\epsilon + r_b - \frac{\alpha_e r_b r_c + r_b^2}{r_c + r_b} = r_\epsilon - (\alpha_e - 1)r_b\frac{1}{1 + r_b/r_c}. \tag{15}$$

This equation shows that the input resistance v_ϵ/i_ϵ can be negative. If $r_b \ll r_c$, then the input resistance is negative when $r_\epsilon < (\alpha_e - 1)r_b$, the condition previously discussed.[5]

[5] J. Bardeen and W. H. Brattain have previously derived essentially the same equation. Their expression is written as $\alpha f' r_b = 1$, where $f' = 1/r_{11}$ so that this is equivalent to $\alpha r_b/r_{11} = 1$. From the definition of α and the formula for α_e given in Figure 2.7, it can be shown that the two conditions are equivalent. I am informed by R. M. Ryder that from the circuit point of view, all of these conditions are equivalent to $\mu\beta = 1$ in feedback terminology.

4.1c. Effects Associated with Transit Time. Two important effects arise from the fact that a finite transit time is required for holes to traverse the r_c side of the filament: during this time the holes recombine with electrons and the modulation effect is attenuated for this reason; also the modulation of the conductivity of the filament at any instant is the result of the emitter current over a previous interval and for this reason there will be a loss of modulation when the period of the a-c signal is comparable with the transit time or less. In accordance with the results given in the problems of Chapter 3, for small injected hole densities the change in conductance of the active branch L_c of the filamentary transistor of Figure 4.1 is dependent on the total number of added holes and not on their distribution along the filament.

In the problem at the end of this chapter, a treatment based on the total number of holes present is given. In this section, the result is obtained by reasoning by analogy using the results of Section 4.1a.

For the small signal theory, the effect of transit time is readily worked out in analytic terms. We shall give a derivation based on the assumption that the lifetime of a hole before it combines with an electron is τ_p. According to this assumption, the fraction of the holes injected at instant t_1 which are still uncombined at time t_2 is $\exp[-(t_2 - t_1)/\tau_p]$. This means that the effect in the filament at any instant t_2 is the average, weighted by this factor, of all the contributions prior to t_2 back to time $t_2 - \tau_t$ where τ_t is the transit time; holes injected prior to $t_2 - \tau_t$ have passed out of the filament by time t_2. If the emitter current is represented by $i_{\epsilon 0}e^{i\omega t}$, the effective average emitter current is

$$i_{\epsilon \text{eff}}(t_2) = i_{\epsilon 0}\int_{t_2-\tau_t}^{t_2} \exp[i\omega t_1 - (t_2 - t_1)/\tau_\epsilon]dt_1/\tau_t. \qquad (16)$$

The term dt_1/τ_t is chosen so that a true average is obtained since the sum of all the dt_1 intervals add up to τ_t. The integral is readily evaluated and gives

$$i_{\epsilon \text{eff}}(t_2) = i_{\epsilon 0}e^{i\omega t_2}\frac{1 - \exp[-i\omega\tau_t - (\tau_t/\tau_p)]}{i\omega\tau_t + (\tau_t/\tau_p)}. \qquad (17)$$

The result so far as the equivalent circuit is concerned is that obtained by taking α_e as[6]

$$\alpha_e = \gamma(1 + b)\beta, \qquad (18)$$

[6] The derivation of Equations (11a) and (11b), describing the equivalent circuit, shows that hole injection enters only through the term $\delta r_c I_c$ in (9). This term leads only to $\alpha_e r_c i_\epsilon = (1 + b)\gamma r_c i_\epsilon$ in (11b) and should be replaced by $(1 + b)\gamma r_c i_{\epsilon\,\text{eff}} = (1 + b)\gamma\beta r_c i_\epsilon$ leading to (18).

where

$$\beta = \frac{1 - \exp\left[-i\omega\tau_t - (\tau_t/\tau_p)\right]}{i\omega\tau_t + (\tau_t/\tau_p)}. \tag{19}$$

β represents the effect of recombination and transit angle, $\omega\tau_t$, in reducing the gain.

We shall consider two limiting cases of this expression. First if $\omega\tau_t$ is very small, the new factor becomes

$$\beta = (\tau_p/\tau_t)(1 - e^{-\tau_t/\tau_p}). \tag{20}$$

If τ_t is much larger than τ_p, so that the holes recombine before traversing the filament, then the exponential is negligible and β becomes simply τ_p/τ_t. This means that the effectiveness of the holes is reduced by the ratio of their

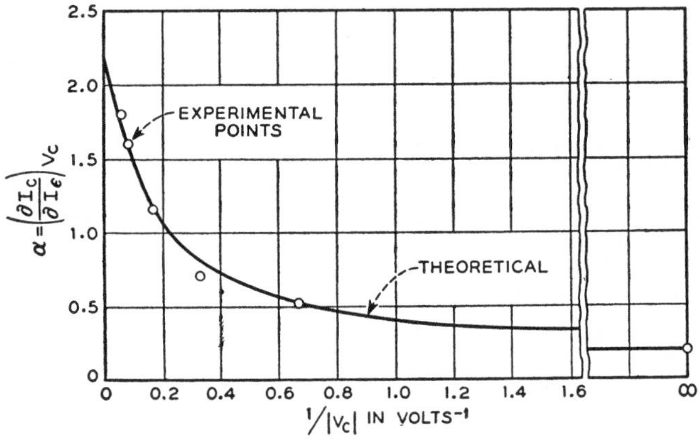

FIG. 4-2—α Versus $1/|V_c|$ Showing Agreement with the Theory for the Value of β.

effective distance of travel to the entire length of the filament, that is, τ_p/τ_t is the ratio of distance traveled in one lifetime to the entire length of the filament. Essentially the holes modulate only the fraction of the filament which they penetrate. The transit time depends on the field in the filament which is $|V_c - V_J|/L_c$, the absolute value being used since V_c is negative. The transit time is thus

$$\tau_t = L_c/[\mu_p|V_c - V_J|/L_c] = L_c^2/\mu_p|V_c - V_J|. \tag{21}$$

For very small emitter currents $V_c - V_J = r_cV_c/(r_c + r_b)$ so that

$$\tau_t = L_c^2(r_c + r_b)/\mu_p r_c|V_c| \tag{22}$$

and τ_t is inversely proportional to V_c. For large values of V_c, τ_t approaches zero and β approaches unity. The dependence of β upon V_c has been investigated by measuring α and plotting it as a function of $|1/V_c|$ as shown in

Figure 4.2. The value of

$$\alpha = -(\partial I_c/\partial I_\epsilon)_{V_c} = -(i_c/i_\epsilon)_{V_c} \tag{23}$$

is readily found from the equivalent circuit, using equation (11b), to be

$$\alpha = \frac{r_b}{r_b + r_c} + \frac{\alpha_e r_c}{r_b + r_c}. \tag{24}$$

For one particular structure investigated, the values of r_b and r_c, obtained at $I_\epsilon = 0$, were in the ratio 1:4. The value of α obtained by extrapolating the data to $|V_c| = \infty$ is 2.2; the value given by the formula for this case with $\beta = 1$ and $b = 1.5$ is

$$\alpha = 0.2 + 0.8 \times 2.5 \times \gamma, \tag{25}$$

from which we find $\gamma = 1.0$, in agreement with the result of Figure 3.5 of Section 3.1 that substantially all of the emitter current is carried by holes.[7] The theoretical curve shown on the figure is

$$\alpha = 0.2 + 0.8 \times 2.5 \times |V_c/10|(1 - e^{10/|V_c|}). \tag{26}$$

This corresponds to

$$\frac{\tau_t}{\tau_p} = \frac{10}{|V_c|} = \frac{L_c^2(r_c + r_b)}{\tau_p \mu_p r_c |V_c|}, \tag{27}$$

from which it was concluded that for the particular bridge studied τ_p was 0.2 microsecond.

If τ_t is much shorter than τ_p, then the holes penetrate the whole filament and β becomes

$$\beta = \frac{1 - \exp(-i\omega\tau_t)}{i\omega\tau_t} = \frac{\exp(-i\omega\tau_t/2)\sin(\omega\tau_t/2)}{(\omega\tau_t/2)}. \tag{28}$$

For small values of $\omega\tau_t$, β approaches unity since $(\sin x)/x$ approaches unity as x approaches zero. For $\omega\tau_t/2 = \pi$, the response is zero. This is the condition that $\tau_t = 2\pi/\omega = 1/f$. For this case the filament is just so long that the modulation is averaged over the time of one cycle of the input signal and since this average includes all phases, the modulation vanishes.

Preliminary experiments with filamentary transistors, made in accordance with the principles discussed above, appear to confirm the general aspects of the theory.[8] Power gains of 15 db have been obtained and frequency responses showing a drop of 3 db in α at 10^6 cycles/sec have been observed. Noise measurements indicate an improvement of 10 to 15 db over the average type-A transistor for comparable conditions of preparation.

[7] This depends on the value of 1.5 for b which is based on the "Hall mobility." See Section 12.9.

[8] W. Shockley, G. L. Pearson, J. R. Haynes, *Bell Syst. Tech. J.*, **28**, 344–366 (1949).

4.2 *p-n* JUNCTIONS AND *p-n* JUNCTION TRANSISTORS

4.2a. The Nature of the Current in *p-n* Junctions.[1] The physical theory of rectifying junctions in transistors may be more simply described for *p-n* junctions than for metal point contacts, and a discussion of the latter is, therefore, postponed until Section 4.3. Both collectors and emitters may be made from *p-n* junctions and it is possible to show how emitter junctions may be designed to have high efficiency for injecting carriers of the sign not normally present in appreciable numbers.

The *p-n* junctions which we shall discuss occur when a piece of germanium or silicon has a variable concentration of donor and acceptor centers so that a transition from *p*-type to *n*-type occurs in a continuous solid specimen. If two separate pieces of germanium of opposite conductivity types are simply placed in contact, however, layers of oxide or other material on the surface, surface states, roughness, etc., in general prevent a true *p-n* junction from being formed.[2]

Good *p-n* junctions may be formed in many ways. They occur naturally in melts of relatively pure silicon because of segregation of the impurities upon solidification.[3] They have been produced in germanium by converting one part of a piece of *n*-type material to *p*-type either by nuclear bombardment or by heating.[4]

In Figure 4.3 we show an idealized *p-n* junction. Part (a) shows pictorially the distribution of donors, acceptors, holes, and electrons; (b) shows the densities of donors and acceptors; and (c) shows the densities of holes and electrons. These densities adjust themselves under the thermal equilibrium situation represented so that there is no current either of holes or electrons and so that the recombination of holes and electrons in any small element of volume just balances the thermal rate of production of holes and electrons. Since the holes are more concentrated in the *p*-type

[1] *P-n* junctions were investigated before the war at Bell Telephone Laboratories by R. S. Ohl. Work on *p-n* junctions in germanium has been published by the group at Purdue directed by K. Lark-Horovitz: S. Benzer, *Phys. Rev.* **72**, 1267 (1947); M. Becker and H. Y. Fan, *Phys. Rev.* **75**, 1631 (1949); and H. Y. Fan, *Phys. Rev.* **75**, 1631 (1949). Similar junctions occur in lead sulfide according to L. Sosnowski, J. Starkiewicz, and O. Simpson, *Nature* **159**, 818 (1947); L. Sosnowski, *Phys. Rev.* **72**, 641 (1947); and L. Sosnowski, B. W. Soole, and J. Starkiewicz, *Nature* **160**, 471 (1947). The theory described here has been discussed in connection with photoelectric effects in *p-n* junctions by F. S. Goucher, Meeting of the American Physical Society, Cleveland, March 10–12, 1949 and by W. Shockley, G. L. Pearson, and M. Sparks, *Phys. Rev.* **76**, 180 (1949). For a general review of conductivity in *p*- and *n*-type silicon see G. L. Pearson and J. Bardeen, *Phys. Rev.* **75**, 865 (1949); J. H. Scaff, H. C. Theuerer, and E. E. Schumacher, *J. of Metals* **185**, 383 (1949); and W. G. Pfann and J. H. Scaff, *J. of Metals* **185**, 389 (1949). The latter two papers also discuss photovoltaic barriers.

[2] For a review of these matters see J. Bardeen, *Phys. Rev.* **71**, 717–727 (1949).

[3] See J. H. Scaff, H. C. Theuerer, and E. E. Schumacher, *J. of Metals* **185**, 383–388 (1949); and W. G. Pfann and J. H. Scaff, *J. of Metals* **185**, 389–392 (1949).

[4] M. Becker and H. Y. Fan, *Phys. Rev.* **75**, 1631 (1949); W. Shockley, G. L. Pearson, and M. Sparks, *Phys. Rev.* **76**, 180 (1949).

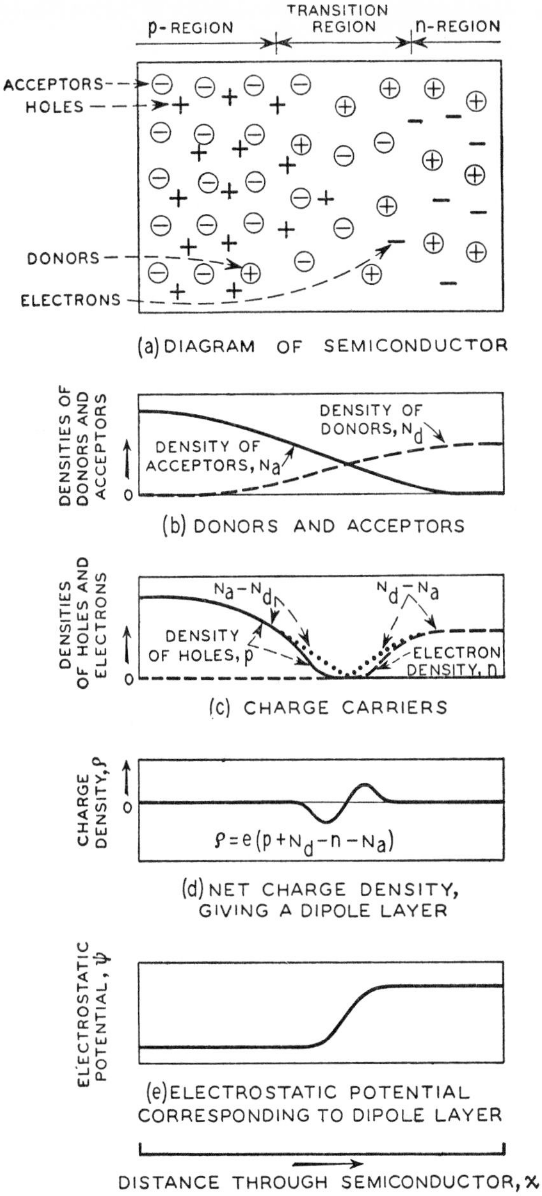

FIG. 4–3—The p–n Junction.

(a) Distribution of donors, acceptors, holes, and electrons
(b) and (c) Densities.
(d) Net charge density producing a dipole layer.
(e) Electrostatic potential corresponding to dipole layer.

material, they tend to diffuse to the n-region. This tendency is prevented by an electrostatic field shown in (e). This field makes the electrostatic potential more positive in the n-type material so that the n-region is thus able to keep holes out and to hold electrons in. A similar situation holds true for electrons. In part (d) the net charge density is shown (neglecting any electrons trapped on donors or holes on acceptors, a legitimate assump-

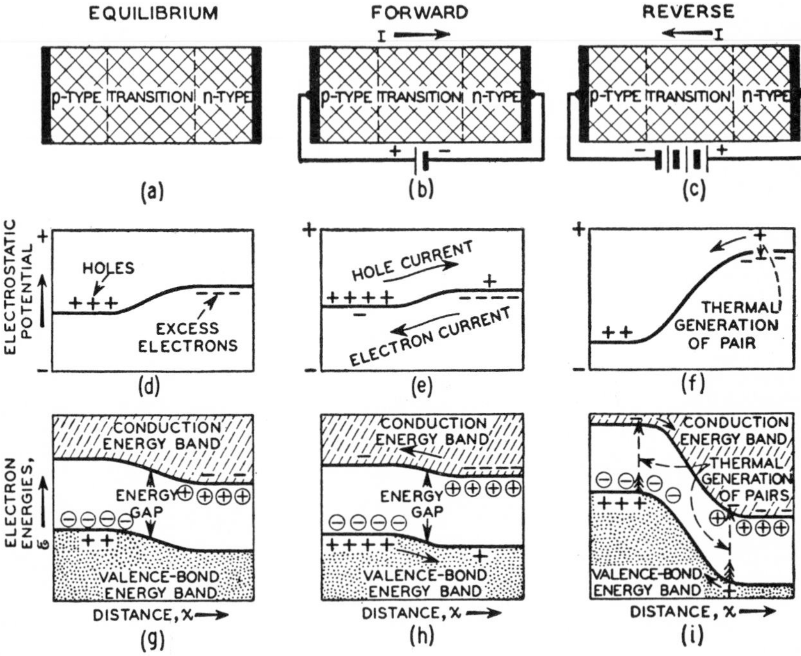

FIG. 4-4—Distribution of Holes and Electrons and Energy as a Function of Position in a p-n Junction under Applied Biases.

tion for germanium, see Chapter 10). This charge density produces a dipole layer which in turn produces the potential difference shown in (e). The determination of the potential distribution under equilibrium conditions is, in principle, a straightforward application of the equilibrium theory discussed in Chapters 10 and 12.[5] (The potential rise is then found to be an inevitable consequence of the requirement that the Fermi level be constant throughout the specimen; when the appropriate analysis is carried out, the situation illustrated in Figure 4.3 is deduced.)

We shall next consider the nature of current flow across the junction when voltages are applied. The situations of interest are shown in the first two rows of Figure 4.4; the bottom row will be discussed later. Under

[5] See W. Shockley, *Bell Syst. Tech. J.* **28**, 435–489 (1949) for mathematical details covering the same ground as this section.

equilibrium conditions some of the holes shown in (d) will acquire enough energy by thermal agitation to climb the potential rise and diffuse into the *n*-region. Once in this region they will combine with electrons. This current is exactly balanced (according to the principle of detailed balancing, see Chapter 11) by holes thermally generated in the *n*-region, as members of hole electron pairs, which diffuse to the transition region and slide down the potential hill into the *p*-region. In a similar way electron currents flow and also exactly balance each other.

Under conditions of *reverse bias*, a negative potential is applied to the *p*-region so that the height of the potential hill is increased, part (f). For biases of few tenths of a volt or more, the hill is so high that a negligible number of holes acquire sufficient thermal energy to climb it, and the hole current *into* the *n*-region substantially vanishes. The hole current *from* the *n*-region, however, is practically the same as for (d); in part (f) the origin of this current is represented as the thermal generation of a hole electron pair followed by the diffusion of the hole to the transition region and its subsequent transit to the *p*-region. This hole current, as remarked previously, is almost unaffected by the applied potential. For this reason the reverse hole current saturates, that is, reaches a limiting value as the reverse voltage increases. Precisely similar comments apply to the electron flow, and it is also evident that the same hill which holds holes in the *p*-region holds electrons in the *n*-region; that is, reverse voltage is reverse for both hole and electron currents.

In the forward direction, the hill is diminished in height and, therefore, a relatively large hole current flows over it and into the *n*-region. Some of these holes diffuse back, the fraction being determined by the lifetime of a hole in the *n*-region. Mathematical theory can be applied to the currents involved, and the results can be expressed in terms of a current voltage relationship, which we shall quote below.

Also shown in Figure 4.4 are the "energy band" diagrams appropriate to this situation. Although the basic theory of energy bands is not developed until the end of Chapter 5, the aspects of the theory needed in connection with Figure 4.4 are sufficiently simple so that they can be described directly from the figure. As we have discussed, an electron can be removed from the valence band and set in motion in the conduction band. Such an electron is in an excited state (like the car on the second floor) and its energy is higher than it would be in a valence bond. These conduction band levels are represented by the shaded region in the upper parts of the bottom row of Figure 4.4. A similar situation holds true for the levels in the valence-bond energy band. These lower levels correspond to holes, which are simply unoccupied levels for electrons. The donors and acceptors are represented as plus and minus charges and are located at energies corresponding to their abilities to bind holes and electrons. For the case of interest in this section, we shall suppose that binding occurs to a negligible extent so that we need not be concerned with these energies. (A further discussion will be found in Chapter 9.) The tendency of electrons to seek the lowest energy is equivalent to holes seeking the highest electronic energy level. Since it is quantum-mechanically impossible to have a half-excited electron (that is, there is no mezzanine or ramp in the garage), there is a gap of energy levels between the valence-bond states and the conduction states. The creation of hole-electron pairs by thermal excitation corresponds to the raising of an electron across the energy gap as shown in part (i). Kinetic energy of motion for the electron carries it to energies higher in the conduction band, and kinetic energy of motion of a hole carries it lower. These topics are discussed on the basis of the quantum theory of energy bands at the end of Chapter 7 (see Figure 7.6, for example). Since, furthermore, the charge on the electron has, unfortunately, been chosen as negative, the energy level diagram for electrons is upside down compared to the electrostatic potential.

The net hole current across the junction can be derived for the model of Figure 4.4 and is found to be given by the following equation, familiar in rectifier theory:[6]

$$I_p = I_{ps}\,[\exp\,(eV/kT) - 1] \tag{1}$$

where V is the voltage (positive values of V being forward) applied across the junction itself, e is the electronic charge, and kT is thermal energy. When V is expressed in volts and T is 300°K, corresponding to room temperature, this equation becomes

$$\begin{aligned} I_p &= I_{ps}\,[\exp\,(39V) - 1] \\ &= I_{ps}(10^{17V} - 1). \end{aligned} \tag{2}$$

The value of I_{ps} depends only on the n-type material and not on the p-type. The value of I_{ps} is

$$I_{ps} = ep_nD_p/\sqrt{D_p\tau_p} \tag{3}$$

where p_n is the concentration of holes in the n-region, τ_p the lifetime of a hole in the n-region, and μ_p and D_p the mobility and diffusion constant for holes. ($D_p = \mu_pkT/e \doteq 1700/39 = 44$ cm^2/sec as shown in Chapter 12.)

For a semiconductor in equilibrium, a mass action law [see Section 10.3, equation (19)] applies to the product of hole and electron concentrations.

[6] W. Shockley, "Theory of *p-n* Junctions", *Bell Syst. Tech. J.* **28**, 435–489 (1949). This equation was first published by C. Wagner, *Phys. Zeits.* **32**, 641–645 (1931), for a different model and represents the maximum nonlinearity attainable for purely electronic rectification. Equations (1) and (3) are derived in Chapter 12.

Hence, if the *n*-type material has a high concentration of electrons, it has a small concentration of holes and I_{ps} will be small.

Similar equations apply to the flow of electrons. Forward flow for electrons corresponds to the same polarity as for holes so that the hole and electron currents are those of two rectifiers in parallel. From the reasoning already given, we see that if the concentration of holes in the *p*-region is much greater than the concentration of electrons in the *n*-region, then I_{ns} will be less than I_{ps} and the current will flow across the junction mainly as holes.

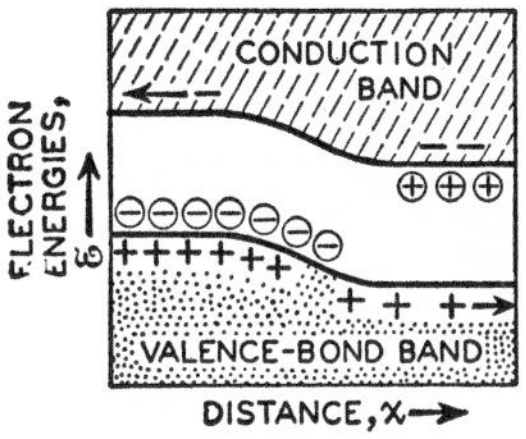

Fig. 4–5—Unsymmetrical *p-n* Junction Showing How More Current (Forward is Shown) Is Carried by More Abundant Type of Carrier (Holes).

There are a number of ways of obtaining this last result and it may be helpful to consider a different one. In Figure 4.5 we show a junction having concentration of holes in the *p*-region greater than that of electrons in the *n*-region. For forward biases both holes and electrons must climb over the same potential energy hill. If there are many more holes than electrons, many more of them will climb over the hill and most of the current will flow in the form of holes. Such a junction will obviously be a good emitter of holes.

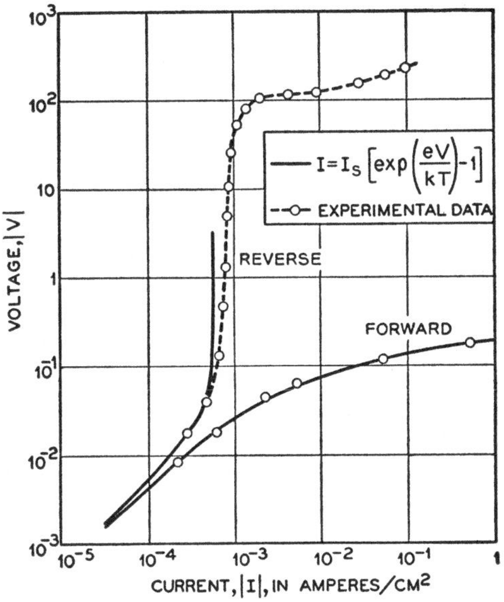

Fig. 4–6—Theoretical Rectification Curve and Experimental Data for a *p-n* Junction.

In Figure 4.6 we compare the theoretical curve with experimental data for a *p-n* junction. The fit is seen to be quite good except at high reverse fields where secondary effects reminiscent of field emission set in. It should be remarked that a 10 or 20% variation of e/kT from the theoretical value of 39 volts^{-1} will materially impair the fit of the data. This fact is a verification that the effective charge on the current carriers is e, the charge of the free electron.

4.2b. A *p-n* Junction Transistor. In Figure 4.7 we show an idealized transistor structure using two *p-n* junctions, which separate the two *p*-type regions P_ϵ and P_c from the *n*-type region N_b. Compared to the Type-A,

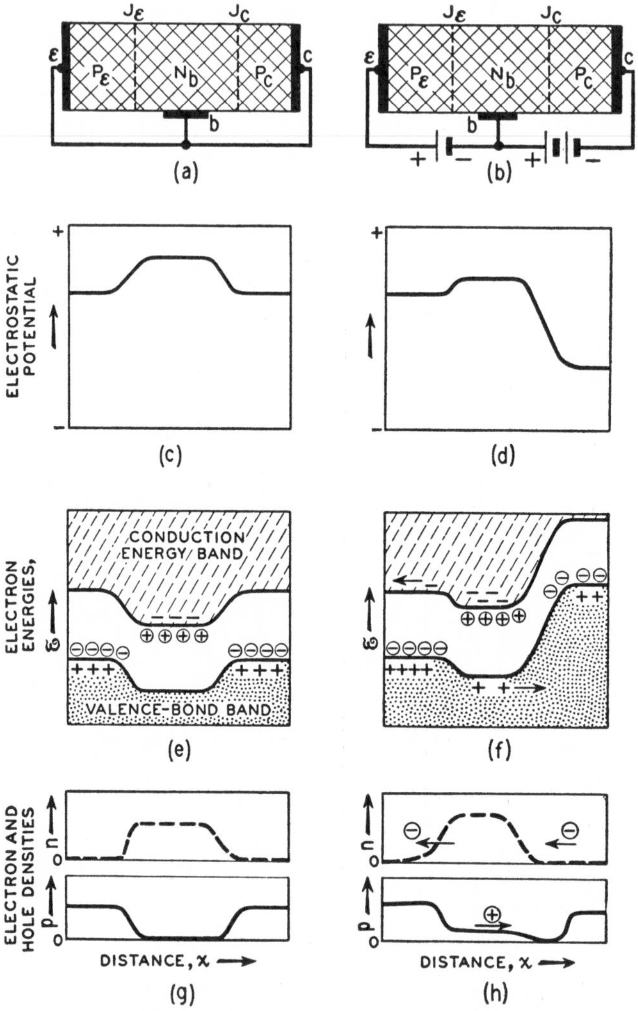

FIG. 4-7—A *p-n-p* Transistor Compared with a Vacuum-tube Triode.

this transistor has the expositional advantage that a detailed mental picture can be formed of the distribution of all the atoms involved. An instructive analogy can be drawn between this structure and a vacuum-tube triode, shown to the right.

In part (g) of the figure, the equilibrium concentration of holes and electrons is indicated together with the electrostatic potential (c). Holes, as suggested schematically in the diagram, tend to seek points of lowest electrostatic potential; electrons, having a minus charge, seek the highest

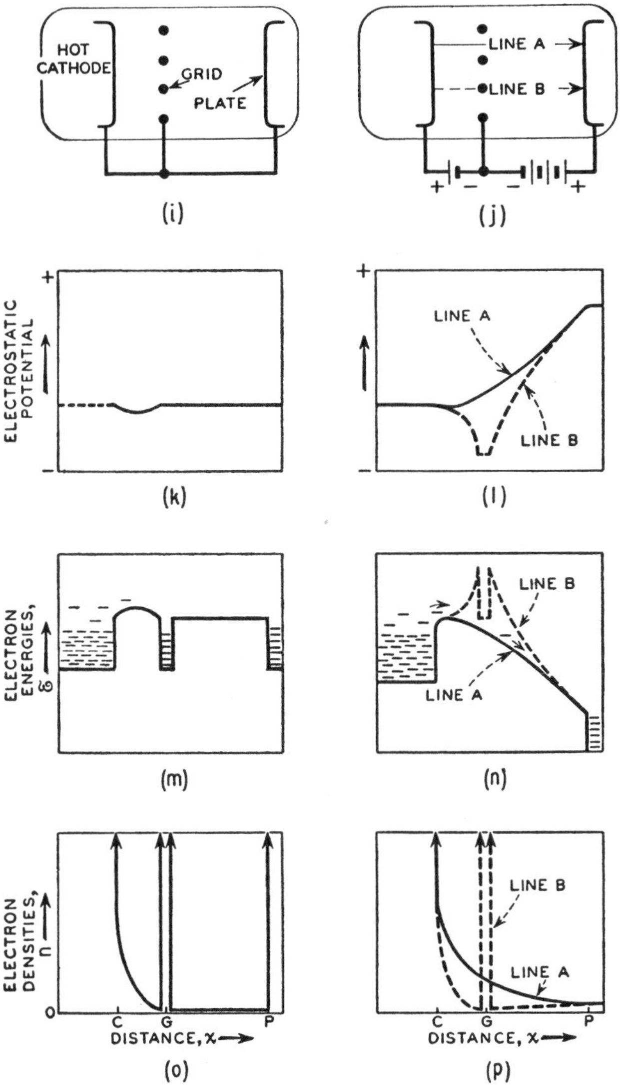

FIG. 4-7—*Continued.*

points. The energy band diagram is shown in (e). In (b) voltage in the reverse direction is applied across J_c and a small forward voltage across J_e. If the *p*-type regions have much higher conductivity than the n-type region, most of the current across the junctions will be in the form of holes. Furthermore, if the n-type region N_b is so narrow that a hole can diffuse across it with a small chance of recombining, the holes in the current across J_ϵ will also flow across J_c.[7] If the electron currents across the junctions are negligible and recombination in N_b is neglected, this transistor will draw no base current. If the emitter is grounded, a-c voltages applied to the base will require negligible currents (at least at low frequencies) and the behavior of the transistor will be like that of the vacuum tube on the right, except that the mobile carriers are positive rather than negative.

An interesting additional comparison between the transistor and the vacuum tube may be based on the ways in which electrons carry current. In fact it appears that *in both cases current can be carried by electrons in two ways which remain substantially separate and it is the control of one form of current by the other that enables amplification to occur.* We shall first discuss the two ways for the vacuum tube and then compare them with the hole and electron processes for a transistor.

In the vacuum tube there are high concentrations of electrons in the cathode, the grid, and the plate as shown in parts (o) and (p) of Figure 3.7. There is a much smaller, but important, concentration of thermionically emitted electrons just in front of the cathode. When operating voltages are applied, the potential energy diagram for an electron will be as shown in (n). At the grid wires themselves the potential energy is high since the grid is negative in respect to the cathode. Between the grid wires the potential energy is relatively low, and the electrons flow through these passes between the peaks. If electrons are withdrawn from the grid wires, so as to reduce its negative charge, the potential energy will become lower in the grid region, and a larger current will flow. In the control region we thus find one electron current controlling another: electrons which flow in and out of the grid wires control electrons flowing in the space. The two currents do not become mixed because (1) the grid is negative with respect to the cathode so that thermionic electrons cannot reach it and (2) electrons in the grid wires cannot escape because they are held in by the work function of the grid.[8]

In the transistor structure the application of voltage between emitter and base has an effect similar to the application of voltage between grid and cathode in the vacuum tube. The transition region corresponds to the grid cathode spacing and has a capacitance analogous to the grid cathode capacitance. For example, if a negative potential is applied to N_b, the

[7] Theory indicates that a negligible number of holes will be generated in the steep field region of J_c for voltages at which satisfactory operation may be obtained.

[8] If the grid is overheated, thermionic emission occurs; we are not interested in such an unfavorable case, however.

flow of electrons into N_b will charge this capacitance (and the collector junction capacitance as well) and will reduce the height of the potential hill over which the holes must drift to reach the collector junction. Thus for the transistor there is a region in which electron flow by one means, that is, excess electrons, controls flow by another means, that is, holes. However, in this case there will always be some recombination, and separation of the current to the same degree as in a vacuum tube will be difficult to attain.

It is evident that for the *p-n-p* transistor structure two quantities γ and β can be introduced as for the filamentary transistor. Since saturation current is being drawn across J_c, we would expect that the change in collector current will be simply $\alpha = \beta\gamma$ times the change in the emitter current. The resistances of J_ϵ and J_c will replace r_ϵ and r_c of the last section. The value of r_b will be more complicated and will be an average of the resistance from the contact on the *n*-type material to the junction J_1. From the approximate power gain formula, $\alpha^2 r_c/r_\epsilon$ we see that, since r_c is much greater than r_ϵ for the applied voltage condition shown, there will be large power gain, provided most of the holes can diffuse across the *n*-type material so that α is not too small. If α is nearly unity, the a-c base current will be very small. In this event the transistor can operate efficiently, in a manner similar to a grounded cathode vacuum tube, with the emitter grounded and the input signal applied to the base.

In Section 4.5 we consider modifications of the *p-n-p* structure which will lead to collectors with high values of α.

4.3 ON THE NATURE OF METAL SEMICONDUCTOR CONTACTS

The type-A and coaxial transistors employ metal contacts for their emitter and collector points. These contacts, like the *p-n* junctions just described, may carry their current either as holes or as electrons, depending on the circumstances. In this section we shall endeavor to describe on an elementary level the phenomena occurring at metal semiconductor contacts. In order to describe these matters in any detail, it is advantageous to make use of the techniques of Fermi-Dirac statistics and energy bands. For this reason, we shall give only a superficial discussion here; the reader is referred to Chapter 12 in which a mathematical formulation of the statistics of "space charge layers" is presented.[1]

The theory of rectification in germanium differs from the conventional theory by taking into account both hole and electron currents. In this section we shall lay particular stress on the hole emission aspect of the forward current in *n*-type germanium and will also indicate in a preliminary

[1] See also Torrey and Whitmer, *Crystal Rectifiers*, McGraw-Hill Book Co., New York, 1948, for a general survey of point contact rectifiers.

way how current amplification may occur at the collector. We shall return to both these topics, using somewhat more advanced concepts, in Section 4.5 where forming and the theory of high values for α are discussed.

The term *emitter* was first introduced by Bardeen and Brattain to emphasize the role of the emitter contact in supplying a hole current for the collector; at that time the evidence indicated that the important hole flow occurred in a surface layer. The phrase *hole injection* was introduced to describe the penetration of holes into the body of the germanium. Hole injection was proposed in connection with the theory of *p-n* junctions and the *p-n-p* transistor and was independently observed by J. N. Shive in connection with the double surface transistor.[2]

In order to explain how hole injection can occur when the emitter is a metal point contact, we must consider the role of the surface states discussed with Figure 2.3. We shall first describe what occurs when the point contact approaches and finally touches the surface. In order to simplify the discussion, we shall suppose that the metal point is electrically connected to the semiconductor. Then between the point and the semiconductor there will be an electric field arising from the contact difference in potential. As the surfaces approach each other, this field becomes more intense, the induced charge on the semiconductor surface arising largely from a change in the number of electrons in the surface states. The tendency of the surface states to stabilize the surface potential is so great that *only a very small change in potential in the interior of the semiconductor is produced by the contact-difference-in-potential field even when the metal is only one or two atomic diameters away from the surface.* Furthermore, it is quite probable that the germanium surface is covered to a depth of one or two atoms with a layer of oxide or adsorbed atoms. Consequently, even after the metal is brought into mechanical contact with the semiconductor, the potential just inside the semiconductor surface will be much the same as it was before contact was established. Furthermore, it is the potential just inside the surface that determines the rectifying properties of the contact, and the argument just presented shows that this potential is only slightly affected by putting the metal contact on the surface.[3] This argument has been used by Bardeen to explain why the rectifying properties of a germanium surface are independent of the work function of the metal used for the point contact.[4]

Although the potential at the surface of the semiconductor is independent

[2] For further historical details see Section 3.1, and J. Bardeen and W. H. Brattain, *Phys. Rev.* **75**, 1208–1225 (1949).

[3] These matters are discussed by J. Bardeen, *Phys. Rev.* **71**, 717–727 (1947), particularly equation (23).

[4] For additional data as germanium rectifiers see S. Benzer, *J. App. Phys.* **20**, 804–815 (1949). References.

of the work function of the metal which makes contact, *the application of voltage to the point contact produces equal changes in the potential inside the semiconductor.* This apparently contradictory result is explained as follows: Under thermal equilibrium conditions, the electrons in the surface states adjust their numbers so as to produce the fixed potential at the semiconductor surface by exchanging electrons with the interior of the semiconductor and also with the metal; the end result, however, like all equilibrium situations, does not depend on the nature of the mechanism which permits it to be established. When a potential is applied to the metal contact, on the other hand, the situation is not an equilibrium one. Accordingly, the surface states acquire a charge which represents a compromise between attempts to get into equilibrium by electron exchange with the metal on the one side and with the interior of the semiconductor on the other side. If the metal is not in contact, so that electron exchange with it is prevented, then the situation of Figure 2.3 prevails and the surface gets into equilibrium with the interior and stays at the fixed potential in respect to the semiconductor. When the metal is in contact, however, the surface states exchange electrons very easily with the metal and maintain the same potential in respect to the metal as they would under equilibrium conditions. Thus the surface of the semiconductor comes to a potential in respect to the metal which is independent of the work function of the metal; and this difference in potential then remains fixed so that the surface of the semiconductor follows the potential changes applied to the metal.

The theory proposed by Bardeen and Brattain for hole injection by emitter points is based on the role of the surface states. According to these ideas a large density of electrons is normally present in the surface states. As a result electrons are repelled from the surface as shown in the Surface States Diagram, Figure 2.3. The effect is supposed to be still more pronounced than shown in Figure 2.3, however, and repulsion from the region near the surface occurs not only for excess electrons (in the conduction band) but for some of the electrons in the valence bands as well. Accordingly a concentration of holes, represented in part (a) of Figure 4.8, is present immediately next to the metal contact. The situation is thus in many ways similar to the *p-n* junctions described in the previous section. If the junction of Figure 4.8 is biased in the forward direction, as shown in (b), a major portion of the current may be carried by holes. Similar considerations to those described in the last section have been used by Bardeen and Brattain as criteria for a good hole emitter. In essence, if the hole concentration near the surface is much higher than the electron concentration in the interior, then most of the current will be carried by holes. We shall return to a further discussion of this criterion in connection with the theory of "forming" proposed by Bardeen and Pfann.

The reverse-current situation is represented in part (c). For this case the applied potential is in the same direction as the potential produced by the surface states and electrons are driven still farther from the point of contact. The reverse current is composed in part of electrons which gain sufficient energy thermally in the metal to get into the conduction band and then flow into the semiconductor. This excitation process is similar to the process of thermionic emission from a hot metal filament. However, the work function for escape into the semiconductor is so much lower than that of a metal that appreciable currents flow in the semiconductor at room temperature. Another portion of the current consists of holes which are

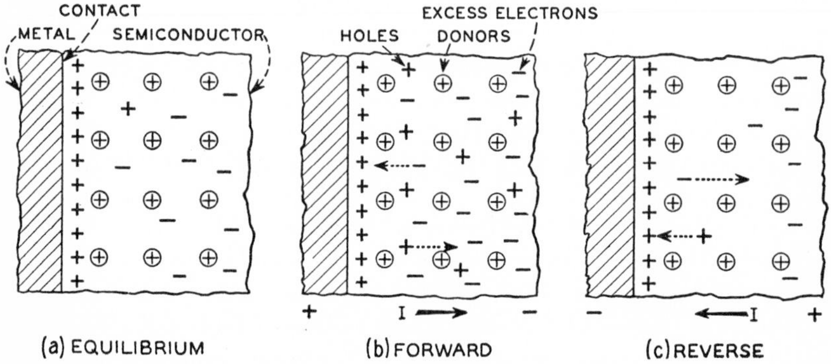

FIG. 4-8—Transistor Contact Diagram Showing the Nature of Rectification as It Occurs in a Transistor.

formed thermally in the interior or on adjacent portions of the surface of the semiconductor. These holes are thus generated in much the same way and play much the same role as they do in a *p-n* junction. There is, however, one very important difference. Since the emission of electrons from the metal is limited by the work function they must surmount, the electron emission will be enhanced by any lowering of the work function. A hole flowing towards the metal attracts electrons and makes it easier for them to escape. Thus the presence of holes promotes the electron flow. We shall return to a further discussion of this process, which is one of the ways of explaining why α_i can be >1, in the section on forming.

Rectification occurs in the metal semiconductor contact of Figure 4.8 for much the same reasons as in the *p-n* junction and we need not repeat the arguments in detail here. The chief difference is that the recombination and generation process taking place in the *p*-region of the *p-n* junction is here replaced by the emission and absorption of electrons over the potential barrier of the metal and by a similar process for holes.

In a practical rectifier, it is usually necessary to consider the voltage drop in other parts of the unit as well as in the rectifying junction itself.

In a point contact rectifier this additional resistance is referred to as the *spreading resistance.* This may be visualized with the aid of Figure 4.9. The barrier layer introduces a resistance between the metal point and the semiconductor. The resistance of this layer is the junction resistance R_J. In series with it is the spreading resistance R_S of the body of the semiconductor through which the current must flow in order to reach the metal base. The latter contact is usually so large that its resistance, even if it does rectify, is negligible; in general, the base contact is designed so as to

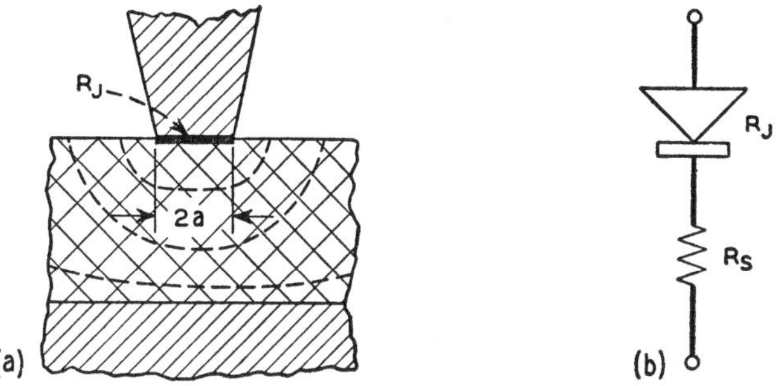

FIG. 4-9—The Equivalent Circuit of a Point Contact Rectifier.

reduce its nonlinearity. Most of the resistance R_S arises from the immediate neighborhood of the emitter point where the lines of flow are constricted; the value of R_S is thus a function of the area of contact. If the contact is regarded as circular a disc of radius a, then the spreading resistance is given by the formula

$$R_S = \frac{1}{4\sigma a} \tag{1}$$

where σ is conductivity of the material. If the disc is taken as a hemisphere, the formula is $R_S = 1/2\pi\sigma a$.

According to "one current" theories of rectification, which do not allow for carrier injection, R_S is a constant. R_S can be determined experimentally by applying large forward voltages so that R_J becomes small; the limiting resistance then becomes R_S. It was found[5] that the limiting resistance in the forward direction for germanium was much smaller than

[5] R. Bray, K. Lark-Horovitz, and R. N. Smith, *Phys. Rev.* **72**, 530 (1947). It was proposed by these authors that this effect was due to a change in conductivity in high fields and this effect was further investigated by R. Bray, *Phys. Rev.* **74**, 1218 (1948). The high field effect was shown to be due to hole injection from the terminals by E. J. Ryder and W. Shockley, *Phys. Rev.* **75**, 310 (1949), and this conclusion is further substantiated by the experiments described in Section 3.1. It has also been proposed that the low spreading resistance may be due to layer of surface conductivity of reversed conductivity type: J. Bardeen and W. H. Brattain, *Phys. Rev.* **74**, 230, 231 (1948).

could be explained by the basis of the values of a and σ. The explanation in most cases is that hole injection in n-type germanium increases the conductivity in R_S so that for forward currents both R_J and R_S are decreased.

An additional effect of hole injection in n-type germanium rectifiers results from the high density of holes present while a large forward current flows: If the applied voltage is suddenly reversed, these holes are withdrawn by the point and produce a current pulse. For some applications, this effect can be large enough to cause damage to the unit.[6]

For purposes of the transistor theory, we have placed major emphasis upon the behavior of the two forms of current in the rectifying contact. There are, however, other features of theoretical interest which have been developed in connection with the "one current" theory of rectifiers. Except for the hole current, which is vital for our discussion, these theories describe most of the features shown in Figure 4.8. These theories are associated chiefly with the names of N. F. Mott[7] and W. Schottky.[8] According to Schottky's theory, interesting information about the space charge layer, which is depleted of electrons, can be obtained by measuring the complex impedance of the junction and interpreting it as a resistance and capacitance in parallel. This capacitance can be shown to be that of a layer of dielectric having the thickness of the space charge layer. As is shown in Figure 4.8, the thickness of this layer varies with applied voltage, being wider at reverse voltages and narrower at forward voltages. Measurements of the capacitance as a function of applied voltage have been reported by a number of workers and the results are generally in excellent agreement with Schottky's predictions.[9] More recently Bardeen has analyzed the frequency effects in more detail and has defined the limits under which Schottky's approximate equivalent circuit should be valid.[10]

It is possible to show that a simple interpretation of the voltage drop across the barrier layer can be made in terms of the local conductivity. For this purpose, the layer is divided into a set of thin parallel sub-layers. These have varying electron densities and thus varying resistivities, and

[6] An effect of this sort has been observed by L. A. Meacham and S. E. Michaels, *Phys. Rev.* **78**, 175–176 (1950) and has been called the "enhancement" effect by M. C. Waltz and R. R. Blair.

[7] N. F. Mott, *Proc. Roy. Soc.* London **171A**, 27–38 (1939); also N. F. Mott and R. W. Gurney, *Electronic Processes in Ionic Crystals*, Oxford University Press, 1940.

[8] W. Schottky and E. Spenke, *Wiss. Veröff. aus die Siemens Werken* **18** (3), **1–67** (1939); W. Schottky, *Zeits. für Physik* **113**, 367–414 (1939); **118**, 539–592 (1942); F. Rose and E. Spenke, *Zeits. für Physik* **126**, 632–641 (1949). For a simple summary of Schottky's theory, see J. Joffé, *Electrical Communication*, **22**, 217–225 (1945).

[9] See the references to Schottky's papers already cited and Torrey and Whitmer, *Crystal Rectifiers*, McGraw-Hill, New York, 1948, and a paper compiled by S. J. Angello, "Semiconductor Rectifiers", *Electrical Engineering* **68** (10), 865–872 (1949).

[10] J. Bardeen, *Bell Syst. Tech. J.* **28**, 428–434 (1949).

the total resistance of the barrier layer (defined as applied voltage divided by current) can be shown to be simply the sum of the individual resistances of the sub-layers. The total width of the barrier and the resistances of the sub-layers both increase with reverse voltage and decrease with forward voltage, thus accounting for the dependence of resistance upon current in the barrier layer. This picture can be put into quantitative form and shown to be essentially equivalent to other approaches.

There is, however, a serious pitfall in the method just described; once the idea of separating the junction into parallel layers has been considered, there is a strong temptation to deal with each layer as a resistance and capacitance in parallel and then to try to synthesize the junction impedance by combining these parallel combinations in series. This proposal can be shown to be entirely incorrect; the arguments, however, are not simple and we refer the reader to the recent analysis of Bardeen from which it may be seen that the correct solution bears no resemblance to such a series synthesis.

4.4 THEORY OF THE TYPE-A TRANSISTOR

4.4a. Relationship Between the Physical Picture and the Equivalent Circuit. We shall now combine several ideas, some of which were developed for simpler cases, in order to obtain a description of processes in the type-A transistor and a derivation of its equivalent circuit. The ideas of chief importance are:

(1) Hole injection by the emitter point;
(2) Current multiplication at the collector point with an intrinsic alpha of α_i;
(3) The relationship of the filamentary transistor to its equivalent circuit.

We shall modify (3) in a suitable way so that it applies to the geometry of the type-A transistor. As a result of this procedure we shall obtain a physical picture of current flow in the type-A transistor together with an interpretation of its equivalent circuit in terms of more fundamental mechanisms. For this purpose we shall refer to Figure 4.10 which illustrates the current flow in this case.

The numerical values on Figure 4.10 are in approximate agreement with typical data for a type-A transistor as represented in the sets of static characteristics of Figure 2.6 except for the set of emitter voltages given in parenthesis. These latter correspond, as is discussed in more detail later, to increasing r_{12} in the equivalent circuit by 100 ohms so that a negative input resistance is developed.

There are two principal differences between this structure and the filamentary type: (1) The high collector resistance in this case is highly localized and arises from the junction resistance at the collector; the major portion of the transit time for a hole is spent in a relatively low resistance part of the unit between the emitter and collector. (2) The electron paths from the collector are not restrained by the geometry and flow over a wide range of directions.

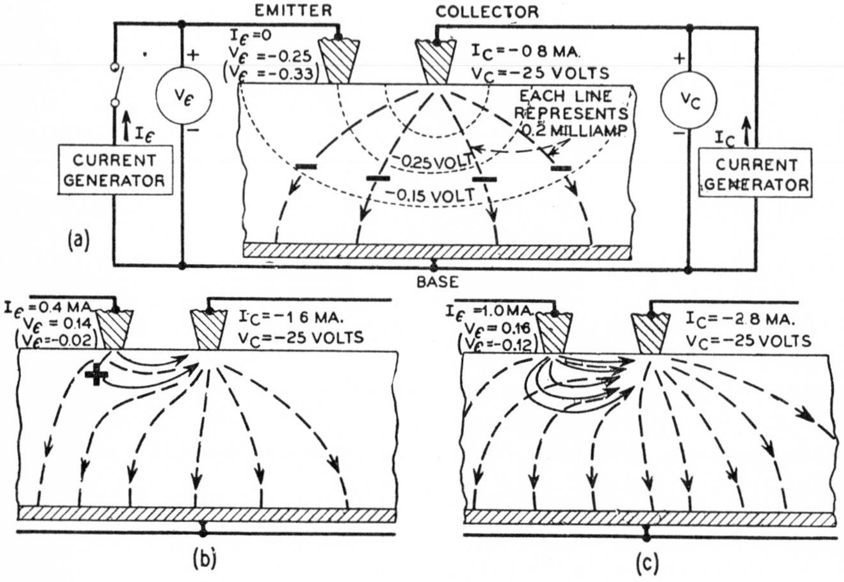

FIG. 4–10—Transistor Currents Diagram.

The figure has been drawn as if all of the emitter current consisted of holes and all of these holes reached the collector. Actually we should introduce factors γ and β:

γ = fraction of the emitter current carried by holes (for a p-type transistor, the fraction carried by electrons);

β = fraction of holes (or electrons for p-type) leaving emitter which arrive at the collector.

For ease of reference we shall also repeat the definitions of the various forms of α for a n-type transistor (for p-type transistor holes and electrons are interchanged):

$\alpha = -(\partial I_c/\partial I_\epsilon)$ with V_c = constant;

α_e = alpha for the current or voltage generators in the equivalent circuit of Figure 2.7 (subscript e for equivalent circuit).

α_i = "intrinsic α" = increase in collector current per unit increase of hole current to the collector point at constant collector voltage;

$1 - \alpha_i =$ evidently, increase electron current from collector per unit increase in hole current.

In terms of β and γ, the hole current reaching the collector will be

$$I_{pc} = -\beta\gamma I_\epsilon. \tag{1}$$

The minus sign results from the sign convention that a *positive* current corresponds to current *into* the unit. If each of these holes provokes the emission of $(\alpha_i - 1)$ extra electrons, an increase of collector current amounting to

$$\begin{aligned} \Delta I_c &= -\beta\gamma I_\epsilon - (\alpha_i - 1)\beta\gamma I_\epsilon = -\alpha_i\beta\gamma I_\epsilon \\ &= -\alpha I_\epsilon \end{aligned} \tag{2}$$

will be produced by the emitter current. Thus the quantity α is really the composite result of several effects and is given by the equation

$$\alpha = \alpha_i\beta\gamma. \tag{3}$$

The total collector current will thus be

$$I_c = I_{c0}(V_c) - \alpha I_\epsilon \tag{4}$$

where $I_{c0}(V_c)$ is the function of V_c which gives the reverse characteristic of the collector point in the absence of emitter current. This equation thus gives one relationship among the three parameters V_c, I_c, I_ϵ, which together with V_ϵ are used to describe the static characteristics.

In order to derive an equation involving V_ϵ, we use methods similar to those for the emitter point of the filamentary transistor, Section 4.1. For that case we found that, according to equations (1) and (4),

$$I_\epsilon = f_\epsilon(V_\epsilon - V_J) \tag{5}$$

where V_J was the potential in the germanium just below the emitter point and was given by

$$V_J = (I_\epsilon - I_c)R_b. \tag{6}$$

For the type-A transistor, R_b is a less clearly defined concept. As Figure 4.10 indicates, when I_ϵ is zero, the emitter will float at a potential corresponding to the equipotential which passes through it. For Figure 4.10, this equipotential is separated from the base by 0.25 volt/0.8 ma = 312 ≐ 300 ohms. When emitter current flows, the conductivity in its neighborhood is modulated and the current paths distorted.

An exact treatment of this situation is very difficult and we shall accordingly proceed by introducing the simplifying *assumption of additivity of potentials*. This additivity assumption would, of course, be exactly fulfilled if the conductivity modulation were negligible. For constant conductivity, the potential, which we shall again call V_J, immediately below the emitter point would be obtained simply by superposing the potentials due to the currents I_c and I_ϵ individually. According to our previous definition of R_b, the collector current will produce a contribution R_bI_c. Similarly I_ϵ will produce a potential $R_b'I_\epsilon$. The spreading of I_c over a large area equipotential surface by the time it passes the emitter point will cause R_b to be less than R_b'. In terms of R_b and R_b' and the additivity assumption we may write

$$I_\epsilon = f_\epsilon(V_\epsilon - V_J) = f_\epsilon(V_\epsilon - R_b'I_\epsilon - R_bI_c). \tag{7}$$

This gives a relationship between V_ϵ and the two currents I_ϵ and I_c and, together with the equation for I_c, completes the set of two equations required for the static characteristics.

The quantity $f_\epsilon(V_\epsilon - V_J)$ is not readily measurable since V_J cannot be directly determined. It is, however, straightforward to measure the current voltage characteristic of the emitter contact by itself. This leads to the relationship

$$I_\epsilon = f_{\epsilon 0}(V_\epsilon) = f_\epsilon[V_\epsilon - R_b'f_{\epsilon 0}(V_\epsilon)] \tag{8}$$

for the case of zero collector current. It is an elementary problem in functional relationships to show that from these two last equations we must have[1]

$$I_\epsilon = f_{\epsilon 0}(V_\epsilon - R_bI_c). \tag{9}$$

This last equation was derived more directly by Bardeen and Brattain by the following argument. If $I_\epsilon = f_{\epsilon 0}(V_\epsilon)$ when $I_c = 0$, then when $I_c \neq 0$ an effective bias of $+R_bI_c$ is applied. To obtain a given emitter current it is thus necessary to add sufficient extra voltage to overcome the bias, and thus V_ϵ is replaced by $V_\epsilon - R_bI_c$. This argument is equivalent to the one given previously and both depend on the assumption of additivity for the voltages due to I_ϵ and I_c. We have presented the more detailed argument to show more completely how the additivity assumption enters the theory.

The additivity assumption will be a good approximation even if considerable conductivity modulation occurs. This modulation will have as its principal effect a modification of the forward current characteristic of the emitter point, and this effect will be most important in the immediate neighborhood of the emitter point where the *spreading resistance*, discussed in Section 4.3, chiefly arises. If the collector does not disturb the modulation in this region greatly, then the contribution of I_ϵ to the emitter voltage will be largely independent of collector current. Furthermore, if the modulation of conductivity takes place over a limited region near the

[1] The first equation states that there is a functional relationship between I_ϵ and $V_\epsilon - R_bI_c$ writing this as $I_\epsilon = g(V_\epsilon - R_bI_c)$, we see that, for $I_c = 0$, we must have $g(V_\epsilon) = f_{\epsilon 0}(V_\epsilon)$.

emitter, the value of r_b will not be greatly altered. If these conditions hold, the additivity assumption will thus be well fulfilled. The additivity assumption is actually found to be adequate only over limited ranges of the currents, and r_b is found for one example to be about 1000 ohms for $I_\epsilon = 0.1$ ma and about 400 ohms for I_ϵ between 0.2 and 0.8 ma.[2] For the purpose of showing semiquantitatively the relationship of the physical picture to the equivalent circuit, however, the additivity assumption is satisfactory.

In terms of the functions already described the equations for the equivalent circuit may be readily derived as follows: The collector resistance r_{22} is obtained from

$$I_c = I_{c0}(V_c) - \alpha I_\epsilon \tag{10}$$

by finding the differential of both sides. We thus obtain

$$\dot{i}_c = \frac{1}{r_{22}} v_c - \alpha \dot{i}_\epsilon, \quad \frac{1}{r_{22}} = \frac{dI_{c0}}{dV_c}, \tag{11}$$

leading to

$$\begin{aligned} v_c &= \alpha r_{22} \dot{i}_\epsilon + r_{22} \dot{i}_c \\ &= r_{21} \dot{i}_\epsilon + r_{22} \dot{i}_c \quad (r_{21} \equiv \alpha r_{22}), \end{aligned} \tag{12}$$

which corresponds to the right-hand branch of the equivalent circuit of Figure 2.7(a). From the formula for emitter current,

$$I_\epsilon = f_{\epsilon 0}(V_\epsilon - R_b I_c), \tag{13}$$

we obtain

$$\dot{i}_\epsilon = \frac{1}{r_{11}} v_\epsilon - \frac{r_b}{r_{11}} \dot{i}_c, \quad \frac{1}{r_{11}} = \left[\frac{\partial f_{\epsilon 0}}{\partial V_\epsilon}\right]_{I_c = \text{const}}, \tag{14}$$

leading to

$$v_\epsilon = r_{11} \dot{i}_\epsilon + r_{12} \dot{i}_c \quad (r_{12} \equiv r_b) \tag{15}$$

which corresponds to the left side of Figure 2.7(a).

These last equations show the relationship of the physical model of Figure 4.10 to the equivalent circuit. The description of the equivalent circuit has been given in some detail and will not be repeated here. We shall, however, review the question of instability and negative input impedance. The decrease in input impedance with increasing current was shown in Section 4.1 to be due to decreasing resistance in the input contact with increasing currents. This same effect occurs in the type-A transistor. The positive feedback is again associated with having α greater than unity and with the presence of a resistance r_b. If the collector is connected to a zero impedance voltage supply, a negative resistance αr_b appears in the

[2] R. M. Ryder, *Bell Laboratories Record* **27**, 89–93 (1949).

input circuit in addition to r_{11}. On Figure 4.10 we show the effect of increasing r_b by 100 ohms while leaving $f_{\epsilon 0}$ unaltered. The result is to change V_ϵ by $100 I_c$ or by -0.08, -0.16, and -0.28 volt for the cases shown, the new values being given in parentheses. This increase in r_b is sufficient to give the unit a negative input impedance in the range $I_\epsilon = 0.4$ to 1.0 ma as shown in going from (b) to (c) of Figure 4.10.

The theory of the double surface or coaxial transistor is, except for the change in geometry, the same as that just described.

4.4b. Transit Time Effects. Transit time plays a very different role in the filamentary and the type-A transistors. The high impedance region of the output of the type-A transistor is the barrier layer which is about 10^{-3} cm thick and has a voltage drop of 25 volts. For such a layer the transit time would be negligible compared to effects of interest in transistors. If all holes followed paths of equal length from emitter to collector and took the same time, the current generator in the collector circuit would be a faithful but delayed replica of the input signal. If, however, holes travel in unequal paths, as shown in Figure 4.10, then the hole current at the collector will represent an average of the emitter currents at various previous times. If the important spread of times is comparable to one cycle of an a-c input signal, there will be a great deal of cancellation and the gain will be reduced.

In Figure 4.11 (a) we indicate a means of estimating the transit time and its spread. The current leaving the emitter is divided into several parts each bounded by flow lines or surfaces. In order to deal with a simple approximation, we neglect diffusion effects and modulation of the conductivity. Under these conditions a simple relationship exists between the volume through which the current flows and the average transit time. This is illustrated in (b). A current I flows through a tube bounded by flow surfaces and having an area $A(l)$ where l is length along the tube. The lines of flow in this case are readily computed from electrostatic theory.[3] The field E and drift velocity v at position l are obtained as follows:

$$\sigma A E = I, \quad E = I/\sigma A, \quad v = \mu E = \mu I/\sigma A. \tag{16}$$

The length of time required to traverse a distance dl is then dt where

$$dt = dl/v = \sigma A dl/\mu I = (\sigma/\mu I) dV_I \tag{17}$$

where dV_I is an element of volume of the tube. From this it is evident that the transit time is simply

$$t = \sigma V_I/\mu I \tag{18}$$

where V_I is the volume used by the current I.

[3] See Jeans, *Mathematical Theory of Electricity and Magnetism*, Cambridge at the University Press, 1927, Figure 19.

It is evident that, for the lines of flow of Figure 4.11, there will be great variations in transit time. As a rough approximation we shall take the spread of transit times as being comparable with the transit time for electrons in the group I_2. This group will contain about half the emitter current. From the figure the volume of its tube, which has the line from ϵ to c as a symmetry axis, is estimated as that of a cylinder S units long, and

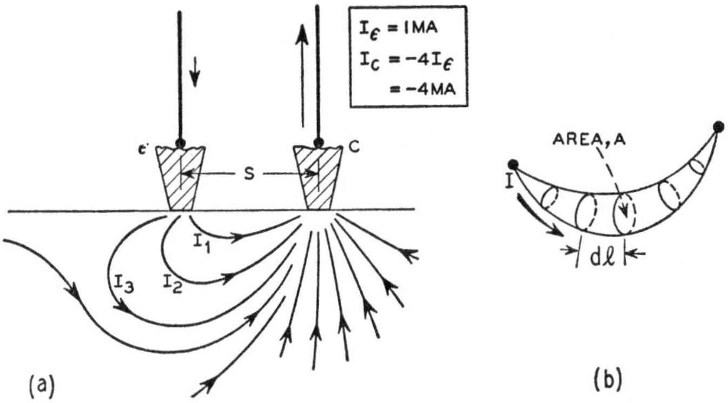

FIG. 4–11—Spread in Emitter Current Paths and Geometry of a Tube of Flow.

about 0.6S units in average radius over half a circular cross section. Its volume would thus be

$$V_I = \frac{\pi}{2}(0.6S)^2 S \doteq S^3/2. \tag{19}$$

The transit time would then be

$$\tau = \frac{\sigma S^3}{2\mu I_2} = \frac{\sigma S^3}{\mu I_\epsilon}. \tag{20}$$

If we insert the typical values:

$$I_\epsilon = 10^{-3} \text{ amp}, \quad \mu = 1700 \text{ cm}^2/\text{sec}, \quad \sigma \doteq 0.2 \text{ (ohm-cm)}^{-1}, \tag{21a}$$

$$S = 5 \times 10^{-3} \text{ cm } (= 2 \text{ mils}), \tag{21b}$$

we obtain

$$\tau = \frac{0.2 \times 125 \times 10^{-9}}{1700 \times 10^{-3}} \doteq 0.015 \text{ microsecond}. \tag{22}$$

If, as supposed, the transit time spread is comparable to this, the output would be nearly zero at 60 mc/sec and the three db point would be about at something less than half this frequency or 30 mc.

This crude calculation, which was given only to illustrate the method, predicts a cutoff frequency, about four times higher than that observed for a spacing of 2 mils, which corresponds to $S = 5 \times 10^{-3}$ cm. The discrepancy is probably due in considerable measure to conductivity modulation near the emitter point, which in effect will raise σ and increase the transit time.

It is evident from this treatment that situations which constrict the current paths between emitter and collector will tend to decrease the transit time. One example of this sort is furnished by effect of applying a magnetic field to a transistor in such a direction as to deflect holes which start downwards in Figure 4.11 towards the collector. This will tend to constrict the tubes of flow and to reduce transit time.[4]

4.5 ON THE THEORY OF "FORMING" AND HIGH VALUES OF α

It was observed by Brattain in some of the earliest transistor experiments that the performance of the transistor could be greatly improved by passing large reverse currents through the collector point. Since that time a considerable art has developed in connection with forming, and a number of techniques for applying controlled currents and voltages for various lengths of time have been used. We shall not attempt to survey these but will restrict the discussion to (a) some general observations on the effect of current polarity and conductivity type on the forming process and to (b) some speculations about the nature of collectors which have high values, say greater than 5, for α.

4.5a. The Forming Process. The forming treatments considered by Bardeen and Pfann[1] involve applications of direct current in either the forward or reverse direction for times of the order of seconds or less. The extent of the changes caused by forming depends in part on the surface treatment of the germanium and in part on the material used for the point contact. For their investigations, Bardeen and Pfann used phosphor bronze points on surfaces which had been ground and etched. They found that the effect of forming can be systematized in tabular form as shown on p. 109.

As Bardeen and Pfann point out, this array of data can be correlated with the aid of one assumption about the effect of forming current:

The forming current leads to a change in height of the potential barrier opposite to the change imposed during forming by the applied voltage. In other words, if the point contact is negative during forming, so that the potential barrier over which an electron must climb in going from the semiconductor to the metal is increased, then after the forming voltage is removed the barrier will be altered so as to be lower than its preformed

[4] This effect was reported by C. B. Brown at the June 1949 IRE Conference on Electron Devices at Princeton University, and in *Phys. Rev.* **76**, 1736–1737 (1949).

[1] J. Bardeen and W. G. Pfann, *Phys. Rev.* **77**, 401–402 (1950)

TABLE 4.1 EFFECT OF FORMING TREATMENT

Unformed	Formed with point + (+ current)	Formed with point − (− current)
	n-TYPE	
(c) *Forward Current* (*point* +): Mostly holes, good emitter *Reverse Current* (*point* −): Small, good rectifier	(d) *Forward Current* (*point* +): Little change, or larger *Reverse Current* (*point* −): Smaller	(e) *Forward Current* (*point* +): Hole current smaller *Reverse Current* (*point* −): Increased; collector formed
	p-TYPE	
(f) *Forward Current* (*point* −): Small, few electrons; poor emitter *Reverse Current* (*point* +): Large, poor rectifier	(g) *Forward Current* (*point* −): Like unformed *Reverse Current* (*point* +): Increased	(h) *Forward Current* (*point* −): Electron current increased, emitter formed *Reverse Current* (*point* −): Large

value. In Figure 4.12 this effect of forming is illustrated. The six cases shown correspond to those of Table 4.1, and for each of these the effect of forming is to alter the potential barrier in a sense opposite to that of the applied forming voltage, which is also shown in the figure.

(c) The normal contact to *n*-type germanium is supposed to produce a barrier layer with a rise in potential φ_s, which is determined largely by the surface states and very little by the nature of the metal contact. It is also supposed that the barrier is not entirely uniform and is higher at some points. At these points, the hole concentration is especially large and when the junction is operated in the forward direction, the current is carried chiefly by holes which flow into the germanium from these regions of high concentration.

(f) Since the potential of the surface states is the same for both *n*-type and *p*-type, the distribution of potential will be altered from (c) as shown in (f) for *p*-type germanium. The low potential barrier for holes leads to easy flow of holes and poor rectification and small electron emission. This is in agreement with the known poor rectification usually found for contacts with *p*-type germanium.

In (a) and (b) the fields present during forming are shown. The resulting forming effects can be interpreted by assuming that under forming conditions the temperature and field are so high that motion of the impurities results. The motion will then produce the changes in concentration and potentials shown below (a) and (b), the result being that φ_s will tend to be changed in a direction opposite to the applied potential.

(e) The desired change for n-type germanium is to cause an increase in collector reverse current. This permits it to set up an electric field sufficiently large so that it can efficiently collect the injected holes.

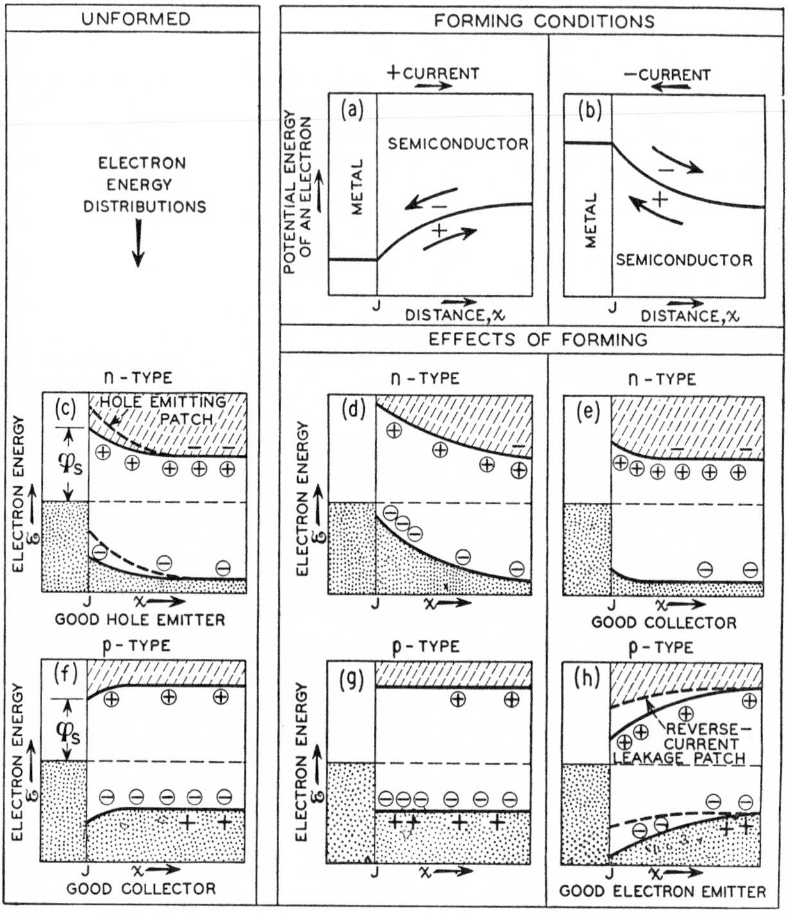

FIG. 4-12—Interpretation of Forming by Bardeen and Pfann.

(h) In p-type germanium the reverse current is large enough without forming, but the emitter current ordinarily contains too small a fraction of electrons. By lowering φ_s as shown in (h), regions of high electron concentration are formed in front of the metal point and these can act as good electron emitters.

4.5b. Interpretation of High Values of α. The hole current collected in a type-A transistor causes an enhanced electron flow from the collector, as is evidenced by the fact that α is greater than unity. It was mentioned in

Section 4.3 that α values greater than one can be explained by the space charge due to holes as they approach the collector point. This space charge will have an effect similar to the positive ions of Figure 4.12(e), whose concentration directly in front of the collector contact tends to lower the work function for electron emission and thus to increase the reverse current.

This theory of α was proposed by Bardeen and Brattain in their first article.[2] As we shall discuss below, it seems possible to explain α values as high as 3 by this means. Values of α much larger than 3 have been observed, however, and in order to explain these it is necessary to make major revisions of the way in which the space charge of the holes is effective in increasing electron emission. After discussing the theoretical limit of the simple space charge theory, we shall describe the other theories.

We shall now show that the maximum value for α based on the simple space charge theory has the same value $(1 + b)$ for a type-A transistor that it has for a filamentary transistor and for much the same reasons. Let us suppose that an increment of hole current ΔI_p produces by its space charge an increment of electron current ΔI_n. Now consider what the total space charge situation would be if $\Delta I_n = b\Delta I_p$, corresponding to $\alpha = 1 + b$. We shall show that ΔI_n cannot be this large. It is evident that ΔI_n will produce a space charge ρ_n, which tends to cancel ρ_p due to ΔI_p; evidently ρ_n cannot quite cancel ρ_p because, if it did, there would be no net positive space charge and hence no enhanced electron emission. However, if $\Delta I_n = b\Delta I_p$, then ρ_n will just equal ρ_p since, for equal charge densities, electron and hole currents would be in the ratio of $b : 1$. From this we conclude that the hole current cannot produce an electron current as great as $b\Delta I_p$ and hence that α, or more precisely α_i, cannot be as great as $1 + b$. This argument depends upon three assumptions, tacitly made in the preceding discussion: (1) The ratio of mobilities is not affected by the high fields near the collector point; (2) holes and electrons flow along the same current paths (which may not be true if the junction is "patchy"); and (3) holes have no trouble entering the metal and thus do not accumulate in front of the metal on their way in. Although these assumptions are probably not exactly fulfilled, it is unlikely that they introduce any significant errors in the argument.[3]

It has also been proposed that the current multiplication may be due to the generation of secondaries by fast-moving holes in the intense field near the collector. There is a large amount of evidence against this theory. In some cases high values of α have been observed at such low voltages, say 5 volts, that secondary generation seems quite unreasonable. Furthermore, if secondary generation were the important process, we should expect

[2] J. Bardeen and W. H. Brattain, *Phys. Rev.* **75**, 1208–1225 (1949).

[3] Similar reasoning has been presented by L. P. Hunter, *Phys. Rev.*, **77**, 558 (1950).

α to increase rapidly with applied voltage; in fact, this process would be similar to avalanche production in dielectric breakdown and the value of α would be expected to vary exponentially with voltage. This has not been observed, and the data are much more consistent with the idea that α_i is independent of applied voltage, as if each hole provoked the emission of a fixed number of electrons regardless of the value of the collector voltage.

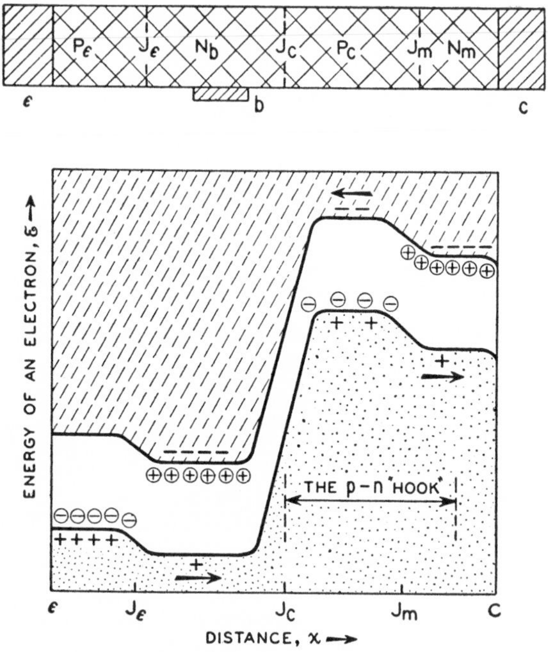

FIG. 4–13—The "*p-n* Hook" as a Current Multiplier.

For these reasons, we shall reject the theory of secondary generation and proceed to other theories based on the space charge of the holes.[4]

The space charge theories which we shall propose have the common feature that in each case the holes are impeded in their progress to the metal contact so that their space charge accumulates, thus having an enhanced effect on the electrons.

In Figure 4.13 a transistor having a current multiplying collector region is shown. This transistor differs from that of Figure 4.7 in having a "multiplying" region of n-type material, denoted by N_m, directly behind the p-type region, P_c. The region P_c is supposed to be so thin that electrons entering it from N_m have a good chance of diffusing through it before combining with a hole. Owing to the operation of the emitter, a hole current

[4] W. Shockley, *Phys. Rev.* **78**, 294–295 (1950).

is injected across J_e, through N_b and across J_c. These holes tend to become trapped in P_c because of the "hook" in potential which is produced by the junction J_m. We next consider the current flowing across J_m. If N_m is much more n-type than P_c is p-type, then most of the current across J_m will be carried by electrons as discussed in Section 4.2. In addition, the small width of P_c permits electrons to escape to the left in Figure 4.13 without recombining; this also tends to reduce diffusion back to N_m and hence further to enhance the electron current. As a result the electron current flowing across P_c to the base will be much larger than the hole current flowing across J_m to N_m. Consequently J_m acts as a current multiplying junction and causes the hole current arriving at the collector to produce a greatly enhanced electron emission. This explanation of high α is referred to as the *p-n hook theory.*

It may be noted that a high degree of symmetry exists between the right and left sides of Figure 4.13, except for the reversal of sign. This symmetry suggests another way of viewing the role of the hook region. This region is substantially the same as an electron emitter N_m and a p-type base P_c. Hole arrival at P_c biases this electron emitter forward and produces electron injection into P_c and thence to J_c, which plays the role of a collector junction in both cases.

The theory of the hook current multiplier is mathematically very similar to the theory of the p-n-p transistor for which the basic mathematical equations have been published elsewhere.[5]

Collector contacts which appear to function by the hook mechanism just discussed have been produced by W. G. Pfann, A. E. Anderson, R. M. Ryder, R. L. Wallace, and L. Valdes at Bell Telephone Laboratories. Values of α as high as 20 or more have been obtained by this means.

From the previous discussion of the simple space charge case, a limiting value of $1 + b$ was set for α_i. If the value of b could be increased, it is evident that the hole space charge would be relatively increased and α would be raised. One possible process which might have this effect is that of "hole trapping." If a concentration of centers which had the property of binding holes tightly could be produced directly in front of the collector, then the holes would tend to accumulate there with a resultant increase in space charge. If we suppose that each hole spends, on the average, a fraction x of its time in these traps and a fraction $1 - x$ of its time moving with normal mobility, then the net effect will be to reduce its mobility by a factor $1 - x$ and to raise the effective value of b to $b_{\text{eff}} = b/(1 - x)$. If $(1 - x)$ is small, this process will lead to very large values of b_{eff} and hence to large possible values of α.

Of the various mechanisms proposed for high values of α, it appears that the p-n hook is in best agreement with the facts and that considerations

[5] W. Shockley, *Bell Syst. Tech. J.* **28**, 435–489 (1949).

based on this idea have good promise of leading to advantageously designed units.

4.6. Phototransistors and Counters. The phenomenon of current multiplication at a collector and the phenomenon of optical generation of holes and electrons have been combined by J. N. Shive in the phototransistor.[1] The phototransistor is somewhat similar to the coaxial transistor,[2,3] which consists of a thin wafer of germanium with emitter and collector terminals on opposite sides. In the phototransistor, the emitter is lacking and holes are generated in the *n*-type germanium by light absorption. These holes then flow to the collector and produce a multiplied photocurrent. Thus if each photon produces one hole-electron pair and the hole in reaching the collector provokes the emission of two electrons, $\alpha_i = 3$, then three electronic units of charge will be produced per photon absorbed. The ratio of number of electronic charges produced per photon absorbed is called the *quantum efficiency;* for the phototransistor this efficiency can be greater than 100% because of the α_i of the collector.

The quantum efficiency of the light absorption process itself has been studied by Goucher.[4] In his experiment the change in conductance of an *n*-type germanium filament due to illumination was measured. This change should be proportional to the total number of holes present, the formula being given in problem 3 of Chapter 3. The number of holes present is equal to the rate of generation times the lifetime, as may readily be established by the methods of Section 12.6. The rate of generation is equal to the quantum efficiency times the rate of photon absorption, which is determined from the energy and wave length of the light and the optical constants of the germanium. In Goucher's experiments the lifetime was measured by hole injection techniques like those discussed in Sections 3.1d and 12.6. By comparing the change in conductance with the rate of photon absorption, Goucher concludes that over a range of about 1.0 to 1.8 μ each photon absorbed produces one hole-electron pair.

Collector junctions have also been used by K. G. McKay[5] to detect holes produced in the germanium by alpha particles. He finds that the sensitive region has a diameter of between 10^{-3} and 10^{-2} cm and that the maximum pulse height corresponds to the passage of 10^6 electrons. The pulse width was less than 0.05 microseconds. McKay has also suggested the use of large area *p-n* junctions for counters and such effects have been observed by Orman, Fan, Goldsmith and Lark-Horovitz.[6]

[1] J. N. Shive, *Phys. Rev.*, **76**, 575 (1949).
[2] J. N. Shive, *Phys. Rev.*, **75**, 689–690 (1949).
[3] R. L. Wallace and W. E. Kock, *Elect. Engg.*, **68**, 222–223 (1949).
[4] F. S. Goucher, *Phys. Rev.*, **78**, 646 (1950).
[5] K. G. McKay, *Phys. Rev.*, **76**, 1537 (1949).
[6] C. Orman, H. Y. Fan, G. J. Goldsmith and K. Lark-Horovitz, *Phys. Rev.*, **78**, 646 (1950).

p-n junctions themselves are quite effective photocells.[7] It is not necessary for the hole-electron pairs to be generated in the junction itself to produce a current but merely near enough so that there is a good probability of the "injected" carriers diffusing to the junction. This probability is determined by the diffusion length discussed in Section 12.5.

Problems

1. Assume that the probability that an injected hole recombines in time dt is dt/τ_p. Show that the fraction of the holes injected at time t' which are uncombined at time t is $\exp (t' - t)/\tau_p$. (See Section 8.3 for a similar analysis of physically different problems.)

2. Suppose that in a filamentary transistor the emitter current is $i_\epsilon(t)$ and consists of a fraction γ of holes. If the transit time is τ_t, show that the number of holes in the filament is

$$p_{\text{tot.}} = (\gamma/e)\int_{t-\tau_t}^{t} i_\epsilon(t') \exp [(t' - t)/\tau_p]dt'$$

and hence, using problem 3 of Chapter 3, that the increase in current is

$$\begin{aligned}\delta i_c(t) &= (V_J - V_c)\gamma(\mu_n + \mu_p)L^{-2}\int_{t-\tau t}^{t} i_\epsilon(t') \exp [(t' - t)/\tau_p]dt' \\ &= \gamma(1 + b)\int_{t-\tau_t}^{t} i_\epsilon(t') \exp [(t' - t)/\tau_p]dt'/\tau_t.\end{aligned}$$

Show that this reduces to equations (18) and (19) of Section 4.1 for $\delta i_c(t)/i_\epsilon(t) = \alpha_e = \gamma(1 + b)\beta$ when $i_\epsilon(t) = i_{\epsilon 0} \exp i\omega t$.

3. A p-n junction has a resistance of $R = 400$ ohms for low voltages and an area of $A = 0.5\ \text{cm}^2$. Assuming that the current is carried chiefly by holes and using (1) of Section 4.2, show that

$$(e/kT)I_{ps}A = 1/R$$

so that

$$I_{ps} = kT/eAR = 1/39AR = 1.28 \times 10^{-4}\ \text{amperes/cm}^2.$$

Assuming that the lifetime of a hole in the n-region is 100 microseconds, show from (3) of Section 4.2 that this leads to

$$p_n = (D_p\tau_p)^{1/2}/ARe\mu_p = 1.2 \times 10^{12}\ \text{cm}^{-3}$$

and that this concentration corresponds to a conductivity for holes of $3.3 \times 10^{-4}\ \text{ohm}^{-1}\ \text{cm}^{-1}$. This value is much smaller than 2.1×10^{-2}

[7] See footnote 1 of Section 4.2.

$\text{ohm}^{-1}\ \text{cm}^{-1}$ corresponding to the conductivity of pure germanium at room temperature and shows that the high resistance of a *p-n* junction arises from the difficulty of forcing hole current to pass through *n*-type material where there are very few holes to carry it. This subject is discussed more fully in Chapter 12.

PART II DESCRIPTIVE THEORY OF SEMICONDUCTORS

PART II

Books on the Energy Band Theory of Solids

W. Hume-Rothery, "Metallic State; Electrical Properties and Theories," *Oxford, Clarendon Press*, 1931. 367p.
A. Sommerfeld and H. Bethe, "Elektronentheorie der Metalle," *Handbuch der Physik*, **24**, pt. 2, 333–622. Berlin, Springer, 1933.
P. P. Ewald, Th. Pöschl and L. Prandtl, "The Physics of Solids and Fluids," *Blackie*, London, 2 ed. 1936. 306p.
H. Fröhlich, "Elektronentheorie der Metalle," *Springer*, Berlin, 1936. 386p.
N. F. Mott and H. Jones, "The Theory of the Properties of Metals and Alloys," *Oxford, Clarendon Press*, 1936. 326p.
A. H. Wilson, "Theory of Metals," *Cambridge Univ. Press*, 1936. 272p.
A. H. Wilson, "Semiconductors and Metals; an Introduction to the Electron Theory of Metals," *Cambridge Univ. Press*, 1939. 120p.
N. F. Mott and R. F. Gurney, "Electronic Processes in Ionic Crystals," *Oxford, Clarendon Press*, 1940. 275p.
F. Seitz, "The Modern Theory of Solids," *McGraw-Hill*, 1940. 698p.
F. Seitz, "The Physics of Metals," *McGraw-Hill*, 1943. 330p.
J. C. Slater, "Introduction to Chemical Physics," *McGraw-Hill*, 1939. 521p.
W. Hume-Rothery, "Atomic Theory for Students of Metallurgy," *Institute of Metals*, London, 1947. 288p.
W. Hume-Rothery, "Electrons, Atoms, Metals and Alloys," *Iliffe*, London, 1948. 377p.
F. O. Rice and E. Teller, "Structure of Matter," *John Wiley and Sons*, 1949. 361p.

Review Articles

F. Seitz and R. P. Johnson, "Modern Theory of Solids," *J. App. Phys.* **8**, 84–97, 186–199, 246–260 (1937).
J. C. Slater, "Electronic Structure of Metals," *Rev. Mod. Phys.* **6**, 209–280 (1934).

CHAPTER 5

QUANTUM STATES, ENERGY BANDS, AND BRILLOUIN ZONES

5.1 INTRODUCTION TO CHAPTERS 5, 6, AND 7

In non-quantum-mechanical treatments of conductivity, the current carriers are thought of as small particles moving in random thermal motions on which is superimposed the effect of electric and magnetic fields. Except for their smallness, the particles are imagined to have much the same attributes as larger solid bodies such as are encountered in ordinary experience, golf balls, for example. Although non-quantum-mechanical theories have failed to account for many aspects of the behavior of electrons, quantum mechanics, which includes wave mechanics, has succeeded. Quantum mechanics brings with it a set of new concepts which include the "wave function", "probability density", "energy level", "spin", "quantum state", and "Pauli exclusion principle" — concepts which are based on the development of the mathematical postulates of the theory.[1] In general, a mastery of this mathematical machinery is achieved only in a postgraduate course in physics. Fortunately, the results of the application of quantum mechanics to the motions of electrons in crystals do not need to be described in terms of the mathematical machinery itself but can be presented, instead, largely in pictorial terms. It is the purpose of Chapters 5, 6, and 7 to develop this picture.

Before encountering the details of the argument, the reader will benefit from a preview of the end result to be reached, together with some idea of the route to be followed. There is, in the author's opinion, a definite value in traversing a route from which one may view, at least superficially, the detailed processes used in developing a theory. The alternative would be to skip these intermediate steps and to present, at once, the final physical picture of the properties of holes and electrons as they emerge from the theory. This alternative seems undesirable for several reasons. It gives no idea of the fundamental concepts on which the behaviors of holes are based and thus opens the possibility of error due to applying the picture beyond its range of validity. Furthermore, from the purely scientific viewpoint, a major theoretical triumph is achieved by showing how the

[1] Chapter 14 is an introduction to some of these ideas. It follows an unconventional course and stresses analogies with transmission line theory.

electrons, through the mechanism of hole conduction, can simulate the behavior of positive particles.

The objective aimed at in this chapter and the next is a description of the behavior of an electron in an ideally perfect crystal, in which the perfection of the atomic arrangement is undistorted by thermal vibrations, impurity atoms, or other causes. The electron is found to have certain permitted modes of motion, quantum states. In these, it has certain energies and velocities. A consequence of Schroedinger's wave equation is, however, that the classical relationship $\mathcal{E} = mv^2/2$ does not apply to electrons in a crystal; and a further consequence is that the momentum is not equal to mv. A "crystal momentum", $\boldsymbol{P}$, defined in Section 5.5 in terms of the wave functions for electrons is introduced. As we find in Chapter 7, this vector quantity satisfies the equation of $d\boldsymbol{P}/dt = \boldsymbol{F}$ when a force $\boldsymbol{F}$ acts on the electron, a relationship which justifies calling it momentum in keeping with Newton's second law of motion. Unlike the momentum of a free particle, the crystal momentum of an electron is, in a certain sense, limited in range. Because of the repetitive or periodic property of the potential field of the crystal in which the electron moves, we need not consider larger values of momenta than those lying in a certain region, called the Brillouin zone, in the three-dimensional $\boldsymbol{P}$-space.

The energy and velocity of an electron are determined by the value of $\boldsymbol{P}$ inside the Brillouin zone. The relationship is multivalued, however, so that for a given value of $\boldsymbol{P}$ there will be a large number of possible quantum states, each with its characteristic energy and velocity. The various sections of this chapter are used to develop these ideas, the mathematical details being given in Part III. As a result we obtain a description of the behavior of an electron in the crystal in terms of energy, velocity, and momentum which are not related in the same way as they are in classical mechanics.

As an aid to the reader in following or reviewing the argument, Figures 5.1, 6.1, and 6.3 have been prepared. These indicate the logical sequence of the argument, and, where the steps involved have been presented in the text, reference numbers are given. These three figures represent the development of the Brillouin zone concepts of electron motion in a perfect crystal[2] under the condition for which no external forces act.

When an external force acts on the electron, then, as a consequence of the wave equation, it is found that the crystal momentum changes according to the law[3] $d\boldsymbol{P}/dt \equiv \dot{\boldsymbol{P}} = \boldsymbol{F}$. For the analysis of the conductivity and Hall effect, forces due to electric and magnetic fields must be considered. The arguments used in deriving the basic acceleration law are shown in

[2] This requirement of perfection excludes thermal vibrations or impurity atoms and thus eliminates the random processes which give rise to resistance.

[3] We shall occasionally use the symbol "·" for "d/dt" or "$\partial/\partial t$".

Figure 7.1 together with some of the consequences of this law. This material is discussed in Chapter 7, which deals with the dynamics of electrons in a perfect lattice.

The relationships of energy $\mathcal{E}$, velocity $\boldsymbol{v}$, and momentum $\boldsymbol{P}$ and the law of force $\dot{\boldsymbol{P}} = \boldsymbol{F}$ lead to a determination of the effects of applied fields on the

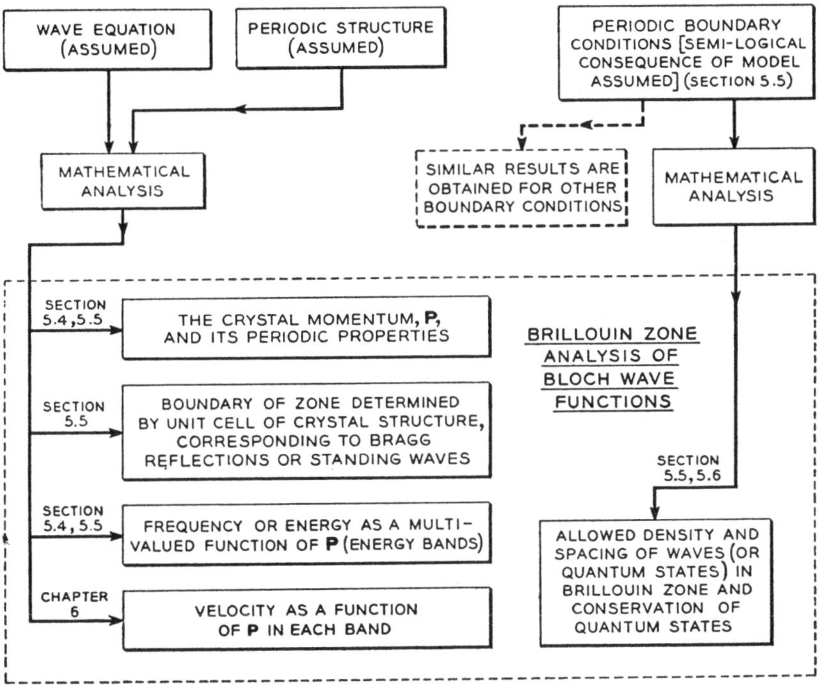

FIG. 5–1—The Brillouin Zone Method of Analyzing Bloch Wave Functions for Electrons in Crystals.

motion of an electron. The situation is quite like that for the classical case except for the different relationships between $\mathcal{E}$, $\boldsymbol{v}$, and $\boldsymbol{P}$. These differences are essential in explaining the processes in electrical conductors. On the basis of the new relationships between $\mathcal{E}$, $\boldsymbol{v}$, and $\boldsymbol{P}$, the behaviors of an excess electron and of a hole are analyzed. The results justify the description of holes and electrons as classical particles given in Chapter 1. This is the principal aim of Chapters 5, 6, and 7.

In later chapters, the effect of random processes which introduce electrical resistance and establish thermal equilibrium is discussed. These processes are caused by disturbances that make the crystal imperfect and are thus automatically excluded from Chapters 5 to 7, which are restricted to a discussion of a perfect crystal in which each atom is free from thermal motion and is exactly located at its proper lattice point.

5.2 QUANTUM STATES, SPIN, AND THE PAULI EXCLUSION PRINCIPLE

5.2a. The Nature of the Solutions of Wave Equations. An electron, as is now generally accepted, is a very small particle whose laws of motion are governed by Schroedinger's wave equation.[1] Schroedinger's equation combined with the Pauli exclusion principle furnishes a satisfactory basis for explaining most of the phenomena occurring in solid state physics and chemistry. We shall discuss how its application leads to the theory of energy bands, Bloch wave functions, and Brillouin zones. But first, as a necessary introduction, we shall comment briefly on the quantum theory for one-electron systems.[2]

When an electron moves in three-dimensional space (with its position described by coordinates x, y, z), under the influence of forces which can be represented by a potential energy $\mathcal{U}(x, y, z)$, its possible energies $\mathcal{E}$ (not a function of x, y, z) and modes of motion are described by Schroedinger's wave equation:

$$\frac{h^2}{8\pi^2 m}\left(\frac{\partial^2\psi}{\partial x^2} + \frac{\partial^2\psi}{\partial y^2} + \frac{\partial^2\psi}{\partial z^2}\right) + (\mathcal{E} - \mathcal{U})\psi = 0 \tag{1}$$

where h = Planck's constant;
m = the mass of the electron;
$\mathcal{E}$ = the energy of the electron;
ψ = $\psi(x, y, z)$, an unknown wave function.

We shall discuss the meaning of the wave function ψ below after we have compared this equation to several other equations of mathematical physics.[3] Thus, the wave equation for the scalar or vector potential of an electromagnetic wave oscillating with frequency ν is

$$\frac{c^2}{4\pi^2}\left(\frac{\partial^2 A}{\partial x^2} + \frac{\partial^2 A}{\partial y^2} + \frac{\partial^2 A}{\partial z^2}\right) + \nu^2 A = 0. \tag{2}$$

In Figure 5.2 (f) to (i) we represent a stretched membrane (drum head)

[1] This equation is known to be inexact but it applies well so long as the speed of the electron is small compared to the velocity of light. It stands in relation to the Dirac equation as Newton's laws of motion do with respect to Einstein's relativistic equations.

[2] This treatment follows in an abbreviated form one presented by the author in "The Quantum Physics of Solids I", *Bell Syst. Tech. J.* **28**, 645–743 (1939), a publication which is available on request to the Publication Department of Bell Telephone Laboratories as Monograph B-1184. Most of the material originally planned for II of the series, which was never written owing to the interruption of the war, has been incorporated in the present book.

[3] A more detailed discussion of the interpretation and solution of Schroedinger's equation, together with a comparison between it and an electrical transmission line, is given in Chapter 14.

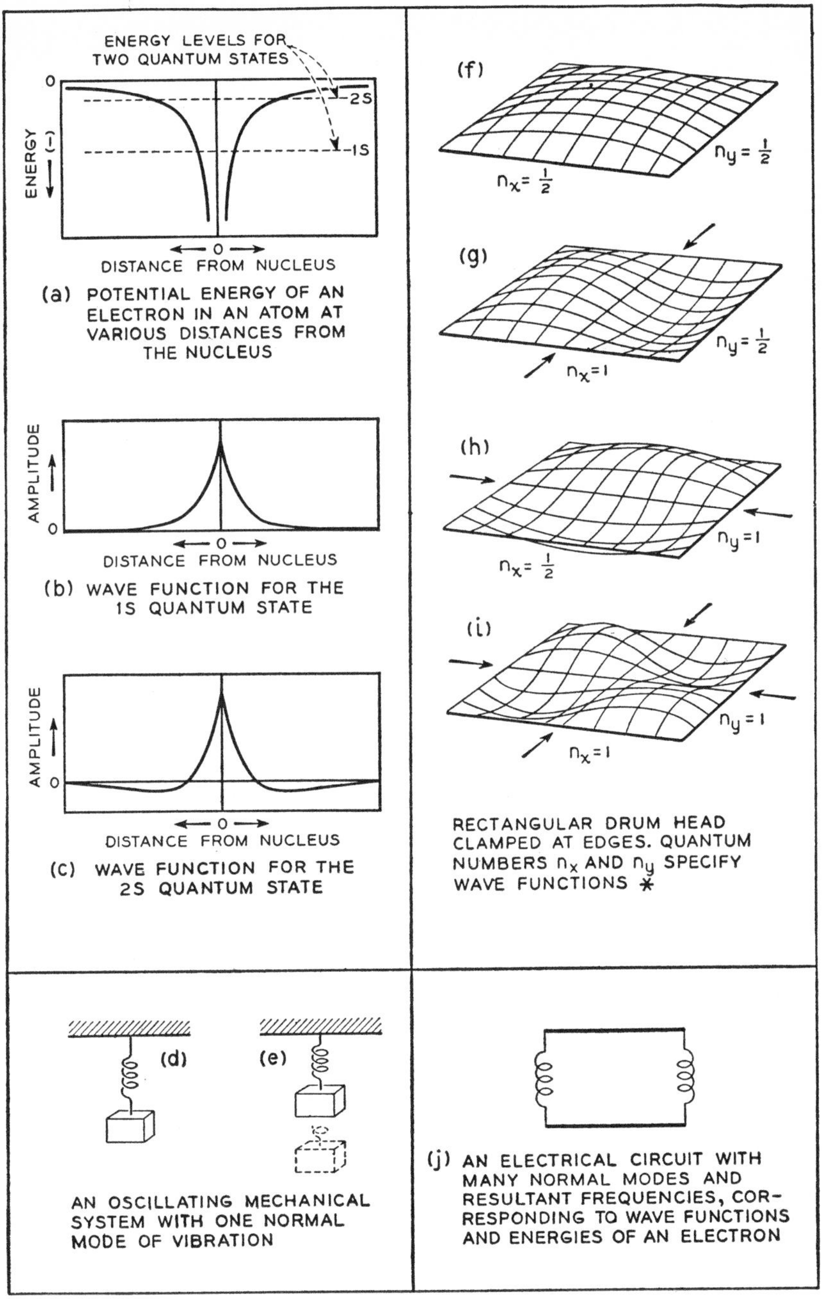

FIG. 5-2—Wave Functions of an Electron With Some Mechanical and Electrical Analogues. [(f) to (i) are from "Vibration and Sound", by P. M. Morse, McGraw-Hill, New York, 1937, Courtesy of the McGraw-Hill Book Co.]

whose vertical displacement ϕ satisfies the wave equation

$$\frac{T}{4\pi^2\sigma}\left(\frac{\partial^2\phi}{\partial x^2} + \frac{\partial^2\phi}{\partial y^2}\right) + \nu^2\phi = 0 \tag{3}$$

where σ is the mass per unit area, T the surface tension, and ν is the frequency. This equation has solutions, subject to the boundary condition that $\phi = 0$ on the rectangular boundary where the membrane is clamped, only for certain frequencies called the "normal modes" or "natural frequencies" of the drum head. These normal modes can be described by a set of "quantum numbers" n_x and n_y as shown in Figure 5.2. There is some choice as to the exact convention most suitable for the quantum numbers, and the ones selected for (f) to (i) have been chosen to conform to those later used for Brillouin zones. These quantum numbers are defined as the number of wave lengths of the vibration in each of the two directions so that each is equal to $\frac{1}{2} + \frac{1}{2}$ (number of nodal lines, heavy arrows, in the figure).

Another system, whose normal modes may be familiar to the reader, is the length of transmission line with choke coils across its ends (j). (This structure and the simple harmonic oscillator of (d) and (e) are introduced for use later in describing what happens when isolated atoms are brought together to form a crystal.)

Part (a) shows the potential energy function $\mathcal{U}$ for an electron in a hydrogen atom. It is defined so that the energy is zero when the electron is infinitely far away from the proton; consequently, $\mathcal{U} = -e^2/r$ where e is the electronic charge and r the distance from the proton. When this potential energy is put in Schroedinger's equation, solutions for ψ which satisfy the proper boundary conditions ($\psi = 0$ when r approaches infinity in this case) can be obtained only for certain values of $\mathcal{E}$. These solutions are called "eigenfunctions" or "proper functions", and the energies are called "eigenvalues" or "proper energies". It is found that these wave functions, two of which are shown in (b) and (c), can be described by a set of quantum numbers like those for the rectangular drum head. For example, the 1 in 1*s* and 2 in 2*s* are each just one more than the number of times the value of ψ crosses zero as a function of r.

A feature to be stressed about Schroedinger's equation is that the quantum numbers are not introduced in it as a special assumption.[4] They arise in just the same way as those shown with the rectangular drum

[4] There may be some choice about the numerical values assigned to the quantum numbers. Thus in Figure 5.2, n_x and n_y might have been chosen to be the number of half wave lengths along each edge, so as to avoid half integers. However, the point we wish to make is that no special assumption is made in order to introduce quantum numbers in the first place. This was not true of the old Bohr theory; in it quantum numbers were introduced as a special assumption.

head of Figure 5.2 or the normal modes of a resonant cavity. The mathematical aspects of all of these problems are very similar and all involve solving a differential equation subject to certain boundary conditions. When the proper wave functions are examined in detail, it is usually found that they are characterized by having a certain number of wave lengths or half wave lengths in the x, y, and z directions (or in three other directions depending on the boundary conditions). These numbers of half wave lengths are thus naturally arising quantum numbers which can be used to characterize the solution of the wave equation. For an electron in a hydrogen atom, the most suitable quantum numbers are simply related to the number of half wave lengths (or nodes in the wave function) in the r, θ, and φ directions in spherical coordinates. For our purposes, however, we do not need to refer to the detailed meaning of quantum numbers of electrons in atoms.[5] We shall be, however, much concerned with the quantum numbers used to describe the motions of electrons in crystals and will give these an interpretation in terms of the wave lengths of the wave functions involved (and thereby introduce the crystal momentum $\boldsymbol{P}$).

In connection with the motion of electrons through crystals, we shall later have occasion to compare the behavior of electron wave functions with those of electromagnetic waves or mechanical waves. For this reason the interpretation of the wave functions in both cases should also be compared.

According to quantum mechanics, the wave function ψ is a "probability amplitude" (a phrase developed in quantum mechanics and not found elsewhere), and the square of its absolute value, denoted by $|\psi|^2 = |\psi^2| = \psi^*\psi$, is a probability density.[6] The probability density has the following interpretation: the electron in its wave function is to be thought of as a very small particle with extent negligible compared to the size of the atom. It itself is not spread out over the volume occupied by the wave function. However, in its motion it traverses the space about the atom spending varying amounts of time in each region. The probability density at any point is simply proportional to the amount of time the electron spends (per unit volume) in a small volume about the point. In fact, the scale factor of the wave function is usually so adjusted that $|\psi|^2 dV$ is simply the fraction of time the electron spends in the volume dV. As a consequence, $|\psi|^2 dV$ is the probability of finding the electron in the volume dV. (ψ

[5] An elementary discussion of this topic is given in the Monograph B-1184. previously referred to.

[6] As discussed in Chapter 14, ψ itself may be regarded as a mathematical aid used in applying quantum mechanics to explain the behavior of electrons. There is even less purpose served in discussing what ψ is a displacement of than in discussing what kind of an ether is being stressed by an electromagnetic wave. For any set of values of x, y, z, and t, ψ is a complex number which can be written as $u + iw$ where u and w are real numbers. $|\psi|$ = the positive square root of $u^2 + w^2$ and $\psi^*\psi = u^2 + w^2$.

itself has thus the dimensions of $cm^{-3/2}$.) When these probabilities are added up for all space, the sum is unity since the electron must be somewhere. A wave function which adds up to unity in this way is said to be normalized.

Consequently, $|\psi^2|$ gives the long-time-average charge-density distribution produced by the electron. It is such average-charge densities that are represented in Figures 1.1, 1.2, 1.4, and 1.14. (These take the place of the Bohr orbits of the old quantum theory.) For many purposes, the electron moves so quickly over its possible positions that this average itself can be used. For other cases, such as for van der Waals' forces, it is necessary to take into account the fluctuations in position of the electron. Other aspects of the electron's behavior, such as its average velocity, can be obtained from the wave function in ways discussed in Chapter 6 and Chapter 14.

The probability density behaves in a way closely analogous to the energy density of electromagnetic or mechanical vibrating systems, provided that these have no loss, so that the energy of oscillation is conserved. For these cases the voltages and currents (or electric and magnetic fields) and the displacement and velocity are the amplitudes involved. The energy densities are proportional to the squares of these amplitudes. There is an important difference, however: In the electromagnetic or mechanical cases, the solutions are usually obtained in complex form, and, for these cases, the real part and the imaginary part of the solution are each individually equally meaningful (in the physical rather than mathematical sense) as solutions of the problem. Each of these solutions shows a transfer back and forth, twice each cycle, of electric to magnetic energy or of kinetic to potential energy. For the quantum mechanical case, on the other hand, the wave function is complex, its time dependence is of the form $\exp(-2\pi i\mathcal{E}t/h)$ where $\mathcal{E}$ is the energy, t is time, and h is Planck's constant, and neither the real nor imaginary part of ψ is alone a solution of Schroedinger's equation. Only the complex wave function itself is. The value of $|\psi^2|$ is thus independent of time and represents an unvarying probability distribution.[7] This is a natural result, since unlike energy density, which can exist in two forms, both for the electromagnetic and for the mechanical case, there is only one form of probability density so that the possibility of exchanging between one form and another is excluded. (There is, however, an exchange between the real and imaginary parts of the wave function ψ, which has a bearing on the wave velocities discussed later.)

In the cases of classical waves and in the quantum mechanical case as

[7] While $|\psi^2|$ is always independent of time for the normal mode solutions or eigenfunction solutions of Schroedinger's equation, there are other solutions in which $|\psi^2|$ varies with time and the probability distribution moves in space. Such cases are described in connection with wave-packets, Section 6.3.

well, it is sometimes of value to get a new wave function by adding together other wave functions. This addition is permitted because of the superposition principle which arises from the linearity of the differential equations involved. Considerations of wave functions built up by superposition are particularly helpful in understanding the bodily motion of the electron through empty space or through a crystal and we shall return to them in Section 6.3.

The important feature about the normal modes of any system, such as the drum head, is that any free oscillation for that system can be expressed in terms of them. Thus if the drum head is struck a sharp blow and then allowed to oscillate, its motion will not be represented by a single one of the normal modes of Figure 5.2 but instead by a combination of the effects of several normal modes, each oscillating with its own proper frequency, independently of the others, the total displacement being given by the sum of the component displacements. A disturbance of this sort does not have a single frequency and satisfies a wave equation containing a derivative with respect to time:

$$\frac{T}{\sigma}\left(\frac{\partial^2\phi}{\partial x^2}+\frac{\partial^2\phi}{\partial y^2}\right)-\frac{\partial^2\phi}{\partial t^2}=0. \tag{4}$$

The mathematical expression for the solution of this equation is obtained by adding together the solutions for the normal modes as follows: If $\phi_1(x, y), \phi_2(x, y)\cdots$ are normal mode solutions with frequencies $f_1, f_2, \cdots$, then a general solution of the time dependent equation is

$$\phi(x, y, t) = A_1\phi_1(x, y)\sin 2\pi f_1(t - t_1) + A_2\phi_2(x, y)\sin 2\pi f_2(t - t_2) + \cdots \tag{5}$$

where $A_1, A_2, \cdots$ and $t_1, t_2 \cdots$ are constants. For each of these normal mode terms individually, the time-dependent equation reduces to the one containing f^2 and no time derivatives.

The corresponding time-dependent Schroedinger equation is

$$\frac{h^2}{8\pi^2 m}\left(\frac{\partial^2\psi}{\partial x^2}+\frac{\partial^2\psi}{\partial y^2}+\frac{\partial^2\psi}{\partial z^2}\right)-\mathcal{U}\psi+\frac{ih}{2\pi}\frac{\partial\psi}{\partial t}=0. \tag{6}$$

It can also be solved by adding together eigenfunction solutions, each multiplied by a time-dependent term.[8] We shall have occasion to consider wave functions constructed in this way in connection with wave-packets, which represent the motion of the electron through the crystal, in Chapter 6. In this chapter, however, we shall deal only with eigenfunction solutions.

[8] Such time-dependent terms are considered in Chapter 14 and are also used in Equations (2) and (5) of Section 5.4.

5.2b. Spin and Quantum States. In addition to its motion through space, the electron also rotates about its own axis. This rotation introduces effectively a fourth degree of freedom and a fourth quantum number — the "spin". For reasons associated with the theory of relativity, the fourth quantum number is found to have only two permitted values, denoted by $+\frac{1}{2}$ and $-\frac{1}{2}$. In considering the cohesive forces between atoms certain "exchange" effects associated with the spin are of great importance. The spin of the electron also produces the magnetic moment which is responsible for ferromagnetism. For the purpose of treating the motions of electrons in semiconductors, however, we are concerned only with the fact that the spin gives the electron an extra degree of freedom so that it may move in accordance with a given wave function either with plus spin or with minus spin.

We are now in a position to introduce the concept of a "quantum state". *A quantum state describes a possible mode of behavior of an electron. It is specified by stating its four quantum numbers, three describing the wave function and the other the spin.*

So far our discussion has been concerned with the behavior of a single electron. When several electrons are simultaneously present and exert forces of electrostatic repulsion on each other, it is no longer exact to think of each electron as moving in a constant force field, and, consequently, it is not exact to speak of quantum states for individual electrons but instead only of the quantum state for the system as a whole. However, when large numbers of electrons are present, certain approximate methods may be used. According to these, when one electron is being considered, the effect of the others may be averaged to give a force field which is then taken as fixed and definite. The motion of the electron in this average force field is then investigated, and its quantum state is determined. The same process is carried out for all the other electrons also. (This process is usually repeated until a "self-consistent" solution is obtained: that is, starting from a set of trial wave functions for the electrons, the force field on each individual electron due to the other electrons is worked out. From these force fields a new set of wave functions is found. This process is repeated until a set of wave functions is obtained which produces a force field which gives the same set of wave functions back again.) To the degree of approximation used in the theory of semiconductors, this averaging of the force field is accepted as satisfactory, and the behavior of each electron is described by specifying the quantum state it occupies.

5.2c. Pauli Exclusion Principle. In terms of the quantum state concept we may now state the "Pauli exclusion principle" which is a foundation stone in the theory of matter.

The Pauli exclusion principle states simply that no two electrons may occupy one and the same quantum state.

If it were not for the existence of a law of nature of this form, all of the electrons in complex atoms would drop to the lowest energy level and the structure of the periodic table, the diversity of chemistry, and the reader of this book would not exist.

In applying the Pauli exclusion principle, care must be taken to distinguish properly between quantum states. Thus the exclusion principle does not require that, if one electron is in the lowest energy quantum state in a particular hydrogen atom, then no other electron in the universe can be in the same quantum state on another hydrogen atom. In fact, in a gas of hydrogen atoms, we should expect half the electrons to be in the lowest quantum state with one spin and the other half in the state of opposite spin. This apparent paradox is resolved by noting that, when we have two atoms, the quantum state scheme is essentially doubled. In fact the wave functions around two nuclei require a new set of quantum numbers. It is found, however, that, if the atoms are far apart, the wave functions themselves are simply what they were for one atom and are obtained twice over, once centered about one atom and next about the other. In other words, two separate atoms do not have the *same* quantum states; each of them has wave functions of exactly the same shape which are located at different points in space and correspond to distinct quantum states, one set for each atom.

The mathematical formulation of the Pauli exclusion principle is expressed by requiring that the wave function representing the system of electrons be "antisymmetrical". The meaning of this term and various other mathematical aspects of the Pauli exclusion principle are discussed in Section 15.7. It is not necessary, however, to study these analytical features in order to understand how the principle is applied in this and the following chapters.

5.3 QUANTUM STATES AND ENERGY BANDS IN CRYSTALS

There are two principal procedures for introducing the systems of quantum states in a solid. In one of these the average field method discussed in the last section is used to determine the individual motions of the electrons in the solid. It is necessary to use this method in describing the momentum, acceleration, and velocity of charge carriers in semiconductors. The other procedure traces the quantum states in the solid back to their origin in the isolated atoms of which the solid is composed. We shall treat the latter method first.[1]

Accordingly, we imagine that the N_s atoms of the solid are arranged perfectly on the crystal lattice in question but that they are separated by

[1] This treatment follows closely, in some respects, that of W. Shockley, "The Quantum Physics of Solids I", *Bell Syst. Tech. J.* **17**, 645 (1939), Monograph B-1184.

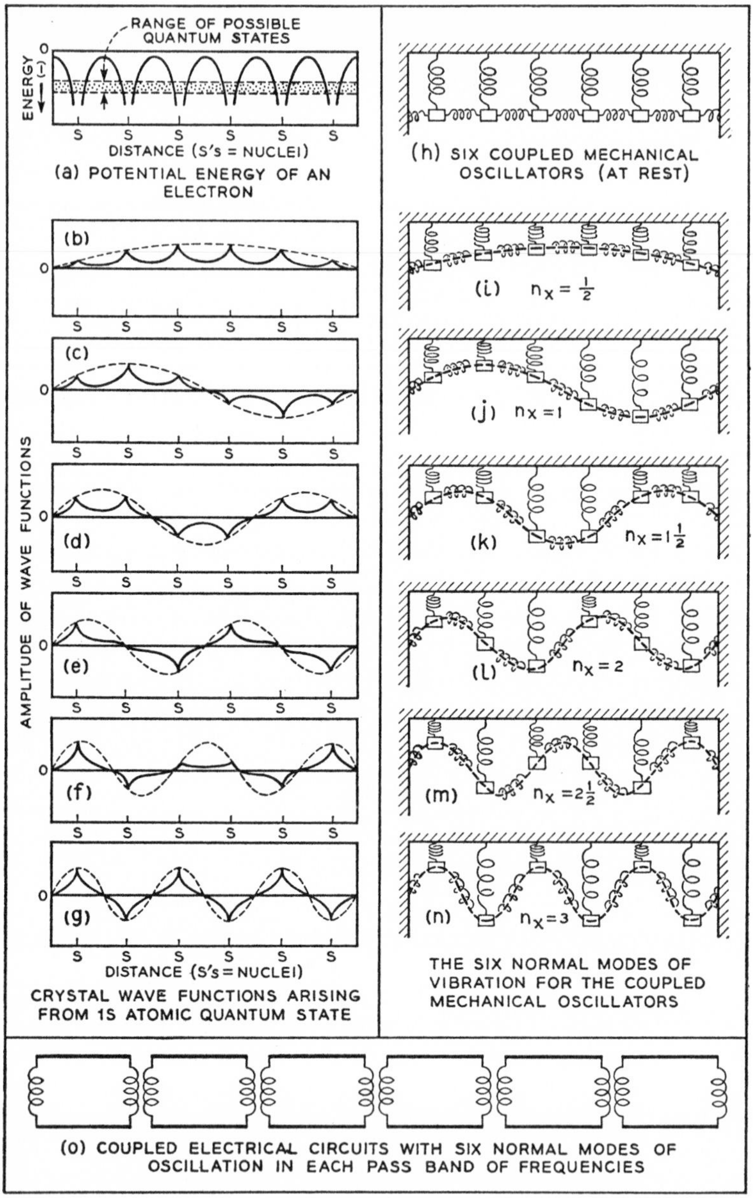

FIG. 5-3—A One-dimensional Crystal and Some Mechanical and Electrical Analogues. [An electron, because of the wave mechanical laws that govern it, can move through a region where its potential energy is greater than its total energy, as shown in (a). In this region its wave function is damped like an electromagnetic field in a wave guide when the frequency is below cutoff.]

many times the normal lattice constant so that there is substantially no interaction between them.[2] The quantum state distribution for the crystal is then essentially that of one atom duplicated N_s times. That is, each energy level corresponding to an atomic quantum state occurs N_s times, once for the corresponding quantum state on each atom. However, as the lattice constant is reduced the situation changes and, when the atoms are so close together that appreciable overlapping of the wave functions between adjacent atoms occurs, the quantum state and energy level schemes are profoundly modified. The result of this modification is shown pictorially in Figure 5.3. In parts (h) to (n) of the figure, the effect of interaction is illustrated in terms of a set of one-dimensional mechanical oscillators. The interaction between atoms is represented here by weak springs coupling the oscillators together. When the coupling is zero, each oscillator has its own normal mode of vibration, all of the same frequency. As soon as coupling occurs, the vibrations can no longer be restricted to one oscillator but extend throughout the system. However, if the coupling is weak, each of the modes (i) to (n) will have almost the same frequency as an isolated oscillator. As the coupling becomes stronger, the band of frequencies spreads over a wider range.

A very similar behavior occurs for the atoms. As soon as the wave functions overlap, the quantum states are no longer restricted to single atoms but instead extend over the entire crystal as is suggested in (b) to (g) of Figure 5.3. The energy levels corresponding to the quantum states split in much the same way as do the frequencies of vibration for the system of oscillators. In this way each atomic energy level leads to a *band* of energy levels or *energy band* in the crystal. This is illustrated in Figure 5.4 which shows qualitatively the results predicted from Figure 5.3.

In the process of splitting, the total number of degrees of freedom of the mechanical oscillators and the number of quantum states for the atoms are invariant. Thus, for example, if the six atoms of Figure 5.3 contained twelve electrons, so that the lowest quantum state of each spin was occupied, then there would still be just enough quantum states to accommodate all the electrons in the band of energy levels produced by the interaction; there would be six wave functions each giving rise, because of spin, to two quantum states.

For systems of larger numbers of atoms, there will be more wave functions. However, the one of lowest energy will have a slowly varying modulation wave like that of the dashed line in Figure 5.3(b) and the one

[2] The crystal lattice describes the relative arrangement of the atoms. The diamond structure of Figure 1.3 is common to carbon, silicon, germanium, and one modification of tin. Ordinarily by "lattice constant" one means the value observed in nature. In this chapter, however, we use "lattice constant" as a variable and "normal lattice constant" for the one observed in nature.

of highest energy will be simply a continuation of (g) to more atoms.[3] Hence the upper and lower energies for the electron wave functions, or frequencies for the normal modes of the analogues, will be independent of the length of the structure, provided that it contains a large number of units. The conclusions just reached can be summarized in two basic theorems for energy bands:

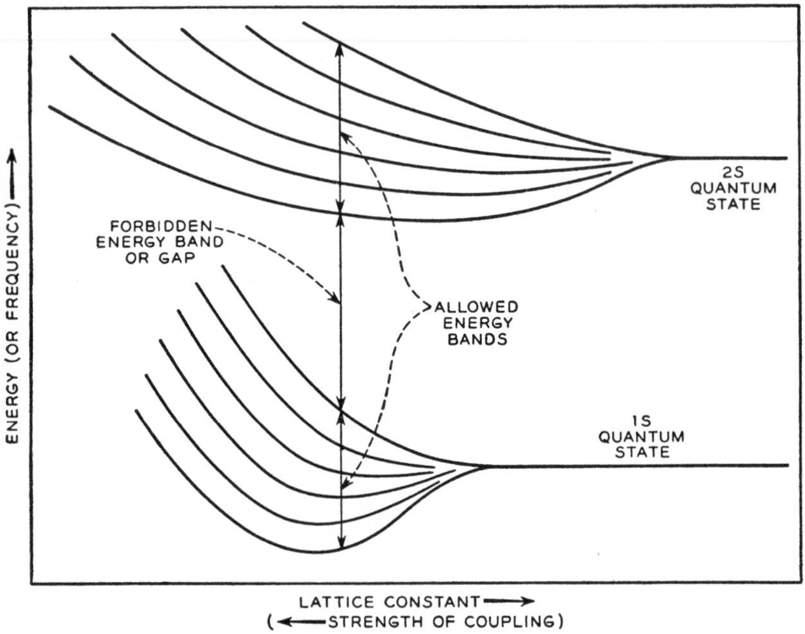

FIG. 5–4—Dependence of Energy Levels upon Lattice Constant, or Frequency of Vibration upon Strength of Coupling.

THE BAND WIDTH THEOREM: *The width of the energy band arising from an atomic energy level will be independent of the number of atoms in the crystal.*[4]

THE THEOREM OF THE CONSERVATION OF QUANTUM STATES: *The number of quantum states in the energy band will be the same as the number of atomic quantum states from which the band was produced.*

There are no quantum states with energies between those shown in the energy bands. For this reason the energy range running from the top of

[3] We shall be intimately concerned with waves like (g), for which the wave length λ is twice the period a of the structure, in the remainder of this chapter. Such waves correspond to the edges of the Brillouin zone.

[4] If we imagine that the lattice constant can be varied as shown in Figure 5.4, then the width of the band depends on the lattice constant. It also depends upon the energy level from which the band arose.

one band to the bottom of the next is referred to as a *forbidden energy band* or *energy gap*. The energy bands containing quantum states are referred to as *allowed energy bands;* although, where confusion is unlikely to arise, they are usually referred to simply as "energy bands", as was done in the two theorems.

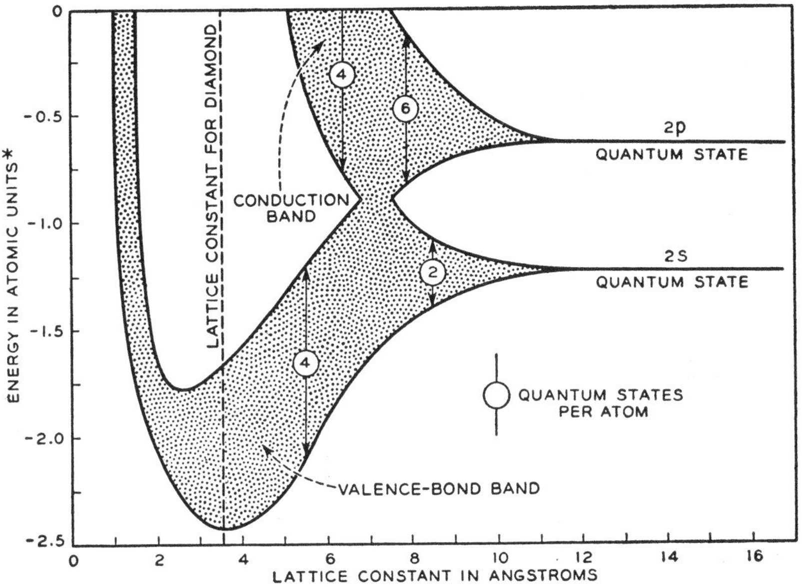

FIG. 5–5—Energy Bands for Diamond versus Lattice Constant.

* The unit in this figure is the Rydberg = 13.5 electron volts = ½ the atomic unit of Appendix A.

For the case of carbon, the important energy bands arise from the quantum states occupied by the valence electrons. These are denoted as the $2s$ and $2p$ states and contain a total of eight quantum states per atom. They are, therefore, half occupied in the isolated atoms. Each atom also contains two electrons in the $1s$ state. However, these electrons have their wave functions so compactly bound to the nucleus that no appreciable overlapping between adjacent atoms occurs. They can be considered, therefore, as unaffected by their being in the crystal. The energy bands for carbon[5] in the diamond structure are shown in Figure 5.5. At large lattice constants, the lower band is filled and the upper partially filled. At smaller distances, however, the eight states per atom are split equally between the two bands so that the lower band is completely filled. This lower band corresponds to electrons in the valence-bond structure. To

[5] G. E. Kimball, *J. Chem. Phys.* **3**, 560 (1935).

excite an electron requires enough energy to raise it to a quantum state in the upper band. It is to be noted that the average energy of an electron in the lower band at the actual lattice constant is much less than for the free atoms. This drop in energy is associated with the "binding energy" of the crystal. A comparison of the total energy of a set of isolated carbon atoms with that of the same atoms arranged in the crystal shows that the loss in energy of the valence electrons makes the latter much more stable. The energy required to pull the crystal apart into free atoms is called the binding energy. For a valence crystal, like diamond, the approximations employed in the energy band method of calculation are probably less accurate than those dealing directly with the electron-pair bond so far as binding energy is concerned. However, our main interest in quantum-mechanical theory centers about the problems of electrical conductivity rather than those of cohesion, and for conductivity the energy band theory is not only the best approximation currently available but apparently an adequate one as well.

5.4 BLOCH WAVE FUNCTIONS IN A PERIODIC POTENTIAL FIELD

In the last section, we concluded that the energy levels of an array of isolated atoms are split into energy bands when the atoms are brought close together to form a crystal. By analogy with the behaviors of classical systems, the theorems on band width and conservation of quantum states were developed. We must next consider the wave functions obtained from the viewpoint of their nature as solutions of the wave equation for an electron moving in a periodic potential. This will form a basis for later discussions of the velocity and acceleration of the electrons.

The solutions obtained for Schroedinger's equation for an electron in a crystal depend upon three attributes of the problem considered:

(1) The form of the potential field in which the electron moves: that is, the value of $\mathcal{U}(x, y, z)$ at each point in one unit cell of the crystal. (Since the crystal is periodic the value will be the same in every unit cell.)
(2) The size of the crystal.
(3) The boundary conditions at the surface of the crystal.

The nature of the problem, as discussed in Section 5.2, leads to the introduction of a set of quantum members which can be used to specify the wave functions and the associated characteristic energies.

In order to deal with running waves which carry current through the crystal, we shall consider a type of boundary condition different from that implicitly introduced in connection with Figure 5.3. We shall see below that this change in boundary conditions does not affect the theorem of the conservation of quantum states.

The *boundary condition* shown in Figure 5.3 is one for which the wave function vanishes at the ends of the one-dimensional crystal. We shall refer to this case as "vanishing" boundary conditions. These boundary conditions are mathematically equivalent to setting up infinite force fields just outside the crystal which reflect the electron inwards. Under these conditions the electron wave is a standing wave which can be regarded as made up of two running waves. However, each of these running waves individually, although it does satisfy Schroedinger's equation for the particular potential $\mathcal{U}$, is not a solution which satisfies the boundary conditions. The normal modes for the coupled oscillators similarly satisfy boundary conditions for which the end points of the half-coupling springs at the ends of the row have no displacement; the dashed modulating waves show this behavior by having nodes at the end. For these boundary conditions, the electron waves (b) to (g) do not carry any probability density along the crystal nor do the mechanical oscillations (i) to (n) carry any mechanical energy. In order to deal with eigenfunctions or normal modes which support a flow, it is most convenient to use periodic boundary conditions.

Periodic or *cyclic* boundary conditions amount to doubling the crystal back on itself or arranging the coupled oscillators in a closed loop. Such an arrangement is shown in Figure 5.6. The six oscillators are now mounted below a rigid ring and are coupled together as for Figure 5.3. Some means of constraint is assumed which restricts their motion to the vertical axis. The lowest frequency of oscillation is that shown in (a). For this case, all the bobs move together and the coupling springs are inactive. In (b) the motion is given by a sine wave function which has one wave length around the loop. In this case, the coupling springs between (2) and its neighbors (1) and (3) exert a restoring force on (2) and cause the frequency of the normal mode of (b) to be higher than that of (a). [In order to simplify the diagram, the coupling springs have been omitted from the figure except in (a).] There are two normal modes for (b), one for each direction of travel of the sine wave. The frequency of these waves would be the same as (j) in Figure 5.3; in fact, the standing wave (j) can be obtained by allowing the right- and left-hand waves of Figure 5.6(b) to interfere. These two waves are given the quantum numbers $n_x = +1$ and $n_x = -1$: $+$ for the wave running towards $+x$ and $-$ for the wave running towards $-x$, and 1 because the wave has one wave length around the crystal. In accordance with this convention, we have the relationship $\pm 1/\lambda = n_x/A_x = n_x/N_x a$, which we shall use later.

Part (c) of Figure 5.6 shows a wave with two wave lengths around the loop, $n_x = \pm 2$. There are again two normal modes, with waves running

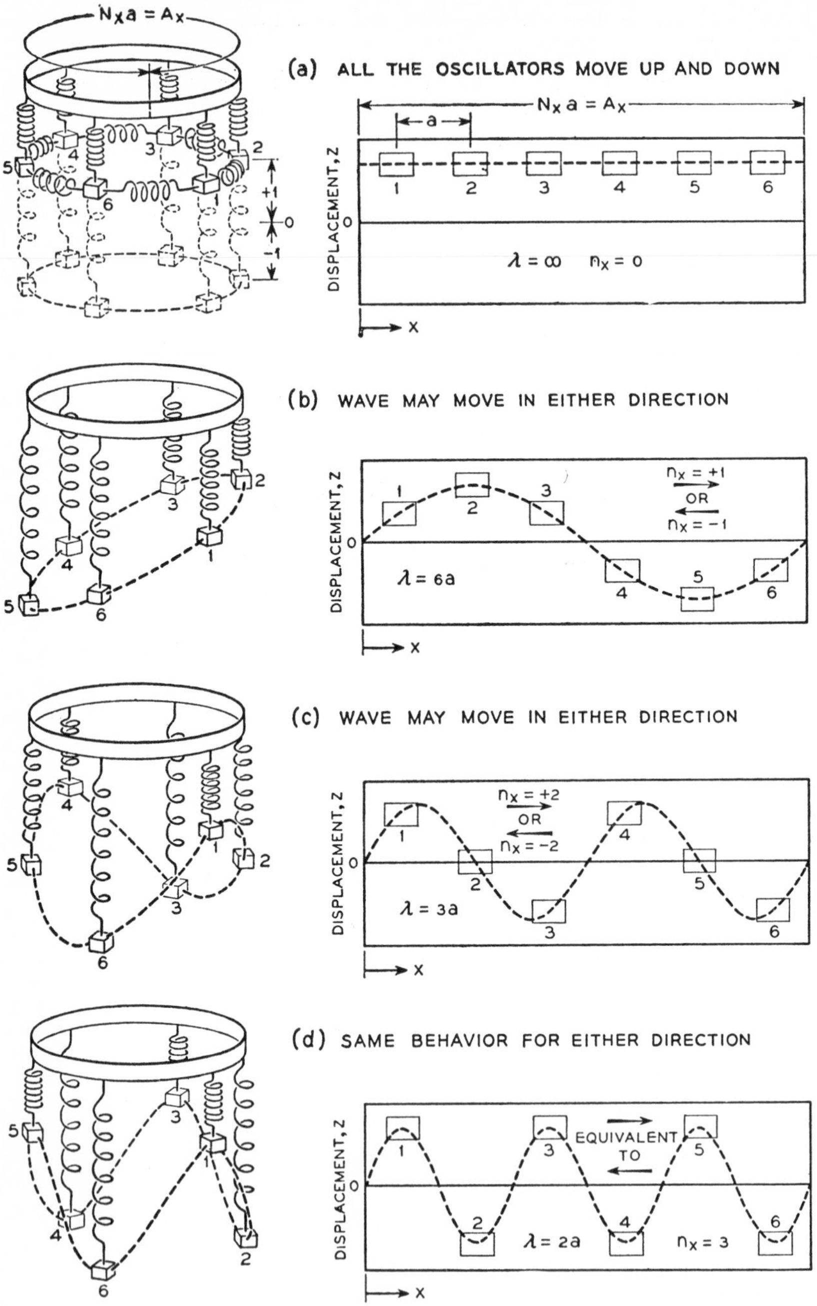

FIG. 5-6—Normal Modes for Coupled Oscillators with Cyclic or Periodic Boundary Conditions.

to the left and to the right each having the same frequency as (1) of Figure 5.3.

A critical condition occurs for $\lambda = 2a$, however. For this case, both the right-going and left-going waves produce exactly the same motion of the bobs, in this instance a very simple one in which the even-numbered bobs have identical harmonic oscillations just 180° out of phase with odd-numbered bobs.

We see that there are altogether six normal modes for the periodic boundary condition of Figure 5.6, just as there were six for the "clamped end" boundary condition of Figure 5.3. However, the proper frequencies are somewhat different in the two cases.[1]

We must next consider the solutions of Schroedinger's equation which are analogous to the running waves for the coupled mechanical oscillators just discussed. Some insight into the relationship between the two problems may be gained by writing the equation for the displacement produced by a sine wave with wave length λ, frequency ν, velocity c (equal to $\lambda\nu$), and unit amplitude in Figure 5.6. The displacement z of the bob at position x, measured along the rim of the circle, may be written in any one of several forms:

$$z = \sin 2\pi\left(\frac{x}{\lambda} - \nu t\right) = \sin (2\pi/\lambda)(x - ct)$$

$$= \sin 2\pi\nu\left(\frac{x}{c} - t\right). \tag{1}$$

If we increase x by a, that is, go from one bob to the next, the phase angle in the wave is increased by $2\pi a/\lambda$. This effect is equivalent to replacing t by $t - a/c$. In other words, the oscillation at one bob is just the same as at the previous one except that it is delayed by a time a/c, the time required for the wave to traverse one period of the structure. The velocity for such waves varies with λ, an example being given in Chapter 15.

It seems logical to expect that the solutions to Schroedinger's equation for waves in the crystal will give the same behavior in every unit cell, except for a lag in phase angle corresponding to the time required for the wave to progress from one cell to the next. The proof that waves of this sort are eigenfunctions is due to F. Bloch[2] and the corresponding wave functions are called *Bloch functions*.[3] For a one-dimensional crystal the Bloch function is

$$\psi_\lambda(x, t) = \mathbf{u}_\lambda(x) \exp\left[2\pi i\left(\frac{x}{\lambda} - \nu_\lambda t\right)\right]. \tag{2}$$

[1] Some problems bearing on these matters are given at the end of this chapter.

[2] F. Bloch, *Z. Physik* **52**, 555 (1928).

[3] These subjects are discussed in more detail in Chapter 14.

In this $\mathbf{u}_\lambda(x)$ is the same function in every unit cell; it is periodic so that

$$\mathbf{u}_\lambda(x + a) = \mathbf{u}_\lambda(x). \tag{3}$$

It is thus seen that the behavior of $\psi_\lambda(x, t)$ is just the same at $x + a$ as at x after a lapse of time given by

$$\frac{a}{\lambda} = \nu_\lambda t \quad \text{or} \quad t = a/\lambda\nu_\lambda; \tag{4}$$

in other words, the wave progresses to the right with phase velocity $\lambda\nu_\lambda$. The periodic function $\mathbf{u}_\lambda(x)$ and the frequency ν_λ both depend upon λ as is indicated by the subscripts. $\mathbf{u}_\lambda(x)$ is, in general, complex, with a varying phase angle, although it becomes real for $\lambda = \infty$. (For $\lambda = 2a$, $\mathbf{u}_\lambda(x)$ varies as $\exp(-2\pi ix/\lambda)$ so that ψ_λ has a special behavior as shown in Figure 5.7(c).) The energy $\mathcal{E}_\lambda$ of the wave function also depends upon λ and is related to the frequency ν_λ by the equation[4]

$$\mathcal{E}_\lambda = h\nu_\lambda, \tag{5}$$

as may be easily verified by taking the partial derivative with respect to time in the Schroedinger time-dependent equation. As is always true for energy eigenfunctions, the value of $|\psi_\lambda(x, t)|^2$ is independent of time. [This follows analytically from the fact that the only time dependent term in (2) is $\exp(-2\pi i\nu_\lambda t)$.]

The remainder of this chapter and the next will be devoted chiefly to examining the behavior of the Bloch wave functions $\psi_\lambda(x, t)$ and interpreting the results. The value of λ, or better that of $\pm h/\lambda = hn_x/A = hn_x/N_x a$, will be found to be a useful "quantum number" to characterize the states in an energy band. Of particular importance is the significance of conditions like those shown in Figure 5.3(g) and (n) and 5.6(d) for which the phase difference is 180° for one period of the lattice or, in other words, $\lambda = 2a$. (These conditions determine the boundary of the Brillouin zone discussed in the next section.)

Figure 5.7 (p. 140) shows some running waves of the Bloch form. In order to give an idea of how the wave function ψ and probability density $|\psi|^2$ depend on time and distance, we have plotted the real and imaginary parts of ψ and also $|\psi|^2$ in the figure. As described in connection with the figure, we see that each wave gives the same probability density at every atom. However, the real and imaginary parts of ψ individually have the type of wave motion along the crystal required by equation (2) of Section

[4] This famous equation, which is now a part of the general formalism of quantum mechanics, was first discussed by Planck in 1900 in connection with thermal equilibrium of electromagnetic radiation in a cavity. In 1905 Einstein used it to interpret photoelectric emission by supposing that light quanta or photons deliver energy in units of $h\nu$ to electrons in the photoelectric effect.

5.4. As we shall see in Chapter 6, this wave motion is associated with electron current, the electron being, in a sense, carried along by the wave.

In terms of these running waves, the condition $\lambda = 2a$ can be given special significance. This is a condition for which the behavior of the running

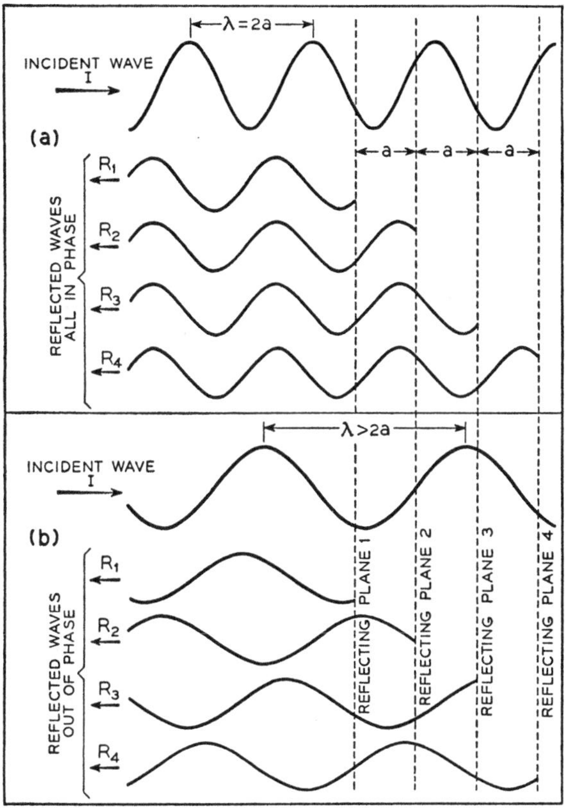

FIG. 5-8—Phase Relations between Reflected Waves, Showing Critical Condition for $\lambda = 2a$.

wave degenerates to that of a standing wave. Furthermore, the same standing wave is obtained whether waves running to the right or to the left are used initially. [Discussed in connection with Figure 5.6(d).]

Because of the fundamental importance of the limiting condition $\lambda = 2a$ in the theory of Brillouin zones, we shall use, in addition to the discussion of the wave behavior of Figure 5.7, another argument leading to the same result. In Figure 5.8(a) we imagine that a wave I whose sinusoidal behavior corresponds to $\lambda = 2a$ is traversing the crystal from left to right. As it passes by each plane of atoms, a certain fraction of it is reflected

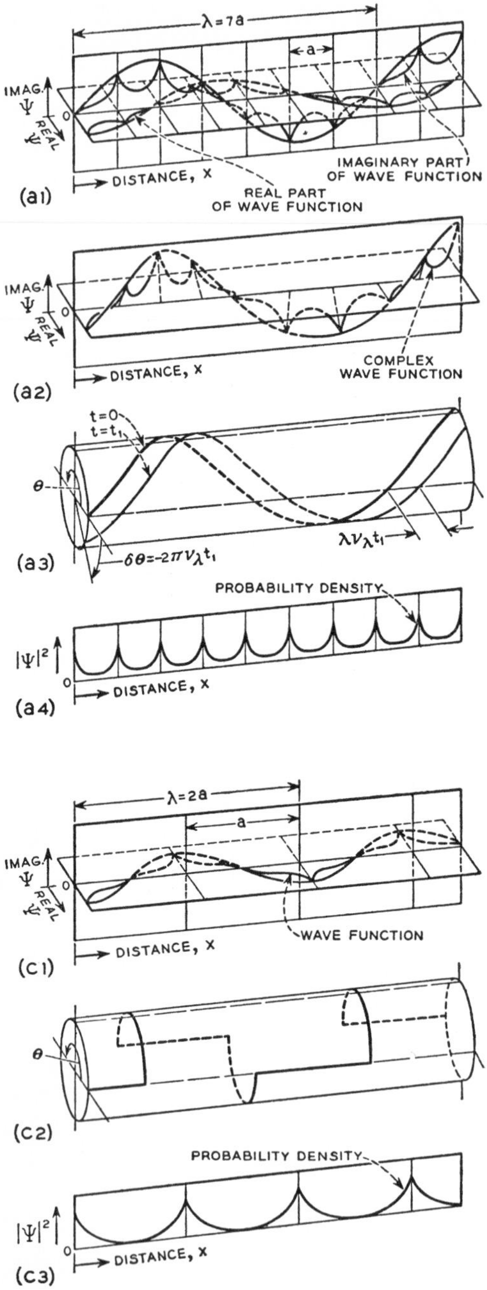

FIG. 5-7—The Wave Function $\psi = \mathbf{u}_P(x) \exp(iPx/\hbar)$.

This figure shows the behavior of wave functions for three quantum states of the lowest energy band and some wave functions for higher energy bands.

(a) shows a relatively long wave length, $\lambda = 7a$. Its real and imaginary parts (a1) consist of a part with period a multiplied by a sinusoidal part. The sinusoidal parts are related as sine and cosine. This results in the same value of $|\psi|^2 = (\text{Real } \psi)^2 + (\text{Imag. } \psi)^2$ in every unit cell (a4), since $\sin^2 + \cos^2 = 1$. Furthermore, the phase angle of the wave function, $\theta = \tan^{-1}(\text{Imag. } \psi/\text{Real } \psi)$, varies linearly with x, (a3), and goes through 360° in 7 lattice constants. As t increases from 0 to t_1, the phase angle for any fixed value of x changes by $\delta\theta = -2\pi\nu_\lambda t_1$, (a3). This is equivalent to moving the wave function along $(\delta\theta/2\pi)$ wave lengths. The phase velocity is, therefore, $\lambda(\delta\theta/2\pi)/t_1 = \lambda\nu_\lambda$.

(b) shows a wave with $\lambda = (7/3)a$, only slightly larger than the critical value, $\lambda = 2a$. The scale is enlarged to show the more rapid variations. Again, the sinusoidal waves for real and imaginary parts of ψ are 90° out of phase at the *centers of the atoms*. However, between atom centers, the phase angle changes irregularly, tending to remain constant near each atom center, then changing abruptly between atoms where the wave function is small. If $\mathbf{u}_\lambda(x)$ were real or had a constant phase angle, the phase angle for ψ would vary linearly with (x). However, $\mathbf{u}_\lambda(x)$

is, in general, complex and can be taken as real only for $\lambda = 0$.

(c) For $\lambda = 2a$, the wave function is multiplied by -1 (a phase change of 180°) between one atom and the next. For the case shown, the wave function is entirely real (at another time it will be entirely imaginary due to the $\exp(-2\pi i \nu_\lambda t)$ time dependence). The transition of phase by 180° occurs discontinuously at the midpoint between atoms where the wave function vanishes so that the phase is indeterminate. These discontinuities have been drawn like a left-handed screw in keeping with (a) and (b). However, exactly the same behavior of ψ would result if the discontinuities were like a right-handed screw. Thus, the same behavior of the wave function for $\lambda = 2a$ results for waves running initially in either direction. In other words, this is a standing wave.

(d) shows some real and imaginary parts of wave functions for electrons in the conduction band for metallic sodium as computed by Slater.* We include them to show that numerical calculations of wave functions of the Bloch type for higher energy bands have been carried out. The heavy sine-wave lines have the wave length λ of the Bloch waves, and the dashed lines show the plane free electron waves into which the Bloch wave functions would transform if the potential energy function $\mathcal{U}$ were diminished to zero.

* J. C. Slater, *Rev. Mod. Phys.* **6**, 209–280 (1934).

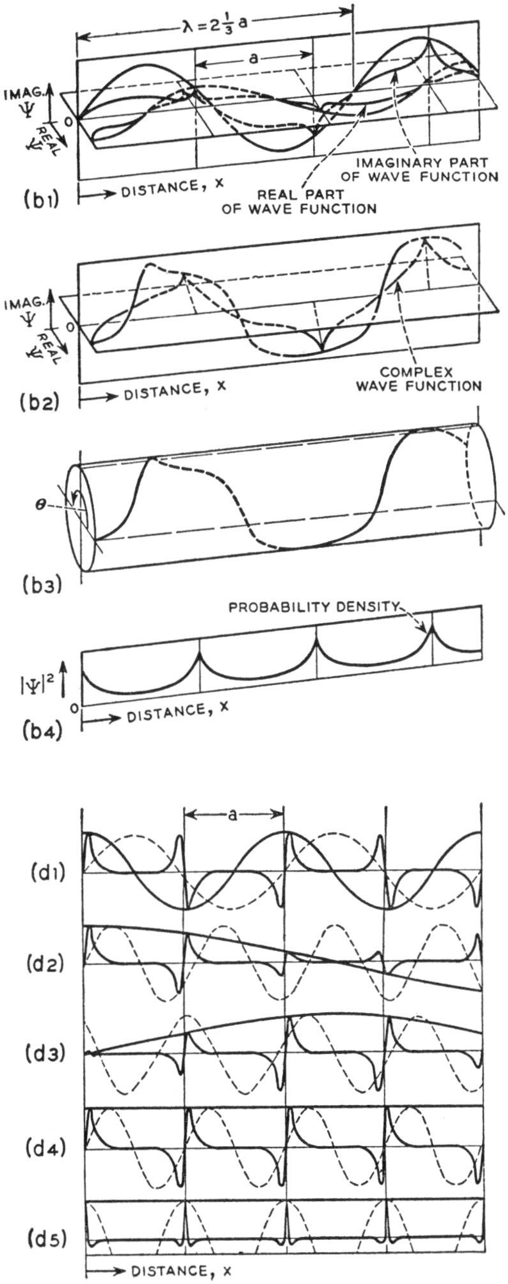

giving rise to reflected waves R_1 to R_4 from planes 1 to 4, which have been arbitrarily selected for study. As is seen, each reflected wave starts 180° out of phase with the wave of the preceding plane. However, by the time it has reached the preceding plane it has lost an additional 180° and is in phase. Hence the waves R_1 to R_4 are all in phase and do not cancel one another. The situation is quite different in (b), where I has a wave length different from two lattice constants. Then the waves R_1 to R_4 are out of phase and cancel out on the average. As a consequence of this, it is possible for the wave in (b) to traverse the crystal with only a small fraction of reflected wave present. However, the wave in (a) will build up a reflected intensity just equal to the incident intensity; the reflected waves will then be converted to the incident wave by additional reflections just as fast as the incident is converted to reflected. However, this situation of waves of equal intensity moving in both directions is mathematically equivalent to a standing wave. Hence it is impossible to transmit energy or electrons through periodic structures such as crystals, or coupled mechanical or electrical oscillators, when the wave length is twice the period of the structure.

It is evident that, if we start with modulation waves of very long wave length and make them shorter and shorter, a natural limit is reached when $\lambda = 2a$. In fact, this range of wave lengths just covers the states in the energy band. However, in order to discuss the extension of the ideas presented here to three dimensions, and to lay the foundation for discussing velocity and momentum, we must introduce a suitable means of describing the modulation waves in three dimensions. The quantum number required for this purpose plays the role of momentum for electrons in crystals and is introduced in the next section.

Before discussing this next topic, the very close analogy between the results discussed here and certain electrical and acoustical problems should be mentioned. In electrical filter systems made of identical networks connected in tandem, there are frequency bands of transmission and bands of attenuation. The frequency at which the transition between these occurs corresponds to the condition $\lambda = 2a$, the currents in adjacent networks being 180° out of phase. Between the transmitting bands the electrical signal decays exponentially from network to network through the filter. The electron waves corresponding to energies lying outside the energy bands are similarly attenuated and cannot be made to fit the periodic boundary conditions we have considered. They are of interest in connection with the surfaces of semiconductors, however, but these are matters not covered in this text. The behavior is entirely similar for acoustical lines having periodic structures.[5]

[5] For a general treatment of such problems see L. Brillouin, *Wave Propagation in Periodic Structures*, McGraw-Hill Book Co., New York, 1946.

5.5 BRILLOUIN ZONES AND THE CRYSTAL MOMENTUM P AS A QUANTUM NUMBER

In this section we shall introduce the crystal momentum in place of the wave length as a means of characterizing the Bloch wave functions. The crystal momentum, which we shall shortly define, has many convenient properties: Used as a quantum number, its allowed values are described by a very simple scheme, and the critical condition $\lambda = 2a$ is simply represented in terms of the crystal momentum. In Chapters 6 and 7 we shall use it in studying the velocity and acceleration of electrons.

We shall first deal with a one-dimensional case. For this we define the crystal momentum P_x along the x-axis as follows:

$$P_x = \pm h/\lambda \tag{1}$$

(+ for waves running to the right, − for waves running to the left). Since h has the dimensions of action, P has the dimensions of momentum (mass × length/time).[1] Furthermore, if the periodic potential were imagined to diminish so that the Bloch functions became simply plane electron waves in free space (which quantum mechanics interprets as electrons moving freely with constant momentum), then equation (1) would give the correct relationship between momentum and wave length. In the crystal, however, the electron is acted upon by forces so that it cannot be considered to have a definite momentum and for this reason we have used the phrase "crystal momentum" to indicate that P_x is not the momentum in the ordinary sense. (We shall use capital P for crystal momentum, small p for ordinary momentum.) However, since P_x has the attribute, discussed later, of changing at a rate equal to the applied force, it may be used as a momentum for many purposes.

In terms of P_x, the Bloch wave function takes the form

$$\psi_{P_x}(x, t) = \mathbf{u}_{P_x}(x) \exp [(2\pi i/h)(P_x x - \mathcal{E}_{P_x} t)]. \tag{2}$$

We must next ask what values of P_x are allowed. The answer to this question depends on the size of the crystal and the boundary conditions. We shall consider a crystal consisting of N_x unit cells of length a and thus being $A_x = N_x a$ long, and shall assume periodic or cyclic boundary conditions equivalent to supposing that the crystal is bent around until the first and last atoms become near neighbors. As a consequence of this boundary condition, the wave function at positions x and $x + A_x$ must be identical because these two values of x are really the same place. In other words, ψ must be periodic with period A_x. Since $\mathbf{u}_{P_x}(x + a) = \mathbf{u}_{P_x}(x)$, we also

[1] This result may be recalled from the Heisenberg uncertainty relation, which specifies that the uncertainties in momentum and position satisfy the inequality $\Delta p \Delta x \geq h/4\pi$ where p represents ordinary momentum.

have $u_{P_x}(x + A_x) = u_{P_x}(x)$. Consequently, for ψ_{P_x} to be periodic with period A_x, we must have

$$\exp[(2\pi i/h)P_x(x + A_x)] = \exp[(2\pi i/h)P_x x], \quad \text{or} \quad \exp[(2\pi i/h)P_x A_x] = 1,$$
$$\text{or} \quad (2\pi i/h)P_x A_x = 2\pi i n_x \tag{3}$$

where n_x is an integer and is, in fact, simply the number of wave lengths of ψ around the cyclic crystal (as shown in Figure 5.6). This leads to the condition

$$P_x = n_x h/A_x \quad \text{or} \quad \delta P_x = h/A_x. \tag{4}$$

Thus the allowed values of P_x are spaced h/A_x apart. This situation is represented in Figure 5.9(a). The P_x-axis extends to $\pm h/2a$ corresponding to the limiting condition $\lambda = 2a$ previously discussed, and, as explained, the wave functions at these two limiting points are identical. As we shall show later, values of P_x lying outside the range $-h/2a$ to $+h/2a$ are equivalent to P_x-values lying in this range. *This region of P_x space, which includes a complete set of non-equivalent points and is centered at $P_x = 0$, is defined as the one-dimensional Brillouin zone.* Its chief importance is that we need not, for most purposes, consider any P_x-values lying outside of the Brillouin zone since all the eigenfunctions for the one-dimensional crystal are contained in the zone. In fact, because of the reciprocal relationships:

(Size of Brillouin zone) $\propto$ (Spacing in crystal)$^{-1}$

(Size of crystal) $\propto$ (Spacing in Brillouin zone)$^{-1}$

there are just as many allowed P_x-values in the Brillouin zone as there are atoms in the crystal, this being the Brillouin-zone statement of the theorem of conservation of quantum states. This theorem is proved by noting that the number of atoms in the crystal is $A_x/a = N_x$ and the number of allowed P_x-values is $(h/a)/(h/A_x) = A_x/a = N_x$ also. Taking into account the two values of spin leads to $2N_x$ quantum states in the Brillouin zone.

The dependence of energy on P_x has also been indicated for two energy bands in Figure 5.9(a). From this it is possible to make an energy level diagram for the allowed P_x-values as is indicated on the right edge of (a). These same energy levels are plotted in (b), which is drawn to a reduced scale to show how the energy bands arise in the Brillouin-zone picture.

The actual computation of energy as a function of P_x can be carried out for idealized one-dimensional crystals, and treatments will be found in any modern text on the electron theory of solids and in some on quantum theory.[2] For electrons in actual crystals, however, laborious numerical calculations are, in general, required and only a limited number of cases have been dealt with.

[2] A simple case is treated in Chapter 14.

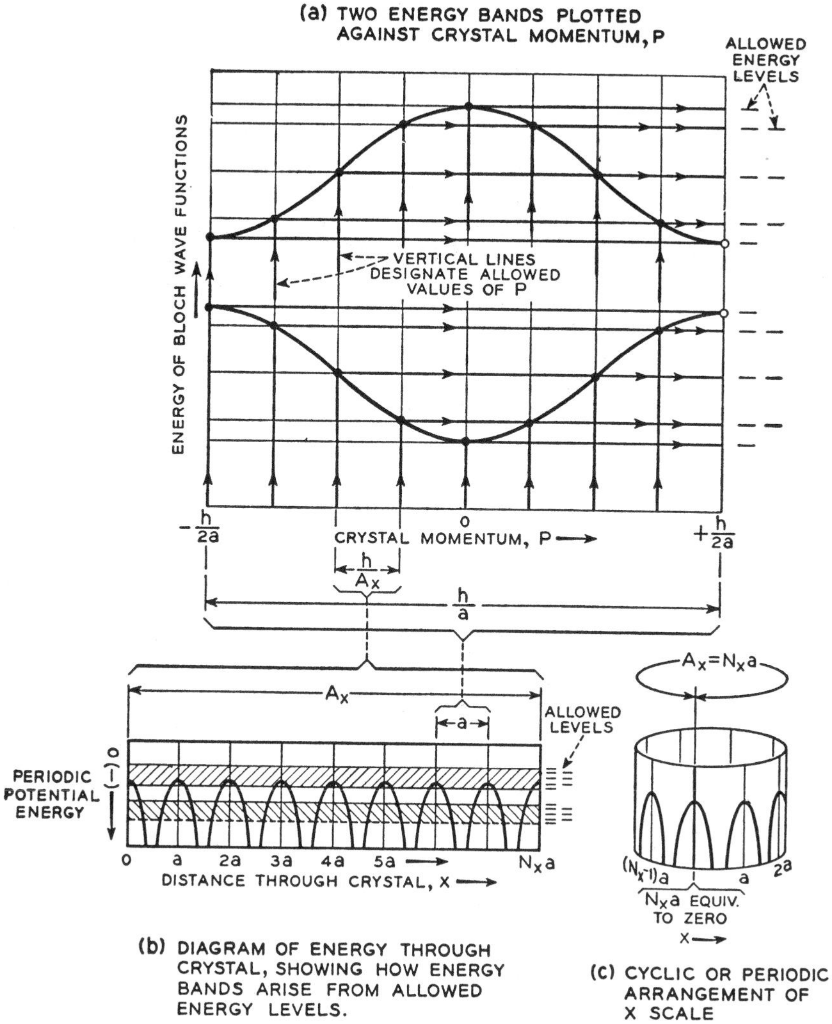

FIG. 5-9—Structure of Brillouin Zone and Energy Bands for a One-dimensional Cyclic Crystal. (The Allowed Values of P_x are Given by the Formula: $P_x = h\,n_x/A_x$.)

We must next extend the Brillouin-zone concept to three dimensions. For this case P_x is replaced by a vector denoted by $\boldsymbol{P}$, having components P_x, P_y, P_z. Again we shall find that there is a limited volume in $\boldsymbol{P}$-space to be considered which is bounded by planes corresponding to $\lambda = 2a$. *This limited volume of* $\boldsymbol{P}$-space, *which may be defined as the most compact volume about* $\boldsymbol{P} = 0$ *containing all the non-equivalent values of* $\boldsymbol{P}$, *is the three-dimensional Brillouin zone.* Only values of $\boldsymbol{P}$ within this zone need be considered and of these only a certain allowed set gives eigenfunctions satisfying the boundary conditions. We shall show that, for this allowed set, the theorem of conservation of quantum states once more applies.

For three dimensions the Bloch function can be written as

$$\psi_{P_xP_yP_z}(x, y, z, t) = u_{P_x,P_y,P_z}(x, y, z) \times \exp [(2\pi i/h)(P_xx + P_yy + P_zz - \mathcal{E}_{P_xP_yP_z}t)] = \psi_P(r, t) = u_P(r) \exp [(2\pi i/h)(P \cdot r - \mathcal{E}_Pt)] \quad (5)$$

where the second form makes use of the vector notation

$$\boldsymbol{P} = \boldsymbol{i}P_x + \boldsymbol{j}P_y + \boldsymbol{k}P_z \quad (6)$$

$$\boldsymbol{r} = \boldsymbol{i}x + \boldsymbol{j}y + \boldsymbol{k}z \quad (7)$$

$$\boldsymbol{P} \cdot \boldsymbol{r} = P_xx + P_yy + P_zz \quad (8)$$

where $\boldsymbol{i}$, $\boldsymbol{j}$, and $\boldsymbol{k}$ are the customary unit vectors along x, y, and z axes (this use of $\boldsymbol{i}$ should not be confused with $i = \sqrt{-1}$). We shall suppose that we are dealing with a simple cubic crystal with lattice constant a. (Later we shall comment on the modifications required for the treatment of a more complex crystal, such as the diamond structure.) By arguments similar to those used in the one-dimensional case,[3] it can be concluded that wave functions with $P_x = h/2a$ correspond to standing waves in the x-direction and that, for any arbitrary fixed values of P_y and P_z, $P_x = h/2a$ and $P_x = -h/2a$ lead to the same wave function. Thus the Brillouin zone in (P_x, P_y, P_z) space or $\boldsymbol{P}$-space is a cube whose bounding planes are $P_x = \pm h/2a$, $P_y = \pm h/2a$, $P_z = \pm h/2a$. For more complex lattices, a more general method of attack must be used to find the Brillouin zone. However, in all cases the boundaries of the zone are determined by standing wave conditions like $\lambda = 2a$. The planes which do the reflecting, however, will not necessarily be perpendicular to the x, y, and z axes and one must consider other possible planes. A procedure for doing this has long been used in studies of X-ray diffraction in crystals. According to this theory, the X rays will produce Bragg reflections from certain planes in the crystal when the wave vector $(\boldsymbol{P}/h)$ satisfies conditions equivalent to $\lambda = 2a$ where a is the spacing between the planes in question. In this way certain boundaries in P-space are found which enclose the Brillouin zone for more complex lattices. We shall not consider these more complicated cases, however, since it suffices for the present purposes to deal only with results which are generally valid for all types of lattices, and these can be illustrated adequately with the simple cubic model.

[3] For details of these arguments, which involve new mathematical complexities rather than new physical ideas, the reader is referred to F. Seitz, *Modern Thoery of Solids*, McGraw-Hill Book Co., New York, 1940; A. H. Wilson, *The Theory of Metals*, Cambridge at the University Press, 1936; and Section 14.8.

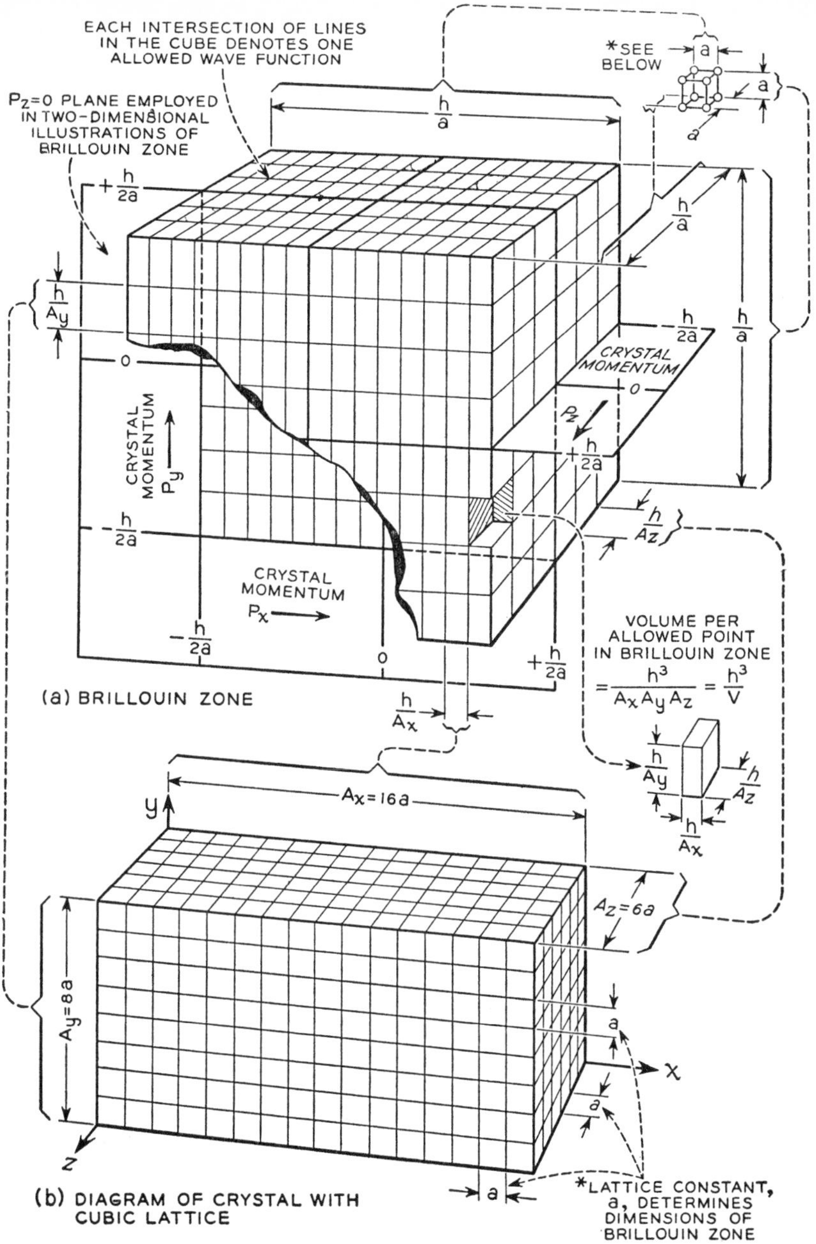

FIG. 5-10—Three-dimensional Coordinates of the Brillouin Zone for a Crystal with Simple Cubic Structure.

If the simple cubic crystal consists of a rectangular parallelepiped with edges of length A_x, A_y, and A_z parallel to the x, y, and z axes, we may find the allowed values for $\boldsymbol{P}$ by using periodic boundary conditions as before. In this case it is hard to imagine bending the crystal around in such a way that the faces on opposite sides are brought together as shown in Figure 5.9(c). However, mathematically, the requirement that the wave function be periodic with periods A_x, A_y, and A_z in the three dimensions can easily be formulated. Experience in problems of this sort in theoretical physics has shown that the results are, in general, independent of the exact boundary conditions used at the surface of the sample.[4] For this reason, we shall not try to justify the periodic boundary condition on physical grounds but will indicate that substantially equivalent results would be obtained if we made the wave functions vanish on the boundary, corresponding to a potential field which holds the electrons inside the crystal. If we require that the wave function take on the same value at x and $x + A_x$, we conclude by the same reasoning as before that

$$P_x = n_x h/A_x \quad \text{or} \quad \delta P_x = h/A_x. \tag{9}$$

Since the values of y and z are left unaltered by this increase in x, they do not affect the reasoning. Similarly, by dealing with y and z individually, we conclude that

$$P_y = n_y h/A_y \quad \text{or} \quad \delta P_y = h/A_y \tag{10}$$

and

$$P_z = n_z h/A_z \quad \text{or} \quad \delta P_z = h/A_z. \tag{11}$$

Thus the allowed points in $\boldsymbol{P}$-space fall on the corners of a rectangular or orthorhombic lattice for which the unit cell has edges h/A_x, h/A_y, h/A_z so that there is a volume of $h^3/A_xA_yA_z$ or h^3/V of $\boldsymbol{P}$-space for each unit cell or allowed wave function. The reciprocal relationship between crystal and Brillouin zone in three dimensions thus is:

(Size of Brillouin zone) $\propto$ (Unit Cell of Crystal)$^{-1}$

(Dimensions of Crystal) $\propto$ (Dimensions in Array of allowed points in Brillouin zone)$^{-1}$.

We shall have occasion to deal with the lattice of points in $\boldsymbol{P}$-space defined by equations (9) to (11) in other connections as well and shall use the phrase *basic **P**-lattice of the Brillouin zone* to describe the array distinct points in the Brillouin zone specified by these equations.

This lattice and the reciprocal relationships are represented in Figure 5.10. The allowed points in the $P_z = 0$ plane are shown in greater detail in Figure 5.11(a). Here each distinct allowed point is shown by a solid

[4] We shall have further occasion in Chapter 7 to discuss the unimportance of the exact boundary conditions employed.

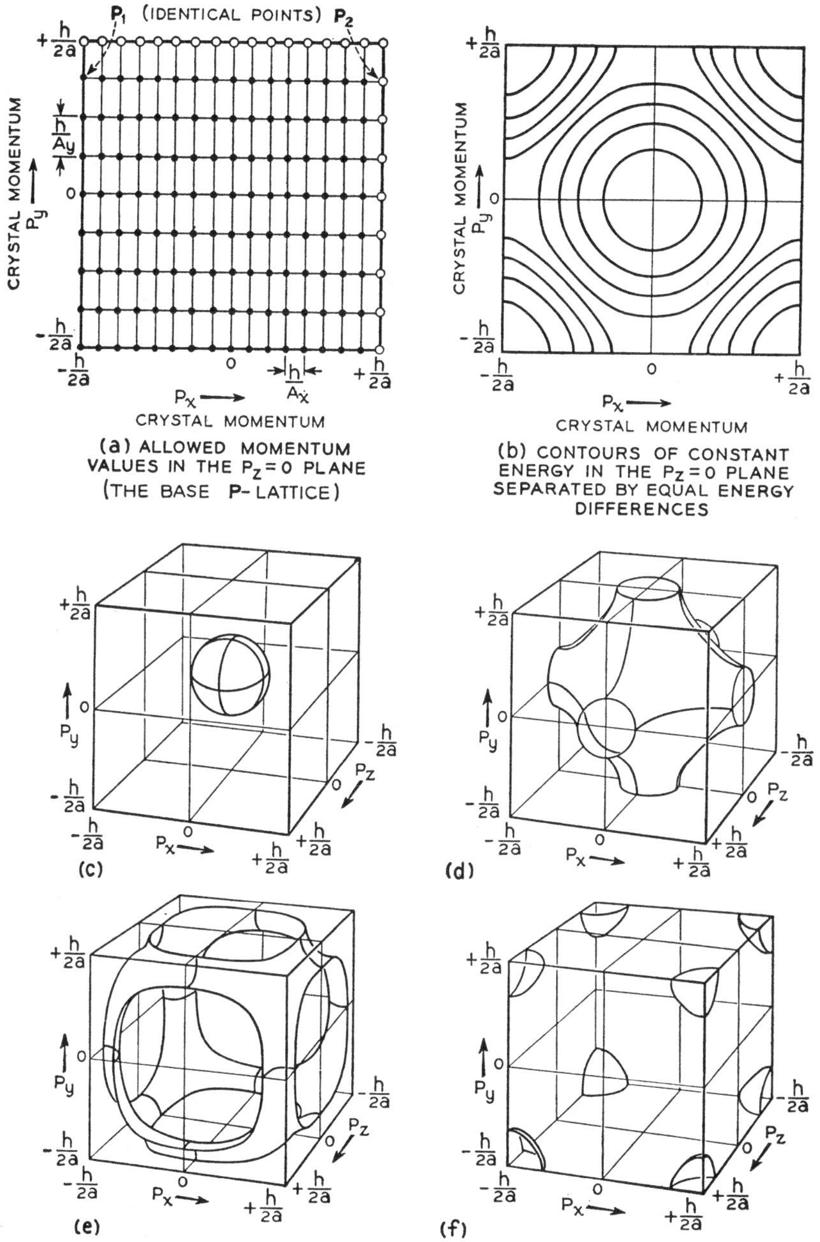

(c), (d), (e) & (f) TYPICAL SHAPES OF CONSTANT ENERGY CONTOUR SURFACES FOR SIMPLE CUBIC LATTICE STRUCTURE

FIG. 5-11—Brillouin Zone, Permitted Quantum States and Energy Contours.

dot; the open circles like $\boldsymbol{P}_2$ on the edges of the Brillouin zone give wave functions identical with those of other points like $\boldsymbol{P}_1$. Thus it is seen that there are just as many allowed points as there are rectangles in the plane (that is, there is one rectangle to the upper right of each allowed point). For three dimensions there are also just as many allowed points as there are elementary volumes h^3/V. This is a general result, as is shown in Section 14.8, and may be expressed in the form

$$\text{density of allowed points in } \boldsymbol{P}\text{-space} = V/h^3. \qquad (12)$$

Since the volume of the Brillouin zone is $(h/a)^3$, the number of allowed points is $(h/a)^3 \times (V/h^3) = V/a^3$. Thus we have the Brillouin-zone formulation of the Theorem of Conservation of Quantum States:

The number of allowed points in the Brillouin zone is equal to the number of unit cells in the crystal.

It is thus easy to see one relationship between the Brillouin zone and the energy band. If the crystal is assembled by varying the lattice constant as discussed in Section 5.3, then each atomic energy level spreads into a band of energy levels; the number of quantum states, usually one per atom, for each spin, then give rise to $2N_s$ quantum states in the band, where N_s is the total number of atoms. These $2N_s$ quantum states correspond to the N_s allowed points in the Brillouin zone combined with the two possibilities for the spin. If there is more than one atom per unit cell, the situation is somewhat more complicated. In this case the energy level will probably split into several overlapping bands each requiring a Brillouin zone to represent it. We have discussed the derivation of the theorem for the case of a simple cubic crystal. The theorem as stated is, however, actually applicable to any crystal lattice, provided that "unit cell" is properly interpreted. The unit cell is the smallest parallelepiped which is typical of the crystal and, if repeated over and over parallel to itself, would generate the entire crystal. For the diamond structure of Figure 1.3, the unit cell is a slanting parallelepiped containing two atoms and having a volume of one fourth the cube shown. The four electrons per atom in the valence-bond band of Figure 5.5 thus correspond to eight electrons per unit cell or enough to fill four Brillouin zones completely, using both spins. Thus the energy bands of interest in diamond-type crystals will correspond to Brillouin zones whose energies overlap in complicated ways. We shall indicate in Section 7.5 that certain approximations are made in treating this case. Some other features relating to overlapping energy bands are discussed in Chapter 12.

The energy of the wave function depends, of course, on the value of $\boldsymbol{P}$. In Figure 5.11(b) the dependence of energy upon $\boldsymbol{P}$ is illustrated for a hypothetical case, lines of equal energy being shown for points on the $P_z = 0$ plane exhibited in Figure 5.10. As is seen, the figure has a high degree of symmetry, consistent with the symmetry of the simple cubic model being considered. The minimum energy lies at the center of the figure and the maximum value at the corners. The contours in Figure 5.11(b) are spaced at approximately equal energy differences. In the three-dimensional diagrams, the constant energy lines are surfaces, nearly spherical ones near the center with more complicated shapes for larger energies. Finally, near the corners of the cube, the surfaces once more become nearly spherical. It does not always happen that the lowest energy occurs at the center. As we shall see later, so far as the behavior of the electrons as current carriers is concerned, the minimum may equally well be at the corners and the maximum at the center.

It has been stated that points on opposite faces of the Brillouin zone, for example $\boldsymbol{P}_1$ and $\boldsymbol{P}_2$ in Figure 5.11, for which the crystal momenta differ by $\Delta P_x = \pm h/a$, $\Delta P_y = \Delta P_z = 0$, have the same wave function. This means that increasing P_x beyond $h/2a$ is equivalent to re-entering the zone on the opposite face and proceeding inwards. This result can be shown to be correct by dealing mathematically with the Bloch wave functions. It can also be illustrated for the analogous case of the coupled oscillators of Figures 5.3 and 5.6. In Figure 5.12, three modulating waves are shown moving past a group of harmonic oscillators according to a scheme of presentation similar to Figures 5.3 and 5.6. However, in this case only the bobs of the oscillators are shown. Part (a) shows the conditions $P_x = \pm h/2a$ or $1/\lambda = 1/2a$. The two waves are drawn as though they progressed to right and left in successive diagrams. In part (b) we show a wave with $P_x = (h/2a) + \frac{1}{6}(h/2a)$, that is, a momentum lying beyond the edge of the zone by $16\frac{2}{3}$ per cent. Its wave pattern is also shown as though it advanced to the right reading down the page. In addition, on part (b) we show $P_x = -(h/2a) + \frac{1}{6}(h/2a)$; that is, a wave moved in from the left edge of the zone by the same amount that the other lies outside to the right. Since it is characterized by a negative value of P_x, this wave is to be considered as moving to the left. As is seen, the motion of the oscillator bobs is the same for the two waves, showing that proceeding beyond the zone boundary is equivalent to entering it on the opposite side. We could thus equally well use for the zone the boundaries $P_x = 0$, $P_x = h/a$, $P_y = 0$, $P_y = h/a$ since the points in this area would give the same set of modulation waves as Figure 5.11(a). Such a rearrangement has advantages for some purposes and is employed in Section 7.5.

$\boldsymbol{P}$-space is periodic in the sense that adding $\pm h/a$ to any component of $\boldsymbol{P}$ gives rise to the same wave function. For this reason points outside of the Brillouin zone are always equivalent to points inside the zone and it is never really necessary to consider $\boldsymbol{P}$ values lying outside the Brillouin zone. One Brillouin zone, however, gives only enough quantum states to accommodate two electrons per unit cell; whereas most crystals contain many more than two electrons per unit cell. This difficulty may be solved by using the same points in the Brillouin zone many times over, simply specifying in each case which energy band is concerned. Another pro-

$P = +\dfrac{h}{2a}$
WAVES MOVE TO RIGHT

$P = -\dfrac{h}{2a}$
WAVES MOVE TO LEFT

(a) THE MOTIONS OF BOBS FOR PROGRESSIVE WAVES WITH $\lambda = 2a$ ARE THE SAME FOR BOTH DIRECTIONS OF MOTION. FOR ELECTRONS, THE WAVE FUNCTIONS ARE THE SAME FOR $P = +\dfrac{h}{2a}$ AND $P = -\dfrac{h}{2a}$

SUCCESSIVE POSITIONS OF POINTS SHOW VERTICAL MOTIONS OF HARMONIC OSCILLATOR BOBS.

$P = \left(\dfrac{h}{2a}\right) + \dfrac{1}{6}\left(\dfrac{h}{2a}\right)$
WAVES MOVE TO RIGHT

$P = -\left(\dfrac{h}{2a}\right) + \dfrac{1}{6}\left(\dfrac{h}{2a}\right)$
WAVES MOVE TO LEFT

(b) WAVES CORRESPONDING TO $P = \left(\dfrac{h}{2a}\right) + \dfrac{1}{6}\left(\dfrac{h}{2a}\right)$ AND $P = -\left(\dfrac{h}{2a}\right) + \dfrac{1}{6}\left(\dfrac{h}{2a}\right)$ PRODUCE THE SAME MOTIONS OF THE BOBS AND, IN THE CRYSTAL, THE SAME WAVE FUNCTIONS FOR ELECTRONS

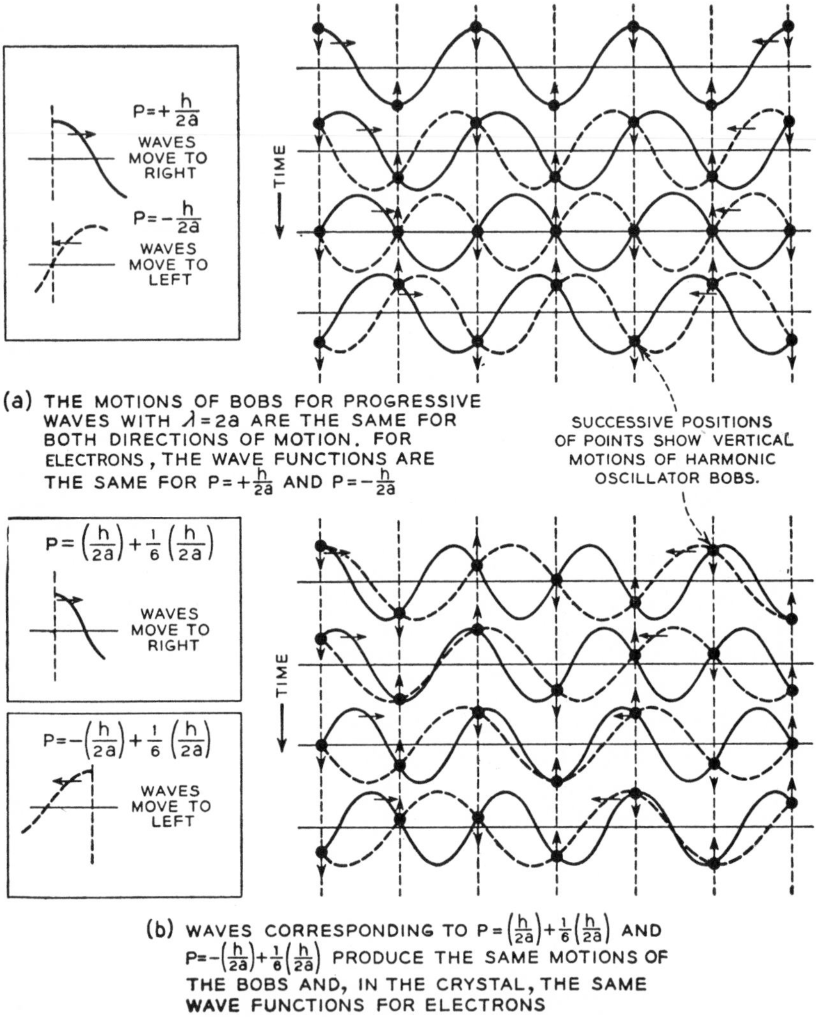

Fig. 5–12—Values of Momentum Lying Outside the Brillouin Zone are Equivalent to Values Lying Inside Obtained by Subtracting h/a from P.

cedure, which is more complicated, is to use the remaining portions of P-space outside of the Brillouin zone for other bands. There is, in fact, a definite procedure for doing this and the more complicated volumes involved in P-space are referred to as the second, third, etc., Brillouin zones. (The dashed lines of Figure 5.7(d) are associated with this procedure.) For our purposes, however, it is simplest to use only the first Brillouin zone, which is the one already described. For a given point in this zone there is a definite modulation wave, exp $(2\pi i \boldsymbol{P} \cdot \boldsymbol{r}/h)$ superimposed on the atomic wave function. By using the same modulation wave but different atomic functions, wave functions in the different energy bands are produced. Some examples of these have been shown in Figure 5.7(d).

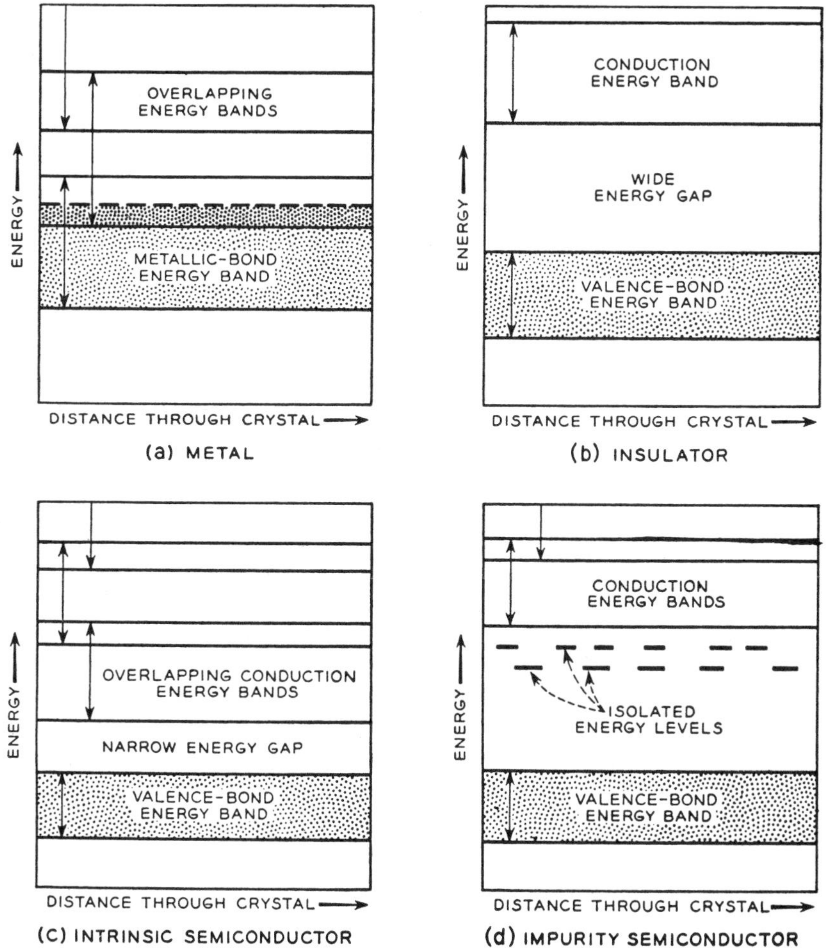

Fig. 5–13—Energy Band Diagrams for Several Crystal Types.

5.6 ENERGY BANDS FOR METALS, INSULATORS, AND SEMICONDUCTORS

There are characteristic differences between the energy bands of metals, insulators, and intrinsic semiconductors in terms of which their electrical conductivity can be understood. As we shall see later, when all the quantum states of a Brillouin zone are occupied, the zone cannot participate in conduction. In the case of a metal, the energy bands overlap in such a way that, as occupied by the available electrons, some Brillouin zones are left partially filled. On the other hand for pure diamond, one set of Brillouin zones is occupied by valence electrons, and the next higher set (the conduction band) is left entirely empty. The energy gap between them is so great that electrons are not thermally excited, and diamond, therefore, has either completely filled or completely empty Brillouin zones and is an insulator when in thermal equilibrium at room temperature. On the other hand, germanium has its valence-bond bands and conduction bands so close together that it would exhibit appreciable conductivity at room temperature even if pure. The diamond form of tin, called gray tin, which is stable below room temperature, is probably either a metallic conductor with overlapping energy bands or at least an intrinsic semiconductor of very high conductivity. The energy gaps for tetravalent elements are thus thought to be:

Carbon (Diamond)	*Silicon*	*Germanium*	*Tin*[1] *Gray*	*Tin Metallic*	*Lead*
6 to 7 ev	1.11 ev	0.72 ev	0.1 ev	(overlapping bands)	

The donors and acceptors, which can bind holes and electrons as described in Chapter 1, give rise to energy levels which lie in the energy gap. A quantum-mechanical treatment of these energy levels and their relationship to energy bands is given in Chapter 9.

The energy level schemes for the various cases just discussed are shown in a conventional form in Figure 5.13. We shall return to a discussion of the statistical distribution, under thermal equilibrium conditions, of electrons among such energy levels at the close of Chapter 10.

Problems

1. Carry out an analysis, similar to that for Figure 5.6, for four oscillators and for eight oscillators. Discuss the differences that would be introduced if the number of oscillators were changed to an odd number.

2. The patterns shown in Figure 5.6 resemble traces seen on oscilloscopes. Show that, if the horizontal plates are driven at frequency f and the vertical plates at nf, patterns like the dashed lines of Figure 5.6 will be obtained. If the y plates have also a small signal at f and 90° out of phase with the horizontal plates, even the perspective will be simulated. What will happen if the y frequency is slightly different from nf?

3. Assume a force constant for the vertical spring and an initial tension

[1] G. Busch, J. Wieland and H. Zoller, Conference on the Properties of Semiconducting Materials, University of Reading, July 10–15, 1950.

in the coupling springs; assume also that the masses of Figure 5.6 slide on frictionless vertical rods. Work out the frequency of the waves for small amplitude disturbances.

4. Compare qualitatively the distribution of allowed frequencies for "clamped end" boundary conditions and periodic boundary conditions, for an odd number of coupled oscillators.

CHAPTER 6

VELOCITIES AND CURRENTS FOR ELECTRONS IN CRYSTALS

This chapter serves to complete the Brillouin zone description of the properties of the Bloch wave functions; with it we conclude the general description of the behavior of electrons moving in the periodic field of a perfect crystal in the absence of external fields. The new aspect added here is the velocity of motion associated with each quantum state in the Brillouin zone. In a text designed exclusively for physicists, this topic would be treated principally on the basis of the postulates and operations of quantum mechanics. While in our treatment we refer to these quantum-mechanical methods, we emphasize primarily the analogy with other problems involving wave motion. At the end of this chapter we are within two steps of the elementary treatment of conductivity: In Chapter 7 we consider the effect of electric and magnetic fields on altering the behavior of electrons, and in Chapter 8 we introduce the idea of random processes which produce resistance.

6.1 THE VELOCITY AND CURRENT CONCEPTS

When an electron is moving in the crystal in the manner prescribed by one of the Bloch wave functions, it has an "average" velocity of motion which gives rise to a current. In fact, for each quantum state in the Brillouin zone there is a certain average velocity, the one which an electron would have if it occupied that state. It is the aim of this chapter to discuss the relationship of this average velocity, which will be denoted by the vector symbol $\boldsymbol{v}$, to other properties of the Bloch wave functions.

As described in connection with Figure 5.7, the complex Bloch waves flow through the crystal. However, the electron probability density is the same in every unit cell so that no apparent flow can be detected by examining the distribution of the probability density for the electron in space. This brings us to the important matter of interpreting the meaning of the Bloch wave function in terms of more familiar descriptions of the electron's behavior. We shall find that the theory indicates that the electron is actually moving through the crystal with a definite average velocity. It should be pointed out at once that this involves no contradiction with the uniform probability distribution. In a number of other cases, the entity which flows may show no shift in its density distribution. For example, a rotating flywheel may look perfectly stationary, yet the iron which is uniformly distributed around the rim is actually moving. The energy density along a power line may be uniform and unchanging, yet power flows. (An example of this sort is presented in Section 15.3.)

The Bloch function represents a state of motion for the electron in which it is equally likely to be in any unit cell of the crystal while at the same time it has an average velocity of motion through the crystal. The reason that this motion does not carry it out of the crystal after a time is a mathematical one associated with the periodic boundary conditions. These correspond (as shown in Figure 5.9(c) for the one-dimensional case) to a cyclic crystal in which flow out of one end is equivalent to flow in at the other. Thus the electron flows indefinitely through the crystal.

This flow can be expressed as a current density in a form useful for later conductivity considerations. Thus, in the one-dimensional case, if the average velocity of motion is v_x, then the electron traverses one cycle about the loop in a time A/v_x. This means that the charge $-e$ of the electron flows around the loop v_x/A times per second and hence that the current is $I = -ev_x/A = (-e/A)v_x$, the second form showing that the current can be expressed as charge per unit length, $(-e/A)$, times average velocity. In three dimensions, the current density vector or charge per unit area per unit time can be expressed similarly as

$$\boldsymbol{I} = (-e/V)\boldsymbol{v}, \tag{1}$$

that is, as the charge per unit volume times the vector velocity.[1] The derivation is as follows: If the electron has an average vector velocity $\boldsymbol{v}$ with components v_x, v_y, v_z, then it makes v_x/A_x traversals in the x-direction per unit time; on each traversal it carries $-e$ across the end face of the crystal giving rise to a current of $-ev_x/A_x$. Since the end face has an area A_yA_z, the current density in the x-direction is given by the formula

$$I_x = (-ev_x/A_x)/A_yA_z = (-e/V)v_x, \tag{2}$$

where $V = A_xA_yA_z$ is the volume of the crystal. Similar calculations for the y and z axes then establish the general formula. Actually, the argument given here, which serves to introduce the formula for current density and to illustrate the idea of the average velocity, is in a sense the reverse of the one logically employed in the theory to find the average velocity. This will become apparent from the method of attack used to evaluate the average velocity.

6.2 THE POYNTING VECTOR OR DENSITY-FLOW METHOD

Since no motion can be detected by computing the change in the probability density, other methods of determining the velocity must be em-

[1] Even before the electron crosses the surfaces, it produces a displacement current given by Equation (1). See, for example, W. Shockley, *J. App. Phys.*, 9, 635–636 (1938).

ployed. These are directly available in the theory of quantum mechanics, and a definite procedure is given for calculating the average current flow from a wave function. We shall not present this method here[1a] but shall

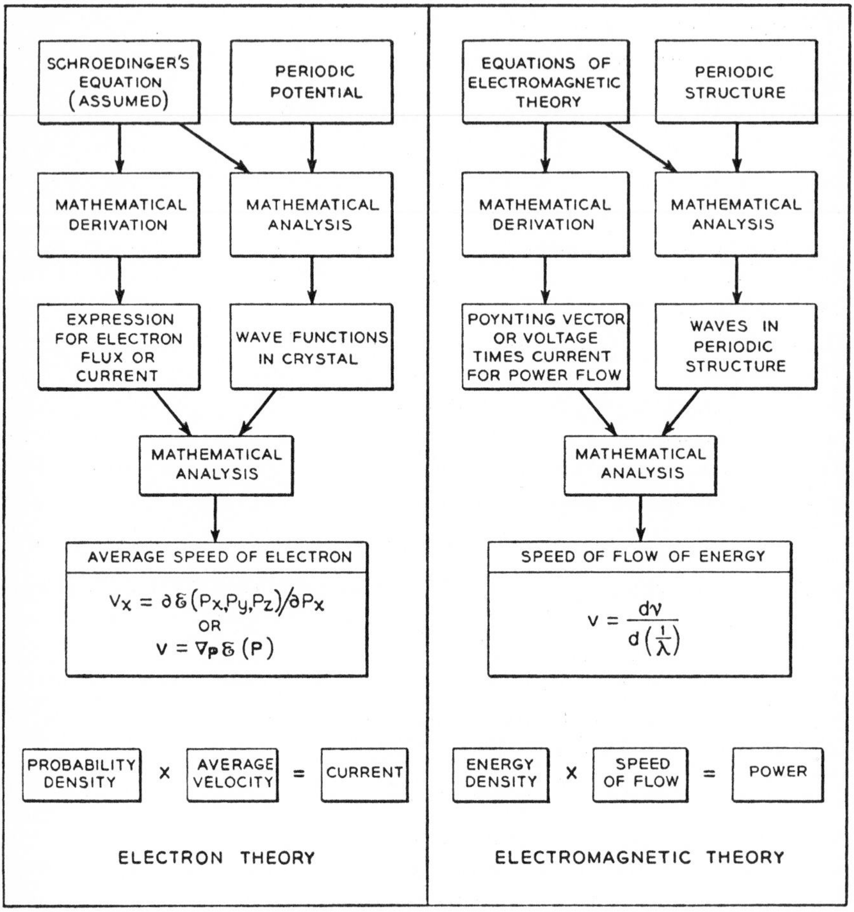

FIG. 6–1—Comparison of the Calculation of Average Speed for an Electron with the Calculation of Flow of Electromagnetic Energy.

describe instead its analogue in terms of power flow in electromagnetic theory. The relationship is shown in Figure 6.1. In Section 15.3 the procedure is actually carried out in detail for an artificial lumped-constant transmission line having no loss. For this case, the energy per unit length is computed for a line with a particular running wave, the running wave being found by the procedure leading to "waves in periodic structure"

[1a] Derivations are given in Sections 14.3 and 15.4.

shown in the figure. The equation for power flow is then simply obtained by evaluating the current flowing from left to right at some point on the line and multiplying this by the voltage and proper phase factor. This familiar expression for power flow is analogous to the Poynting vector for electromagnetic waves or to the quantum mechanical expression for average current. We next argue that the power flow along the line is due to the flow of its energy density. Consequently, if its velocity of flow is v, all the energy in a length v of line flows past a point in unit time, and, therefore, power is energy density times speed of flow as shown in Figure 6.1. If the line has no phase distortion, that is, all frequencies travel at the same speed, or, in other words, there is no dispersion, then this rate of flow always turns out to be the velocity of the waves. However, if the speed of the waves depends on the wave length, then the power does not flow at the same speed as the waves but instead at a speed given by the formula

$$v = \frac{d\nu}{d(1/\lambda)}. \tag{1}$$

This expression equals the speed of the waves only for the case of constant wave speed for which

$$\nu = c/\lambda \quad \text{so that} \quad \frac{d\nu}{d(1/\lambda)} = \frac{dc(1/\lambda)}{d(1/\lambda)} = c, \tag{2}$$

where c is the speed of the waves.

When the analogous procedure using the expressions for electron current density are employed as indicated in Figure 6.1, precisely the same formula is obtained for electron probability density as for electrical energy for the *one-dimensional case.* However, since the relationships

$$h\nu = \mathcal{E} \quad \text{and} \quad P_x = h/\lambda \tag{3}$$

apply, we can rewrite the formula in the form

$$v = \frac{d\nu}{d(1/\lambda)} = \frac{d\mathcal{E}}{dP_x}. \tag{4}$$

This formula is identical in form with that obtained classically, for which $p = mv$ and $\mathcal{E} = mv^2/2 = p^2/2m$ so that $d\mathcal{E}/dp = p/m = v$. However, the physical meaning is different, since, for electrons in a periodic potential, $\mathcal{E}$ does not equal $p^2/2m$.

There is a fundamental explanation for the fact that the same formula $v = d\nu/d(1/\lambda)$ is obtained for both electromagnetic waves and electron waves. This explanation is based on the use of wave-packets and group velocity, topics which are discussed in the next section.

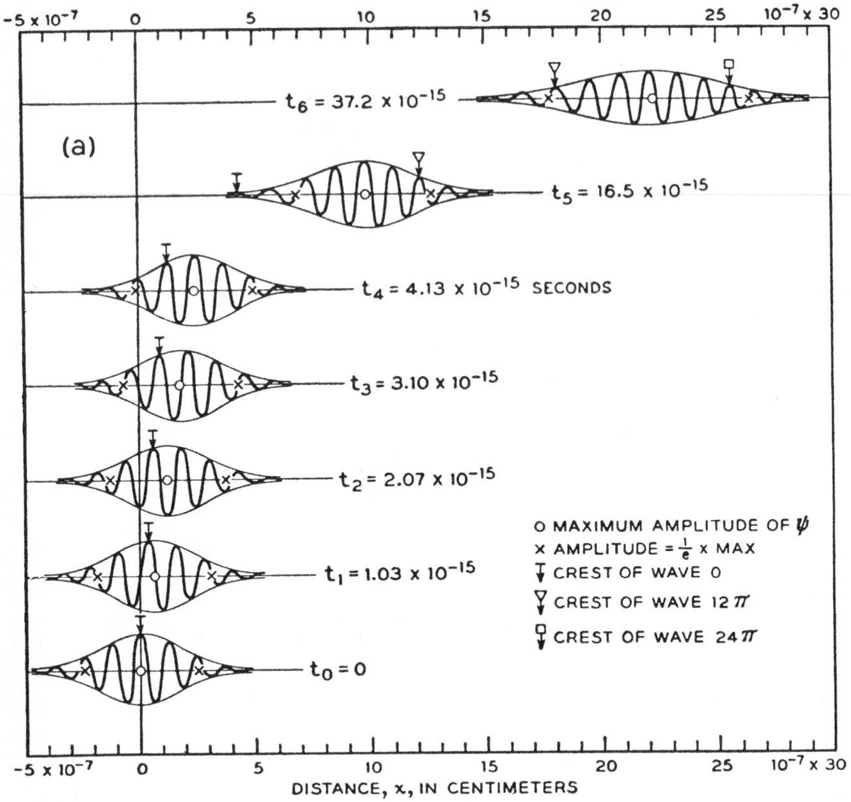

FIG. 6–2—Motion of a Minimum-uncertainty Wave-Packet in Free Space.

The product of uncertainty in position and momentum at $t = 0$ for this packet has the minimum value $\Delta p\, \Delta x = h/4\pi$ permitted by the Heisenberg uncertainty relation.

(a) Plots of the real part of the wave function $\psi(x,t)$ for several instants of time. Since this is a one-dimensional wave-packet, ψ has the dimension of $\text{cm}^{-1/2}$ The imaginary part of the wave function is similar in shape and $\psi\,(x,t)$ is shown by the envelope lines. The velocity of the group is twice as great as the phase velocity of the waves which compose it: while the crest of the waves advance one wave length between t_0 and t_4, the group advances two wave lengths. The packet is composed of momenta varying by about $\pm 8\%$ from the mean value. This spread in component velocities causes the packet to spread appreciably in length during the time from t_0 to t_6, the higher momentum components gathering near the front of the packet and giving the wave function a shorter wave length while the lower components similarly gather at the rear.

(b) The continuous progression of the lines of constant complex phase angle is

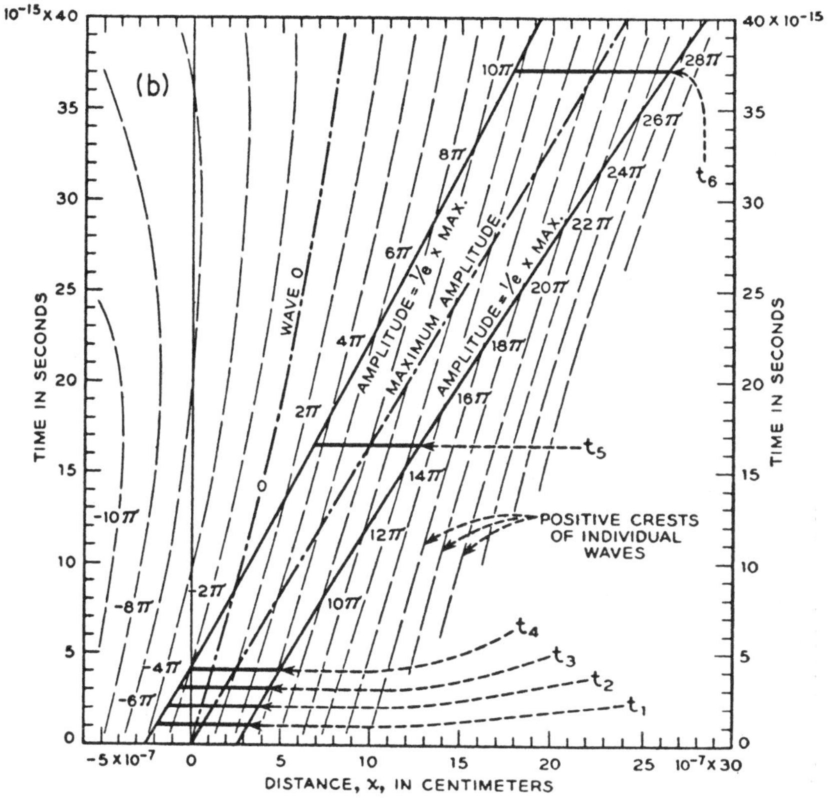

FIG. 6–2—*Continued.*

shown by the dashed lines. When the angle is a multiple of 2π, the wave function is real and positive, corresponding to the crests in (a). The line for maximum probability amplitude advances with twice the phase velocity and the lines at which the probability amplitude falls to $1/e^2$ of the maximum are seen to spread out. The bending back of the lines of constant phase near $x = 0$ for the larger times corresponds to motion to the left, which results from the fact that there is some probability that the electron will have a negative momentum.

The average velocity corresponds to 1 electron volt of energy with a velocity 5.92×10^7 cm/sec and a wave length of 12.2×10^{-8} cm. The value of Δx at $t = 0$ was arbitrarily chosen as 12.2 Å in computing the packet. This gives $\Delta p = h/4\pi\Delta x = p_1/4\pi$ where p_1 corresponds to 1 ev. The spread in momentum increases the energy from $p_1^2/2m$ to $(p_1^2 + \Delta p^2)/2m = p_1^2(1 + (1/16\pi^2))/2m = 1.0063$ ev.

For earlier times than $t = 0$, the packet also spreads out in a way symmetrical to that for later times. The theory of this wave-packet is discussed in Section 15.5.

6.3 WAVE-PACKETS AND GROUP VELOCITY

We shall next interpret the velocity formula in terms of wave-packets and group velocities, both of which are useful concepts in wave mechanics. Accordingly, we suppose that instead of sending a single frequency down the transmission line we send a pulse signal at a certain carrier frequency. This pulse may be regarded as the result of superimposing a number of single-frequency waves. The pulse localization results from the constructive interference of these waves at the location of the pulse and the destructive interference elsewhere. In general, these waves travel at varying speeds, and thus the interference pattern changes. As a result of the shifting of the interference pattern of the group of single-frequency waves, the pulse travels at a speed different from that of the individual waves. The terminology used to describe this situation is as follows: The pulse is said to be a *wave-packet*, which is made up by the constructive interference of a *group* of waves. Each of these individual waves moves with its own velocity called the *phase velocity*, whereas the packet moves with the *group velocity*. If all the waves of the group travel with the same phase velocity, then the wave-packet moves with them with unaltered shape. However, if the phase velocity varies with the wave length, then the interference pattern shifts and the wave-packet may go either faster or slower than the waves which build it up. When the mathematics of this interference problem is analyzed (see Chapter 15), it is found that the group velocity is given by the formula

$$v = \frac{d\nu}{d(1/\lambda)}. \tag{1}$$

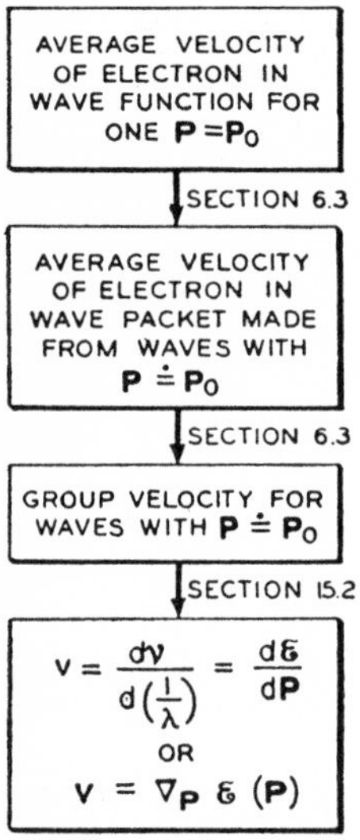

FIG. 6–3—Outline of Argument Used to Derive Velocity of Electron from Group-velocity Concept.

The reason that the same formula results from both the power flow calculation and the group velocity calculation may be understood from the following argument: In the case of the wave-packet the energy is localized in a certain region of the transmission line, hence it must flow at the same speed as the wave-packet. On the other hand, when the wave-packet is passing a certain point on the line, the situation locally is much the same as for the state of steady power flow, and energy will be flowing past at the speed determined from the steady flow calculation. Hence the speed of flow of the steady-state power must be given by the group-velocity formula. (The equality for the artificial line is given in detail in Section 15.3.)

This same procedure may be used to analyze the behavior of the Bloch waves. The argument is outlined in Figure 6.3. In order to determine the velocity in one Bloch function, we combine it with others of approximately the same value of P_x. In this way we get a localized wave-packet. (Such a wave-packet for an electron is shown in Figure 6.2, and its properties are discussed in the associated text.) We then argue that since all the wave functions are very similar (that is, only a small difference in their P_x values), they correspond to about the same average velocity for the electron, and thus the packet which represents the average of their effects must also have the same velocity. It is then a straightforward mathematical problem to evaluate the group velocity from a study of the interference patterns. The result is quite general, and, as derived in Section 15.2, it applies to any set of interfering waves. It agrees, of course, with that obtained by the density-flow method. We shall next discuss the consequences of the formula for a one-dimensional case.

6.4 VELOCITIES FOR QUANTUM STATES IN BRILLOUIN ZONES

As pointed out previously, the relationships $\mathcal{E} = h\nu$ and $P_x = h/\lambda$ enable us to transform the group velocity equation to

$$v = d\mathcal{E}/dP_x \tag{1}$$

for the one-dimensional case. This result is in agreement with the classical Hamiltonian expression. In the classical theory of analytical dynamics the energy is written in the form $\mathcal{H}(p_i, q_i)$ where the quantities q_i are coordinates and the quantities p_i are momenta. For a single particle the coordinates and momenta are conveniently chosen as

$$q_1 = x, \qquad q_2 = y, \qquad q_3 = z$$

$$p_1 = m\frac{dx}{dt}, \quad p_2 = m\frac{dy}{dt}, \quad p_3 = m\frac{dz}{dt}.$$

Other choices can be made for special problems, such as the problem of planetary motion. The relationship between classical and quantum mechanical Hamiltonian expression is discussed in Section 14.3. Hamilton's equations of motion are

$$q_i = \frac{\partial\mathcal{H}}{\partial p_i} \quad \text{and} \quad \dot{p}_i = -\frac{\partial\mathcal{H}}{\partial q_i}.$$

For the case of an electron moving in one dimension in a constant potential energy $\mathcal{V}_0$, both quantum and classical mechanics give kinetic energy $= p^2/2m$, so that the total energy is $\mathcal{E} = \mathcal{V}_0 + p^2/2m$, and the group velocity is $v = p/m$ or $p = mv$, the classical result.[1] The value $\mathcal{V}_0$, although it contributes to the frequency of the Bloch wave, through $\nu = \mathcal{E}/h$, does not affect the group velocity. In other words, the group velocity is independent of the zero selected for the energy scale whereas the phase velocity is not. Since only the group velocity contributes to the current, this variability of the phase velocity is, however, not important.

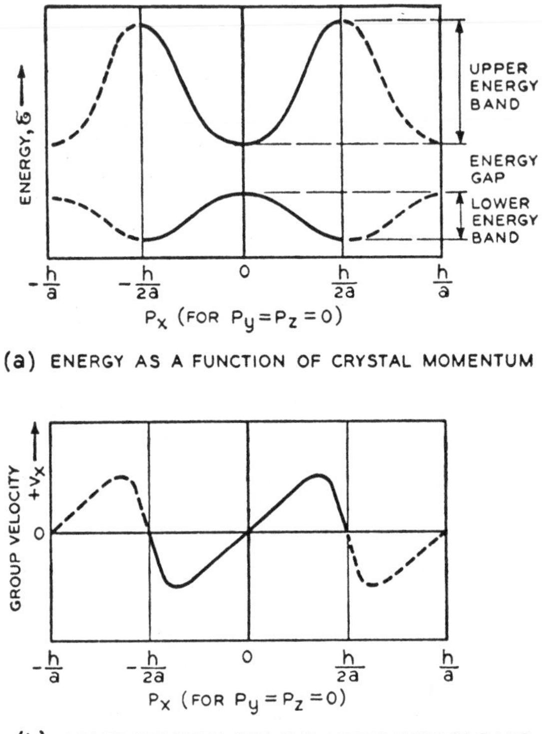

FIG. 6-4—Energy and Velocity in Brillouin Zones.

For definiteness, some of these results are shown graphically in Figure 6.4. Here energy and group velocity are plotted for points along the P_x-axis. The upper diagram shows the energy for two energy bands, and the lower diagram shows the velocity for the upper energy band. In both cases dotted lines show the periodic dependence of energy and velocity on P_x, points outside the zone being equivalent to points inside obtained by adding or subtracting h/a from P_x.

[1] We use p for momentum in the ordinary classical or quantum-mechanical sense and P for the crystal momentum.

In three dimensions, it is necessary to find a vector velocity with components similar to $v = d\mathcal{E}/dP_x$ for one dimension. These are shown in Chapter 15 to be

$$v_x = \partial\mathcal{E}(P_x, P_y, P_z)/\partial P_x \tag{2a}$$

$$v_y = \partial\mathcal{E}(P_x, P_y, P_z)/\partial P_y \tag{2b}$$

$$v_z = \partial\mathcal{E}(P_x, P_y, P_z)/\partial P_z \tag{2c}$$

or, in vector notation

$$\boldsymbol{v} = \nabla_{\boldsymbol{P}}\mathcal{E}(\boldsymbol{P}). \tag{3}$$

Geometrically, this relates $\boldsymbol{v}$ to the energy surfaces in the Brillouin zone as follows: To find the velocity corresponding to $\boldsymbol{P}$, construct in $\boldsymbol{P}$-space a line perpendicular to the surface of constant energy passing through $\boldsymbol{P}$. Proceed along this line in the direction of increasing energy and evaluate $d\mathcal{E}/d|\boldsymbol{P}|$ along this line. The group velocity corresponding to $\boldsymbol{P}$ has the direction and magnitude so obtained, it being, of course, supposed that the P_x, P_y, P_z axes of the Brillouin zone are parallel to x, y, z axes of the crystal.[2] This process will lead to a vector distribution of velocities in the P_xP_y plane of a Brillouin zone like that shown in Figure 6.5. Here the lengths of the arrows indicate the velocity corresponding to the point in $\boldsymbol{P}$-space from which they originate. Since $\boldsymbol{P}$ and $\boldsymbol{v}$ have different dimensions, only the directions and the relative magnitudes of the arrows are significant.

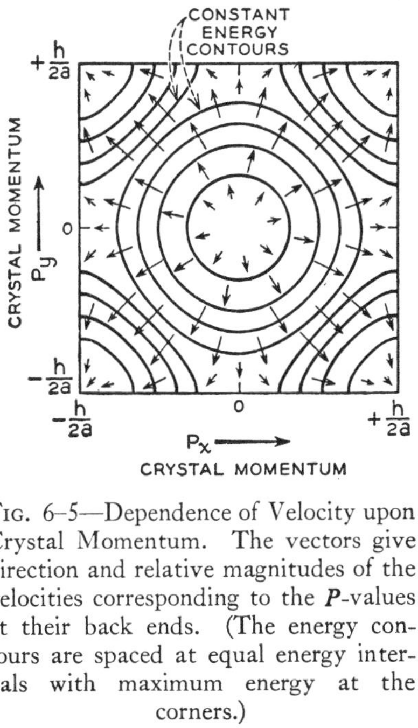

FIG. 6–5—Dependence of Velocity upon Crystal Momentum. The vectors give direction and relative magnitudes of the velocities corresponding to the $\boldsymbol{P}$-values at their back ends. (The energy contours are spaced at equal energy intervals with maximum energy at the corners.)

The evaluation of the velocity completes the development of the Brillouin zone description of the properties of Bloch wave functions as outlined in Figure 5.1. To summarize briefly the present aspect, we find that an electron in a Bloch function moves so that it is equally likely to be anywhere in the crystal. It has an average velocity of motion $\boldsymbol{v}$ which carries it cyclically through the crystal over and over again. In this connection, it should be noted that as the electron moves through the crystal in accordance with the velocity shown in Figure 6.5, its $\boldsymbol{P}$ value does not change. That is,

[2] We have merely defined in words the meaning of $\nabla_{\boldsymbol{P}}\mathcal{E}$, the gradient of $\mathcal{E}$ in $\boldsymbol{P}$-space.

the vector in Figure 6.5 shows a velocity in ordinary space, not in $\boldsymbol{P}$ space. The value of $\boldsymbol{P}$ is changed by externally applied forces or scattering processes, which are considered later.

We can reach a conclusion bearing on conductivity from the matters just discussed, which we shall present here as a prelude to the following chapters, where we shall again derive it as a limiting case: From the properties of the Brillouin zone, we conclude that a full zone can carry no current. This result follows simply from symmetry. For all lattices it is found that the energy takes on equal values for $\boldsymbol{P}$ and $-\boldsymbol{P}$ and, as a consequence of this, the velocities for $\boldsymbol{P}$ and $-\boldsymbol{P}$ are equal and opposite. In a full zone, since every allowed state is occupied, the electrons always cancel off their velocities in pairs so that there can never be any velocity unbalance resulting in an average velocity for the electrons as a whole. (A further discussion of the additivity of the charge and current densities is given in Sections 15.7 and 15.8.)

CHAPTER 7

ELECTRONS AND HOLES IN ELECTRIC AND MAGNETIC FIELDS

The chief aim of this chapter is to show that an electron in an otherwise empty conduction band, and a hole in an otherwise full valence-bond band behave, respectively, much like negative and positive particles in free space, when acted upon by electric and magnetic fields. The treatment is founded on the basic law for the rate of change of crystal momentum when the electron is subjected to a force. This law, although in close analogy with Newton's second law $\boldsymbol{F} = m\boldsymbol{a}$, must be derived on the basis of quantum mechanical arguments such as those given in Chapter 15. In Section 7.3 we give a derivation, which, although not rigorous, is helpful in making the law seem reasonable. The method of derivation followed in Chapter 15 is indicated in general terms in Figure 7.1, together with some consequences of the law which are treated in various sections of this chapter as indicated on the figure.

7.1 THE EFFECT OF A FORCE $\dot{\boldsymbol{P}} = \boldsymbol{F}$

When the field in which the electron moves in the crystal is modified by adding, to the periodic potential $\mathcal{U}$ of the crystal, terms representing applied electric and magnetic fields, the Bloch wave functions are no longer solutions of the time-dependent Schroedinger's equation. However, it is found that if the vector $\boldsymbol{P}$ in the Bloch functions is varied with time according to the law

$$\frac{d\boldsymbol{P}}{dt} \equiv \dot{\boldsymbol{P}} = \boldsymbol{F}, \tag{1}$$

then the resulting wave function is a good solution. The expression to be used for $\boldsymbol{F}$ is the classical one. The electric field $\boldsymbol{E}$ exerts a force $-e\boldsymbol{E}$ and the magnetic field $\boldsymbol{H}$ exerts a force at right angles to both $\boldsymbol{H}$ and the average velocity $\boldsymbol{v}$ of the electron. This leads to the expression for the force on the electron

$$\boldsymbol{F} = -e(\boldsymbol{E} + \boldsymbol{v} \times \boldsymbol{H}/c) \tag{2}$$

where c is the speed of light.[1]

[1] For the meaning of $\boldsymbol{v} \times \boldsymbol{H}$ for a particular case, see Figure 8.7.

The similarity between equation (1) and Newton's second law of motion $\boldsymbol{F} = m\boldsymbol{a} = d(m\boldsymbol{v})/dt$ is the basis for calling $\boldsymbol{P}$ the crystal momentum.

If the force is steadily maintained, the point in the Brillouin zone representing the electron (that is, "representative point") moves steadily in the direction prescribed by the force. When it reaches the surface of the zone, it becomes equivalent to a point on the opposite side and continues into the zone from the equivalent point.

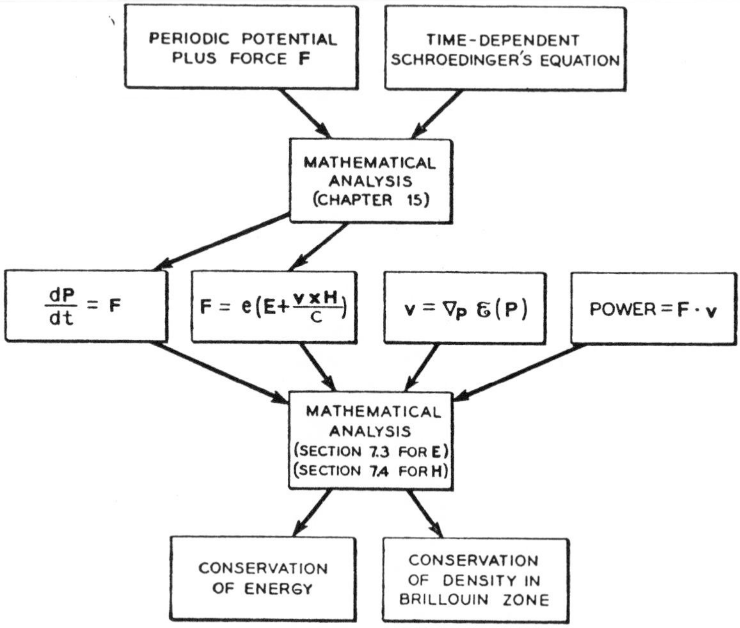

FIG. 7–1—Outline of Derivation of $\dot{\boldsymbol{P}} = \boldsymbol{F}$ and the Consequences $\dot{\boldsymbol{P}} = \boldsymbol{F}$.

The result $\dot{\boldsymbol{P}} = \boldsymbol{F}$ implies that all points in the Brillouin zone represent possible quantum states instead of just the basic $\boldsymbol{P}$-lattice of equations (9) to (11) of Section 5.5 which arose as a consequence of the periodic boundary conditions. The reason for this difference is that, in general, the application of electric and magnetic fields alters the periodic character of the boundary conditions. Thus in a straight wire carrying a current the voltage at one end of the wire will not be the same as at the other, and it is meaningless to require periodic boundary conditions to apply between the two points. However, it is actually possible to bend a wire into a circle, in which case periodic boundary conditions should apply to the coordinate which runs around the circle. It is also possible to apply an electric field to such a wire by using it as a short-circuited secondary winding on a transformer. A model of this sort is treated in Section 15.9 and it is found that, although the induced electric field produces changes in the wave function, it does not change the number of wave lengths of the wave function around the circle. However, it does change the energy of the electron and the current carried by it in precisely the way they would change if the periodic requirements on the boundary condition were disregarded and the wave function altered according to the formula $\dot{\boldsymbol{P}} = \boldsymbol{F}$.

The details of the argument involve mathematical procedures which cannot well be described in simpler terms. However, the result is that, for the periodic model, all energies and velocities vary just as if the lattice of points representing allowed quantum states shifted under an electric field in accordance with the law $\dot{\boldsymbol{P}} = \boldsymbol{F}$. We shall accordingly treat the effect of fields in this way.

The same result may also be obtained by considering a wave-packet composed of Bloch functions all having approximately the same value of $\boldsymbol{P}$. When electric and magnetic fields act on the packet, it is altered so that the average value of $\boldsymbol{P}$ changes according to the law $\dot{\boldsymbol{P}} = -e(\boldsymbol{E} + \boldsymbol{v} \times \boldsymbol{H}/c)$. Corresponding results are obtained if the wave-packet is made to represent a hole. (Sections 15.6 to 15.8.)

As a consequence of these facts, we are justified in taking the law $\dot{\boldsymbol{P}} = \boldsymbol{F}$ as being the proper generalization of Newton's Second Law of Motion for electrons in a crystal. Since the application of this law to all electrons in the states of a Brillouin zone produces a flow of all the occupied states in the zone, we may also consider it to apply to the unoccupied states as well. No error is introduced by this process, since, if the states are empty, we are not concerned with them directly and, if they are occupied, the behavior is correctly given. The procedure of considering that the empty states also vary according to the same law, however, has advantages in analyzing the behavior of holes and is used in Section 7.6.

Finally it may be remarked that in the presence of a magnetic field the relationship between classical momentum $\boldsymbol{p}$ and velocity $\boldsymbol{v}$, and between the corresponding quantum mechanical operators, is no longer $\boldsymbol{p} = m\boldsymbol{v}$. The correct relationship is discussed in Section 15.6. It is shown, however, that the relationships

$$\mathcal{E} = \mathcal{E}(\boldsymbol{P}) \quad \text{and} \quad \boldsymbol{v} = \nabla_{\boldsymbol{P}}\mathcal{E}(\boldsymbol{P}) \tag{3}$$

are unaltered by a magnetic field so that equations (1), (2), and (3) constitute a complete set of equations for determining the behavior of an electron in applied fields, at least for linear terms in $\boldsymbol{E}$ and $\boldsymbol{H}$.

7.2 THE THEOREM OF THE CONSERVATION OF QUANTUM STATES

The flow of points in the Brillouin zone is compatible with the theorem of the conservation of states for the case of electric and magnetic fields.

We shall treat first the electric field, returning to magnetic fields later.

In the case of the electric field the result is obvious. Since the force is $-e\boldsymbol{E}$ and the same for all values of $\boldsymbol{P}$, the lattice of allowed quantum states simply moves as a solid with translational motion through the Brillouin zone. It is evident that this leaves the total number of allowed points in the Brillouin zone unchanged and also has no effect on the density of their distribution.[1]

7.3 THE PRINCIPLE OF THE CONSERVATION OF ENERGY FOR AN ELECTRIC FIELD

If a particle is accelerated by a force $\boldsymbol{F}$, we should expect its gain in energy to be equal to the work done on it by the force. The power furnished to the particle by the force is $\boldsymbol{v} \cdot \boldsymbol{F}$ and the rate of change of energy is $\dot{\mathcal{E}} = \nabla_{\boldsymbol{P}}\mathcal{E} \cdot \dot{\boldsymbol{P}}$. According to the laws we have discussed $\dot{\boldsymbol{P}} = \boldsymbol{F}$ and $\boldsymbol{v} = \nabla_{\boldsymbol{P}}\mathcal{E}$, so that these two powers are equal, and conservation of energy applies to the behavior of an electron subjected to a force.

Although energy is conserved when the electron is acted upon by a steady electric field, its behavior in other respects is strikingly different from that of a classical particle in free space. To a large degree, the particularly unusual features, which we shall shortly describe, are suppressed by the random processes of scattering which are always active in real crystals (except metals in the superconducting state). However, it is worth while to present these results because they are direct, logical consequences of the theory presented in Figures 6.4 and 6.5 and, in addition, they illustrate principles which we shall later use.

Accordingly, we shall consider the motion of an electron in a perfect crystal under the influence of a uniform force $\boldsymbol{F}$ in the x-direction. If this electron is initially at the lowest state in the upper band at $P_x = 0$ in Figure 7.2, then its value of P_x will increase toward the right at a constant rate $\dot{P}_x = F_x$. Its velocity will initially increase to the right also, and, in fact, directly in proportion to the momentum so that, as long as the curve of part (b) may be well approximated by a straight line, the velocity increases uniformly in time as it would for a classical particle, such as a falling body. For these conditions the behavior of the electron can actually be treated in a classical fashion, as will be discussed later. However, after a certain length of time the velocity will reach the maximum of the curve in part (b) and, thereafter, will decrease. Thus in the region where dv_x/dP_x is negative, the electron behaves as if it had a *negative* mass; that is, in this region the continued application of a force to the right actually produces a deceleration to the right and the particle slows down.

[1] It also leaves the wave functions orthogonal.

This behavior has no classical analogue in the dynamics of particles and is due to the wave-mechanical laws governing the electron. If the field is maintained long enough, P_x reaches the value $h/2a$, after which its velocity becomes negative as indicated by the dotted line, and the electron moves against the force. (This behavior is, of course, equivalent to moving P_x

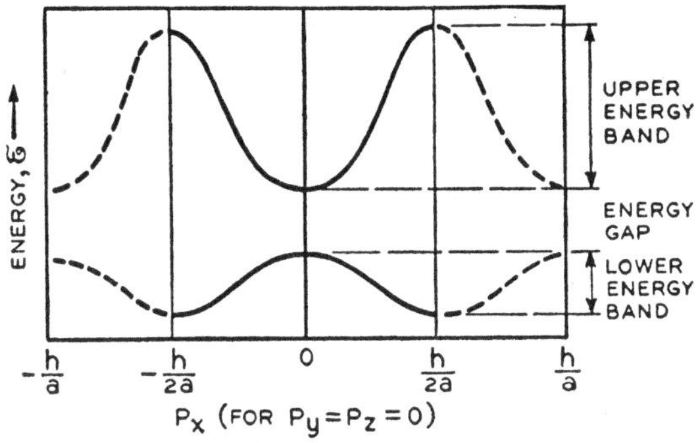

(a) ENERGY AS A FUNCTION OF CRYSTAL MOMENTUM

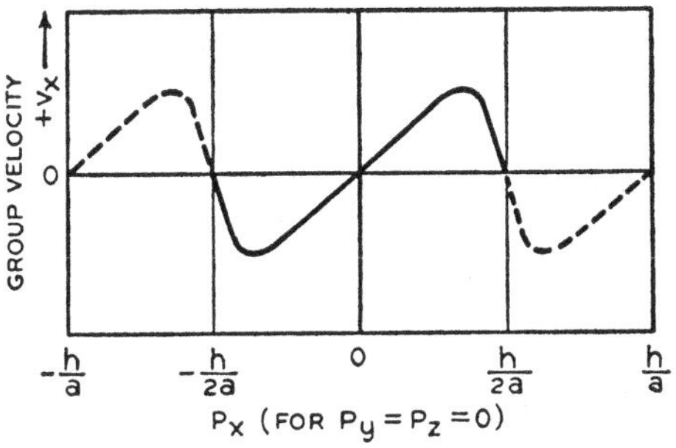

(b) GROUP VELOCITY FOR THE UPPER ENERGY BAND

FIG. 7–2—Energy and Velocity in Brillouin Zones.

back to an interior point on the opposite side of the zone in accordance with the periodic property of P_x.) Finally, the velocity reaches a maximum value in the negative direction, corresponding to the minimum of curve (b), and then decreases in magnitude to zero once more, after which the cycle repeats itself. In other words, the particle oscillates back and forth. If a wave-packet were made up using wave functions, all having nearly the same momentum, this wave-packet would oscillate back and forth in the crystal.

In fact, the position of the wave-packet as a function of time would be a curve of exactly the same shape as that which gives its energy as a function of momentum, a result which can be seen from the following equations:

$$x = \int v_x dt = \int (\partial \mathcal{E}/\partial P_x) dt = \int (\partial \mathcal{E}/\partial P_x) dP_x/F_x$$

$$= \mathcal{E}(P_x)/F_x + \text{const} = \mathcal{E}(F_x t)/F_x + \text{const} \tag{1}$$

since if F_x is applied at $t = 0$, $P_x = F_x t$ and $dt = dP_x/F_x$. Since the work done on the particle is xF_x, this equation, in addition to giving the position of the electron as a function of time, is also an expression of the conservation of energy. The period of the motion is h/aF_x, the time required for P_x to traverse the Brillouin zone.

The oscillatory aspect of the electron's behavior in this case can be profitably considered in calculating the probability of exciting an electron from one band to the other under the influence of a strong applied field[1]—a subject not within the scope of this presentation, however. Instead, the discussion just given should be regarded as intended to illustrate features of the theory of Brillouin zones. As we shall see, collision processes introduce random transitions of the electron from one state to another in times less than the period of the motion so that the anomalous behaviors considered here play no important role in ordinary conduction processes.

7.4 THE EFFECT OF A MAGNETIC FIELD

We shall next consider the forces exerted by a magnetic field upon a moving electron and the consequent changes in momentum. We shall treat the restricted case of a crystal momentum in the $P_z = 0$ plane for a magnetic field in the z-direction. In Figure 7.3(a) we represent the P_xP_y plane of a Brillouin zone with the energy contours and velocity shown as in Figure 6.5. The velocity, it will be recalled, is proportional to the gradient of $\mathcal{E}$ in the zone and is thus perpendicular to the energy contours; the direction and relative magnitude of the velocity are indicated by arrows in the figure.

When a magnetic field is applied in the $+z$-direction, the vector force $-e\boldsymbol{v} \times \boldsymbol{H}/c$ on the electron will result in a changing momentum. Since the force is perpendicular to $\boldsymbol{v}$, the momentum will change as indicated by the arrows in (b). This results in a flow of representative points along the energy contours. Thus, the electrons gain no energy from the magnetic field—a result consistent with the fact that the force produced by $\boldsymbol{H}$ is perpendicular to the motion of the electron and hence does no work on it. As discussed in connection with the Pauli principle, this flow should not change the density of the points and should thus be incompressible. This necessary result comes directly from the formula for the change of momen-

[1] C. Zener, *Proc. Roy. Soc.* **145A**, 523–529 (1934).

tum. Thus if we consider the lines L_1 and L_2 in Figure 7.3(b), we can show that the flow of particles across L_1 is just the same as across L_2; for example, if L_1 is half as long as L_2, then the velocity of an electron at L_1 will be twice that at L_2, and consequently the force and rate of change of momentum will be doubled. Hence, equal volumes of momentum space will flow across L_1 and L_2 in equal times. Accordingly, if the quantum states are initially uniformly distributed, there will be no tendency

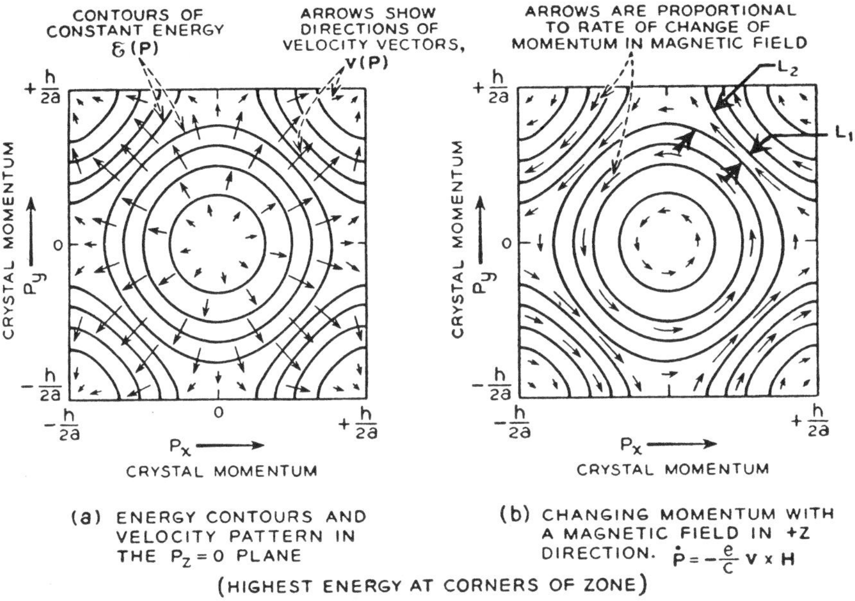

FIG. 7-3—Energy, Velocity and Effect of a Magnetic Field in a Brillouin Zone.

for them to accumulate between L_1 and L_2 since the flow across L_1 and L_2 are equal. (For three dimensions this result can be put in the form that divergence of $\dot{\boldsymbol{P}}$ is zero,[1] so that the general conclusion is that a magnetic field by itself serves only to move the points about on their energy surfaces without producing any accumulations or affecting the properties of the equilibrium distribution.)

These general results for the change in $\boldsymbol{P}$ for an electron in a Brillouin zone under the influence of electric and magnetic fields will next be applied to the behavior of a single electron in an otherwise empty conduction band and a single hole in an otherwise full valence-bond band—still moving in a perfect crystal, however, so that random processes are prevented.

[1] $\nabla_P \cdot (\nabla_P \mathcal{E} \times H) = H \cdot (\nabla_P \times \nabla_P \mathcal{E}) + \nabla_P \mathcal{E} \cdot (\nabla_P \times H) = 0 - 0.$

7.5 THE BEHAVIOR OF AN EXCESS ELECTRON: EFFECTIVE MASS AND REARRANGEMENT OF THE BRILLOUIN ZONE

In order to apply the theory just discussed to problems of conductivity and Hall effect, it is necessary to deal with specific properties of the energy surfaces in the Brillouin zones. The calculation of energy surfaces has actually been carried out for a number of metals by using approximate methods. However, no detailed calculations are as yet available for semiconductors. For this reason, a certain amount of guess-work is involved in dealing with these cases. Fortunately, however, there are general principles, like those involved in determining the shape and size of the Brillouin zone, which can be used as guides. These principles are based largely on symmetry conditions which the surfaces must satisfy in order to be consistent with the basic symmetry of the crystal. In addition there are some quantitative calculations on certain features of the energy surfaces which help to establish numerical magnitudes. Finally there are experimental data which can be used to test the reasonableness of the assumptions concerning the shape.

On the basis of these arguments, we shall proceed by assuming that the diamond-structure semiconductors have approximately spherical energy surfaces at the highest energies of the valence-bond band and the lowest energy of the conduction band, these being the regions of interest for conductivity.[1] Furthermore, we are interested only in energies quite near these edges of the bands, the reason for this being, as discussed in more detail in connection with the Fermi-Dirac statistics, that thermal energies can excite electrons by only about 0.03 electron volt of energy, whereas the band may be several electron volts wide. Hence the states of interest are within a few per cent of the edge of the energy band.

In order to deal with electrons and holes, we must consider both the bottom of the conduction band and the top of the valence-bond band. Furthermore, for each of these, the energy of interest may occur at either the center or the corner of the Brillouin zone. Fortunately, all of these cases may be treated on an equivalent basis, as we shall show. We shall, therefore, treat first the simplest case and later show the relationship of the other cases to it. Accordingly, we suppose that the bottom of the conduction band comes at an energy $\mathcal{E}_0$ at the center of the zone; this is the situation shown in Figures 6.5 and in 5.11 if the highest energy is at the corners. General theoretical arguments show that the spherical surfaces must have the equation

$$\mathcal{E} = \mathcal{E}_0 + K(P_x^2 + P_y^2 + P_z^2) = \mathcal{E}_0 + KP^2 \tag{1}$$

[1] See, for example, F. Seitz, *Phys. Rev.* **73**, 549–564 (1948). I am informed by C. Herring, however, that there are some strong theoretical reasons for believing that one of the bands is degenerate at its limiting energy with consequent complications in the shape of the energy surfaces. We shall return to this question in Chapter 12.

where P is the magnitude of $\boldsymbol{P}$. This gives rise to a velocity

$$v_x = \partial\mathcal{E}/\partial P_x = 2KP_x \tag{2a}$$

$$v_y = \partial\mathcal{E}/\partial P_y = 2KP_y \tag{2b}$$

$$v_z = \partial\mathcal{E}/\partial P_z = 2KP_z \tag{2c}$$

or in vector notation,

$$\boldsymbol{v} = 2K\boldsymbol{P}. \tag{3}$$

If a force $\boldsymbol{F}$ is applied to the particle, the velocity changes according to the law

$$\dot{\boldsymbol{v}} = 2K\dot{\boldsymbol{P}} = 2K\boldsymbol{F}. \tag{4}$$

This is analogous to Newton's second law in the form

$$\boldsymbol{F} = m_n\boldsymbol{a} \quad \text{or} \quad \dot{\boldsymbol{v}} = \boldsymbol{F}/m_n, \tag{5}$$

if we interpret $1/2K$ as an effective mass m_n for an excess electron. In other words, the application of a force changes the velocity of the electron just as if it had a mass of $1/2K$.

The quantity $1/2K$ for states in the conduction band (or $1/2K'$ for the valence-bond band) is of the same order of magnitude as the mass of the free electron, a result which we shall derive by general considerations of the wave functions in Section 14.7. From the experimental data on silicon and germanium it appears to be quite near the electron mass. For most of the purposes of this chapter not enough use is made of the difference between the effective mass and the mass of the free electron to warrant emphasizing the effective mass at this point in the treatment by giving it a special symbol. We shall, therefore, write $1/2m$ for K and proceed accordingly.[2]

On the basis of this assumption, we see that the response of an electron at the minimum energy point of a Brillouin zone to a force is the same as for a free electron with

$$\dot{\boldsymbol{v}} = \boldsymbol{F}/m. \tag{6}$$

We shall use this relationship later in connection with conductivity.

What happens if the lowest energy of the zone comes at the corner of the zone? To illustrate this it will suffice to consider a two-dimensional example. This is shown in Figure 7.4, where, for emphasis, the higher energy contours have been omitted. In (a) the effect of a force in accelerating an electron from $\boldsymbol{P}_0$ to $\boldsymbol{P}_5$ is indicated. Each time the momentum reaches a point on the zone boundary, the wave function becomes equivalent to a point on the opposite boundary, as shown. (It may be noted in

[2] A wave-packet electron made from quantum states where the effective mass is very small can be accelerated more readily in a crystal than it can be in vacuum. The way in which interaction with the crystals helps to accelerate the electron is discussed in Section 14.7.

this case that when the point reaches a boundary, $\boldsymbol{v}$ is parallel to the boundary so that, as previously discussed, the wave function carries no current in a direction perpendicular to the boundary.) If the zone is rearranged for convenience as shown in part (b), the change in momentum from $\boldsymbol{P}_0$ to $\boldsymbol{P}_5$ becomes a continuous straight line. If we now describe the momentum of the electron by the vector $\boldsymbol{P}'$ which is drawn from the center of (b), we once more have the relationship that

$$\mathcal{E} = \mathcal{E}_0 + K'(P_x'^2 + P_y'^2 + P_z'^2) = \mathcal{E}_0 + K'\boldsymbol{P}'^2 \tag{7}$$

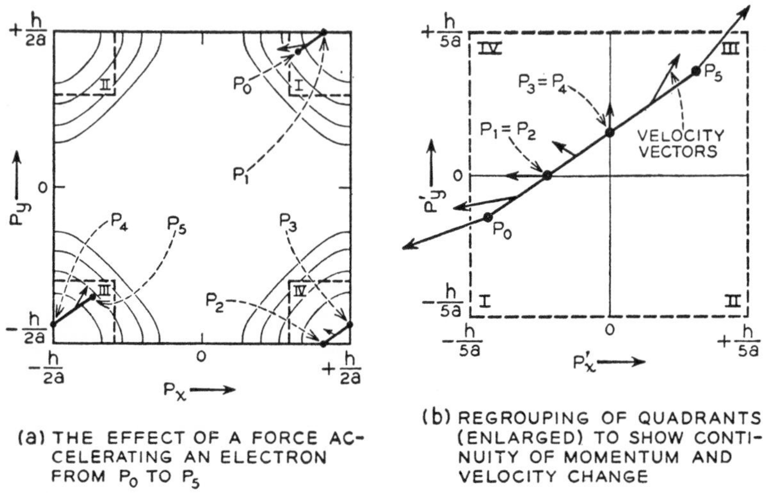

FIG. 7-4—Rearrangement of Brillouin Zone for the Case Where the Lowest Energy Comes at the Corners.

as a consequence of general theoretical arguments about the energy surfaces near the corners of the zone. Since the behavior of $\boldsymbol{P}'$ under a force in (b) is just the same as that of $\boldsymbol{P}$ itself at the center of the zone in the case previously considered, it is evident that similar results will be obtained here and $1/2K'$ can be regarded as an effective mass. As Figure 7.4 illustrates, the velocity of the electron varies continuously with $\boldsymbol{P}'$ and shows no discontinuous behavior at the edge of the Brillouin zone.

Thus the effect of a force on the electron is to change its crystal momentum according to the equation

$$\dot{\boldsymbol{P}} \text{ or } \dot{\boldsymbol{P}}' = \boldsymbol{F} \tag{8}$$

and, since its velocity is related to $\boldsymbol{P}$ or $\boldsymbol{P}'$ by the equation

$$\boldsymbol{v} = 2K\boldsymbol{P} = \boldsymbol{P}/m$$

or

$$\boldsymbol{v} = 2K'\boldsymbol{P}' = \boldsymbol{P}'/m, \tag{9}$$

the change in velocity is given by

$$\dot{\boldsymbol{v}} = \boldsymbol{F}/m \quad \text{or} \quad \boldsymbol{F} = m\boldsymbol{a} \tag{10}$$

where $\boldsymbol{a} \equiv \dot{\boldsymbol{v}}$ is the acceleration.

In terms of $\boldsymbol{P}$, $\boldsymbol{P}'$ and $\boldsymbol{v}$ the energy can be written as

$$\mathcal{E} = \mathcal{E}_0 + (\boldsymbol{P}^2 \quad \text{or} \quad \boldsymbol{P}'^2)/2m = \mathcal{E}_0 + m\boldsymbol{v}^2/2. \tag{11}$$

Thus the energy has the form of a constant plus a term which is formally identical with a kinetic energy associated with velocity $\boldsymbol{v}$. Actually, the term $m\boldsymbol{v}^2/2$ represents a combination of changing potential energy of the electron in the periodic potential and changing kinetic energy as the Bloch wave function for the electron changes with changing $\boldsymbol{P}$. Thus although the expression for the energy may be used as if it arose from changing kinetic energy of motion of the excess electron, this feature should be considered a mathematical simplification rather than a statement of a basic physical fact. As will be seen in Section 7.6, a similar simplification may be made in the case of a hole.

We next must verify that under the influence of electric and magnetic fields, the Brillouin zone electron behaves like a classical one so far as its acceleration is concerned. This result follows directly from the $\boldsymbol{F} = m\boldsymbol{a}$ together with equation for the force, which as we discussed in Section 7.1, is taken as

$$-e(\boldsymbol{E} + \boldsymbol{v} \times \boldsymbol{H}/c) = \boldsymbol{F} = m\dot{\boldsymbol{v}} = m\boldsymbol{a}. \tag{12}$$

Here e is a positive number so that $-e$ is the charge on the electron. This equation states that the rate of change of $\boldsymbol{v}$ in electric and magnetic fields is related to $\boldsymbol{E}$, $\boldsymbol{H}$, and $\boldsymbol{v}$ in just the same way as it is for a free classical electron.

This result seems so obvious that it may be worth while to review briefly the route taken in reaching it. When we treat the case of the hole in the next few pages, we shall use the same machinery but will conclude that a plus sign should be used in the equation expressing $\boldsymbol{a}$ in terms of $\boldsymbol{E}$, $\boldsymbol{H}$, and $\boldsymbol{v}$. In essence, we have indicated that it is a consequence of quantum theory that an electron may have a state of motion through a periodic potential field which gives it a crystal momentum $\boldsymbol{P}$ and velocity $\boldsymbol{v}$ which remain constant in time if no external forces act (the crystal being assumed perfect). This in itself is a surprising result; a classical electron moving in the crystal field would be deflected so often that its vector velocity would take on practically random values and would, on the average, be zero. It is, thus, the Schroedinger wave equation which leads us to the idea that the electron moving in the crystal can have values for crystal momentum and average velocity which are independent of time. Next, we find that from the same quantum mechanical framework we can deduce that $\dot{\boldsymbol{P}} = \boldsymbol{F}$. Finally, as an approximation, we conclude that near the minimum energy point of the zone $\boldsymbol{v} = 2K\boldsymbol{P}$, and as a further approximation, that $\boldsymbol{v} = \boldsymbol{P}/m$.[3] (In later chapters we shall introduce m_n and m_p for effective masses and in Chapter 12 we shall consider a more complicated possibility.) Finally we equate $\boldsymbol{F}$ to the classical expression for force on an electron of charge $-e$ and velocity $\boldsymbol{v}$ moving in a field $\boldsymbol{E}$ and $\boldsymbol{H}$; this last step represents in a sense an additional assumption in the exposition as presented here. Actually this expression for the force is contained in Schroedinger's equation, and thus does not really represent a new assumption.

Thus we reach the end of our treatment of the behavior of an excess electron in the conduction band. The Brillouin zone theory indicates that it behaves in the same way as a free classical electron, provided its energy is always so low that it stays near the bottom of the conduction band. In the next paragraphs we shall verify that the behavior of a hole in the valence-bond band is similarly like that of a positively charged electron.

7.6 THE BEHAVIOR OF A HOLE

As we shall show, the hole is really an abstraction which gives a convenient way of describing the behavior of the electrons. An essential feature in making this abstraction is the fact that a full Brillouin zone with every allowed state occupied can carry no net current. This feature permits the behavior of the hole to be found directly from the behavior which the missing electron would have if it were present.[1]

We indicated at the end of Chapter 6 that the average vector velocity for all the electrons in a Brillouin zone was zero because their individual velocities cancelled off in pairs. However, this conclusion was based on symmetry arguments and followed a discussion of the basic $\boldsymbol{P}$-lattice of the Brillouin zone. (Figure 5.10.) After that discussion, we permitted the array of allowed points to be translated through the Brillouin zone keeping the density constant. This does not invalidate the conclusion that the full zone can carry no current, for when the velocities are added up for all the electrons, they are found to cancel so accurately that what is left over is far less than that of one electron; in fact, the bigger the crystal and the more electrons in the zone, the more perfect is the cancellation. Thus the cancellation failure is negligible compared to the effect produced by removing an electron so as to leave a hole in the band. This is the key to the procedure for computing the effect of the hole.

We reason as follows: Consider first a full Brillouin zone. Each electron, as discussed in Section 6.1, contributes a current density $(-e/V)\boldsymbol{v}$ to the crystal. We single out a particular quantum state for consideration;

[3] When we apply this theory to practical cases, the unlimited variations of $\boldsymbol{P}$ produced by an electric field are reduced by the effect of collisions so that only small $\boldsymbol{P}$ values need be considered and the approximation applies.

[1] See Sections 15.7 and 15.8 for an analytical treatment of the same subject matter.

call it state s, with velocity $\boldsymbol{v}_s$. We now add up all the currents due to all the electrons, but we split the sum into two parts with the electron in state s separated from the others. Since the zone is full, this sum is zero, and we have

$$\sum_{i \neq s} (-e/V)\boldsymbol{v}_i + (-e/V)\boldsymbol{v}_s = 0, \tag{1}$$

from which we obtain another equation:

$$\sum_{i \neq s} (-e/V)\boldsymbol{v}_i = (e/V)\boldsymbol{v}_s. \tag{2}$$

Now the left side of this equation is simply the net current due to a Brillouin zone with every state occupied except s. Hence we see that this current is just what we would get if we had an empty zone with one *positively* charged electron moving with $\boldsymbol{v}_s$. This is a lemma in the proof that holes act like positively charged electrons. Stated in words:

The current due to a hole corresponds to a charge $+e$ moving with the velocity associated with the vacant quantum state:

$$\boldsymbol{I}\text{ (hole)} = (+e/V)\boldsymbol{v}_s. \tag{3}$$

Our next problem is to find how the current in the zone varies under applied electric and magnetic fields. For this purpose we must consider the dependence of energy upon momentum near the top of the energy band. As we did for the case of the bottom of the energy band, we disregard the possibility that this upper energy may be degenerate and conclude that the surfaces must be spherical and that the energy must vary as P^2; the problem of rearranging the zone if the highest energy occurs at the corners is handled in the same way; and we also use the electron mass m for $1/2K$. However, there is a vital difference from the case of the bottom of the conduction band, since now we must have an energy which decreases as $\boldsymbol{P}$ or $\boldsymbol{P}'$ varies from the maximum energy point. Consequently, we have

$$\begin{aligned} \mathcal{E} &= \mathcal{E}_0 - K(P_x^2 + P_y^2 + P_z^2) \\ &= \mathcal{E}_0 - P^2/2m_p \end{aligned} \tag{4}$$

where m_p is the effective mass for a hole. As for m_m of (5) in Section 7.5, we shall not distinguish between m_p and m in this chapter. From this we conclude that the relationship between velocity and momentum is

$$\boldsymbol{v} = -\boldsymbol{P}/m, \tag{5}$$

and the relationship between velocity and energy is

$$\mathcal{E} = \mathcal{E}_0 - mv^2/2. \tag{6}$$

Hence if we had an isolated electron in this part of the Brillouin zone, it would behave in a very anomalous way, as discussed in connection with the periodic motion in Section 7.3. For example, if the isolated electron were at the top of the band, the application of an electric field would decrease its energy rather than increase it. However, we do not have an isolated electron but instead a vacant quantum state at the top of an otherwise filled band. This situation obviously represents a minimum energy condition for the band as a whole since the electrons are all in lower states of energy than the vacant state. When the hole acquires velocity, it does so at the expense of displacing an electron from a lower state into the higher state. We must now see that the resulting changes in energy and current are those to be expected for a charge of $+e$ and a mass of $+m$.

We shall, for purposes of exposition, take a preliminary step in the proof by showing the behavior of a hole initially at rest corresponds to a positive and not a negative mass. Accordingly, we suppose that the vacant quantum state s is at the highest energy of the valence-bond band so that $\mathcal{E} = \mathcal{E}_0$ and $\boldsymbol{v}_s = 0$. The application of an electric field $\boldsymbol{E}$ causes the quantum states to change according to the law $\dot{\boldsymbol{P}} = -e\boldsymbol{E}$. Hence, the velocity of the vacant state, which bears a negative mass relationship to the momentum, will vary according to the equation

$$\dot{\boldsymbol{v}}_s = -\dot{\boldsymbol{P}}/m = +e\boldsymbol{E}/m, \tag{7}$$

and the current will vary according to

$$\dot{\boldsymbol{I}} = (+e/V)\dot{\boldsymbol{v}}_s = (+e^2/m)\boldsymbol{E}. \tag{8}$$

This relationship shows that the current increases in the direction of the applied field. If $\boldsymbol{v}_s = 0$ at $t = 0$, then the power delivered to the specimen would be given by the equation

$$\boldsymbol{E} \cdot \boldsymbol{I} = (e^2/m)\boldsymbol{E}^2 t \tag{9}$$

so that energy would be absorbed in accelerating the hole. Now suppose we try to interpret the behavior of the hole as corresponding to a charge e_p and a mass m_p. Then we would conclude that

$$\dot{\boldsymbol{v}} = e_p\boldsymbol{E}/m_p \tag{10}$$

and

$$\boldsymbol{E} \cdot \boldsymbol{I} = (e_p{}^2/m_p)\boldsymbol{E}^2 t. \tag{11}$$

It is evident that the only choice for e_p and m_p which will cause these two equations to give the same results as (7) and (9) is $e_p = e$ and $m_p = m$. The choice $m_p = -m$ is excluded by the sign of the power input equation—attributing a negative mass to the hole would cause it to deliver energy to the field rather than to absorb it.

We shall next derive a more general expression which includes the effects of magnetic fields as well. For this case

$$\dot{\boldsymbol{P}} = \boldsymbol{F} = -e(\boldsymbol{E} + \boldsymbol{v}_s \times \boldsymbol{H}/c) \tag{12}$$

and $m\dot{\boldsymbol{v}}_s = -\dot{\boldsymbol{P}}$ as before. From these equations we obtain

$$m\dot{\boldsymbol{v}}_s = +e(\boldsymbol{E} + \boldsymbol{v}_s \times \boldsymbol{H}/c). \tag{13}$$

This is just the equation for $\boldsymbol{F} = m\boldsymbol{a}$ for a positive charge. Furthermore, we have shown that the current density is given, as in equation (3), by

$$\boldsymbol{I} = (e/V)\boldsymbol{v}_s. \tag{14}$$

From these two equations we may obtain an equation relating $\dot{\boldsymbol{I}}$ to $\boldsymbol{E}$, $\boldsymbol{H}$, and $\boldsymbol{I}$:

$$\dot{\boldsymbol{I}} = (e^2/m)[(\boldsymbol{E}/V) + (\boldsymbol{I} \times \boldsymbol{H}/ec)] \tag{15}$$

for a hole in the valence-bond band where e is the positive magnitude of electronic charge.

A similar treatment for the case of an electron at the bottom of the conduction band, using equation (12) of Section 7.5 instead of (13) of Section 7.6 and $(-e/V)\boldsymbol{v}$ instead of $(+e/V)\boldsymbol{v}_s$, leads to

$$\dot{\boldsymbol{I}} = (e^2/m)[(\boldsymbol{E}/V) - (\boldsymbol{I} \times \boldsymbol{H}/ec)] \tag{16}$$

where again e is the positive magnitude of the electronic charge so that the charge on the electron is $-e$.

It is evident that these equations differ in just the way expected for a change in sign of e. An interpretation involving a negative mass would lead to a current which was accelerated opposite to $\boldsymbol{E}$ for the case of $\boldsymbol{H} = 0$ and is thus inadmissible. (See Problem 7.) Equations (15) and (16) are in a form readily adapted to treating conductivity and Hall effect and are employed for the latter purpose in Section 8.7.

The equation of conservation of energy may be established for the case of a hole in the valence-bond band by addition for all the electrons in the zone. Since, for each of them, $d\mathcal{E}/dt = \boldsymbol{F} \cdot \boldsymbol{v}$, the rate of change of the total energy in the band must be equal to the rate at which power is being furnished by the electric field.

It is more illuminating, however, to interpret the energy changes as due to a "pseudo kinetic energy of motion of the hole". This idea is approached by calculating how the energy of the electrons depends on the energy of the empty quantum state. If the hole changes its velocity from $\boldsymbol{v}_s = 0$ to $\boldsymbol{v}_s = \boldsymbol{v}_1$, it is evident that the redistribution of the electrons has a net result of shifting one electron from energy

$$\mathcal{E}_0 - m{v_1}^2/2 \quad \text{to} \quad \mathcal{E}_0 \tag{17}$$

or an increase in energy of $m{v_1}^2/2$. Of course, the redistribution takes place by the flow of the quantum states[2] according to $\dot{\boldsymbol{P}} = \boldsymbol{F}$, so that the

[2] It is evident from the discussion of Section 7.2 that the quantum states should be considered to flow, whether occupied by an electron or not.

electron formerly at $\boldsymbol{v}_1$ is not the one raised to $\boldsymbol{v} = 0$. However, the net increase in energy for all the electrons is this amount. In other words, the energy gained by the electrons is equal to the kinetic energy calculated for a hole of mass $+m$ moving with velocity $\boldsymbol{v}_1$.

Finally, it may be added that the charge density in the crystal due to one hole in the valence-bond energy band may be determined from the wave functions occupied by the electrons. When this is done, it is found that the hole contributes a deficit electron density. Furthermore, the wave functions may be combined to produce a wave-packet for the hole and this wave-packet is found to have the group velocity $\boldsymbol{v}_s$ corresponding to the velocity of the vacant quantum states combined in making the wave-packet (Section 15.8). Wave-packets for the positive charge of holes may thus be dealt with in the same way as the wave-packets for electrons. It is such wave-packets for holes that should be used to describe the flow of holes in n-type germanium in transistors or the Haynes-Shockley experiment of Section 3.1.

7.7 SUMMARY

Starting with the law derived from quantum mechanics that $\dot{\boldsymbol{P}} = \boldsymbol{F}$ and certain approximations about the behavior of the energy surfaces in a Brillouin zone, it has been shown that the behaviors of one electron in the conduction band and of one hole in the valence-bond band are given by the classical equations for an electron with minus charge and plus charge, respectively. The argument has been long, and the behavior of the hole has been shown to be essentially a shorthand way of describing the behavior of all the electrons.

Some of the essential conclusions about the behavior of the electron and the hole are shown in Figure 7.5. Part (a) represents the Brillouin zone for the conduction band, a small portion near the minimum energy having been singled out for attention. In this region, the velocity and the acceleration of an excess electron by a magnetic field will be as shown near the center portion of Figure 7.3. (It may have been necessary to rearrange the zone and use $\boldsymbol{P}'$ in Figure 7.5.) Part (b) similarly represents the highest energy in the valence-bond zone, in which the velocity and the acceleration in a magnetic field of a hole will be as shown near the corners of Figure 7.3. If an electric field E_x alone were present, all the points in the zone would move to the left in accordance with the equation $\dot{P}_x = -eE_x$, and the velocity of the electron and hole would change as $\boldsymbol{v}$ (electron) $= \boldsymbol{P}/m$, $\boldsymbol{v}$ (hole) $= -\boldsymbol{P}/m$. These differences are indicated in the four parts of Figure 7.5 for an electron and a hole, each initially in the state with $\boldsymbol{v} = 0$. The energy for the quantum state occupied by the electron would increase with time in (a) and (c) and that of the state occupied by the hole (that is,

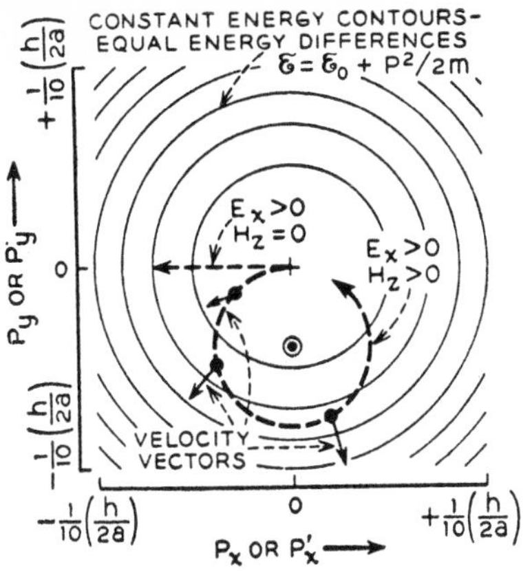

(a) CHANGE IN QUANTUM STATE OF AN ELECTRON INITIALLY AT THE STATE OF LOWEST ENERGY

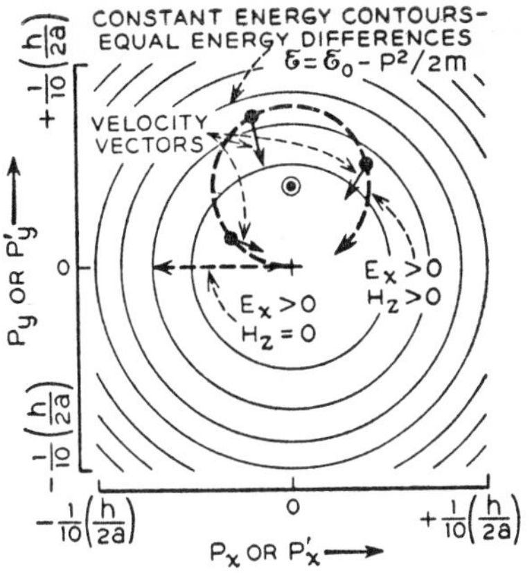

(b) CHANGE IN QUANTUM STATE OF AN ELECTRON INITIALLY AT THE STATE OF HIGHEST ENERGY

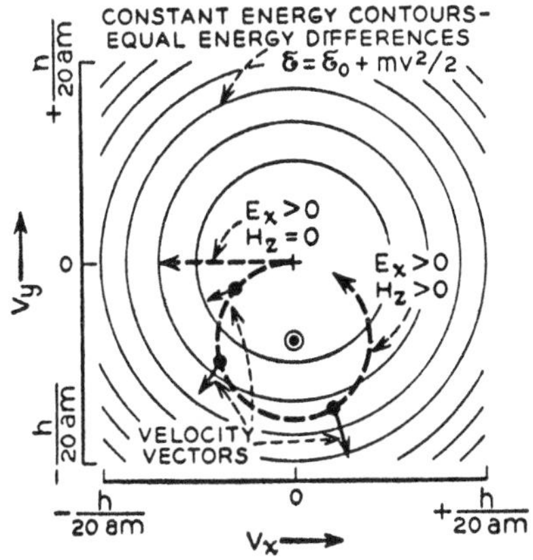

(C) CHANGE IN VELOCITY AND ENERGY OF ONE ELECTRON IN THE CONDUCTION BAND

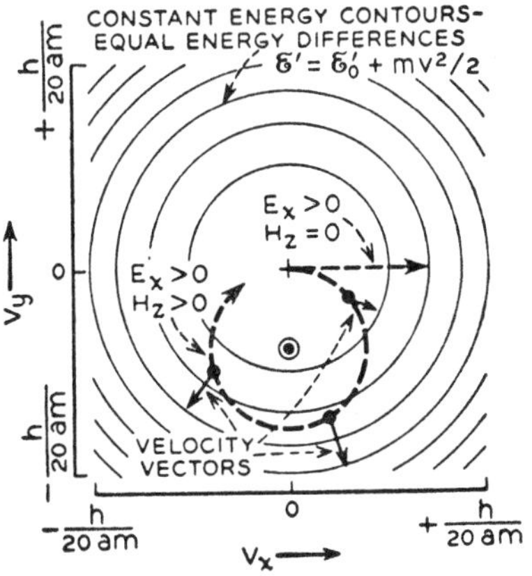

(d) CHANGE IN VELOCITY FOR HOLE IN VALENCE-BOND BAND AND CHANGE IN ENERGY, $\mathcal{E}'$, OF ALL ELECTRONS IN BAND

FIG. 7–5—Summary of the Behavior of an Excess Electron or a Hole under Influence of Combined Electric and Magnetic Fields. Dashed Arrows Show Variation with Time.

left empty) would decrease with time in (b); the total, however, of all the electrons in the valence-bond band would increase as shown in (d).

If a magnetic field alone were applied, all of the quantum states would simply rotate as a rigid body about the P_z-axis, the angular velocity being (as discussed in the next chapter; also see footnote below) $\pm eH/mc$, the directions of rotation being those shown in Figure 7.3.

If both E_x and H_z are present, the dependence of the quantum states upon time will be as shown[1]: The momenta and velocities of the quantum states involved move on circles with constant angular velocity. The two fields result in a bodily rotation of all points in P-space (not merely the one initially at $\boldsymbol{v} = 0$) about the point $\odot$ in the diagrams—a displacement

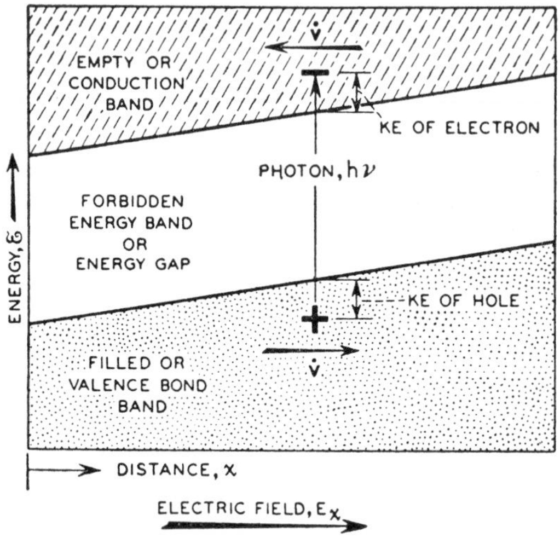

FIG. 7–6—Conventional Energy Band Picture, Showing Interpretation of Electron and Hole Energies.

which obviously conserves the density of quantum states in the Brillouin zone.

A conventional scheme for representing the motions and energies of holes and electrons is shown in Figure 7.6. Here we consider the production of a hole-electron pair by a photon as discussed in Chapter 1. The photon excites an electron from a state in the lower band to one in the higher band. In general, neither of the states is at the edge of the forbidden band, so that both the hole and the electron have velocities of motion and contribute their apparent kinetic energies to the energy of the crystal. If the location of the exciting photon has been closely controlled (by focusing the light with a lens, for example), then the wave functions for the hole and the electron will be represented by localized wave-packets. The effect of the electric

[1] This interesting case is not essential to the later development—the formulae of the Hall effect being obtained in a simpler way. It is instructive, however, to work it out in detail. The solution can be compactly obtained in complex notation as follows: Letting $w = x + iy$, the equation $m\ddot{w} = \pm e(E_x - i\dot{w}H_z/c)$ is solved by $w = -icE_x/H_z +$ const exp $(\pm ieHt/mc)$ so that $\odot$ corresponds to $v_y = -cE_x/H_z$. This can easily be integrated to give $w(t)$.

field is to tilt the energy bands since an electron in a wave-packet state at the bottom of the conduction band will have a lower energy at the positive side of the crystal. In the absence of an electric field, the lowest energy state for the excited pair is that for which the electron is at the bottom and the hole at the top of their respective bands. The general rule that holes tend towards high levels on energy diagrams is, of course, just the same as saying that electrons tend towards low levels. The effect of the electric field will be to accelerate the wave-packets in the indicated directions. We shall return to a more detailed treatment of such energy level diagrams in Chapter 12 where they are used in analyzing the rectification of *p-n* junctions and related topics.

In order to proceed further to the discussion of conductivity and Hall effect we must consider the effect of random processes upon large numbers of holes and electrons behaving as shown in Figure 7.5. This we do in the following chapter.

7.8 CONNECTION BETWEEN QUANTUM AND CLASSICAL MECHANICS

From the material presented in this chapter, it can be suggested how the results of quantum mechanics may merge with those of classical mechanics in the limiting case. As we have discussed, it is possible to build up wave-packets which move with certain group velocities. If the electron is not in a crystal (so that the periodic potential energy is eliminated), then quantum mechanics leads to just the same formula for energy, as a function of momentum, as does the classical theory for an electron. Consequently, the group velocity, momentum, and energy are related as in the case of a free particle. Furthermore, in electric and magnetic fields, the electron is subjected to a force which changes its momentum as if it were a classical particle. As a result, the wave-packet will behave in just the same way as a free electron.

Analysis along the lines presented above can be carried out in detail. When this is done, it is concluded that so long as the dimensions of the environment in which the electron moves are very large compared to its wave length, the wave-packet may be regarded as traveling like a classical free particle. The requirement that the dimensions be large is not very stringent. For example, wave-packets for electrons in the smallest of vacuum tubes would be a thousand times smaller than the grid wire spacing and thus could be dealt with as particles.

In this sense, classical mechanics may be regarded as a limiting case of quantum mechanics just as geometrical optics is a limiting case of physical optics, which deals with diffraction phenomena for light. In both cases, when the size of the structures involved is very large compared to the wave length, the appropriate limiting methods may be employed. The same differences in treatment are used in connection with both radio waves and microwaves. For antenna design and small-scale effects, diffraction phenomena are vital; for large-distance effects such as ionosphere reflections and the target end of radar systems, geometrical optics suffices.

There is, of course, far more to the philosophy of quantum mechanics as applied to large-scale phenomena than is involved in showing that the electron wave-packet behaves like a classical electron. However, a treatment of these important questions belongs not in a treatise on a limited phase of the role of electrons in semiconductors, but in texts on quantum mechanics and the philosophy of physics to which the interested reader is here referred.[1]

Problems

1. Verify, by considerations of symmetry, from Figure 7.3 that, if an electric field $E_x = E_y$ is applied, the electron undergoes periodic motions in the x and y directions in space with the same period, so that it periodically retraces the same path. What can be said of the motion if $E_x = 2E_y$?

2. By the method of the footnote of Section 7.7, find the trajectories in fields E_x and H_z for holes and electrons with initial velocities v_x and v_y.

3. Prove that, if the initial velocities at $t=0$ are $v_x = v_0 \cos\theta$, $v_y = v_0 \sin\theta$, the displacement at time t when averaged over θ is independent of v_0. (This shows that the effect of random velocities in Section 8.6 can be disregarded.)

4. For some crystals, such as bismuth and zinc, the important parts of the Brillouin zones have energies of the form

$$\mathcal{E} = \mathcal{E}_0 + K_x P_x^2 + K_y P_y^2 + K_z P_z^2.$$

Find the equation of motion which replaces $m\boldsymbol{a} = \boldsymbol{F}$ for this case and integrate these equations for $\boldsymbol{F}$ due to H_z.

5. What happens to the mass of the crystal as a whole when the hole and electron of Figure 7.6 recombine?

6. Verify by using in part equations (15) and (16) of Section 7.6 that the power expended by the electric field $\boldsymbol{E}\left(=\int \boldsymbol{I}\cdot\boldsymbol{E}\,dV\right)$ can be accounted for by the changing energy of the electron or hole.

7. Show that a particle of negative mass initially at rest so that $\boldsymbol{I} = 0$ at $t = 0$ produces a negative resistance and delivers during time t an energy

$$e^2E^2t^2/2|m|.$$

Show that a particle of positive mass absorbs an amount of energy given by the same formula.

[1] P. W. Bridgman, *The Logic of Modern Physics*, Macmillan, New York, 1927. R. B. Lindsay and H. Margenau, *Foundations of Physics*, John Wiley & Sons, New York, 1936.

CHAPTER 8

INTRODUCTORY THEORY OF CONDUCTIVITY AND HALL EFFECT

8.1 INTRODUCTION

In this chapter the laws of motion for electrons and holes derived in Chapter 7 are applied to the analysis of conductivity and Hall effect. The exposition presented here is based on a simplified treatment of the "mean free time". In Chapter 11, a more detailed discussion of the mean free time is presented. The purpose of presenting the simplified treatment at this point in the text is to illustrate the connection between experimental results and the abstract matters discussed in Chapters 5, 6, and 7 before proceeding with the theoretical topics of the next three chapters, which treat energy level diagrams, the Fermi-Dirac Statistics, and collision processes. In addition, the simplified treatment, in anticipating the results to be derived later, serves as an introduction to the following chapters.

In this chapter, we shall introduce, in addition to the mean free time, the concepts of an equilibrium thermal distribution of electrons and holes and of random processes which bring this equilibrium about. However, these latter two ideas will enter the equations only through the mean free time, denoted by $\boldsymbol{\tau}$.

There are two important ways of thinking about the current density in a semiconductor. Although mathematically equivalent for most purposes, the mental images associated with these are somewhat different and for this reason both methods are described. According to one method, the electrons or holes are thought to be in definite quantum states in the Brillouin zone and thus to be carrying currents of $(\mp e/V)\mathbf{v}$ in accordance with equations (1) of Section 6.1 and (3) of Section 7.6. This method is well adapted to calculating currents on the basis of diagrams showing how applied electric and magnetic fields disturb the distribution of electrons in the momentum space of the Brillouin zone. According to the other method, the holes and electrons are treated as classical particles having at any instant definite locations and velocities. This treatment is justified by the analysis presented in Chapters 6 and 7 which shows that wave-packets for holes and electrons behave like classical particles having effective masses which may differ from the free electron mass. According to this particle treatment, the electrons and holes move in the presence of electric and magnetic fields in curved paths (like the parabola for a falling body) between collisions. (In Chapter 11, a more general method of attack is used which is applicable to the case in which the effective mass concept must be modified.)

There are cases, however, in which the wave-packet method is appropriate to the physical situation whereas the method of using individual

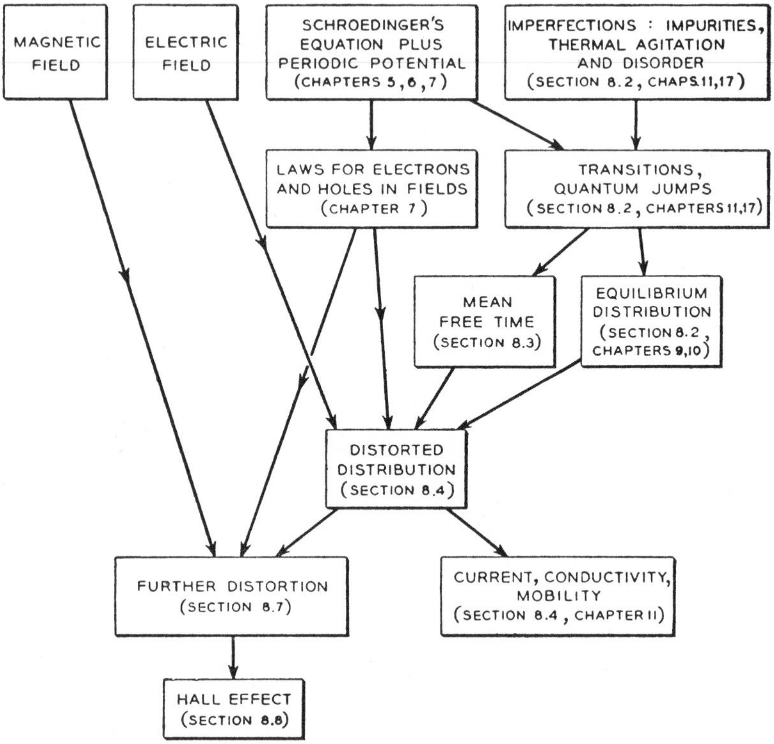

Fig. 8–1—Outline of Arguments Used in Developing Expressions for Conductivity and Hall Effect.

states in the Brillouin zone is not. The latter is adapted to the case in which the wave function gives equal probability that the electron (or hole) is in any unit cell of the crystal. In the experiment of Haynes and Shockley, Section 3.1, on the other hand, the holes are injected at a definite point in the germanium and when first injected have negligible probability of being anywhere except near that point. After a measurable lapse of time, the probability that there are holes at the collector point first rises appreciably above zero. The appropriate description, in this case, is that of wave-packets or particles, and one may think of the holes as being injected and pulled along the germanium as described in Chapter 3. The justification for this procedure is the mathematical one that wave-packets of electrons or holes have the behavior of classical particles.

Figure 8.1 has been prepared to indicate the logical connection among the various ideas discussed in this chapter. The section numbers given in the figure indicate where the topics are discussed. This figure will probably be of most value after Sections 8.2 to 8.7 have been read.

8.2 RANDOM PROCESSES AND TRANSITION PROBABILITIES

As stated in the introduction to this chapter, the theory of the mean free time is based on the idea of random processes. As its name suggests, the mean free time, which has already been referred to several times without being defined, is related to the time during which an electron moves without being affected by a random process. We shall consider first the nature of random processes, and then show how these processes can lead to an equilibrium distribution. In the following section we shall define the mean free time and relate it to these random processes.

The random processes referred to occur because the crystal is imperfect. If the crystal were ideal so that the potential field in which the electron moves were perfectly periodic, then the electron would remain in one quantum state indefinitely; or, if acted upon by electric and magnetic fields, it would change its momentum according to the law $\dot{\boldsymbol{P}} = \boldsymbol{F}$. In a real crystal, however, the potential field is not perfectly periodic and, as indicated in Figure 8.1, there are three main classes of imperfections. The heat stored in the crystal is present in the form of vibrations of the atoms. If temperature vibrations are lacking, imperfection may be present in the form of impurity atoms, such as donors or acceptors as discussed in Chapter 1, or in the form of places where atoms are missing from their normal positions and are perhaps squeezed into places in the crystal where they do not belong. (Additional imperfections, such as "dislocations", may be present also.)

A description of the motion of electrons in imperfect crystals is based on the quantum-mechanical idea of transition probabilities. When the crystal is imperfect, the electron does not stay in one state of the Brillouin zone indefinitely; instead, after a time, it will make a quantum jump to another state in the Brillouin zone. This process is illustrated in Figure 8.2(a) which shows the allowed states near the lowest energy of a Brillouin zone, with an excess electron occupying one of them. As a consequence of the imperfection of the periodic field, there is a transition probability that the electron will jump from the state which it occupies to any one of a number of other states. This probability depends in a complicated way upon the values of $\boldsymbol{P}$ for the initial and final states and is not restricted to such a small group of states as that shown in Figure 8.2(a).[1] Figure 8.2(b) shows a corresponding picture for a hole near the top of the valence-bond band.

[1] A discussion of the transitions is given in more detail in Chapter 11.

The transition of the hole comes about as a consequence of an electron jumping into it as indicated by the arrows. In this case the Pauli exclusion principle exerts an influence on the transitions, since an electron can make only those jumps which carry it to a state not already occupied by another electron. This means that the only jumps possible for electrons are those which end in the hole; and whenever such a jump occurs, the result is equivalent to a jump for the hole. Although the processes illustrated in (a) and (b) have this great difference, the net result is that the transition probability for the hole is approximately the same as that for an electron. This may be seen as follows:

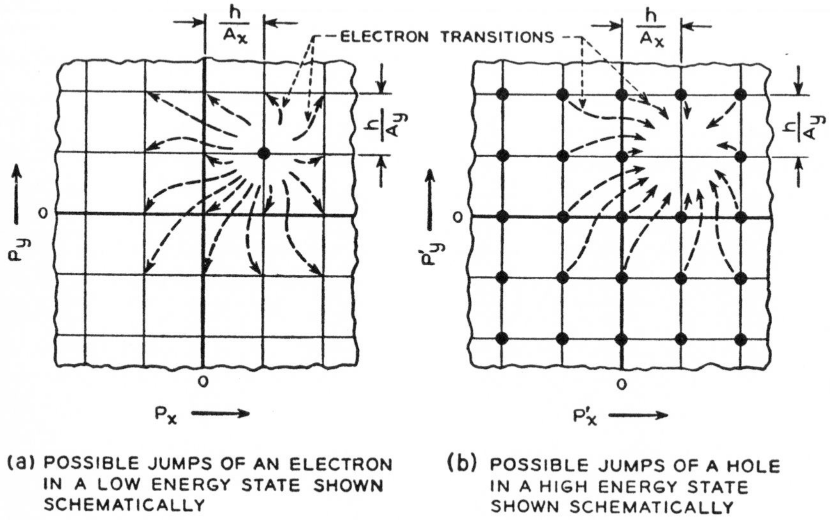

FIG. 8–2—Transitions for an Electron and for a Hole.

In Figure 8.2(a) the one electron can make jumps to any one of, say, A states. In Figure 8.2(b) each electron can make a jump to only one state; however, there are now A electrons which may jump. Furthermore, the probability of a jump to a vacant state is about the same for an electron near the top as for one near the bottom of a band. (Discussed in Chapter 11.) As a consequence, the probability that the hole in (b) makes a jump corresponds to A electrons each having one possible transition whereas the probability that the electron in (a) makes a jump corresponds to one electron with A possible transitions. These two over-all transition probabilities are thus comparable, a result which, as we shall later show more fully, accounts for the fact that holes and electrons have approximately the same mobility.[2]

[2] In Chapter 1 we discussed the similar dynamics of holes and electrons. We here point out the similarity of the effects of random processes upon them. In Chapter 10 we shall show that their thermal equilibrium statistics are equivalent.

For the purposes of this chapter, we shall suppose that only a small fraction of the states in the conduction band are occupied by electrons, so that we may disregard the possibility that a transition is prevented because the end state is occupied. A similar assumption is made for holes. Actually the removal of these assumptions is readily accomplished in the analytical theory; it would add unnecessary complications, however, to treat the more general case in this chapter.

If the jump which the electron makes is due to thermal vibrations, it will (according to the theory of transition probabilities) gain or lose energy each time it jumps. In this way, it can exchange thermal energy with the vibrating atoms of the crystal and come to equilibrium. It is, of course, meaningless to speak of one electron being at thermal equilibrium since thermal equilibrium is a statistical situation in which some electrons have large energies and some have small energies. However, if a large number of electrons are present in a conduction band, their equilibrium distribution is a definite thing which we shall discuss in the next two chapters. It is worth mentioning that the thermal equilibrium distribution of electrons in a crystal, or of atoms in gas or any similar example, is independent of the exact nature of the transition probabilities which permit it to arise. In other words, no matter what the nature of the transitions is, so long as they provide a means of exchanging energy, the system will eventually reach one and the same end state of thermal equilibrium. The rate, however, at which the end state is approached depends in detail upon the nature of the transitions—an idea which brings us to the consideration of the mean free time.

8.3 THE MEAN FREE TIME

As stated before, the theory of transitions or collisions presented in this chapter is based on certain simplifying assumptions. In particular we assume:

(1) *The probability that the electron (or hole) makes a transition in any small interval dt of time is* dt/τ *where* τ *is a constant.*

This assumption has two independent implications: (a) The probability of transition does not depend on the elapsed time since the last transition. (b) The probability of transition does not depend upon the quantum state occupied by the electron or hole. Both of these implications come from the statement that the "probability parameter τ", which has the dimensions of time (seconds), is a constant so that it does not depend on the factors mentioned in (a) or (b). By "a constant" we mean, of course, constant under given conditions of temperature and composition. Also τ will not be the same for electrons as for holes.

As we shall show τ is the mean free time; the natural definition, however, for the mean free time is stated in words very different from those employed in (1); for this pedagogical reason, we shall not call τ the mean free time until, with the aid of equation (1) of Section 8.3, we have defined the mean free time, denoted by $\bar{t}$, and proved it equal to τ in equation (5) of Section 8.3.

The second assumption is:

(2) *The end state of the transition is independent of the initial state, and the probability of arriving at any particular end state is proportional to the probability that the end state would normally be occupied in the thermal equilibrium distribution.*

What assumption (2) means is that no matter how the electrons are distributed among the quantum states at $t = 0$, as soon as each one has made a transition, they will be distributed in the thermal equilibrium distribution. (In other words, they cannot remember what they were doing before the collision.)

Assumptions (1) and (2), together with the equations of Chapter 7, furnish the basis from which we derive the equations for conductivity and Hall effect. These final equations correspond to a physical picture, which we shall now describe, inaccurately, in order to suggest the significance of (1) and (2) in the later development. According to (1), the electrons are accelerated by the applied fields for approximately a time τ; then they collide. If these collisions produced only small deflections of the velocity of the particles, then the condition after collision would be influenced by the field acting before collision. According to (2), however, the collision wipes out all memory of the pre-collision state of the particle. Thus the behavior of the particles under the influence of fields is substantially that which would occur τ seconds after an instantaneous application of the fields.[1]

Before defining the mean free time and proving its equality to the probability parameter τ in assumption (1), we shall attempt to make (1) appear to be reasonable by giving a discussion of collision processes on the basis of a particle model. According to this model, the electron moves as a particle in a straight line path (or a curved path if electric and magnetic fields are present) for a length of time and is then abruptly deflected, after which the process repeats. These abrupt deflections are analogous to collisions between molecules in a gas and may also be referred to as collisions. In particle language, assumption (1) is equivalent to saying that

[1] As we shall discuss in Chapter 11, neither assumption (1) nor (2) is in good agreement with the best theories: the time between collisions depends in an important way on the speed of the electron (or hole) and, after a transition, energy and direction will not be, on the average, random as suggested by (2). If the time between collisions is properly averaged, however, by procedures discussed in Chapter 11 (taking into account the variation of time between collisions with velocity and the dependence of velocity after collision upon velocity before collision), *then* the so obtained average τ may be inserted in assumption (1) and employed with assumption (2) *and the analysis presented in this chapter will give the correct results:* that is, the same as those obtained by the more involved procedures of Chapter 11.

the probability that the electron will suffer a collision is independent of how long it has traveled since its last previous collision. This is intuitively seen to be a reasonable result for the motion of molecules in a gas. Consider a molecule selected at random, and call it molecule (a); although molecule (a) has had a collision at some time in the past, the region of space into which it is moving contains molecules whose paths were unaffected by the previous collision of molecule (a). Thus the environment into which molecule (a) moves is unaffected by the length of time molecule (a) has been moving since its last collision, and, consequently, the probability that it suffers a collision in the next increment dt of time should be independent of how long it has been moving freely.[2]

The quantum mechanical theory of transition probabilities leads one to the same conclusion: the probability that a quantum jump may take place is also dt/τ regardless of how long the electron has stayed in the quantum state.

We are now in a position to define and discuss the mean free time. We shall do this first in the case of molecule (a) in a gas so as to have a specific picture in mind. We follow the course of molecule (a) over a *long period of time* T which starts and ends at instants of collision. During time T, molecule (a) has C collisions (not counting the starting one), the intervals between them being $t_1, t_2, \cdots, t_C$. These intervals are called "free times", during which the molecule moves freely uninfluenced by random processes, and their average is the mean free time. This leads to the definition:

The mean free time, $\bar{t}$, *is the (unweighted) average of the free times between collisions:*

$$\bar{t} = \frac{t_1 + t_2 + \cdots + t_C}{C} = T/C. \tag{1}$$

[The word (unweighted) is inserted in the definition to suggest to the reader that some other averaging procedure might be employed. We meet such a case in Section 8.7; and a paradox encountered there is explained as being due to a difference in averaging procedures.]

We must next study the distribution of the free times t_1 to t_C; as we shall see their average is τ; however, all values from zero to many times τ occur in the distribution.

The analysis is most easily visualized by considering not C successive free times for one molecule, but instead one free time each for C molecules. By assumption (1), any free time, no matter how it originates, has the same statistical behavior so that the same distribution of free times will be obtained for either case. Suppose that by chance all C of these molecules

[2] This problem is closely analogous to the coin which is about to be tossed. No matter how many times it has come up heads in succession, the probability is always $\frac{1}{2}$ that it will come up heads on the next throw.

have collisions simultaneously (in pairs) with other molecules at $t = 0$;[3] we then follow each of these molecules until it has its next collision, record its free time in our list of t_1, t_2 to t_C; thereafter, we may disregard its subsequent behavior since we have selected for study only the first free time for each molecule after its $t = 0$ collision. The significant statistical features of this situation can be derived by consideration of the number of molecules, denoted by $C(t)$, which have not collided up to time t after the

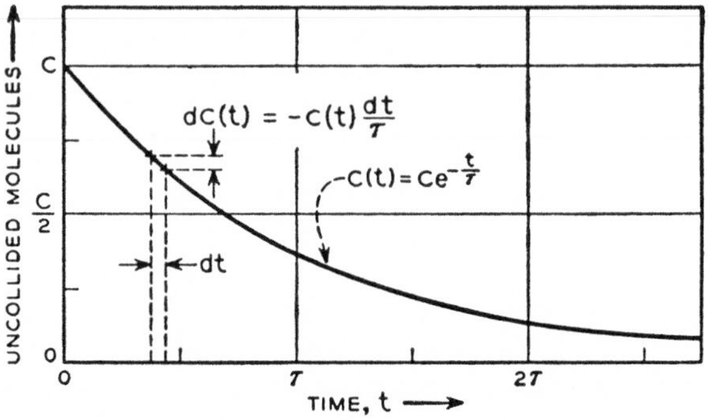

FIG. 8-3—Dependence of the Number of Uncollided Molecules upon Time.

$t = 0$ collision. How does $C(t)$ vary with time? Figure 8.3 shows how the dependence is derived: at time t there are $C(t)$ molecules which are uncollided; in an interval dt the probability that each of these will collide is dt/τ. [Assumption (1).] Hence, the number colliding in dt will be simply $C(t)dt/\tau$, and during dt the number of uncollided molecules will decrease by this amount. Expressed as an equation

$$dC(t) = -C(t)dt/\tau. \tag{2}$$

This is a differential equation for $C(t)$ which can be integrated as follows:

$$-dt/\tau = dC(t)/C(t) = d\ln C(t)$$
$$\ln C(t) = -t/\tau + a$$
$$C(t) = e^a e^{-t/\tau} = Ce^{-t/\tau} \tag{3}$$

the constant a being evaluated so that $C(t) = C$ when $t = 0$. From this expression we can find the number of mean free paths which end in dt at t. As discussed this number is $C(t)dt/\tau$ and is thus:

$$\left.\begin{array}{c}\text{number of molecules}\\ \text{colliding}\\ \text{between } t \text{ and } t+dt\end{array}\right\} = Ce^{-t/\tau}dt/\tau. \tag{4}$$

[3] Or, which is entirely equivalent, consider C collisions selected at random and measure time for each molecule from the instant of collision. Or consider one molecule and C of its free times, measuring time for each one from the instant of the collision which initiates it.

Hence, among all the C collisions considered, the above number have a free time of t and, therefore, contribute t times $Ce^{-t/\tau}dt/\tau$, to the sum $t_1 + t_2 + \cdots + t_C$. Adding this up over all collisions leads to a mean free time given by

$$\bar{t} = \frac{t_1 + t_2 + \cdots + t_C}{C} = \frac{1}{C}\int_0^\infty tCe^{-t/\tau}dt/\tau = \tau, \tag{5}$$

(the integral

$$\int_0^\infty te^{-t/\tau}dt/\tau = \tau\int_0^\infty e^{-t/\tau}(t/\tau)d(t/\tau) = \tau\int_0^\infty e^{-x}xdx = \tau \tag{6}$$

being readily integrated by parts or found in the tables).

It is worth while to review briefly what the argument has been in arriving at equation (5). First assumption (1) was introduced: (For a model like a gas, arguments were presented to show that (1) was reasonable with the probability parameter τ being constant and independent of the time after the previous collision.) It was not at that point stated that τ was the mean free time. Second, a mean free time $\bar{t}$ was defined. Third, the distribution of free times resulting from assumption (1) was evaluated and from this distribution the value of $\bar{t}$ was computed. It was then found that the value of $\bar{t}$ is τ.

The same mathematical results would follow from assumption (1) no matter what model was used. We could equally well have used electrons described either by the Brillouin zone scheme or the particle scheme. If we had started with C uncollided electrons at $t = 0$, we would have found formulae (2) to (5) from just the same analysis. The same results would apply for holes.

So far in this discussion, no use has been made of assumption (2). The results in the next section, however, will depend upon it. As stated earlier, detailed consideration of the processes leading to collisions will be discussed in a later chapter.

8.4 BRILLOUIN ZONE TREATMENT OF AVERAGE VELOCITY, MOBILITY, AND CONDUCTIVITY

We must now apply the concept of the mean free time and the distribution of free times given by equation (4) of Section 8.3 to a calculation of the current produced by an electric field. There are several ways of viewing this problem and, since all involve ideas which are useful in gaining insight into the conduction process, we shall compare them.

First is what we may call the Brillouin zone or distribution-in-momentum method. This is the one which best adapts itself to detailed analytic calculation when variations of τ with the initial state are taken into account. We shall, however, deal with it here in accordance with assumption (1) that τ is constant. Before an electric field is applied, the electrons occupy a set of quantum states near the lowest energy in a manner described in detail in Chapters 9 and 10. This situation is represented schematically in Figure 8.4(a) for electrons and 8.4(b) for holes. For the cases considered here, only a few electrons or holes are imagined present so that they

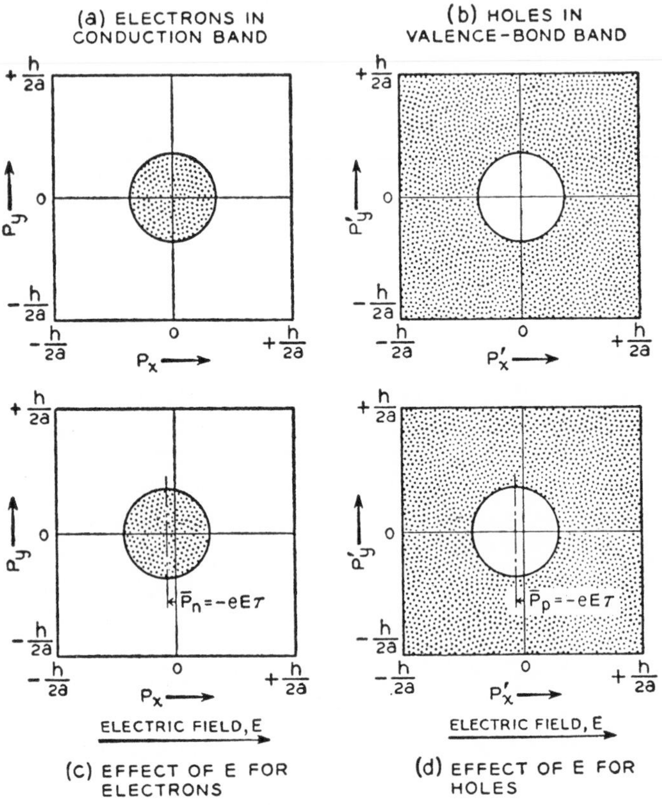

FIG. 8-4—Effect of Electric Field, E_x, on the Distribution of Electrons and Holes in the Brillouin Zone.

occupy only a very small fraction of the states, inside the shaded areas, and a negligible number have energies outside the shaded areas. Under thermal equilibrium, this distribution is symmetrical so that just as many electrons are moving to the right as to the left and the average values of $\boldsymbol{P}$ and $\boldsymbol{v}$ are zero.

We shall now specifically consider the case of electrons assuming that there are n of them present in the crystal, whose volume V we shall later take equal to unity so that n will be the density or concentration of electrons (that is, number per unit volume). We shall find it convenient to deal with the total crystal momentum of the n electrons which is

$$\boldsymbol{P}_{\text{tot}} = \boldsymbol{P}_1 + \boldsymbol{P}_2 + \cdots + \boldsymbol{P}_n. \tag{1}$$

The average momentum of an electron is evidently $\boldsymbol{P}_{\text{tot}}/n$ and its average velocity $\bar{\boldsymbol{v}}$ is

$$\bar{\boldsymbol{v}} = (\boldsymbol{v}_1 + \boldsymbol{v}_2 + \cdots + \boldsymbol{v}_n)/n = \boldsymbol{P}_{\text{tot}}/nm \tag{2}$$

because of the relationship $\boldsymbol{P} = m\boldsymbol{v}$ which holds near the bottom of the conduction band, as discussed in Section 7.5. The current density of each electron, as discussed in Section 6.1, is $(-e/V)\boldsymbol{v}$ so that the current due to the n electrons is

$$\begin{aligned} I &= (-e/V)(\boldsymbol{v}_1 + \boldsymbol{v}_2 + \cdots + \boldsymbol{v}_n) = (-en/V)\bar{\boldsymbol{v}} \\ &= (-ne)\bar{\boldsymbol{v}} \quad (\text{for } V = 1). \end{aligned} \tag{3}$$

Under the influence of an electric field, which we take to be in the x-direction with component E_x, all the electrons change their momenta according to $\dot{\boldsymbol{P}} = -e\boldsymbol{E}$. This shifts the distribution to the left as shown in Figure 8.4(c). At the same time, collisions tend to restore the electrons to the random distribution shown in (a). The steady state is reached when these two tendencies cancel with a net shift of the distribution to the left.

We shall, therefore, calculate the change in time dt produced in $\boldsymbol{P}_{\text{tot}}$ due to the field and due to collisions and by equating these find an equation describing the steady state. We need, of course, consider only the x-component. This gives

$$dP_{x\text{tot}} = -neE_x dt \tag{4}$$

for the change due to the field. During the same time, a fraction dt/τ of the electrons suffer a collision according to assumption (1), and return to the random state according to assumption (2). Furthermore, according to assumption (1), the electrons colliding are a random sample of all the electrons; they, therefore, eliminate just their proportional share of the total momentum and reduce $P_{x\text{tot}}$ by an amount

$$dP_{x\text{tot}} = -P_{x\text{tot}} dt/\tau. \tag{5}$$

Equating the sum of these two changes to zero gives the steady state condition

$$\begin{aligned} -P_{x\text{tot}} dt/\tau - neE_x dt &= 0 \\ P_{x\text{tot}} &= -neE_x\tau \end{aligned} \tag{6}$$

from which we readily infer for the vector field E that

$$P_{\text{tot}} = -neE\tau \tag{7a}$$

$$\bar{P} = P_{\text{tot}}/n = -eE\tau \tag{7b}$$

$$\bar{v} = -(e\tau/m)E \tag{8a}$$

$$I = (-ne)\bar{v} = (ne)(e\tau/m)E. \tag{8b}$$

These equations evaluate the current in terms of the constants τ and m which describe the behavior of the electron in the conduction band. As is indicated in Figure 8.4(c) they correspond to a displacement of the distribution of electrons in the Brillouin zone as shown.

Equations (8a) and (8b) can be re-expressed in terms of the *conductivity* and the *mobility*. By definition the mobility is the ratio of average velocity to electric field *without regard to algebraic sign*. Thus we have

$$\mu_n = e\tau_n/m_n \quad \text{and} \quad \bar{v}_n = -\mu_n E \tag{9}$$

where the subscript n shows that the mobility has been evaluated for negative carriers, that is excess electrons. The conductivity is by definition the ratio of I to E and is

$$\sigma = ne^2\tau_n/m_n = ne\mu_n \tag{10}$$

and is thus the total charge density, again without regard to algebraic sign, times the mobility.

Precisely similar relationships apply for holes. We shall use p, for positive carriers, to represent the concentration or density of holes and also as a subscript. The shift in momentum is again derived from the law $\dot{P} = F$. The sum P_{tot} of P over the unoccupied states, or holes, will change by $-epEdt$ owing to the field and will change because of collisions by $-P_{\text{tot}}dt/\tau_p$. Hence the steady state will be given by a formula like equation (5) for electrons and, as the reader may verify, the relationships $P = -m_p v$ and $I = (+e/V)v$ for holes (discussed in Section 7.6) lead to

$$\mu_p = e\tau_p/m_p \qquad \bar{v}_p = +\mu_p E \tag{11}$$

$$\sigma = pe\mu_p. \tag{12}$$

When both holes and electrons are present, as occurs in the intrinsic range, the total current density is the sum of the hole and electron currents and the conductivity is

$$\sigma = ne\mu_n + pe\mu_p \tag{13}$$

where we have used n and p for densities as follows:

$$n = \text{density of negative carriers} = \text{electrons/cm}^3; \tag{14a}$$

$$p = \text{density of positive carriers} = \text{holes/cm}^3. \tag{14b}$$

At this point in the exposition the reader may find it advantageous to consider Figure 8.1 and, with its aid, review the argument up to this point.

The units used in this discussion are absolute electrostatic units. It may be instructive to convert them to practical units and to consider some typical values.[1] We shall use the subscript L (for laboratory) to denote practical units since p has already been required for holes. In terms of the absolute electrostatic units, the mobility has the dimensions of cm/sec per electrostatic volt/cm. Thus we have

$$\mu = e\tau/m \quad \text{cm}^2/\text{e.s. volt sec.} \tag{15}$$

For practical work fields are measured in volts/cm and since 1 e.s. volt = 300 volts and the value of μ in practical units is[2]

$$\begin{aligned}\mu_L \text{ cm}^2/\text{volt sec} &= \mu(\text{cm}^2/\text{e.s. volt sec})(1 \text{ e.s. volt}/300 \text{ volts}) \\ &= (\mu/300)\text{cm}^2/\text{volt sec} = e\tau/300m \text{ cm}^2/\text{volt sec.}\end{aligned} \tag{16}$$

For $\mu_L = 10^3$ cm^2/volt sec or $\mu = 10^3 \cdot 300$, the mean free time τ is

$$\begin{aligned}\tau = 10^3 \cdot 300 \cdot m/e &= 3 \times 10^5 \times 9.1 \times 10^{-28}/4.8 \times 10^{-10} \\ &= 5.7 \times 10^{-13} \text{ sec.}\end{aligned} \tag{17}$$

This time is so short that relaxation effects could not be observed even with waves in the millimeter range.

A typical value of conductivity (for germanium) is 1 ohm^{-1} cm^{-1}. If we express e in practical units $e_L = 1.60 \times 10^{-19}$ coulomb, then the conductivity in practical units will be

$$\sigma_L = ne_L\mu_L \tag{18}$$

corresponding to

$$n = \sigma_L/e_L\mu_L = 1/1.6 \times 10^{-19} \times 10^3 = 6 \times 10^{15} \text{ cm}^{-3}. \tag{19}$$

It may be worth mentioning that a conductivity of 1 ohm^{-1} cm^{-1} is so high that, in the body of the semiconductor, displacement currents (due to changing dielectric displacement) may be neglected compared to conduction currents (due to the motion of holes and electrons) at least up to microwave frequencies. (This is not true in the high resistance layers of rectifying junctions, however.) In absolute e.s.u., the displacement current is

$$(1/4\pi)\kappa\omega E = \kappa f E/2 \tag{20}$$

where $\omega = 2\pi f$ is the angular frequency and κ the dielectric constant; for germanium κ is 19. (A better value is 16; see p. 224.) The conductive current is

$$\sigma E = 9 \times 10^{11}\sigma_L E. \tag{21}$$

[1] A more complete list of formulae are given in equations (8) to (17) of Section 8.8.

[2] Some rules for changing units are given in an appendix.

Hence the ratio

$$\frac{\text{displacement current}}{\text{conduction current}} = \kappa f/2 \times 9 \times 10^{11}\sigma_L = 1.05 \times 10^{-11}f. \quad (22)$$

Hence the displacement current is negligible up to $f = 10^{10}\ \text{sec}^{-1}$ corresponding to a 3-cm wave length in vacuum.

[The same result can be obtained more compactly in M.K.S. units for which, as discussed in an appendix, the conductivity σ_m is obtained from σ_L as follows:

$$\begin{aligned}\sigma_L\ \text{ohm}^{-1}\ \text{cm}^{-1} \times (100\ \text{cm}/1\ \text{meter}) &= 100\sigma_L\ \text{ohm}^{-1}\ \text{meter}^{-1}\\ &= \sigma_M\ \text{ohm}^{-1}\ \text{meter}^{-1}. \quad (23)\end{aligned}$$

The ratio of displacement current to conduction current is

$$(\partial D/\partial t)/\sigma_M E = \kappa\varepsilon_0\omega/\sigma_M. \quad (24)$$

For $\sigma_M = 100\ \text{ohm}^{-1}\ \text{meter}^{-1}$, corresponding to $1\ \text{ohm}^{-1}\ \text{cm}^{-1}$, and $\kappa = 19$, and $\varepsilon_0 = 8.85 \times 10^{-12}$ this leads to equation (22). The value of ω for which the ratio of currents is unity is equal to the reciprocal of the *relaxation time* for dissipation of charge density, as may be seen by calculating the rate of change of charge density M.K.S. units:

$$\partial\rho/\partial t = -\nabla \cdot \boldsymbol{I} = -\nabla \cdot \sigma_M \boldsymbol{E} = -(\sigma_M/\kappa\varepsilon_0)\nabla \cdot \boldsymbol{D} = -(\sigma_M/\kappa\varepsilon_0)\rho \quad (25)$$

which leads to a rate of decrease in $\ln \rho$ of $(\partial\rho/\partial t)/\rho = -\sigma_M/\kappa\varepsilon_0$ in agreement with the result stated above.]

8.5 PARTICLE TREATMENT OF DRIFT VELOCITY, MOBILITY, AND CONDUCTIVITY

In this section we shall derive the same relationships as in Section 8.4 but on the basis of a wave-packet or classical particle model. We shall see that proper care must be taken in averaging over the free times in order to avoid error by a factor of 2.

We shall again deal with electrons and simply indicate how the same results may be obtained for holes. Thus we consider n electrons acting as mobile particles in the semiconductor. According to assumption (1), each electron has a probability dt/τ of colliding with an imperfection in the lattice (not with another electron; electron-electron collisions can be neglected for most purposes in the theory of semiconductors). Each collision, in accordance with assumption 2, brings the electron into the thermal equilibrium distribution, statistically speaking, so that its direction and speed of motion are uncorrelated with those before collision. As a consequence, its path is an irregular random motion like that shown in Figure 8.5(a). Each leg of this motion is a "free path", and the average of them is the mean free path. There is a relationship between the mean free path and the mean free time. In fact, if the average thermal velocity of motion is v_θ, we have

$$\text{mean free path} = v_\theta \tau. \tag{1}$$

We shall return to a discussion of mean free paths in Chapter 11; in this chapter, as already stated, we shall emphasize the mean free time.

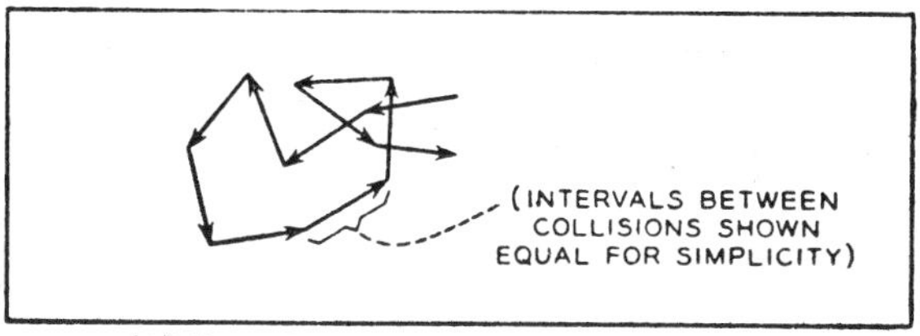

(a) RANDOM MOTION OF AN ELECTRON IN A CRYSTAL

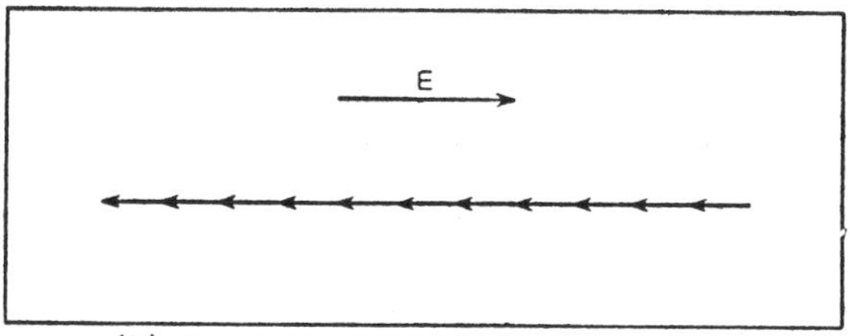

(b) COMPONENT OF ELECTRON MOTION ADDED BY ELECTRIC FIELD

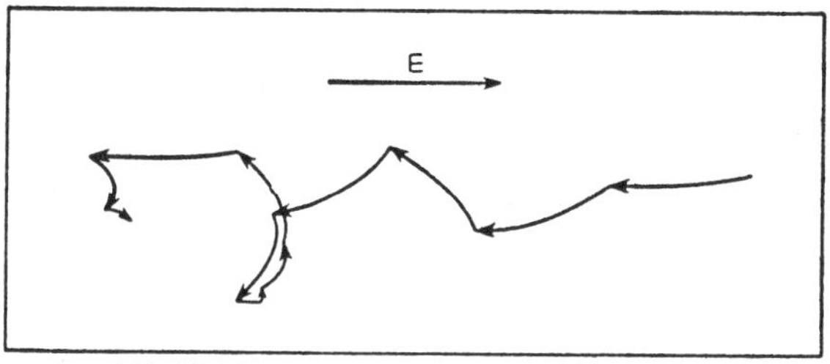

(c) COMBINATION OF (a) AND (b): ELECTRON MOTION IN AN ELECTRIC FIELD

FIG. 8-5—Effect of Electric Field in Superimposing Drift on Random Motion.

Under the influence of an electric field, the free paths become curved; in fact, they are segments of parabolas, like the trajectory of a falling body. During a path of duration t_1, the net displacement due to the acceleration ($a = -eE/m$) produced by the field is $\frac{1}{2}at_1^2$, as for a falling body. Thus after C collisions, the path would be deformed from (a) to (c) in Figure 8.5 by a displacement of $\frac{1}{2}at_1^2$ for each path or $(C/2)at_1^2$ for all the paths if the free time for each path were t_1. However, the free times are distributed according to the statistics given in equation (4) of Section 8.3 so that the free time varies. The total displacement is thus

$$(a/2)(t_1^2 + t_2^2 + \cdots + t_C^2). \tag{2}$$

According to equation (4), a number $C \exp(-t/\tau)dt/\tau$ of free times have durations between t and $t + dt$, and these contribute an amount

$$t^2 Ce^{-t/\tau}dt/\tau \tag{3}$$

to the sum in (2). Adding up for all the possible free times gives

$$\begin{aligned} t_1^2 + t_2^2 + \cdots + t_C^2 &= \int_0^\infty t^2 Ce^{-t/\tau}dt/\tau \\ &= C\tau^2 \int_0^\infty (t/\tau)^2 e^{-(t/\tau)} d(t/\tau) \\ &= C\tau^2 \int_0^\infty x^2 e^{-x} dx \\ &= 2C\tau^2. \end{aligned} \tag{4}$$

Thus the average value of t^2 for each collision is $2\tau^2$ and $\overline{t^2}$ is not simply the square of $\bar{t}$ which would be τ^2. The fact that the average of the square of a quantity is greater than the square of its average is a generally valid result. For example, on a 60-cycle line the average square of voltage is $(110)^2$, but the average voltage is zero and so is the square of the average. (Further examples are discussed in Section 14.3.) Hence the displacement due to the C collisions is

$$(a/2)2C\tau^2 = Ca\tau^2. \tag{5}$$

These C collisions will require, on the average, a time $C\tau$, since this is the definition (1) of Section 8.3 of the mean free time. Consequently, the average velocity due to the field is

$$\mathbf{v}_d = \mathbf{a}C\tau^2/C\tau = \mathbf{a}\tau \tag{6}$$

where $\mathbf{v}_d$ is the average velocity with which the particles "drift" down the field.

If the acceleration is due to an electric field $\mathbf{E}$, its value is $\mathbf{a} = -e\mathbf{E}/m$ and

$$\mathbf{v}_d = -(e\tau/m)\mathbf{E} \tag{7}$$

which is equivalent to expression (8a) obtained in Section 8.4 for $\bar{\mathbf{v}}$.

The same expression for the current may be obtained for the particle picture as may be seen with the aid of Figure 8.6. Here we consider the number of electrons flowing across area A in time t. All the electrons in the volume $A|\mathbf{v}_d|t$ will cross the area carrying a charge of $(-e)$ from right to left or $+e$ from left to right. There will be some flow of electrons in and out of the volume owing to their random motions; this flow will cancel out on the average, however, since the random motions alone produce no net current. Thus if the density of electrons is n per unit volume, the net charge transported in the direction of the field will be $en|\mathbf{v}_d|At$ giving rise to a current density which is parallel to $\mathbf{E}$ and has the value (current density is charge per unit time per unit area)

$$\mathbf{I} = (en\mathbf{v}_d At)/At = en\mathbf{v}_d = (ne^2\tau/m)\mathbf{E} \tag{8}$$

again identical with the result of Section 8.4. Similar results will obviously be obtained for holes.

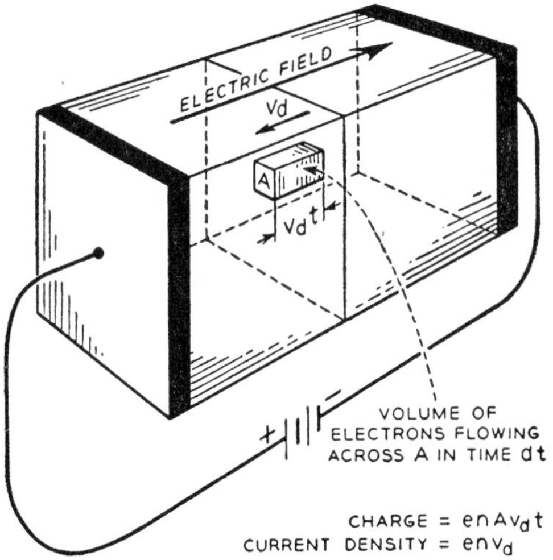

FIG. 8-6—Relationship Between Drift Velocity and Current.

(We shall here, parenthetically, discuss a difficulty which can arise in applying the mean free time concept carelessly to the computation of drift velocity. We argue, erroneously, as follows: on the average, a particle has a free time of τ. Suppose we observe it at a random time; on the average we find it midway between collisions and, consequently, at a time $\tau/2$ since its last collision. Thus a random observation will find it with velocity $\mathbf{a}\tau/2$ and not $\mathbf{a}\tau$.

(The argument is wrong. If we observe the particle at a random time it will, on the average, have been free for a time τ and will remain free for an additional time τ, so that it will have, on the average, a drift velocity $\mathbf{a}\tau$, as derived in equation (6) above. But the last sentence, which is correct, appears to imply that the mean free time is 2τ. The factor of 2 comes from a difference in averaging procedures: in (1) of Section 8.3, which gives the definition of the mean free time, each mean free time is given equal weight. However, the selection of a sample of mean free times at random instants, automatically weights the choice in favor of the long mean free times, and, as a result, the average so obtained comes out 2τ. This weighting process is mathematically closely related to computing $\overline{t^2}$, which also leads to "2" by turning out to be $2\tau^2$.)

8.6 DISCUSSION OF HALL EFFECT IN PARTICLE LANGUAGE

The Hall effect has played a decisive role in revealing the mechanism of conduction in semiconductors. The reason is that when data from measurements of both the Hall effect and the conductivity of an impurity semiconductor are combined, it is possible to determine both the concentration and mobility of the carriers; whereas, a measurement of the conductivity alone gives only a product $n\mu_n$ or $p\mu_p$.

The Hall effect, discovered in 1879, occurs when a transverse magnetic field is applied to a conductor carrying current. Under these conditions it is found that in addition to the longitudinal electric field normally present a transverse field is produced so that the current (which continues to flow longitudinally) and the electric field are no longer parallel. The effect thus produces a measurable transverse voltage across the specimen. The cause of the effect is that the magnetic field exerts a force on the electrons and deflects them sidewise. Before the development of the band theory of solids, it was a mystery why the electrons which carried the current were sometimes deflected so as to produce the effects expected of positive charges. However, as we shall show, this is just the effect to be expected for holes. We shall give first an introductory pictorial description of the Hall effect in terms of particles. For this purpose, we neglect the random velocities of the particles (which are here regarded as classical non-wave mechanical particles) by supposing that after each collision they are brought to rest so that all of the motion they acquire is produced by the applied fields. (It can be shown that the random velocities cancel out on the average here, as they do for the case of conductivity just discussed in Section 8.5.) Under these assumptions, the motions of positive and negative particles in combined electric and magnetic fields will be as shown in Figure 8.7. A positive particle starting at the origin will be accelerated by the electric field. As it gathers speed, it is subjected to a sidewise thrust by the magnetic field according to the vector formula

$$\mathbf{F} = (e/c)\mathbf{v} \times \mathbf{H}. \qquad (1)$$

This deflects it along the curved path until it suffers a collision and starts over. As a consequence, the current and electric field deviate by an angle θ_p, as shown. An electron is deflected to the same side: it acquires a velocity opposite to a hole but since the thrust of the magnetic field is proportional to velocity times charge (which is also reversed compared to a hole), the sidewise thrust is in the same direction. The current due to electrons deviates in the opposite direction at an angle of θ_n from the electric field. The angles of deviation can readily be obtained on the basis of this model, and are given by the equations

$$\theta_p = e\tau_p H/mc = \mu_p H/c, \qquad \theta_n = e\tau_n H/mc = \mu_n H/c \tag{2}$$

when absolute e.s.u. and e.m.u. are employed. We shall omit the derivation here since we shall obtain the same formula more simply in the next

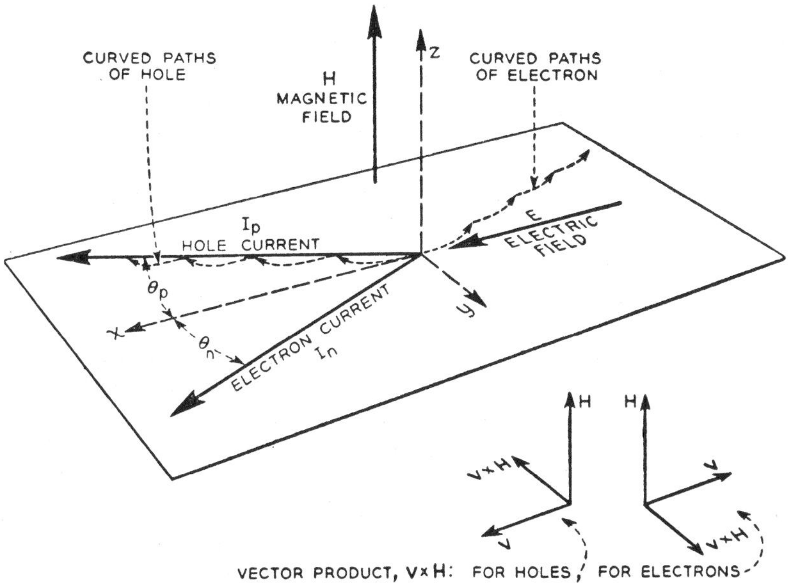

FIG. 8-7—Electron and Hole Current in Relation to Electric and Magnetic Fields.

section. This formula is familiar in connection with magnetron theory, eH/mc being the angular velocity of an electron in a magnetic field.

As will be described later, the angle θ (or its equivalent), together with the direction of the deviation between electric field and current, can be found in a Hall effect experiment. This leads to a determination of the sign of the carriers and an evaluation of μ from (2), which can then be combined with the conductivity data to find the concentration of carriers.

We shall next show that the Brillouin zone theory also leads to a deviation between current and electric field and to the same formula for θ. After that the Hall coefficient will be defined, the experimental procedure for measuring it will be described and the results re-expressed in practical units.

8.7 BRILLOUIN ZONE TREATMENT; HALL, DRIFT, AND MICROSCOPIC MOBILITIES

We next consider the effect of applying electric and magnetic fields simultaneously. This is shown in Figure 8.8. We consider in (a) and (b) the case where only a few electrons are present in the Brillouin zone, corresponding to an *n*-type semiconductor. Under equilibrium conditions, with no field present, these electrons will occupy states distributed near the center of the zone. As discussed in connection with Figure 8.4, we have indicated the electrons by shading and have not attempted to show the gradual dropping off of the concentration from the lowest energy values at the center of the zone to higher energy values, but have instead drawn the distribution as though all energies up to a certain state were equally occupied and all those beyond that were unoccupied. For the purposes of the qualitative exposition in this paragraph and the next, no error is introduced by this procedure. (For the analytic purposes, employed later, we do not refer to the figure.) In part (a) of this figure we show the effect of an electric field alone. The electric field is supposed to act in the plus x-direction so that the force on the electron acts in the minus direction. As previously discussed, this results in a shift of electron distribution so that the steady state value will be as shown in the shaded area. If all the quantum states in the shaded area were filled and those outside were empty, this shift would effectively add the electrons corresponding to region A and subtract those corresponding to region B. Since the distribution of electrons is diffuse, however, the net result is merely to increase the number of electrons in quantum states in the Brillouin zone near region A and to decrease the number near region B. Now, as the figure indicates, the velocities near A have a component to the left and those near B, to the right, so that the net result of these changes is to unbalance the velocity distribution and produce an average momentum $\boldsymbol{P}$ to the left for the electrons, as is shown. Since the electron has a negative charge, this is equivalent to a current to the right in the direction of the applied field.

When a magnetic field is also applied, all of the quantum states move in accordance with Figure 7.3(b), as previously discussed. This motion shifts the group of points in A counterclockwise on the diagram as indicated, and the distribution will be distorted into the form shown in Figure 8.8(b); the average velocity vector will, thus, make an angle with the field, as is shown on the figure. From this figure, it is easy to understand why the angle between the current and the electric field is given by the magnetron frequency times the mean free time. If an electron suffered no collisions, then in the presence of crossed electric and magnetic fields its velocity would run through a complete cycle of directions in the period corresponding to the magnetron frequency. This means that if the electric field were suddenly removed and collisions were prevented, the shaded circle would rotate about the center of the diagram with the magnetron frequency, as has been described in connection with Figure 7.5. This process is terminated in a mean life time τ so that, as we shall show analytically below, the

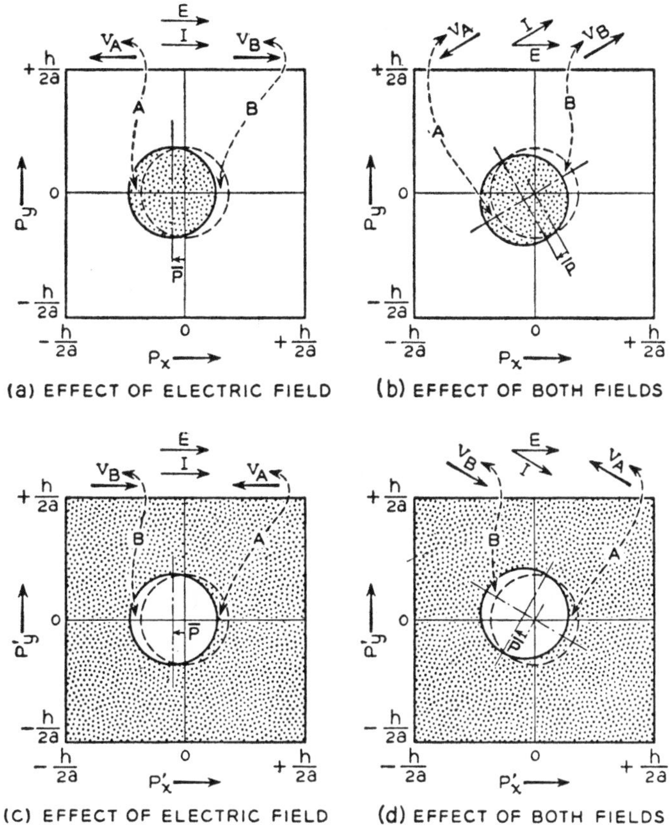

FIG. 8-8—Effect of Electric Field, E_x, and Magnetic Field, H_z, on Electron and Hole Distributions in the Brillouin Zone.

effective angle through which the distribution is rotated is τ times the magnetron frequency.

The formula for the angle between current and electric field can readily be derived analytically from assumptions (1) and (2) of Section 8.3. According to these, if the total current at any instant is I, then it will decrease in time dt due to collisions by an amount Idt/τ, since during dt a random or representative fraction dt/τ of the electrons is restored to the thermal equilibrium state. At the same time, each electron is changing its current because of the accelerations produced by the applied fields.

We shall calculate the change in current due to these fields by summing the changes for each electron. (The procedure is essentially the same as that used in deriving equation (4) of Section 8.4.) If we multiply equation (16) of Section 7.6, which expresses $\dot{\boldsymbol{I}}$ in terms of $\boldsymbol{I}, \boldsymbol{E}$, and $\boldsymbol{H}$, by dt so as to find $d\boldsymbol{I}$, and suppose that this $d\boldsymbol{I}$ corresponds to electron "1" of the n electrons we are dealing with, we obtain

$$d\boldsymbol{I}_1 = (e^2/mV)\boldsymbol{E}dt - \boldsymbol{I}_1 \times \boldsymbol{H}(e/mc)dt. \tag{1}$$

Summing over the n electrons and taking V as unit volume, we obtain

$$d\boldsymbol{I} = d\boldsymbol{I}_1 + d\boldsymbol{I}_2 + \cdots + d\boldsymbol{I}_n = (ne^2/m)\boldsymbol{E}dt - \boldsymbol{I} \times \boldsymbol{H}(e/mc)dt \tag{2}$$

for the change in $\boldsymbol{I}$ due to the acceleration of the fields during time dt. (The reader should note that the effect of random motions is not apparent in this equation, since only the total current $\boldsymbol{I}$ occurs in it.) The change due to collisions is

$$d\boldsymbol{I} = -\boldsymbol{I}dt/\tau. \tag{3}$$

For the steady state the sum of these changes must be zero, giving

$$(ne^2/m)\boldsymbol{E}dt - \boldsymbol{I} \times \boldsymbol{H}(e/mc)dt - \boldsymbol{I}dt/\tau = 0$$

or

$$\boldsymbol{I} = (ne^2\tau/m)\boldsymbol{E} - \boldsymbol{I} \times \boldsymbol{H}(e\tau/mc). \tag{4}$$

In order to simplify the algebra involved in interpreting this vector equation, we shall take the x-axis as parallel to $\boldsymbol{I}$ (rather than to $\boldsymbol{E}$) and shall compute the direction and magnitude of $\boldsymbol{E}$. For these conditions the vector product $\boldsymbol{I} \times \boldsymbol{H}$ is thus parallel to y with magnitude $-I_xH_z$, $\boldsymbol{H}$ being parallel to the z-axis as before. Thus the x and y components of equation (4) of Section 8.7 are:

$$I_x = (ne^2\tau/m)E_x = \sigma E_x \tag{5a}$$

$$0 = \sigma E_y - (-I_xH_z)(e\tau/mc) \tag{5b}$$

or

$$E_x = I_x/(ne^2\tau/m) = I_x/\sigma \tag{6a}$$ [1]

$$E_y = -(I_x/\sigma)(e\tau H_z/mc). \tag{6b}$$

This shows that the $\boldsymbol{E}$ vector is turned clockwise in the x-y plane in respect

[1] Since no approximations have been involved in dealing with the consequences of equation (16) of Section 7.6 and assumptions (1) and (2) in deriving (6a), we conclude that these assumptions lead to the result that the power dissipation I_xE_x due to current I_x is independent of H. This shows that magnetoresistance (an increase in resistance in magnetic fields) cannot be derived from these assumptions. Magnetoresistance, theory shows, is connected with variations in mean free time of the sort excluded by assumption (1) or with variations of the energy surfaces from spherical form. See Chapter 12.

to the $\boldsymbol{I}$ vector. For small angles this will give

$$\theta = E_y/E_x = e\tau H_z/mc = \mu H_z/c. \tag{7}$$

This is the same result which was discussed in connection with the particle treatment and shown in Figure 8.7.

The only effect of using holes instead of electrons is, as pointed out at the end of Section 7.6, to change the coefficient of the H_z term and thus reverse the rotation. Of course, the values of $\boldsymbol{\tau}$ and m appropriate to the carriers in question should be used. The diagrams showing the displacement of electrons in the Brillouin zone, however, have quite a different appearance and are shown in Figure 8.8(c) and (d). As explained for Figures 7.3 and 7.5, the rotation produced near the top of the energy band is opposite to that near the bottom, thus accounting for the opposite rotations for electrons and holes of (b) and (d) of Figure 8.8.

In this treatment we have stressed the *Hall angle* θ rather than the *Hall constant* R_H. The reason is that the Hall constant, which we shall discuss in the next section, measures the number of carriers, whereas θ, or rather θ/H, measures the mobility; and the latter quantity is much more fundamental in semiconductors, in which the number of carriers is a highly structure-sensitive property, depending as it does on the impurity content. Furthermore, θ/H is the more readily measured quantity. (It will have the same value even if the yz cross-section of Figure 8.9 discussed in the next section is irregular or unknown in shape.) In order to describe the mobility as measured by the Hall effect, we introduce the term *Hall mobility* and the symbol

$$\mu_H = c\frac{\theta}{H} = \text{Hall mobility}. \tag{8}$$

In general, the Hall mobility will not be equal to the mobility required in the expression for conductivity. It has been possible only recently to measure the mobility in terms of drift velocity by using the techniques described in Section 3.1. We shall introduce the term *drift mobility* defined as the drift velocity of injected carriers divided by electric field.

$$\mu_D = \frac{\text{average drift velocity}}{\text{electric field}} = \text{drift mobility}. \tag{9}$$

The drift mobility as measured by the transmission velocity of a hole pulse in a germanium filament may actually be different in concept from the drift velocity discussed in earlier sections of this chapter; if the semiconductor contains levels capable of binding holes tightly enough so that they are free to move for only a fraction of the time, then the drift mobility will be only a fraction of the *microscopic mobility*, which we define as the mobility of an untrapped particle. The drift of color centers in alkali halide crystals is an example of this sort, and the drift mobilities are very low, being of the order of 10^{-4} cm/sec per volt/cm.[2] Evidently such trapping effects will reduce the drift mobility but will not affect the microscopic mobility or the Hall mobility.[3] Trapping by donors and acceptors in germanium at room

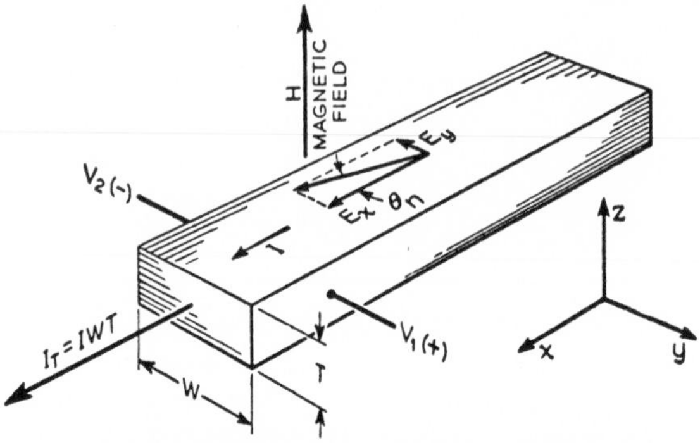

(a) VOLTAGE COMPONENTS FOR CONDUCTION BY ELECTRONS IN A MAGNETIC FIELD

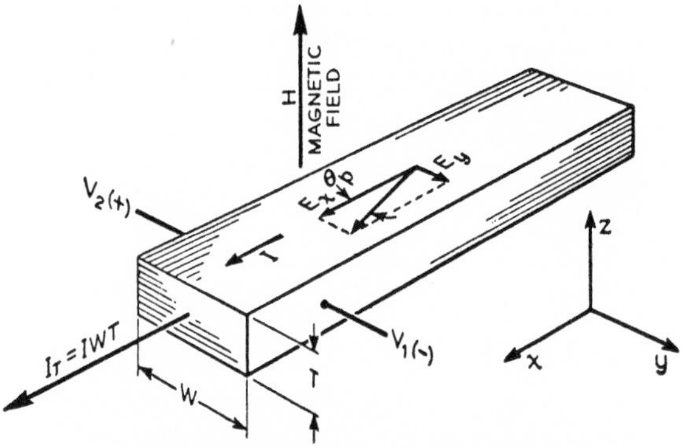

(b) VOLTAGE COMPONENTS FOR CONDUCTION BY HOLES IN A MAGNETIC FIELD

FIG. 8–9—Transverse Field Due to Hall Effect.

temperature will be negligible, as is discussed in Chapter 10; in silicon at

[2] See N. F. Mott and R. W. Gurney, *Electronic Processes in Ionic Crystals*, Oxford University Press, 1940, p. 141, for a review of the work of R. W. Pohl and his colleagues on this subject. See also F. Seitz, "Color Centers in Alkali Halide Crystals", *Rev. Mod. Phys.*, **18**, 384–408 (1946).

[3] The earliest experiments of which the author is aware in which a comparison of drift and Hall mobilities were made for the same particles were carried out during some unpublished work with photoelectrons in silver chloride by J. R. Haynes, using techniques described by J. R. Haynes and W. Shockley, "Report of a Conference on Strength of Solids", *The Physical Society*, London, 1948, p. 151. No evidence for trapping which reduced the mobility was found at room temperature.

room temperature and in germanium at lower temperatures, however, trapping should produce distinguishable differences between the drift and Hall mobilities. In Chapter 12, a review of mobility data is presented.

It may be worth while to point out the relationship of drift and microscopic mobility to conductivity. Suppose we have an n-type semiconductor with a certain density of excess electrons which are continuously and frequently becoming trapped and then free again. Under these conditions the conductivity will correspond to the untrapped density so that we may write

$$\sigma = e \times \text{(microscopic mobility)} \times \text{(density of untrapped excess electrons)}$$

$$= e\mu_D \times \text{(total density of excess electrons)}.$$

Thus either of the mobilities may be used equally well provided that it is properly paired with the appropriate density. This sort of a determination of μ from conductivity and total donor density is discussed in Section 12.9.

In the next section we shall describe the standard technique for carrying out a Hall effect experiment and shall convert the formulae from the absolute electromagnetic system employed here to other units. For this purpose, we shall not distinguish between the different definitions of mobility. In the following section we shall discuss the difference between microscopic mobility, denoted simply as μ, and Hall mobility and quote results, derived later in Chapter 11, relating the two. A further discussion of this difference is given in Section 12.9.

8.8 MEASUREMENT OF THE HALL EFFECT—PRACTICAL AND M.K.S. UNITS

The Hall effect is measured by using samples similar to those illustrated in Figure 8.9; in these the current flows parallel to the long dimension, which is supposed to be large compared to the transverse dimensions.[1] When a magnetic field is applied to such a sample, it deflects the current carriers to one side and causes the current to deviate from the direction of the electric field. During the initial transient phase, while the magnetic field is building up, a transverse current actually flows in the specimen and carries charge across it from one side to the other. This charge accumulates on the two opposite faces, making one positive and the other negative, and sets up a transverse electric field. After steady-state conditions have been established, this transverse field just balances the deflecting effect of the magnetic field and no further charge accumulation on the surfaces takes place. The current then flows parallel to the long direction (x-axis) in Figure 8.9. The electric field is not parallel to the x-axis and has a y-com-

[1] For a discussion of the Hall effect in short samples see I. Isenberg, B. R. Russell, and R. F. Greene, *Rev. Sci. Inst.* **19**, 685–688 (1948).

ponent. In accordance with the results of the last section, the angle θ between the current vector $\boldsymbol{I}$ and the electric field vector $\boldsymbol{E}$ is given by equation (7) of Section 8.7. Part (a) of Figure 8.9 shows relative orientations of $\boldsymbol{E}, \boldsymbol{I}, \boldsymbol{H}$ and the specimen for the case of electrons and part (b) for the case of holes. Because θ is a small angle,[2] we may replace its cosine by unity and its sine and tangent by θ. (*For brevity we shall omit the subscripts y and z from I_y and H_z.*) As discussed previously, $\theta = e\tau H/mc = \mu H/c$. Also because θ is small, we may write $I = \sigma E_x$ for the current density. Accordingly, the expression for transverse field may be simplified to

$$E_y = \theta E_x = \theta I/\sigma = (\pm \mu/c\sigma)HI \tag{1}$$

where the plus sign holds for holes and the minus sign for electrons.

The Hall coefficient R_H is by definition the coefficient of HI in the foregoing equation. It gives the ratio of the transverse electric field to the longitudinal current density times the magnetic field. It is given by the equation

$$R_H = \theta/\sigma H = \pm\mu/c\sigma = 1/n(\pm e)c \tag{2}$$

and is thus seen to give a direct measure of the number of carriers present and also their sign. (If both holes and electrons are present the more complicated expression, equations (10) of Section 8.9, is obtained.)

Usually the Hall coefficient is measured in terms of the transverse voltage produced by the transverse field between two directly opposite points on the sample. Also the total current through the slab, denoted by I_T, rather than the current density I, is used. These quantities are related to those in the previous equation as follows:

$$I_T = IWT, \quad V_2 - V_1 = E_y W \tag{3}$$

where W and T are the transverse dimensions of the sample. Combining these with the previous equation we obtain

$$V_2 - V_1 = R_H H I_T/T \quad \text{or} \quad R_H = (V_2 - V_1)T/HI_T. \tag{4}$$

These equations are expressed in the absolute c.g.s. system of units so that currents and voltages are to be measured in electrostatic units. Except for this, the form given in the last equation is that used to measure R_H in practical units. As defined in connection with the directions shown in Figure 8.9, the sign of the Hall effect determines whether or not the conduction is by electrons or holes. We shall discuss the case where both holes and electrons are present in Section 8.9.

Practical Units. We must next convert the Hall Effect formulae to the

[2] At least so far as first-order effects alone are concerned.

units ordinarily used in the laboratory:

cm, gram, sec *for mechanical quantities,*
volts, coulombs *for electrical quantities,*
gauss or oersted *for magnetic field.*

The subscript L, for laboratory, will be used for quantities measured in these units. Summarizing the previous results, which are expressed in c.g.s. units and absolute e.s.u. and e.m.u., we have

$$R_H = \theta/\sigma H = 1/n(\pm e)c \tag{5}$$

$$c|R_H|\sigma = \mu = e\tau/m \tag{6}$$

$$\tau = mc|R_H|\sigma/e. \tag{7}$$

The Hall coefficient in practical units, denoted by R_L, is expressed in cm^3/coulomb, that is, the reciprocal of a charge density expressed in coulombs per cubic centimeter. This can be obtained from equation (5) by eliminating the speed of light, c, and multiplying by a factor to convert absolute electrostatic units of charge to coulombs. Denoting by e_L the charge on the electron in coulombs we have $e = ce_L/10$ in accordance with the customary definition of the ratio of these units. This leads to the following relationship between practical and electrostatic Hall coefficients:

$$R_L = \frac{1}{n(\pm e_L)} = \frac{c}{n(\pm e)10} = \frac{c^2}{10} \cdot \frac{1}{n(\pm e)c} = 9 \times 10^{19} R_H \tag{8}$$

where we have used the customary approximation that $c = 3 \cdot 10^{10}$ cm/sec. In order to determine R_L from practical measurements, we shall first see how R_H is obtained when currents and voltages are measured in practical units. Denoting the total current through the specimen by I_L we then have

$$(V_2 - V_1) = (V_2 - V_1)_L/300, \quad I_T = 3 \times 10^9 I_L \tag{9}$$

and

$$R_H = \frac{(V_2 - V_1)T}{HI_T} = \frac{(V_2 - V_1)_L T}{9 \times 10^{11} I_L H} \tag{10}$$

so that we get

$$R_L = 10^8 (V_2 - V_1)_L T/HI_L. \tag{11}$$

Equation (11) defines the value of R_L in cm^3/coulomb in terms of measured quantities. Using the relationship $E_{yL} = (V_2 - V_1)_L/W$ and $E_{xL} = I_L/\sigma_L WT$, one may rearrange equation (10) to obtain the angle between electric field and current

$$\theta = E_{yL}/E_{xL} = \frac{(V_2 - V_1)_L T\sigma_L}{I_L} = 10^{-8} R_L \sigma_L H. \tag{12}$$

Since the conductivity in practical units is equal to the charge density of carriers in practical units times the mobility in practical units, we have an easy way of obtaining the mobility in practical units as follows:

$$\sigma_L = ne_L\mu_L = \mu_L/|R_L| \text{ ohm}^{-1}\text{ cm}^{-1};\ e_L = 1.60 \times 10^{-19} \text{ coulombs} \quad (13)$$

$$\mu_L = \sigma_L|R_L| \text{ cm}^2/\text{volt sec.} \quad (14)$$

This result for μ_L, which leads to the useful formula

$$\theta = 10^{-8}\mu_L H, \quad (15)$$

can be converted to an expression for the mean free time by using $\mu = e\tau/m$. Equation (14) must then be multiplied by 300 to convert it to mobility per electrostatic volt, and we thus obtain

$$\tau = (300m/e)\mu_L = 5.7 \times 10^{-16}\mu_L \text{ sec} = 5.7 \times 10^{-16}R_L\sigma_L \text{ sec.} \quad (16)$$

At room temperature, the thermal velocity of an electron is 10^7 cm/sec (provided its effective mass is that of a free electron). Hence the average distance it travels between collisions is

$$l = 10^7\tau = 0.57 \times 10^{-8}R_L\sigma_L \text{ cm} = (0.57\mu_L) \text{ A} \quad (17)$$

where A = 1 angstrom = 10^{-8} cm. This gives a simple relationship between mobility and mean free path applicable at room temperature. (Mean free paths for silicon are discussed further in Chapter 11.)

M.K.S. Units. In M.K.S. units the basic force equation becomes

$$\mathbf{F}_M = e_M[\mathbf{E}_M + \mathbf{v}_M \times \mathbf{B}_M] \quad (18)$$

where the units are

$$\mathbf{E}_M = \text{volts/meter} \quad (19)$$

$$\mathbf{B}_M = \text{webers/meter}^2 \quad (20)$$

$$e_M = 1.60 \times 10^{-19} \text{ coulombs.} \quad (21)$$

A field of one weber/cm^2 is equal to 10^4 gauss. In terms of these units the basic equations become

$$\sigma_M = n_M e_M \mu_M \text{ ohm}^{-1}\text{ meter}^{-1} \quad (22)$$

$$n_M = \text{carriers/meters}^3 \quad (23)$$

$$\mu_M = \text{meters}^2/\text{volt sec} \quad (24)$$

$$R_M = 1/n_M e_M = \text{meters}^3/\text{coulomb} \quad (25)$$

$$\theta \text{ radians} = \mu_M B_M. \quad (26)$$

That the ratio of transverse electric field to longitudinal current density times magnetic flux density gives the M.K.S. Hall constant may be verified as follows:

$$R_M = \frac{E_{yM}}{I_M B_M} = \frac{E_{xM}\theta}{E_{xM}\sigma_M B_M} = \frac{\mu_M}{\sigma_M} = \frac{1}{n_M e_M}. \tag{27}$$

Equation (26) shows that mobility has the dimensions of $1/B_M$ = meters^2/weber or meters^2/volt sec because of the definition of the weber. The expressions for $\boldsymbol{\tau}$ and for l corresponding to (16) and (17) become

$$\boldsymbol{\tau} = 5.7 \times 10^{-12}\mu_M \text{ sec} \tag{28}$$

$$l = 5.7 \times 10^{-7}\mu_M \text{ meters} = 5700\mu_M \text{ angstroms.} \tag{29}$$

8.9 MODIFICATIONS OF THE HALL EFFECT FORMULA

As stated earlier, assumptions (1) and (2) of Section 8.3 are only an approximation to the actual situation, in which the probability of a transition depends in a complicated way on both the initial and the end state. When the variation of this probability is taken into account, it is found that a different average mean free time, call it $\boldsymbol{\tau}_H$, occurs in the Hall angle

$$\theta = e\boldsymbol{\tau}_H H/mc \tag{1}$$

and in the mobility (call this average $\boldsymbol{\tau}_\mu$):

$$\mu = e\boldsymbol{\tau}_\mu/m. \tag{2}$$

For electrons in semiconductors at temperatures so high that thermal scattering predominates, the times and mobilities are related by the formula, derived as (28c) in Section 11.4,

$$\boldsymbol{\tau}_H = \frac{3\pi}{8}\boldsymbol{\tau}_\mu \quad \text{and} \quad \mu_H = \frac{3\pi}{8}\mu. \tag{3}$$

As a consequence, the Hall coefficient, as given by equation (5) of Section 8.8 in terms of measured values θ, σ, and H, is

$$R_H = \frac{\theta}{\sigma H} = \frac{\pm e\boldsymbol{\tau}_H H}{mc}\cdot\frac{m}{ne^2\boldsymbol{\tau}_\mu H} = \frac{\boldsymbol{\tau}_H}{\boldsymbol{\tau}_\mu}\cdot\frac{1}{n(\pm e)c} = \frac{3\pi}{8}\cdot\frac{1}{n(\pm e)c}. \tag{4}$$

This is the formula usually used for Hall effect in semiconductors. For simple metals and for very impure semiconductors, which have the electron statistics of metals, $\boldsymbol{\tau}_H$ turns out to be the same as $\boldsymbol{\tau}_\mu$ and the formula is $R_H = 1/nec$. These results are derived in Chapter 11.

When both electrons and holes are present, the Hall constant is a complicated average of the Hall constant and conductivity for each type of carrier. In Figure 8.10 we show the construction necessary to evaluate the Hall constant when both holes and electrons are present. The magnetic field comes out of the paper, parallel to z as for Figure 8.7. Under these conditions the hole current I_p and electron current I_n deviate in opposite directions from the electric field. The net current $\boldsymbol{I}$ must, of course, have no transverse component across the specimen since such a component would quickly set up charges and produce a transverse field which would reduce the transverse current to zero. Although the net transverse current is zero, the transverse electron and hole currents individually will not vanish and there will be a tendency for both holes and electrons to be swept to one side of the filament as discussed in Section 3.2.

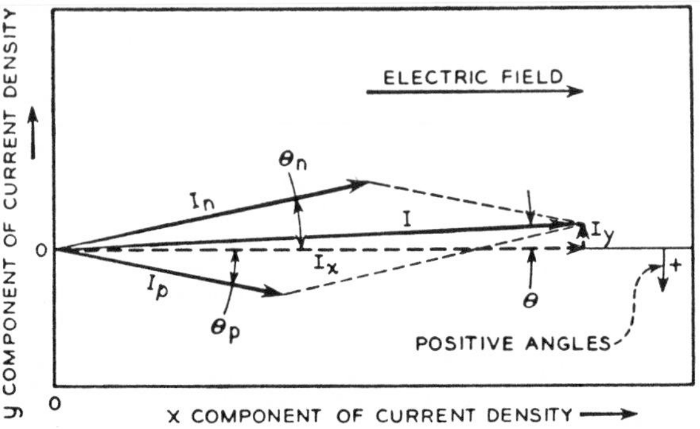

FIG. 8–10—Hall Effect for Both Holes and Electrons. (Magnetic Field in +z Direction, toward the Reader.)

For electric fields so small that no conductivity modulation is produced by this effect, the densities of holes and electrons will have their equilibrium values and the currents will be as shown in Figure 8.10. Our problem is thus to evaluate θ from the construction shown in the figure and to use this value to determine the Hall constant in accordance with $R_H = \theta/\sigma H$, equation (2) of Section 8.8. By convention, we shall regard an angle like θ_n as negative. When both types of carriers are present, we have from equation (13) of Section 8.4

$$\sigma = e(n\mu_n + p\mu_p). \tag{5}$$

Using the approximation employed for the Hall effect that $\cos\theta_n = \cos\theta_p = 1$, we obtain a current parallel to $\boldsymbol{E}$ of

$$I_x = I_n + I_p = (ne\mu_n + pe\mu_p)E_x. \tag{6}$$

The y-component of current is obtained by the approximation $\sin\theta_n = \theta_n$ and $\sin\theta_p = \theta_p$ and by recalling that the Hall effect convention gives $I_y/I_x = -\theta$ for Figure 8.10. Thus we obtain

$$I_y = -(I_n\theta_n + I_p\theta_p). \tag{7}$$

The values for θ_n and θ_p are defined in equation (2) of Section 8.6 as

$$\theta_n = -\mu_n H/c, \qquad \theta_p = +\mu_p H/c. \tag{8}$$

From these equations we find that the Hall angle for the total current is

$$\theta = \frac{-I_y}{I_x} = \frac{-(ne\mu_n E_x \mu_n H/c) + (pe\mu_p E_x \mu_p H/c)}{(ne\mu_n + pe\mu_p)E_x}$$
$$= \frac{(-n\mu_n^2 + p\mu_p^2)(H/c)}{n\mu_n + p\mu_p}. \tag{9}$$

Substituting this in the formula for R_H gives (see also Prob. 11 p. 219)

$$R_H = \frac{\theta}{\sigma H} = \frac{(-n\mu_n^2 + p\mu_p^2)(H/c)}{(n\mu_n + p\mu_p)(en\mu_n + ep\mu_p)H}$$
$$= \frac{-n\mu_n^2 + p\mu_p^2}{ec(n\mu_n + p\mu_p)^2} = \frac{-nb^2 + p}{(nb + p)^2 ec} \tag{10}$$

where (using Haynes' drift mobility data for Ge, see Section 12.9)

$$b = \mu_n/\mu_p (= 2.1 \text{ for Ge and } = 3.0 \text{ for Si}) \tag{11}$$

is the ratio of electron to hole mobility, n is the density of electrons (that is, number/cm^3), p the density of holes, e the charge on the electron, and c the speed of light. The corresponding values of R_H in practical and M.K.S. units are

$$R_L = \frac{-nb^2 + p}{(nb + p)^2 e_L}; \quad R_M = \frac{-n_M b^2 + p_M}{(n_M b + p_M)^2 e_L} \tag{12}$$

where $e_L = 1.6 \times 10^{-19}$ coulombs. The measured value of R_L is defined in equation (11) of Section 8.8. It may be readily verified that if conduction is by electrons only (that is, $p = 0$) or holes only (that is, $n = 0$) this formula reduces to equation (8) of Section 8.8 or $R_L = 1/n(\pm e_L)$. Making allowance for the difference in τ_H and τ_μ, modifies (12) to the form

$$R_L = -\frac{3\pi}{8} \cdot \frac{nb^2 - p}{(nb + p)^2 e_L}. \tag{13}$$

This is the form frequently quoted in the literature. For an example of the use of this formula in interpreting conductivity data, the reader is referred to the paper by Pearson and Bardeen.[1]

One simple consequence of (12) should be mentioned in closing. A p-type sample, which has $n \ll p$ and, consequently, $R_L > 0$, will, upon heating, become intrinsic with $n \cong p$. Since b is >1, this will cause a reversal in sign of the Hall coefficient at the temperature for which $n = p/b^2$. This point of reversal is used as discussed in Problem 7 in determining b.[1]

[1] G. L. Pearson and J. Bardeen, *Phys. Rev.* 75, 865–883 (1949).

Problems

1. From the distribution given by equation (4) of Section 8.3 compute the average values of t, t^2, t^3, t^4.

2. Obtain the general formula $\overline{t^n} = n\,!\,\tau^n$ by considering the integral $I(a)$ and its derivatives:

$$I(a) = \int_0^\infty e^{-at}dt = 1/a; \quad \overline{t^n} = (-1)^n a \frac{d^n I(a)}{da^n}.$$

3. In the long interval T of equation (1) of Section 8.3 find the total time occupied by paths with free times between t and $t + dt$. This time divided by T is the probability that an observation at a random instant will select a mean free time with duration between t and $t + dt$. Prove that the average of the mean free times selected in this way is 2τ. (Compare with the discussion at the end of Section 8.5.)

4. Show that if the free times all had exactly the same length t' (this would be a violation of assumption (1)), then the drift velocity would be $vd = -e\boldsymbol{E}t'/2m$ for electrons.

5. The lowest energy bands in a solid are very narrow since there is hardly any overlap of the wave functions; on the other hand, the high bands are relatively wide. Supposing that the mean free time is about the same for the valence-bond band and the conduction band, which would you expect to have the higher mobility, an electron or a hole?

6. Suppose a magnetic field is instantaneously applied to a semiconductor which is subjected to a constant $\boldsymbol{E}$. There will be a short time lag before the current reaches its new steady state value. What process will determine this time?

7. The ratio of mobilities b is sometimes obtained from data on a p-type specimen by using the temperature T_0 for which $R_H = 0$, so that $nb^2 - p = 0$. For somewhat lower temperatures, all of the acceptors are ionized and p has a "saturation" value p_s equal to the excess of acceptors over donors and n is negligible. As the intrinsic range is approached, holes and electrons are produced in equal numbers by the breaking of valence bonds so that $p = p_s + n$. If the saturation conductivity line is extrapolated to T_0, a conductivity σ_e is obtained, whereas the actual conductivity is σ_0. Show that

$$\frac{\sigma_0}{\sigma_e} = \frac{b}{b-1} \quad \text{or} \quad b = \frac{\sigma_0}{\sigma_0 - \sigma_e}.$$

8. Verify the practical unit formulae of Section 8.8 by deriving them from the M.K.S. expressions.

9. This problem is intended to illustrate some fundamental features of magnetoresistance effects. Suppose for this purpose that the electrons in a crystal can be separated into two classes, with densities n and n', relaxation times τ and τ' and both with the same mass. Show from (6a) and (6b) of Section 8.7 that the Hall angles are

$$\theta = \tan^{-1}(e\tau H_z/mc) = \tan^{-1}\omega\tau$$

$$\theta' = \tan^{-1}(e\tau' H_z/mc) = \tan^{-1}\omega\tau'.$$

Carry out a calculation like that shown for Figure 8.10 and show that although there is no net transverse currents there are equal and opposite transverse currents due to the individual classes. Show that if $n'\tau' = n\tau$, the resistance, defined as power dissipation per unit volume for unit current density, varies as

$$\frac{4 + 2(\omega\tau)^2 + 2(\omega\tau')^2}{4 + (\omega\tau + \omega\tau')^2}.$$

Show that if $\tau = \tau'$ this ratio is unity for all values of H_z. However, if $\tau' = 10\tau$ it varies from unity to $202/121 = 1.67$ as H_z varies from 0 to ∞.

10. Analyze the side thrust on an n-type filament carrying current in a magnetic field. Consider the fact that the net force per unit volume due to the transverse field E_y upon the electrons must be equal and opposite to that on the donors, which are mechanically attached to the filament. The electrons also transfer momentum to the filament by collisions. Show that the condition $\tau_\mu = \tau_H$ corresponds to the case in which collisions transfer no net transverse momentum to the filament. The problem of visualizing the details of momentum transfer for the electrons of an almost full band so as to establish the mechanism for the force on holes is more difficult but can be done by methods like those of Section 7.6. This problem also shows that the procedure sometimes employed to obtain the Hall angle formula by equating eE_y to ev_xH_z/c or to ev_xB_{Mx} is not fundamental since it neglects momentum transfer by collisions.

11. Show that if b_D is defined as the ratio of drift mobilities (Section 8.7) and b_H as the ratio of Hall mobilities:

$$b_D = \mu_{Dn}/\mu_{Dp}, \quad b_H = \mu_{Hn}/\mu_{Hp}$$

then the Hall coefficient [equation (10) Section 8.9] becomes

$$R_H = (\mu_{Hp}/\mu_{Dp})(p - nb_Db_H)/(p + nb_D)^2ec.$$

Show that in Problem 7, the result becomes

$$\sigma_0/\sigma_e = b_D(1 + b_H)/(b_Db_H - 1).$$

CHAPTER 9

DISTRIBUTIONS OF QUANTUM STATES IN ENERGY

9.1 ENERGY LEVEL DIAGRAMS FOR PURE CRYSTALS

For the purpose of dealing with the equilibrium distribution at various temperatures of the electrons among the quantum states of the crystal, it is necessary to consider the *distribution in energy* of the quantum states. This quantity, denoted by $N(\mathfrak{E})$, is defined in the following way: Consider a crystal of *unit volume*. Suppose that a complete list of the quantum states and their energies were prepared for all the allowed quantum states in all the Brillouin zones of the crystal. From this list, select and count all the quantum states whose energies lie in some particular interval $\mathfrak{E}$ to $\mathfrak{E} + d\mathfrak{E}$ of energy; call this number dS. Then $N(\mathfrak{E})$ is defined by

$$N(\mathfrak{E}) = dS/d\mathfrak{E} \text{ or } dS = N(\mathfrak{E})d\mathfrak{E}. \tag{1}$$

Thus we have

DEFINITION OF $N(\mathfrak{E})$: *$N(\mathfrak{E})$ is the number of quantum states per unit energy per unit volume of the crystal.*

In accordance with the theorem of the conservation of quantum states discussed in Sections 5.3 and 5.5, the number of quantum states in the Brillouin zone is simply proportional to the size of the crystal while the width of the energy bands is independent of the size of the crystal (that is, the energy band width theorem, Section 5.3). Thus, the number of quantum states per unit energy range for a crystal of volume V is simply $VN(\mathfrak{E})$.

If the energy $\mathfrak{E}$ falls in a forbidden band of energies, that is in a gap between the allowed energy bands (see Sections 5.3 and 5.6), the value of $N(\mathfrak{E})$ is zero. If $\mathfrak{E}$ falls within an energy band, then the value of $N(\mathfrak{E})$ may be obtained in principle as follows: Corresponding to the energy band there will be one or more Brillouin zones. In each of these the values $\mathfrak{E}$ and $\mathfrak{E} + d\mathfrak{E}$ define two closely neighboring energy surfaces. In any particular zone there will thus be a certain volume $dV_{\boldsymbol{P}}$ of $\boldsymbol{P}$-space between the two surfaces. The volume per allowed point in this space is h^3/V, where V is the volume of the crystal (see Sections 5.2, 5.5, and 14.8, and Figure 5.10), and each such point gives rise to two quantum states, one with each spin. Hence, the number of quantum states in the range $d\mathfrak{E}$

per unit volume of crystal will be

$$dS = \frac{2dV_P/(h^3/V)}{V} = (2/h^3)dV_P. \tag{2}$$

If several zones have overlapping energies, then the total number of states in $\mathcal{E}$ to $\mathcal{E} + d\mathcal{E}$ is obtained by adding the contributions of the individual zones. In connection with the theory of conductivity, it is found advantageous to re-express dV_P in terms of the element of area of the energy surfaces, which has the dimensions of $\boldsymbol{P}^2$ and is denoted by $d\Omega$, and the energy difference $d\mathcal{E}$ between two neighboring surfaces. The formula is

$$dV_P = d\mathcal{E}d\Omega/v \tag{3}$$

where $v = |\boldsymbol{v}| = |\nabla_P\mathcal{E}(\boldsymbol{P})|$ is the speed associated with $\boldsymbol{P}$ at the point of interest. The derivation of (3) will be found in Section 11.2 in connection with equation (5), in which the formula is used; it is repeated here for ease of reference.

As an example, we consider the number of quantum states per unit energy near the bottom of an energy band. As discussed in Sections 7.5 and 7.6, we may use the approximation that near the top or bottom of an energy band, the energy varies as P^2 or P'^2 where $P = |\boldsymbol{P}|$ and $P' = |\boldsymbol{P}'|$. For simplicity, we suppose the lowest energy is at the center of the zone so that according to equation (11) of Section 7.5

$$\mathcal{E} = \mathcal{E}_0 + P^2/2m. \tag{4}$$

From this we obtain

$$d\mathcal{E} = (P/m)dP. \tag{5}$$

The volume of the spherical shell between P and $P + dP$ in P-space is $4\pi P^2dP$ and it, therefore, contains

$$dS = (2/h^3)4\pi P^2dP \tag{6}$$

quantum states. This can be expressed in terms of the energy by solving (4) for P and (5) for PdP obtaining

$$P = [2m(\mathcal{E} - \mathcal{E}_0)]^{1/2} \tag{7a}$$

and

$$PdP = md\mathcal{E}. \tag{7b}$$

These may be substituted in equation (6) to give

$$dS = (8\pi/h^3)P^2dP = (4\pi/h^3)(2m)^{3/2}(\mathcal{E} - \mathcal{E}_0)^{1/2}d\mathcal{E} = N(\mathcal{E})d\mathcal{E} \tag{8}$$

leading to the desired value for $N(\mathcal{E})$:

$$N(\mathcal{E}) = (4\pi/h^3)(2m)^{3/2}(\mathcal{E} - \mathcal{E}_0)^{1/2}. \tag{9}$$

This equation gives the number of quantum states per unit volume of the crystal per unit energy range for energies slightly above $\mathcal{E}_0$, the minimum energy of the Brillouin zone. The total number of states having energies less than a certain maximum value $\mathcal{E}_m$ is obtained by integrating $N(\mathcal{E})d\mathcal{E}$ from $\mathcal{E}_0$ to $\mathcal{E}_m$:

$$\int_{\mathcal{E}_0}^{\mathcal{E}_m} N(\mathcal{E})d\mathcal{E} = \frac{8\pi}{3h^3}(2m)^{3/2}(\mathcal{E}_m - \mathcal{E}_0)^{3/2} = 4.5 \times 10^{21}(\mathcal{E}_m - \mathcal{E}_0)^{3/2}, \quad (10)$$

the numerical value giving *number per cm*^{3} for energy in electron volts. This formula is frequently used in the theory of metals in which it is assumed that the energy band has the form (9) and that all the states below a certain energy are occupied. Equation (10) can then be used to calculate the energy of the highest occupied state.

Entirely similar results will be obtained for the distribution of states near the top of a band. In this case, equation (4) of Section 7.6 or $\mathcal{E} = \mathcal{E}_0 - P^2/2m$ is used, and equation (9) is modified by replacing $(\mathcal{E} - \mathcal{E}_0)^{1/2}$ by $(\mathcal{E}_0 - \mathcal{E})^{1/2}$. It should be mentioned that these results are dependent on the same assumptions about the simplicity of the energy surfaces as those employed in Section 7.5.

In Figure 9.1 the behavior predicted from equation (9) and certain general considerations have been combined to show the qualitative features to be expected for the distribution of quantum states in energy for diamond. We consider a unit volume of crystal containing N_s atoms. Three energy bands are shown. The lowest corresponds to the 1*s* atomic level and is occupied by the two core electrons per atom, shown for the carbon atom in Figure 1.2. Since, as discussed in Section 5.5, the diamond structure has two atoms per unit cell, two Brillouin zones are required to accommodate the four core electrons per unit cell. The eight valence electrons per unit cell similarly require four Brillouin zones which overlap to give rise to the valence-bond energy band. If we assume that equation (9) holds near the edges of the bands [a form similar to equation (9) should always hold, provided m is replaced by a suitable effective mass], then the distribution should appear as in Figure 9.1(a). Energy is plotted upward in this figure so as to correspond to the energy band picture in (b). The number of states in the valence-bond band is

$$S = \int_{\mathcal{E}_1}^{\mathcal{E}_2} N(\mathcal{E})d\mathcal{E}, \quad (11)$$

an element of the integration being shown by the shaded area. The core electron band is narrow in energy and thus has a high value of $N(\mathcal{E})$.

The distribution of quantum states in the conduction band has been shown as continuing indefinitely to higher energies. This is in keeping with the idea that the higher atomic levels, which produce the conduction

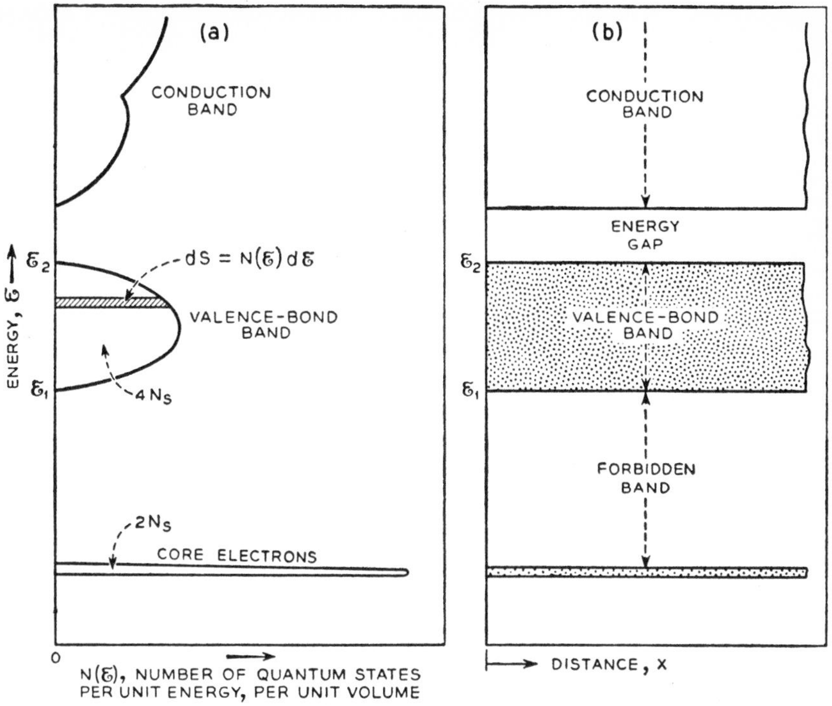

Fig. 9-1—Distribution in Energy of the Quantum States.

energy bands, interact so strongly that the resulting energy bands cover all possible higher energies, leaving no gaps.

Figure 9.1 is appropriate, so far as the valence-bond band and bottom of the conduction band are concerned, to pure silicon and germanium. However, it must be modified when impurities are present.

9.2 ENERGY LEVELS DUE TO IMPURITIES—DONOR AND ACCEPTOR LEVELS

We shall consider first the case of a donor impurity, for example, a phosphorous atom inserted in silicon. As discussed in Chapter 1, an atom like arsenic or phosphorus has one too many plus charges for its share of the four valence bonds. Therefore, it constitutes a positive charge in the lattice. It also brings with it an extra electron which neutralizes this charge but cannot be a part of the valence-bond structure. This extra electron uses wave functions in the conduction band. As we have seen, an electron near the bottom of this band behaves in the crystal in much the same way as an electron behaves in free space, except that the electron in the crystal may have an effective mass different from that of an electron in empty space. In accordance with this picture, we might expect it to behave in the presence of the impurity much as a free electron would behave in the presence of a positive charge. Detailed considerations of the wave mechanical nature of this problem shows that this picture is essentially correct.[1] The electron moving in the conduction band has modes of motion around the impurity atom essentially like those which an electron in free space has around a proton in a hydrogen atom. There are two important differences: In the case of the hydrogen atom, the electron is attracted by the charge of one proton. In the case of the phosphorous impurity, the electron is attracted by the same charge. However, this charge is embedded in a dielectric medium. Because of the polarizability and dielectric constant of the semiconductor, the attraction of the donor impurity for the electron is reduced from e^2/r^2 to $e^2/\kappa r^2$ where κ is the dielectric constant, e the electronic charge, and r the distance of separation. The other essential difference is that the effective mass is not that of a free electron. This effect is probably somewhat less important. As a consequence, the wave functions for the electron about the donor impurity are like hydrogen wave functions but for an atom with a much smaller effective nuclear charge. The well-known formula for the binding energy of a single electron to a nucleus is given by

$$2\pi^2 m^* e^4 Z^2/h^2 = 13.6Z^2 \text{ electron volts.} \tag{1}$$

In this formula e is the charge on the electron, m^* its effective mass which we take as the free electron mass for purposes of calculation, h is Planck's constant, and Z is the nuclear charge. For the semiconductor case, Z is equal to $1/\kappa$. Since κ is 11.9 for silicon[2] and 16.1 for germanium,[2] the binding energy is 0.10 electron volt for silicon and 0.05 electron volt for germanium. The radius of the wave function or electron orbit is also increased by a factor κ and thus is so large that it overlaps many atoms as shown in Figure 1.14.

The formula for the hydrogen-like wave function is

$$\Psi = \frac{1}{\sqrt{\pi}}\left[\frac{1}{\kappa a_0}\right]^{3/2} e^{-r/\kappa a_0}, \quad a_0 = \frac{\hbar^2}{me^2} \tag{2}$$

where r is the distance from the ion and a_0 is the radius of the first Bohr orbit and has the value 0.529×10^{-8} cm = 0.529 A (angstroms).

This bound state is removed from the other states in the conduction band

[1] G. Wannier, *Phys. Rev.* **52**, 191–197 (1937); S. R. Tibbs, *Trans. Faraday Soc.* **35**, 1471–1484 (1939); S. Peckar, *J. Phys. U. S. S. R.* **10**, 431–433 (1946); H. M. James, *Phys. Rev.* **76**, 1602–1610, 1611–1624 (1949); J. C. Slater, *Phys. Rev.* **76**, 1592–1601 (1949); of which the last is most pertinent.

[2] H. B. Briggs, *Phys. Rev.* **77**, 287 (1950). See also J. F. Mullaney, *Phys. Rev.* **66**, 326–339 (1944), H. B. Briggs and W. H. Brattain, *Phys. Rev.* **75**, 1705–1710 (1949), M. Becker and H. Y. Fan, *Phys. Rev.* **76**, 1530–1531 (1949).

by about the interval of 0.10 electron volt for silicon. As a result, each donor impurity removes a state from the conduction band and establishes it as a quantum state of lower energy. For most practical cases, the total amount of impurity present is very small compared to the total number of atoms, so only a very small fraction of the quantum states in the conduction band is reduced to lower energies. The hydrogen atom has, in addition to its lowest state, a series of higher energy levels, and these will be present around the donor impurity also. However, these levels will lie very close to the conduction band, and electrons will be so readily excited from them into the conduction band that it is not necessary to consider these higher levels further for the applications with which we are concerned.

The donor impurity atoms will have a similar effect upon states in the valence-bond band, and one state at the bottom of the valence-bond band will be moved to a lower level for each donor atom present. States at the bottom of this band are, however, without interest because, at all temperatures met with in semiconductors, these states are fully occupied.

Acceptor impurities, however, do have an important effect upon the states in the valence-bond band. These impurities represent a negative charge since their atomic cores are insufficiently charged to neutralize the four electrons in their share of the valence bonds. Consequently, they can attract a positive charge. As we have seen, a hole in the valence-bond band has many of the attributes of a positive charge. Quantum-mechanical theory shows that a hole in the valence-bond band, in addition to behaving as a positive charge for Hall effect and conductivity, will have modes of motion around an acceptor impurity of a type entirely analogous to the motion of an electron around a donor impurity. As a consequence of this, some of the states in the valence-bond band are raised to higher levels by the presence of acceptor impurities.

In Figure 9.2 we show the effect of acceptor and donor impurities on the quantum-state distribution. We consider unit volume of material so that the numbers of quantum states are, in effect, densities (number per unit volume). The valence-bond band is represented as having $4N_s$ quantum states, where, as before, N_s is the number of atoms per unit volume of the crystal. There are N_d donor atoms present and these depress N_d states from the valence-bond band and from the conduction band to lower levels as described above. Owing to the two values for the spin, these levels actually are $2N_d$ in number. However, we must take into account the fact that, if an electron is in a quantum state around one of the donor atoms, it neutralizes this atom so that another electron cannot be trapped there with the same binding energy. This means in effect that only N_d rather than $2N_d$ of the states are available. (This feature of the effect of spin upon the statistics of electrons in quantum states is described only approximately by the foregoing method. However, for purposes of this exposition it is not necessary to use a more exact treatment. See Problem 2, Chapter 16.) Similarly, the N_a acceptor atoms remove N_a states from the valence-bond band, leaving a total of $4N_s - N_d - N_a$ quantum states in the band. As is shown in the figure, N_d states are pushed out of the valence-bond band to the lower level, as previously discussed. The notation for the energy levels uses the subscripts corresponding to conduction, donor, acceptor and valence-bond.

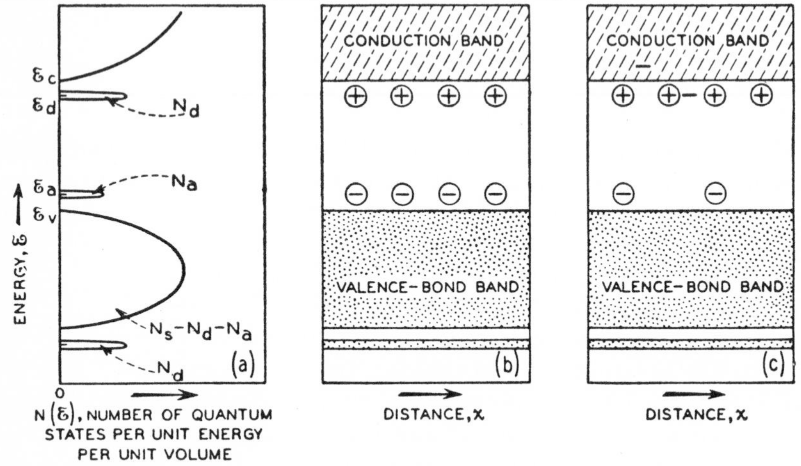

FIG. 9–2—Energy Level Diagrams.

(a) Quantum state distribution including effects of donors and acceptors for an *n*-type semiconductor.

(b) Energy level diagram showing locations and charges of impurities for a sample with balanced impurities.

(c) As for (*b*) for an *n*-type sample at low temperature.

Diagrams like Figure 9.2(b) and (c) are frequently employed to describe and analyze situations in which the distribution in space as well as energy is important. In these diagrams the ⊕ and ⊖ symbols represent energy levels at donors and acceptors and also the net charge. The charge is defined for the situation in which the valence-bond structure is complete and no excess electrons are left over. For this case each donor represents an excess positive charge. The donor symbol is located at the energy with which the donor can bind an electron and is thus somewhat below the edge of the conduction band. Electrons in the conduction band and trapped in donors are represented as in Figure 9.2(c). A hole, due to an electron missing from the valence-bond band, would similarly be represented by a plus sign, and, since holes tend towards higher levels in diagrams like Figure 9.2(c), a hole trapped to an acceptor would be represented by a plus sign next to the acceptor symbol. It is evident from these definitions that electrical neutrality for diagrams like Figure 9.2(c) requires the vanishing of the sum of the charges shown. Diagrams of this sort are used in Chapters 4 and 12.

For high densities of impurities, new effects set in, which are associated with pronounced overlapping of the wave functions of the impurity atoms. This situation is closely analogous to those discussed in Section 5.3 in con-

Table 9.1 Energies in Electron Volts

		Silicon	Germanium
$\mathcal{E}_c - \mathcal{E}_v$	Experiment	$1.12 - 3 \times 10^{-4}T$ [a]	0.72 [b]
$\mathcal{E}_c - \mathcal{E}_d$	Theory	0.08 [c] − 0.054 [d]	0.05 [c]
$\mathcal{E}_a - \mathcal{E}_v$	Theory	0.08 [c]	0.05 [c]
$\mathcal{E}_c - \mathcal{E}_d$	Experiment	0.06 [e]	0.013 [f]
$\mathcal{E}_a - \mathcal{E}_v$	Experiment	0.075 [e]	0.013 [f]
$\mathcal{E}_a - \mathcal{E}_v$	Bombarded Centers	. . .	0.046 ± 0.01 [g]
$\mathcal{E}_a - \mathcal{E}_v$	Heat-produced Centers	. . .	0.04 [g]

[a] G. L. Pearson and J. Bardeen, *Phys. Rev.* **75**, 865–883 (1949). This value was deduced from an extensive analysis of the temperature dependence of pn. See Section 16.3 for a discussion.

[b] Unpublished estimate by J. Bardeen.

[c] Using m = free electron mass, $\kappa = 13$ for Si, $\kappa = 16.1$ for Ge in eq. (1).

[d] Reference (a), based on an electron mass $m_n = 0.67m$.

[e] Reference (a). See also Figure 9.3 and text.

[f] Preliminary analysis of early data by G. L. Pearson and W. Shockley, *Phys. Rev.* **71**, 142 (1947).

[g] Preliminary results by G. L. Pearson.

nection with bringing atoms together to form a solid. As was shown in Figure 5.4, when the atoms approach close enough, appreciable splitting of the atomic energy levels occurs, and it becomes appropriate to deal with the electrons as being in wave functions corresponding to states in the Brillouin zone. A similar effect will occur with the wave functions around the impurities. As the impurity density becomes higher, the wave functions overlap more and finally form a band of energies. Electrons in these states will then conduct in the same way as do electrons in a metal. In fact, the situation closely corresponds to that of an alkali metal which may be thought of as an array of small positive ions neutralizing a space charge of electrons moving in free space. In the semiconductor there is a random array of donor (or acceptor) ions neutralizing a space charge of electrons (or holes) moving, not through free space, but in the manner appropriate to their states in the Brillouin zone. Evidence for this type of behavior is found in both n- and p-type silicon and germanium. Samples of high impurity content have been observed to have substantially metallic behavior with numbers of carriers which are independent of temperature at temperatures below the intrinsic range. Examples of this behavior are shown in Figures 1.10 and 1.11 and are discussed further in Chapter 11.

On the basis of analyses of Hall effect and conductivity as a function of temperature, values have been determined for a number of the energy parameters discussed in this section. These are quoted in Table 9.1 together with explanatory remarks and references. The data for germanium is of a preliminary character.

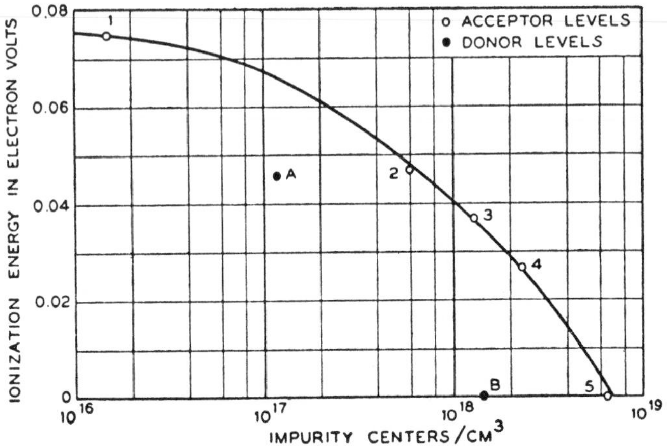

FIG. 9–3—Dependence of Ionization Energy of Impurity Centers upon Density.

Pearson and Bardeen find that for silicon the ionization potentials $\mathcal{E}_c - \mathcal{E}_d$ and $\mathcal{E}_a - \mathcal{E}_v$ vary with impurity concentration in the way shown in Figure 9.3. They interpret the decrease with increasing concentration as being due to the electrostatic attraction of other donors for an electron which has escaped from its own donor. This attraction reduces the average energy of an electron which has escaped from its own donor so that the ionization potential for its escape is reduced. They find that the data for holes in Figure 9.3 can be fitted by the formula

$$\mathcal{E}_a - \mathcal{E}_v = 0.08 - 4.3 \times 10^{-8} N_a^{1/3} \quad \text{electron volt} \tag{3}$$

from which the solid curve is drawn. The value of the coefficient is of the right order of magnitude as may be seen by calculating the electrostatic potential due to one ion at a radius equal to half the average distance between ions. This radius will be approximately $\frac{1}{2}N_a^{1/3} = r$. The potential at that radius is

$$\begin{aligned} \frac{e}{\kappa r} \text{ electrostatic volts} &= 300 \, \frac{4.8 \times 10^{-10}}{11.9r} \text{ volts} \\ &= 1.2 \times 10^{-8} r^{-1} = 2.4 \times 10^{-8} N_a^{1/3} \text{ volts.} \end{aligned} \tag{4}$$

In their calculation a more detailed procedure is used which takes into account the average potential of the electron at points nearer the other ion and a value of $3 \times 10^{-8} N_a^{1/3}$ is obtained.

It has been pointed out by Seitz[3] that there are objections to the procedure used by Pearson and Bardeen due to the fact that the wave functions implied in their analysis are not orthogonal. Seitz's investigations indicate that the interpretation of Figure 9.3 may involve quite subtle effects.

Problems

1. Obtain formula (10) of Section 9.1 by considering a sphere in momentum space with radius P_m given by $\mathcal{E}_m = P_m^2/2m$.

2. Calculate the first and second ionization potentials for substitutional impurities differing in valence by two from silicon and germanium taking 24.5 electron volts as the first ionization potential of helium. Compute the radii of the wave functions and compare with the lattice constant for the case of one bound electron.

[3] Personal communication.

CHAPTER 10

FERMI-DIRAC STATISTICS FOR SEMICONDUCTORS

10.1 THE FERMI-DIRAC DISTRIBUTION FUNCTION

In this section, the effects of thermal agitation in distributing the electrons among the various quantum states will be described. We shall consider first the state of affairs at the absolute zero of temperature and then see how this is modified by the excitation of electrons from quantum states of lower energy to those of higher energy. The population of quantum states at various energies can be described in terms of the Fermi-Dirac distribution function, which is discussed in the first part of this section. This distribution function is generally applicable to electrons in the sense that it is independent of the detailed nature of the distribution of quantum states in energy. In the latter parts of the section, the Fermi-Dirac distribution function is applied first to a hypothetical example with only 36 quantum states and 20 electrons in order to illustrate the general features in their simplest form. This example is followed by an application of the methods to the distribution of quantum states in energy appropriate to an impurity semiconductor.

At the absolute zero of temperature, the electrons fall to the lowest energy states available to them and fill these up subject to the limitations imposed by the Pauli exclusion principle. As a result all the quantum states below a certain energy are filled and those above that energy are empty. If thermal agitation is present, however, electrons will tend to be excited to higher states. On the basis of statistical mechanical theory, this tendency is expressed in terms of energy by the quantity kT which represents the average thermal energy of a one-dimensional oscillator at temperature T. Accordingly, electrons which at absolute zero have energies only about kT less than empty states may be excited to these higher states. However, electrons having energies many times kT less than empty states have an almost negligible probability of being excited. Thus for the equilibrium distribution, the sharp division between filled and empty states becomes smeared out into a transition region about $2kT$ wide at higher temperatures.

This situation is described in detail by the Fermi-Dirac distribution function. This function, denoted by the letter $\mathbf{f}$, depends upon $\mathcal{E}$ and T in a way shortly to be discussed. The values of $\mathbf{f}$ range from zero to unity and are interpreted as follows:

The value of the Fermi-Dirac function **f** *is the probability that under thermal equilibrium a quantum state of energy* $\mathcal{E}$ *is occupied by an electron. It is, therefore, the average value of the fraction of all the quantum states of energy* $\mathcal{E}$ *occupied by electrons.*

The significance of the properties just described will become more apparent in terms of the examples discussed below.[1]

The function **f** is plotted in Figure 10.1 and is seen to vary f om zero for high energies to unity for low energies. It makes the tra sition from values nearly equal to zero to values nearly equal to unity over a range of energy equal to about $4kT$. The energy $\mathcal{E}_F$, for which **f** takes on the value $\frac{1}{2}$, is called the *Fermi level.*

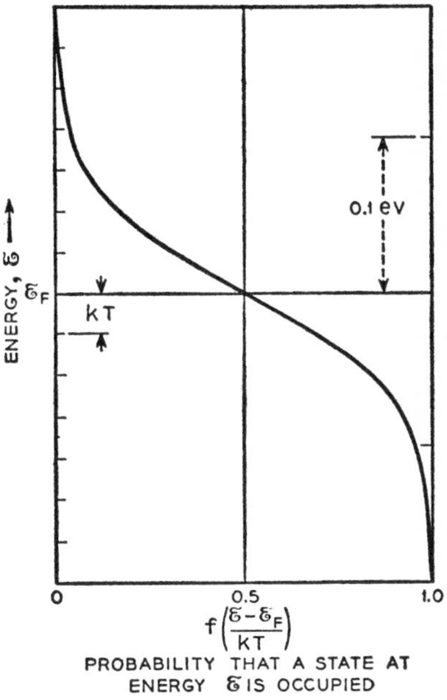

FIG. 10–1—Fermi-Dirac Distribution Function.

For purposes of comparison with energies at room temperature, a scale in electron volts is also shown. At room temperature, $T = 300°K$, kT corresponds to an energy of 0.026 electron volt. It is sometimes also convenient to compare these energies with the energy of light quanta. In Figure 10.2, we compare the energy scales in terms of degrees Kelvin, electron volts, wave lengths in angstroms, and frequency ν of the light quanta.

The statistical nature of the Fermi-Dirac distribution function should be pointed out. What **f** prescribes is the *probability* that a quantum state of given energy be occupied. If, for example, the system being considered has six quantum states at the energy for which $\mathbf{f} = \frac{1}{2}$, then under equilibrium conditions there will be on the average three electrons in these states. However, this number will fluctuate as electrons make transitions in and out of the states as a result of thermal agitation, and in fact for $\frac{1}{64}$ of the time the quantum states will be filled with six electrons; for $\frac{6}{64}$ of the time they will contain five electrons; for $\frac{15}{64}$, four; for $\frac{20}{64}$, three; for $\frac{15}{64}$, two; for $\frac{6}{64}$, one and for $\frac{1}{64}$ of the time all the states will be empty. For this case, with only a small number of states involved, the statistical

[1] An outline of the statistical mechanical theory of the Fermi-Dirac distribution function is presented in Chapter 16.

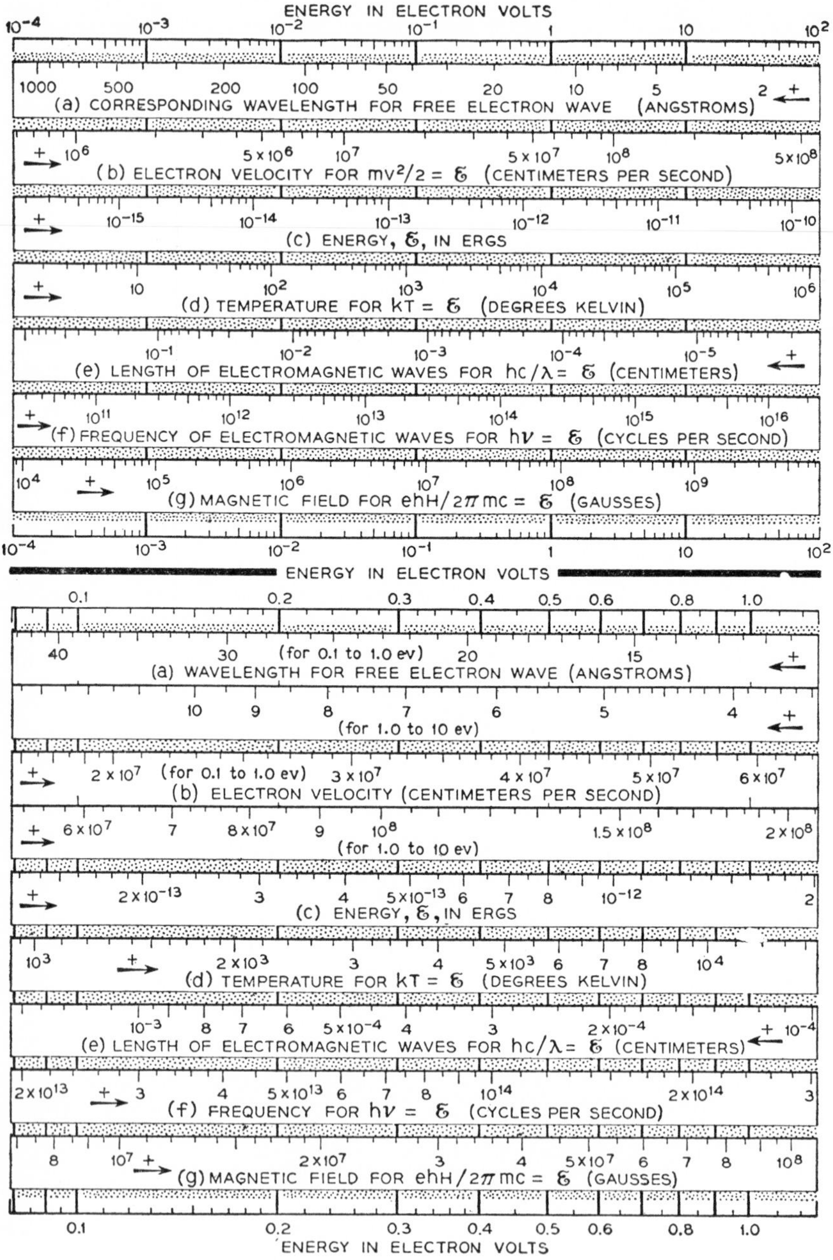

FIG. 10–2—Comparison of Energy Scales.

fluctuations are large. In a semiconductor, however, we are generally concerned with groups containing 10^{10} or more states. For such large numbers as these, the statistical fluctuations are negligible on a percentage basis for the purposes of this chapter.

The mathematical form of the Fermi-Dirac distribution function gives complete symmetry to holes and electrons as will be shown below by consideration of its mathematical form. If, as is indicated on Figure 10.1, the energy level for which $\mathbf{f} = \frac{1}{2}$ is denoted by $\mathcal{E}_F$, called the Fermi level, then the value of $\mathbf{f}$ at energy $\mathcal{E}$ is given by the formula[2]

$$\mathbf{f} \equiv \mathbf{f}[(\mathcal{E} - \mathcal{E}_F)/kT] = \frac{1}{1 + \exp[(\mathcal{E} - \mathcal{E}_F)/kT]}. \tag{1}$$

As explained previously, the value of $\mathbf{f}$ gives the probability that a quantum state at energy $\mathcal{E}$ is occupied. If $\mathcal{E}$ is larger than $\mathcal{E}_F$ by several times kT, then the exponential in the denominator is large compared to unity and $\mathbf{f}$ takes the approximate form

$$\mathbf{f} \cong \exp[-(\mathcal{E} - \mathcal{E}_F)/kT] \quad \text{for} \quad \mathcal{E} > \mathcal{E}_F. \tag{2}$$

This approximation gives the same distribution of electrons as do the "classical" statistical mechanical distributions of Maxwell and Boltzmann. We shall refer to it and to the corresponding approximation for holes, discussed below, as the *classical approximation.*

On the other hand, if $\mathcal{E}$ is less than $\mathcal{E}_F$, the exponential in the denominator is small and the approximation $(1 + \varepsilon)^{-1} \cong 1 - \varepsilon$ may be used giving

$$\mathbf{f} \cong 1 - \exp[(\mathcal{E} - \mathcal{E}_F)/kT] \quad \text{for} \quad \mathcal{E} < \mathcal{E}_F. \tag{3}$$

Thus for energies much above $\mathcal{E}_F$, $\mathbf{f}$ approaches zero as $\exp(-\Delta\mathcal{E}/kT)$ where $\Delta\mathcal{E}$ is the height above $\mathcal{E}_F$. Below $\mathcal{E}_F$, $\mathbf{f}$ approaches unity, the difference between $\mathbf{f}$ and unity being given by the expression $\exp(-\Delta\mathcal{E}/kT)$ where $\Delta\mathcal{E}$ is now the energy *below* $\mathcal{E}_F$.

Now, if $\mathbf{f}$ is the fraction of quantum states occupied by electrons, then obviously $1 - \mathbf{f}$ is the fraction left vacant or, in other words, the fraction occupied by holes. The approximation (3) thus shows that the probability of finding that a quantum state is occupied by a hole at $\Delta\mathcal{E}$ below $\mathcal{E}_F$ is just equal to the probability of finding a quantum state occupied by an electron at $\Delta\mathcal{E}$ above $\mathcal{E}_F$. This result is valid in general and not merely for approximations (2) and (3). If we write $\mathbf{f}_p = 1 - \mathbf{f}$ for the distribution function for holes, we find by straightforward algebraic manipulation

$$\mathbf{f}_p \equiv \mathbf{f}_p[(\mathcal{E} - \mathcal{E}_F)/kT] = 1 - \mathbf{f} = \frac{1}{1 + \exp[(\mathcal{E}_F - \mathcal{E})/kT]} \tag{4}$$

[2] As (1) shows, $\mathbf{f}$ is a function of three variables $\mathcal{E}$, $\mathcal{E}_F$ and T. In cases where T and $\mathcal{E}_F$ do not vary, we shall sometimes write $\mathbf{f}$ as simply $\mathbf{f}$ or $\mathbf{f}(\mathcal{E})$.

which shows on comparison with (1) that $\mathbf{f}_p$ is the same function of energy measured downward from $\mathcal{E}_F$ as $\mathbf{f}$ is for energies measured upwards. In other words, the function $\mathbf{f}$ of Figure 10.1 has a center of symmetry at $\mathcal{E} = \mathcal{E}_F$ and $\mathbf{f} = \frac{1}{2}$.

The functions $\mathbf{f}$ and $\mathbf{f}_p$ are each functions of the single variable $(\mathcal{E} - \mathcal{E}_F)/kT$. In many cases, however, $\mathcal{E}_F$ and T are constant and the only variable of interest is $\mathcal{E}$ or $\boldsymbol{P}$ upon which $\mathcal{E}$ depends. In such cases we may for brevity write the functions in one of the following forms:

$$\mathbf{f} = \mathbf{f}(\mathcal{E}) = \mathbf{f}[\mathcal{E}(\boldsymbol{P})] = \mathbf{f}(\boldsymbol{P}) \tag{5a}$$

$$\mathbf{f}_p = \mathbf{f}_p(\mathcal{E}) = \mathbf{f}_p[\mathcal{E}(\boldsymbol{P})] = \mathbf{f}_p(\boldsymbol{P}), \tag{5b}$$

the latter forms being specially adapted for use with quantum states in the Brillouin zone. This notation is not mathematically consistent but should cause no confusion.

One of the principal problems in determining the equilibrium distribution of the electrons in the quantum states arises in connection with the location of the Fermi level $\mathcal{E}_F$ on the energy scale. As will be discussed below, the location of $\mathcal{E}_F$ is determined by: (1) the distribution of quantum states for the system under consideration; (2) the total number of electrons in the system; and (3) the absolute temperature. The temperature is an important variable in determining the location of the Fermi level for semiconductors, the situation here being quite different from that in metals for which the location of the Fermi level is substantially the same at all temperatures. Once the Fermi level has been determined for a given temperature, however, the distribution of electrons in energy can be obtained directly by combining the values of $N(\mathcal{E})$, the number of quantum states per unit energy range, with $\mathbf{f}[(\mathcal{E} - \mathcal{E}_F)/kT]$, the fraction of these that are occupied. Detailed examples of these procedures are given below.

Under conditions in which $\mathcal{E}_F$ is substantially independent of temperature, the temperature dependence of $\mathbf{f}$ for values of $\mathcal{E}$ which differ from $\mathcal{E}_F$ by several times kT can be approximated by equations (2) or (3). For each of those the number of electrons or holes varies simply as $\exp(-\Delta\mathcal{E}/kT)$ as discussed previously. This type of behavior leads to a straight line plot for $\ln \mathbf{f}$ vs $1/T$ as mentioned in Chapter 1, with $\Delta\mathcal{E}$ giving the slope or activation energy. However, in order to determine what $\Delta\mathcal{E}$ means in terms of the energy-level diagram, it is necessary to know the location of $\mathcal{E}_F$. This subject is investigated below.

10.2 FERMI-DIRAC STATISTICS FOR A SIMPLE MODEL

Rather than discuss initially the case of a semiconductor whose quantum state distribution is represented by Figure 9.2, we consider a small system with only 36 states and 20 electrons. The advantage of dealing with a small system is that illustrative distributions of electrons can be shown in detail rather than in analytical form. For the small numbers involved, statistical fluctuations are important and should be considered; we disregard them here, however, since the general features to be brought out are applicable for large systems where fluctuations are unimportant. The

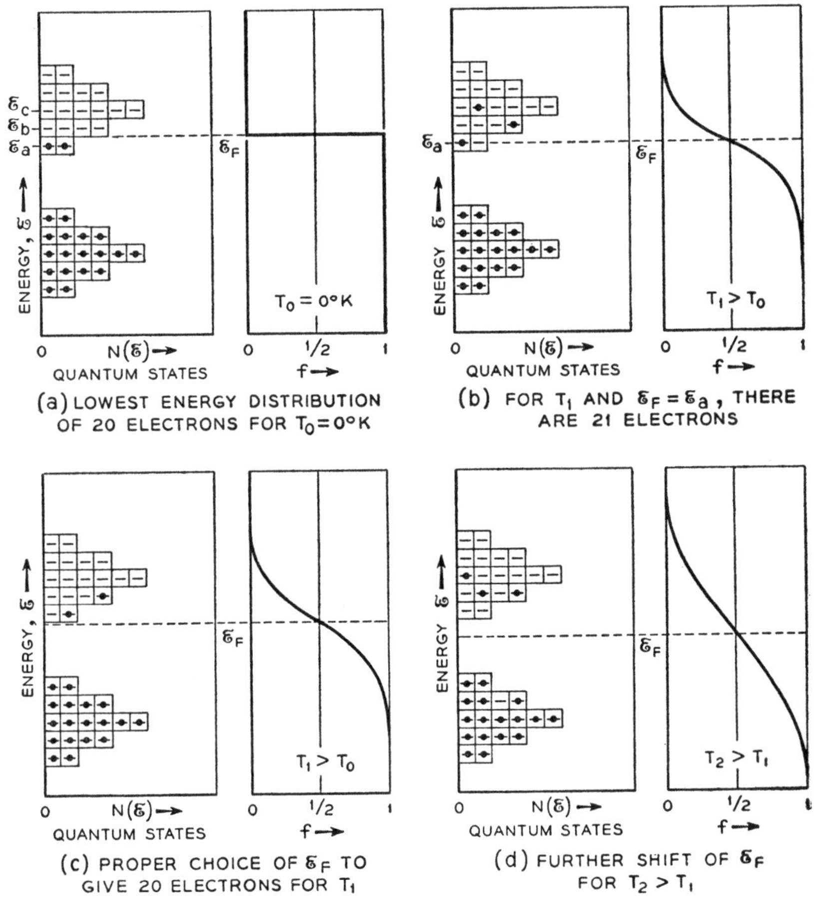

Fig. 10–3—Quantum State Distribution and Fermi-Dirac Distribution Combined to Give Equilibrium Distribution.

quantum state distribution, Figure 10.3, has been drawn to represent approximately two energy bands, so that it has some similarity to the distribution in a semiconductor.

In Figure 10.3(a) we represent the situation at the absolute zero of temperature. At this temperature the electrons will attempt to occupy the lowest quantum states; however, because of the Pauli principle that not more than one electron may occupy a given state, they will fill up states to relatively high energies, as is shown. At the absolute zero of temperature, the transition of the Fermi function from zero to unity will be infinitely sharp, as a result of which all of the states above the level $\mathcal{E}_F$ are empty and all of those below are filled. Since the system considered contains 20 electrons as shown, the Fermi level must lie between the energy levels $\mathcal{E}_a$ and $\mathcal{E}_b$. If it were drawn below $\mathcal{E}_a$, the system would contain a maximum of 18 electrons, and if it were drawn above $\mathcal{E}_b$, it would contain a minimum of 24 electrons.

For higher temperatures, however, it is found that the Fermi level must lie below $\mathcal{E}_a$. This can be understood by considering (b), for which kT is comparable to the spacing between levels in the upper group. For this case, the Fermi level is drawn as though it were precisely at the energy $\mathcal{E}_a$. Under this condition, on the average, half of the states at $\mathcal{E}_a$ will be occupied. Since electrons are excited well above the Fermi level for the case shown, there is an appreciable probability of finding electrons at energies $\mathcal{E}_b$ and $\mathcal{E}_c$, and the expected numbers are approximately one at each of these higher energy values. Of course, while the situation represented in (b) is only one of a large number of possibilities which are simultaneously represented by the Fermi function, it has been drawn so as to represent correctly the average behavior to be expected so far as numbers of electrons at each energy are concerned. Accordingly, we will find an average of three electrons in the upper group. On the other hand, the lower group of quantum states lie so far below the Fermi level that the probability of finding a hole in them is negligible for the situation represented in (b). Consequently, we see the total number of electrons for this case is 21 or one higher than the required number. In order to produce the desired number of 20 electrons, it is necessary to shift the Fermi level downwards as is shown in (c). This downward shift of the Fermi level with respect to the distribution of quantum states decreases the probability that any state is occupied by an electron, as may be easily seen from the shape of the Fermi-Dirac distribution curve. As a consequence, the shift downwards of (c) compared to (b) decreases the probability that the states in the upper group are occupied; and if the correct shift is made, the probability will be such that on the average only two electrons will be found in the upper group and thus the required number, 20, will be achieved.

As mentioned before, the shift of Fermi level with temperature is a characteristic feature in the behavior of semiconductors. It is quite different from the situation in metals where the Fermi level at higher temperatures comes at very nearly the same level as it does at absolute zero.

At still higher temperatures, the situation for our simple example will be as shown in (d). In this case kT is comparable to the spacing between the two groups of quantum states. Under these conditions, some electrons will be excited from the lower group to the higher group, and in order to keep the total number of electrons equal to 20, the Fermi level has been moved still farther down and nearer the center position between the two groups.

We must next consider the extension of these results in analytical form to the quantum state distribution of a semiconductor.

10.3 FERMI-DIRAC STATISTICS FOR A SEMICONDUCTOR

An essential condition to be met in the interior of a semiconductor is that of electrical neutrality. This condition follows from the well-known theorem in the theory of conductivity that the net charge density within a conductor at equilibrium must be zero. As applied to a semiconductor this requirement is equivalent to that used in Section 10.2, namely, that there must be a fixed total number of electrons in the quantum states. We shall first consider the factors which determine this number and next see how the Fermi level must be adjusted to meet the requirement.

Since we are interested in relatively small deviations from the perfect valence-bond structure, it is convenient to deal not with the total number of electrons, but instead only with the net excess over the number required exactly to fill the valence-bond structure. This net excess number of electrons per unit volume is made up of four contributions:

n = number of excess electrons in the conduction band (that is, not bound to donors) per unit volume of crystal;

n_d = number of excess electrons in bound states about donors per unit volume of crystal;

p = number of holes in the valence-bond band (that is, not bound to acceptors) per unit volume of crystal;

p_a = number of holes in bound states about acceptors per unit volume of crystal.

If all these quantities are zero, the valence bonds will be complete and there will be no excess electrons. If they are not zero, then the net excess number of electrons is clearly

$$n + n_d - p - p_a. \tag{1}$$

In addition to the four densities just defined, a number of others will be required in discussing the statistics of this chapter. For convenience we tabulate them below, although N_c and N_v will not be precisely defined until later. For brevity we use "density" to replace "number per unit volume".

If the valence-bond structure were perfect and there were no excess electrons, the crystal would not, in general, be neutral because of the localized charges of the donors and acceptors. As discussed in Chapters 1 and 9, when the valence-bond structure is perfect around a donor atom,

TABLE 10.1

Positive Charges		
	p	= density of holes in valence-bond band
	p_a	= density of holes bound to acceptors
	p_t	$= p + p_a$
	N_d	= density of donors
Negative Charges		
	n	= density of excess electrons in conduction band
	n_d	= density of excess electrons bound to donors
	n_t	$= n + n_d$
	N_a	= density of acceptors
Other Densities		
	N_s	= number of atomic sites per unit volume
	$N_c \equiv N_c(T)$	= effective density of states in conduction band
	$N_v \equiv N_p(T)$	= effective density of states in valence-bond band

the latter represents a localized positive charge and an acceptor represents similarly a net negative charge. For donor and acceptor densities of N_d and N_a this means that there would be a net charge density in the semiconductor of $e(N_d - N_a)$ if the valence-bond structure were complete. Since (1) represents the excess electrons over a complete valence-bond structure, the net charge density, which must vanish, is given by

$$e(N_d - N_a) - e(n + n_d - p - p_a) = 0. \tag{2}$$

This can be rewritten as

$$n + n_a - p - p_a = N_d - N_a \tag{3}$$

which states that the net excess of electrons is just equal to the excess of donors over acceptors (a situation discussed in connection with Figure 1.9).

The values of n, n_a, p and p_a are determined by (1) the distribution of quantum states in energy, (2) the value of $\mathcal{E}_F$, and (3) the temperature. The appropriate equations are obtained as follows: In the energy range $d\mathcal{E}$ there are $dS = N(\mathcal{E})d\mathcal{E}$ quantum states. A fraction $\mathbf{f}[(\mathcal{E} - \mathcal{E}_F)/kT]$ of these are occupied, leading to

$$\mathbf{f}[(\mathcal{E} - \mathcal{E}_F)/kT]dS = \mathbf{f}[(\mathcal{E} - \mathcal{E}_F)/kT]N(\mathcal{E})d\mathcal{E} \tag{4}$$

electrons in states in the range $d\mathcal{E}$. If this quantity is integrated over a range of energies, such as $\mathcal{E}_1$ to $\mathcal{E}_2$ shown in Figure 10.4, which covers the donor states and the conduction band, it will give n_d plus n. Thus we shall have

$$n_t = n + n_d = \int_{\mathcal{E}_1}^{\mathcal{E}_2} \mathbf{f}[(\mathcal{E} - \mathcal{E}_F)/kT]N(\mathcal{E})d\mathcal{E}. \tag{5}$$

The exact value of $\mathcal{E}_2$ is unimportant since $\mathbf{f}$ approaches zero rapidly at high energies. In fact, **f** is negligible except near the very bottom of the energy band. This is illustrated in part (b) of the figure in which $N(\mathcal{E})\mathbf{f}$ and $N(\mathcal{E})\mathbf{f}_p$ are plotted, and it is seen that the important parts of the integrals come near the edges of the bands. This subject will be dealt with in more detail later. A similar calculation gives the number of holes:

$$p_t = p + p_a = \int_{\mathcal{E}_3}^{\mathcal{E}_4} \mathbf{f}_p[(\mathcal{E} - \mathcal{E}_F)/kT]N(\mathcal{E})d\mathcal{E} \tag{6}$$

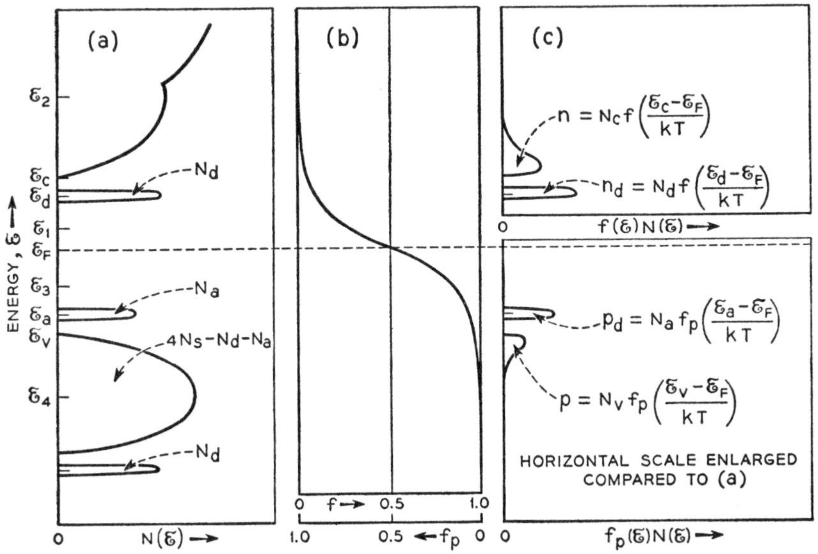

FIG. 10–4—Distribution of Quantum States and Densities of Electrons and Holes for an n-Type Semiconductor.

(a) Distribution of Quantum States in Energy for an n-Type Semiconductor.
(b) Fermi-Dirac Distribution.
(c) Electrons and Holes per Unit Volume, per Unit Energy, and Formulae for Densities.

where $\mathcal{E}_3$ to $\mathcal{E}_4$ includes the full band and the acceptor levels and $\mathbf{f}_p = 1 - \mathbf{f}$ (given by [equation (4) of Section 10.1]) is the probability that a quantum state of energy $\mathcal{E}$ is vacant (that is, occupied by a hole).

For a given temperature T, both n_t and p_t of equations (5) and (6) are functions only of $\mathcal{E}_F$. Hence the equation

$$n_t - p_t = N_d - N_a \tag{7}$$

may be regarded as containing only one unknown, namely $\mathcal{E}_F$, since N_d, N_a and the distribution $N(\mathcal{E})$ may be regarded as specified by the nature of the semiconductor considered, the situation being similar to that discussed in connection with simple model containing 20 electrons of Figure 10.3. In Chapter 16, some detailed examples of solving for $\mathcal{E}_F$ are given. In this section we shall indicate only how n_t and p_t vary in general and then describe the final statistical results.

In the integral for n_t, the integration extends over the N_d donor levels. These are all supposed to lie at one energy $\mathcal{E}_d$ for which the **f** has the value $\mathbf{f}[(\mathcal{E}_d - \mathcal{E}_F)/kT]$; hence, the integral gives for the number of electrons in the donor levels the value

$$n_d = N_d\mathbf{f}[(\mathcal{E}_d - \mathcal{E}_F)/kT] \tag{8}$$

as is indicated on Figure 10.4.

The number of electrons in the states in the conduction band is obtained by an integration over $N(\mathcal{E})$ near the bottom of the conduction band. The situation is simplified by the fact that, for most cases, $\mathcal{E}_F$ lies well below the bottom of the conduction band so that approximation (2) of Section 10.1 may be used. This enables us to write

$$n = \mathrm{e}^{\mathcal{E}_F/kT}\int_{\mathcal{E}_c}^{\mathcal{E}_2} N(\mathcal{E})\mathrm{e}^{-\mathcal{E}/kT}\,d\mathcal{E}. \tag{9}$$

Since the exponential decreases by e^{-1} for each increment kT of $\mathcal{E}$, we may limit, as a rough approximation, the integration to the interior of the energy surface in the Brillouin zone for which $\mathcal{E} - \mathcal{E}_c = kT$ or $P^2/2m = kT$. Since kT is very small compared to the width of the energy band, the approximation [equation (11) of Section 7.5] $\mathcal{E} = \mathcal{E}_c + P^2/2m$ may be used so that the energy surface is a sphere containing a volume $4\pi P^3/3$ where $P^2/2m = \mathcal{E} - \mathcal{E}_c = kT$. This sphere contains $(2/h^3)(4\pi P^3/3) = 8\pi P^3/3h^3$ quantum states per unit volume of the crystal. Since in this volume $\exp(-\mathcal{E}/kT)$ is approximately $\exp(-\mathcal{E}_c/kT)$, we may approximate the integral by $\exp(-\mathcal{E}_c/kT)$ times $8\pi P^3/3h^3$. Writing $P=(2mkT)^{1/2}$,

$$n \cong \mathrm{e}^{-(\mathcal{E}_c-\mathcal{E}_F)/kT}(8\pi/3)(2mkT/h^2)^{3/2}. \tag{10}$$

The exact integration carried out in Chapter 16 gives

$$n = \mathrm{e}^{-(\mathcal{E}_c-\mathcal{E}_F)/kT}2(2\pi mkT/h^2)^{3/2} = N_c\mathrm{e}^{-(\mathcal{E}_c-\mathcal{E}_F)/kT}, \tag{11}$$

a result about one-third larger. Comparing this with expression (8) above for n_d and using approximation (2) of Section 10.1 for **f**, we see that the conduction band behaves like a group of states localized at energy $\mathcal{E}_c$, the number in the group being not N_d but instead

$$N_c = 2(2\pi mkT/h^2)^{3/2} = 4.82 \times 10^{15}T^{3/2}\ \mathrm{cm}^{-3}. \tag{12}$$

[A factor $(m_n/m)^{3/2}$ is to be introduced if the effective mass is not m.] N_c may be referred to as the *effective density of states in the conduction band*.

The expression for n is valid only when $\mathcal{E}_F$ is less than $\mathcal{E}_d$ by several times

kT. At room temperature, with m taken as the mass of the free electron, $N_c = 2.41 \times 10^{19}$ cm^3 or about 1/2000 the atomic density. N_c varies slowly with T compared to the exponential functions involved in the theory and can be regarded, for many purposes, as substantially constant.

Precisely similar calculations may be made for p_a and p giving

$$p_a = N_a \mathbf{f}_p[(\mathcal{E}_a - \mathcal{E}_F)/kT] = N_a \mathbf{f}[(\mathcal{E}_F - \mathcal{E}_a)/kT] \tag{13}$$

$$p = 2(2\pi mkT/h^2)^{3/2} \mathrm{e}^{-(\mathcal{E}_F - \mathcal{E}_v)/kT} = N_v \mathrm{e}^{-(\mathcal{E}_F - \mathcal{E}_v)/kT}. \tag{14}$$

N_v and N_c will differ if the electrons and holes have different effective masses.

The use of N_c and N_v is limited to cases for which $\mathbf{f}$ and $\mathbf{f}_p$ can be approximated by the simple exponential functions of equations (2) and (3) of Section 10.1. If the concentration of carriers is increased sufficiently, these approximations are no longer valid and other formulae must be employed. We shall illustrate this for the case of electrons.

(The reader may skip the next three paragraphs and Figure 10.5 on a first reading.)

If the number of electrons in the conduction band is very large, say one per atom as in the case of a monovalent metal like sodium or copper, then the Fermi level will lie very far above the bottom of the conduction band. For the case of sodium, which has been most extensively investigated, there is experimental and theoretical evidence that the energy is given by $P^2/2m$ up to the highest occupied levels so that the equation (9) of Section 9.1 for $N(\mathcal{E})$ may be used. Calculation then shows that the levels must be occupied up to a maximum energy $\mathcal{E}_m$ of about 3.16 electron volts above the bottom of the conduction band. This energy is so much greater than kT that the transition region of $\mathbf{f}$ is negligible in comparison, and the value of $\mathcal{E}_F$ may be calculated simply by solving equation (10) of Section 9.1 for $\mathcal{E}_F$:

$$n = \int_{\mathcal{E}_c}^{\mathcal{E}_F} N(\mathcal{E})d\mathcal{E} = (8\pi/3h^3)[2m(\mathcal{E}_F - \mathcal{E}_c)]^{3/2} \tag{15a}$$

$$\mathcal{E}_F - \mathcal{E}_c = \mathcal{E}_m = h^2(3n/8\pi)^{2/3}/2m = 21.6 \times 10^{-16} n^{2/3} \text{ electron volts} \tag{15b}$$

where $\mathcal{E}_F = \mathcal{E}_c + \mathcal{E}_m$, and m is taken as the free electron mass. Cases of this sort, for which the temperature no longer has an appreciable effect, are said to be *degenerate* and the distribution of electrons is called a *degenerate electron gas*.

For any given temperature, the statistics of the electrons in the band will vary from classical to degenerate as the number is increased. The dividing line is conventionally described by the degeneracy concentration n_{deg} or the degeneracy temperature T_{deg}. These two quantities are related by the equation,

$$T_{\text{deg}} = \left[\frac{3}{\pi}\right]^{2/3} \frac{h^2}{8km} n_{\text{deg}}^{2/3}$$
$$= 4.2 \times 10^{-11} n_{\text{deg}}^{2/3}. \tag{16}$$

This equation is interpreted as follows: At absolute zero the electrons are condensed into the states of lowest energy and require all the states up to a certain energy $\mathcal{E}_m$ given by (15). The degeneracy temperature defined by the equation satisfies the equation $kT_{\text{deg}} = \mathcal{E}_m$. In other words, if T is high enough to raise electrons from the bottom of the band to levels higher than $\mathcal{E}_m$, the degeneracy will be largely destroyed. Equation (16) serves

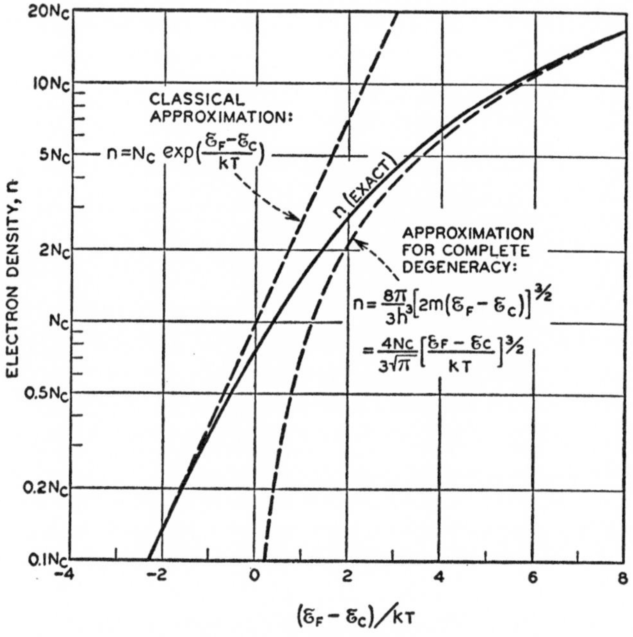

FIG. 10–5—Exact Dependence of n Upon $(\mathcal{E}_F - \mathcal{E}_c)$ and T Compared to Limiting Forms.

equally well to give the degeneracy temperature for a given concentration or the concentration for a given temperature.

When n is comparable to n_{deg}, there is no simple approximate formula for n in terms of T and $\mathcal{E}_F - \mathcal{E}_c$, and the value must be obtained by an exact integration[1] of $N(\mathcal{E})\mathbf{f}(\mathcal{E})d\mathcal{E}$. The result of the exact integration is shown in Figure 10.5. On this plot the limiting forms (11) and (15) are

[1] J. McDougall and E. C. Stoner, *Roy. Soc., London Trans.* **237A**, 67–104 (1938); Müller-Pouillet, *Lehrbuch der Physik* **4** (4), pp. 271 ff.; A. Sommerfeld, *Zeits. für Physik* **47**, 1–32 (1928); L. Nordheim, *Annalen der Physik* **9**, 607–678 (1931).

also shown. The point corresponding to $n = n_{\text{deg}}$ is given by $\mathcal{E}_m = kT$ and thus corresponds to the value of n for $\mathcal{E}_F - \mathcal{E}_c = kT$ on the completely degenerate curve, this value of n being about $0.75N_c$. The exact solution for $n = 0.75N_c$ corresponds to $\mathcal{E}_F - \mathcal{E}_c$ slightly negative so that even the states at the bottom of the band are less than half-filled and the distribution is not very degenerate. The distribution will be quite degenerate, however, for situations in which $\mathcal{E}_F - \mathcal{E}_c$ is $2kT$ or more. This corresponds to values of n about three times larger than the n_{deg} corresponding to a given T, or conversely to temperatures of $3^{-2/3} = 0.48$ as large as the T_{deg} corresponding to a given n.

We shall have occasion in Chapter 11 to discuss degenerate cases. For the purposes of illustrating the method of finding the equilibrium distribution in a semiconductor, however, it is simpler and equally instructive to consider a non-degenerate case. Accordingly, we return to the expressions for n, p, n_d, and p_a which give these quantities as simple functions of $\mathcal{E}_F$. These are substituted in the equation (3) for electrical neutrality, which then becomes an equation which may be solved for $\mathcal{E}_F$. As mentioned previously, graphical procedures for obtaining the solution are shown in Chapter 16. The description of the results obtained for a particular example is as follows:

We consider a case for a germanium sample which, like the silicon sample A of Figures 1.9 and 1.12, has $N_d > N_a$. In particular we have chosen $N_d = 10^{15}$ cm^{-3} and $N_a = 10^{14}$ cm^{-3}. The energy scale has been chosen with zero midway between the energy bands and with $E_c = 0.36$, $E_d = 0.32$, $E_a = -0.32$, and $E_v = -0.36$ electron volt. The dependence of $\mathcal{E}_F$ upon T for this case is shown in Figure 10.6, the calculations being described in Chapter 16. Qualitative descriptions of a number of the features of Figure 10.6 are given below.

At very low temperatures we have

$$n_d = N_d - N_a \tag{17}$$

since there will be no holes and all the extra electrons are in bound states on the donors. Hence, a fraction n_d/N_d of the donor levels will be filled. Consequently, $\mathcal{E}_F$ must be within a few times kT of $\mathcal{E}_d$; for otherwise, the states would be either all full or all empty in accordance with the behavior of **f** at low temperatures shown in Figure 10.3(a). Since $\mathcal{E}_F$ is near $\mathcal{E}_d$, the number of electrons in the conduction band will vary as $\exp[-(\mathcal{E}_c - \mathcal{E}_d)/kT]$ in accordance with equation (11). Thus, for this case, the activation energy obtained from a $\log n$ vs $1/T$ plot will be the binding energy $\mathcal{E}_c - \mathcal{E}_d$ of an electron in a donor state rather than half this value. The temperature range over which this approximation is valid extends only up to about 10°K as may be seen from the plot of $\mathcal{E}_F$ versus T given in Figure 10.6.

As the temperature is increased still further, an appreciable number of electrons are excited to the conduction band, and $\mathcal{E}_F$ tends to move down so as to keep n_t constant, by reducing the number of electrons in the donor levels. Over the temperature range for which there is substantially complete ionization of the donor levels, the equation

$$n \cong N_d - N_a \tag{18}$$

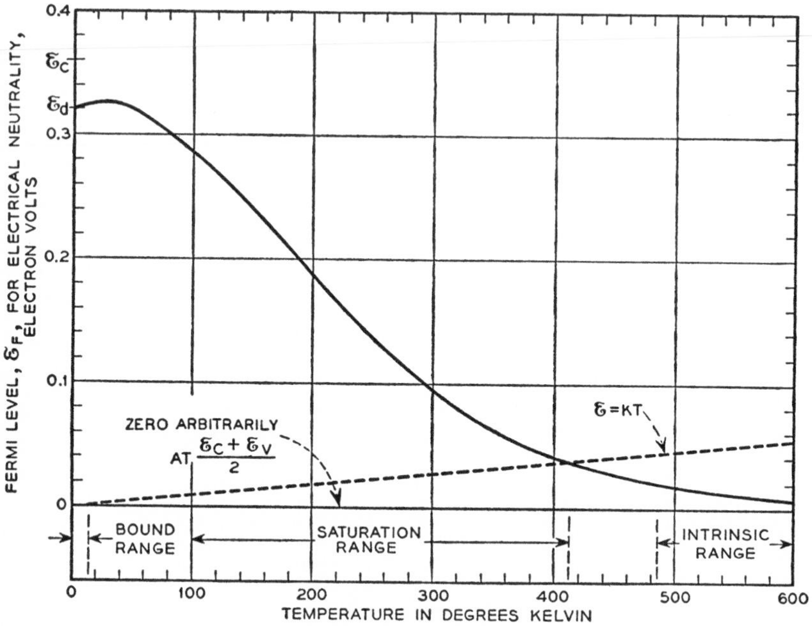

FIG. 10–6—The Fermi Level for Electrical Neutrality Plotted Against Temperature, for Germanium with $N_d = 10^{15}$ cm^{-3} and $N = 10^{14}$ cm^{-3} and $\mathcal{E}_c - \mathcal{E}_d = 0.04$ ev.

holds. This temperature range is referred to as the *saturation range* since all available excess electrons are in the conduction band. The saturation range is entered as soon as $\mathcal{E}_F$ has fallen several times kT below $\mathcal{E}_d$ so that the donor levels are substantially empty. On Figure 10.6, the value of kT in electron volts is shown as a function of temperature; it is seen that at about 100°K, $\mathcal{E}_d - \mathcal{E}_F > 2kT$ so that for temperatures above this the donors are fully ionized. In this range, increasing the temperature requires a decrease in $\mathcal{E}_F$ for the same reasons shown in Figure 10.3(b) and (c).

This decrease finally brings $\mathcal{E}_F$ to a position a few kT above the midpoint between the bands. As soon as this occurs, an appreciable number of holes is produced. At still higher temperatures, $\mathcal{E}_F$ approaches the midposition more closely, for otherwise either the holes or the electrons would greatly outnumber the other, and the condition, equation (3), of electrical neutrality could not be satisfied. As a consequence of this stabilization of $\mathcal{E}_F$, both n and p increase exponentially as $\exp[-(\mathcal{E}_c - \mathcal{E}_F)/kT]$ and $\exp[-(\mathcal{E}_F - \mathcal{E}_v)/kT]$; since $\mathcal{E}_F$ is midway between $\mathcal{E}_c$ and $\mathcal{E}_v$, these both reduce to $\exp[-(\mathcal{E}_c - \mathcal{E}_v)/2kT]$ so that, as mentioned in Chapter 1, the activation is half the forbidden energy gap. The intrinsic range has been selected for Figure 10.6 as the range for which $n/p < \mathbf{e}$, that is, it is the range for which $\mathcal{E}$ lies less than $\frac{1}{2}kT$ above the midpoint of the energy gap.

In the event that N_a is greater than N_d, there is always a net excess of holes. Because of the general symmetry between hole densities and electron densities, as shown in equations (3) of Section 10.1 and (7) to (14) of this section, an entirely similar behavior occurs, with $\mathcal{E}_F$ starting at $\mathcal{E}_a$ at low temperatures and rising to $(\mathcal{E}_c + \mathcal{E}_v)/2$ at high temperatures.

The description in this section thus puts in somewhat more formal terms the physical picture described in Chapter 1. In Chapter 16, additional details are given showing more fully the trend for $\mathcal{E}_F$ and the way in which the electron and hole densities vary.

This section may be summed up by saying that, once the distribution of quantum states in energy is known and the temperature specified, a definite procedure can be stated for finding the equilibrium distribution of electrons and holes. This procedure requires the Fermi level, $\mathcal{E}_F$, to be adjusted to give electrical neutrality [equation (3) of Section 10.3]. Certain simplifications can be introduced in the procedure, such as the use of N_c and N_v, the effective numbers of quantum states in the conduction band and valence-bond band; the general trend of the behavior can be seen to be that discussed in Chapter 1 in connection with Figures 1.9 to 1.12. However, there are few simplifying features, and only for the low temperature and intrinsic ranges can a simple interpretation be given of the slopes of the plots of the logarithm of n versus $1/T$.

The fact that there is complete statistical symmetry between holes and electrons, as shown by equation (4) of Section 10.1 and its consequences, should be noted. This symmetry, together with the results of Chapters 7 and 8, shows that the only significant feature distinguishing the behavior of holes from that of electrons in semiconductors is the sign of the effective charge.[2]

An important consequence of the form of the approximations (11) and (14) for n and p is that their product is a function of T only and is independent of n and p individually:

$$\begin{aligned} np &= N_cN_v \exp[(\mathcal{E}_c - \mathcal{E}_v)/kT] = N_cN_v \exp(-\mathcal{E}_G/kT) \\ &= 2.33 \times 10^{31} T^3 \exp(-\mathcal{E}_G/kT) \text{ cm}^{-6}. \end{aligned} \tag{19}$$

For intrinsic material, n and p are equal to each other and each is equal to

[2] The mathematical investigation of the equivalence takes major portions of Chapters 15 and 17.

the square root of equation (19). We shall use this constancy in the next section and shall discuss it further in Sections 12.4 and 16.4 and in connection with its use by Pearson and Bardeen in analyzing silicon alloys. Equation (19) can also be derived on the basis of the statistical theory of detailed balancing, which requires that the rate of recombination of holes and electrons (proportional to np) should equal their rate of generation, which should depend on T as does the right side of equation (19) but not on n or p.

10.4 APPLICATION AT ROOM TEMPERATURE

Because of the practical interest in high resistivity germanium at room temperature, we shall present a simplified analysis and describe the results. In Section 16.3 the same problem is analyzed by the general methods of Section 10.3 making use of illustrative values of 0.72 electron volt for $\mathcal{E}_c - \mathcal{E}_v$ and 0.04 electron volt for $\mathcal{E}_a - \mathcal{E}_v$ and $\mathcal{E}_c - \mathcal{E}_d$. We shall use here a somewhat more direct procedure which utilizes different experimental data. If the intrinsic line for germanium is extrapolated to room temperature it yields a value of 48 ohm-cm for the resistivity.[1] On the other hand the resistance of a relatively pure n-type sample may be 2.4 ohm-cm. (Using 2600 cm^2/volt sec for the mobility for electrons in germanium, this gives a concentration $n = \sigma_L/e_L\mu_L = 1/2.4 \times 1.6 \times 10^{-19} \times 2600 = 1.0 \times 10^{15}$ electrons/cm^3 in agreement with the example of Chapter 16.) The ratio of mobilities for germanium is estimated[2] to be $b = 1.5$. From this we can determine directly the ratio of n' (single prime for the 2.4-ohm sample) to n'' (double prime for the intrinsic sample) as follows: For the intrinsic sample $p'' = n''$, almost by definition. Therefore, its conductivity is

$$\sigma'' = e(n''\mu_n + p''\mu_p) = e\mu_p(1 + b)n''. \tag{1}$$

For the 2.4-ohm sample, p' is negligible (as we shall verify below) and

$$\sigma' = e\mu_n n' = e\mu_p b n'. \tag{2}$$

Dividing the second by the first gives

$$\frac{\sigma'}{\sigma''} = \frac{48}{2.4} = 20 = \frac{bn'}{(1 + b)n''} \tag{3}$$

whence

$$\frac{n'}{n''} = \left(\frac{1 + b}{b}\right) 20 = 33 = \mathrm{e}^{3.5}. \tag{4}$$

[1] Since this section was written, better germanium specimens have become available and a new estimate of 60 ohm-cm is considered best. The values of μ_n and μ_p have also been revised as a result of Haynes' experiments. It has not been practical to revise all examples in the text accordingly, and this section, together with Chapter 16, is now somewhat inaccurate. They are internally consistent, however, and illustrate the principles involved.

[2] See Section 12.9 for a review of the best available data.

From equation (11) of Section 10.3, which states that

$$n = N_c \exp\left[-(\mathcal{E}_c - \mathcal{E}_F)/kT\right], \tag{5}$$

we see at once that $\mathcal{E}_F$ must be $3.5kT$ higher for the 2.4-ohm-cm sample. Since, for the intrinsic sample, $\mathcal{E}_F$ is midway between the bands, $\mathcal{E}_F$ for the 2.4-ohm sample must be $3.5kT$ or 0.09 electron volt above the midpoint. This situation is summarized in Figure 10.7 which also shows a possible impurity distribution. The same situation (except for small differences in rounding off numbers) is represented more analytically in Chapter 16, Figure 16.2.

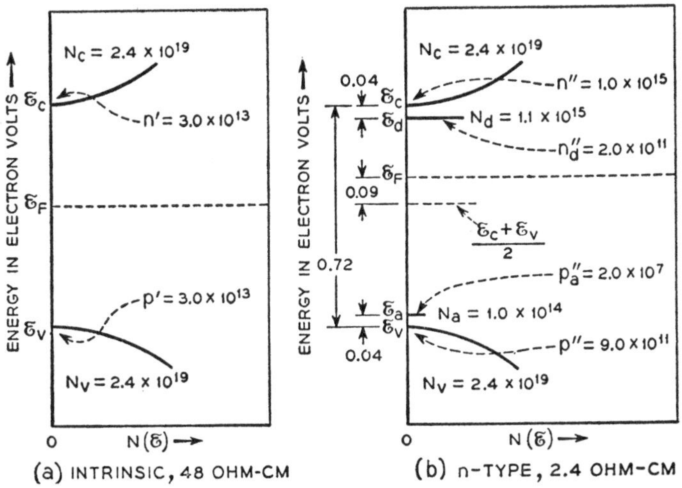

FIG. 10-7—Theoretical Distributions of Quantum States in Energy, Effective Densities, and Densities of Electrons and Holes for Germanium. (The values shown are illustrative and not exact.)

It should be noted that $n'p' = n''p''$ in Figure 10.7. This is a direct consequence of equation (19) of the last section.

A plot for 2.4-ohm-cm silicon would look much the same as Figure 10.7, and the Fermi level would come at about the same distance (about 0.27 electron volt) from the conduction band. However, since the energy gap for silicon is 1.12 volts, the Fermi level would lie 0.29 electron volt above the midpoint, and the number of holes would be very small.

The changes from n-type to p-type are obvious.

At room temperature, diamond is an insulator. Its conductivity is, therefore, 10^{-13} ohm^{-1} cm^{-1} or less. The mobility of electrons and holes is probably about one tenth as much for diamond as for germanium, hence the density of carriers will be only about 10^3 cm^{-3}. Taking N_c as 10^{19}, this means that $\mathbf{f} = 10^{-17}$ and, consequently, $\mathcal{E}_F$ is not within 1 electron volt of either the conduction or valence-bond band. This energy difference of 1 electron volt is larger than that predicted for the binding energy of donor levels (0.35 electron volt) and suggests that there are levels of other sorts lying at least 1 electron volt away from the edges of the energy gap. Further evidence for such levels is furnished by data on the ultraviolet absorption of diamond which varies from sample to sample from around 7 to 4 electron volts, indicating levels lying several volts from the edges of the energy gap. Further evidence for such deep lying levels has been obtained from studies of the release of trapped electrons and holes in experiments on electron bombardment induced conductivity.[3]

Applications of the statistical methods discussed here to cases in which the distribution of holes and electrons vary from one part of the semiconductor to another have been carried out in various connections. Some of the basic equations required are discussed in Section 12.4. Some of the applications of particular interest apply to metal semiconductor contacts and are presented in the references quoted in Section 4.3. The distribution of potential and its dependence on temperature for a semiconductor with surface states has been treated[4] and similar calculations have been carried out in some detail for *p-n* junctions.[5] An excellent review of the problem with emphasis on metal-semiconductor contacts is contained in a paper by Slater.[6]

Problems

1. The equilibrium condition for temperatures below the intrinsic range may be regarded as resulting from the "pseudo-chemical" reaction

$$\text{excess electron} + \text{ionized donor} \rightleftarrows \text{neutral donor}.$$

The three concentrations involved are respectively n, $N_d - n_d$, and n_d. The mass action law would then give

$$\frac{n(N_d - n_d)}{n_d} = K.$$

Derive this equation employing the approximation (10) of Section 10.3 and show that

$$K = N_c \exp[(\mathcal{E}_d - \mathcal{E}_c)/kT].$$

(This result corresponds to equations (22) and (23) for G. L. Pearson and J. Bardeen, *Phys. Rev.*, 865–883 (1949), except that our value of K is twice theirs. The difference arises from the fact that an electron trapped on a donor may have either plus or minus spin and, therefore, the correct expression for n_d is $N_d\mathbf{f}[(\mathcal{E}_d - kT\ln 2 - \mathcal{E}_F)/kT]$; see Problem 2, Chapter 16.)

[3] K. G. McKay, *Phys. Rev.* **77**, 816–825 (1950).

[4] J. J. Markham and P. H. Miller, *Phys. Rev.* **75**, 959–967 (1949).

[5] W. Shockley, *Bell Syst. Tech. J.* **28**, 435–489 (1949), references.

[6] J. C. Slater, *Phys. Rev.* **76**, 1592–1601 (1949), references.

If the mass action law is combined with the condition for electrical neutrality, show that

$$\frac{n(n + N_a)}{(N_d - N_a - n)} = K$$

and that

$$(n + K)(n + N_a) = N_d K.$$

This quadratic may be solved for n so that n may be plotted as a function of T. This formula is used by Pearson and Bardeen to fit Figure 1.12; they use $N_d = 12 \times 10^{16}$, $N_a = 1.5 \times 10^{16}$, $\mathcal{E}_c - \mathcal{E}_d = 0.045$ electron volt, $m_n = 0.33$ free electron mass and the factor of $(\frac{1}{2})$, mentioned above, in K. Calculate some values of n from this formula and compare with Figure 1.12.

Refer to Chapter 16, equations (10) and (11) of Section 16.4, and compute the values of p shown in Figure 1.12.

2. Obtain the formula for N_c or N_v by integrating **f** directly over $\boldsymbol{P}$-space. See Section 16.2 for a hint. What is the significance of the $\frac{3}{2}$ power in equation (12) of Section 10.3?

3. Two groups of quantum states N_1 and N_2 in number at energies $\mathcal{E}_1$ and $\mathcal{E}_2$ are occupied by n_1 and n_2 electrons. Suppose electrons in N_1 can make transitions to any vacant state in N_2 with transition probability T_{12} per unit time. The rate of transitions from N_1 to N_2 is then

$$n_1 T_{12}(N_2 - n_2).$$

From general statistical theory, the transition probability T_{21} from 2 to 1 is

$$T_{21} = T_{12}\mathbf{e}^{(\mathcal{E}_2 - \mathcal{E}_1)/kT}.$$

Show that if n_1 and n_2 are given by the Fermi-Dirac distribution, then the rate of transition from 1 to 2 is just balanced by that from 2 to 1:

$$n_2 T_{21}(N_1 - n_1).$$

Show that if a third group of quantum states, described by N_3, $\mathcal{E}_3$, and n_3, is in equilibrium with $\mathcal{E}_1$, it must also be occupied in accordance with the same Fermi-Dirac distribution.

CHAPTER 11

MATHEMATICAL THEORY OF CONDUCTIVITY AND HALL EFFECT

11.1 INTRODUCTION

In this chapter, the basic formulae for mobility and Hall effect are derived. The procedure is divided into two parts. In the first part it is shown that any disturbance of the distribution of electrons from the equilibrium value tends to decay, owing to the "scattering" of electrons, and the effect of this decay upon the current which may be present is evaluated. In the second part, expressions are derived for the rate at which current builds up due to the application of electric and magnetic fields. Since only a linear theory is desired, the rate of build-up due to an electric field can be calculated from the undistorted or equilibrium distribution, the effect of the electric field upon the distorted part of distribution being quadratic in the electric field. For the effect of a magnetic field, however, it is necessary to consider disturbances produced by the electric field in order to get any effects. The steady-state solution is then found by equating the rate of build-up of current to the rate of decay. The problem is more complicated than that of Chapter 8 because the rate of decay is not the same for all classes of electrons, and suitable averaging processes must be carried out.

In Sections 11.2 and 11.3 the decay process is considered. Section 11.2 is concerned with a general formulation of the decay process and with the derivation of formulae useful for application to specific cases. In this treatment it is shown that the rate of decay is not affected by the fact that the Pauli exclusion principle prevents electrons from making transitions to already occupied states. In this section, reference is made to parts of Chapter 17 in which the quantum-mechanical theory of transition probabilities is given in detail.

In Section 11.3 we treat the two most important transition-producing mechanisms: scattering by charged ions, similar to the Rutherford scattering of α-particles by atoms which first gave evidence of the nuclear structure of atoms, and scattering by thermal vibrations of the atoms. The first mechanism is treated on the customary basis of an approximate model. The thermal scattering, which is treated in detail in Chapter 17, is given a brief and simplified treatment in Section 11.3 which serves to illustrate the principles involved.

In Section 11.4, we treat the effect of the combined interaction of the applied fields and the decay process. The effect of an electric field upon the equilibrium distribution is evaluated and combined with the laws of decay derived in Sections 11.2 and 11.3, and the steady-state distribution is determined. The effect of a magnetic field upon the steady state due to the electric field is next considered, and a modified steady-state distribution is determined. From these steady-state distributions, general expressions for mobility, conductivity, and Hall effect are found. These general expressions are worked out in detail for a number of cases of particular importance in semiconductors.

The formulae for intrinsic semiconductors, in which holes and electrons carry comparable portions of the current, are given in Chapter 8 in terms of the mobilities of holes and electrons. These formulae are repeated in the notation of this chapter in Section 11.4.

In Section 11.5 a comparison between theory and experiment is presented.

11.2 THE RELAXATION TIME

In Chapter 8 we gave a simplified treatment of the scattering process based on two assumptions:

(1) The probability that an electron (or hole) makes a transition in any small interval dt of time dt/τ where τ is a constant.

(2) The probability of a transition to any particular end state is independent of the initial state and is directly proportional to the probability that the end state would normally be occupied in the thermal-equilibrium distribution.

In Chapter 17, an analytical examination of these assumptions is carried out on the basis of Schroedinger's equation. It is found that, for cubic semiconductors such as silicon and germanium and the customary assumption of spherical energy surfaces, the probability of transition depends neither on the direction of motion nor the time after the last transition but, in general, does depend on the speed of motion. Thus τ in assumption (1) must be replaced by $\tau(v)$ where $v = |\mathbf{v}|$ is the speed, or by $\tau(\mathcal{E})$ where $\mathcal{E}$ is the energy.

It is also found that assumption (2) is in error and that the electron can gain or lose only a small fraction of its energy in a given collision. These gains and losses are important in converting to heat the electrical power delivered to the electrons in the form of I^2R or V^2/R. However, these effects are quadratic in the field and can be disregarded in developing the linear theory of mobility and Hall effect. The reason that the change in energy is small is discussed in detail in Section 17.6 in connection with equation (31). In brief, the electron can gain or lose in one transition an amount of energy $h\nu$ where ν is the frequency of the atomic vibration with which it interacts. Such a quantum of energy is called a phonon. The vibrations with which the electron interacts are those with wave lengths comparable to the electron's wave length. This condition leads to the relationship

$$mv = h/\lambda \text{ for the electron,} \tag{1}$$

$$\nu = c/\lambda \quad \text{or} \quad h\nu = ch/\lambda = mvc \text{ for the phonon,} \tag{2}$$

where c is the speed of sound in the crystal. The ratio of phonon energy to electron energy is thus

$$\frac{h\nu}{mv^2/2} = 2c/v = (10^6 \text{ cm/sec})/(10^7 \text{ cm/sec}) = 10^{-1} \tag{3}$$

for typical values at room temperature. Thus in one collision the electron may change its energy by about 10 per cent. This change is relatively small; furthermore, there will be balancing effects, and electrons falling to lower energies will be replaced by other electrons excited to higher energies. For these reasons negligible errors will be introduced by making the customary approximation of conductivity theory that energy is conserved in the transitions. This result is in disagreement with assumption (2) which supposes that the energy after transition is independent of the energy before transition.

Assumption (2) also fails in regard to the distribution of directions after collision. As we shall show in Section 11.3a, for scattering by ionized donors and acceptors there may be a high degree of correlation between the direction of motion before and after collision. We shall derive in this section a general formula which takes into account such correlations and will apply it to impurity scattering in later sections.

Figure 11.1 represents the case with which we shall be chiefly concerned. It represents in momentum space two equal energy surfaces separated by a small energy difference $\delta\mathcal{E}$. An element of volume in the shell between the two surfaces may be described by the corresponding surface area $\delta\Omega$, which has the dimensions of $\boldsymbol{P}^2$ or (momentum)^2. (We shall use the symbol δ for these infinitesimal areas in this section so as to reserve d for changes occurring during a time interval dt.) An electron in a state $\boldsymbol{P}_i$ may make transitions to other states of almost exactly equal energy. As a result of these transitions, the electrons (more exactly, the states occupied by electrons) become uniformly distributed in the shell so that the probability that any given state be occupied by an electron becomes the same for all states and acquires the value $\mathbf{f}$ given by the Fermi-Dirac distribution function of Chapter 10, evaluated for the energy of the shell considered.

The number of states in the range $\delta\mathcal{E}\delta\Omega_i$ is obtained as follows: The volume in momentum space of the element is $d\Omega$ times the shell thickness,

$\delta P_\perp$. Evidently,

$$\delta\mathcal{E} = \left|\nabla_P \mathcal{E}(P)\right|_i \delta P_\perp = v_i \delta P_\perp, \tag{4}$$

the last equality following from the group-velocity formula. Hence the volume of the element is $\delta P_\perp \delta\Omega_i = \delta\mathcal{E}\delta\Omega_i/v_i$. The density of states in

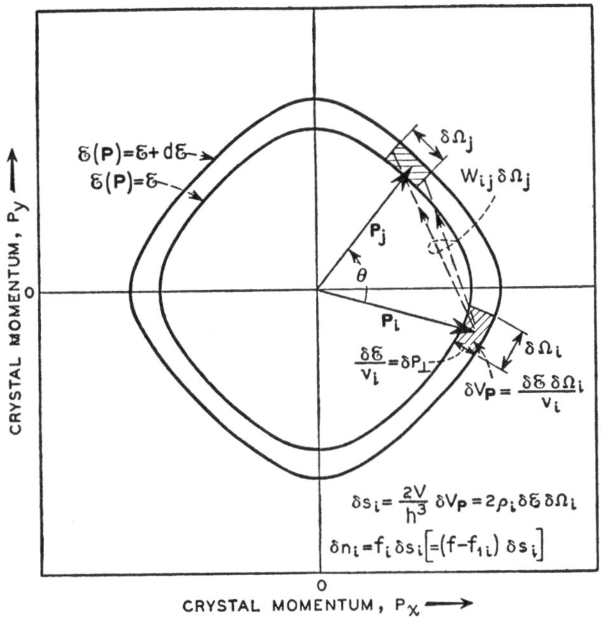

FIG. 11-1—Notation Used for Describing Transitions from States in Range $\delta\mathcal{E}\delta\Omega_i$ to States in Range $\delta\mathcal{E}\delta\Omega_i$. The Change in Direction, θ, Is Useful for Spherical Energy Surfaces.

$\boldsymbol{P}$-space is $2V/h^3$ where V is the volume of the crystal, as for equation (12) of Section 5.5. Hence the number of states δs_i in the element is

$$\delta s_i = (2V/h^3)\delta\mathcal{E}\delta\Omega_i/v_i = 2\rho_i \delta\mathcal{E}\delta\Omega_i \tag{5}$$

where

$$\rho_i \equiv V/h^3 v_i \tag{6}$$

is the density of states of each spin per unit energy per unit surface area.

For the case in which the band is nearly empty, so that the effect of the Pauli exclusion principle may be neglected, the transitions may be described in terms of a probability parameter W_{ij}. This parameter is used to specify the total probability of transition from a state i to all of the states j lying in a region $d\Omega_j$ of the energy shell. In these transitions there may be energy losses or gains. For reasons discussed above we neglect these changes in energy and assume that all the states to which state $\boldsymbol{P}_i$ can make transitions lie in the shell $\delta\mathcal{E}$ so that we may speak of transition probability per unit area of the shell. Accordingly the probability that an electron makes a transition from a state near $\boldsymbol{P}_i$ to a state in range $\delta\Omega_j$ near $\boldsymbol{P}_j$ in time dt is

$$W_{ij}\delta\Omega_j dt. \tag{7}$$

The total number of transitions of this sort is obtained simply by multiplying this probability by the number of electrons in the element $d\Omega_i$; and this number is simply ds_i times the probability $\mathbf{f}_i$ that these states are occupied. Hence the total rate of transition from $d\mathcal{E}d\Omega_i$ to $d\mathcal{E}d\Omega_j$ is

$$(\delta\mathcal{E}\delta\Omega_i \rightarrow \delta\mathcal{E}\delta\Omega_j) = \mathbf{f}_i\delta s_i W_{ij}\delta\Omega_j \tag{8}$$

$$= \mathbf{f}_i 2\rho_i W_{ij}\delta\mathcal{E}\delta\Omega_i\delta\Omega_j. \tag{9}$$

Similarly the transitions from j to i will be given by

$$(\delta\mathcal{E}\delta\Omega_j \rightarrow \delta\mathcal{E}\delta\Omega_i) = \mathbf{f}_j 2\rho_j W_{ji}\delta\mathcal{E}\delta\Omega_i\delta\Omega_j. \tag{10}$$

These expressions are found to be consistent with the *principle of detailed balancing*[1] which states that each process and its reverse occur with equal frequency under equilibrium conditions. For thermal equilibrium $\mathbf{f}_i = \mathbf{f}_j = \mathbf{f}(\mathcal{E})$, the Fermi-Dirac distribution function for the energy $\mathcal{E}$ of the shell. Furthermore, as discussed in Section 17.2, the quantum-mechanical formulae for W_{ij} and W_{ji} are such that

$$\rho_i W_{ij} = \rho_j W_{ji}; \tag{11}$$

hence the two rates previously evaluated are equal. We shall make use of this result later in this section in order to determine the effect of the Pauli principle upon the relaxation process when a large fraction of the states are occupied.

The total probability of transition from the state $\boldsymbol{P}_i$ is simply the sum of all the transition probabilities to all areas $d\Omega_j$. We shall write it as W_i where

$$W_i = \int_{\Omega(\mathcal{E})} W_{ij}d\Omega_j = 1/\tau_c. \tag{12}$$

This probability is essentially the reciprocal of the mean free time between collisions, denoted by τ_c since, by the arguments of Chapter 8, the probability of collision should be dt/τ_c and this is evidently equal to $W_i dt$. Since assumption (2) is invalid, however, τ_c cannot be related to the rate of decay of a current, and a new formula must be derived for the relaxation time.

We shall next consider the law of decay for any current produced by a

[1] P. W. Bridgman, *Phys. Rev.*, **31**, 101 (1928).

disturbance of the equilibrium distribution. We shall deal first with the case of principal interest, in which the energy surfaces are taken as spheres and all directions of motion as equivalent. We shall later mention briefly another case in which τ_c depends on direction. For the spherical case, it is evident from symmetry that for the equilibrium condition there is no net current. We shall next suppose that we are dealing with a perturbed situation. Let us then consider the current due to the group of electrons δn_i in an element of volume $d\mathcal{E}d\Omega_i$. These will produce a current density

$$\delta \boldsymbol{I} = (-e\boldsymbol{v}_i/V)\delta n_i. \tag{13}$$

In time dt, a fraction $W_{ij}d\Omega_j dt$ will be scattered into a new direction with velocity $\boldsymbol{v}_j$ so that the change in current will be

$$d\delta\boldsymbol{I} = dt \int_{\Omega(\mathcal{E})} [-e(\boldsymbol{v}_j - \boldsymbol{v}_i)/V]\delta n_i W_{ij} d\Omega_j \tag{14}$$

where $\Omega(\mathcal{E})$ is the energy surface. For the cases of particular interest, the transitions are of such a nature that W_{ij} is a function $W(\theta)$ only of θ, the angle between initial and final velocities. Under these conditions, we may choose a set of spherical coordinates with $\theta = 0$ parallel to $\boldsymbol{v}_i$. In terms of this coordinate system, we may write

$$\boldsymbol{v}_i = \boldsymbol{i}_z v_i \tag{15a}$$

$$\boldsymbol{v}_j = (\boldsymbol{i}_z \cos\theta + \boldsymbol{i}_x \sin\theta\cos\varphi + \boldsymbol{i}_y \sin\theta\sin\varphi)v_i \tag{15b}$$

$$d\Omega_j = (m_n v_i)^2 \sin\theta\, d\theta d\varphi \tag{15c}$$

where $\boldsymbol{i}_x$, $\boldsymbol{i}_y$, $\boldsymbol{i}_z$ are mutually orthogonal unit vectors. With the aid of these we find

$$d\delta\boldsymbol{I} = dt\delta n_i(-e/V)\boldsymbol{i}_z v_i \int_0^\pi (\cos\theta - 1)W(\theta)(m_n v_i)^2 \sin\theta d\theta 2\pi \tag{16}$$

since the components along $\boldsymbol{i}_x$ and $\boldsymbol{i}_y$ integrate to zero. This expression may be rewritten as

$$\begin{aligned} d\delta\boldsymbol{I} &= -dt\delta\boldsymbol{I}\int (1 - \cos\theta_j)W(\theta_j)d\Omega_j \\ &= -dt\delta\boldsymbol{I}W_0\langle 1 - \cos\theta\rangle = -dt\delta\boldsymbol{I}/\tau \end{aligned} \tag{17}$$

where

$$W_0 = \int W(\theta)d\Omega_j = 1/\tau_c \tag{18}$$

and

$$\langle 1 - \cos\theta\rangle = (1/W_0)\int (1 - \cos\theta)W(\theta)d\Omega. \tag{19}$$

These equations lead, as we shall shortly show, to the result that the current $\delta\boldsymbol{I}$ decays with a "relaxation time" equal to $\tau = \tau_c/\langle 1 - \cos\theta\rangle$ where

$\langle 1 - \cos\theta \rangle$ is the average value of $1 - \cos\theta$ for all the collisions. It may be helpful to note that, if collisions were equally likely to all areas of the sphere, then $\langle \cos\theta \rangle$ would be zero so that τ_c and τ would be equal. On the other hand, if all transitions involved only small changes in direction, $\langle \cos\theta \rangle$ would be close to unity and relaxation time τ would be much larger than τ_c.

In order to relate these equations to the relaxation of the current, we shall consider the current due to an arbitrary distribution $\mathbf{f}_i$ over the energy shell, supposing still that $\mathbf{f}_i \ll 1$. Each element of area $\delta\Omega_i$ will then produce a current $\delta \boldsymbol{I}_i$, and in time dt the electrons in these elements will suffer transitions such that

$$d\delta \boldsymbol{I}_i = -\delta \boldsymbol{I}_i dt/\tau. \tag{20}$$

The change in the total current $\boldsymbol{I} = \sum \delta \boldsymbol{I}_i$ will then change by

$$d\boldsymbol{I} = \sum d\delta \boldsymbol{I}_i = -(\sum \delta \boldsymbol{I}_i) dt/\tau = -\boldsymbol{I} dt/\tau, \tag{21}$$

or

$$\dot{\boldsymbol{I}} = -\boldsymbol{I}/\tau. \tag{22}$$

This shows that the total current decays according to a simple relaxation equation for which the solution is

$$\boldsymbol{I} = \boldsymbol{I}_0 e^{-t/\tau}. \tag{23}$$

Thus τ is the relaxation time for the decay of the current and is, as we shall see, the correct quantity to use in calculating the mobility. We shall refer to it as the *relaxation time* or the *mean free time*.

For the case of thermal vibrations, the transition probability is independent of θ and $\langle \cos\theta \rangle = 0$ so that $\tau = \tau_c$. For impurity scattering, $\langle \cos\theta \rangle$ approaches unity for large energies so that $\tau \gg \tau_c$; for this case τ_c enters the theory only as an intermediate step.

Another model, which we mention for completeness, has been considered by various writers. For this model, τ_c is a function of $\boldsymbol{P}_i$ and $\langle \cos\theta \rangle = 0$, for spherical energy surfaces. For more complicated surfaces, it is supposed that scattering is of such a nature that after one collision an electron is equally likely to be scattered into any state in the energy shell. The rate of decay of current in this case is given by the formula

$$\dot{\boldsymbol{I}} = -\sum \delta \boldsymbol{I}_i/\tau_i, \tag{24}$$

summed over all the elements of the energy shell, where τ_i is the τ_c appropriate to the state near $\boldsymbol{P}_i$.

The Effect of the Pauli Exclusion Principle. If a large fraction of the states in the energy shell are occupied, some electron transitions will be prevented because the end states will already be occupied. This prevention may be taken account of very simply, and the surprising result is obtained that *the relaxation process is not affected by having an appreciable proportion of the states occupied.*

The simplest way of seeing this result is as follows: If a transition from a state $\boldsymbol{P}_i$ to a state $\boldsymbol{P}_j$ is prevented because $\boldsymbol{P}_j$ is occupied, then at the same time a transition from $\boldsymbol{P}_j$ to $\boldsymbol{P}_i$ is also prevented. Suppose we disregard the restrictions imposed by the Pauli exclusion principle and permit both transitions to occur; then on the average the excluded collisions will cancel out in pairs and the relaxation process will proceed at the same rate as it would if the exclusion principle were enforced. Thus the relaxation of the current will proceed at the same rate whether the Pauli exclusion principle is enforced or not and, consequently, the relaxation time for a given energy shell will be unaffected by the number of electrons occupying states in the shell. We shall verify this conclusion analytically below.

We consider once more the transitions from $\delta\mathfrak{E}\delta\Omega_i$ to $\delta\mathfrak{E}\delta\Omega_j$. Since a fraction $\mathbf{f}_j$ of the states in $\delta\mathfrak{E}\delta\Omega_j$ are occupied, the number of possible transitions is reduced by a factor $(1 - \mathbf{f}_j)$ and we obtain

$$(d\mathfrak{E}d\Omega_i \rightarrow d\mathfrak{E}d\Omega_j) = \mathbf{f}_i(1 - \mathbf{f}_j)2\rho_i W_{ij} d\mathfrak{E}d\Omega_i d\Omega_j. \tag{25}$$

(In Section 17.7 we examine this problem from the point of view of antisymmetrical electron wave functions and conclude that the semi-intuitive introduction of the factor $(1 - \mathbf{f}_j)$ can be justified by a rigorous procedure.) The rate of transition from $\delta\mathfrak{E}\delta\Omega_j$ to $\delta\mathfrak{E}\delta\Omega_i$ will be given by a similar expression with $\mathbf{f}_j(1 - \mathbf{f}_i)$. As pointed out above, the quantum-mechanical treatment leads to *the principle of detailed balancing* by having

$$\rho_i W_{ij} = \rho_j W_{ji} \tag{26}$$

so that the net rate of transition from $\delta\mathfrak{E}\delta\Omega_i$ to $\delta\mathfrak{E}\delta\Omega_j$ is

$$\begin{aligned}(\delta\mathfrak{E}\delta\Omega_i \leftrightarrows \delta\mathfrak{E}\delta\Omega_j) &= [\mathbf{f}_i(1 - \mathbf{f}_j) - \mathbf{f}_j(1 - \mathbf{f}_i)]2\rho_i W_{ij}\delta\mathfrak{E}\delta\Omega_i\delta\Omega_j \\ &= (\mathbf{f}_i - \mathbf{f}_j)2\rho_i W_{ij}\delta\mathfrak{E}\delta\Omega_i\delta\Omega_j. \end{aligned} \tag{27}$$

Thus the net rate depends only on $\mathbf{f}_i - \mathbf{f}_j$ and not on $\mathbf{f}_i$ and $\mathbf{f}_j$ individually. From this it is evident that, if the equilibrium represented by $\mathbf{f}$ is disturbed so that $\mathbf{f}_i = \mathbf{f} + \mathbf{f}_{1i}$, then the rate of transitions will depend only on $\mathbf{f}_{1i}$. Hence the rate of decay of a disturbance is unaffected by the degree to which the states are occupied and, consequently, the relaxation time is unaltered.

In a semiconductor, in which the number of electrons varies with varying impurity content, the change in number of electrons will be accompanied by changes in the scattering mechanism. For such conditions, of course, the relaxation time will depend upon the impurity concentration and it will be impractical to change the electron concentration without changing the relaxation time. This does not invalidate the result obtained previously in the form in which we wish to employ it. For this purpose we need the result that, for a given set of values of W_{ij}, the relaxation time does not depend on **f**.

Similarity of Holes and Electrons. The similarity of the behavior of holes and electrons follows at once from the result that the rate of electron transitions between $\delta\mathfrak{S}\delta\Omega_i$ and $\delta\mathfrak{S}\delta\Omega_j$ depends on $\mathbf{f}_i - \mathbf{f}_j$. Evidently each electron transition may be described equally well as a hole transition in the opposite direction. Furthermore, the fraction of the states occupied by holes are $1 - \mathbf{f}_i = \mathbf{f}_{pi}$ and $1 - \mathbf{f}_j = \mathbf{f}_{pj}$. Consequently, the rate of hole transitions is

$$\begin{aligned}(\delta\mathfrak{S}\delta\Omega_i \leftrightarrows \delta\mathfrak{S}\delta\Omega_j)_p &= -(\delta\mathfrak{S}\delta\Omega_i \leftrightarrows \delta\mathfrak{S}\delta\Omega_j)\\ &= -(\mathbf{f}_i - \mathbf{f}_j)2\rho_i W_{ij}\delta\mathfrak{S}\delta\Omega_i\delta\Omega_j\\ &= (\mathbf{f}_{pi} - \mathbf{f}_{pj})2\rho_i W_{ij}\delta\mathfrak{S}\delta\Omega_i\delta\Omega_j. \qquad (28)\end{aligned}$$

This last form is formally identical, except for the subscript p, to the expression for electrons. Thus we conclude that the relaxation of a given energy shell is the same whether it is occupied by only a few electrons, or only a few holes, or is partially filled.

In Section 17.7, we deal with the case of a few holes in more detail and evaluate the transition probabilities directly. The conclusion is the same as that presented in this section, that the holes may be dealt with formally on the same basis as electrons. It is, furthermore, shown that their transition probabilities will be of the same order of magnitude.

11.3 TRANSITIONS DUE TO THERMAL AGITATION AND DUE TO SCATTERING BY IMPURITIES

In this section we shall treat the transitions produced by thermal vibrations and by ionized impurities in the lattice.[1] We shall treat the latter case first as it is conceptually somewhat simpler. For it we shall use a classical model and compute the scattering in terms of the electrostatic deflection of an electron as it moves by a charged ion. The problem of transitions induced by thermal vibrations of the atoms is essentially a wave-interference phenomenon and is treated as such from the outset. For both cases, expressions for $\boldsymbol{\tau}(v)$ are obtained for use in the next section in formulae for mobility and Hall effect.

11.3a. Scattering by Ions; the Conwell-Weisskopf Formula. The problem of scattering by charged donors or acceptors or other impurity ions may be dealt with by purely classical methods, the holes and electrons being treated as wave-packets. This procedure permits us to calculate the scattering on the basis of the deflections of charged particles moving in

[1] For a discussion in somewhat simpler terms with more application to metals, see V. Weisskopf, *American Journal of Physics* **11**, 1–12 (1943).

field of charged ions. It is necessary, however, to introduce simplifying assumptions in order to obtain the results in analytic form. The results obtained classically are a good approximation to the quantum-mechanical ones and, for the case of scattering by an isolated ion, the scattering formulae are identical.

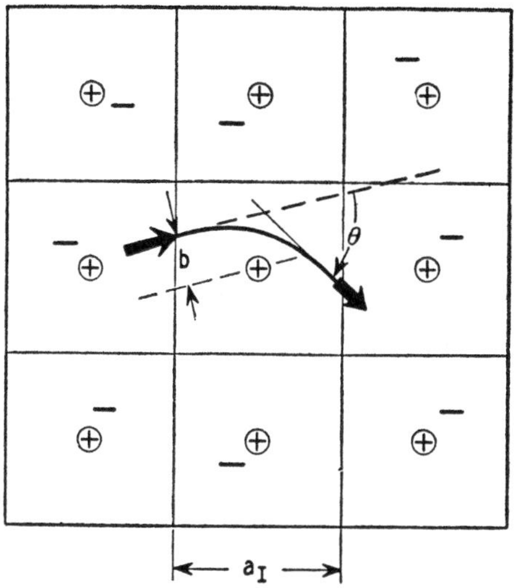

FIG. 11–2—Simplified Representation of the Scattering of an Electron by a Donor Ion.

We shall discuss first the model shown in Figure 11.2, which represents an n-type semiconductor with a density N_d of fully ionized donors. The deflection of an electron passing by one of the donors is shown. The model has been simplified by placing the donors on a simple cubic lattice with lattice constant a_I. Evidently

$$N_d = 1/a_I^3. \tag{1}$$

Furthermore, the other electrons have been uniformly spread about around all the other donors so that they are effectively shielded and the electron sees only the field of one donor. We can then consider that its deflection arises solely from the single ion in the center of the cell. Actually, the electron will not be perfectly shielded from the other ions and will be deflected to some extent by forces coming from other cells. For purposes of calculation we shall define the passage through each cell as a collision and will, furthermore, take the average time between collisions as $\tau_c = a_I/v$. Even so the problem is still too difficult and we shall make the further approximation that the deflection θ produced by passing through one cell is the same as it would be if the electron traversed an infinite hyperbolic path in the presence of the same charge. This last approximation leads us to overestimate the deflection due to the central ion and tends to compensate for the neglect of the effect of the other ions.

Before the analytical consequences of the model are derived, it may be worth while to estimate the order of magnitude of the effects for a typical case: n-type germanium, with a density $N_d = 10^{15}$ cm^{-3}; the value of a_I is 10^{-5} cm or 10^3 A (angstroms), and $\kappa = 19$.[1a] The potential energy at a distance $a_I/2$ is

$$\mathcal{E}_I = \frac{e^2}{\kappa(a_I/2)} = e\,\frac{4.80 \times 10^{-10}}{19(a_I/2)_A 10^{-8}} \text{ electrostatic volts}$$
$$= \frac{0.76}{(a_I/2)_A} \text{ electron volts} = \frac{30}{(a_I/2)_A} kT \text{ at } 300°\text{K}$$
$$= 0.06kT \text{ at } 300°\text{K}, \tag{2}$$

the intermediate expressions being included so that energies in electron volts and units of kT may be easily estimated for distances in angstroms. This calculation shows that, when the electron is at the center of the face of one of the cubes, its interaction with the ion is only $\frac{1}{16}$ of the thermal energy kT. Thus the interactions are relatively weak and the deflections will be large only for paths which carry the electrons into a radius of about $(\frac{1}{16})$ of $a_I/2$. On the other hand, at low temperatures, such as 20°K for example, the energy at the edge of the cell will be comparable to kT, and an electron which happens to lose energy near an ion may move around it in an elliptical orbit before escaping or becoming bound. For these low temperature ranges, the present theory will be a rather poor approximation. However, it is the best available at the time of writing and improvements will probably await the acquiring of better experimental data than is now available. There is an interesting range, however, at which the approximation is valid and at which impurity scattering at the same time is the dominant process. We shall discuss the comparison between theory and experiment in this range in later sections of this chapter.

We next consider the deflection of an electron which starts at infinity with velocity v on a line such that it would miss the ion by a distance b. Figure 11.3 shows its hyperbolic trajectory. The diagram also shows the trajectory of a hole starting with the same velocity and also directed to miss by b. It is a consequence of the equations for the orbits, as we shall show, that, if the effective masses are equal, so are the deflections θ in spite of the difference in the sign of the charges. The equation for an orbit in

[1a] Since this section was written, a revised value of 16 has been obtained by H. B. Briggs. See Chapter 9.

a central electrostatic field is given in terms of the radius r and angle ϕ by the equation

$$\frac{1}{r} = \pm \frac{e^2 m}{\kappa J^2} + A \cos \phi \tag{3}$$

where the + sign holds for attractive forces (electron in Figure 11.3) and the − sign for repulsive force (hole). J is the angular momentum which may be written as

$$J = mr^2\dot{\phi} = xp_y - yp_x = mbv. \tag{4}$$

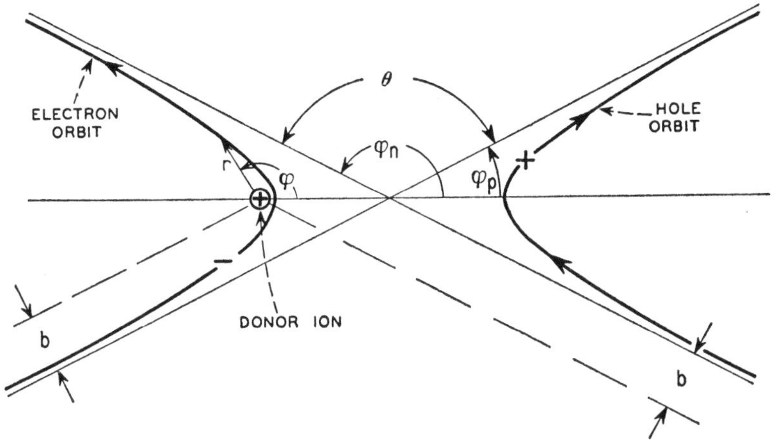

Fig. 11–3—Positive and Negative Particles of Equal Masses and Energies are Deflected through the Same Angle θ in Traversing Infinite Hyperbolic Orbits.

The last equation follows from the fact that the angular momentum is conserved and has the magnitude mbv when the particles are moving on the asymptotes; since only J^2 enters the formula, the algebraic sign of J has been disregarded.

Our problem is to evaluate θ in terms of b. It is evident that $\theta = 2\phi_n - \pi$ and $\theta = \pi - 2\phi_p$ for the two cases where ϕ_n and ϕ_p are the angles of the asymptotes. The values of ϕ_n and ϕ_p are found as follows: As the particles approach infinity, ϕ approaches its maximum magnitude and we have

$$0 = \frac{e^2 m}{\kappa J^2} + A \cos \phi_n; \quad 0 = -\frac{e^2 m}{\kappa J^2} + A \cos \phi_p. \tag{5}$$

Near the maximum, the right side may be expanded as a series in $\phi - \phi_n$ or $\phi - \phi_p$, leading to

$$\frac{1}{r} = -(\phi - \phi_n) A \sin \phi_n; \quad \frac{1}{r} = -(\phi - \phi_p) A \sin \phi_p. \tag{6}$$

These last may be solved for $(\phi_n - \phi)r$ and $(\phi_p - \phi)r$, quantities which are evidently both equal to b for large values of r. This gives

$$b = 1/A \sin \phi_n, \quad b = 1/A \sin \phi_p. \tag{7}$$

A may now be eliminated from the equations giving

$$\begin{aligned} \cot \phi_n &= -\frac{e^2 mb}{\kappa J^2}, & \cot \phi_p &= \frac{e^2 mb}{\kappa J^2} \\ &= -\frac{e^2}{\kappa m v^2 b} & &= \frac{e^2}{\kappa m v^2 b}. \end{aligned} \tag{8}$$

For $v^2 b = \infty$, $\cot \phi_n$ and $\cot \phi_p$ are zero corresponding to $\phi_n = \phi_p = \pi/2$ and to $\theta = 0$ and thus to no deflection. For finite values of $v^2 b$, ϕ_n and ϕ_p differ from $\pi/2$ by equal amounts with $\phi_n > \pi/2$ and $\phi_p < \pi/2$ thus leading to equal deflections θ. The relationship between θ and ϕ_n and ϕ_p leads for both cases to

$$\tan \theta/2 = e^2/\kappa m v^2 b. \tag{9}$$

In order for an electron to be deflected from its initial direction through an angle between θ and $\theta + d\theta$, corresponding to a solid angle $2\pi \sin \theta d\theta$, it must be directed towards an annular ring lying in a range b to $b + db$ where

$$db = \frac{e^2}{\kappa m v^2} d \cot \frac{\theta}{2} = -\frac{e^2}{2\kappa m v^2} \operatorname{cosec}^2 (\theta/2) d\theta. \tag{10}$$

The projected area $2\pi b db$ of this annular ring at which the hyperbolic asymptote must be pointed in order to produce deflections in the solid angle $2\pi \sin \theta d\theta$ is defined as $\sigma(\theta) 2\pi \sin \theta d\theta$, where $\sigma(\theta)$ is called the *differential cross section.* From this definition we obtain

$$\sigma(\theta) = \frac{2\pi b db}{2\pi \sin \theta d\theta} = \left(\frac{e^2}{2\kappa m v^2}\right)^2 \operatorname{cosec}^4 (\theta/2). \tag{11}$$

This is the Rutherford scattering law. The same relationship is also obtained by solving and interpreting Schroedinger's equation.[2]

The formula indicates that for small angles of scattering, σ approaches infinity; this corresponds to the fact that there is an infinite cross section for large misses which produce negligible deflections. For θ approaching π, corresponding to reversal of motion, $\sigma = r_\pi^2/16$ where r_π is the radius at which the potential energy equals $mv^2/2$.

For our application, the quantity of particular importance is $1 - \cos \theta$

[2] See, for example, L. I. Schiff, *Quantum Mechanics*, McGraw-Hill Book Co., New York, 1949, equation (20.11).

and we have

$$1 - \cos\theta = 1 - \cos(2\phi_n - \pi) = 1 + \cos 2\phi_n = 2\cos^2\phi_n$$
$$= \frac{2}{1 + \tan^2\phi_n} = \frac{2}{1 + \left(\dfrac{\kappa m v^2 b}{e^2}\right)^2}, \tag{12}$$

which expresses the quantity $1 - \cos\theta$ in terms of b and v. The formula applies to both holes and electrons.

We must next average $1 - \cos\theta$ over all collisions. To do this we give each "radius of miss" b a weight proportional to its projected area and for simplicity we replace the square of edge a_I by a circle of radius $a_I/2$. We then say that the probability that the asymptote is directed so as to pass in a range db is simply $(2\pi b db)/\pi(a_I/2)^2$. The value of $\langle 1 - \cos\theta\rangle$ then becomes

$$\langle 1 - \cos\theta\rangle = \frac{4}{\pi a_I{}^2}\int_0^{a_I/2}(1 - \cos\theta)2\pi b\,db$$
$$= 2\left(\frac{\mathcal{E}_I}{2\mathcal{E}}\right)^2 \ln\left[1 + \left(\frac{2\mathcal{E}}{\mathcal{E}_I}\right)^2\right] \tag{13}$$

where

$$\mathcal{E} = mv^2/2, \quad \mathcal{E}_I = \frac{2e^2}{\kappa a_I}. \tag{14}$$

For an electron with $\mathcal{E} = kT$ in the example discussed with equation (2)

$$\langle 1 - \cos\theta\rangle = 2(\tfrac{1}{32})^2 \ln[1 + (32)^2] \doteq \tfrac{1}{65}. \tag{15}$$

Thus an electron is in effect only $\frac{1}{65}$ scattered in traveling a distance a_I so that the mean free path for the example chosen will be $65a_I$ or 6.5×10^{-4} cm. From formula (17) of Section 8.8 we see that this mean free path is about 40 times the mean free path of a thermal electron. At room temperature, therefore, impurity scattering is unimportant in semiconductors with low impurity contents.

In terms of τ_c and $\langle 1 - \cos\theta\rangle$ the relaxation time τ_I for impurity scattering is found to be

$$\tau_I = \tau_c/\langle 1 - \cos\theta\rangle = \frac{a_I}{2v}\left(\frac{2\mathcal{E}}{\mathcal{E}_I}\right)^2 \frac{1}{\ln\left[1 + \left(\dfrac{2\mathcal{E}}{\mathcal{E}_I}\right)^2\right]}. \tag{16}$$

This formula was introduced by Conwell and Weisskopf[3] in calculating the

[3] E. Conwell and V. F. Weisskopf, *Phys. Rev.* **69**, 258A (1946); **77**, 388–390 (1950). The problem is mathematically identical with that of the scattering of electrons in an ionized gas which has been solved to the same approximation by S. Chapman, *Monthly Notices, R. A. S.* **82**, 292–297 (1922).

effect of impurity scattering upon the mobility. Owing to differences in the approximations their expression differs by a factor $4/\pi$ leading to

$$\tau_I = \frac{2a_I}{v\pi} \frac{(2\mathcal{E}/\mathcal{E}_I)^2}{\ln [1 + (2\mathcal{E}/\mathcal{E}_I)^2]}. \tag{17}$$

We shall return to this formula in the next section and average it appropriately over the energy distribution, thus obtaining the Conwell-Weisskopf formula for mobility.

If several types of impurity are present, with charges varying both in sign and magnitude, substantially the same approximations may be carried out. For this case, a_I should be computed from the total density of scattering centers without regard to sign or magnitude so as to divide the crystal in this way into one cube for each center. The contribution of each center, however, is approximately proportional to $\mathcal{E}_I^2$ as may be seen from the formula for $\langle 1 - \cos\theta \rangle$. Since $|\mathcal{E}_I|$ is proportional to e^2/κ for single charged ions, $\mathcal{E}_I^2$ will be proportional to Z^2 for an ion of charge Z_e. Thus the average charge to use in the formula will be approximately the root mean square charge of the scattering ions and the number to use will be the total number.

There is strong theoretical and experimental evidence that donors and acceptors contribute in an important way to scattering even when their charges are neutralized by bound holes and electrons. This possibility is discussed by Pearson and Bardeen[4] who conclude that in silicon equal contributions will be made by an ionized and a neutral center at about 100°K. At lower temperatures, the ionized center may be more effective. The scattering is associated with exchange effects between the incident and the bound electron and is similar to problems already investigated for the collision cross section of hydrogen atoms.[5] The application of these important theories to semiconductors is not at present so well developed, however, as to justify a discussion of the other forms of impurity scattering at this point.

11.3b. Scattering by Thermal Agitation; the Deformation Potential Formula. Thermal vibrations of the atoms distort the periodic potential field in which the electrons move. As a consequence, the Bloch wave functions are no longer exact solutions, and an electron started with one value of $\boldsymbol{P}$ makes transitions to other states and is, in effect, scattered. This situation is closely analogous to the reflections and attenuations which may occur in a non-uniform but lossless transmission line. At the locations where the impedance changes, mismatches of impedance occur and the

[4] G. L. Pearson and J. Bardeen, *Phys. Rev.* **75**, 865–883 (1949).

[5] H. S. W. Massey and C. B. O. Mohr, *Proc. Roy. Soc., London* **132A**, 605–630 (1931); **136A**, 289–311 (1932).

wave is reflected. We shall give a treatment in this section based on the idea of mismatch and reflection and thus show how the various important quantities enter the formula for the mean free path.

As a result of thermal agitation, the atoms in the crystal are vibrating in a random fashion with respect to their neighbors. The description of this pattern and a determination of what aspects of it are most important in scattering the electrons are treated in detail in Chapter 17. We shall here summarize and simplify the physical picture which results from that analysis. In the first place, the important distortions of the atomic arrangement are those which produce uniform effects over a distance of about $(\frac{1}{4})\lambda$ for the electron wave. This is a familiar result from wave theory. In a material like a glass or a liquid, for example, there is a great amount of disorder on a scale of atomic dimensions. A wave length of visible light, however, extends over several thousand atomic spacings and as a result the effects of the atoms on the waves are averaged over groups containing of the order of $(1000)^3$ atoms. For such large numbers, the statistical fluctuations are negligible, the material appears to be homogeneous, and no appreciable scattering of light occurs. On the other hand, if the transmission properties of the medium change by only a small fraction of themselves in a given wave length, then the situation is essentially that of a tapered transmission line or a tapered waveguide. For such cases the traveling wave can adjust itself as it progresses and almost no reflection occurs. In quantum mechanics such cases are the ones best adapted to treatment by the Wentzel-Kramers-Brillouin approximation, which is discussed in most quantum-mechanics texts. If, however, there is a change in property of the material in a region which extends over $(\frac{1}{4})\lambda$, there will be a relatively strong reflection: Under these conditions, the waves reflected from the entrance side and the exit side will combine constructively since the reflection conditions at the two faces give reflections differing by 180° and the round trip through the region adds another phase lag of 180°, making 360° altogether.

In the correct quantum-mechanical treatment, the problem of determining the important scale for distortions is eliminated by a mathematical device. Essentially what is done is to make a Fourier analysis of the deformation. Each component corresponds to a possible normal component or standing wave pattern for the atomic vibrations, like the waves of Figure 5.3 or 5.6. Each of these vibrations is thermally excited and has, at the temperatures of most interest, an average energy of kT. The scattering of the electrons is then computed by adding the effects of the individual waves. From this it is found that the most important waves are those whose wave length is somewhat less than the electrons' wave lengths. A longitudinal wave of this sort will produce alternating regions of compression and dilation somewhat less than $\frac{1}{2}\lambda$ (λ for the electron wave) wide, in general agreement with our conclusion that disturbances about $\frac{1}{4}\lambda$ wide are the most important.

Furthermore, it should be noted that the speed of a thermal electron is about 10^7 cm/sec whereas that of the sound wave is about 5×10^5 cm/sec, or 20 times less. Consequently the deformation in the crystal may be regarded as static while the electron moves by. The relatively slow movement of the pattern will, however, produce a "Doppler effect" on the

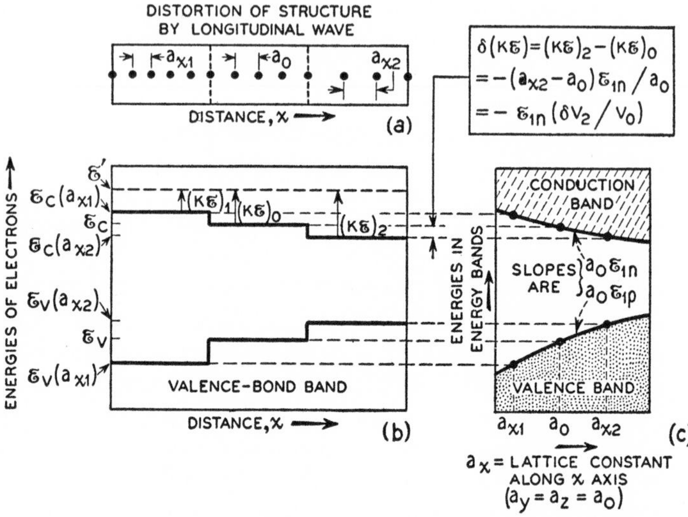

FIG. 11–4—The Deformation Potentials for Electrons and Holes. (The variation of atomic spacing produced by a longitudinal wave has the effect of introducing a varying potential energy.)

scattered electron wave and thus change the energy of the electron. As discussed in Section 11.2, this effect is small and can be neglected for our purposes. (See Problems for a further discussion.)

We must next ask what effect a deformation of the lattice will have upon its transmission properties for an electron wave. The answer is found by considering the dependence of the energy boundaries upon lattice constant as illustrated in Figure 11.4. Here a one-dimensional section of the crystal is shown. In it, a longitudinal wave progressing in the x-direction is represented, for purposes of illustration, as changing the degree of compression abruptly so that the crystal is divided, momentarily, into three parts having lattice constants along the x-axis of a_1, a_0, and a_2 respectively. We now imagine that this pattern is frozen and consider its effect on the motion of an electron. Part (c) shows the dependence of the energy-band boundaries upon a_x, the lattice constant along the x-axis, it being supposed that

$a_y = a_z = a_0$, where a_0 is the equilibrium value or true lattice constant. (This plot differs from Figures 5.4 and 5.5 for which the variable was $a = a_x = a_y = a_z$.) The changes of lattice constant will thus change the energy-band structure in the crystal as shown in (b). The changes in energy-band boundaries may be seen to be substantially equivalent to the introduction of a varying potential energy for an electron, and we shall refer to them as *deformation potentials*. An electron of energy $\mathcal{E}'$, as it moves through the crystal, will have variations in its kinetic energy (energy in excess of $\mathcal{E}_c$) as shown. The discontinuities in kinetic energy will give rise to reflection coefficients at the transition boundaries which are closely analogous to those for a discontinuous transmission line. Entirely similar considerations apply to the case of the hole and we need not consider it separately.

We shall illustrate the reflection process for an electron by considering a simplified one-dimensional problem in which an electron wave in region a with kinetic energy $(K.E.)_a = P_a{}^2/2m$ strikes a region b in which its kinetic energy is $(K.E.)_b = P_b{}^2/2m$ and is partially transmitted and partially reflected. We shall take the boundary as occurring at $x = 0$ and shall write the wave functions as follows:

$$\left.\begin{array}{ll}\text{Incident wave:} & e^{iP_ax/\hbar} \\ \text{Reflected wave:} & Re^{-iP_ax/\hbar}\end{array}\right\} \text{region } a, \quad x < 0 \qquad (18a), (18b)$$

$$\text{Transmitted wave:} \quad Te^{iP_bx/\hbar} \qquad \text{region } b, \quad x > 0. \tag{18c}$$

The electron current density (as the reader may verify from the equations of Section 15.4) is proportional to $1 - |R|^2$ so that $|R|^2$ is to be taken as the fraction of incident current which is reflected. The boundary conditions at $x = 0$ require continuity of ψ and $d\psi/dx$ and thus lead to

$$1 + R = T, \quad (iP_a/\hbar)(1 - R) = (iP_b/\hbar)T \tag{19}$$

and hence to

$$|R|^2 = \left(\frac{P_a - P_b}{P_a + P_b}\right)^2, \tag{20}$$

a formula strictly analogous to the reflection formula $(Z_1 - Z_2)^2/(Z_1 + Z_2)^2$ at the junction of transmission lines of impedance Z_1 and Z_2. (For the transmission-line problem, continuity requirements for voltage V and current $I = \pm(1/Z)V$ replace those for ψ and $\pm P\psi$.) For a small change in energy $\delta\mathcal{E}_c$, we may write

$$\delta\mathcal{E}_c = P\delta P/m_n \tag{21}$$

and

$$|R|^2 = \frac{(\delta P)^2}{4P^2} = \frac{m_n{}^2(\delta\mathcal{E}_c)^2}{4P^4}. \tag{22}$$

We must next consider the value of the deformation potential $\delta\mathcal{E}_c$ to be expected for a region about $\frac{1}{4}\lambda$ in size. In Figure 11.5 we have arbitrarily divided the crystal into cubes $b\lambda$ along the edge having volume $(b\lambda)^3 = V_0$. We shall assume initially that each cube contracts and expands independently of the others, considering later how this assumption may be corrected. Our problem is then to find the probability that, if the incident wave falls on a particular cube, it will be reflected. To estimate this effect, we shall re-express $\delta\mathcal{E}_c$ in terms of a change in volume. For a crystal with

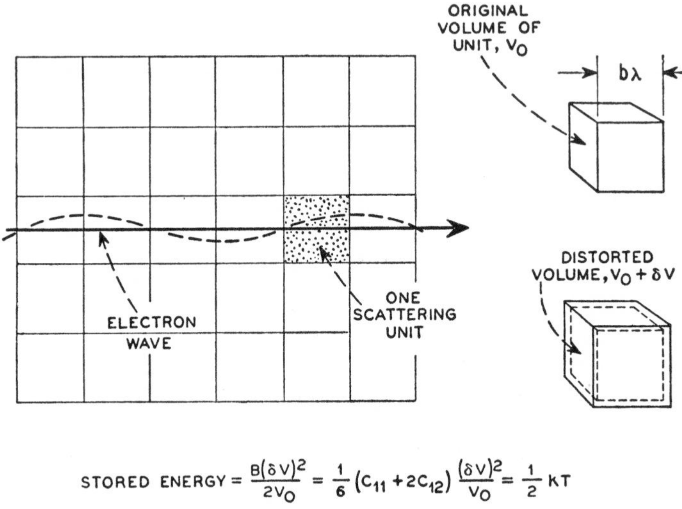

FIG. 11–5—For Purposes of Calculation the Crystal Is Divided into Cubes Which Expand and Contract Due to Thermal Agitation. Each Block Acts Independently in Scattering the Electron Wave.

cubic symmetry it is evident that $\delta\mathcal{E}_c$ must depend symmetrically upon a_x, a_y, and a_z. The only linear symmetric form may be written as

$$\delta\mathcal{E}_c = (a_x + a_y + a_z - 3a_0)\mathcal{E}_{1n}/a_0 \tag{23}$$

where $\mathcal{E}_{1n}$ is a constant for small changes of the a's. But this form is equivalent to

$$\delta\mathcal{E}_c = \mathcal{E}_{1n}\frac{\delta V}{V_0} \tag{24}$$

where V_0 is the original volume of the unit considered and $V_0 + \delta V$ the distorted volume. (For the case of holes we similarly conclude that

$$\delta\mathcal{E}_v = \mathcal{E}_{1p}\frac{\delta V}{V_0}, \tag{25}$$

a more complete discussion of both cases being given in Chapter 17.)

The volume considered in Figure 11.5 has many degrees of freedom, in fact three times as many as atoms, and it contains kT of thermal energy for each of these modes. Of all the possible modes of distortion, however, the one of most importance is that which produces a dilation (or contraction) of the whole unit and thus contributes to the average value of $\delta\mathcal{E}_c$. All other independent modes will give zero average dilation and will correspond to fine-scale fluctuations of potential of the sort which we decided to neglect. For the deformation of importance the stored elastic energy is

$$-\tfrac{1}{2}p\delta V = \tfrac{1}{2}B\frac{\delta V}{V_0}\delta V \tag{26}$$

where $p = -B\delta V/V_0$ is the pressure and B is the bulk modulus or the reciprocal of the compressibility. This energy must, on the average, be $\frac{1}{2}kT$ so that

$$(\delta\mathcal{E}_c)^2 = \mathcal{E}_{1n}{}^2\left(\frac{\delta V}{V_0}\right)^2 = \mathcal{E}_{1n}{}^2\frac{kT}{V_0B}. \tag{27}$$

This leads to a probability of reflection of

$$|R|^2 = \frac{m_n{}^2\mathcal{E}_{1n}{}^2kT}{4P^4V_0B} = \frac{m_n{}^2\mathcal{E}_{1n}{}^2kT}{4P^4(b\lambda)^3B}. \tag{28}$$

In terms of $|R|^2$ we estimate the mean free path l as follows: By definition, the probability that an electron be scattered while moving a distance δx is $\delta x/l$. In traversing the cube the probability is about $2|R|^2$, that is, $|R|^2$ from front and back sides. Thus we have

$$\frac{b\lambda}{l} = 2|R|^2$$

or

$$l = b\lambda/2|R|^2 = \frac{2(b\lambda)^4P^4B}{m_n{}^2\mathcal{E}_{1n}{}^2kT} = \frac{2b^4h^4B}{m_n{}^2\mathcal{E}_{1n}{}^2kT}, \tag{29}$$

the last equality following from $\lambda P = h$.

The exact expression for l derived in Chapter 17 is

$$l = \frac{\pi\hbar^4c_{ll}}{m_n{}^2\mathcal{E}_{1n}{}^2kT} \tag{30}$$

where c_{ll} is the elastic constant for a longitudinal wave propagating in the [110] direction and is about 25 per cent larger than c_{11}. The ratio of the two values of l is

$$\frac{l\text{ (approx.)}}{l\text{ (exact)}} = \frac{2(2\pi b)^4B}{\pi c_{ll}} \doteq 2 \tag{31}$$

for $B = \frac{1}{3}(c_{11} + 2c_{12}) \doteq \frac{1}{2}c_{11}$ and $b = \frac{1}{4}$. This agreement is as good as could be expected from our simple model, which was discussed for the purpose of illustrating the principles rather than for calculating the exact formula.

The exact treatment leads to the result, which we shall not attempt to show from the simple model, that the scattering is substantially spherical so that the direction of motion after collision is independent of the direction of motion before collision. This leads, as discussed in Section 11.2, to the result that $\langle \cos \theta \rangle = 0$ for this case. In accordance with these results we have for this case

$$\tau = \tau_c = l/v \text{ and } \langle 1 - \cos \theta \rangle = 1. \tag{32}$$

Certain flaws in the model should be mentioned. It is incorrect to treat the blocks as independent since the surrounding material exerts a restraining force which approximately doubles the stiffness and thus tends to increase l. The scattering treatment of one block is so very crude, however, that it would not be worthwhile to try to improve the treatment of B without adding further refinements. Adding further refinements would soon lead to the path followed in Chapter 17, however, to which the reader who is interested in the basic theory is here referred.

In the next section, formula (30) for the mean free path will be used to obtain an equation for the mobility, equation (30) of Section 11.4. An important test of the theory can be made from values of $\mathcal{E}_{1n}$ and the corresponding quantity for holes, and $\mathcal{E}_{1p}$. These values can be determined from the measured mobilities, elastic constants, and assumed values of the effective mass. The sum of the two values may then be compared with

$$\mathcal{E}_{1G} = d(\mathcal{E}_c - \mathcal{E}_v)/d \ln V \tag{33}$$

where V is the volume of the crystal and $\mathcal{E}_{1G}$ is the change in energy gap per unit dilation. In Chapter 12 we shall discuss various ways in which $\mathcal{E}_{1G}$ may be determined directly. If $\mathcal{E}_{1n}$ and $\mathcal{E}_{1p}$ have opposite signs, as shown in Figure 11.4, then the relationship

$$|\mathcal{E}_{1G}| = |\mathcal{E}_{1n}| + |\mathcal{E}_{1v}| \tag{34}$$

should hold. The comparison of $\mathcal{E}_{1G}$ with $|\mathcal{E}_{1n}| + |\mathcal{E}_{1v}|$ presented in Chapter 12 shows that the theory of scattering by deformation potentials is probably correct, at least in its general features.

11.4 AVERAGE MOBILITIES, MEAN FREE TIMES, AND HALL CONSTANTS

As a result of the analysis in Sections 11.2 and 11.3, we can suppose that there is a relaxation time $\tau(\mathcal{E})$ associated with the states in an energy shell bounded by surfaces $\mathcal{E}$ and $\mathcal{E} + d\mathcal{E}$ and that this time is independent of the degree to which the states are occupied. According to the development of the concept of the relaxation time, any current produced by a disturbance of the equilibrium among states will decrease by a fraction $dt/\tau(\mathcal{E})$ in time dt. We shall now use this result to calculate the steady-state currents.

The conductivity and other quantities of interest are determined by calculating the current due to the electrons in a partially filled Brillouin zone when their equilibrium is disturbed by applied fields. A magnetic field alone, as discussed in Section 7.4, causes an incompressible flow of points in the Brillouin zone along the equal energy surfaces; along a given energy surface the density of occupied states, being $\mathbf{f}(\mathcal{E})$ times the density of states, is constant; consequently, the flow produced by $\boldsymbol{H}$ does not disturb the equilibrium. An electric field does alter the distribution and the non-equilibrium distribution so produced can then be affected by magnetic fields in the manner discussed in Chapter 8.

We shall introduce a symbol $\mathbf{f}'(\boldsymbol{P})$ to represent the fraction of the quantum states, having crystal momenta near $\boldsymbol{P}$, which are occupied by electrons under non-equilibrium conditions. For small fields, $\mathbf{f}'$ will differ by a very small amount from the Fermi-Dirac distribution function $\mathbf{f}(\boldsymbol{P}) = \mathbf{f}[\mathcal{E}(\boldsymbol{P})]$ and we may write

$$\mathbf{f}'(\boldsymbol{P}) = \mathbf{f}[\mathcal{E}(\boldsymbol{P})] + \mathbf{f}_1(\boldsymbol{P}) \tag{1}$$

where $\mathbf{f}_1$ represents the small disturbance produced by the field.

We shall use Figure 11.6 in deriving the effect of the applied field. The coordinates are $\boldsymbol{P}$-space and two energy surfaces corresponding to $\mathcal{E}$ and $\mathcal{E} + d\mathcal{E}$ are shown. The total number of quantum states in the shell between the two surfaces is denoted by dS and the number in any small element by ds. The number of electrons occupying states in ds is $\mathbf{f}'(\boldsymbol{P})ds$. These electrons contribute a current density, evaluated according to equation (1) of Section 6.1, of

$$d\boldsymbol{I}_s = \frac{-e}{V}\boldsymbol{v}\mathbf{f}'ds, \tag{2}$$

since each of them has the velocity $\boldsymbol{v} = \nabla_P\mathcal{E}(\boldsymbol{P})$ evaluated at the $\boldsymbol{P}$ corresponding to ds. The total contribution of the shell to the current density is

$$d\boldsymbol{I}_S = \frac{-e}{V}\int_{\text{shell } d\mathcal{E}} \boldsymbol{v}\mathbf{f}'ds. \tag{3}$$

Under equilibrium conditions, this integral is zero because of the symmetry of the energy surfaces: For all Brillouin zones, regardless of the crystal structure, $\mathcal{E}(\boldsymbol{P}) = \mathcal{E}(-\boldsymbol{P})$; consequently the surfaces have a center of symmetry and for each element ds located at $\boldsymbol{P}$ and having velocity $\boldsymbol{v}$

there is always an equal element ds' located at $-\boldsymbol{P}$ which has an opposite velocity $\boldsymbol{v}' = -\boldsymbol{v}$ and these elements will cancel in the integral. Hence, the current arises entirely from $\mathbf{f}_1$.

The disturbance in equilibrium due to the electric field is represented by $\dot{\boldsymbol{P}} = -e\boldsymbol{E}$ on the figure. In time dt, the flow produced by $\dot{\boldsymbol{P}}$ moves elec-

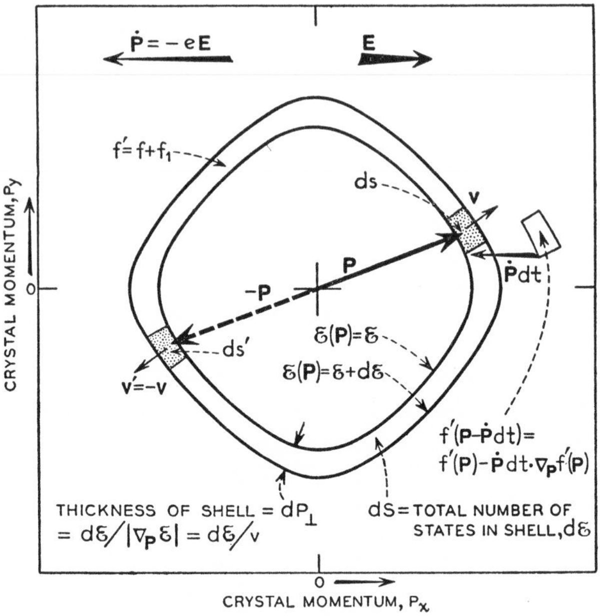

FIG. 11-6—Disturbance of Equilibrium by Electric Field.

trons so as to replace the $\mathbf{f}'(\boldsymbol{P})ds$ electrons in ds by the $\mathbf{f}'(\boldsymbol{P} - \dot{\boldsymbol{P}}dt)ds$ electrons in the volume displaced by $-\dot{\boldsymbol{P}}dt$ from ds. This causes the number of electrons in ds to change at a rate (due to $\boldsymbol{E}$) of

$$\left[\frac{d}{dt}\mathbf{f}'ds\right]_{\text{due to } \boldsymbol{E}} = \left[\frac{d}{dt}\mathbf{f}'\right]_{\text{due to } \boldsymbol{E}} ds = -\dot{\boldsymbol{P}} \cdot \nabla_{\boldsymbol{P}}\mathbf{f}'(\boldsymbol{P})ds$$
$$= -[\dot{\boldsymbol{P}} \cdot \nabla_{\boldsymbol{P}}\mathbf{f}(\boldsymbol{P}) + \dot{\boldsymbol{P}} \cdot \nabla_{\boldsymbol{P}}\mathbf{f}_1(\boldsymbol{P})]ds. \tag{4}$$

The second term in this expression involves a product of $\dot{\boldsymbol{P}}$, which is proportional to $\boldsymbol{E}$, and $\mathbf{f}_1$, which is also proportional to $\boldsymbol{E}$, and thus represents a quadratic effect, which we can neglect in a linear theory. We shall, therefore, calculate the rate of increase of current due to $\nabla_{\boldsymbol{P}}\mathbf{f}(\boldsymbol{P})$ and shall later equate this to the rate of decay of current due to collisions in order to determine the steady state. Because $\mathbf{f}(\boldsymbol{P})$ depends only on $\mathcal{E}(\boldsymbol{P})$, we may re-express $\nabla_{\boldsymbol{P}}\mathbf{f}[\mathcal{E}(\boldsymbol{P})]$ as follows:

$$\nabla_{\boldsymbol{P}}\mathbf{f}[\mathcal{E}(\boldsymbol{P})] = \nabla_{\boldsymbol{P}}\mathcal{E}(\boldsymbol{P})d\mathbf{f}(\mathcal{E})/d\mathcal{E} = \boldsymbol{v}\frac{-1}{kT}(1 - \mathbf{f})\mathbf{f}, \tag{5}$$

where we have used the useful relationship

$$\frac{d\mathbf{f}(\mathcal{E})}{d\mathcal{E}} = \frac{d}{d\mathcal{E}}\{[\exp(\mathcal{E} - \mathcal{E}_F)/kT] + 1\}^{-1} = -\frac{1}{kT}\mathbf{f}(1 - \mathbf{f}). \tag{6}$$

From these we obtain

$$\left(\frac{d}{dt}\mathbf{f}'\right)_{\text{due to } \boldsymbol{E}} = -\dot{\boldsymbol{P}} \cdot \nabla_P \mathbf{f}(\boldsymbol{P}) = \frac{(-e)}{kT}\boldsymbol{E} \cdot \boldsymbol{v}\mathbf{f}(1 - \mathbf{f}). \tag{7}$$

The rate of change of $d\boldsymbol{I}_s$ due to $\boldsymbol{E}$ is evidently

$$\left[\frac{d}{dt}\,d\boldsymbol{I}_s\right]_{\text{due to } \boldsymbol{E}} = \left[\frac{-e}{V}\right]\boldsymbol{v}\left[\frac{d}{dt}\mathbf{f}'\right]_{\text{due to } \boldsymbol{E}} ds,$$

and by using the foregoing relationships, this may be reduced to

$$\left[\frac{d}{dt}\,d\boldsymbol{I}_s\right]_{\text{due to } \boldsymbol{E}} = \frac{e^2}{kTV}\boldsymbol{v}\,(\boldsymbol{E} \cdot \boldsymbol{v})\mathbf{f}(1 - \mathbf{f})ds. \tag{8}$$

It should be pointed out that the situation shown in Figure 11.6 has been exaggerated for purposes of exposition by making the vector $\dot{\boldsymbol{P}}dt$ very large. Actually we have used the figure to calculate a derivative and this corresponds to the case for which $\dot{\boldsymbol{P}}dt$ approaches zero. Under these conditions the rate of change of $\mathbf{f}'ds$ arises from the value of $\mathbf{f}'$ just on the two surfaces of the shell. (We have also shown in the figure a formula for the thickness of the shell. This is for use in the problems at the end of this section.)

The fact that rate of change of equation (8) depends upon the degree to which the states are occupied through the simple factor $\mathbf{f}(1 - \mathbf{f})$ has the following interpretation: When $\mathbf{f}$ is very small, as it is for a high resistance n-type semiconductor, $\mathbf{f}(1 - \mathbf{f})ds$ is approximately $\mathbf{f}ds$, which is simply the number of electrons in the states ds; the contribution of the states ds is, therefore, simply proportional to the number of electrons in them. When an appreciable fraction of the states are occupied, however, the efficiency is reduced with which they can use their electrons in producing changes in current; in particular if the states ds were completely occupied, no change in their current could be produced and the effect of the electric field would be simply to replace one electron by another. The proper efficiency factor is evidently $(1 - \mathbf{f})$ since this gives the probability that an accelerated electron will produce a change in the situation rather than simply replacing another electron. Furthermore, a similar situation applies for holes; thus as discussed in Chapter 10, $(1 - \mathbf{f}) = \mathbf{f}_p$ is the probability that a state is occupied by a hole; hence the expressions $\mathbf{f}(1 - \mathbf{f})$ and $(1 - \mathbf{f}_p)\mathbf{f}_p$ are equal and equation (10) is symmetrical in holes and electrons. The same reasoning may be applied to holes, in which case the effectiveness factor is $(1 - \mathbf{f}_p) = \mathbf{f}$. We shall use the interpretations of $\mathbf{f}ds$ as number of electrons and $(1 - \mathbf{f})ds = \mathbf{f}_p ds$ as number of holes in later integrated expressions for the current and the conductivity.

The rate of change of the total current due to the shell $d\mathfrak{E}$ is obtained by integrating (8) over the quantum states in the shell. (The general results are more complicated than needed for our purposes: Thus if the electric field is in the x-direction, for example, the change in current may have a component in the y-direction; it will have such a component if the average value of $E_x v_x \boldsymbol{v}$ has a y-component. We shall give the general formula for such cases in the Problems and restrict the treatment here to the symmetrical case of spherical energy surfaces.) For the case of spherical energy surfaces, the average value of $v_x v_y$ over the surface of the sphere vanishes by symmetry, and the average values of $v_x{}^2$, $v_y{}^2$, and $v_z{}^2$ are equal by symmetry; furthermore, since $v_x{}^2 + v_y{}^2 + v_z{}^2 = v^2$, where $v = |\boldsymbol{v}|$, the average value of $v_x{}^2 = v^2/3$. For this case integration over the shell $d\mathfrak{E}$ leads to

$$\left[\frac{d}{dt} dI_{Sx}\right]_{\text{due to } \boldsymbol{E}} = E_x \frac{e^2}{kTV} \cdot \frac{v^2}{3} \cdot \mathbf{f}(1 - \mathbf{f})dS \tag{9}$$

for the x-component of current due to a field E_x.

Under steady-state conditions the increase in current in shell $d\mathfrak{E}$ due to $\boldsymbol{E}$ must be just balanced by the rate of decay. This latter is simply $(d\boldsymbol{I}_S)/\tau(\mathfrak{E})$ so that we have

$$\begin{aligned} 0 &= \left[\frac{d}{dt} dI_{Sx}\right]_{\text{due to } \boldsymbol{E}} + \left[\frac{d}{dt} dI_{Sx}\right]_{\text{collisions}} \\ &= E_x \frac{e^2}{kTV} \frac{v^2}{3} \mathbf{f}(1 - \mathbf{f})dS - \frac{1}{\tau} dI_{Sx} \end{aligned}$$

leading to

$$dI_{Sx} = E_x \frac{e^2}{3kTV} \tau v^2 \mathbf{f}(1 - \mathbf{f})dS, \tag{10}$$

where $d\boldsymbol{I}_S$ is the contribution of the dS states in shell $d\mathfrak{E}$ and $d\boldsymbol{I}_{Sx}$ is its x-component. The total current is obtained by integrating this equation (10) over all energies.

11.4a. Non-degenerate Semiconductors. For the case of semiconductors, the element of integration can conveniently be changed by using the approximations of equations (2) and (4) of Section 10.1:

$$n\text{-type:} \qquad \frac{\mathbf{f}(\mathbf{f} - 1)dS}{V} \doteq \mathbf{f}\frac{dS}{V} = dn_{\mathfrak{E}} \tag{11a}$$

$$p\text{-type:} \qquad \frac{\mathbf{f}(\mathbf{f} - 1)dS}{V} \doteq \mathbf{f}_p\frac{dS}{V} = dp_{\mathfrak{E}} \tag{11b}$$

where $dn_{\mathcal{E}}$ and $dp_{\mathcal{E}}$ are respectively the contributions to the electron and hole densities arising from the energy shell $d\mathcal{E}$.

Before carrying out the integration for current and conductivity for the semiconductor case, we shall discuss the Hall angle θ and obtain a corresponding expression for it so that both integrations may be carried out at the same time. We shall deal with the case of a small magnetic field H_z, at right angles to E_x, and shall neglect any effects proportional to H_z^2. Under the influence of E_x alone the shell $d\mathcal{E}$ acquires a current $d\boldsymbol{I}_S$. For the spherical approximation, this current is parallel to $\boldsymbol{E}$, its only component being

$$dI_{Sx} = E_x \frac{e^2}{3kT} v^2 \tau(\mathcal{E}) dn_{\mathcal{E}}. \tag{12}$$

The magnetic field rotates the entire pattern of states at rate $\omega = eH_z/mc$ and thus tends to build up a current dI_{Sy} at a rate equal to ωdI_{Sx}. This component of current, like any component of current produced in shell $d\mathcal{E}$, decays at a rate equal to its steady-state value times $1/\tau$ and is thus

$$dI_{Sy} = \omega\tau dI_{Sx}. \tag{13}$$

(The effect of H_z upon the x-component of current is quadratic in H_z and may be neglected.) The total x and y components of current are thus:

$$I_x = \frac{E_x e^2}{3kT} \int v^2 \tau(\mathcal{E}) dn_{\mathcal{E}} \tag{14a}$$

$$I_y = \frac{E_x e^2 \omega}{3kT} \int v^2 \tau^2(\mathcal{E}) dn_{\mathcal{E}}. \tag{14b}$$

The integrals represent the sums of $v^2\tau$ and $v^2\tau^2$ over all the electrons in unit volume of the crystal and are thus simply n times the average values. Denoting the average over all the electrons of the distribution by $\langle v^2\tau\rangle$ and $\langle v^2\tau^2\rangle$, we may write for the conductivity σ and the mobility μ (omitting the subscript n for brevity, it being understood that electrons are concerned).

$$\sigma = \frac{I_x}{E_x} = ne\frac{e\langle v^2\tau\rangle}{3kT} = ne\mu. \tag{15}$$

The Hall mobility μ_H and Hall mean free time τ_H, defined in Section 8.9, are

$$\mu_H = \frac{c\theta}{H_z} = \frac{cI_y}{H_z I_x} = \frac{\omega\langle v^2\tau^2\rangle}{H_z\langle v^2\tau\rangle} = \frac{e\langle v^2\tau^2\rangle}{m\langle v^2\tau\rangle} \tag{16a}$$

$$\tau_H = \frac{\theta}{\omega} = \frac{mc\theta}{eH_z} = \frac{\langle v^2\tau^2\rangle}{\langle v^2\tau\rangle}. \tag{16b}$$

We shall next carry out the necessary averages for (15) and (16) by integrating over the electron distribution. The number of electrons having speeds in the range dv about v is given by the approximate expression (2) of Section 10.1 for **f** and by the volume $4\pi P^2 dP (\propto v^2 dv)$ of momentum space. Hence the relative probability that an electron has its velocity in the range v to $v + dv$ is

$$\mathrm{e}^{-mv^2/2kT} v^2 dv \tag{17}$$

and the average value of any function $g(v)$ is

$$\langle g(v) \rangle = \frac{\int_0^\infty \mathrm{e}^{-mv^2/2kT} g(v) v^2 dv}{\int_0^\infty \mathrm{e}^{-mv^2/2kT} v^2 dv}, \tag{18}$$

the denominator serving to normalize the probabilities. If $g(v) = v^n$, the integrals become standard forms and we have

$$\langle v^n \rangle = v_T{}^n \frac{\left(\frac{n+1}{2}\right)!}{\frac{1}{2}!}, \tag{19}$$

where v_T is the thermal velocity (actually root mean square along one axis):

$$v_T = \sqrt{2kT/m}, \tag{20}$$

and $x! = \Gamma(x + 1)$ for non-integral values of x. The integrals with which we shall be concerned can be evaluated with the aid of a general formula and two particular values:

$$(x + 1)! = (x + 1)x!, \quad 1! = 1, \quad \tfrac{1}{2}! = \sqrt{\pi}/2. \tag{21}$$

We shall first test these formulae by finding the average value of $\frac{1}{2}mv^2$; this is known to be $\frac{3}{2}kT$ from the general equipartion theorem which gives $\frac{1}{2}kT$ to each component of motion; thus we should find that $\langle v^2 \rangle = 3kT/m$. Our formula checks this by giving

$$\langle v^2 \rangle = v_T{}^2 \frac{\frac{3}{2}!}{\frac{1}{2}!} = \frac{2kT}{m} \cdot \frac{\frac{3}{2} \cdot \frac{1}{2}!}{\frac{1}{2}!} = \frac{3kT}{m}. \tag{22}$$

Inserting $m\langle v^2 \rangle$ for $3kT$ in (15), we obtain

$$\sigma = ne \frac{e\langle v^2 \tau \rangle}{m\langle v^2 \rangle} = ne \frac{e\tau_\mu}{m}, \quad \mu = \frac{e}{m} \frac{\langle v^2 \tau \rangle}{\langle v^2 \rangle} = \frac{e}{m} \tau_\mu \tag{23}$$

where τ_μ is the appropriate mean free time to use for mobility. We see that τ_μ is a weighted average of the relaxation time over the particles of the distribution, each particle being given weight v^2. It is similarly seen that τ_H is the quotient of the weighted average of τ^2 divided by the weighted average of τ. The Hall-effect formula, derived in Chapter 8, can, therefore, be written as

$$R_H = \frac{\tau_H}{\tau_\mu}\frac{1}{n(\pm e)c} = \frac{1}{n(\pm e)c}\frac{\langle v^2\tau^2\rangle\langle v^2\rangle}{\langle v^2\tau\rangle^2} \tag{24}$$

since our formula shows that τ_H/τ_μ is

$$\frac{\mu_H}{\mu} = \frac{\tau_H}{\tau_\mu} = \frac{\langle v^2\tau^2\rangle\langle v^2\rangle}{\langle v^2\tau\rangle^2} = \frac{\langle v^2\tau^2\rangle}{\langle v^2\rangle} \div \left(\frac{\langle v^2\tau\rangle}{\langle v^2\rangle}\right)^2. \tag{25}$$

The second form of equation (25) shows that τ_H/τ_μ is the weighted average of τ^2 divided by the square of the weighted average of τ. Since a mean squared value is always greater than the square mean value, as is shown in Section 14.3 for a different example, the Hall mobility is always greater than the drift mobility. This conclusion depends, however, on the assumption of spherical energy surfaces and a mean free time dependent on energy only. In Section 12.9 we shall quote data which throws doubt upon the validity of these simplifying assumptions and shall discuss possible explanations.

We shall next evaluate these averages for several dependencies of $\tau(\mathcal{E})$ on the energy and shall then derive expressions for degenerate cases.

FORMULAE FOR NON-DEGENERATE SEMICONDUCTORS

τ = constant. The simplest case is that for which the mean free time is constant with value τ_0. For this case we have

$$\tau_\mu = \tau_H = \tau_0 \tag{26a}$$

$$\mu = e\tau_0/m \tag{26b}$$

$$R_H = \frac{\tau_H}{\tau_\mu n(\pm e)c} = \frac{1}{n(\pm e)c} \tag{26c}$$

Thermal Vibrations. The next case is that appropriate to scattering by thermal vibrations. For this case the mean free path is constant and

$$\tau = l/v. \tag{27}$$

Using the notation of equation (15) that $\langle\ \rangle$ denotes an average of the bracketed property over all the electrons in the distribution giving each electron equal weight, we find

$$\tau_\mu = \frac{\langle lv\rangle}{\langle v^2\rangle} = \frac{lv_T 1!}{v_T{}^2\cdot\frac{3}{2}!} = \frac{l}{v_T\dfrac{3\sqrt{\pi}}{4}} = \frac{4l}{3\sqrt{2\pi kT/m}} \tag{28a}$$

$$\mu = \frac{e\tau_\mu}{m} = 0.753\frac{el}{mv_T} = \frac{4el}{3\sqrt{2\pi mkT}} \tag{28b}$$

$$\frac{\mu_H}{\mu} = \frac{\tau_H}{\tau_\mu} = \frac{\langle l^2\rangle\langle v^2\rangle}{\langle lv\rangle^2} = \frac{\frac{1}{2}!\frac{3}{2}!}{(1!)^2} = \frac{3\pi}{8} = 1.18 \tag{28c}$$

$$R_H = \frac{3\pi}{8n(\pm e)c} = \frac{1.18}{n(\pm e)c}. \tag{28d}$$

The formula for l derived on the basis of deformation potentials and quoted in Section 11.3b is for electrons

$$l = \frac{\pi\hbar^4 c_{ll}}{\mathcal{E}_{1n}{}^2 kTm_n{}^2}, \tag{29}$$

where c_{ll} is the elastic constant for [110] longitudinal acoustical waves, $\mathcal{E}_{1n}$ is the derivative of $\mathcal{E}_c$ with respect to the logarithm of the volume of the unit cell, and m_n is the effective mass. Inserting this in the formula for μ gives

$$\mu_n = \frac{2\sqrt{2\pi}}{3} \frac{\hbar^4 c_{ll}}{\mathcal{E}_{1n}{}^2 m_n{}^{5/2} k^{3/2}} T^{-3/2}. \tag{30a}$$

Expressed in cm^2/volt sec for μ, dynes/cm^2 for c_{ll}, electron volts for $\mathcal{E}_{1n}$ and °K for T, this becomes

$$\mu = 3.2 \times 10^{-5} c_{ll}(m/m_n)^{5/2}/\mathcal{E}_{1n}{}^2 T^{3/2}. \tag{30b}$$

Similar expressions apply to holes.

Impurity Scattering; the Conwell-Weisskopf Formula. For this case the dependence of τ upon v is given by the complicated relationship (17) of Section 11.3,

$$\tau_I = \frac{\dfrac{2a_I}{\pi v}\left(\dfrac{2\mathcal{E}}{\mathcal{E}_I}\right)^2}{\ln\left[1 + \left(\dfrac{2\mathcal{E}}{\mathcal{E}_I}\right)^2\right]} \tag{31}$$

where the symbols are defined in terms of the density N_I of scattering impurities and the dielectric constant κ as follows:

$$a_I = N_I{}^{-1/3}, \quad \mathcal{E}_I = 2e^2/\kappa a_I, \quad \mathcal{E} = m_n v^2/2 \tag{32}$$

where m_n is the effective mass for an electron. τ_I can be rewritten as

$$\tau_I = \frac{A\mathcal{E}^{3/2}}{\ln\left[1 + \left(\dfrac{2\mathcal{E}}{\mathcal{E}_I}\right)^2\right]}. \tag{33}$$

The quantity $(2\mathcal{E}/\mathcal{E}_I)$ was found to be about 30 for a typical high resistivity sample at room temperature so that the logarithm term has a value of 6.8. This term varies slowly with $\mathcal{E}$, in fact a change of $e^{\pm 1}$ in $\mathcal{E}$ will make the logarithm vary from 4.8 to 8.8. For this reason, no great error will be made by replacing the logarithm term by a constant corresponding to its value at the most important part of the range while averaging $v^2\tau$. Now the average of $v^2\tau$ is the same as the average of $2m_n\mathcal{E}\tau$, and the number of electrons in a range $d\mathcal{E}$ is proportional to $\mathcal{E}^{1/2}\exp(-\mathcal{E}/kT)d\mathcal{E}$. Thus the integrand in $\langle v^2\tau\rangle$ involves the factors

$$(\mathcal{E}^{3/2} \text{ from } \tau_I)(\mathcal{E} \text{ from } v^2)\mathcal{E}^{1/2}e^{-\mathcal{E}/kT} = \mathcal{E}^3 e^{-\mathcal{E}/kT}; \tag{34}$$

this has its maximum at $\mathcal{E} = 3kT$. Hence we replace $\mathcal{E}$ by $3kT$ in the logarithm term and write τ_I in the form

$$\tau_I = Bv^3, \quad B = \frac{m_n{}^2\kappa^2/2\pi N_I e^4}{\ln\left[1+\left(\dfrac{6kT}{\mathcal{E}_I}\right)^2\right]}. \tag{35}$$

Since τ is now proportional to a power of v, the previously derived formulae are applicable and we obtain[1]

$$\tau_\mu = \frac{B\langle v^5\rangle}{\langle v^2\rangle} = Bv_T{}^3\frac{3!}{\frac{3}{2}!} = \frac{8}{\sqrt{\pi}}Bv_T{}^3; \tag{36a}$$

$$\mu = \frac{e}{m_n}\tau_\mu = \frac{\dfrac{8}{\pi}\sqrt{\dfrac{2}{\pi}}\dfrac{\kappa^2(kT)^{3/2}}{N_I e^3 m_n{}^{1/2}}}{\ln\left[1+\left(\dfrac{3\kappa kT}{e^2N_I{}^{1/3}}\right)^2\right]}; \tag{36b}$$

$$\frac{\mu_H}{\mu} = \frac{\tau_H}{\tau_\mu} = \frac{B^2\langle v^8\rangle\langle v^2\rangle}{B^2\langle v^5\rangle^2} = \frac{\frac{9}{2}!\,\frac{3}{2}!}{3!\,3!} = \frac{315\pi}{512} = 1.93; \tag{36c}$$

$$R_H = \frac{1.93}{n(\pm e)c}. \tag{36d}$$

Intrinsic or Near-Intrinsic Range. In Chapter 8, the formulae for conductivity and Hall coefficient for the case in which both holes and electrons are present were derived in terms of μ and θ. We shall repeat the formulae here modified in keeping with the notation of this chapter:

$$\sigma = ne\mu_n + pe\mu_p \tag{37}$$

$$R_H = \frac{\theta}{\sigma H} = \frac{-(ne\mu_n e\tau_{Hn}H/m_nc) + (pe\mu_p e\tau_{Hp}H/m_pc)}{\sigma^2 H}$$

$$= \frac{-n\tau_{\mu n}\tau_{Hn}/m_n{}^2 + p\tau_{\mu p}\tau_{Hp}/m_p{}^2}{ec(n\tau_{\mu n}/m_n + p\tau_{\mu p}/m_p)^2} \tag{38}$$

[1] The formula for μ is due to E. Conwell and V. F. Weisskopf, *Phys. Rev.* **69**, 258A (1946); **77**, 388–390 (1950).

If the ratio of τ_H/τ_μ is the same for holes and electrons, this formula becomes

$$R_H = \frac{\tau_H}{\tau_\mu}\frac{-nb^2 + p}{(nb + p)^2ce} \tag{39}$$

where $b = \mu_n/\mu_p$. For thermal scattering, $\tau_H/\tau_\mu = 3\pi/8$, and the formula reduces to that quoted in Chapter 8.

11.4b. The General Case. For the case of a degenerate semiconductor, similar expressions can be obtained, but, as discussed previously, the effectiveness of electrons must be weighted by $1 - \mathbf{f} = \mathbf{f}_p$ and holes by $(1 - \mathbf{f}_p) = \mathbf{f}$. The corresponding general expressions derived from equation (10) and slight generalizations of (14a) and (14b) become for n-type:

$$\sigma = ne\frac{e\langle \mathbf{f}_p v^2 \tau\rangle}{3kT} = ne\frac{e\tau_\mu}{m}, \tag{40}$$

and

$$\tau_H = \langle \mathbf{f}_p v^2 \tau^2\rangle/\langle \mathbf{f}_p v^2 \tau\rangle, \tag{41}$$

where again each conduction electron is given equal weight in taking the average. The expressions suggest that for the degenerate case τ_μ is a weighted average of τ over the electrons with each electron having a weight $v^2\mathbf{f}_p$. For spherical energy surfaces this is found to be the case, for then $3kT$ is $m\langle v^2\mathbf{f}_p\rangle$ (see Problems) and then we may write

$$\tau_\mu = \langle \tau v^2\mathbf{f}_p\rangle/\langle v^2\mathbf{f}_p\rangle; \quad \tau_H/\tau_\mu = <\tau^2v^2\mathbf{f}_p><v^2\mathbf{f}_p>/<\tau v^2\mathbf{f}_p>^2. \tag{42}$$

It should be noted that (40) is the general expression of which (42) is only a special case. Precisely similar expressions are obtained for holes by replacing $\mathbf{f}_p$ by $\mathbf{f}$ and averaging over the holes.

11.4c. Simple Metals and Degenerate Semiconductors. For the case of a highly degenerate semiconductor, or a metal, most of the states in the band of interest are either full with $\mathbf{f} = 1$ and $\mathbf{f}(1 - \mathbf{f}) = 0$ or else they are empty with $\mathbf{f} = 0$ and again with $\mathbf{f}(1 - \mathbf{f}) = 0$. The important contributions in the sum over the energy shells thus come from a relatively narrow energy region, about $2kT$ wide, centered at $\mathcal{E} = \mathcal{E}_F$, in which $\mathbf{f}(1 - \mathbf{f})$ is approximately $(\frac{1}{2})^2 = 0.25$. In this narrow range, the other functions of $\mathcal{E}$ may be regarded as constant. Accordingly, the integral for σ, obtained from the integrated form of (8) by dividing by E_x, is

$$\sigma = \frac{e^2}{3kTV}\tau v^2\int \mathbf{f}(1 - \mathbf{f})dS \tag{43}$$

where the approximation of spherical energy surfaces has been used in taking v^2 out of the integral. The integral can be converted directly to an integration over energy by using equation (9) of Section 9.1:

$$dS = (4\pi/h^3)(2m)^{3/2}(\mathcal{E} - \mathcal{E}_0)^{1/2}d\mathcal{E}. \tag{44}$$

It is simpler, however, to use another procedure. If we let[2] $\mathcal{E}_1$ represent $\mathcal{E} - \mathcal{E}_0$, then the number of states S in the Brillouin zone with energy less than $\mathcal{E}_1$ is proportional to P^3 or to $\mathcal{E}_1^{3/2}$. As a consequence of this, we have at the highest occupied states, that is, for $\mathcal{E}_1 = \mathcal{E}_F - \mathcal{E}_0$,

$$\frac{dS}{d\mathcal{E}_1} = \frac{3}{2}\frac{S}{\mathcal{E}_1} = \frac{3}{2}\frac{nV}{\mathcal{E}_1} \quad \text{or} \quad dS = \frac{3nV}{2\mathcal{E}_1}d\mathcal{E}_1 \tag{45}$$

since the total number of occupied states is the electron density n times V. The integration of $\mathbf{f}(1 - \mathbf{f})d\mathcal{E}$ is then carried out, using equation (6), as follows:

$$\int \frac{1}{kT} d\mathcal{E}_1\mathbf{f}(1 - \mathbf{f}) = \int -d\mathbf{f} = 1 \tag{46}$$

when integrated across the Fermi level $\mathcal{E}_F$ since $\mathbf{f}$ varies from unity to zero over this range. Thus σ becomes

$$\sigma = \frac{e^2}{3V}\tau v^2 \frac{3nV}{2\mathcal{E}_1} = ne\frac{e\tau}{m}; \quad \mu = \frac{e\tau}{m}, \tag{47}$$

the second equality following from $\mathcal{E}_1 = mv^2/2$. This expression is just that obtained on the basis of a single mean free time in Chapter 8. The physical reason for this simple result is that all the electrons are accelerated by the field so that $\dot{I} = ne^2\mathbf{E}/m$ as discussed in Chapter 5. The decay of the current is all brought about by electrons in the outermost energy shell however, since all inner shells are full. Thus, the conductivity involves the density n of all the electrons, but the relaxation time τ of only those of highest energy. The transverse component of current produced by the magnetic field also decays with the time constant τ and consequently $\tau = \tau_H$. Since τ_μ and τ_H are equal, the formula for the Hall constant is simply

$$R_H = -1/nec. \tag{48}$$

The Conwell-Weisskopf formula has been extended to the degenerate range by Johnson and Lark-Horovitz.[3] For this case only a single energy is involved and it is not necessary to average τ_I over the energy. The energy $\mathcal{E}$ required is the highest occupied state of the degenerate electron gas; it is evaluated as follows:

$$n = \frac{2}{h^3}\frac{4\pi}{3}P^3, \quad P = \frac{h}{2}\left(\frac{3n}{\pi}\right)^{1/3}, \quad \mathcal{E} = \frac{h^2}{8m}\left(\frac{3n}{\pi}\right)^{2/3}. \tag{49}$$

[2] $\mathcal{E}_1$ in these manipulations is not to be confused with $\mathcal{E}_1$ used for thermal scattering.

[3] V. A. Johnson and K. Lark-Horovitz, *Phys. Rev.* **71**, 374–375 (1947). They also give a general integral for intermediate cases.

The formula for the relaxation time discussed in (17) of Section 11.3 is

$$\tau_I = \frac{4}{\pi} \frac{\tau_c}{\langle 1 - \cos\theta \rangle} = \frac{4}{\pi} \frac{a_I}{v} \frac{(2\mathcal{E}/\mathcal{E}_I)^2}{2 \ln [1 + (2\mathcal{E}/\mathcal{E}_1)^2]} \tag{50}$$

The quantity $2\mathcal{E}/\mathcal{E}_I$ has the value[4]

$$\frac{2\mathcal{E}}{\mathcal{E}_I} = 2\mathcal{E}\frac{\kappa a_I}{2e^2} = \frac{h^2}{8m}\left(\frac{3n}{\pi}\right)^{2/3} \frac{\kappa}{e^2 n^{1/3}} = \frac{(3/\pi)^{2/3}}{8} \cdot \frac{\kappa h^2 n^{1/3}}{me^2}$$

$$= 3.4 \times 10^{-7} n^{1/3} \quad (\text{for } S_i)$$

$$= 5.0 \times 10^{-7} n^{1/3} \quad (\text{for } G_e), \tag{51}$$

with m as the electron mass and $\kappa = 13$ and 19 respectively.[4a] In the foregoing equation we have set $n = n_I$ since in a "doped" degenerate sample the electrons and donors must be very nearly equal in number. From this equation we see that $2\mathcal{E}/\mathcal{E}_I = 1$ for $n \doteq 3 \times 10^{19}$. From a limiting form for the average of the mean free time, Johnson and Lark-Horovitz conclude that $2\mathcal{E}/\mathcal{E}_I$ will be small.[5] On this basis they simplify their expression by what is equivalent to expanding the logarithm term, leading to a cancellation of the term $(2\mathcal{E}/\mathcal{E}_I)^2$ in the numerator, a result equivalent to setting $\langle 1 - \cos\theta \rangle$ equal to 2. The physical meaning is that, for $\mathcal{E} \ll \mathcal{E}_I$, the electron will follow an almost parabolic orbit around the ion and shoot back out of the cell with its velocity completely reversed, that is, the collision will have $\theta = 180°$ or $\cos\theta = -1$. For this case the formula for mobility becomes

$$\mu = \frac{e}{m}\tau_I = \frac{e}{m}\frac{2a_I}{\pi v} = \frac{2en^{-1/3}}{\pi P} = \frac{4}{\pi}\left[\frac{\pi}{3}\right]^{1/3} \frac{e}{hn^{2/3}} \tag{52}$$

and the resistivity becomes

$$\rho = \frac{1}{ne\mu} = \frac{\pi}{4}\left[\frac{3}{\pi}\right]^{1/3} \frac{h}{e^2} n^{-1/3}. \tag{53}$$

The value obtained by Johnson and Lark-Horovitz is π times smaller due to differences in the approximations. In practical units their value becomes

$$\rho = 6270 n^{-1/3} \text{ ohm-cm} \tag{54}$$

with n in cm^{-3}.

11.5 COMPARISON WITH EXPERIMENT

The most extensive presentation of a comparison of theory with experiment for semiconductors of the type covered by the theory of this chapter

[4] $2\mathcal{E}/\mathcal{E}_I$ is the same function of n as is y of equation (62) in G. L. Pearson and J. Bardeen, *Phys. Rev.* **75**, 865–883 (1949).

[4a] See Chapter 9 for later values of κ.

[5] This conclusion does not seem to be valid unless $n < 3 \times 10^{19}$ as discussed previously.

is found in the paper of Pearson and Bardeen, to which we have so often referred.[1] The specimens studied by them may be broadly divided into two classes: degenerate and non-degenerate, divided by the condition [see equation (16) of Section 10.3]

$$T = \left[\frac{3}{\pi}\right]^{2/3} \frac{h^2}{8km_n} n^{2/3} = 4.2 \times 10^{-11} \left[\frac{m}{m_n}\right] n^{2/3}\ {}^\circ\mathrm{K} \tag{1a}$$

$$kT = 21.6 \times 10^{-16}(m/m_n)n^{2/3} \text{ electron volts} \tag{1b}$$

for n-type, and a similar equation for p-type.

For the degenerate cases, the important electron velocity corresponds to the highest occupied electron level and does not change appreciably with temperature. For these cases, the impurity density is so high that impurity scattering dominates lattice scattering. As a result, the resistivity and Hall coefficient are both practically constant over a wide temperature range. Samples 6, 7, 8 in Figures 11.7 and 11.8 (taken from Pearson and Bardeen) exhibit this behavior. These are p-type samples containing added densities of boron to the amount indicated on the curves. The fact that the added boron content for sample 8 exceeds the number of carriers is interpreted, on the basis of other evidence, as due to the precipitation of a second boron-rich phase on the grain boundaries of this alloy, thus diminishing the effective boron concentration. The samples become non-degenerate at higher temperatures, the line of demarcation being indicated as T_{deg} on Figure 11.8 where m/m_n has been set equal to unity for simplicity. (At still higher temperatures they would become intrinsic.) The mobilities for 6, 7, 8 are respectively 31, 30, 54 cm^2/volt sec. This increase of mobility with increasing impurity content is in the opposite direction from the trend predicted by extending the Conwell-Weisskopf formula into this range even if the complete expression rather than the Johnson–Lark-

[1] Very extensive and early investigations of the properties of germanium along the lines discussed here have been reported from the group at Purdue working under the direction of Professor K. Lark-Horovitz. Partial reports of their findings are given in the following references: K. Lark-Horovitz, A. E. Middleton, E. P. Miller, and I. Walerstein, *Phys. Rev.* **69**, 258A (1946).

Additional investigations on germanium are reported in the following references: J. Bardeen and W. H. Brattain, *Phys. Rev.* **75** (8), 1208–1225 (1949); *Bell Syst. Tech. J.* **28** (2), 239–277 (1949).

Additional references of general interest in this connection are: K. Lark-Horovitz and V. A. Johnson, *Phys. Rev.* **69**, 258A (1946); V. A. Johnson and K. Lark-Horovitz, *Phys. Rev.* **71**, 374–375 (1947); **72**, 531 (1947); K. Lark-Horovitz, A. E. Middleton, E. P. Miller, W. W. Scanlon, and I. Walerstein, *Phys. Rev.* **69**, 259 (1946); V. A. Johnson and K. Lark-Horovitz, *Phys. Rev.* **69**, 259 (1946); W. Ringer and H. Welker, *Zeits. für Naturforschung* **3A**, 20–29 (1948); W. C. Dunlap, Jr., *G. E. Review* **52**, 9, February 1949. Much of the foregoing material is reviewed by G. L. Pearson, K. Lark-Horovitz, and N. C. Jamison, *Electrical Engineering* **68**, 1047–1056 (1949).

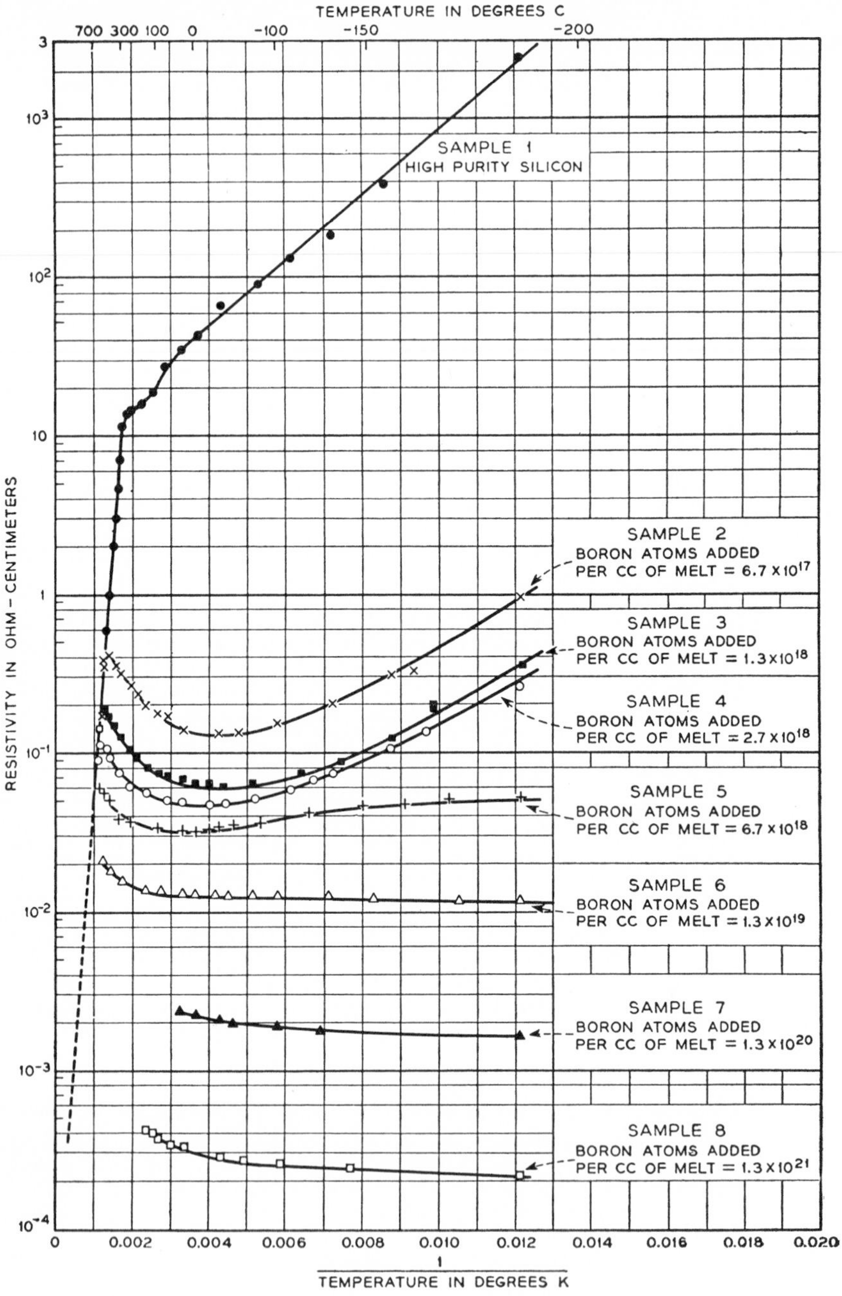

FIG. 11-7—Resistivity of Silicon-boron Alloys versus Temperature.

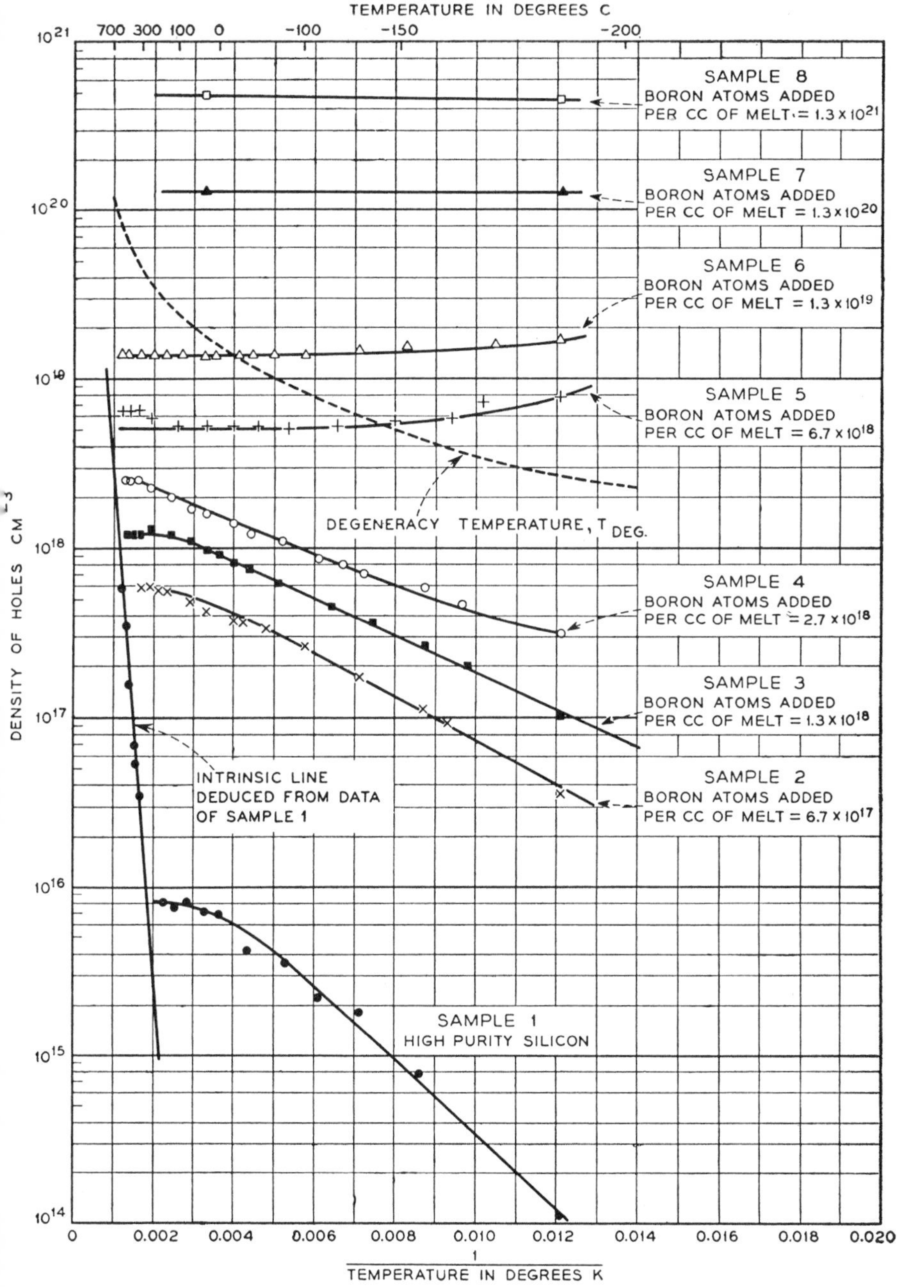

FIG. 11–8—Density of Holes in Silicon-boron Alloys versus Temperature.

Horovitz approximation is used. Pearson and Bardeen have considered in addition a formula derived by Mott for application to alloys. Although all these formulae give general agreement with experiment as to order of magnitude of the mobilities, none appears to agree in detail in giving the correct dependence upon n_I. Sample 5, which is expected to be degenerate below $-150°$C, is anomalous in its behavior and does not have a constant mobility at low temperatures.

Johnson and Lark-Horovitz have also made a comparison between their formula $\rho = 6270n^{-1/3}$ and some germanium samples having resistivities in the range of 2.9 to 5.1×10^{-3} ohm-cm, obtaining good agreement.

There are difficulties in connection with applying the Conwell-Weisskopf procedure to the degenerate samples. The wave length of the fastest electrons of the degenerate gas is given by the formula

$$\lambda = \frac{h}{P} = 2\left[\frac{\pi}{3n}\right]^{1/3} = 2\left[\frac{\pi}{3}\right]^{1/3} a_I. \tag{2}$$

Hence the unit cube considered in the scattering model is only $\frac{1}{2}$ wave length wide and its "radius" is about $\frac{1}{4}\lambda$. For this relationship between wave length and spacing, the approximation of using the Rutherford formulae becomes completely untenable on the classical basis and very dubious on the quantum-mechanical basis since the individual scattering centers are so close together. Thus in addition to problems of determining the effective potential field in which the electrons move, there will be problems of coherence in the scattering.

For the non-degenerate samples, the formulae of the last section should be applicable. However, in certain ranges of temperature both scattering processes may be important and it is necessary to consider the combined effects of impurity and thermal scattering. For electrons of a given energy, the two relaxation processes will be additive. Hence, a perturbation of the equilibrium will decay in time dt by a fraction

$$\frac{dt}{\tau_I} + \frac{dt}{\tau_T} = \left[\frac{1}{\tau_I} + \frac{1}{\tau_T}\right] dt = \frac{1}{\tau}\, dt, \tag{3}$$

where

$$\tau = \tau_I \tau_T / (\tau_I + \tau_T). \tag{4}$$

This quantity should then replace τ in the averages used in computing τ_μ and τ_H. If τ_μ and τ_L depended upon v in the same way, this would lead to the conclusion that

$$\tau_\mu = \frac{\tau_{\mu I} \tau_{\mu T}}{\tau_{\mu I} + \tau_{\mu T}} \tag{5}$$

and a similar expression for τ_H. The errors introduced by this process, due to the fact that τ_I and τ_T depend differently on v, may amount to 50 per cent or more. We shall not analyze this question further at this point but will use the approximation generally employed in comparing theory and experiment and take the two relaxation times as additive.

We thus write as an approximation,

$$\frac{1}{\tau_\mu} = \frac{1}{\tau_{\mu I}} + \frac{1}{\tau_{\mu T}}, \tag{6}$$

and hence

$$\frac{1}{\mu} = \frac{1}{\mu_I} + \frac{1}{\mu_T}, \tag{7}$$

where μ_I and μ_T are the mobilities corresponding to the effects of impurities and thermal agitation individually. For these the formulae are:

$$\mu_I = \frac{\dfrac{8}{\pi}\sqrt{\dfrac{2}{\pi}}\,\dfrac{\kappa^2 (kT)^{3/2}}{N_I e^3 m_n^{1/2}}}{\ln\left[1 + \left(\dfrac{3\kappa kT}{e^2 N_I^{1/3}}\right)^2\right]} \tag{8}$$

giving for silicon with $\kappa = 13$,

$$\mu = 5.5\times 10^{17} T^{3/2}/N_I \ln(1+x^2)\ \text{cm}^2/\text{volt sec} \quad (\text{for } N_I\ \text{cm}^{-3},\ T^\circ\text{K}) \tag{9a}$$

$$x = 2.3\times 10^4 T/N_I^{1/3} \tag{9b}$$

and similarly for germanium with $\kappa = 19$,

$$\mu = 1.2 \cdot 10^{18} T^{3/2}/N_I \ln(1 + x^2) \tag{10a}$$

$$x = 3.4 \cdot 10^4 T/N_I^{1/3}. \tag{10b}$$

The formula for μ_T is

$$\mu_T = \frac{2\sqrt{2\pi}}{3}\,\frac{\hbar^4 c_{ll}}{\mathcal{E}_{1n}^2 m_n^{5/2} k^{3/2}} = \text{const}\ T^{-3/2} \tag{11}$$

and the experimentally determined values are

$$\mu_{nT} = 15 \cdot 10^5 T^{-3/2}\ \text{cm}^2/\text{volt sec for silicon}; \tag{12a}$$

$$\mu_{pT} = 5 \cdot 10^5 T^{-3/2}\ \text{cm}^2/\text{volt sec for silicon}; \tag{12b}$$

$$\mu_{nT} = 19 \cdot 10^6 T^{-3/2}\ \text{cm}^2/\text{volt sec for germanium}; \tag{12c}$$

$$\mu_{pT} = 8.9 \cdot 10^6 T^{-3/2}\ \text{cm}^2/\text{volt sec for germanium}. \tag{12d}$$

The temperature dependencies of μ_I and μ_T are quite different, their variations being approximately as $T^{3/2}$ and as $T^{-3/2}$ respectively. In Figure 11.9 these mobilities are plotted for holes and electrons in silicon and germanium.[2]

[2] Values of κ and of drift velocities in germanium obtained since this figure was prepared differ slightly. See Chapters 9 and 12.

Pearson and Bardeen find that they can fit the mobility data for non-degenerate samples, Figure 11.10, reasonably well with a formula of the form

$$\frac{1}{\mu} = \frac{a}{T^{3/2}} + bT^{3/2} \tag{13}$$

which corresponds to replacing the μ_I curves of Figure 11.9 by straight lines. The values of a which they obtain are within about a factor of 2

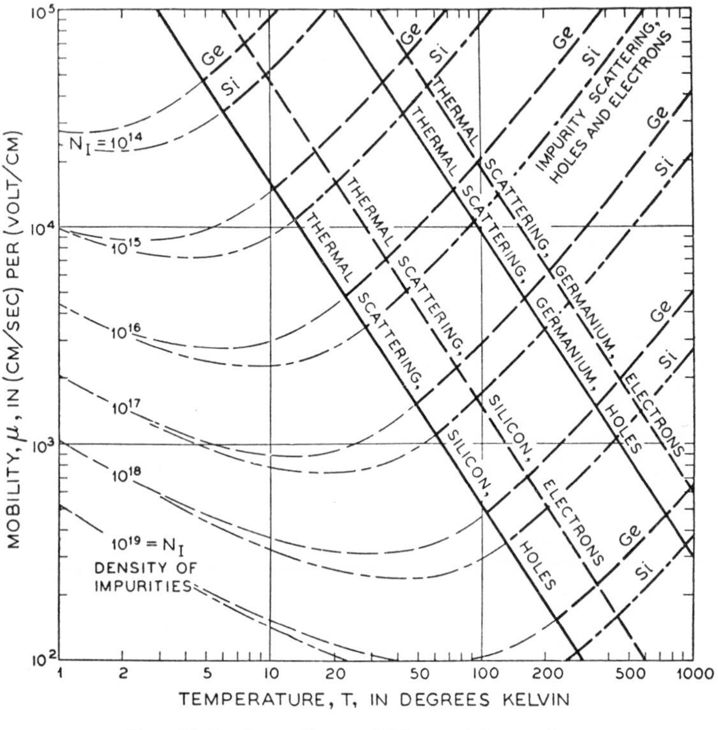

FIG. 11-9—Impurity and Thermal Scattering.

of those predicted from the equation for μ_I and the impurity content, the impurity content being estimated from the saturation behavior. The same degree of agreement is obtained for the data quoted in Chapter 1.

Curve 1 deserves special comment; it represents a highly purified sample of silicon obtained from E. I. du Pont de Nemours & Company. The anomalous behavior of very high resistivity samples has often been observed in silicon and germanium and is attributed to inhomogeneity. For such samples, small density differences from place to place will make large proportional differences in conductivity. Under these conditions, the current paths become tortuous, and for such paths an artificially low Hall effect and mobility will be obtained.

$$l = 3\sqrt{2\pi m_n kT}\,\mu/4e = 4.12 \times 10^{-10}(m_n T/m)^{1/2}\mu_L cm \qquad (14)$$

The value of the mean free path in the thermal agitation range is from (28a) of Section 11.4. For $T = 300°K$ and $m_n = m$, $l = 7.1 + 10^{-9}\mu_L cm$.

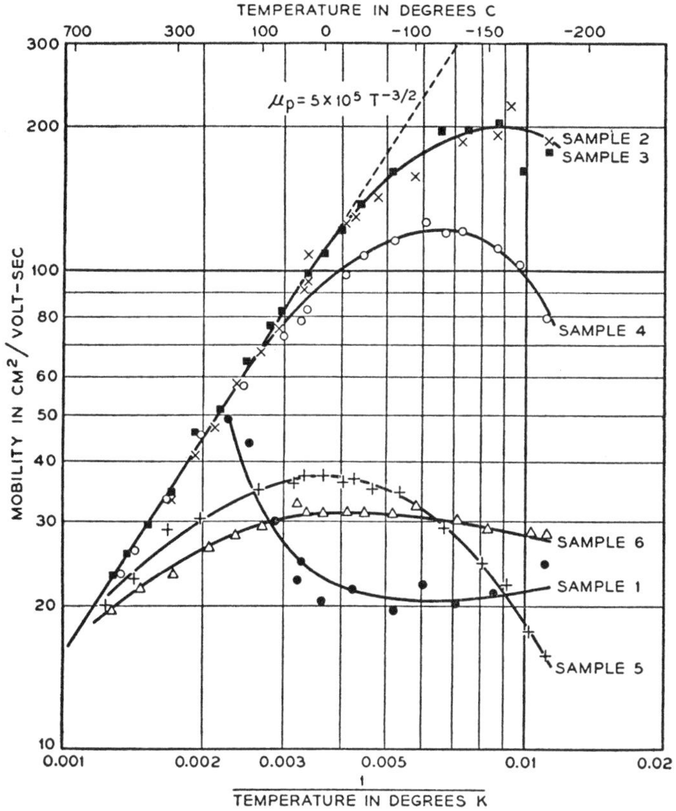

FIG. 11-10—Hole Mobility in Silicon-boron Alloys versus Temperature.

Values of the mean free path based on this formula with $m_n/m = 1$ and $m_p/m = 1$ are shown in Figures 11.11 and 11.12. The values for low temperatures, where lattice scattering predominates, correspond to inserting (36b) of Section 11.4 into (14) thus obtaining

$$l = \frac{\dfrac{4\kappa^2(kT)^2}{\pi N_I e^4}}{\ln\left[1 + \left(\dfrac{3\kappa kT}{e^2 N_I^{1/3}}\right)^2\right]} = \frac{a_I 16\pi\left(\dfrac{kT}{\mathcal{E}_I}\right)^2}{\ln\left[1 + \left(\dfrac{6kT}{\mathcal{E}_I}\right)^2\right]} \qquad (15)$$

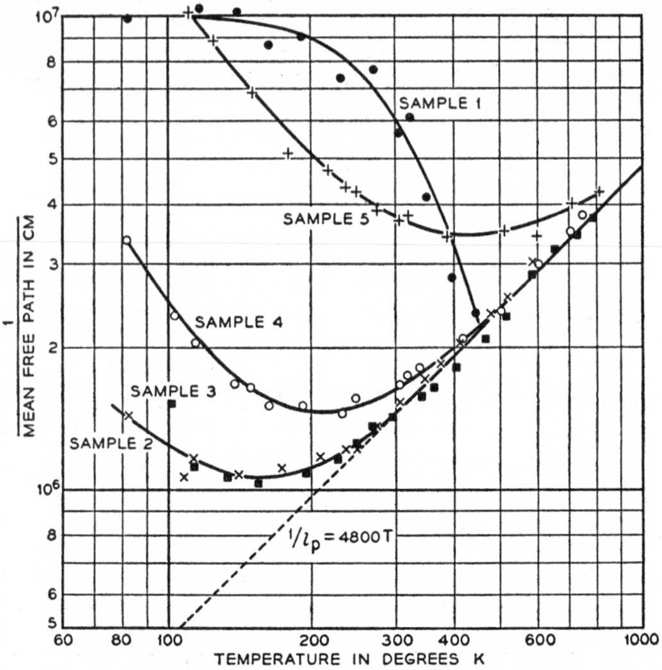

FIG. 11–11—Mean Free Path in Silicon-boron Alloys versus Temperature.

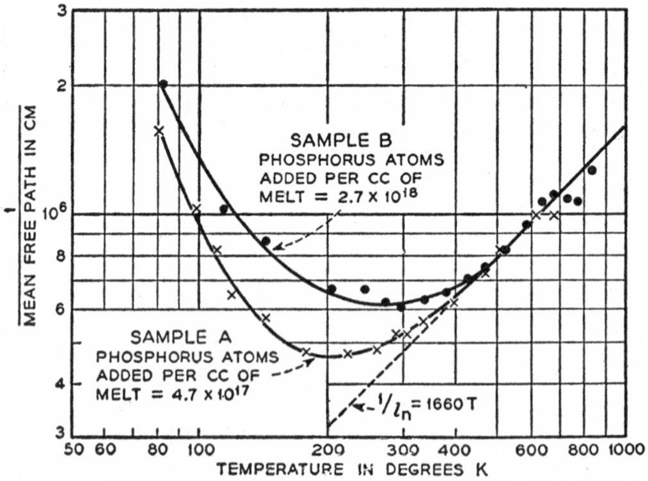

FIG. 11–12—Mean Free Path in Silicon-phosphorous Alloys versus Temperature.

where, in the notation of Section 8.3a,

$$a_I = N_I^{-1/3} \quad \text{and} \quad \mathcal{E}_I = 2e^2/\kappa a_I. \tag{16}$$

Formula (15), as well as the curves of Figure 11.10, may be used to estimate ion densities from mobility data.

Problems

1. Derive the equation of the hyperbolic orbits. This may be done by setting the radial acceleration equal to the centrifugal force minus the attractive force

$$m\ddot{r} = mr\dot{\theta}^2 - Ze^2/\kappa r^2,$$

and eliminating $\dot{\theta}$ by the principle of conservation of angular momentum in a central field:

$$mr^2\dot{\theta} = J.$$

The resulting equation for $\ddot{r}$ may be converted to an equation for $d^2u/d\theta^2$, where $u = 1/r$, by two steps:

$$\frac{dr}{dt} = \frac{dr}{d\theta}\dot{\theta} = \frac{J}{m}\frac{1}{r^2}\frac{dr}{d\theta} = -\frac{J}{m}\frac{du}{d\theta}$$

followed by a similar procedure for $(d/dt)(dr/dt)$. The resulting equation may then be integrated to find u as a function of θ, the result being:

$$u = \frac{1}{r} = \frac{Ze^2m}{\kappa J^2} + A\cos(\theta + \theta_0)$$

where A and θ_0 are integration constants.

For the derivation of the $\ddot{r}$ equation and other steps, see standard introductory texts to theoretical physics, such as L. Page, *Introduction to Theoretical Physics*, D. Van Nostrand Co., New York, 1935.

2. Apply the expression for current density of Chapter 15 to the transmitted wave of equation (18) and verify that the transmitted current is

$$4P_aP_b/(P_a + P_b)^2 \sim |4Z_1Z_2/(Z_1 + Z_2)^2|,$$

the latter form applying to electrical transmission lines. Verify that the sum of reflected and transmitted currents equals the incident current.

3. Estimate the change in energy of an electron when it interacts with a phonon by considering the Doppler effect produced by the motion of the deformation pattern. Consider first the case of motion along the x-axis. Let the incident electron wave function be $\exp i(P_1x - \mathcal{E}_1t)/\hbar$ and the reflected wave be $\exp i(-P_2x - \mathcal{E}_2t)\hbar$. Let the pattern move with speed c so that its position is given by $x - ct = 0$. Show that the requirement of constant phase relation between incident and reflected waves at the location $x - ct$ leads to $P_2 - P_1 = 2mc$ if $\mathcal{E} = P^2/2m$ and hence for $v_1 = P_1/m > c$ that $\Delta\mathcal{E} \doteq mvc$.

Show further that, if the moving pattern has a periodic form such as $\sin 2\pi(x - ct)/\lambda_c$, the requirement of reinforcement of the waves reflected from successive planes of equal deformation is $(P_2 + P_1)\lambda/\hbar = 2\pi$ and hence that $\mathcal{E}_1 - \mathcal{E}_2 = 2\pi\hbar c/\lambda_c = h\nu$, independently of the form of the functional dependence of $\mathcal{E}$ on P.

A more difficult problem is to generalize these results to three dimensions and compare them with the procedure of Section 17.6, Figure 17.6, and associated text.

4. Show that in Section 11.4 equation (40) is equivalent to (43) when applied to a metal.

5. Establish the formulae required for calculating $\langle v^n \rangle$ for integral values of n by letting $1/v_T^2 = \alpha$ and writing

$$I_n = \int_0^\infty e^{-\alpha v^2} v^{n+2} dv = \frac{1}{\alpha^{(n+3)/2}} \int_0^\infty e^{-x^2} x^{n+2} dx$$

and evaluating $dI_n/d\alpha$ by differentiating under the integral sign. This leads to equation (21) of 11.4. Evaluate I_{-1} directly, and I_{-2} by considering

$$(I_{-2})^2 = \int_0^\infty e^{-\alpha x^2} dx \int_0^\infty e^{-\alpha y^2} dy = \int_0^{\pi/2} d\theta \int_0^\infty e^{-\alpha r^2} r dr = \pi/4\alpha.$$

6. Prove that, if the spherical approximation holds, so that $dS = av^2dv$ where a is a constant, then

$$\langle v^2 \mathbf{f}_p \rangle = 3kTm$$

when averaged over the electrons in the distribution. This may be proved from the definition of $\langle v^2\mathbf{f}_p \rangle$ which is

$$\langle v^2 \mathbf{f}_p \rangle = \int_0^\infty v^2 \mathbf{f}_p \mathbf{f} a v^2 dv \bigg/ \int_0^\infty \mathbf{f} a v^2 dv$$

by integrating the numerator by parts with the aid of equation (6) of Section 11.4.

7. Obtain the formal expression for the conductivity tensor for a non-cubic crystal in which $\boldsymbol{\tau}$ is a function of $\mathcal{E}$ only. The rate of change of current in $d\Omega$ in shell $d\mathcal{E}$ due to $\boldsymbol{E}$ is then

$$\left[\frac{d}{dt} d\boldsymbol{I}_s\right]_E = \frac{2e^2}{h^3kT} \boldsymbol{v}(\boldsymbol{E} \cdot \boldsymbol{v})\mathbf{f}(1 - \mathbf{f}) d\frac{\Omega d\mathcal{E}}{v}$$

where $v = |\boldsymbol{v}|$. Hence show that

$$\sigma = \frac{2e^2}{h^3kT} \int \boldsymbol{v}\boldsymbol{v}\boldsymbol{\tau}\mathbf{f}(1 - \mathbf{f}) \frac{d\Omega d\mathcal{E}}{v}$$

where $\boldsymbol{vv}$ is a tensor operator such that

$$(\boldsymbol{vv})\boldsymbol{E} = \boldsymbol{v}(\boldsymbol{v} \cdot \boldsymbol{E}).$$

8. There is an apparent paradox associated with the occurrence of $\langle v^2\tau\rangle$ in the expression for τ_μ. From the viewpoint of an individual electron, the reasoning of Chapter 8 suggests that $\mu = e\tau/m$. Hence, if different electrons have different values for τ, we should expect to have $\mu = e\langle\tau\rangle/m$. The error in the latter expression can be attributed to the neglect of the variation of mean free time of the electron due to the change in its velocity while it is being accelerated. The problem is to show that when this effect is taken into account the correct formula for conductivity may be obtained by considering the trajectories of the individual electrons as was done in Section 8.5.

Suppose that after a collision an electron is equally likely to end in any state in an energy shell $d\mathcal{E}$ but that $\tau(\boldsymbol{P})$ may vary from point to point in the energy shell. Show from the principle of detailed balancing and the symmetry condition $\mathcal{E}(\boldsymbol{P}) = \mathcal{E}(-\boldsymbol{P})$ that the rate at which electrons are scattered into a volume $dV_{\boldsymbol{P}i}$ is unaffected to the first order by applied electric fields and has a value

$$\delta g_i = (2/h^3)\mathrm{f}[\mathcal{E}(\boldsymbol{P}_i)]dV_{\boldsymbol{P}i}/\tau(\boldsymbol{P}_i).$$

If the average displacement of each of these electrons before its next collision is $\bar{\boldsymbol{r}}_i$, show that the average value of $\sum\boldsymbol{v}$ for all the electrons is

$$\sum\boldsymbol{v} = \sum\delta g_i\bar{\boldsymbol{r}}_i = \frac{2}{h^3}\int\frac{1}{\tau_i}\,\bar{\boldsymbol{r}}_i\mathrm{f}[\mathcal{E}(\boldsymbol{P}_i)]dV_{\boldsymbol{P}i}.$$

Owing to the effect of the field, which is taken as $\boldsymbol{i}_xE_x$ for simplicity, each electron is accelerated by a force F_x. As a result, its mean free time and velocity vary as

$$\tau = \tau_i + \frac{\partial\tau}{\partial P_x}F_xt$$

$$\boldsymbol{v} = \boldsymbol{v}_i + \frac{\partial\boldsymbol{v}}{\partial P_x}F_xt.$$

Show that the probability that the electron has not collided after time t is

$$c(t) = \left(1 + \frac{\partial\tau}{\partial P_x}F_x\frac{t^2}{2\tau_i^{\,2}}\right)\mathrm{e}^{-t/\tau_i}$$

and hence that its average displacement is

$$\bar{\boldsymbol{r}}_i = \boldsymbol{v}_i\tau_i + \left(\boldsymbol{v}_i\frac{\partial\tau}{\partial P_x} + \tau_i\frac{\partial\boldsymbol{v}}{\partial P_x}\right)F_x\tau_i.$$

(Note that the Pauli exclusion principle does not invalidate the integrated forms of these expressions.) Integrate this expression over $\boldsymbol{P}$-space by writing $dV_P = dP_x dP_y dP_z$ and integrating dP_x by parts so as to obtain

$$\sum \boldsymbol{v} = \frac{2F_x}{kTh^3} \int \boldsymbol{v} v_x \tau \mathbf{f}(1 - \mathbf{f}) dV_p$$

Show that this is equivalent to the result of Problem 7.

9. For spherical energy surfaces and τ a function of v only show from the equation for $\boldsymbol{r}_i$ in Problem 8 that the "mobility" of the electrons in an energy shell is

$$[e\tau(v)/m][1 - \tfrac{1}{3}(d \ln \tau / d \ln v)].$$

Average this over a Maxwellian velocity distribution taking $\tau(v) = l/v$ and obtain (28b) of Section 11.4.

CHAPTER 12

APPLICATIONS TO TRANSISTOR ELECTRONICS

12.1 INTRODUCTION

In the earlier chapters of Part II the theory of the behavior of holes and electrons is presented largely in pictorial and descriptive terms. In Chapter 11, however, the statistical treatment of the influence of electric and magnetic fields is given in detail. The quantum-mechanical foundation for all these subjects is presented in Part III in terms of the wave functions for electrons in crystals.

There remain, however, certain steps to be taken before the attributes of holes and electrons discussed in the earlier chapters of Part II can be embodied in an analytic theory for use in treating transistors and other semiconductor devices of the sort discussed in Part I. These additional steps are taken in this chapter, and some detailed examples of application are presented.

This chapter serves the additional purpose, as has been indicated at various points in the text, of giving a critique of the data presented earlier. In particular the closing sections are concerned with values of $\mathcal{E}_{1G}$ and the mobilities of holes and electrons. Much of the best data on these important topics was being accumulated as this book was being written and the values chosen for examples were later shown in many instances to be inaccurate. Since these values were used to illustrate principles rather than to compute definitive results, the task of revising them to keep them up to date while the book was in preparation has not been undertaken. Instead the best values available at the latest date at which proof could be revised have been incorporated in the last sections of this chapter.

The basic equation which is required for treating many problems involving the flow of particles is the *equation of continuity*. This equation states the principle of conservation of matter in the following form: The rate of increase of the particles in an element of volume is equal to the net inward flow across the surfaces of the element. In the case of holes and electrons this must be generalized, as discussed in Section 12.2, to allow for recombination and generation of hole electron pairs in the element of volume.

In order to deal with most semiconducting devices it is necessary to consider the random diffusive motions of the charge carriers provoked by thermal agitation as well as the systematic drifts produced by electric fields and magnetic fields. (See Figure 8.5 for an illustration of both types of motion.) The diffusive motion is described by the *diffusion constant* which is related to the mobility by the Einstein relation $D_n = kT\mu_n/e$ for electrons of $D_p = kT\mu_p/e$ for holes. This relationship can be derived directly by the statistical methods of Chapter 11 or from a consideration of the principle of detailed balancing. We shall give the latter presentation in Section 12.3.

In Section 12.4 we shall introduce the *quasi Fermi levels*, concepts which are of considerable aid in analyzing processes in semiconductors when hole and electron densities are not in equilibrium.

In Section 12.5 we shall apply diffusion theory to obtain the characteristics of a simplified model of a *p-n* junction including the rectification characteristic and the dependence of admittance on frequency.

In Section 12.6 we shall deal with diffusion and recombination in a germanium filament. In many cases the recombination takes place chiefly on the surface of the filament and is limited by diffusion to the surface.

In Section 12.7 a treatment due to C. Herring of the motion of a pulse of holes (or electrons) in a body of opposite conductivity type will be presented indicating how deformation of the wave shape may arise.

In Section 12.8 various ways of determining $\mathcal{E}_{1G}$, the rate of change of energy gap with volume, are reviewed and compared with values obtained from mobility measurements.

In Section 12.9, data relating to mobility and the effect of magnetic fields upon conduction are reviewed and the best values obtainable at the time of going to press for drift and Hall mobilities in germanium are quoted.

In Section 12.10, some experimental results and speculations on the origin of noise are presented.

The concluding section expresses in general terms an appraisal of semiconductor theory and experiment and analyzes future prospects. This section belongs in some ways at the end of the book but since it most closely relates to the material of Parts I and II it is placed at the end of Part II.

12.2 NOTATION, CONTINUITY EQUATION, USE OF *q* FOR *e*

In the remainder of this chapter we shall use the following notation. Some of the terms are defined in Section 12.3.

$$\psi = \text{electrostatic potential} \tag{1}$$

$$E = -\nabla\psi = \text{electric field} \tag{2}$$

$$\varphi = \text{Fermi level} \tag{3a}$$

$$\varphi_n = \text{quasi Fermi level for electrons} \tag{3b}$$

$$\varphi_p = \text{quasi Fermi level for holes} \tag{3c}$$

$$+q = \text{charge of one hole} = e \tag{4}$$

$$\mathbf{e} = \text{base of Napierian logarithms.} \tag{5}$$

The notation q for the electronic charge has been found convenient by the author when dealing with problems in which both applied potentials and temperature appear explicitly, as they do frequently in transistor electronics. The use of q replaces Boltzmann factors like

$$\mathrm{e}^{-e\psi/kT} = \exp(-e\psi/kT) \text{ by } \mathrm{e}^{-q\psi/kT}.$$

The alternative use of ϵ for the base of Napierian logarithms is undesirable since ϵ and ϵ_0 can occur in the dielectric constant. The choice of ψ = psi for potential and φ = phi for Fermi level is made so that the phonetic relationship may act as an aid to memory.

We shall also assume that any deviations of hole or electron densities from their thermal equilibrium values tend to decay with time constants, or lifetimes, denoted by

$$\tau_p = \text{lifetime for holes in } n\text{-type material};[1] \tag{6a}$$

$$\tau_n = \text{lifetime for electrons in } p\text{-type material.} \tag{6b}$$

In connection with p-n junctions we shall introduce for the thermal equilibrium densities the following notation:

$$n_n \text{ and } p_n, \text{ electrons and holes in } n\text{-type}; \tag{7a}$$

$$n_p \text{ and } p_p, \text{ electrons and holes in } p\text{-type.} \tag{7b}$$

The basic equation which is required to treat the behavior of holes and electrons under conditions in which the concentrations are functions of time and space is the *continuity equation.* This may be written, for the hole density p, in the form

$$\frac{\partial p}{\partial t} = (g - r) - \frac{1}{q} \nabla \cdot \boldsymbol{I}_p. \tag{8}$$

The derivation is as follows: Consider a small region dx, dy, dz centered at x, y, z. The rate of change of holes in $dxdydz$ is

$$\frac{\partial p}{\partial t} dxdydz. \tag{9}$$

These holes arise from the excess of generation over recombination in the volume and from a net flow across the surfaces. The first process is represented by

$$(g - r)dxdydz \tag{10}$$

where g is the net rate of generation of holes per unit volume (due to light, X rays, thermal excitation, etc.), and r the net rate of decay by recombination with electrons. Since p represents untrapped holes, r may include

[1] Tau appears in the form $\boldsymbol{\tau}$ in Chapters 8 and 11 to represent the relaxation time, the dependence upon $\boldsymbol{v}$ being indicated by the fact that it is bold face. The symbols for lifetime used in this chapter are of lighter weight so that no confusion should result.

holes which become trapped without recombining and g may include their re-emission. The term containing $\nabla \cdot \boldsymbol{I}_p$ represents hole flow across the boundaries and is derived as follows: The current density in the x-direction at the midpoint of the $dydz$ face located at $x - (dx/2)$ is

$$I_{px}(x, y, z) - \frac{\partial I_p(x, y, z)}{\partial x} \frac{dx}{2}. \tag{11}$$

This midpoint value will be the average value for the face so that the hole flow across the face **into** $dxdydz$ will be

$$\frac{1}{q}\left[I_{px} - \frac{\partial I_{px}}{\partial x}\frac{dx}{2}\right] dydz.$$

The flow **out of** $dxdydz$ across the face at $x + (dx/2)$ differs only by replacing the $(-)$ by $(+)$. The net flow across the $dydz$ pair of faces into $dxdydz$ is thus

$$\frac{1}{q}\left[I_{px} - \frac{\partial I_{px}}{\partial x}\frac{dx}{2}\right] dydz - \frac{1}{q}\left[I_{px} + \frac{\partial I_{px}}{\partial x}\frac{dx}{2}\right] dydz$$
$$= -\frac{1}{q}\frac{\partial I_{px}}{\partial x}\, dxdydz. \tag{12}$$

Proceeding similarly for the $dxdy$ and $dxdz$ faces leads to a **net flow into** $dxdydz$ of

$$-\frac{1}{q}\left[\frac{\partial I_{px}}{\partial x} + \frac{\partial I_{py}}{\partial y} + \frac{\partial I_{pz}}{\partial z}\right] dxdydz \equiv -\frac{1}{q}\nabla \cdot \boldsymbol{I}_p dxdydz \tag{13}$$

where the quantity in square brackets is by definition $\nabla \cdot \boldsymbol{I}_p$, the *divergence* of $\boldsymbol{I}_p$. Equating the net rate of change in $dxdydz$ to the net generation plus the net flow in across the surfaces leads to the *continuity equation* (8) times a common factor $dxdydz$, which may be divided out, thus establishing (8).

In general it will be assumed that deviations for injected carrier densities p or n from the equilibrium values p_n and n_p tend to decay with characteristic lifetimes which are independent of the concentration of the carriers or the way in which they are introduced. For this purpose we consider that *due to recombination and spontaneous generation the net rate of change* of injected holes in n-type is

$$(p_n - p)/\tau_p \tag{14a}$$

and for electrons in p-type the corresponding contribution is

$$(n_p - n)/\tau_n. \tag{14b}$$

The continuity equation for holes is then

$$\frac{\partial p}{\partial t} = \frac{p_n - p}{\tau_p} - \frac{1}{q}\nabla \cdot \boldsymbol{I}_p + g_p' \tag{15a}$$

where g_p' is the net rate of generation of holes in the volume due to photons or other **external** exciting agency. The similar equation for electrons in p-type material is

$$\frac{\partial n}{\partial t} = \frac{n_p - n}{\tau_n} + \frac{1}{q} \nabla \cdot \boldsymbol{I}_n + g_n'. \tag{15b}$$

We shall use these equations for the case in which $g_p' = g_n' = 0$ to treat the flow of holes and electrons in p-n junctions and filaments. (Sections 12.5 and 12.6.)

12.3 THE EINSTEIN RELATIONSHIP BETWEEN MOBILITY AND DIFFUSION CONSTANT

We shall consider the diffusion process only for the case in which the density of carriers is so small that the effect of the Pauli exclusion principle can be neglected. Under these conditions we can suppose that each carrier drifts in the electric field and undergoes random diffusion independently of all the others. We shall also suppose that the electric field is small enough so that it does not appreciably disturb the equilibrium distribution of velocities of the carriers, this being the assumption which permitted the neglect of terms higher than the first power in $\boldsymbol{E}$ in Chapter 11. Under these conditions it is justifiable to consider that the hole current (we shall treat holes for our example) arises from superposing a conduction and a diffusion current density. The latter is directly proportional[1] to the gradient of hole density ∇p. Thus we write

$$\boldsymbol{I}_p = q\mu_p p\boldsymbol{E} - qD_p \nabla p. \tag{1a}$$

$$\boldsymbol{I}_n = q\mu_n n\boldsymbol{E} + qD_n \nabla n. \tag{1b}$$

The minus sign occurs in (1a) because, in the absence of an electric field, the net diffusion of holes will be away from the direction in which p increases. Since the charge on the electron is minus, the corresponding sign in (1b) is plus.

If the electrostatic field is produced under conditions of thermal equilibrium, as for example in the case of the transition region of a p-n junction as discussed in Section 4.2 or a metal semiconductor contact as in Section 4.3, then the statistical mechanical principle of detailed balancing[2] requires that for every hole which is crossing a given area in one direction per unit time there is another which is crossing it in the reverse direction with the same speed. This last statement is true even if magnetic fields

[1] This can be shown by a statistical treatment of the individual holes by methods like those of Chapter 11.

[2] R. C. Tolman, *The Principles of Statistical Mechanics*, Oxford at the Clarendon Press, 1938, p. 165.

are present,[3] a fact which we shall soon use. Thus under equilibrium conditions I_p must vanish. Under equilibrium conditions, however, the hole density is governed by the *Boltzmann factor*, exp $(-q\psi/kT)$, which represents the tendency of holes to avoid regions of high potential energy.[3a] The formula for p is given in equation (14) of Section 10.3,

$$p = A\mathrm{e}^{-q\psi/kT} \tag{2}$$

where A is a constant involving N_v and $\mathcal{E}_F$, which we shall discuss more fully in connection with the quasi Fermi levels. Inserting equation (2) into equation (1), evaluating ∇p and requiring that $I_p = 0$, we find that

$$\begin{aligned} 0 &= q\mu_p p\boldsymbol{E} - qD_p\left(-\frac{q}{kT}\nabla\psi\right)A\mathrm{e}^{-q\psi/kT} \\ &= q\left(\mu_p - \frac{q}{kT}D_p\right)p\boldsymbol{E} \end{aligned} \tag{3}$$

which can be satisfied only if the

Einstein relationship $$\mu_p = \frac{q}{kT}D_p \tag{4}$$

holds.[4] A precisely similar calculation establishes the Einstein relationship

$$\mu_n = \frac{q}{kT}D_n \tag{5}$$

for electrons. (In treating electrons, it should be remembered that μ_n is defined as positive and so is D_n.)

The dimensions of D_p are (length)^2/time. In the c.g.s. system D_p is expressed in cm^2/sec. If μ_{pL} is expressed in cm^2/volt sec (the L system of Section 8.8), then equation (4) leads to

$$\mu_{pL}\ \mathrm{cm}^2/\mathrm{volt\ sec} = \frac{q}{kT}D_p\ \mathrm{cm}^2/\mathrm{sec}, \tag{6}$$

which is consistent because q/kT is charge/energy which has the dimensions of volts^{-1}; at 300°K, $kT/q = 0.026$ volt so that for the L system and the M.K.S. system we have

$$D_p = D_{pL} = 0.026\mu_{pL} \tag{7a}$$

$$D_{pM} = 0.026\mu_{pM}. \tag{7b}$$

[3] See J. H. VanVleck, *The Theory of Electric and Magnetic Susceptibilities*, Oxford at the Clarendon Press, 1932, pp. 94–98.

[3a] The relationship of $q\psi$ to $\mathcal{E}_v$ is discussed more fully in the next section.

[4] A. Einstein, *Annalen der Physik*, **17**, 549–560 (1905); Müller-Pouillet, *Lehrbuch der Physik*, Braunschweig, 1933, IV (3), pp. 316–319.

The quantity $D_p\tau_p$, which occurs in the analysis of later sections, has the dimensions of (length)^2 and is denoted by the symbol L_p^2 where L_p is called the *diffusion length*. The diffusion length, as discussed in Sections (12.5) and (12.6) and the Problems, is the average distance along any given direction, x-direction for example, which a hole will diffuse before recombining, if its lifetime is τ_p.

Under non-equilibrium conditions the current may be written as the sum of a conduction term and a diffusion term as given by equation (1), but the two terms in general will not cancel.

When a magnetic field is present, the conduction current does not usually flow parallel to the electric field and the diffusion current does not flow parallel to ∇p. The conductance and diffusion constants both become tensors which we may derive from the resistivity tensor ρ. We shall treat the simplest case: For a cubic crystal with $\boldsymbol{H}$ along the z-axis, [001] direction, we may write the electric field in the form

$$E_x = \rho_t(I_x - I_y \tan\theta) \tag{8a}$$

$$E_y = \rho_t(I_x \tan\theta + I_y) \tag{8b}$$

$$E_z = \rho_l I_z \tag{8c}$$

where ρ_t is the resistivity transverse to $\boldsymbol{H}$ [defined by setting power dissipation density equal to $\rho_t(I_x^2 + I_y^2)$], and ρ_l (l for longitudinal) is the resistivity parallel to $\boldsymbol{H}$, and θ is the Hall angle, the angle between $\boldsymbol{E}$ and $\boldsymbol{I}$ for transverse currents. This equation is analogous to the Hall effect formulae of Chapters 8 and 11 except that we have written it in a form adaptable to large values of H and θ. In this form θ is plus for holes, and negative for electrons. Equations (8a) and (8b) may be written compactly as a complex equation:

$$E_x + iE_y = (\rho_t e^{i\theta}/\cos\theta)(I_x + iI_y). \tag{8ab}$$

In order to define the diffusion constant we shall solve equation (8) for the currents and express these as $I_x = qpv_x$, etc., where the v's are the drift velocities. We introduce

$$\mu_t = 1/qp\rho_t, \quad \mu_l = 1/qp\rho_l \tag{9}$$

for this purpose. We then obtain

$$v_x + iv_y = e^{-i\theta}\mu_t \cos\theta(E_x + iE_y) \tag{10ab}$$

or

$$v_x = \mu_t \cos^2\theta(E_x + E_y \tan\theta) \equiv \mu_{xx}E_x + \mu_{xy}E_y \tag{10a}$$

$$v_y = \mu_t \cos^2\theta(-E_x \tan\theta + E_y) \equiv \mu_{yx}E_x + \mu_{yy}E_y \tag{10b}$$

$$v_z = \mu_l E_z \equiv \mu_{zz}E_z. \tag{10c}$$

The diffusion current is then found from the principle of detailed balancing and the Boltzmann distribution (2), neither of which is affected by a magnetic field. The resulting relationships are

$$I_x = -q\left(D_{xx}\frac{\partial p}{\partial x} + D_{xy}\frac{\partial p}{\partial y}\right) \tag{11a}$$

$$I_y = -q\left(D_{yx}\frac{\partial p}{\partial x} + D_{yy}\frac{\partial p}{\partial y}\right) \tag{11b}$$

$$I_z = -qD_{zz}\frac{\partial p}{\partial z} \tag{11c}$$

where each D is (kT/q) times the corresponding μ of equations (10). A similar relationship holds for electrons.

12.4 ENERGY BAND DIAGRAMS, ELECTROSTATIC POTENTIAL, AND QUASI FERMI LEVELS

In Figure 12.1 we represent the energy level scheme for a *p-n* junction. We have discussed similar diagrams on a largely intuitive basis in Chapter 4 and briefly in connection with Figures 5.13 and 7.6. We shall now lay an analytical foundation for treating them. The diagram implies that an electron in the lowest state of the conduction band, or better a localized wave-packet made up of wave functions near the lowest state, will have different energies in different parts of the crystal. This situation represents a macroscopic electric field since the energy of interaction of the electron with the local atomic field is the same no matter where the electron is in the crystal, provided that its wave function distributes it in the same way with respect to its immediate environment in all cases. Thus if the electrostatic potential at one point is ψ_A and at another is ψ_B, then, since the charge on the electron is $-q$, the energy $\mathcal{E}_c$ of the lowest state in the conduction band will have a value of

$$\mathcal{E}_{cB} = \mathcal{E}_{cA} + q(\psi_A - \psi_B) \tag{1}$$

or, if B is varied while A is held fixed,

$$\mathcal{E}_{cB} = -q\psi_B + \text{constant.} \tag{2}$$

A similar relationship will hold for a wave-packet at the top of the valence-bond band, and the energy of an electron in such a wave-packet will also vary as $-q\psi$. From this we can conclude that the energy of the missing wave-packet, corresponding to a hole, varies as $q\psi$. We shall illustrate this more fully as follows:

The potential energy of a hole may be defined in terms of the electrostatic work done in moving the hole. We may evaluate this energy by the following process: Suppose we have a hole at B. We can get the hole to A by exciting an electron at A from the valence-bond band to the conduction band, which costs us $\mathcal{E}_G$; we can then move the electron to B putting in work $-q(\psi_B - \psi_A)$ and then let it fall into the hole at B regaining $\mathcal{E}_G$. Thus moving the hole from B to A requires work to the amount $q(\psi_A - \psi_B)$, which is just that required to move a charge $+q$ from B to A. Accordingly, we may take the energy of a hole to vary as $+q\psi$. This result is entirely consistent with the conclusions reached in Chapters 7, 8, and 11 that a hole could be considered to behave as a positive particle with positive mass and to be accelerated by a force $q\boldsymbol{E}$. These same

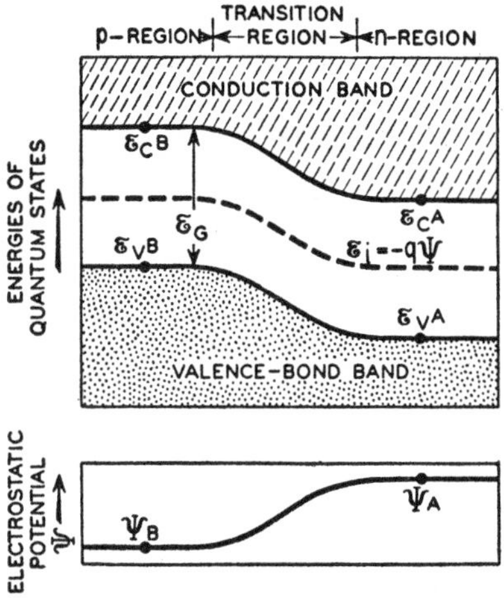

FIG. 12–1—The Relationship between Energy Levels and Electrostatic Potential.

conclusions are reached on the basis of more detailed arguments in Section 15.8, especially as shown in equations (10), (11), and (12).

It is convenient for our purposes to define the electrostatic potential in terms of a *level of energy* (not called an energy level because there need be no quantum states at that energy) approximately midway between the bands. We shall give the mathematical definitions in this paragraph and shall enlarge below upon the physical interpretation. We shall denote the level as $\mathcal{E}_i$ (i for intrinsic for reasons which will soon become evident) and define it as

$$\mathcal{E}_i = \tfrac{1}{2}[\mathcal{E}_c + \mathcal{E}_v + kT \ln (N_v/N_c)], \tag{3}$$

the symbols having the meanings of Chapter 10. The electrostatic potential will be defined as

$$\psi = -\mathcal{E}_i/q \quad \text{or} \quad \mathcal{E}_i = -q\psi. \tag{4}$$

The energies $\mathcal{E}_c$ and $\mathcal{E}_v$ are then

$$\mathcal{E}_c = -q\psi + \tfrac{1}{2}kT \ln (N_c/N_v) + \tfrac{1}{2}\mathcal{E}_G \tag{5}$$

$$\mathcal{E}_v = -q\psi + \tfrac{1}{2}kT \ln (N_c/N_v) - \tfrac{1}{2}\mathcal{E}_G \tag{6}$$

where

$$\mathcal{E}_G = \mathcal{E}_c - \mathcal{E}_v. \tag{7}$$

We shall also define a Fermi level in terms of an electrostatic potential by the equation

$$\varphi = -\mathcal{E}_F/q \quad \text{or} \quad \mathcal{E}_F = -q\varphi. \tag{8}$$

We shall refer to φ, as well as to $\mathcal{E}_F$, as the Fermi level. The reason for these choices is that, when φ is equal to ψ, the densities of holes and electrons are equal as may be seen by substituting into equations (11) and (14) of Section 10.3, which give in terms of ψ and φ

$$n = N_c \mathrm{e}^{-(\mathcal{E}_c - \mathcal{E}_F)/kT} = n_i \mathrm{e}^{q(\psi - \varphi)/kT} \tag{9}$$

$$p = N_v \mathrm{e}^{-(\mathcal{E}_F - \mathcal{E}_v)/kT} = n_i \mathrm{e}^{q(\varphi - \psi)/kT} \tag{10}$$

where

$$n_i = \sqrt{N_v N_c}\, \mathrm{e}^{-\mathcal{E}_G/2kT} = \sqrt{pn}, \tag{11}$$

in accordance with the results of equation (19) of Section 10.3. The quantity n_i represents the density of holes and of electrons in an intrinsic sample for which the condition of zero space charge requires the equality $p = n$.

We shall next describe in words the content of the equations presented above. For this purpose we consider a semiconductor containing a p-n junction so that the electrostatic potential and the values of p and n vary with position. There will evidently be a point in the transition region at which p and n are equal and thus each is equal to n_i, the value in an intrinsic sample. At this point the Fermi level will be approximately midway between the bands. However, if the Fermi level were exactly midway then $\mathbf{f} = \exp[(\mathcal{E}_F - \mathcal{E}_c)/kT]$ and $\mathbf{f}_p = \exp[(\mathcal{E}_v - \mathcal{E}_F)/kT]$ would be equal and the ratio of p to n would be N_c/N_v. For the case of N_c greater than N_v, the midpoint of the energy band must lie above $\mathcal{E}_F$ so as to reduce n and increase p until $n = p$. An upward shift of the band by $\Delta\mathcal{E}$ decreases n by $\exp(-\Delta\mathcal{E}/kT)$ and increases p by the reciprocal ratio $\exp(\Delta\mathcal{E}/kT)$ so that n/p is reduced by $\exp(-2\Delta\mathcal{E}/kT)$. Equating this last factor to N_v/N_c, the required reduction in n/p, leads to

$$\Delta\mathcal{E} = (kT/2) \ln N_c/N_v.$$

Consequently, at the point where $n = p = n_i$, the level of energy which lies $\Delta\mathcal{E}$ below the midpoint of the energy gap will be equal to $\mathcal{E}_F$. This level of energy is $\mathcal{E}_i$, defined in (3), since it lies at

$$\begin{aligned}\tfrac{1}{2}(\mathcal{E}_c + \mathcal{E}_v) - \Delta\mathcal{E} &= \tfrac{1}{2}(\mathcal{E}_c + \mathcal{E}_v) - \tfrac{1}{2}kT \ln N_c/N_v \\ &= \tfrac{1}{2}(\mathcal{E}_c + \mathcal{E}_v + kT \ln N_v/N_c).\end{aligned}$$

At points in the p-n junction where $\mathcal{E}_i$ rises above $\mathcal{E}_F$, n is decreased from

n_i by a factor exp $(\mathcal{E}_F - \mathcal{E}_i)/kT$ and p is increased from n_i by the reciprocal factor. If we define an electrostatic potential by $-q\psi = \mathcal{E}_i$ and a Fermi potential by $-q\varphi = \mathcal{E}_F$, the preceding reasoning leads to equations (9) and (10) for n and p. Equations (9) and (10) can be remembered by the following scheme: When $\varphi = \psi$, the densities correspond to the intrinsic case. Deviations of ψ from φ introduce factors $\exp \pm q(\psi - \varphi)/kT$. The plus sign corresponds to electrons since increases in ψ correspond to more positive potentials and thus to lower potential energies for the negative electrons and hence to higher densities; the minus sign corresponds to holes, which are repelled by positive potentials.

Differences in the Fermi potential φ are the quantities measured by a voltmeter. In order to illustrate this conclusion, we imagine that two conductors A and B are connected by a battery but that the circuit is open or of such high resistance that negligible currents flow. Under these conditions each conductor is substantially in thermal equilibrium and has a Fermi-Dirac distribution of electrons in its energy level scheme. The system as a whole is not in thermal equilibrium, however, and if metals A and B were brought in contact, currents would flow and irreversible heating effects would take place. The source of energy is, of course, the battery which is itself not in equilibrium. If metals A and B are identical and have identical work functions, then in the space between the two there will evidently be an electric field which produces a voltage difference between points just outside each metal equal to the battery voltage. Since the two metals have identical properties, each will have the same difference between the Fermi level and a wave-packet state representing an electron at rest just outside the metal.[1] Hence the difference in Fermi levels will be the battery potential. This result will also hold if the conductors are different, being A' and B', as may be seen as follows: Suppose A' is placed in contact with A and B' with B. Then since A' and A are in local equilibrium, they will have the same Fermi level and so will B and B'. Hence the difference in Fermi potentials between A' and B' will also be equal to the battery potential. Another example of the theorem that the change in Fermi potential is equivalent to a change in applied voltage is given at the end of this section. A statistical mechanical treatment relating the Fermi level to a *chemical potential* for electrons is given at the end of Section 16.1.

We shall next consider the application of equations (9) and (10) to the problem of the determination of the thermal equilibrium distribution of potential in a semiconductor of non-uniform impurity content, the densities of donors and acceptors being functions $N_d(\mathbf{r})$ and $N_a(\mathbf{r})$ of position $\mathbf{r}$. We shall also assume that the donors and acceptors are fully ionized, as is

[1] This difference is the "true work function" of the surface. For a detailed discussion see C. Herring and M. H. Nichols, *Rev. Mod. Phys.*, **21**, 185–270 (1949), Section I.2.

the case for germanium at room temperature when the Fermi level lies more than 0.1 volt from the edges of the energy gap. Under these conditions the charge density ρ is

$$\begin{aligned}\rho &= q(p - n + N_d - N_a) \\ &= q[N_d(\mathbf{r}) - N_a(\mathbf{r}) + 2n_i \sinh q(\varphi - \psi)/kT]. \end{aligned} \quad (12)$$

Here φ may be regarded as specified, and ψ as an unknown function of position $\mathbf{r}$. [The reader may find it instructive to compare equation (12) with Figures 16.2 to 16.6 in which the difference between heavy solid and heavy dotted lines is essentially equal to ρ plotted as a function of $q(\psi - \varphi)$.] The dependence of ψ upon $\mathbf{r}$ is determined from Poisson's equation, in e.s.u. and M.K.S. respectively,

$$\nabla^2\psi = -4\pi\rho/\kappa \quad \text{or} \quad \nabla^2\psi = -\rho_M/\kappa\epsilon_0 \quad (13)$$

which thus becomes an equation with one unknown ψ when (12) is inserted for ρ, it being assumed that $N_d(\mathbf{r})$ and $N_a(\mathbf{r})$ and the boundary conditions on ψ are specified.

It is not difficult to show that this equation always has solutions. The one-dimensional case is treated in detail in the literature.[2] We shall not give an analytical treatment here but shall discuss the implications of equations (12) and (13) in terms of physical arguments. Suppose we have a solution for the case of uniform values of N_d and N_a. (Such solutions are obtained in Chapters 10 and 16.) Then suppose we increase N_d in some small region R, holding n and p arbitrarily constant with values n_0 and p_0. This change will increase the electrostatic potential in R from ψ_0 to ψ_1 where ψ_1 is more positive since the change in N_d represents an added positive charge. We next imagine that ψ is held fixed with value ψ_1 and that the hole and electron densities are allowed to readjust themselves with a resultant flow of holes out and electrons into R. These flows will make a negative contribution to the charge density inside R, consistent with the sign of the term containing ψ in equation (12), which becomes more negative as ψ becomes more positive. The change in charge density due to carrier flow thus tends to counteract the original change produced by increasing N_d. This tendency to counteract is really the essential point. If we change N_d, the change in n and p is such as to tend to compensate the changed charge density. As a result the flow of holes and electrons tends to stabilize the potential and to neutralize the effect of changes in impurity density. The foregoing remarks are supposed to furnish food for thought and to improve the reader's insight into the behavior of the charge densities. A rigorous argument can be built up from this beginning to show that a

[2] W. Shockley, *Bell Syst. Tech. J.* **28**, 435–489 (1949), Appendix VII; J. C. Slater, *Phys. Rev.*, **76**, 1592–1601 (1949). References are given in both papers.

solution for ψ, and consequently n and p, can be found for arbitrary values of $N_d(\mathbf{r})$ and $N_a(\mathbf{r})$.

We shall next apply the arguments just given to the p-n junction shown in Figure 12.2. To the left of the junction in the p-region, we shall suppose that $N_a - N_d$ is positive and constant with a value $p_p - n_p$. In this region, the electron density n_p is very small compared to the hole density so that the space charge condition leads with a negligible error to

$$N_a - N_d = p_p \tag{14}$$

or, letting ψ_p represent the value of ψ in the p-region,

$$p_p = n_i e^{-q(\psi-\varphi)/kT} \tag{15}$$

so that by dividing by n_i and taking the logarithm, we obtain

$$\psi_p = \varphi - \frac{kT}{q} \ln (p_p/n_i). \tag{16}$$

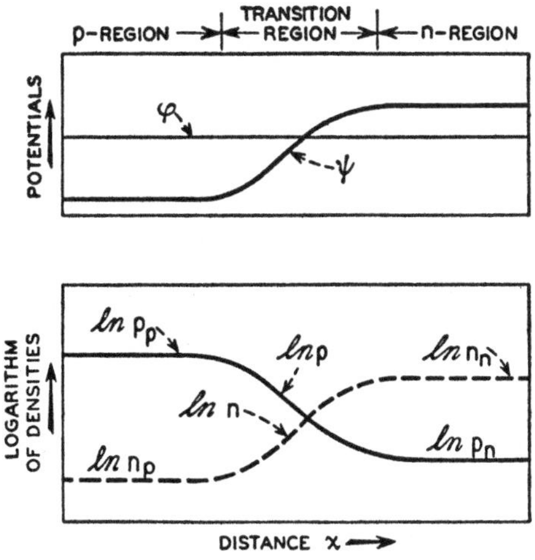

FIG. 12–2—The Dependence of n and p Upon Potential ψ and Fermi Level φ for a p-n Junction with Higher Conductivity in the p-Region.

In the n-region, we shall suppose that $N_d - N_a$ is positive and uniform with a value denoted by n_n. The value of ψ in the n-region is thus

$$\psi_n = \varphi + \frac{kT}{q} \ln (n_n/n_i). \tag{17}$$

Thus ψ_n is more positive than ψ_p for the equilibrium condition of constant φ. This situation is similar to the one discussed in Chapter 4. We shall next express in analytic form the conclusion presented in Chapter 4 that the tendency for holes to diffuse out of the p-region, where their density is greatest, is canceled by the tendency for them to flow back owing to the electric field. We shall introduce the *quasi Fermi levels* for this purpose.

The combined action of diffusion and conduction is conveniently treated with the aid of the *quasi Fermi levels* φ_n for electrons and φ_p for holes. These are defined in terms of n, p, and ψ for non-equilibrium conditions by the equations

$$n = n_i e^{q(\psi - \varphi_n)/kT}, \quad \varphi_n = \psi - \frac{kT}{q} \ln (n/n_i) \tag{18}$$

$$p = n_i e^{-q(\psi - \varphi_p)/kT}, \quad \varphi_p = \psi + \frac{kT}{q} \ln (p/n_i). \tag{19}$$

It is evident that for thermal equilibrium, which is the only condition for which a Fermi level φ is really defined, both φ_p and φ_n reduce to φ.

In terms of φ_n and φ_p, the total current densities may be rewritten using the Einstein relationship to eliminate D_p in the simple forms

$$\begin{aligned} \boldsymbol{I}_p = q\mu_p p\boldsymbol{E} - qD_p\nabla_p &= -q\mu_p p\nabla\psi - q\frac{kT}{q}\mu_p p\nabla \frac{-q(\psi - \varphi_p)}{kT} \\ &= -q\mu_p p\nabla\varphi_p \end{aligned} \tag{20}$$

and similarly

$$\boldsymbol{I}_n = -q\mu_n n\nabla\varphi_n. \tag{21}$$

These equations are formally the same as those for *currents produced by electric fields* $-\nabla\varphi_p$ *and* $-\nabla\varphi_n$ *in materials of conductivity* $q\mu_p p$ *and* $q\mu_n n$.

The cancellation of the competing tendencies discussed in connection with Figure 12.2 is thus simply interpreted as the condition that φ_n and φ_p are both constant, being equal to φ, so that diffusion and conduction currents are equal and opposite. In the next section we shall discuss *p-n* junctions which are not in thermal equilibrium and show how the quasi Fermi levels may be used in analyzing the processes involved.

As an additional example of the behavior of φ_n, φ_p and ψ we consider a long, uniform rod of semiconductor carrying a current in the x-direction. We assume that ohmic contacts are made at the ends so that the densities of the holes and electrons are unperturbed. For this case, the differences $\psi - \varphi_n$ and $\psi - \varphi_p$, which determine the densities n and p, are independent of position and have the same values as for the thermal equilibrium state, for which $\varphi_n = \varphi_p = \varphi$. The plot of φ_n, φ_p and ψ versus distance is the same as for the equilibrium case except that a potential $-E_x x$ will be added to φ_n, φ_p, and ψ. Equations (20) and (21) for the current densities will reduce to the conventional form $I_x = (q\mu_p p + q\mu_n n)E_x$ and the change in electrostatic potential, or voltage, and changes in quasi Fermi levels will be equal.

12.5 THE THEORY OF A *p-n* JUNCTION

In this section we shall illustrate the use of the quasi Fermi levels in analysis and at the same time give a treatment of a simplified model of a p-n junction.[1] Figure 12.3 represents the distribution of φ_p, φ_n, and ψ in a p-n junction under equilibrium conditions and also biassed in the forward direction by a potential $\delta\varphi$, which is represented as if applied to the p-region on the left while the **right side** of the n-region is held at **constant potential.** Under these conditions holes are injected into the n-region, and electrons into the p-region. We suppose that the uniform regions are so wide that injected carriers recombine before reaching the positions x_a and x_b. As a consequence, the total current[2] I is carried almost entirely by holes at the left edge of the figure and almost entirely by electrons at the right edge. The hole current I_p must thus equal I at $x = x_a$ and must gradually drop to zero at $x = x_b$, while $I_n = I - I_p$ correspondingly increases. The mechanism of change in current type can be understood by considering the hole current flowing past x_{Tn}: This current decreases because the holes have a lifetime τ_p in the n-region and combine with electrons so that the hole current attenuates as x increases; a similar effect takes place for the electron current entering the p-region. We shall assume in the analysis below that the transition region is so narrow that recombination in it is negligible; consequently, I_p has the same value at x_{Tp} and x_{Tn}.

At x_a and x_b, the hole and electron densities thus have the values characteristic of uniform material and consequently at x_a, for example, φ_p and φ_n are equal and differ from ψ by the same amount as for the equilibrium case. This explains why the figure shows

$$\varphi_p(x_a) = \varphi_n(x_a) = \varphi_p(x_b) + \delta\varphi = \varphi_n(x_b) + \delta\varphi \tag{1}$$

where $\delta\varphi$ is the applied potential and why $\delta\psi = \delta\varphi_p = \delta\varphi_n$ at $x = x_a$ and $\delta\psi = 0$ at $x = x_b$. We shall next suppose that $\delta\varphi$ is small compared to kT/q so that only small disturbances in n and p are produced by the applied potentials.

On the basis of the foregoing assumptions we shall prove the important result for p-n junctions that *φ_p is nearly constant in the p-region and drops by $\delta\varphi$ in the n-region while φ_n is nearly constant in the n-region and changes by $\delta\varphi$ in the p-region.* To prove this statement we shall suppose that the two homogeneous regions have conductivities substantially larger than intrinsic, so that the densities of the less abundant carriers, denoted by n_p and p_n, are very much smaller than the more abundant carriers p_p and n_n. This

[1] The treatment follows that of W. Shockley, *Bell. Syst. Tech. J.* **28**, 435–489 (1949) but omits the more detailed considerations. See this article and Section 4.2 for additional references.

[2] In the analytic development below, the scalar symbols I, I_p, and I_n will represent current densities in the x-direction.

relationship follows from

$$n_p = n_i^2/p_p \quad \text{and} \quad p_n = n_i^2/n_n. \tag{2}$$

[See Section 12.2 and equation (11) of Section 12.4 for details.] We shall next obtain the ratio of $\nabla\varphi_p$ at x_{Tp} and x_{Tn} in terms of these densities by

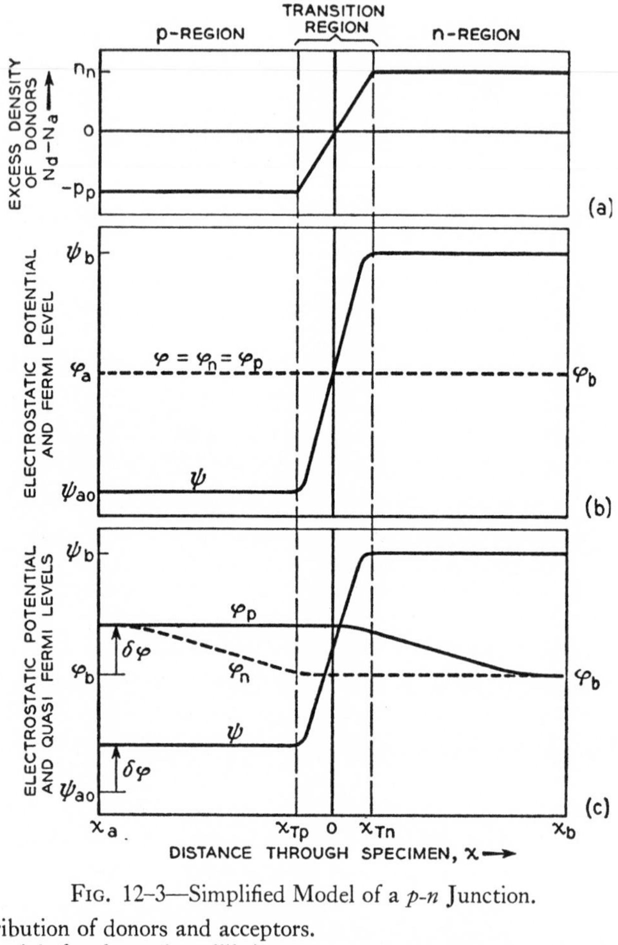

FIG. 12–3—Simplified Model of a *p-n* Junction.

(a) Distribution of donors and acceptors.
(b) Potentials for thermal equilibrium.
(c) Effect of a potential $\delta\varphi$ applied in the forward direction.

using the expression for current density (20) of Section 12.4. At x_{Tp} we obtain

$$I_p = -q\mu_p p_p \frac{d\varphi_p(x_{Tp})}{dx}. \tag{3a}$$

Since we neglect recombination in the transition region, this same current flows at x_{Tn} and we have

$$I_p = -q\mu_p p_n \frac{d\varphi_p(x_{Tn})}{dx}. \tag{3b}$$

From these we obtain

$$\frac{d\varphi_p(x_{Tn})/dx}{d\varphi_p(x_{Tp})/dx} = \frac{p_p}{p_n} = \frac{p_p n_n}{n_i^2}. \tag{3c}$$

From (3c) we conclude that the change in φ_p is much larger in the n-region than in the p-region, at least in the neighborhood of the junction. Similar arguments apply to φ_n. Consequently $\nabla\varphi_n$ and $\nabla\varphi_p$ will have to be very large wherever I_p is appreciable in the n-region or I_n is appreciable in the p-region. Hence we conclude that the important changes in φ_p and φ_n occur as shown in Figure 12.3, where injected currents flow.

Another way of stating this significant result is that *the hole current is carried by material of high hole conductivity in the p-region and of low hole conductivity in the n-region;* hence the gradient must be much greater in the n-region.

The last statement brings out an important point: *The resistance of a p-n junction is much greater than the integrated resistivity of the material composing the junction.* This point follows from the fact that the hole current, for example, must flow a certain distance in regions where the density of the carriers is much less than n_i, and in such a region $\nabla\varphi_p$ is much greater than it would be even in intrinsic material. To be more specific, the junction resistance may be much greater than the resistance of a layer of intrinsic material as thick as the transition region.[3]

Until the injected carrier density becomes comparable with the density normally present, the arguments presented previously will be valid, and the important changes in φ_p and φ_n will occur in the regions in which the corresponding carriers are present only in relatively small densities. In the p-region, p itself is nearly constant and equal to p_p; this follows from the condition for electrical neutrality which requires that the changes in p be equal to the changes in n produced by injection, and these latter are small compared to p_p. Hence by equation (19) of Section 12.4, ψ as well as φ_p is nearly constant in the p-region; similarly, ψ and φ_n are nearly constant in the n-region.

The purpose of establishing the conclusions just discussed, which are illustrated by the curves for ψ, φ_p, and φ_n in Figure 12.3, is to obtain a set of boundary conditions which can be applied to hole flow into the n-region across the plane at x_{Tn}. Similar reasoning may be used to obtain expressions for electron flow into the p-region across x_{Tp}. In carrying out the

[3] See Shockley, *loc. cit.*, for a general analysis of this subject.

analysis we shall assume that all quantities are independent of y and z. As a consequence, all currents flow parallel to the x-axis, and we need consider only the x-component. For brevity we shall continue to omit the subscript x and replace

$$I_{px} \quad \text{by} \quad I_p \tag{4}$$

in this section. We shall neglect the contribution of recombination in the transition region itself to the conversion, as x increases, of I_p to I_n, the transition region being assumed very narrow compared to other dimensions such as the diffusion lengths L_p and L_n which will be discussed; consequently, I_p may be taken to have the same value from $x = x_{Tp}$ to $x = x_{Tn}$. On the basis of this assumption we may take the total conduction current per unit area across $x = 0$ to be

$$I = I_n(x_{Tp}) + I_p(x_{Tn}). \tag{5}$$

We shall calculate $I_p(x_{Tn})$ using the boundary conditions already discussed, and obtain $I_n(x_{Tp})$ by analogy. In the discussion, the description in words will be given in terms of forward current, $\delta\varphi > 0$; the equations, however, will be equally applicable to $\delta\varphi < 0$, the case of reverse currents.

At the right-hand boundary, x_{Tn}, of the transition region, the relationship

$$\varphi_p = \varphi_n + \delta\varphi \tag{6}$$

will hold. For convenience we repeat the following definitions of quantities in the n-region:

$n_n = N_d - N_a$, normal density of electrons;
$p_n = n_i^2/n_n \ll n_n$, normal density of holes;
τ_p, lifetime of a hole in the n-region.

Since we have assumed that the injected carrier density is small compared to the type normally present, the boundary conditions at x_{Tn} are

$$n(x_{Tn}) \doteq n_n = n_i e^{q(\psi-\varphi_n)/kT} \tag{7a}$$

$$p(x_{Tn}) = n_i e^{q(\varphi_p-\psi)/kT}. \tag{7b}$$

Replacing $\exp(-q\psi/kT)$ in (7b) by (n_i/n_n), $\exp(-q\varphi_n/kT)$ obtained from (7a) gives

$$p(x_{Tn}) = (n_i^2/n_n)e^{q(\varphi_p-\varphi_n)/kT} = p_n e^{q\delta\varphi/kT}. \tag{8}$$

Equation (8) is the key equation of the rectification theory. It states that the hole density at x_{Tn} varies exponentially with the applied voltage. From the linear theory, which we will give in detail soon, it follows that the hole current must vary in proportion to the deviation of p from its equilibrium value and hence as $p - p_n = p_n[\exp(q\delta\varphi/kT) - 1]$. This leads at once to the correct form for the d-c characteristic given in (22).

We shall suppose that $\delta\varphi$ consists of d-c and a-c components so that

$$\delta\varphi = V_0 + v_1 e^{i\omega t} \tag{9}$$

where v_1 is an a-c signal, assumed so small that linear theory may be employed (that is, $v_1 \ll kT/q$). Expanding (8) in series gives

$$p(x_{Tn}) = (p_n e^{qV_0/kT})[1 + (qv_1/kT)e^{i\omega t}].$$

We resolve this density into its equilibrium value p_n, a d-c component p_0, and an a-c component $p_1 \exp(i\omega t)$:

$$p(x_{Tn}) = p_n + p_0 + p_1 e^{i\omega t} \tag{10}$$

where

$$p_0 = p_n(e^{qV_0/kT} - 1) \tag{11}$$

$$p_1 = (qp_n v_1/kT)e^{qV_0/kT}. \tag{12}$$

So long as $p(x_{Tn}) \ll n_n$, the normal density of electrons in the n-region, the lifetime τ_p, and diffusion constant D_p for a hole will be substantially unaltered by $\delta\varphi$.

Our problem is to determine the value of $p(x, t)$ for $x > x_{Tn}$ and from this to obtain the resultant hole current $I_p(x_{Tn}, t)$ flowing across x_{Tn}. The total current $I(t)$ is the sum of this hole current, a similar electron current, and a displacement current $\dot{\boldsymbol{D}}/4\pi = (\kappa/4\pi)\dot{\boldsymbol{E}}$ or $\dot{\boldsymbol{D}} = \kappa\epsilon_0\dot{\boldsymbol{E}}$ in M.K.S. units. For p-n junctions the displacement current is, in general, small compared to the conduction current except at high reverse biases or low temperatures.[4]

As discussed previously, the hole current injected into the n-region dies away with a relaxation time, or hole lifetime, τ_p. In accordance with the discussion of Section 12.2, equation (15), the *continuity equation* for this case is

$$\frac{\partial p}{\partial t} = \frac{p_n - p}{\tau_p} - \frac{1}{q}\nabla \cdot \boldsymbol{I}_p. \tag{13}$$

Since ψ is assumed substantially constant compared to φ_p, the current density is largely diffusion current with a value of

$$I_p = -qD_p \frac{\partial p}{\partial x}. \tag{14}$$

The insertion of this in equation (13) leads to

$$\frac{\partial p}{\partial t} = \frac{p_n - p}{\tau_p} + D_p \frac{\partial^2 p}{\partial x^2}. \tag{15}$$

[4] See W. Shockley, *Bell Syst. Tech. J.* **28**, 435–489 (1949), Section 4.4.

The solution of this equation subject to the boundary conditions at x_{Tn} is

$$p = p_n + p_0 e^{-x'/L_p} + p_1 e^{i\omega t - x\,(1+i\omega\tau_p)^{1/2}/L_p} \tag{16}$$

where

$$x' = x - x_{Tn}, \quad x' = 0 \text{ at } x = x_{Tn} \tag{17}$$

$$L_p = \sqrt{D_p\tau_p} = \textit{diffusion length} \text{ for holes in } n\text{-region.} \tag{18}$$

The physical meaning of the diffusion length is as follows: Suppose at some value of $x = x_1$ the hole density is $p(x_1)$ and the electric field is negligible. Then the d-c part of equation (16) shows that, if there are no hole sources to the right of x_1, p attenuates by a factor e^{-1} in each interval L_p. Some additional interpretations will be found in the problems.

The condition that $\nabla\psi$ is negligible compared to $\nabla\varphi_p$, so that diffusion current dominates, can be given a simple interpretation if $p \gg p_n$. For this condition we may write $p = p(x_1) \exp[-(x - x_1)/L_p]$ which leads to

$$\varphi_p = -\frac{kT(x - x_1)}{qL_p} + \frac{kT}{q}\ln[p(x_1)/n_i] + \psi \tag{19}$$

according to the definition of φ_p of equation (19) of Section 12.4. The diffusion term will dominate the current if in

$$\frac{\partial\varphi_p}{\partial x} = -\frac{kT}{qL_p} + \frac{\partial\psi}{\partial x} \tag{20}$$

the term kT/qL_p is much greater than $\partial\psi/\partial x$. (See problems.)

The solution just obtained for p gives rise to a current per unit area at x_{Tn} of

$$\begin{aligned} I_p(x_{Tn}) &= -qD_p\frac{\partial p}{\partial x} \\ &= qp_0D_p/L_p + qp_1D_pe^{i\omega t}(1 + i\omega\tau_p)^{1/2}/L_p. \end{aligned} \tag{21}$$

The d-c part is obtained by substituting (11) for p_0:

$$\begin{aligned} I_{p0}(x_{Tn}) &= (qp_nD_p/L_p)(e^{qV_0/kT} - 1); \\ &\equiv I_{ps}(e^{qV_0/kT} - 1). \end{aligned} \tag{22}$$

The forward direction corresponds to $V_0 > 0$ and for it the current increases exponentially. In the reverse direction, the exponential term vanishes and I_{p0} approaches a *saturation current* $(qp_nD_p/L_p) \equiv I_{ps}$. The comparison of equation (22), or rather its generalization (33), with experiment is discussed in Section 4.2 and at the end of this section.

The a-c part is similarly obtained from the p_1 term of (21) combined with (12):

$$\begin{aligned} I_{p1}(x_{Tn}) &= (qp_n\mu_p/L_p)e^{qV_0/kT}(1 + i\omega\tau_p)^{1/2}v_1e^{i\omega t} \\ &\equiv (G_p + iS_p)v_1e^{i\omega t} \equiv A_pv_1e^{i\omega t} \end{aligned} \tag{23}$$

where A_p is called the admittance (per unit area) for holes diffusing into the n-region; its real part G_p and imaginary part S_p are the conductance and susceptance. For $\omega\tau_p$ small, the real term G_p is simply the conductance per unit area of a layer L_p thick with conductance

$$qp_n e^{qV_0/kT}\mu_p/L_p = q(p_n + p_0)\mu_p/L_p; \tag{24}$$

for the case of zero bias this confirms the result stated earlier that the resistance to hole current flow arises from the low hole conductivity $qp_n\mu_p$ of the n-region. Equation (24) is also the differential conductance obtained by differentiating the d-c characteristic (22) with respect to V_0.

In this section we have treated τ_p as arising from body recombination. In a sample whose y and z dimensions are comparable to L_p or L_n, surface recombination may play a dominant role. It can be shown that the same theory can be applied to a good approximation in this case, provided that for τ_p we use the actual lifetime in the sample.[5] We shall discuss lifetimes in filaments in the next section.

Precisely similar expressions may be derived for electron currents.

The values of the coefficients $qp_n\mu_p/L_p$ and $qn_p\mu_n/L_n$ which depend on the properties of the material may be expressed in terms of the conductivity σ_n of the n-region, σ_p of the p-region, σ_i the conductivity of intrinsic material, and the ratio

$$b = \mu_n/\mu_p. \tag{25}$$

In terms of these quantities we have

$$\sigma_i = q\mu_p n_i(1 + b), \tag{26}$$

and if $p_n \ll n_n$ and $n_p \ll p_p$, we find

$$q\mu_p p_n = b\sigma_i^2/(1 + b)^2\sigma_n \tag{27}$$

$$q\mu_n n_p = b\sigma_i^2/(1 + b)^2\sigma_p. \tag{28}$$

Using these equations, we may rewrite (22) and a corresponding equation for electron current into the p-region so as to express their dependence on d-c bias V_0 and the properties of the regions:

$$\begin{aligned} I_{p0}(V_0) &= \frac{b\sigma_i^2}{(1 + b)^2\sigma_n L_p} \cdot \frac{kT}{q} (e^{qV_0/kT} - 1) \\ &\equiv G_{p0} \frac{kT}{q} (e^{qV_0/kT} - 1) \\ &\equiv I_{ps}(e^{qV_0/kT} - 1) \end{aligned} \tag{29}$$

[5] See W. Shockley, *loc. cit.*, Appendix V.

$$
\begin{aligned}
I_{n0}(V_0) &= \frac{b\sigma_i^2}{(1+b)^2\sigma_p L_n}\cdot\frac{kT}{q}\,(e^{qV_0/kT}-1)\\
&\equiv G_{n0}\frac{kT}{q}\,(e^{qV_0/kT}-1)\\
&\equiv I_{ns}(e^{qV_0/kT}-1).
\end{aligned}
\tag{30}
$$

The values of G_{p0} and G_{n0} [which are readily seen to be the values of the low-frequency, low-voltage ($V_0 \ll kT/q$) conductances] and the saturation reverse currents are given by

$$G_{p0} \equiv \frac{b\sigma_i^2}{(1+b)^2\sigma_n L_p} \equiv \frac{q}{kT} I_{ps} \tag{31}$$

$$G_{n0} \equiv \frac{b\sigma_i^2}{(1+b)^2\sigma_p L_n} \equiv \frac{q}{kT} I_{ns}. \tag{32}$$

The expression for direct current then becomes

$$
\begin{aligned}
I_0(V_0) &= (G_{p0}+G_{n0})\frac{kT}{q}\,(e^{qV_0/kT}-1)\\
&= (I_{ps}+I_{ns})(e^{qV_0/kT}-1).
\end{aligned}
\tag{33}
$$

This is the expression plotted in Figure 4.6 and compared with data there.

In terms of G_{p0}, the complex admittance per unit area for hole flow may be written in the form

$$A_p = G_p + iS_p = (1+i\omega\tau_p)^{1/2} G_{p0} e^{qV_0/kT}. \tag{34}$$

For low frequencies, such that ω is much less than $1/\tau_p$, we can expand $G_p + iS_p$ as follows:

$$G_p + iS_p = G_{p0}e^{qV_0/kT} + i\omega(\tau_p/2)G_{p0}e^{qV_0/kT}. \tag{35}$$

Hence $(\tau_p/2)G_{p0}\exp(qV_0/kT)$ behaves like a capacitance per unit area.

It is instructive to interpret this capacitance for the case of zero bias, $V_0 = 0$, for which we find:

$$C_p = \tau_p G_{p0}/2 = \tau_p q p_n \mu_p/2L_p = q^2 p_n L_p/2kT. \tag{36}$$

The last formula, obtained by noting that $\tau_p\mu_p = q\tau_p D_p/kT = qL_p^2/kT$, has a simple interpretation: qp_nL_p is the total charge of holes in a layer L_p thick. For a small change in voltage v, this density should change by a fraction qv/kT so that the change in charge divided by the change in v is $(q/kT)(qp_nL_p)$ which differs from C_p only by a factor of 2, which arises from the nature of the diffusion equation.

Although A_p simulates a conductance and capacitance in parallel at low frequencies, its high-frequency behavior is quite different. In Figure 12.4 the behavior of $(1+i\omega\tau)^{1/2} = A_p/G_{p0}$, is shown. For high frequencies

G_p and S_p are equal:

$$G_p = S_p = \sqrt{\tau_p/2}\, G_{p0} \sqrt{\omega} = \frac{b\sigma_i^2 \sqrt{\omega}}{(1+b)^2 \sigma_n \sqrt{2D}}. \tag{37}$$

Thus for high frequencies the admittance is independent of τ_p and is determined by the diffusion of holes in and out of the n-region. The three

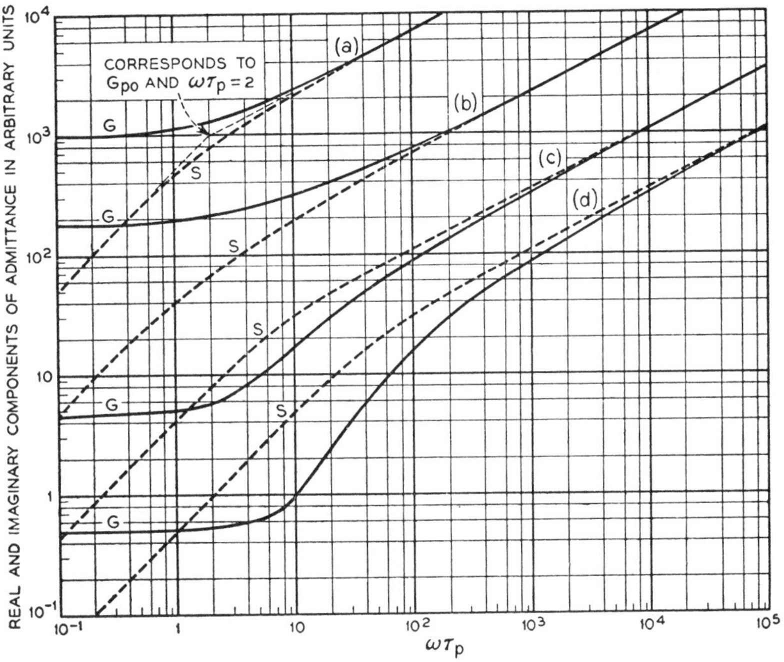

FIG. 12-4—Real and Imaginary Parts, G and S, of the Admittance for Hole Flow into the n-Region.

(a) Shows $10^3\, A_p/G_{po} = 10^3\,(1 + \omega\tau_p)^{1/2}$, and corresponds to equation (34).

(b), (c) and (d) correspond to cases in which the n-region has varying properties.*

* See W. Shockley *Bell Syst. Tech. J.* **28**, (1939), Figure 7 and associated text.

straight asymptotes have a common intersection at the point G_{p0}, $\omega\tau = 2$ on Figure 12.4, a fact which is useful in estimating the value of τ from such data.

The values of $I_{ps} + I_{ns}$ deduced from the known conductivities of the material composing the junction and values of τ, which is a sort of average of τ_p and τ_n, are found to be in general agreement with the values found from the d-c rectification curve within limits imposed by the approximations of the model. At the time of writing, experiments were being undertaken to determine L_p and L_n directly by photoconductive experiments with a view to obtaining separately the I_p and I_n components. The fit for the rectification formula itself, equation (33), is discussed in connection with Figure 4.6.[6]

12.6 LIFETIME IN FILAMENTS, AND THE SUHL EFFECT

12.6a. Introduction. A number of experiments have shown that the recombination process in germanium filaments often takes place largely on the surface and that this recombination may be enhanced by applying simultaneously a longitudinal electric field and a transverse magnetic field. This latter effect arises from the increased recombination on the surface due to the density there of the injected carriers, the Suhl effect discussed in Section 3.2.

In this section we shall solve the continuity equation for a rectangular filament for the case in which a longitudinal field is present and show how surface and body recombination enter into the equations for lifetime and decay constant (= 1/lifetime). We shall treat only certain limited aspects of the Suhl effect because the analysis, although straightforward, is somewhat lengthy.

Throughout this section we shall deal with the case of *holes injected into n-type material in small densities.* For this case arguments similar to those of the last section apply, and we conclude that, since the changes in n are small,

$$\psi - \varphi_n = \frac{kT}{q} \ln n_n/n_i = \text{constant}. \tag{1}$$

The transverse currents in the filament are equal and opposite for holes and electrons, since otherwise a net change would build up on the surface; consequently, the gradients of φ_n and φ_p must be in the ratio of p to n [see equations (20) and (21) of Section (12.4)]. Accordingly, φ_n and ψ must change by negligible amounts across the filament, and the hole current must flow largely by diffusion.

This same conclusion may be reached less analytically as follows: Suppose a few holes are injected. They will set up a slight electric field. This field will draw in electrons to neutralize the charge of the holes. However, the electron density is much larger than the hole density so that the required field need alter the electron distribution only very slightly. The field must, therefore, be so small that the electrons move almost as if no holes were present at all. Hence to a first approximation the holes can be considered to move in the same field as would exist if no holes were present. The effect of the electrons, so far as the holes are concerned, is simply to neutralize the charge density of the holes so that they are shielded from the mutual repulsion of their space charge.

We shall consider the effect of surface recombination upon the lifetime

[6] These results were planned for publication, probably in the *Physical Review*, for late 1950.

for two cases. In the first case we shall imagine that the distribution along the filament is uniform, a situation which might be produced by illuminating the filament by a flash of X rays, for example. After this, we shall show that the hole density decays as

$$\exp(-t/\tau_{p1}), \quad \tau_{p1} = \text{lifetime in filament,} \tag{2}$$

independently of the presence of an electric field. The lifetime is by definition τ_{p1} or the time required for decay by a factor of $\mathbf{e}^{-1}$ to take place.

In the second case, we shall imagine that holes are being steadily injected at some point and are drawn down the filament, decaying by recombination as they go. For this case an attenuation constant a will be introduced, the hole density varying as

$$\mathbf{e}^{-ax}, \quad a = \text{attenuation constant.} \tag{3}$$

We shall define a lifetime as the time required for a hole to drift under the field a distance $1/a$ corresponding to a decay of $\mathbf{e}^{-1}$ in the density. This procedure leads to

$$\tau_{p2} = (1/a)/\mu_p E_x = 1/a\mu_p E_x. \tag{4}$$

We shall show that, for large electric fields, so that motion along the filament arises chiefly from E_x, the values of τ_{p2} and τ_{p1} are equal. We shall find it less convenient to deal with the lifetimes than with the *decay constants*, defined as $1/\tau_{p1}$ and $1/\tau_{p2}$. In terms of decay constants, the effects of surface and volume recombination are found to be additive so that we may write

$$\frac{1}{\tau_{p1}} = \frac{1}{\tau_p} + \nu_s \tag{5}$$

where $1/\tau_p$ is the volume decay constant and ν_s is the surface decay constant.

12.6b. Continuity Equation and Boundary Conditions. We are, therefore, led to investigating the solution of the *continuity equation* [(15) of Section 12.2] for holes in a germanium filament in the presence of a uniform electric field, which we take to lie in the x-direction, along the filament. We shall take the boundaries to lie at $y = \pm B$ and $z = \pm C$ so that the cross section is rectangular with area $4BC$. The hole current has components

$$I_x = q\mu_p p E_x - qD_p \frac{\partial p}{\partial x} \tag{6a}$$

$$I_y = -qD_p \frac{\partial p}{\partial y} \tag{6b}$$

$$I_z = -qD_p \frac{\partial p}{\partial z}. \tag{6c}$$

These must satisfy the *continuity equation*

$$\begin{aligned}\frac{\partial p}{\partial t} &= \frac{p_n - p}{\tau_p} - \frac{1}{q}\nabla \cdot \boldsymbol{I}_p \\ &= \frac{p_n - p}{\tau_p} - \mu_p E_x \frac{\partial p}{\partial x} + D_p\left(\frac{\partial^2 p}{\partial x^2} + \frac{\partial^2 p}{\partial y^2} + \frac{\partial^2 p}{\partial z^2}\right).\end{aligned} \tag{7}$$

Evidently $p = p_n$ is the solution of this equation corresponding to no injected holes. We shall introduce the symbol

$$p_1 = p - p_n \tag{8}$$

to represent the disturbance due to hole injection.

In terms of p_1, equation (7) is a homogeneous linear partial differential equation since each term involves p_1 linearly. Furthermore, it is separable[1] and has solutions that can be written as products of separate functions of t, x, y and z. One such solution is

$$p_1 = \mathbf{e}^{-\nu t - ax} \cos by \cos cz \tag{9}$$

where in order to satisfy (7) the constants ν, a, b and c must satisfy the relationship

$$\nu - \frac{1}{\tau_p} + \mu_p E_x a + D_p(a^2 - b^2 - c^2) = 0. \tag{10}$$

The most general solution of (7) corresponding to arbitrary initial conditions at $t = 0$ can be expressed as the sum of a series of terms of the form of (9), in general including sine terms as well as cosine terms for y and z and complex or imaginary values for a.

For the τ_{p1} case described above, if the filament is illuminated equally on all sides so that the same value of p_1 is produced on all sides, only the symmetrical cosine terms will be needed. A number of such terms will be needed, however, in order that their sum at $t = 0$ will represent the hole density. Each term in the series will thereafter decay with its own characteristic lifetime so that the decay of the total number of holes in the filament will not be represented by a simple exponential decay but by a sum of decaying terms having different lifetimes. This reasoning leads naturally to the question: What is meant by the lifetime in a filament when the decay is made up of a mixture of lifetimes?

The answer is that one of the solutions of the form (9) has a lifetime so much longer than the others that after the others have decayed to negligible proportions it will still persist. It is this longest lifetime with which we shall be chiefly concerned. The other terms can be considered to be initial

[1] A solution can be written as the product of separate functions of the variables, as may be seen from the following development.

transients which die away rapidly leaving only the slowest decay to be observed.

A similar argument applies to the τ_{p2} case. For this case we may imagine that at $x = 0$ the hole current distribution in the cross-section of the filament requires a superposition of many terms of the form $\cos by \cos cz$ and sine terms as well. All but one of these decay very rapidly as the holes drift down the filament under the field and the remaining one, as we shall show, gives the same value for lifetime as τ_{p1}.

[The problem of expanding an arbitrary function of x, y and z in a series of *eigenfunctions* satisfying (7) and the boundary conditions on the surface is similar to many other problems of mathematical physics.[2] A detailed solution for flow across a p-n junction into a square filament has been published.[3]]

At the boundaries $y = \pm B$ we shall suppose that the excess of recombination over generation is directly proportional to the disturbance p_1 with proportionality constant s. Consequently, the current toward the surface satisfies

$$(1/q)I_y = \pm sp_1 \qquad y = \pm B \tag{11a}$$

$$(1/q)I_z = \pm sp_1 \qquad z = \pm C. \tag{11b}$$

The quantity s has the dimension of a velocity or length per unit time. We shall call it the *surface recombination velocity*. Equation (11a) states that the rate of recombination is the same as if a current of holes of density p_1 were drifting with an average velocity s into the surface and being removed.

The boundary condition for B and $-B$ are equivalent when p_1 is given by (9); and (11a) and (11b) reduce to

$$bB \tan bB = sB/D_p \equiv \chi_y = \eta \tan \eta \tag{12a}$$

$$cC \tan cC = sC/D_p \equiv \chi_z = \zeta \tan \zeta \tag{12b}$$

where we have introduced the symbols

$$\eta \equiv bB \text{ and } \zeta \equiv cC. \tag{13}$$

For any value of χ_y there is one solution for η in the range $0 \leq \eta \leq \pi/2$ and also one solution in every interval of the form $n\pi$ to $(n + \frac{1}{2})\pi$ and the same is true for χ_z and ζ. The term having the longest lifetime, as we shall show below, is that corresponding to the smallest solutions, denoted by η_0 and ζ_0, of (12a) and (12b). In terms of these b and c become η_0/B and

[2] See Section 14.5 for an analogous quantum mechanical case and the following references, all from McGraw-Hill Book Co., Inc.: L. I. Schiff, *Quantum Mechanics*, 1949, Section 10; P. M. Morse, *Vibration and Sound*, 1936; E. C. Kemble, *The Fundamental Principles of Quantum Mechanics*, 1937.

[3] W. Shockley, *Bell Syst. Tech J.*, **28**, 435–489 (1949), Appendix V.

ζ_0/C respectively. Equation (10) may then be regarded as an equation in two undetermined parameters a and ν and may be rewritten with these terms gathered together in the form

$$\nu + D_p a^2 + \mu_p E_x a = \frac{1}{\tau_p} + D_p \left(\frac{\eta_0^2}{B^2} + \frac{\zeta_0^2}{C^2}\right) \equiv \nu_f. \tag{14}$$

The terms on the right of the first equals sign are determined by the constants of the filament, i.e., τ_p, B, C and s. We shall show below that this combination, denoted by ν_f, is the decay constant for the filament and is equal to $1/\tau_{p1}$ and $1/\tau_{p2}$. Because of this equality we may introduce a *filament lifetime*

$$\textit{filament lifetime} = \tau_f = 1/\nu_f. \tag{14b}$$

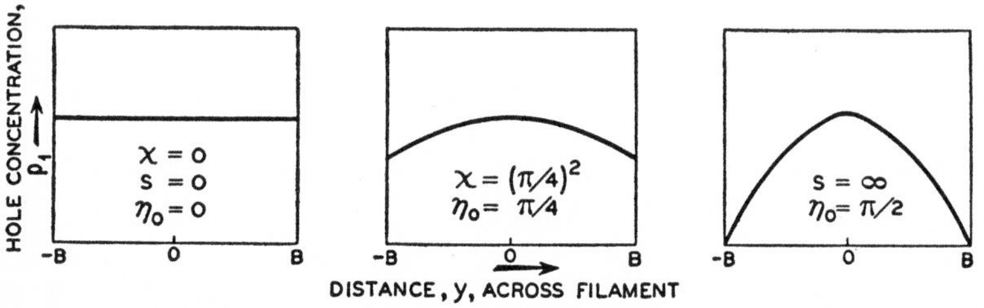

FIG. 12–5—Variation of the Density of Injected Holes across a Filament.

In Figure 12.5 we show the distribution of injected hole density across the filament for the η_0, ζ_0 solution on the line $z = 0$. This distribution shows how p_1 varies with y for three values of χ. For $s = 0$, the distribution is uniform since no diffusion to the surface is required to balance surface recombination. For $s = \infty$, the hole density is zero at the surface, and a high diffusion gradient produces a flow of holes to the surface where they recombine. For an intermediate value of χ, the value of p_1 is finite at the surface, and the gradient is just large enough to account for recombination.

We shall discuss the τ_{p1} case first, for which the density is independent of x so that $a = 0$. For this case equation (14) reduces to

$$\nu = \nu_f \tag{15}$$

independently of the value of E_x. The solution p_1 can then be written as

$$p_1 = \mathrm{e}^{-\nu_f t} \cos by \cos cz \tag{16}$$

so that the hole density decays exponentially with a decay constant ν_f which is equal to $1/\tau_{p1}$ in accordance with equation (2) so that

$$\frac{1}{\tau_{p1}} = \nu_f = \frac{1}{\tau_p} + D_p \left(\frac{\eta_0^2}{B^2} + \frac{\zeta_0^2}{C^2}\right) = \frac{1}{\tau_p} + \nu_s. \tag{17}$$

This equation shows that the decay constant ν_f for the filament is the sum of the "bulk material" decay constant $1/\tau_p$ plus a surface decay constant ν_s defined by equation (17).

We can now see why terms involving η_n and ζ_n with n greater than zero can be disregarded for many purposes. For these terms the values of η_n^2 and ζ_n^2 will be at least four times larger than for η_0^2 and ζ_0^2; consequently, such terms will decay rapidly compared to the η_0, ζ_0 term and will become negligible in comparison with it, at least in all cases where ν_s is important. The distribution of hole density across the filament for these higher order terms is alternating in sign and the rapid decay is a result of flow of excess holes in regions of hole deficit. The distribution of hole density for the case $s = \infty$, corresponding to $p = 0$ at the boundary, is identical with the displacement of the rectangular drum head of Figure 5.2. Part (f) corresponds to the $\eta_0\zeta_0$ term. The other parts of the figure correspond to using sine as well as cosine functions in p_1 and have values of η and ζ for the sine functions which satisfy equations of the form $\eta \cot \eta = \chi$. We shall neglect these higher order terms in the following discussion.

The values of ν_s for the η_0, ζ_0 terms is

$$\nu_s = D_p\left[\frac{\eta_0^2}{B^2} + \frac{\zeta_0^2}{C^2}\right] \tag{18}$$

and this is a function of D_p, B, C, and s with s entering through the dependence of η_0 and ζ_0 upon $\chi_y = sB/D_p$ and $\chi_z = sC/D_p$. Simple forms are obtained for large and small values of s. For $s \rightarrow \infty$, both η_0 and ζ_0 become $\pi/2$ corresponding to $p_1 = 0$ on the boundary. For this case

$$\nu_s = \frac{\pi^2 D_p}{4}\left(\frac{1}{B^2} + \frac{1}{C^2}\right), \quad s \rightarrow \infty. \tag{19a}$$

For small values of s, equations (12) and (18) reduce so that ν_s again takes a simple form [see Problems for a physical argument that also leads to (19b)]:

$$\nu_s = s\left(\frac{1}{B} + \frac{1}{C}\right), \quad s \rightarrow 0. \tag{19b}$$

For a filament of square cross section ν_s can be expressed as a function of the three variables s, B, and D_p by a single curve in terms of the following equation in which $\eta_0(\chi_y)$ is the solution of (12a):

$$\frac{\nu_s B^2}{D_p} = 2[\eta_0(\chi_y)]^2 = 2[\eta_0(sB/D_p)]^2. \tag{20}$$

This relationship is plotted in Figure 12.6.

Figure 12.6 may be used in combination with equation (17) to calculate values of τ_f from known dimensions B, body lifetime τ_p, and surface recombination constant s. It may also be used to determine s from known values of body lifetime τ_p and filament lifetime τ_f. Experiments

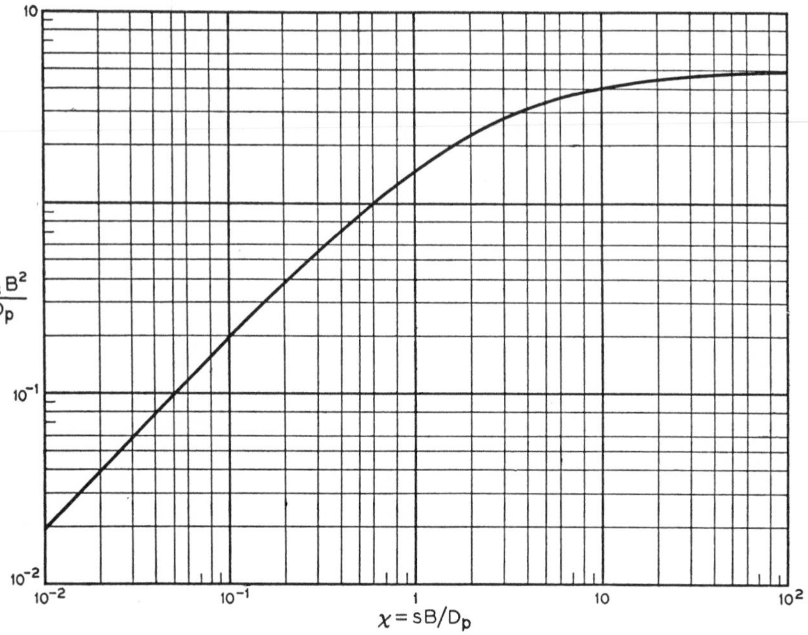

Fig. 12-6—The Dependence of the Surface Decay Constant ν_s upon the Surface Recombination Velocity s.

of this sort have been carried out by a number of workers and approximate values for s, obtained for various surface treatments of germanium, are as shown in Table 12.1.

Table 12.1 Surface Recombination Constants for Germanium

sand blasted surface	~10^4 cm/sec
etched[a]	~2000 cm/sec
X[a]	~200 cm/sec

[a] Details of these treatments are planned for publication in 1950–51.

We shall now show that the same value of τ_f is obtained by studying the dependence of hole density upon distance along a filament in the presence of a high electric field and by converting this distance to time using the drift velocity $E_x\mu_p = v_x$. For this case the solution is a *steady-state* solution independent of time so that ν in equation (14) should be set equal to zero.

The value of a is then found to be

$$a = \frac{-v_x \pm \sqrt{v_x^2 + 4D_p\nu_f}}{2D_p}. \tag{21}$$

The solution of interest corresponds to the plus sign; the minus sign corresponds to decay towards the left ($-x$-direction) or diffusion against the field. (See Problems.) For large values of $v_x^2/D_p\nu_f$, such as are encountered in filaments used experimentally, we obtain by expanding the radical of equation (21) in series or by neglecting the D_p term in (14)

$$a \doteq \nu_f/v_x \tag{22}$$

which leads to the following value of τ_{p2}, as defined by equation (4):

$$\tau_{p2} = 1/av_x = 1/\nu_f = \tau_f. \tag{23}$$

Thus we find that τ_{p2}, as well as τ_{p1}, is equal to τ_f as defined in equation (14).

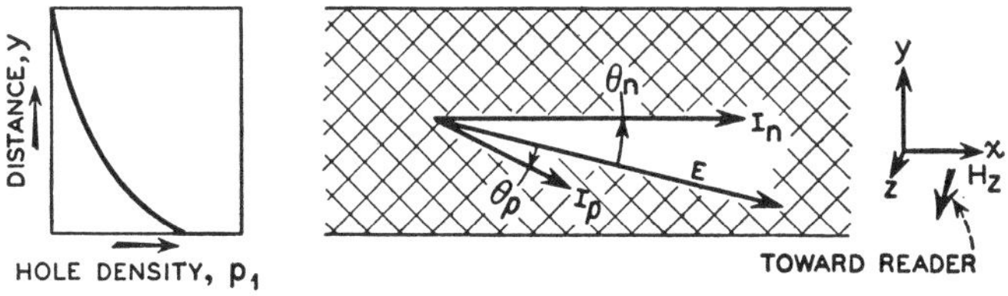

FIG. 12–7—Concentration of Holes near the Surface by the Suhl Effect.

12.6c. The Suhl Effect. We shall next consider the effect of combined electric and magnetic fields upon the lifetime of injected holes. For this case we shall again neglect the influence of the injected holes upon the electric field; there are, however, two differences from the case treated previously. In this case there is a transverse field due to the Hall effect. We shall restrict the treatment to small values of H_z so that only linear terms need be considered. The transverse field E_y is then equal to $-\theta_n E_x$ for the small H_z approximation discussed in Chapter 8. Furthermore, for the coordinate system shown in Figure 12.7, the drift velocity is turned at a Hall angle θ_p from $\boldsymbol{E}$. Finally, the diffusion current is turned at an angle θ_p in respect to the density gradient. [See Section 12.3, equation (11).] For the case treated below, the rotation of the diffusion current produces an effect dependent on $\cos(\theta_p + \theta_n)$. For the small angle approximation used here, this effect is negligible.

If E_y is large enough, the field across the filament will be so strong that substantially all the holes will be concentrated on one side. We shall restrict the treatment to this case, referred to as the strong Suhl effect, and shall choose the origin of the y-coordinate as the side of the filament towards which holes are deflected. This is the case shown in Figure 12.7.

We shall for brevity restrict the treatment to the case analogous to τ_{p1} and assume that p_1 depends only on y and t; the neglect of z corresponds to increasing the z-dimension indefinitely. These assumptions permit us to neglect the x-component of the hole current since this component is independent of x and does not contribute to the divergence. The y-component of hole current is then

$$I_y(y) = -q\mu_p p_1 E_x(\theta_n + \theta_p) - qD_p \frac{\partial p_1}{\partial y}; \tag{24}$$

and at $y = 0$, corresponding to the lower boundary towards which holes are driven, this must satisfy the boundary condition

$$-I_y(0) = qsp_1(0). \tag{25}$$

We shall show that there is a solution for the injected hole density of the form

$$p_1 = \mathrm{e}^{-\nu t - by} \tag{26}$$

which attenuates in the $+y$-direction, as shown in Figure 12.7, and represents hole density near one surface of the filament. In order to abbreviate the algebra and bring out the physics we introduce the symbol v_t where

$$v_t = \mu_p E_x(\theta_p + \theta_n) = \mu_p E_t. \tag{27}$$

v_t is the transverse drift velocity which a hole would have in the field of Figure 12.7. E_t is thus an "effective transverse field". Substituting p_1 and v_t into the preceding equations leads to the following expression for the boundary condition (25):

$$v_t - D_p b = s \tag{28}$$

or

$$b = (v_t - s)/D_p. \tag{29}$$

If $v_t > s$, the holes tend to accumulate near $y = 0$ since they are brought down faster than they can recombine; we shall treat this case which represents a strong Suhl effect; cases with $v_t = s$ and $v_t < s$ represent transition cases between $H_z = 0$ and H_z large and have a more complicated behavior.

The density gradient in Figure 12.7 tends to make holes diffuse away from the surface and is thus in contrast with Figure 12.5 in which holes diffuse to the surface. The situation of Figure 12.7 occurs because we have taken $v_t > s$. The quantity s has the dimensions of a velocity and, for $v_t > s$, holes are swept towards the surface by E_t so rapidly that they tend to accumulate and set up a back diffusion current.

The value of ν is determined by inserting the expression (26) for p_1 into the *continuity equation* (15) of Section 12.2:

$$\frac{\partial p_1}{\partial t} = -\frac{p_1}{\tau_p} - \frac{1}{q}\nabla \cdot I_p, \tag{30}$$

which then reduces to

$$\begin{aligned}\nu &= \frac{1}{\tau_p} + v_t b - D_p b^2 = \frac{1}{\tau_p} + (v_t - D_p b)b \\ &= \frac{1}{\tau_p} + sb \equiv \frac{1}{\tau_p} + \nu_s.\end{aligned} \tag{31}$$

The contribution $\nu_s = sb$ of the surface recombination to the decay constant is thus seen to be directly proportional to b, which, in turn, increases rapidly with

$$E_t = (\theta_p + \theta_n)E_x = 10^{-8}(\mu_{Hp} + \mu_{Hn})H_z E_x, \tag{32}$$

in the notation of Section 8.8 *for practical units.*

The relationship $\nu_s = sb$ of equation (31) has a relatively simple interpretation. The solution

$$p_1 = e^{-\nu t - by} \tag{33}$$

leads to a total number of holes per unit area of surface on the xz plane of

$$\int_0^{2B} e^{-\nu t - by}\,dy = \frac{e^{-\nu t}}{b}(1 - e^{-2Bb}). \tag{34}$$

The quantity $2Bb$ is an effective transverse potential due to the magnetic field measured in units of kT/q. For $v_t \gg s$ it may be written as

$$2Bb = 2B\mu_p E_t/D_p = 2qBE_t/kT. \tag{35}$$

For realizable experimental conditions, the contribution of the v_t term to $2bB$ may be so large that the term $\exp(-2Bb)$ is negligible. [If $\exp(-2Bb)$ is not negligible, the effect of the boundary at $y = 2B$ must be included.] The rate of recombination on the surface is

$$p_1 s = se^{-\nu t} \tag{36}$$

and this is simply sb times the number of holes given by (34). Since the decay constant is the ratio of rate of decrease of holes to total number, this shows that $sb = \nu_s$.

To put the same result in simpler language: *For strong transverse fields, the holes flow in a layer with an effective thickness of $kT/qE_t \doteq 1/b$* so that the total number per unit area is $p_1(0) \times (1/b)$ and the rate of recombination is $p_1(0) \times s$ and the decay constant is $s \div (1/b) = sb$.

The treatment just presented corresponds to the τ_{p1} treatment of Section 12.6a since for it p_1 was independent of x.

A treatment like that given to τ_{p2} can be carried out also when magnetic fields are present. The answer is again that $\tau_{p1} = \tau_{p2}$, provided that θ_p and θ_n may be treated as small angles. When they are large, magneto-resistance effects set in and the complete mobility and diffusion tensors of Section 12.3 must be employed.

Theories of the sort discussed here have been applied in evaluating the surface recombination constant s as discussed in Section 3.2.[4]

12.7 HERRING'S SOLUTION FOR LARGE, INJECTED CARRIER DENSITIES

The formulation of the continuity equation of Section 12.2, the Einstein relationship, and the equations for the quasi Fermi levels are applicable to hole and electron densities of arbitrary magnitudes (provided degeneracy is not encountered). In the applications presented in Sections 12.5 and 12.6, however, only hole densities which were very small compared to electron densities were considered. Under these conditions, the differential equations for p were linear, and separable and simple solutions could be obtained including the effects of recombination, diffusion, and applied electric and magnetic fields. When the hole density is comparable to the electron density, the equations become nonlinear and it is much more difficult to obtain solutions and to visualize the processes involved.

An elegant solution for the nonlinear problem has been obtained by C. Herring[1] for the case of large-amplitude disturbances of the densities. We shall discuss his simplest case which is subject to the following assumptions:

I. $n = p + n_e = p + N_d - N_a$ (1)

which implies that the net charge density is zero and that the donors and acceptors are fully ionized.

II. Recombination is negligible.

III. Conditions are uniform in the y and z directions so that all quantities are functions of x and t.

IV. Diffusion currents can be neglected compared to currents produced by the field E_x.

Assumption II puts the treatment of this section on a very different basis from those of the last two sections in which recombination plays a major role in causing the injected carrier concentrations to decay to their equilibrium values. For this section, the equilibrium values are without

[4] See H. Suhl and W. Shockley, *Phys. Rev.* **76**, 180 (1949).

[1] C. Herring, *Bell. Syst. Tech. J.*, **28**, 401–427 (1949).

significance since Assumption II prevents an approach to them. This is the reason that in (1) we have used not n_n but instead n_e, subscript e for excess of donors over acceptors. Similarly the conductivity $\sigma(0) = q\mu_n n_e$ is not the equilibrium conductivity but simply that corresponding to $p = 0$. (The limiting case of intrinsic material with $n_e \to 0$ leads to $\sigma(0) = 0$ and $E_0 = \infty$ but in such a way, as pointed out in a footnote, that the expression for the densities velocity is properly behaved.)

Under these conditions the hole and electron currents are

$$I_p = q\mu_p pE \tag{2}$$

$$I_n = q\mu_n nE, \tag{3}$$

where we have omitted the subscript x from the current densities and from E. The fraction of the current carried by holes is a function of p and can be written[2] as

$$f(p) = \frac{I_p}{I} = \frac{I_p}{I_n + I_p} = \frac{\mu_p \dfrac{p}{n_e}}{\mu_n\left[1 + \left(1 + \dfrac{1}{b}\right)\dfrac{p}{n_e}\right]}, \tag{4}$$

where n has been eliminated by equation (1) and

$$b = \mu_n/\mu_p. \tag{5}$$

(These equations are discussed more fully in Section 3.1.) In terms of $f(p)$, the continuity equation [(15) of Section 12.2] becomes

$$\begin{aligned}\frac{\partial p}{\partial t} &= -\frac{1}{q}\nabla \cdot \boldsymbol{I}_p = -\frac{I}{q}\frac{\partial f(p)}{\partial x} = -\frac{I}{q}\frac{df(p)}{dp}\frac{\partial p}{\partial x} \\ &\equiv -V(I, p)\frac{\partial p}{\partial x}\end{aligned} \tag{6}$$

where the *densities velocity* $V(I, p)$, defined by this equation, will be shown later to be the velocity with which the hole density propagates. $V(I, p)$ may be written in the form

$$V(I, p) \equiv \frac{I}{q}\frac{df(p)}{dp} = \frac{I}{q\mu_n n_e}\frac{\mu_p}{\left[1 + \left(1 + \dfrac{1}{b}\right)\dfrac{p}{n_e}\right]^2}. \tag{7}$$

In some cases we shall consider I to be independent of time and shall write V simply as $V(p)$. For small densities of injected holes, $V(I, p)$ reduces to the drift velocity $\mu_p E_0$ of holes in a field

$$E_0 = I/qn_e\mu_n = I/\sigma(0) \tag{8}$$

[2] $f(p)$, which is used only in this section, is not to be confused with the Fermi-Dirac functions f and f_p.

corresponding to I in the n-type material. For large values of p, $V(I, p)$ approaches zero.

It is instructive to compare $V(I, p)$ with the drift velocity $v_p = E\mu_p$ of a hole. E is obtained from the conductivity

$$\sigma(p) = q(\mu_p p + \mu_n n) = \sigma(0)\left[1 + \left(1 + \frac{1}{b}\right)\frac{p}{n_e}\right], \tag{9}$$

and the current I by using the equation $E = I/\sigma$ so that

$$v_p(p) = \mu_p E = \frac{I}{qn_e\mu_n} \frac{\mu_p}{\left[1 + \left(1 + \frac{1}{b}\right)\frac{p}{n_e}\right]}. \tag{10}$$

For $p = 0$, this equation, like equation (7), reduces to $\mu_p E_0$. For larger values of p, we have the relationships

$$v_p(p) = v_p(0)\frac{\sigma(0)}{\sigma(p)} = \frac{v_p(0)}{\left[1 + \left(1 + \frac{1}{b}\right)\frac{p}{n_e}\right]} \tag{11a}$$

$$V(I, p) = v_p(0)\left[\frac{\sigma(0)}{\sigma(p)}\right]^2 = \frac{v_p(0)}{\left[1 + \left(1 + \frac{1}{b}\right)\frac{p}{n_e}\right]^2} \tag{11b}$$

which shows that the reduction factor for $V(I, p)$ is the square of that for v_p.

Since n and p are related linearly according to equation (1), n may be determined directly from p and need not be considered separately. We shall also omit a discussion of electrons injected into p-type material since that case can be obtained from the one treated by interchanging subscripts and writing $1/b$ in place of b.

We shall next consider an extreme case corresponding to hole injection into intrinsic material. For this case we shall show, by direct physical arguments rather than by equation (7), that the hole density remains stationary in space although the drift velocity of holes does not vanish.[3] Accordingly we suppose that the hole density is some function $p_0(x)$ at $t = 0$; since $n_e = 0$, n also equals $p_0(x)$ at $t = 0$. Thus for all values of x, the ratio of hole current to electron current is constant and has the value $1/b$, and consequently, the hole current is independent of x. For this condition, the net hole current flowing into one side of any interval δx is equal to the hole current flowing out of the other side, and the hole and electron densities remain constant although both hole and electron

[3] After considering the meaning of V as discussed in connection with Figure 12.8, the reader should verify that the statement above corresponds to showing that as $n_n \rightarrow 0$, V vanishes while v_p remains finite.

flows occur. We may imagine a condition of this sort to be produced by producing photoconductivity by a pulse of light shining on a limited length of a filament in the intrinsic condition. The region of high density would then remain fixed in space while holes flowed through it in one direction and electrons in the other direction.

The mathematical expression of the physical picture described in the preceding paragraph is that $df(p)/dp = 0$ so that the same fraction of current is carried by holes at all values of x and consequently $\nabla \cdot \boldsymbol{I}_p = 0$. For small values of p in an n-type sample the hole density progresses with the hole drift velocity as we shall discuss.

Actually diffusion and recombination would cause the region of high density to spread and decay, and the stationary situation already discussed would represent only an approximation. However, it serves to illustrate that it is not the hole drift velocity which causes the distribution of density to shift but instead the divergence of hole current as required by the continuity equation.

If, on the other hand, the hole density p is so small compared to n that the electric field is altered negligibly so far as hole motion is concerned, then the hole current is simply

$$I_p = E_0 q \mu_p p = q v_p p \tag{12}$$

where v_p is the drift velocity of a hole in the field E_0. For this case the continuity equation becomes

$$\frac{\partial p}{\partial t} = -v_p \frac{\partial p}{\partial x}, \tag{13}$$

which is solved by any function of $x - v_p t$,

$$p = p_0(x - v_p t),$$

which represents translation of the hole density in the $+x$-direction with velocity v_p.

The reader may verify that, for $p/n_n \ll 1$, equations (6) and (7) reduce to (13), whereas for the intrinsic case they reduce to $\partial p/\partial t = 0$.

For intermediate cases, Herring points out that a solution to (7) can be obtained by a geometrical construction which we shall illustrate in terms of a pulse as shown in Figure 12.8. In this figure the pulse has a hole density at $t = 0$ represented by an isosceles trapezoid. In part (b), the density velocity is plotted as a function of p/n_e on the same density scale as for part (a), it being assumed that I is constant in time. Herring's construction is as follows: Take any point in (a) for which p has the value p_1; draw a horizontal line from this point with length $V(p_1)t$; the point so obtained is a point on the translated pulse. That this process satisfies equation (7),

$$\frac{\partial p}{\partial t} = -V(p)\frac{\partial p}{\partial x}, \tag{14}$$

may be seen from (c) which shows that, in time dt, the increase in p at constant x is

$$(dp)_x = [V(p)dt] \cdot \left(-\frac{\partial p}{\partial x}\right) \tag{15}$$

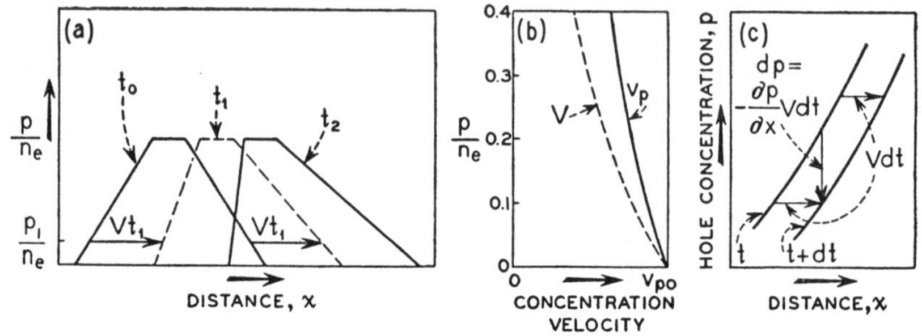

FIG. 12-8—Deformation of a Pulse of Holes Due to Conductivity Modulation. (Drawn for $b = 2$.)

from which (7) is obtained by dividing by dt. A more general expression of the functional relationship between p, x, and t is

$$x(p, t) = x(p, 0) + \int_0^t V[I(t'), p]dt', \tag{16}$$

which is discussed further in the problems.

It is to be noted that the increase in density velocity for low hole density causes the lower densities to go faster than the higher densities so that the trailing edge of the pulse becomes infinitely steep, a shock-wave phenomenon. As the steepness increases, assumption IV (that is, diffusion currents are negligible) becomes untenable and the steep trailing edge tends to widen owing to diffusion. As a result, the width of the trailing edge reaches a limiting value of the order of

$$\Delta x = \frac{2kT}{qE_0} \div \left[1 - \frac{1}{1 + \left(1 + \frac{1}{b}\right)\frac{p_{\max}}{n_e}}\right]^2 \tag{17}$$

where $p_{\max}$ is the maximum of p at the top of the trailing edge. This equation, due to Herring, is obtained by setting the rate of widening due to diffusion equal to the rate of narrowing due to the difference in velocity between the limiting hole densities in the pulse. (See Problems.)

Herring has also considered the effect of recombination, which can be taken account of in a relatively simple way (see Problems) and diffusion. When nonlinear effects due to conductivity are combined with the diffusion mechanism, the problem becomes much more complicated. Steady-state solutions of the one-dimensional case in which the only approximation made is that of assuming negligible failure of the neutrality equation $n = p + n_e$, assumption I, have been obtained by W. van Roosbroeck.[4]

12.8 BAND EDGE ENERGIES VERSUS VOLUME: $\mathcal{E}_{1p}$, $\mathcal{E}_{1n}$, $\mathcal{E}_{1G}$

The method of *deformation potentials* discussed in Sections 11.3 and 17.6 makes use of the quantities $\mathcal{E}_{1p}$ and $\mathcal{E}_{1n}$, which were defined as the change in the band edges $\mathcal{E}_v$ and $\mathcal{E}_c$ per unit dilatation. These quantities enter the expressions for mobility through equations of the form

$$\mu_n = \frac{2\sqrt{2\pi}}{3} \frac{\hbar^4 c_{ll}}{\mathcal{E}_{1n}{}^2 m_n{}^{5/2} (kT)^{3/2}} \tag{1a}$$

which was derived as equation (30) of Section 11.4. For μ_n in cm^2/volt sec c_{ll} in dynes/cm^2, electron volts for $\mathcal{E}_{1n}$ and T in °K this becomes

$$\mu_n = 3.2 \times 10^{-5} c_{ll} (m/m_n)^{5/2} / \mathcal{E}_{1n}{}^2 T^{3/2}. \tag{1b}$$

As discussed in Section 11.3, equation (34), the relationship

$$|\mathcal{E}_{1G}| = |\mathcal{E}_{1n}| + |\mathcal{E}_{1p}| \tag{2}$$

should apply if the slopes of the band edges have opposite signs as shown in Figure 11.4, which has been drawn to agree with the theoretical band structure computed by Kimball for diamond and discussed in connection with Figure 5.5.

This relationship has been investigated by Bardeen[1] for diamond, silicon, germanium, and tellurium. The methods used to determine how $\mathcal{E}_G$ depends on volume are based on the statistical relationship between $\mathcal{E}_G$ and the product np of electron and hole densities given in equation (19) of Section 10.3:

$$\begin{aligned} np &= N_c N_v \mathrm{e}^{(\mathcal{E}_v - \mathcal{E}_c)/kT} \\ &= 4(2\pi mkT/h^2)^3 (m_n m_p/m^2)^{3/2} \mathrm{e}^{-\mathcal{E}_G/kT} \\ &= 2.33 \times 10^{31} T^3 (m_n m_p/m^2)^{3/2} \mathrm{e}^{-\mathcal{E}_G/kT} \mathrm{cm}^{-6}. \end{aligned} \tag{3}$$

The most direct way to study $\mathcal{E}_G$ is simply to change the volume at constant temperature by applied pressure and to study resulting changes in np.

[4] Planned for publication in *Bell Syst. Tech. J.*, **29** (4) (1950).

[1] J. Bardeen and W. Shockley, *Phys. Rev.* **77**, 407–408 (1950). This section gives in abbreviated form the analysis presented in a second paper planned for publication in the fall of 1950.

The volume also changes due to thermal expansion, and this affords another method of studying the variation of $\mathcal{E}_G$ with volume. These and other methods will be discussed.

The results of these investigations are given in Table 12.2 as the bottom row. Values of $|\mathcal{E}_{1n}|$ and $|\mathcal{E}_{1p}|$ are obtained by solving (1), setting $m_n = m_p = m$; their sum is shown in the row next to bottom.

It should be mentioned that the signs of the terms in the last row are in agreement with the predictions of band theory calculations. According to Kimball's calculations for diamond[2] the bands cross at an atomic spacing larger than the equilibrium value, so that $\mathcal{E}_G$ should decrease with dilation and $\mathcal{E}_{1G}$ should be negative; presumably the same will be true for silicon and germanium. Tellurium, on the other hand, is composed of long chains of atoms, and the chains are not held very tightly together. For this reason a relatively small amount of overlapping may occur at the equilibrium lattice constant and the bands may not have crossed, the situation being like that shown in Figure 5.4 or that for large values of lattice constant in Figure 5.5.

TABLE 12.2 COMPARISON OF $\mathcal{E}_{1G}$ DETERMINED BY OTHER MEANS WITH $|\mathcal{E}_{1n}| + |\mathcal{E}_{1v}|$

	Diamond	*Silicon*	*Germanium*	*Tellurium*
$\rho c^2 = c_{ll}(10^{12}$ dynes/cm$)^2$	10.8[a]	2.0[c]	1.55[b]	0.5[c]
μ_n (electrons) (cm²/volt sec, 295°K)......	900[d]	300[e]	3500[f]	530[g]
μ_p (holes) (cm²/volt sec, 295°K)......	>200[h]	100[e]	1700[f]	530[g]
$\mu_n T^{3/2}$	45×10^5	15×10^5	180×10^5	27×10^5
$\mu_p T^{3/2}$	$>10 \times 10^5$	5×10^5	86×10^5	27×10^5
$\|\mathcal{E}_{1c}\|$ (ev)..................	8.8	6.5	1.7	2.4
$\|\mathcal{E}_{1v}\|$ (ev)..................	<30	11.3	2.4	2.4
$\|\mathcal{E}_{1c}\| + \|\mathcal{E}_{1v}\|$	<39	17.8	4.1	4.8
$\mathcal{E}_{1G}$ (ev)..................	?	−30[i]	−5[i]	+5.0[i]

[a] Longitudinal velocity in [110] direction calculated from elastic constants. See R. F. S. Hearmon, *Rev. Mod. Phys.* **18**, 409–440 (1946); (1950).

[b] Calculated from elastic constants; W. L. Bond, W. P. Mason, H. J. McSkimin, K. M. Olsen, and G. K. Teal, *Phys. Rev.* **78**, 176 (1950).

[c] Estimated from compressibility.

[d] C. C. Klick and R. J. Maurer, *Phys. Rev.* **76**, 179 (1949).

[e] G. L. Pearson and J. Bardeen, *Phys. Rev.* **75**, 1777–1778 (1949).

[f] Unpublished data of J. R. Haynes.

[g] Vivian A. Johnson, *Phys. Rev.* **74**, 1255 (1948).

[h] K. G. McKay, personal communication.

[i] See Text.

[2] G. E. Kimball, *J. Chem. Phys.*, **3**, 560–564 (1937).

The value of $\mathcal{E}_{1G}$ may be obtained by the following procedures:

(1) *Intrinsic Conductivity versus Pressure.* In the intrinsic temperature range, n and p are each equal to

$$n_i = 4.82 \times 10^{15} T^{3/2} (m_n m_p / m^2)^{3/4} e^{-\mathcal{E}_G/2kT} \text{cm}^{-3}. \tag{4}$$

Since the mobility varies as $T^{-3/2}$, the explicit dependence of conductivity arises solely from the exponential factor and

$$\sigma_i = \sigma_{i0} e^{-\mathcal{E}_G/2kT}. \tag{5}$$

Using Bridgman's data on tellurium,[3] Bardeen[4] concludes that changes in σ_{i0} with volume are unimportant compared to changes produced in $\mathcal{E}_G$. By a similar procedure Miller and Taylor find $\mathcal{E}_{1G} \sim -5.0$ electron volts for germanium.[5]

(2) *p-n Junction Resistance versus Pressure.* According to the theory of Section 12.5, the current in a p-n junction is

$$I = q\left[\frac{D_p p_n}{L_p} + \frac{D_n p_n}{L_n}\right][e^{qv/kT} - 1]. \tag{6}$$

For temperatures near room temperature in germanium, p_p and n_n are nearly constant, each being equal to the appropriate difference between donor and acceptor densities. Consequently, since pn is given by equation (3), both n_p and p_n must vary as $T^3 \exp(-\mathcal{E}_G/kT)$. If it is assumed that the variations of D_p, D_n, L_p, and L_n with pressure are small compared to the effect of variations in $\mathcal{E}_G$, then equation (6) can be used to interpret the effect of pressure upon resistance. The value obtained in this way for germanium[6] is in agreement with the value quoted under (1).

(3) *Density versus Temperature.* As discussed later in Section 16.4, it is found that the constant coefficient in the product np is found experimentally for silicon to be 32.5 times larger than the value of equation (3). Pearson and Bardeen have proposed that this is due to the temperature dependence of $\mathcal{E}_G$, which they write as

$$\mathcal{E}_G(T) = \mathcal{E}_G(0) - \beta T. \tag{7}$$

This equation leads to

$$e^{-\mathcal{E}_G(T)/kT} = e^{\beta/k} e^{-\mathcal{E}_G(0)/kT} \tag{8}$$

so that a multiplying factor $\exp(\beta/k)$ occurs in np. Setting this factor equal to 32.5 gives

$$\beta = k \ln 32.5 = 3 \times 10^{-4} \text{ electron volt/°K} \tag{9}$$

[3] P. W. Bridgman, *Proc. Am. Acad. Sci.* **68**, 95–123 (1933); **72**, 157–205 (1938); **74**, 21–51 (1940).

[4] J. Bardeen, *Phys. Rev.* **75**, 1777–1778 (1949).

[5] P. H. Miller and Julius Taylor, *Phys. Rev.* **76**, 179A (1949).

[6] Personal communication from J. Bardeen based on unpublished experiments of H. H. Hall and G. L. Pearson.

and combining this with the volume coefficient of expansion α_v gives

$$\mathcal{E}_{1G} = \beta/\alpha_v = 3 \times 10^{-4}/9 \times 10^{-6} = 33 \text{ electron volts.} \qquad (10)$$

For germanium the corresponding factor in np is 3.5 leading to $\beta = 1.0 \times 10^{-4}$ electron volt/°K and since $\alpha_v = 19 \times 10^{-6}$/°K, the value of $\mathcal{E}_{1G}$ is 5.25 electron volts, in good agreement with the values obtained by the other methods.

It also possible to obtain information on the temperature dependence of $\mathcal{E}_G$ from measurements of the dependence of photoelectric effects upon wave length as a function of temperature. Such effects have been observed by F. S. Goucher and H. B. Briggs.[7]

12.9 DRIFT AND HALL MOBILITIES

During the course of the preparation of this book, great improvements were made in the techniques for studying the behaviors of holes and electrons in germanium. By pulsing the sweeping field in experiments like those described in Section 3.1, J. R. Haynes was able to lower the duty cycle and thus raise the permissible sweeping fields by a large factor.[1] In this way, the wave shapes of pulses of injected carriers were greatly sharpened compared to those shown in Figure 3.9. As a result this technique became a precision method which permitted measurement of the drift mobilities with an error probably less than 5 per cent. The values obtained by this improved technique are shown in Table 12.3.

An extended series of Hall effect measurements have been carried out by G. L. Pearson. He finds that the Hall mobility varies more from sample to sample than does the drift mobility. This variation is not fully understood. Pearson's values obtained using the same samples employed by J. R. Haynes together with the spread in value for a number of samples are also shown in Table 12.3. As discussed in Section 8.9 and Section 11.4, equations (25) and (28c), if the mean free path is independent of the velocity of the electrons or holes, as is the case for the deformation potential theory of scattering by thermal agitation, then the drift mobility should be $8/3\pi$ times the Hall mobility. As is seen from Table 12.3, the drift mobility values do not fall within the same range as those obtained from the Hall mobility, and the discrepancy is apparently larger than the probable error. We shall discuss a possible explanation for this discrepancy below.[1a]

All of the measurements just described were carried out using single crystals of germanium, mostly those grown by drawing single crystals from the melt by the method described by Teal and Little.[1b] The importance

[7] Unpublished investigations.

[1] Planned for publication in the *Physical Review* for late 1950.

[1a] This analysis was presented in brief by G. L. Pearson, J. R. Haynes and W. Shockley, *Phys. Rev.* **78**, 295–296 (1950).

[1b] G. K. Teal and J. B. Little, *Bulletin of the 298th Meeting of the American Physical Society.*

TABLE 12.3 MOBILITIES IN CM^2/VOLT SEC IN GERMANIUM

	$(8/3\pi)$ *times*[a] *Hall Mobility*	*Drift*[b] *Mobility*	*Conductivity*[c] *Mobility*
Electrons	2600 ± 300	3600 ± 180	3350 ± 400
Holes	$1700 + 500$ $- 100$	1700 ± 90	

[a] G. L. Pearson, *Phys. Rev.* **76**, 179 (1949).

[b] Reported by J. R. Haynes at the Chicago Meeting of the American Physical Society, Nov. 26, 1949. G. L. Pearson, J. R. Haynes, and W. Shockley, *Phys. Rev.* **78**, 295 (1950).

[c] G. L. Pearson, J. D. Struthers, and H. C. Theuerer, *Phys. Rev.* **75**, 344 (1949).

of using single crystals arises from the added resistance of grain boundaries, a subject to which we return in the next section.

An estimate of electron mobility can be obtained from the experiments of Pearson, Struthers, and Theuerer, mentioned briefly in Chapter 1, dealing with the determination of antimony content in n-type germanium by the use of radioactive antimony. These data show that the conductivity varies linearly with the antimony content so that the conductivity can be written as

$$\sigma = N_d' q\mu_n + \sigma_0$$

where σ_0 is the conductivity of a control sample without added antimony, and N_d' is the antimony content determined by radioactivity. By solving this equation for μ_n for a number of samples, the value of *conductivity mobility* of Table 12.3 is obtained. There is a considerable spread in the data due in part, at least, to the fact that polycrystalline samples were used. The conductivity mobility is in much better agreement with the drift mobility values of Haynes than with the Hall mobility values of Pearson. This agreement is, of course, the result to be expected because, in principle, both the drift experiment and the conductivity experiment measure the average speed with which the carriers travel down the filament under the influence of an applied field. So long as trapping and possibly electron hole collisions do not cause the drift mobility to deviate appreciably from the *microscopic mobility* (see Section 8.7), the two measurements should be concerned with the same physical quantity. It should, of course, be stressed that from the operational viewpoint, discussed in Section 14.1, the two measurements are quite distinct and the fact that they agree increases the reliance to be placed upon the additional physical concepts associated with the theory of holes and electrons.

Hall mobility values for germanium have been found to fluctuate markedly from sample to sample. Figure 12.9 shows a plot of the values obtained in various ways over a period of time. Values differing appreciably from those quoted in Table 12.3 have recently been reported by Dunlap[2] who gives values for $(8/3\pi)\mu_H$ as high as 3400 cm^2/volt sec for electrons and 2650 cm^2/volt sec for holes.

We shall next present a possible interpretation of the apparent fact that for n-type germanium the Hall mobility μ_{Hn}, which we shall take as

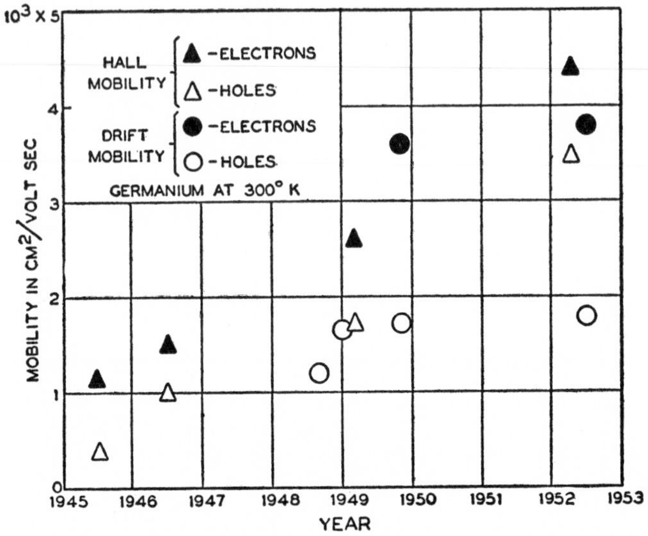

FIG. 12–9—Historical Data on Mobility in Germanium. (Since some of the data represent unpublished "best values" current at the time, no references are given.)

$(3\pi/8)\ 2600 = 3100/\text{cm}^2$ volt sec, is less than the drift mobility μ_n, taken as 3600 cm^2/volt sec measured for electrons injected into p-type germanium. From equation (25) of Section 11.4 we find that

$$\frac{\mu_H}{\mu} = \frac{(\text{weighted average of } \tau^2)}{(\text{weighted average of } \tau)^2} \geq 1.$$

For the case of a constant mean free path and non-degenerate statistics, the value of μ_H/μ is $3\pi/8 = 1.18$, whereas the ratio $3100/3600 = 0.86$ is smaller by a factor of 0.73. Thus the experimental values cannot be reconciled with theory by any assumed dependence of $\tau(v)$ upon the velocity v of the electrons. It is possible, however, to have values of

$$\frac{\mu_H}{\mu} < 1 \quad \text{for re-entrant energy surfaces}$$

in the Brillouin zone in place of the spherical energy surfaces with constant effective mass on which the formulae of Section 11.4 were based.

[2] W. C. Dunlap, Jr., *Phys. Rev.* **77**, 759A (1950).

In Figure 12.10, we show how energy surfaces of a peculiar shape can give rise to a great decrease in μ_H/μ. This figure represents the $P_z = 0$ plane in the Brillouin zone. For this case the velocity vector for a contour of constant energy will vary in magnitude, being greatest where the contour comes close to the origin in $\boldsymbol{P}$-space. If an electric field is applied in the x-direction, electrons for the n-type sample considered will be accelerated

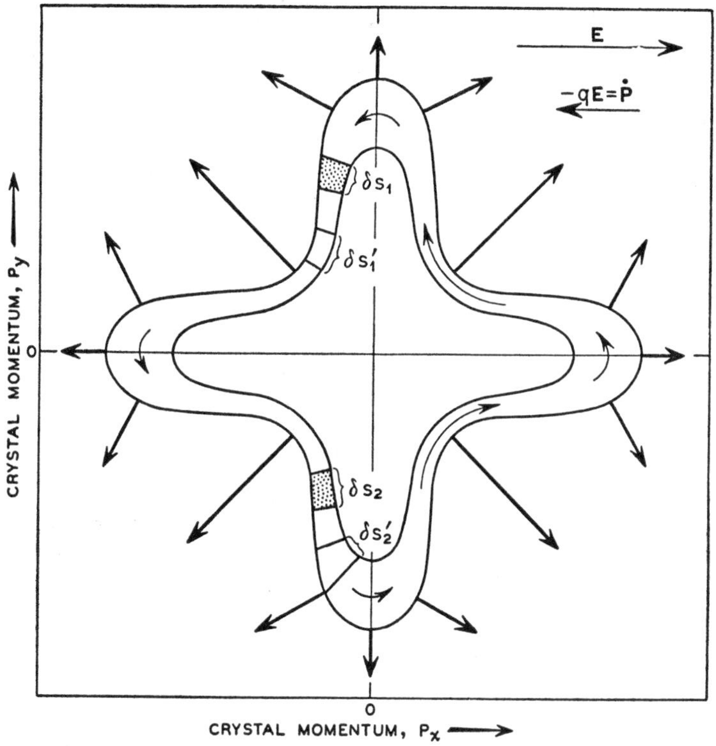

FIG. 12–10—Re-entrant Energy Contours which Will Lead to Low Values of μ_H/μ.

to the left in Figure 12.10 and will alter the number of electrons in states between the energy surfaces. The most important regions will be those for which the component of velocity v_x is large since $\mathbf{f}'$, the fraction of the quantum states occupied by electrons, increases as

$$\left(\frac{d}{dt}\mathbf{f}'\right)_{\text{due to } E_x} = \frac{-q}{kT}\mathbf{f}(1 - \mathbf{f})v_xE_x$$

as shown in Section 11.4, equation (7). Two groups of quantum states with large negative values of v_x are represented as δs_1 and δs_2 in Figure 12.10. For these, E_x produces a rapid increase in the number of electrons.

If a magnetic field is applied, a flow of points along the energy surfaces will take place in the direction shown by the arrows, and the extra electron in δs_1 and δs_2 will be displaced towards $\delta s_1'$ and $\delta s_2'$. This displacement will result in a net increase of the $+y$-component of average velocity since both $\delta s_1'$ and $\delta s_2'$ have more positive v_y values than δs_1 and δs_2. Thus the result[3] is to rotate the average velocity vector in the clockwise direction rather than the counterclockwise direction, as would be the case for spherical energy surfaces, which would have circular intersections in Figure 12.10. Hence the sign of the Hall effect for the case of a few excess electrons at lowest energy states of the conduction band can be that normally associated with holes.

It should be stressed that the electrons near the bottom of a conduction band like that of Figure 12.10 are not equivalent to holes. If one extra electron were present in the band of Figure 12.10, it would constitute a net negative charge and it would move towards the positive terminal in a drift experiment. In these respects it would act like the excess electrons considered in earlier chapters. Its drift velocity, however, would be deflected by a magnetic field so as to give the Hall effect of a hole. Similarly holes in an almost filled band with surfaces like Figure 12.10 would act like holes except that they would give the Hall effect of electrons. We shall introduce the terms *anomalous electron* and *anomalous hole* to describe the behaviors associated with these new concepts. Anomalous holes will give the Hall effect expected for electrons, usually called normal Hall effect, whereas anomalous electrons will give the Hall effect associated with positively charged carriers, usually called anomalous Hall effect. In a filament containing both anomalous holes and anomalous electrons, the carriers would be concentrated on the opposite side of the filament from that shown in Figure 3.10. This would lead to an *anomalous Suhl effect*, and data on the impedance of point contacts, like that of Figure 3.12, would be reversed regarding the sign of the magnetic field.

For a less drastic case than that shown in Figure 12.10, a reduction rather than a reversal of sign will occur for μ_H/μ. This is the proposed explanation for the observed ratio of about 0.86.

That a reversal of sign due to re-entrant surfaces does not occur in germanium is shown directly by the sign of the Suhl effect and by the mutual consistency of Hall effect, direction of motion of injected carriers, sign of the field effect, dependence of conductivity type upon valence of impurity, direction of rectification, and thermoelectric effects. That it is not unreasonable to expect modifications of the sort described previously

[3] The formula for Hall effect for complex surfaces is given in A. H. Wilson, *Theory of Metals*, Cambridge at the University Press, 1936, equation 236. The geometrical interpretation of Figure 12.10 has not in the author's knowledge been previously published.

which might lead to a value 0.86 for μ_H/μ can be seen by considering a case in which the energy contours have been drawn in three dimensions, Figure 12.11. Although the surfaces shown are computed for another crystal, there is no reason to suppose that similar surfaces do not occur for germanium.[4]

Further evidence that the energy surfaces in the Brillouin zone are not spherical is obtained from magnetoresistance effects. It is found that much

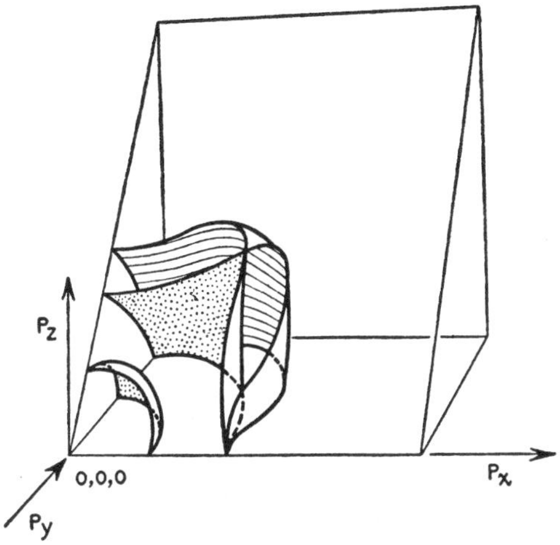

FIG. 12–11—Energy Surfaces Near Edge of an Energy Band. [For sodium chloride after Shockley, *Phys. Rev.* **50**, 754 (1936).]

larger effects are obtained by applying magnetic fields than can be accounted for by the theory of spherical energy surfaces presented in Section 11.4. Recent results of magnetoresistance experiments have been reported by Pearson[5] and Suhl.[6] They find for single crystals of germanium that the magnetoresistance is highly anisotropic which is further evidence for the lack of spherical energy surfaces.[7]

If the energy surfaces are not spherical, the deformation potential theory of the mean free path will need major revisions since scattering by shear modes can then be important. It is reasonable to suppose that scattering by shear waves will probably be less important than by the dilatation waves considered in Chapters 11 and 17 so that the agreement between $\mathcal{E}_{1G}$ and $|\mathcal{E}_{1n}| + |\mathcal{E}_{1p}|$ of Section 12.8 will probably still be valid.

[4] W. Shockley, *Phys. Rev.* **78**, 173 (1950).
[5] G. L. Pearson, *Phys. Rev.* **78**, 646 (1950).
[6] H. Suhl, *Phys. Rev.* **78**, 646 (1950).
[7] W. Shockley, *Phys. Rev.* **79**, 191 (1950).

12.10 ON A THEORY OF NOISE

Considerable progress has been made in understanding the nature of noise in germanium. It has been found by H. C. Montgomery[1] that noise having the typical (1/frequency) or contact-noise spectrum[2] described in Section 2.2 occurs in germanium filaments.

These experiments employ specimens having ends so large that any

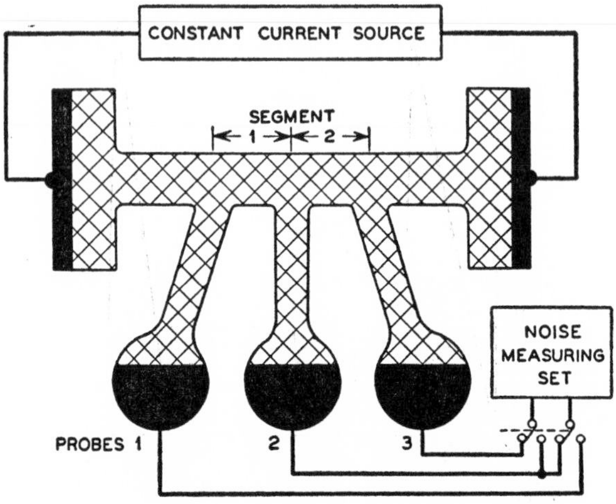

FIG. 12–12—Montgomery's Noise Transmission Effect.

effects due to metal semiconductor contacts are eliminated. Figure 12.12 represents one of the various types used in these experiments. Shapes like that shown, and others of similar complexity for other purposes, have been produced with a magnetostrictively driven cutter designed by W. L. Bond,[3] some examples being shown in the frontispiece. Under typical experimental conditions the filament is supplied by a constant-current source so that fluctuations in voltage between any two points of the filament may be attributed to a fluctuating resistance. In order to measure the noise voltage, arms are cut out of the same single crystal as shown in Figure 12.12 and are used as probes. This method eliminates the instability and additional noise which often occurs in connection with metal semiconductor contacts. The (noise voltage)^2 is found, to a first approximation, to consist of a thermal or Johnson noise component plus a term

[1] H. C. Montgomery and W. Shockley, *Phys. Rev.* **78**, 646 (1950).

[2] The term *contact-noise* arises from the fact that noise of this type was first encountered in connection with contacts: "Resistance Fluctuations in Carbon Microphones and Other Granular Resistances", C. J. Christensen and G. L. Pearson, *Bell Syst. Tech. J.* **15**, 197–223 (1936).

[3] W. L. Bond, *Phys. Rev.* **78**, 646 (1950).

which is directly proportional to I^2 and thus corresponds to a fluctuating resistance. Montgomery has obtained strong evidence that for *n*-type germanium this resistance modulation noise is due to a fluctuating hole current. The preliminary theory of his effect is as follows:

In the body of the germanium or on its surface there are regions which tend to have higher hole density than other regions. The presence of such regions has been observed in unpublished experiments of Haynes which show that the pulsed application of a sweeping field sometimes causes pulses of holes to pass by the collector point although no emitter contact is used at all. Haynes has been able to locate the sources of holes as being in certain definite regions of the filament which are not visibly different from the surrounding material. It is not known in all cases whether these regions are actually *p*-type inclusions or simply regions which are less *n*-type than the adjoining parts of the crystal. We shall for the present purpose assume that the regions are simply less *n*-type and thus under equilibrium conditions have a density of holes higher than the average. In the presence of an electric field, these higher densities are swept away. As a result, the regions, depleted of their normal hole density, attempt to re-establish the density by the emission of holes which are also swept away so that by this mechanism the region acts as a source of holes. If the hole current from this source were purely random in character, the noise in the filament would be much smaller than that observed and would have a "white" frequency spectrum (noise power per cycle independent of frequency) rather than the $1/f$ spectrum actually observed in the filaments as well as in transistors, as discussed in Section 2.2. Accepting for the moment the idea, which we shall discuss further below, that such fluctuating sources of hole current exist in the filament, let us examine the consequences of this picture.

The noise produced by a given fluctuating source of holes will be proportional to the lifetime of the holes in the filament, in fact the noise power will be directly proportional to the square of this lifetime. As we have explained earlier in connection with the Suhl effect, the lifetime of holes in the filament may be materially decreased by the simultaneous application of magnetic fields and electric fields. Reduction of noise in the filament has actually been observed by Montgomery and is in fair quantitative agreement with the theory based on these ideas. From his experiments one finds that the data can best be fitted by assuming that the noise originates in sources of holes located on the surfaces of the filament. This result is consistent with the finding that the surface is the principal locus of recombination for holes with electrons. In fact under conditions of thermal equilibrium, the principle of detailed balancing requires that any region which acts as a locus of recombination also acts as a locus of emission to an exactly balancing degree. There is a somewhat similar symmetry between emission and recombination under nonequilibrium conditions; thus a region having a fluctuating ability to effect the recombination of holes and electrons can introduce noise into an otherwise uniform flow of holes in just the same way as can a region of fluctuating emission.

In addition to the effect of noise reduction by applied magnetic fields, Montgomery has observed a *noise transmission effect* in using a structure, like that of Figure 12.12, having three side arms so that three noise voltages, V_{12}, V_{23}, and V_{13}, may be measured by connecting various pairs of terminals to the noise-measuring set. If the resistance modulation in each of the two segments were independent of the other, then the noise powers V_{12}^2 and V_{23}^2 should add to give the noise power V_{13}^2. Under

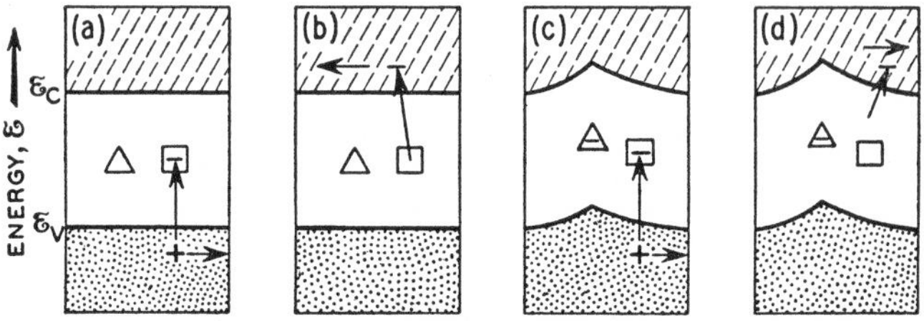

FIG. 12–13—Modulation of the Recombination-emission Process bya Trapped Charge.

the experimental conditions, however, the lifetime of a hole in the filament is so long that most of the holes entering Segment 1 also pass through Segment 2 before recombining. Furthermore, the transit time is a small fraction of a cycle of the measuring frequency. Consequently, the modulation due to the hole current should be approximately the same in both segments and the noise voltages, rather than the noise powers, should add. Confirmation of this prediction is furnished by the fact that Montgomery finds

$$V_{13}^2 = V_{12}^2 + V_{23}^2 + 2CV_{12}V_{23} \tag{1}$$

with a value of 0.85 for the correlation coefficient C. If the surface is treated so that the lifetime is less than the transit time between segments, then the correlation between resistance modulation in one segment and the other is largely eliminated. Combined with the experiment with magnetic fields these experiments with the noise transmission effect constitute very strong evidence that the noise in these filaments is due to a fluctuating hole density.

In Figure 12.13, we show a possible mechanism which will give rise to noise of the sort discussed. We represent a region of the semiconductor containing two imperfections having the hypothetical properties described below. One of these is a *recombination-emission center*, referred to as the *center.* It is supposed that this center has an energy level to which an electron may be excited thermally from the valence band. After an electron is so excited, as shown in (a), the hole left behind moves off so that in this way the center has produced a hole. The electron may then be further excited to the conduction band, after which it moves away leaving the center in a neutral condition. This process may be repeated so that the center can act indefinitely as a source of holes and electrons. Under equilibrium conditions, the reverse process, in which an electron falls into the center from the conduction band and later from there into a hole which happens to diffuse into its neighborhood, will occur with equal frequency. If the region shown in (a) and (b) is one which normally has a relatively high hole concentration and if the holes are swept out by an applied electric field, then the action of the traps can be shown to be such that the region acts as a natural source of holes.

In Figure 12.13 we show a different type of imperfection which can also trap an electron. We shall suppose that this trap is much more stable so that the electron remains in it for a time very much longer than that required for the center to go through a cycle of the sort shown in (a) and (b). We shall refer to this trap as a *modulating trap* or simply a *trap.*[4]

Now the ability of the center to absorb or emit hole-electron pairs can be modulated by the charge of the trap. If the modulating trap shown in the figure becomes charged by trapping an electron, this will alter the shape of the potential energy bands as shown in (c) and (d). As a result of this change, it is somewhat harder for the hole to escape and somewhat easier for the electron to escape. In general, these two effects will not cancel and one process will be modulated in a more important way than the other. Consequently, the source strength of the centers in the neighborhood of the modulating trap will fluctuate as the trap gains or loses an electron. If the energy levels of the traps are distributed over a considerable energy range, then the frequency spectrum characterizing their charge and discharge by capturing and emitting electrons will be of the $1/f$ type.[5]

The case represented in Figure 12.13 is only one of a number of similar possibilities. Furthermore, Montgomery's experiments with magnetic fields indicate that the noise sources are on the surface of the filament. For this case the centers may well be surface states and the traps may also be surface states of a different sort or, alternatively, the modulating role of the traps may be played by ions or polar molecules moving on the surface.

[4] Traps play a major role in the theory of phosphors. For discussion and references see H. W. Leverenz, *Luminescence of Solids*, John Wiley and Sons, Inc., 1950.

[5] I am indebted to J. Bardeen for calling my attention to this conclusion.

It should be stressed that the sources of noise discussed here, like the other imperfections mentioned earlier in this section, are apparently *not necessary attributes of n- and p-type* germanium. In the idealized model of *n*- and *p*-type germanium, consisting of a perfect lattice except for substituted atoms of valences three and five, there appears to be no important source for noise of the type actually observed. It is to be hoped that with improvement in the art, real germanium specimens will be produced so close to the idealized model that for them the noise will be reduced towards the minima set by Johnson noise and space charge limited shot noise.

12.11 CONCLUDING COMMENTS

In this section we shall review briefly the status of transistor electronics and shall indicate areas in which it is expected that scientific and technical progress may occur in the future.

As mentioned many times in the text, the invention of the transistor by Bardeen and Brattain has opened up a new area of research in semiconductor phenomena. The experiments of Haynes on drift mobility, of Pearson on conductivity modulation, of Suhl on magnetic concentration of holes and electrons constitute, the author believes, foundation stones in the entire theory of electronic conductivity in semiconductors. These experiments show that many aspects of the conceptual pictures of the behavior of electrons and holes have a real operational meaning, quite apart from that previously explored by measurements of conductivity, Hall effect, and thermoelectric effects.

In transistor electronics the germanium filament (see Frontispiece) plays the role of the vacuum or gas discharge tube of conventional electronics. The analogue of the thermionic cathode is the emitter contact junction, and its efficiency is the fraction γ of its forward current carried by injected carriers. In Chapter 3 data were quoted for several cases in which γ was measured. Since the time at which those studies were made, the estimate of b has been revised upwards from 1.5 to 2.1, germanium side arms have been used as probes, and a-c techniques for injection and measurement have been employed. W. H. Brattain, who introduced the last two improvements in the studies of γ, has made measurements on an *n*-type germanium. He finds values of γ varying from as low as 0.2 to nearly unity with a tendency for γ to increase for a given point with increasing emitter current. The value of γ is dependent upon surface treatment but not in a readily reproducible way. The apparent value of γ, which is measured as the ratio of small a-c values i_p/i_e, increases with increasing sweeping field and Brattain's values are extrapolated to infinite field.

In addition to the measurements of γ and the new observations of fundamental phenomena made possible by filament techniques, other detailed information has been acquired concerning the behavior of holes and electrons. It has been shown that usually recombination takes place chiefly on the surface of the filament. It is also known that a hole in general will have many thousands or tens of thousands of collisions with the surface before it is finally lost. It appears that this surface recombination process is structure-sensitive to the chemistry and potential fields at the surface.

It is also known that other impurities and imperfections in the crystal besides simply elements of columns 3 and 5 of the periodic table, such as were discussed in connection with Table 1.1, are involved in the recombination process in the body of the germanium. It is found that this body recombination is relatively insensitive to the presence of ordinary donors and acceptors but can be markedly increased by the same sort of heat treatment which tends to convert *n*-type germanium to *p*-type.[1] It has also been found that the *p*-type centers produced by bombardment with nuclear particles are very effective in aiding the recombination of holes and electrons.

It is to be hoped that the nature of these imperfections will be established with the aid of low temperature studies, lifetime measurements, and studies of noise and of trapping phenomena.

Grain boundaries in germanium introduce electrical discontinuities, as mentioned in Section 12.9. Their principal features have been reported by Pearson.[2] He finds that some grain boundaries act like *p*-type regions so that the electrical characteristics are like those of two *p-n* junctions in series; this picture is further supported by the photovoltaic effects, which are also consistent with this view. It is not certain whether the *p*-region at the boundary is due to segregation of chemical impurities or to disorder of a type similar to that produced by nucleon bombardment, but arising from atomic misfit between crystals on opposite sides of the grain boundary. If the germanium is converted to *p*-type by heat treatment, the grain boundary apparently acts simply like a very thin layer of somewhat different conductivity and thus has no appreciable effect upon the resistance of the specimen.

Great progress has been made in the germanium material itself under the pressure of demands of the transistor program. As a result, relatively large crystals of high perfection and chemical purity are now available, and experiments in transistor electronics which would have been impos-

[1] Unpublished investigations carried out by M. Sparks and J. R. Haynes.

[2] G. L. Pearson, *Phys. Rev.* **76**, 459 (1949). See also N. H. Odell and H. Y. Fan, and W. E. Taylor and H. Y. Fan, *Phys. Rev.* **78**, 334–335 (1950). The theory of the structure of grain boundaries for some of the simpler crystals appears to be in accord with experiment. See W. T. Read and W. Shockley, *Phys. Rev.* **78**, 275–289 (1950). References.

sible in the past can now be performed. A preliminary report of crystal growing has been made by G. K. Teal and J. B. Little.[3]

* * * * * *

At this point we shall briefly summarize the aspects of the energy band theory of conduction which have been tested in connection with the transistor program. According to the energy band picture an excess electron

(1) has a charge of $-e$
(2) undergoes random diffusive motions
(3) drifts under an electric field with a characteristic mobility
(4) is deflected sidewise by a magnetic field.

A hole has similar attributes except for the sign of its charge. The new experiments have put all of these aspects on a firm operational footing. The sign of the charge and the drift velocity have been verified by carrier injection experiments. The magnitude of the charge has been confirmed by the rectification formula for *p-n* junctions which depends critically upon e/kT. The sidewise deflection has been directly observed in the Suhl effect experiments. The diffusive motion has been verified by experiments on *p-n* junctions in which the frequency characteristic is found to be in accord with the diffusion theory; it has been still more directly verified in an experiment carried out by Goucher[4] in which carrier injection by light absorption is followed not by drift in a sweeping field but by diffusion in a negligible electric field.

These attributes are the principal consequences of the physical picture of holes and electrons discussed in Part I. In a quantitative form this picture furnishes the mathematical discipline required for the theoretical side of transistor electronics. It is, therefore, of importance to appreciate that as a result of the new experiments growing out of the transistor program, *the important features of the physical picture are now based on direct experimental evidence* and need no longer be considered to be deductions from an involved quantum mechanical analysis.

Further evidence for the validity of the band theory is furnished by the correlation, discussed in Section 12.8, of band spacing or energy gap with size of the unit cell. An additional piece of evidence is the fact that approximately unit quantum efficiency, that is one hole-electron pair per photon, is obtained in photoconductivity experiments[5]; this result is just what should be expected on the basis of the physical picture of excitation by photons described in Chapter 1.

[3] G. K. Teal and J. B. Little, *Phys. Rev.* **78**, 647 (1950).
[4] F. S. Goucher, planned for publication in the *Phys. Rev.* in 1950.
[5] F. S. Goucher, *Phys. Rev.* **78**, 476, 816–817 (1950).

It may be appropriate to speculate at this point about the future of transistor electronics. Those who have worked intensively in the field share the author's feeling of great optimism regarding the ultimate potentialities. It appears to most of the workers that an area has been opened up comparable to the entire area of vacuum and gas discharge electronics. Already several transistor structures have been developed and many others have been explored to the extent of demonstrating their ultimate practicality, and still other ideas have been produced which have yet to be subjected to adequate experimental tests. It seems likely that many inventions unforeseen at present will be made based on the principles of carrier injection, the field effect, the Suhl effect, and the properties of rectifying junctions. It is quite probable that other new physical principles will also be utilized to practical ends as the art develops.

The preparation of this book was motivated by the conviction that electronic conduction in semi-conductors was about to take its place as a subject of major importance in the communications art and other branches of electrical engineering as well. Workers in transistor electronics had expressed a need for a new treatment of the physical principles in descriptive terms with special emphasis upon the features likely to be used as the basis for engineering design. Part II has had as its chief aim the supplying of this description and some of the mathematical tools useful in design considerations. A re-examination of the foundations of the basic theory constitutes one of the major aims of Part III, the other aims being concerned with specially adapted presentations of certain well-known branches of theoretical physics. The more fundamental investigations, because of their strongly mathematical flavor, may be inaccessible to most readers. It is hoped, however, that they will convey the important impression, which the author believes to be true, that the substratum of basic theory is sound for transistor electronics.

Problems

1. Show that if at $t = 0$ there is a distribution of holes on the plane at $x = 0$ with N_p holes per unit area, then owing to diffusion (neglecting all other processes) the value of p at later times is

$$p(x, t) = \frac{N_p}{\sqrt{4\pi Dt}} e^{-x^2/4Dt}.$$

Consider the three-dimensional case and show that if N_p holes are concentrated at $r = 0$ at $t = 0$, then the density at time t is

$$p(r, t) = \frac{N_p}{(4\pi Dt)^{3/2}} e^{-r^2/4Dt}.$$

(This may most easily be done by testing the solution in the continuity equation using rectangular coordinates. See problems of Chapter 11 for the integrals.)

2. Show that, for the one-dimensional diffusion of Problem 1, the value of $\langle x^2 \rangle$, defined as

$$\langle x^2 \rangle = \int_{-\infty}^{+\infty} x^2 p(x, t) dx / N_p$$

is

$$\langle x^2 \rangle = 2Dt.$$

Verify that for any function $p(x, t)$ satisfying

$$\frac{\partial p}{\partial t} = D \frac{\partial^2 p}{\partial x^2}$$

and appropriate boundary conditions

$$\frac{d}{dt} \langle x^2 \rangle = 2D.$$

Extend these results to two and three dimensions.

3. Show that if a plane of holes as in Problem 1 is generated (by the photoconductive effect for example) at $t = 0$ in a medium of hole lifetime τ_p, the solution of the continuity equation for the extra holes is

$$p(x, t) = \frac{p_0}{\sqrt{4\pi Dt}} \mathrm{e}^{-x^2/4Dt - t/\tau_p}.$$

Show that the relative number of holes which recombine in time dt is $[\exp(-t/\tau_p)] dt/\tau_p$. Show that the value of x^2 averaged for each hole from generation to recombination is

$$\langle x^2 \rangle = 2D\tau_p.$$

Obtain this same result by averaging x^2 over the d-c part of the solution (16) of Section 12.5. Show by a physical argument that the two results should be the same. Show that the average distance, not (distance)^2, diffused by a hole is

$$L_p = \sqrt{D\tau_p}$$

for both solutions.

4. Consider a pulse of holes formed at the point $r = 0$ at $t = 0$ in a field $\boldsymbol{E}$ and then allowed to spread and drift owing to diffusion in the field. Show that the distribution at t is like that of Problem 1 but centered at $r = t\mu_p \boldsymbol{E}$ and that the ratio of distance spread sidewise to drift distance is approximately

$$\frac{\text{sidewise spread}}{\text{drift distance}} = \sqrt{\frac{2kT}{eV_0}}$$

where $V_0 = \boldsymbol{E} \cdot$ (drift distance) is the difference in voltage between the initial position and the position at time t. This formula is useful for estimating how far down filament holes must flow before filling the cross section and for other similar applications.

5. Compare the relative importance of the conduction and diffusion terms for hole flow in the n-region as mentioned in connection with equation 20 of Section 12.5. To do this consider a plane, near the transition region, at which we may consider I_p and I_n to be approximately equal. Show that this leads to

$$\frac{d\psi}{dx} \doteq \frac{p_p}{n_n} e^{qV_0/kT} \frac{kT}{qL_p}$$

and hence that diffusion currents are the dominant process for hole flow unless V_0 is so large that the injected hole density is comparable to n_n. Show that for holes in germanium with $\mu_p = 1700\ \text{cm}^2/\text{volt sec}$ kT/qL_p is equal to $3.90\ (10^6\tau_p)^{1/2}$ volts/cm or 3.90 volts/cm for $\tau_p = 1\mu$ sec.

6. Consider surface recombination for the case of $s \to 0$. Show that the formula (19b) of Section 12.6

$$\nu_s = s\left[\frac{1}{B} + \frac{1}{C}\right]$$

can be obtained by calculating the rate of surface recombination assuming that p = const. across the filament and evaluating (total rate of recombination) $\div$ (total number of holes present).

7. Consider the neglected solution of equation (21) of Section 12.6 which was said to represent diffusion against the field. Show that, if recombination is neglected, this solution reduces to the Boltzmann concentration factor discussed in Section 12.3, equation (2).

8. Show that in considering the Suhl effect of Section 12.6 the continuity equation (30) can be separated if a factor $\cos bz$ is included, the effect being to add a term $D_p\varphi_0{}^2/C^2$ to ν in equation (31), the boundary conditions for φ_0 being the same as for $H_z = 0$.

9. Calculate the transverse field E_t and transverse voltage for a filament 20 mils square (that is, 0.05 cm on each edge) with $E_x = 10$ volts/cm and $H_z = 5000$ oersteds for p- and n-type germanium.

10. Assume an effective mass equal to the electron mass and show that the number of electrons having velocities in a range $dv_xdv_ydv_z$ and lying in a volume of space dV is according to approximation (2) of Section 10.1

$$dn = \frac{(2\pi kT/m)^{3/2}}{n} e^{-m(v_x^2+v_y^2+v_z^2)/2kT} dv_xdv_ydv_zdV.$$

Show that the electrons with velocity components $v_xv_yv_z$ striking an element of area $dxdy$ in time dt come from a volume $dxdyv_zdt$. Integrate this over all velocity classes and obtain for the rate at which electrons strike $dxdy$ the formula

$$n\sqrt{\frac{kT}{2\pi m}} = nv_T/\sqrt{4\pi} = 0.282nv_T$$

where v_T is the velocity used in Section 11.4, equation (20), and is related to T as for Figure 10.2.

11. Apply the result of the previous problem to show that the number of collisions with the surface before recombination occurs is given by

$$\text{Number of collisions per recombination} = 0.282v_T/s$$

where s is the surface recombination velocity of Section 12.6. Show that this formula will be valid for values of $s < 10^5$ cm/sec. Show that for $s = 10^3$ cm/sec there are ~2800 per recombination.

12. Show by a geometrical argument that for the construction of Figure 12.8 the total number of holes is independent of time.

13. Derive the partial differential relationship

$$\left(\frac{\partial p}{\partial t}\right)_x = -\left(\frac{\partial x}{\partial t}\right)_p \Big/ \left(\frac{\partial x}{\partial p}\right)_t$$

which must hold whenever x is a function of p and t. Apply this to the relationship

$$x(p, t) = x(p, 0) + \int_0^t V[I(t'), p]dt'$$

and verify that this satisfies equation (6) of Section 12.7 for the case for which the current is $I(t)$.

14. Derive the equation

$$\frac{\partial E}{\partial t} = \frac{E}{I}\frac{dI}{dt} - V(E)\frac{\partial E}{\partial x}$$

where according to equation (9) of Section 12.7

$$E = I/\sigma(p)$$

and consequently

$$V(E) = E\mu_p\frac{E}{E_0}\cdot$$

Show that if E is plotted as a function of x a pulse with an infinitely sharp leading edge at $t = 0$ has a parabolic leading edge for $t > 0$.

15. Consider as an approximation that the trailing edge of the shock wave corresponds to the solution for $x < 0$ of Problem 1,

$$p(x, t) = p_0(4\pi Dt)^{-\frac{1}{2}}e^{-x^2/4Dt}.$$

Show that this tends to spread to the rear at a rate of approximately

$$\frac{d\langle x\rangle}{dt} = \sqrt{2D/t} = 2D/\langle x\rangle.$$

Equate this to the rate at which the concentration $p = 0$ tends to catch up with $p = p_{\max}$ and obtain the formula for the width of the trailing edge quoted at the end of Section 12.7.

16. Show that a solution to the equations of section 12.7 can be obtained for I = constant and constant lifetime τ_p by moving each point $p(x_0, 0)$ of the initial distribution, like that of Figure 12.8, along a curve defined by t such that

$$p(x_0, t) = p(x_0, 0)\mathrm{e}^{-t/\tau_p}$$

$$x = x_0 + \int_0^t V[I, p(x_0, t')]dt'.$$

17. Internal Contact Potentials.[1] Consider a filament of n-type germanium with $s = 0$ (no surface recombination). Suppose that a contact is made to this with a probe of n-type germanium but that the two are separated by a thin layer of p-type germanium so that the probe is separated from the main body by two p-n junctions. Suppose that holes are injected into the filament so that $\varphi_p > \varphi_n$. Then by applying reasoning like that of the first pages of Section 12.5, show that the probe tends to acquire a potential between φ_p and φ_n. In particular, if φ' is the Fermi level deep in the probe where $\varphi_p' = \varphi_n'$ show that $\varphi_n < \varphi' < \varphi_p$. Show that if the p-layer is converted to n-type, by imagining that N_d is increased for example, then $\varphi' \rightarrow \varphi_n$.

18. The treatment of the p-n-p transistor[1] constitutes a further application of the method of this section.

19. The collector currents used in the experiments on which Figure 3.6 is based are sufficiently large so that diffusion currents are negligible compared with conduction currents. Show that the ratio, I_n/I_p, of the electron current to the hole current flowing to the contact is then equal to bn/p. Assuming $b = 2$, and using the data of Figure 3.6 calculate the "intrinsic" α for each of the four contacts. (This method is due to Bardeen.)

[1] See W. Shockley, *Bell Syst. Tech. J.* **28**, 435–489 (1949) for a further discussion.

PART III QUANTUM-MECHANICAL FOUNDATIONS

PART III

Books on Quantum Theory

J. Frenkel, "Wave Mechanics; Elementary Theory," *Oxford, Clarendon Press,* 2 ed. 1936. 309p. "Wave Mechanics, Advanced General Theory," *Oxford, Clarendon Press,* 1934. 524p.

R. W. Gurney, "Elementary Quantum Mechanics," *Cambridge, University Press,* 1934. 157p.

*L. Pauling and E. B. Wilson, "Introduction to Quantum Mechanics," *McGraw-Hill,* 1935. 468p.

E. C. Kemble, "Fundamental Principles of Quantum Mechanics," *McGraw-Hill,* 1937. 611p.

*S. Dushman, "Elements of Quantum Mechanics," *John Wiley and Sons,* 1938. 452p.

*V. Rojansky, "Introductory Quantum Mechanics," *Prentice-Hall,* 1938. 544p.

H. Eyring, J. Walter, G. E. Kimball, "Quantum Chemistry," *John Wiley and Sons,* 1944. 394p.

W. Heitler, "Elementary Wave Mechanics," *Oxford, Clarendon Press,* 1945. 136p.

P. A. M. Dirac, "The Principles of Quantum Mechanics," *Oxford, Clarendon Press,* 3rd ed. 1947. 311p.

L. I. Schiff, "Quantum Mechanics," *McGraw-Hill,* 1949. 404p.

N. F. Mott and I. N. Sneddon, "Wave Mechanics and its Applications," *Oxford, University Press,* 1948. 393p.

Books on Statistical Mechanics

L. Brillouin, "Statistiques Quantiques et Leurs Applications," *Les Presses Universitaires de France,* Paris, 2v. 1930. 404p.

R. H. Fowler, "Statistical Mechanics; the Theory of the Properties of Matter in Equilibrium," *Cambridge, University Press,* 2 ed. 1936. 864p.

L. Landau and E. Lifshitz, "Statistical Physics," Translated from the Russian by D. Shoenberg, *Oxford, Clarendon Press,* 1938. 234p.

R. C. Tolman, "Principles of Statistical Mechanics," *Oxford, Clarendon Press,* 1938. 661p.

R. H. Fowler and E. A. Guggenheim, "Statistical Thermodynamics; a Version ot Statistical Mechanics for Students of Physics and Chemistry," *Macmillan,* 1939. 693p.

J. C. Slater, "Introduction to Chemical Physics," *McGraw-Hill,* 1939. 521p.

J. E. Mayer and M. G. Mayer, "Statistical Mechanics," *John Wiley and Sons,* 1940. 495p.

E. Schroedinger, "Statistical Thermodynamics," *Cambridge, University Press,* 1946. 88p.

R. W. Gurney, "Introduction to Statistical Mechanics," *McGraw-Hill,* 1949. 268p.

G. S. Rushbrooke, "Introduction to Statistical Mechanics," *Oxford, Clarendon Press,* 1949. 334p.

* Recommended for beginners.

CHAPTER 13

INTRODUCTION TO PART III

The topics dealt with in this book are in principle, although not in actuality, directly derivable from the basic laws of quantum mechanics. If complete calculations were available, a strictly logical development would start by stating these basic laws, probably with accompanying references to their historical origin and experimental verification in simple cases. For the case of germanium, the laws would then be applied to element number 32. From this starting point the lowest energy state for a large number, say 10^{22}, of germanium atoms would be determined. This lowest state would correspond to a crystal having the diamond structure and, by comparing its energy with the energy of isolated neutral atoms, the strength of the Ge-Ge covalent bond would be found.

Next all the excited states would be calculated and classified in terms of electron motions and atomic vibrations.

The effect of temperature upon the system would then be obtained by applying the principles of quantum statistics to the excited states. Combining these results with an investigation of the effects of applied electric and magnetic fields would lead to a theory of the conductivity and Hall effect of a pure or intrinsic germanium sample. If we had started with an admixture of 10^{15} atoms of number 51 (antimony), the results would apply to an *n*-type sample. If, in addition, we considered cases in which the impurities were spatially distributed in special ways, this fundamental approach would lead to interpretations including the rectification of *p-n* junctions and the amplification of *p-n* junction transistors.

A procedure of the sort just described is neither practical nor, for that matter, desirable. Calculations of the lowest and some of the excited states have been made for diamond and for silicon. The numerical results, however, are not in very gratifying accord with experiment and could not very satisfactorily be used for design purposes. The important contribution which the theory makes is to provide a *conceptual framework on which to hang experimental facts so as to give them a more fundamental interpretation and to explain and predict relationships among them.* For this purpose an *exact* theory, which specified the correct quantum states or eigenfunctions for the entire system, would be, owing to the complexity of its results, far inferior to an *approximate* theory, which uses much simpler pictures and takes into account their inexactness in terms of various corrections which can also be simply understood. It is chiefly to show how these approximate solutions arise from fundamental quantum mechanics and how they are interpreted that this part of the book is written.

In the following chapters we shall deal with almost all the problems involved in the fundamental or first-principles treatment. The emphasis and level of presentation, however, differ greatly from one topic to another.

The treatment of quantum mechanics in Chapter 14 is at a very practical level and will probably be of value chiefly to readers with engineering training. Unlike most treatments, the emphasis is laid from the outset upon the application of quantum mechanics as a tool for calculating results. This emphasis is particularly appropriate since in the field of transistor electronics, the design of useful devices may be considered to be based on the quantum physics of semiconductors.

In addition to presenting the fundamental laws and formulae of quantum mechanics, Chapter 14 contains applications to some simple problems and compares their mathematical forms with the mathematical forms of corresponding circuit theory analogues. The Bloch wave functions for a general form of a one-dimensional crystal are derived. The chapter closes with an indication of how the method is extended to a general three-dimensional crystal. The problem of constructing the Brillouin zones for general crystals is somewhat lengthy and does not bring out any particularly important new physical ideas. Furthermore, extensive treatments of the topic are found in several texts. For these reasons, no illustrative examples of Brillouin zones for the diamond structure or other crystal structures are presented. The reader interested in additional details will find a list of references at the end of Chapter 14.

The kinematical and dynamical properties of electrons and holes, which were discussed in Chapters 6 and 7 on a descriptive level, are derived as the consequences of quantum mechanics in Chapter 15. The first three sections of that chapter are concerned with deriving the group velocity formula and employ relatively simple mathematical techniques. The later sections are on a more advanced level. Two methods of dealing with holes and electrons are used. The first of these, given in Section 15.6, deals with a wave-packet and shows that the packet simulates the behavior of a particle obeying the laws discussed in Chapter 7. The following two sections show how wave-packets for holes are constructed and interpreted, the results derived being in agreement with the treatment of Chapter 7. In Section 15.9 the other method of dealing with the acceleration of an electron is presented. In this latter case true periodic boundary conditions, along one direction at least, are introduced by imagining that a thin crystal is bent into a loop. Acceleration of the electron around the loop is produced by employing the loop as a short-circuited secondary of a transformer. This case has the conceptual advantage of representing a situation in which the system can be discussed, in principle, in fundamental quantum-mechanical terms while steady currents are flowing around the loop. This model thus represents a step along the fundamental path in which the behavior of the semiconductor would be deduced directly from quantum mechanics applied to a particular system. The treatment of the Hall effect on the basis of this model is omitted since it is treated for the wave-packet case of Section 15.6.

The effects of thermal agitation, used in Chapter 7, are discussed more fully in Chapter 16. Statistical mechanics in general is the subject of many texts, and the treatment here is not intended to be exhaustive. Simple derivations of the Fermi-Dirac distribution for electrons and for the energy distribution of harmonic oscillators, the Planck distribution, are presented. This treatment is based on the use of approximate rather than exact quantum states and furnishes an example of the principle stated earlier that exact solutions may not always be necessary or even desirable. This treatment is followed by a detailed application of the statistics to the semiconductor problems of particular interest in this text.

Chapter 17 deals with the interaction of holes and electrons with vibrating atoms in the crystal. The quantum-mechanical theory of transition probabilities is applied to this situation and it is found that the mechanism of scattering by thermal agitation is closely related to the dependence upon lattice constant of the energy gap between conduction and valence bands. Section 17.7, the closing section, extends the results derived for one electron to the case of a hole and to the intermediate case of many holes and many electrons.

As indicated by the title of this part of the book, the analysis presented here furnishes a foundation upon which the descriptive theory of Part II is based. The most extensive use of this foundation is required for the mathematical theory of conductivity and Hall effect presented in Chapter 11. It may be helpful to point out here how the chapters of Part III contribute to Chapter 11. In Chapter 11, the electrons (or holes) are considered initially to be distributed in the Fermi-Dirac distribution; we may imagine for this purpose that the model of the crystal bent into a secondary transformer turn is used. Under the influence of an electric field, the distribution is disturbed, each electron being accelerated as discussed in Section 15.9. As a result, the principle of detailed balancing (Section 17.2) no longer applies and equilibrium tends to be re-established. The details of the way in which equilibrium is re-established constitutes the principal problem of Chapter 11.

The foundation which supports Chapter 11 is not quite complete. In Chapter 15, the effect of fields in the absence of thermal agitation is presented while in Chapter 17 the effect of thermal agitation in the absence of fields is presented. In the application in Chapter 11, it is assumed that these effects can be directly added. The additivity can be proved by extending the theory of transitions of Section 17.2 to the case in which the electron is being accelerated. The results for an electric field are somewhat different in form from the Weisskopf-Wigner treatment; the physical content, however, is equivalent to adding the two effects just as is done in Chapter 11. Thus it is possible to furnish this missing portion of the foundation structure. The treatment, however, would add little which was new from a conceptual point of view and is omitted in the interest of brevity.

Similar objections may be raised to the use of the scattering formulae of Chapter 17 to wave-packets being accelerated as discussed in Section 15.6. Again the answer is that it is quite possible to show that the scattering results of Chapter 17 could be extended to cover an accelerated wave-packet and thus to prove the desired additivity result. However, the added complexity and length of exposition would probably be of very limited interest to most readers; and interested readers, on the other hand, would be more benefited by carrying out the process themselves.

CHAPTER 14

ELEMENTARY QUANTUM MECHANICS WITH CIRCUIT THEORY ANALOGUES

14.1 ON THE NATURE OF AN APPLIED MATHEMATICAL THEORY[1]

Quantum mechanics is recognized as one of the more difficult branches of theoretical physics, chiefly because it employs certain abstract mathematical tools which in themselves have no direct physical interpretation. Thus one is told, for example, that applying an operator to a wave function in the proper way will lead to predicting the outcome of an experiment. Neither the operator, the method of applying it, nor the wave function has a direct physical interpretation. This lack of a straightforward interpretation is a definite disadvantage from a pedagogical viewpoint—it makes the theory harder to present—but it does not indicate any lack of validity of the theory in its applications. In so far as its applications are concerned, a mathematical theory of an experimental science must satisfy one supreme requirement: *It must agree with experiment.* Although this requirement can be simply stated, we shall give several examples to illustrate its meaning before discussing the agreement with experiment achieved by quantum-mechanical theory and the procedures used to test it.

We shall employ the phrase *applied mathematical theory* to indicate the procedure which the theory specifies in order to predict relationships between measured quantities. We are not here concerned with the modus operandi of the theoretician, or with how theory may be used to plan experiments and, thereby, to improve theory, although we shall comment briefly on such subjects at the end of this section. We are here concerned with the formalism of a well-developed *applied mathematical theory*, such as the quantum-mechanical theory of electrons in solids, and how it is related to experiment. The general nature of the connection between the *formalism of the theory* and the *application of the theory* is indicated in Figure 14.1. We shall explain the various features by means of some examples.

[1] For a more thorough discussion of the subject matter of this section, see in particular P. W. Bridgman, *The Logic of Modern Physics*, Macmillan, 1931, and *The Nature of Physical Theory*, Princeton University Press, 1936, and R. B. Lindsay and H. Margenau, *Foundations of Physics*, John Wiley & Sons, 1936. The discussion in this section is somewhat similar to the chapter "Mathematics in Application" of *The Nature of Physical Theory.*

As the figure indicates, the application concerns measured quantities which are defined by physical operations, such as, for example, the voltages, currents, lengths and times, with which we are chiefly concerned in connection with the properties of semiconductors. The first step in the application of a physical theory is always the same and involves a change of viewpoint. Instead of thinking of the position of a pointer on the scale

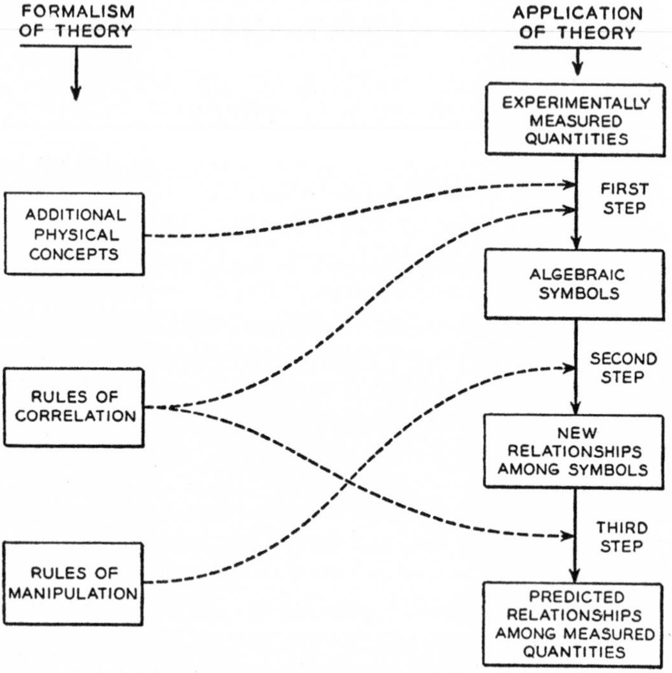

FIG. 14-1—The Nature of an Applied Mathematical Theory.

of an ammeter, one thinks of the symbol I which the theory uses to represent any number which may be obtained as a result of reading the ammeter. This procedure is carried out in accordance with the *rules of correlation* between the "operationally defined" physical quantities and the mathematical symbols.

For the case of a very simple theory, such as the application of Ohm's law to a cylindrical wire, the *rules of correlation* would be

voltmeter reading	$\rightarrow$	V
ammeter reading	$\rightarrow$	I
length in centimeters	$\rightarrow$	L
diameter in centimeters	$\rightarrow$	D
resistivity in ohm-cm	$\rightarrow$	ρ

In this case there would be the additional concept of an area of cross section which would be given the symbol A. This listing would complete the FIRST STEP and bring us to the second box, ALGEBRAIC SYMBOLS, in the figure.

The SECOND STEP would require the insertion of these symbols in the algebraic equations

$$I = AV/\rho L \text{ for Ohm's law} \tag{1}$$

and

$$A = \pi D^2/4 \text{ for area} \tag{2}$$

followed by the use of all the rules of algebraic manipulation, from which one might obtain the *new relationship among the symbols*

$$L = \pi D^2 V/4\rho I,$$

an example of the contents of the third box in APPLICATION OF THE THEORY.

According to the *rules of correlation*, we could relate these symbols to measured quantities, thus taking the THIRD STEP to the fourth box, and could then use this relationship to predict the length of a wire wound on a coil.

Frequently, *additional physical concepts* are introduced describing quantities which are not directly measured. Thus, in the theory of conductivity and Hall effect, we introduce the ideas of electrons and holes and their mobilities and concentrations. In dealing with a sample of germanium containing radioactive antimony, we may make predictions about the transverse voltage observed in a Hall effect experiment. This prediction will involve a relationship between voltage, current density, magnetic field, linear dimensions and counting rate on a Geiger counter, time since the radioactive sample was calibrated, and so forth. However, in making this prediction we introduce many symbols, having quantitative values and associated with definite conceptual entities (like electrons and holes), which are not directly measured. All of these quantities are manipulated according to formulae discussed in Chapters 5 to 11 before the *predicted relationships* are finally cranked out.

In quantum-mechanical theory the role of the experimentally measured quantities is frequently played not by quantities which are actually measured but by ones which could in principle, at least, be measured if enough pains were taken.[2] Thus the position of an individual electron is seldom measured. The path, however, of a single electron can be observed in a cloud chamber, and the time at which it passes through a Geiger counter can be measured as well. The position of the electron can thus be measured in principle. Also, its speed of motion can be measured by measur-

[2] In Section 14.3, however, we shall discuss a case in which actually measured positions are strongly influenced by quantum effects.

ing the time in which it travels a given distance. Quantum mechanics is concerned with the correlations between such measurable quantities as position, velocity, and energy of the electron.[3] In making predictions about the relationships of these quantities, the additional physical concept of a wave function is added. Manipulations are performed upon this wave function in accordance with the *rules of manipulation*. (In quantum mechanics it is hard to specify the difference between a rule of manipulation and a new physical concept; for our purpose, it is not worth while to attempt it.) As a result of these manipulations with the wave function, predictions about the behavior of the electron are made. However, as will be discussed in Section 14.3, no direct physical interpretation of the wave function itself is given, nor is one required. In Section 14.3 we shall write the equations governing the wave function and discuss the rules of correlation. We shall precede this by a discussion in Section 14.2 of the mathematically similar subject of an electrical transmission line. But first we shall briefly discuss some additional aspects of an applied mathematical theory in general.

It should be pointed out that the practice of applying a mathematical theory to experimental science is not, in the hands of a theorist, simply a matter of manipulating the symbols of the theory in a routine way. The sets of mental images employed by the theorist, as he carries out the manipulations, differ greatly from one worker to another and are intellectually the most engrossing aspects of theoretical work. These images and analogies permit the theorist to summarize the theory in a satisfying and convenient way. For example, a physicist with one type of viewpoint may prefer to visualize processes in terms of particles moving and colliding, whereas another may think of algebraic symbols and their patterns of relationship.[4] Frequently, with the aid of these mental pictures, the theorist can arrive quickly at conclusions which could be reached only more laboriously by symbolic manipulation. This procedure has its pitfalls, for the processes used may well be forms of reasoning by analogy and can easily be carried too far. This use of physical intuition may perhaps be called the art of theoretical physics.

One of the difficulties of mastering an applied mathematical theory is closely related to the employment of several sets, rather than one set, of mental images: There are, on the one hand, the physical quantities themselves and, on the other, the symbols which represent them. (These latter are in a sense equally real, consisting of marks on paper or sound

[3] Quantum mechanics embraces, of course, a far larger field, but, for our present illustration, the treatment of an electron suffices.

[4] We are here discussing topics connected with the use of "models", the meaning of an "explanation", and the requirement of "simplicity" in a theory. For a further discussion of these subjects see the references given to the heading of Section 14.1.

waves in air.) Manipulation of the symbols alone (at least to the extent usually needed for physics) does not pose intellectual problems of the same magnitude as those of mathematical physics; neither does experimentation in a laboratory. The difficulty of mathematical physics arises in the translation of the mental images by the rules of correlation. This difficulty is first encountered by students in high school algebra in connection with word problems. These problems are generally recognized as being harder than the problems which are stated in algebraic form at the outset. Still more difficult are the processes discussed in the previous paragraph, in which a mental image of the physical situation is examined from changing viewpoints consistent with the changing relationships among the symbols during the course of the manipulations.

14.2 THE MATHEMATICAL THEORY OF A TAPERED LOSSLESS TRANSMISSION LINE

We shall consider here a lossless transmission line with distributed constants which may vary from one position to another. Applying the appropriate rules of correlation, we say that, at time t and position x on the line, a voltage $V(x, t)$ and a current $I(x, t)$ may be specified. There are, in addition, the inductance per unit length, $L(x)$, and the capacitance per unit length, $C(x)$. The mathematical theory requires these six symbols to satisfy certain differential equations. In the discussion that follows, these equations are manipulated so that new and useful relationships between the measurable quantities, represented by the symbols, are derived. For example, we shall show that the expressions for energy stored per unit length on the line and power flow are consistent: The loss in energy for any segment of the line can be accounted for by outward flow of power at the ends.

Having derived a number of equations, we next ask what sort of a theory we would have if it were impossible to measure current and voltage in the line, but instead only energy density. If this were the case, then V and I would have to be regarded as additional physical concepts introduced in the theory. Looked at in this way, the mathematical theory of the transmission line bears a close relationship to that of quantum mechanics in which two functions, like $V(x, t)$ and $I(x, t)$, are introduced to describe the behavior of the electron, which itself is only observed in connection with its probability density.

In Table 14.1 are given a set of equations referring to this section and to the next. They are written in parallel columns to permit easy comparison. In this section we shall derive and discuss the relationships in column (A) leaving (B) for the next section.

If distance is measured along the line from left to right by the variable x, with positive current flowing to the right, then equations ($1A$) apply to the line. We shall use these to establish the consistency of the power flow and energy density expression. Thus, suppose we start with the physical idea, which may be regarded as an *additional physical concept* in Figure 14.1, that power flow to the right at any point of the line is VI evaluated at that point. Then the net power into the line segment between x_1 and x_2 ($x_1 < x_2$) is

$$P(x_1, x_2) = V(x_1, t)I(x_1, t) - V(x_2, t)I(x_2, t). \tag{1}$$

But the right side is mathematically equal to

$$\begin{aligned} -\int_{x_1}^{x_2} \frac{\partial}{\partial x}(VI)dx &= -\int_{x_1}^{x_2}\left(\frac{\partial V}{\partial x} I + V\frac{\partial I}{\partial x}\right)dx \\ &= -\int_{x_1}^{x_2}\left(-L\frac{\partial I}{\partial t} I - VC\frac{\partial V}{\partial t}\right)dx \\ &= +\int_{x_1}^{x_2}\frac{\partial}{\partial t}(\tfrac{1}{2}LI^2 + \tfrac{1}{2}CV^2)dx \\ &= \frac{\partial}{\partial t}\frac{1}{2}\int_{x_1}^{x_2}(LI^2 + CV^2)dx. \end{aligned} \tag{2}$$

This proves that the net flow of power into segment (x_1, x_2) can be accounted for by the rate of change of the integral of $\frac{1}{2}(LI^2 + CV^2)$ along the line. This is valid even when L and C are functions of x (but not of t) and is, of course, what we should expect on the basis of our previous knowledge that $\frac{1}{2}LI^2$ and $\frac{1}{2}CV^2$ are energies per unit length.

If the line extends to $x = \pm\infty$ and the current-voltage wave is localized (both V and I vanish at $x = \pm\infty$), then the theorem considered above shows that the total energy in the line is independent of time and is equal to the integral

$$\int_{-\infty}^{+\infty} [\tfrac{1}{2}LI^2 + \tfrac{1}{2}CV^2]dx = \mathcal{E}, \tag{3}$$

which is constant [since $P(x_1, x_2)$ vanishes for $x_1 = -\infty$, $x_2 = +\infty$]. The same conclusion is reached if the line has finite length A and is open-circuited ($I = 0$) at both ends, or short-circuited ($V = 0$), or closed on itself so that power flowing out at $x = A$ is simultaneously flowing in at $x = 0$.

Equations (1A) may readily be manipulated to give equation (4A) and a similar equation for $\partial^2 V/\partial t^2$. If we suppose further that L may vary while C does not, and consider a wave of frequency ω so that

$$I(x, t) = f_1(x)\cos\omega t + f_2(x)\sin\omega t, \tag{4}$$

then $I(x, t)$ and f_1 and f_2 must satisfy equation (5A). This equation is formally similar to (5B) which we shall discuss later.

TABLE 14.1 COMPARISON OF EQUATIONS FOR A LOSSLESS TRANSMISSION LINE WITH THE ONE-DIMENSIONAL WAVE EQUATION FOR A PARTICLE

		Transmission Line (A)	Particle ($\hbar = h/2\pi$) (B)
Differential equations	(1)	$\frac{\partial I}{\partial t} = -\frac{1}{L}\frac{\partial V}{\partial x}$ $\frac{\partial V}{\partial t} = -\frac{1}{C}\frac{\partial I}{\partial x}$	$\frac{\partial \mathbf{u}}{\partial t} = -\frac{\hbar}{2m}\frac{\partial^2 \mathbf{w}}{\partial x^2} + \frac{\mathcal{U}\mathbf{w}}{\hbar} = \frac{\mathcal{H}\mathbf{w}}{\hbar}$ $\frac{\partial \mathbf{w}}{\partial t} = +\frac{\hbar}{2m}\frac{\partial^2 \mathbf{u}}{\partial x^2} - \frac{\mathcal{U}\mathbf{u}}{\hbar} = \frac{-\mathcal{H}\mathbf{u}}{\hbar}$
Density	(2)	Energy per unit length $\frac{1}{2}LI^2 + \frac{1}{2}CV^2$	Probability per unit length (normalized) $\mathbf{u}^2 + \mathbf{w}^2$
Density flow	(3)	Energy per unit time = power to right VI	Probability per unit time = probability current to right $\frac{\hbar}{m}\left(\mathbf{u}\frac{\partial \mathbf{w}}{\partial x} - \mathbf{w}\frac{\partial \mathbf{u}}{\partial x}\right)$
Elimination of one variable	(4)	$\frac{\partial^2 I}{\partial t^2} = \frac{1}{LC} \times \left[\frac{\partial^2 I}{\partial x^2} - \frac{d \ln C}{dx}\frac{\partial I}{\partial x}\right]$	$\frac{\partial^2 \mathbf{u}}{\partial t^2} = -\frac{1}{\hbar^2}\mathcal{H}^2\mathbf{u}$
Equation for a single frequency $f = \omega/2\pi$	(5)	For the case of $\frac{dC}{dx} = 0$, I satisfies $\frac{\partial^2 I}{\partial x^2} = -LC\omega^2 I$	$\frac{\partial^2 \mathbf{u}}{\partial x^2} = -\frac{2m}{\hbar^2}[\omega\hbar - \mathcal{U}]\mathbf{u}$

Now suppose that the only measuring instruments we had available were some combination of an electrostatic voltmeter and a-c ammeter so that we could measure only the energy density (2A). If we were able to set up disturbances in the line, or if these were somehow introduced by the experimental conditions, we should be able to study a number of the features of the line. If our measuring devices could operate on a sufficiently fine scale, we could shunt the line and measure standing-wave patterns and in this way learn something about the wave length of the waves. Finally, if we were ingenious enough, we might be able to invent a mathematics which would describe all the properties of the line so far as the measurements of average energy densities were concerned. Now, if this mathematical invention led us to two equations such as (1A), with further consequences such as (2A) and (3A), we could, in the theory, talk about I and V without having any physical way of knowing about them per se or, for that matter, without ever having had any experience with inductances or capacitances except as they occurred in inseparable combination in the transmission line. This lack of direct experience would not, however, hinder us in the application of the theory to predict results of power reflection, standing wave ratios, and so forth. Some additional physical concepts about V and I would enter the theory, but would go out again at the point where the *rules of correlation* were applied (in the third step of Figure 14.1) to predict relationships among observed quantities.

As we shall now discuss, the roles of the real and imaginary parts of Schroedinger's wave function are analogous to I and V of our hypothetical example. However, there is an important difference: Quantum-mechanical theory implies strongly that there is no underlying substratum of experience in terms of which ψ can be interpreted in a manner similar to that in which I and V can be.

14.3 THE MATHEMATICAL PHYSICS OF SCHROEDINGER'S EQUATION

In Table 14.1 in column B we give a number of equations related to the one-dimensional Schroedinger equation. The particle in question moves along the x-axis under the influence of forces which give it a potential energy $\mathcal{U}(x)$. The two wave functions $\mathbf{u}(x, t)$ and $\mathbf{w}(x, t)$ occurring in equations (1B) are required to be real just as I and V in (1A) must be real.

The quantity $\mathbf{u}^2 + \mathbf{w}^2$, which is thus real and positive, is called a probability density. The rule of correlation for its interpretation is as follows: Suppose we have an experiment in which we detect the position of an electron at some instant t. For example, we find where the electron strikes and thus exposes the photographic emulsion of a plate in an electron diffraction camera.[1] In this case, the position of the electron in the

[1] The modern electron diffraction camera, which is the descendant of the early electron diffraction apparatus with which Davisson and Germer first observed that electrons were diffracted like waves, is built like a cathode ray tube. The electron beam (perhaps 10^{-8} amp at 40 KV) strikes the sample, is diffracted and strikes the photographic plate, which usually replaces the fluorescent screen.

emulsion does not change with t for large t's. We repeat this experiment over and over with many electrons, each of which is started down the camera in the identical way, and presumably with the same initial values of **u** and **w** (which in this case will be functions of y and z as well as of x and t). The first rule of correlation is:

(RC. 1)
probability that electron be found between x_1 and x_2

$$= \int_{x_1}^{x_2} (\mathbf{u}^2 + \mathbf{w}^2)\,dx \Big/ \int_{-\infty}^{+\infty} (\mathbf{u}^2 + \mathbf{w}^2)\,dx. \qquad (1)$$

That is, at any instant t, the value of $\mathbf{u}^2 + \mathbf{w}^2$ is proportional to the probability per unit length that the electron is at a given place. In the repetitions of the experiment, this probability gives the fraction of the electrons found in each interval. The theory can thus be compared with experiment. When this is done, it is found that the darkening on the plate is precisely as predicted from the theory. Furthermore, the predictions are far from trivial: From the nature of the interference with the crystal, the wave length of the electron waves can be determined just as accurately as the wave length of the electromagnetic waves of the last section could be found. Furthermore, from the values of the potential energy function, $\mathcal{U}$, in the crystal, intensities of the waves can be calculated. In all of these respects, the mathematical theory represented by equations (1B) of Table 14.1 and *rule of correlation* (RC. 1) appears to be in agreement with experiment.

In order to simplify the probability calculations, **u** and **w** are usually multiplied by a scale factor (normalizing factor) so that $\mathbf{u}^2 + \mathbf{w}^2$ is normalized:

$$\int_{-\infty}^{+\infty} (\mathbf{u}^2 + \mathbf{w}^2)\,dx = 1. \qquad (2)$$

Then the denominator can be left out of (RC. 1). In general we shall use *normalized* wave functions.

As shown in Table 14.1, a density flow (3B) can be found for this case analogous to the power or energy flow for (3A). If we consider a length of line between x_1 and x_2, then the rate of change of probability that the electron be in this interval is

$$\frac{d}{dt}\int_{x_1}^{x_2} [\mathbf{u}^2(x, t) + \mathbf{w}^2(x, t)]\,dx = 2\int_{x_1}^{x_2} \left(\mathbf{u}\frac{\partial \mathbf{u}}{\partial t} + \mathbf{w}\frac{\partial \mathbf{w}}{\partial t}\right) dx. \qquad (3)$$

Substituting (1B) for the time derivatives and integrating by parts gives

$$2\int_{x_1}^{x_2}\left[\mathbf{u}\left(-\frac{\hbar}{2m}\frac{\partial^2\mathbf{w}}{\partial x^2}+\frac{\mathcal{U}}{\hbar}\mathbf{w}\right)+\mathbf{w}\left(\frac{\hbar}{2m}\frac{\partial^2\mathbf{u}}{\partial x^2}-\frac{\mathcal{U}}{\hbar}\mathbf{u}\right)\right]dx$$

$$=\frac{\hbar}{m}\int_{x_1}^{x_2}\left(\mathbf{w}\frac{\partial^2\mathbf{u}}{\partial x^2}-\mathbf{u}\frac{\partial^2\mathbf{w}}{\partial x^2}\right)dx$$

$$=\frac{\hbar}{m}\left(\mathbf{w}\frac{\partial\mathbf{u}}{\partial x}-\mathbf{u}\frac{\partial\mathbf{w}}{\partial x}\right)\Bigg|_{x_1}^{x_2}-\frac{\hbar}{m}\int_{x_1}^{x_2}\left(\frac{\partial\mathbf{w}}{\partial x}\frac{\partial\mathbf{u}}{\partial x}-\frac{\partial\mathbf{u}}{\partial x}\frac{\partial\mathbf{w}}{\partial x}\right)dx$$

$$=\frac{\hbar}{m}\left(\mathbf{u}\frac{\partial\mathbf{w}}{\partial x}-\mathbf{w}\frac{\partial\mathbf{u}}{\partial x}\right)_{x_1}-\frac{\hbar}{m}\left(\mathbf{u}\frac{\partial\mathbf{w}}{\partial x}-\mathbf{w}\frac{\partial\mathbf{u}}{\partial x}\right)_{x_2}, \qquad (4)$$

which shows that the change in probability between x_1 and x_2 can be accounted for by a flow $(\hbar/m)(\mathbf{u}\partial\mathbf{w}/\partial x - \mathbf{w}\partial\mathbf{u}/\partial x)$ to the right as shown in (3B). This result is derived for three dimensions in Section 15.4. It is given here to illustrate the procedure required in later proofs in this section.

In addition to explaining electron diffraction results, predictions based on the same interpretation of $\mathbf{u}^2 + \mathbf{w}^2$, when used for electrons in atoms, give rise to probability densities like those shown for H, H_2, C, Si, and Ge in Chapter 1, and these probability densities lead also to predictions of experimental results, somewhat less directly, by means of other methods of correlation.

In Table 14.1, we have deliberately written Schroedinger's equation in an inconvenient and unfamiliar form. The purpose has been to show that there is no greater cause for mental turmoil about the fact that the wave function

$$\psi(x, t) = \mathbf{u}(x, t) + i\mathbf{w}(x, t) \qquad (5)$$

has real and imaginary components than about the fact that energy density on the transmission line can exist in two forms. The quantities **u** and **w** must be real numbers. They are called the real and imaginary components of the wave function, ψ. ψ is also called the "probability amplitude", to distinguish from "intensity" which is proportional to (amplitude)^2 and which occurs in expressions for density and density flow, especially in theories of mechanical and electrical systems. For our one-dimensional case, **u** and **w** have dimensions of length$^{-1/2}$ (i.e. cm$^{-1/2}$)—cm$^{-3/2}$ in three dimensions—since when squared and integrated over length (cm) they give unity, a pure number with no dimensions. In the rules of correlation, **u** and **w** never occur singly but only in products, like $\mathbf{u}^2$ or $\mathbf{u}d\mathbf{w}/dx$. For this reason there is no direct physical significance to be attached to **u** or **w** or ψ individually. Although called amplitudes, there is no physical entity the displacement of which they represent.[2] In other words, ψ is something put on paper by a theoretical physicist and manipulated according to certain rules as aid to predicting and interpreting experimental data [the most important rule of correlation being (RC. 3), given on p. 375]. However, as a result of experience and practice, the manipulator becomes so facile in applying the rules of correlation that it may appear to the onlooker that the wave functions themselves are considered to be the ultimate reality—an erroneous impression which is frequently refuted in quantum-mechanical texts.

The reader may well ask: If the wave function is so ephemeral a thing, why do more than thirty illustrations of it appear in this text? The answer is that as an aid to prediction of electronic behavior it is at present unexcelled, and that, with familiarity in its use, one gains ability usefully to predict, or at least to interpret, observable results.

The reason that ψ is written in complex notation with an imaginary component is simply a matter of convenience. The rules of correlation employed in combining the two real numbers u and w can be expressed compactly and conveniently by use of the notation of complex numbers, with the formal rule that, wherever i^2 occurs, it may be replaced by -1. Thus, for example, the *complex conjugate* of ψ is ψ^*:

$$\text{if} \quad \psi = u + iw \quad \text{then by definition} \quad \psi^* = u - iw, \tag{6}$$

and

$$\psi^* \psi = (u - iw)(u + iw) = u^2 - i^2 w^2 = u^2 + w^2 \tag{7}$$

and by definition the *absolute value* of ψ is

$$|\psi| = (\psi^*\psi)^{1/2} = \sqrt{u^2 + w^2}, \quad |\psi|^2 = \psi^*\psi. \tag{8}$$

It is an obvious result that $\psi^*\psi$ is always real and positive unless $\psi = 0$.[3]

Proceeding with the use of complex number notation, we convert equations (1B) of Table 14.1 to the customary form as follows: We multiply the first of equations (1B) by $i\hbar$ and the second by $-\hbar$ and add, obtaining in this way

(from the left side)

$$i\hbar \frac{\partial u}{\partial t} - \hbar \frac{\partial w}{\partial t} = i\hbar \frac{\partial}{\partial t}(u + iw) = i\hbar \frac{\partial \psi}{\partial t} = \tag{9a}$$

(from the right side)

$$-\frac{\hbar^2}{2m}\frac{\partial^2 \psi}{\partial x^2} + \mathcal{U}\psi = \left[\frac{1}{2m}\left(\frac{\pm\hbar}{i}\frac{\partial}{\partial x}\right)^2 + \mathcal{U}\right]\psi \equiv \mathcal{H}\psi, \tag{9b}$$

[2] It appears to be as fruitless to be concerned about what ψ is a displacement of as it has been to invent models of the ether in which electromagnetic fields constitute some form of mechanical stress.

[3] Another example in which a complex notation makes for easier manipulation of real quantities is given in the footnote of Section 7.7 and in Section 12.3.

so that

$$i\hbar \frac{\partial \psi}{\partial t} = \mathcal{H}\psi, \tag{9c}$$

where, by definition, $\mathcal{H}$ is the linear operation in the square brackets. [As we shall see later the interpretation of $(\hbar/i)\partial/\partial x$ as momentum indicates that the $+$ sign in $\pm$ is the more appropriate.] We may now state:

The time dependent Schroedinger equation for a particle moving in one dimension in a potential field $\mathcal{U}(x)$ *is*

$$i\hbar \frac{\partial \psi}{\partial t} = \mathcal{H}\psi, \quad \mathcal{H} \equiv -\frac{\hbar^2}{2m}\frac{\partial^2}{\partial x^2} + \mathcal{U}(x) \tag{10}$$

where m *is the mass of the particle and* $\hbar$ *is Dirac's notation for Planck's constant* h *divided by* 2π.

For the pedagogical purposes previously mentioned, we introduced this equation by the less direct procedure of Table 14.1.

So far we have discussed only one rule of correlation, namely, (RC. 1), equation (1), Section 14.3, relating to the probability density $\mathbf{u}^2 + \mathbf{w}^2 = |\psi|^2$. From this we derived the density flow formula (3B) of Table 14.1. We shall next draw further conclusions from (RC. 1) and equation (10) in preparation for a statement of the general rule of correlation (RC. 3), equation (19) of Section 14.3, which is one of the fundamental postulates of quantum theory. In these derivations we shall suppose, for the sake of discussion, that $\psi(x, t)$ has the form of a pulse or wave-packet, although the formulae apply equally well in more general cases.[4] We shall calculate the average position of the electron in the wave-packet, the mean square deviation of its position, and, finally, the average velocity of motion of the wave-packet. The manipulations are given in some detail so that the nature of the averaging processes will be obvious.

We suppose, therefore, that there is an experimental arrangement, like that discussed in connection with the electron diffraction camera, with the aid of which we can study the statistical behavior of many (say N) electrons, all of which have behaviors so much alike that they may be represented by the same wave function $\psi(x, t) = \mathbf{u}(x, t) + i\mathbf{w}(x, t)$, where t is measured from the start of the experiment. (If the experiment were repeated at successive times for the different electrons, then the t's would be measured from different zeros.) Now suppose that at time t, for each electron, we make an observation of the electron's position. Then by (RC. 1), assuming ψ normalized in accordance with equation (2), we find a fraction $|\psi(x, t)|^2 dx$ of them with positions between x and $x + dx$, so that, on the average, we obtain values of x between x and $x + dx$ just

[4] The mathematics of such wave functions is considered in Sections 15.5 and 15.6, a particular wave-packet being shown in Figure 6.2.

$N|\psi(x, t)|^2dx$ times. Consequently, the set of all the measured values $x_1, x_2, \cdots, x_N$ will give an average value of

$$\bar{x} = \frac{(x_1 + x_2 + \cdots + x_N)}{N} = \frac{N\int_{-\infty}^{+\infty} x|\psi(x, t)|^2dx}{N}$$
$$= \int_{-\infty}^{+\infty} x|\psi(x, t)|^2dx. \tag{11}$$

This quantity is, by definition of an average, the average value at time t after each experiment started of the position of the electrons in the N repeated experiments. If our experiment is set up in such a way as to give us a well-defined pulse of probability density, then the x-values will cluster about the mean value $\bar{x}$, with a certain spread about this value. It is instructive and useful for later applications to define a mean value of this spread. This may be done as follows: From each value of x in the list of N individual observations,[5] subtract $\bar{x}$ so as to obtain the deviations $x_1 - \bar{x}$, $x_2 - \bar{x}$, etc.; square these deviations and average them.[6] By definition of the averaging process, this average, called the mean square deviation, is denoted by $\overline{(x - \bar{x})^2}$ and is given by

$$\overline{(x - \bar{x})^2} = \frac{(x_1 - \bar{x})^2 + (x_2 - \bar{x})^2 + \cdots + (x_N - \bar{x})^2}{N}. \tag{12}$$

This can be written in the form

$$\overline{(x - \bar{x})^2} = \overline{(x^2 - 2x\bar{x} + \bar{x}^2)} = \overline{x^2} - 2\bar{x}\bar{x} + \bar{x}^2$$
$$= \overline{x^2} - \bar{x}^2. \tag{13}$$

Now, by the same reasoning as that used for the specification of $\bar{x}$, we conclude that $\overline{x^2}$ is given by

$$\overline{x^2} = \frac{x_1{}^2 + x_2{}^2 + \cdots + x_N{}^2}{N} = \int_{-\infty}^{+\infty} x^2|\psi(x, t_1)|^2dx. \tag{14}$$

The principal reason for carrying out these manipulations in detail at this point is to illustrate the use of the probability density $|\psi|^2$ in computing averages of x and functions of x (that is, the simple function x^2). We shall have occasion to discuss other mean squared values in a later section. It is evident from equations (12) and (13) that, only if $|\psi(x, t)|^2$ con-

[5] In principle we did the experiment N times and recorded the result N times although, in practice, the averaging of the x's may be accomplished by the measuring instrument itself which might, for example, display the total charge induced on a capacitor plate by the N electrons.

[6] The average of the deviations themselves is zero, since the plus and minus deviations cancel out.

centrates all its probability at one particular value of x (so that every $x - \bar{x} = 0$), does $\overline{x^2} = \bar{x}^2$. Otherwise $\overline{x^2}$ always exceeds $\bar{x}^2$.

We shall next make use of the value of $\bar{x}$ to describe one way in which an average velocity may be measured in principle. The value of $\bar{x}$, which we now denote by $\bar{x}(t)$, depends upon t through the time dependence of $\psi(x, t)$. The mean position of the electron thus changes at the rate

$$\frac{d\bar{x}(t)}{dt} = \frac{d}{dt}\int_{-\infty}^{+\infty} x\psi^*(x, t)\psi(x, t)dx$$

$$= \int_{-\infty}^{+\infty}\left[x\frac{\partial\psi^*}{\partial t}\psi + x\psi^*\frac{\partial\psi}{\partial t}\right]dx$$

$$= \int_{-\infty}^{+\infty} 2x\left[\mathbf{u}\frac{\partial\mathbf{u}}{\partial t} + \mathbf{w}\frac{\partial\mathbf{w}}{\partial t}\right]dx \tag{15}$$

where the last form can be derived directly from the $\mathbf{u}^2 + \mathbf{w}^2$ expression. The time-derivative terms can be removed from the $\mathbf{u}$, $\mathbf{w}$ form with the aid of equations (1B) of Table 14.1 or from the ψ form with the aid of equation (10), and the resulting equations integrated by parts (provided ψ vanishes at $\pm\infty$, as it must in accordance with our packet assumption), the manipulations finally yielding the result

$$\frac{\partial}{\partial t}\bar{x} = \int_{-\infty}^{\infty}\psi^*\frac{1}{m}\frac{\hbar}{i}\frac{\partial}{\partial x}\psi dx$$

$$= \frac{\hbar}{m}\int_{-\infty}^{\infty}\left(\mathbf{u}\frac{\partial\mathbf{w}}{\partial x} - \mathbf{w}\frac{\partial\mathbf{u}}{\partial x}\right)dx. \tag{16}^7$$

The last two expressions are equal because

$$\int_{-\infty}^{+\infty}\mathbf{u}\frac{\partial\mathbf{u}}{\partial x}dx = \tfrac{1}{2}\mathbf{u}^2\Big|_{-\infty}^{+\infty} = 0 \tag{17}$$

and a similar result holds for $\mathbf{w}$. Equation (16), written in the ψ form, is an example of one of the general rules of correlation which we shall now express: The symbol

(RC. 2) $$p = \frac{\hbar}{i}\frac{\partial}{\partial x} \tag{18}$$

is called an operator since $p\psi$ performs the operation[7] of differentiating ψ. The operator p is correlated with the classical momentum of the particle and when properly averaged (as described below) gives the mean value of

[7] It may be noted that, whereas $\bar{x} = \int x|\psi|^2, dx$ can equally well be written as $\int \psi^* x\psi dx$ in accordance with (16); $\bar{p} = \int \psi^* p\psi dx$ is not equal to $\int p\,|\psi|^2 dx$.

mdx/dt. There are other operators which are likewise correlated with attributes of the electron's behavior which are, at least in principle, measurable. For all of these operators we have:

(RC. 3) THE AVERAGE VALUE: *Select the operator, $\mathcal{R}$, corresponding to the quantity of interest, operate upon the wave function ψ with it, multiply by ψ^* and integrate. The result is the expected average value which would be obtained from a series of measurements of the quantity made in experiments in each of which the behavior of the electron is described by the wave function ψ:*

$$\langle \mathcal{R} \rangle = \int \psi^* \mathcal{R} \psi dx. \tag{19}$$

We have seen two examples of the use of such operators: The position of the particle has the operator x; and the momentum p, or mdx/dt, has the operator $(\hbar/i)\partial/\partial x$. The averages[8] are computed by

$$\langle x \rangle = \int_{-\infty}^{+\infty} \psi^* x \psi dt \tag{20}$$

$$\langle p \rangle = \int_{-\infty}^{+\infty} \psi^* p \psi dx = \int_{-\infty}^{+\infty} \psi^* \frac{\hbar}{i} \frac{\partial}{\partial x} \psi dx \tag{21}$$

The average value of p, it may be remarked, is, at least in some experiments, a directly measurable quantity. Thus, in velocity-modulated tubes, pulses of many electrons flow down the drift-space and give detectable signals as they cross the output circuit. From the solutions of the time-dependent Schroedinger equation, it is possible to make predictions of the average value of dx/dt for an electron which moves in a potential field $\mathcal{U}$. It can be shown, by methods similar to those discussed in Chapter 7, that such an electron wave-packet will move with precisely the velocity expected for an electron on the basis of Newton's laws of motion. If this same electron beam strikes a crystal at a point along its path, diffraction effects in keeping with its wave length will occur.

Let us next consider the average values and operators for potential and kinetic energy for an electron. Since, when the electron is at x, it has potential energy $\mathcal{U}(x)$, the average potential energy is, by reasoning similar to that used to evaluate $\langle x^2 \rangle$,

$$\langle \mathcal{U} \rangle = \int_{-\infty}^{+\infty} \mathcal{U}(\mathbf{u}^2 + \mathbf{w}^2) dx = \int_{-\infty}^{+\infty} \mathcal{U}|\psi|^2 dx = \int_{-\infty}^{+\infty} \psi^* \mathcal{U} \psi dx. \tag{22}$$

[8] We shall use interchangeably $\langle x \rangle$ and $\bar{x}$ to denote averages, depending upon readability and economy of type setting, in this section.

We shall next show that the logical operator to use for the kinetic energy is

$$K.E. = \frac{p^2}{2m} = \frac{1}{2m}\left(\frac{\hbar}{i}\frac{\partial}{\partial x}\right)\left(\frac{\hbar}{i}\frac{\partial}{\partial x}\right) = -\frac{\hbar^2}{2m}\frac{\partial^2}{\partial x^2} \tag{23}$$

which gives

$$\overline{K.E.} = \int \psi^* \left(-\frac{\hbar^2}{2m}\right)\frac{\partial^2\psi}{\partial x^2}\,dx. \tag{24}$$

The reason for this choice is that we can show that the sum $\langle K.E.\rangle + \langle \mathcal{U}\rangle$ remains constant in time [that is, prove that $d[\langle K.E.\rangle + \langle \mathcal{U}\rangle)/dt = 0]$ no matter what solution we have for equations (1B) of Table 14.1. This result can be established by taking the time derivatives under the integral signs, substituting from equations (1B), and integrating by parts. (We shall do this in the next section, making use of the manipulation procedures for operators.) The result is that

$$\frac{d}{dt}(\langle K.E.\rangle + \langle \mathcal{U}\rangle) = 0. \tag{25}$$

This result can be expressed more simply by noting that the operator for total energy is simply the operator $\mathcal{H}$

$$K.E. + \mathcal{U} = \frac{p^2}{2m} + \mathcal{U} = -\frac{\hbar^2\partial^2}{2m\partial x^2} + \mathcal{U} \equiv \mathcal{H} \tag{26}$$

so that $\langle K.E.\rangle + \langle \mathcal{U}\rangle = \langle \mathcal{H}\rangle$ and $d\langle \mathcal{H}\rangle/dt = 0$.

The presentation given so far has been intended to exhibit quantum mechanics as an applied mathematical theory which is used to make predictions about the behavior of electrons. The exposition has stressed the similarity to the transmission line problem in order to show that it is not necessary to have a further interpretation of $\psi = \mathbf{u} + i\mathbf{w}$ in order to carry out calculations according to rules of manipulation and make predictions on the basis of rules of correlation. The order of presentation has been quite different from that used in most texts on quantum mechanics since it has been assumed that many readers will have some intuitive feeling about transmission lines and thus could be helped by considering the analogy between energy density and probability density. The pair of real functions $\mathbf{u}$ and $\mathbf{w}$ were introduced in place of the mathematically equivalent complex function ψ for the same reason. On the basis of the relatively simple interpretation of $\psi^*\psi$ as a probability density, the average of x was introduced, which led, for example, to the interpretation of the operator p as a means of computing $\dot{x}$. The formal procedure of quantum mechanics introduces these rules of correlation compactly as postulates from the outset and deals conveniently with the integrations by parts by means of the symbolism of Hermitian operators.

We shall introduce this symbolism briefly in the next section, and use it later to help establish the idea of energy eigenfunctions.

14.4 OPERATOR METHODS

Of fundamental importance in quantum mechanics is the fact that the important operators are *Hermitian*. An operator $\mathcal{R}$ is Hermitian if, for any two wave functions ϕ and ψ, the following equation is satisfied:

$$\int_a^b \phi^* \mathcal{R} \psi dx = \int_a^b (\mathcal{R}\phi)^* \psi dx. \tag{1}$$

The result depends upon using wave functions which behave properly a the limits of integration, which satisfy the proper *boundary conditions.* We shall consider two cases: (I) a and b are finite and ψ must vanish at both limits.[1] This corresponds, as we shall show in Section 14.6, to infinite potential energy outside the interval $a - b$ so that electrons are reflected at the boundaries. (II) $a = 0$ and $b = A$ represent the same point of a cyclic or periodic model (see Figures 5.6 and 5.9), so that ψ is the same at $x = a$ and $x = b$. In the remainder of Chapter 14 it will be assumed that the individual wave functions, or sets of wave functions, consistently satisfy one or the other of these two boundary conditions.

That the following operators are Hermitian is evident, almost by inpection:

$\mathcal{R}$ = a positive or negative real number, like $\hbar/m$. (Although the operation is trivial, calling a number an operator is required by the general symbolic procedure.) (2)

$\mathcal{R} = x, x^2$ or any real function of x, like $\mathcal{U}(x)$. (3)

The result is less evident for

$$p = (\hbar/i)\partial/\partial x \tag{4}$$

$$K.E. = (\tfrac{1}{2}m)p^2 = (\hbar^2/2m)\partial^2/\partial x^2. \tag{5}$$

We shall prove the Hermitian property of p by direct manipulation, and the Hermitian property of $K.E.$ from that of p. By substituting p for $\mathcal{R}$ in (1) and integrating by parts, we obtain

$$\begin{aligned} \int_a^b \phi^* p \psi dx &= \frac{\hbar}{i}\int_a^b \phi^* \frac{\partial \psi}{\partial x} dx \\ &= \frac{\hbar}{i}\phi^*\psi\Big|_a^b - \frac{\hbar}{i}\int_a^b \frac{\partial \phi^*}{\partial x}\psi dx \\ &= \int_a^b \left(\frac{\hbar}{i}\right)^* \frac{\partial \phi^*}{\partial x}\psi dx = \int_a^b \left(\frac{\hbar}{i}\frac{\partial \phi}{\partial x}\right)^* \psi dx = \int_a^b (p\phi)^*\psi dx \end{aligned} \tag{6}$$

[1] The case of $a = -\infty$, $b = +\infty$ can be considered as a limiting case of (I).

since the integrated term vanishes for boundary conditions (I) and (II). Before treating *K.E.*, we note that if $\mathcal{R}$ and $\mathcal{S}$ are Hermitian and r and s any two real numbers, then

$$r\mathcal{R} + s\mathcal{S} \text{ is Hermitian.} \tag{7}$$

The proof is simple. Also, if $\mathcal{R}$ is Hermitian so is $\mathcal{R}^2$; since we can write $\chi_1 = \mathcal{R}\psi$ and $\chi_2 = \mathcal{R}\phi$, we have

$$\int_a^b \phi^* \mathcal{R}^2 \psi dx = \int_a^b \phi^* \mathcal{R} \chi_1 dx = \int_a^b (\mathcal{R}\phi)^* \chi_1 dx = \text{etc.}$$
$$= \int_a^b (\mathcal{R}^2 \phi)^* \psi dx. \tag{8}$$

Applying (8) to p^2 and (7) to $(\frac{1}{2}m)p^2$, we conclude that *K.E.* is Hermitian. By applying (7) to *K.E.* and $\mathcal{U}$, we conclude that $\mathcal{H}$ is Hermitian.

The procedure of integration by parts used with p can be used to establish the instructive theorem that the average value of the *K.E.* is never negative: If the state of behavior of the electron is described by a wave function ψ, the value of $\langle K.E.\rangle$ can be transformed as follows:

$$\langle K.E.\rangle = \frac{1}{2m}\int_a^b \psi^* p^2 \psi dx = \frac{1}{2m}\int_a^b \left(\frac{\hbar}{i}\frac{\partial\psi}{\partial x}\right)^* \left(\frac{\hbar}{i}\frac{\partial\psi}{\partial x}\right) dx$$
$$= \frac{\hbar^2}{2m}\int_a^b \left|\frac{\partial\psi}{\partial x}\right|^2 dx. \tag{9}$$

As we shall show next, it is possible to find an operator $\mathcal{S}$ whose average value, computed by the average value rule, gives the time derivative of the average value of another operator $\mathcal{R}$. To rephrase this symbolically,[2]

$$\langle \mathcal{S}\rangle \equiv \int \psi^* \mathcal{S} \psi dx = \frac{d}{dt}\langle \mathcal{R}\rangle \equiv \frac{d}{dt}\int \psi^* \mathcal{R} \psi dx. \tag{10}$$

We shall derive the following equation for $\mathcal{S}$, to which we give the symbol $d\mathcal{R}/dt$,

$$\mathcal{S} \equiv \frac{d\mathcal{R}}{dt} = \frac{1}{i\hbar}(\mathcal{R}\mathcal{H} - \mathcal{H}\mathcal{R}). \tag{11}$$ [3]

We shall first illustrate the meaning of the symbolic operator form on the right for the case $\mathcal{R} = x$; to do this, we let $\mathcal{S}$ operate on any ψ and simplify the resulting expressions:

$$\mathcal{S}\psi = \frac{1}{i\hbar}(\mathcal{R}\mathcal{H} - \mathcal{H}\mathcal{R})\psi$$
$$= \frac{1}{i\hbar}\left[x\left(-\frac{\hbar^2}{2m}\frac{\partial^2}{\partial x^2} + \mathcal{U}\right)\psi - \left(-\frac{\hbar^2}{2m}\frac{\partial^2}{\partial x^2} + \mathcal{U}\right)x\psi\right]; \tag{12}$$

[2] We shall in future equations omit writing the limits of integration; it is to be understood that the integration extends over the range specified by the boundary conditions.

[3] The reader may verify that if A and B are Hermitian, $i(AB - BA)$ is also.

since

$$\frac{\partial^2}{\partial x^2} x\psi = \frac{\partial \psi}{\partial x} + \frac{\partial}{\partial x}\left(x \frac{\partial \psi}{\partial x}\right) = 2\frac{\partial \psi}{\partial x} + x\frac{\partial^2 \psi}{\partial x^2}, \tag{13}$$

we obtain

$$\mathcal{S}\psi = \frac{1}{i\hbar}\left[+\frac{\hbar^2}{2m} 2 \frac{\partial \psi}{\partial x}\right] = \frac{\hbar}{mi}\frac{\partial \psi}{\partial x} = \frac{p}{m}\psi, \tag{14}$$

the result previously obtained in equations (16), (18) and (RC. 3) of Section 14.3, by evaluating $\dot{x}$. To prove (11) in general, we write:

$$\begin{aligned} \langle \mathcal{S} \rangle \equiv \frac{d}{dt}\langle \mathcal{R} \rangle &= \int \left[\frac{\partial \psi^*}{\partial t}\mathcal{R}\psi + \psi^*\mathcal{R}\frac{\partial \psi}{\partial t}\right] dx \\ &= \frac{1}{i\hbar}\int [(-\mathcal{H}\psi)^*\mathcal{R}\psi + \psi^*\mathcal{R}\mathcal{H}\psi]dx \\ &= \frac{1}{i\hbar}\int \psi^*(-\mathcal{H}\mathcal{R} + \mathcal{R}\mathcal{H})\psi dx \end{aligned} \tag{15}$$

which is equivalent to equations (10) and (11).

From this result it easily follows that $\int \psi^*\psi dx$ is independent of time as it must be for $|\psi|^2 dx$ to represent a probability.

The form of (11) and the results just considered for $\dot{x}$ show that two operators do not always commute; by definition:

$$\text{If } \mathcal{R}\mathcal{S} = \mathcal{S}\mathcal{R}, \quad \text{then } \mathcal{R} \text{ and } \mathcal{S} \text{ commute.} \tag{16}$$

The failure of quantum-mechanical operators to commute is what causes quantum mechanics to lead to results different from those of classical mechanics.

If the operator $\mathcal{R}$ in equation (11) is $\mathcal{H}$ itself, then the rate of change of $\langle \mathcal{H} \rangle$ has the operator

$$\frac{d\mathcal{H}}{dt} = \frac{1}{i\hbar}(\mathcal{R}\mathcal{H} - \mathcal{H}\mathcal{R}) = \frac{1}{i\hbar}(\mathcal{H}\mathcal{H} - \mathcal{H}\mathcal{H}) = 0 \tag{17}$$

so that the average value of $\mathcal{H}$ remains constant. This is a quantum-mechanical expression of the principle of conservation of energy. In general any operator which commutes with $\mathcal{H}$ has a constant average value.

In Chapter 5, great emphasis is placed upon the eigenfunctions of Schroedinger's equation. These satisfy the equation

$$\mathcal{H}\psi = \mathcal{E}\psi, \tag{18}$$

which we shall discuss in the next section. In operator language, *the energy $\mathcal{E}$ and wave function ψ which satisfy this equation and the boundary conditions are said to be an eigenvalue and an eigenfunction, respectively,* *of the operator* $\mathcal{H}$. All the operators of quantum mechanics have eigenfunctions and eigenvalues. For example, the operator p has eigenfunctions for the case of cyclic boundary conditions [that is, $\psi(x + A) = \psi(x)$] which satisfy the equation

$$p\psi_s = p_s\psi_s \tag{19}$$

where ψ_s equals exp $(2\pi isx/A)$ where s is any integer and the number p_s equals $2\pi\hbar s/A = hs/A$, as may be verified by substitution in equation (19) using (RC. 2), equation (18) of Section 14.3, for p.

The formulae which give the relationships between operators and the results of their operations upon various wave functions form a major part of quantum theory. The brief discussion which has been presented of this topic has been intended chiefly to show that $\mathcal{H}$ and p are Hermitian and to derive the operator for $d\mathcal{R}/dt$.

We have developed a description of the operators of quantum mechanics in terms of the manipulations used in dealing with wave functions. In closing this section, we shall point out that a very different procedure can be used which is more fundamental but at the same time more abstract and less easily grasped on first contact. In the more fundamental procedure the relationships among the operators themselves, without specific reference to wave functions, are made the central part of the theory. In fact, starting with the relationship

$$px - xp = \hbar/i, \tag{20}$$

which is taken as fundamental in such treatments [whereas we would derive it here from (RC. 2), equation (18) of Section 14.3], and the expression for the energy of a harmonic oscillator of frequency ω

$$\mathcal{H} = (\tfrac{1}{2}m)(p^2 + m^2\omega^2x^2). \tag{21}$$

Dirac[4] derives all of the observable properties of the oscillator by dealing symbolically with the operators without introducing $\psi(x, t)$ at all. This typifies the emphasis which Dirac places upon the symbolic method and is a result which further illustrates the lack of fundamental significance of the wave functions, which, he writes, are introduced in his exposition "merely as an aid to practical calculation". Another treatment in which operators and their symbolic relationships are taken as fundamental is that of J. von Neumann.[5]

14.5 ENERGY EIGENFUNCTIONS

Of particular value in the theory of quantum mechanics are the energy eigenfunctions. These are the particular wave functions for which the

[4] P. A. M. Dirac, *The Principles of Quantum Mechanics*, Oxford University Press, 1930, p. 123, or second edition, 1939, p. 133.

[5] J. von Neumann, *Mathematische Grundlagen der Quantenmechanik*, Julius Springer, 1932.

energy has a definite value. This requires that, if the energy is measured in N similar experiments (just as x was measured in N experiments in Section 14.3), the same value for $K.E. + \mathcal{U} = \mathcal{H}$ will be obtained in each measurement. Now (RC. 3), equation (19) of Section 14.3, does not tell what values will be obtained in a large number of measurements but only what their average $\langle\mathcal{H}\rangle$ will be. However, (RC. 3) can also predict the average of $\mathcal{H}^2$, denoted by $\langle\mathcal{H}^2\rangle$, and by the same reasoning as was used with $\langle x^2\rangle$ and $\langle x\rangle$, we conclude that, if $\langle\mathcal{H}^2\rangle$ is equal to $\langle\mathcal{H}\rangle$, there can be no spread in the measured values for $\mathcal{H}$. In other words, if the application of (RC. 3) to ψ for $\mathcal{H}$ and $\mathcal{H}^2$ gives

$$\langle\mathcal{H}\rangle^2 = \langle\mathcal{H}^2\rangle, \tag{1}$$

then every observation will yield $\langle\mathcal{H}\rangle$. We shall find that we can prove from (1) that

$$\mathcal{H}\psi - \mathcal{E}\psi = 0, \quad \text{where } \mathcal{E} \equiv \langle\mathcal{H}\rangle = \int \psi^*\mathcal{H}\psi dx \tag{2}$$

which is recognized as Schroedinger's equation for the eigenfunctions discussed in Section 5.2. This result is proved for the one-dimensional case considered in this chapter by making use of the Hermitian properties of $\mathcal{H}$, $\mathcal{E}$, and $\mathcal{H} - \mathcal{E}$ ($\mathcal{E}$ is simply a number in these manipulations) in evaluating the integral:

$$\begin{aligned}\int |(\mathcal{H} - \mathcal{E})\psi|^2 dx &= \int [(\mathcal{H} - \mathcal{E})\psi]^*(\mathcal{H} - \mathcal{E})\psi dx \\ &= \int \psi^*(\mathcal{H} - \mathcal{E})[(\mathcal{H} - \mathcal{E})\psi] dx \\ &= \int \psi^*[\mathcal{H}^2 - \mathcal{H}\mathcal{E} - \mathcal{E}\mathcal{H} + \mathcal{E}^2]\psi dx \\ &= \langle\mathcal{H}^2\rangle - \langle\mathcal{H}\rangle\mathcal{E} - \mathcal{E}\langle\mathcal{H}\rangle + \mathcal{E}^2 = \langle\mathcal{H}^2\rangle - \mathcal{E}^2. \end{aligned} \tag{3}$$

Now the extreme right-hand side is zero by (1) and the definition of $\mathcal{E}$ as $\langle\mathcal{H}\rangle$. Hence the integral in the extreme left-hand side is also zero, and the integrand $|(\mathcal{H} - \mathcal{E})\psi|^2$, which is always positive or zero, must be zero too, for all values of x. This establishes (2).

The time dependence of an eigenfunction of Schroedinger's equation, satisfying (2), is very simple. We have, in fact, from equation (10) of Section 14.3 and (2),

$$i\hbar \frac{\partial\psi(x, t)}{\partial t} = \mathcal{H}\psi(x, t) = \mathcal{E}\psi(x, t) \tag{4}$$

which can be integrated at once to give

$$\psi(x, t) = \psi(x, 0)e^{-i\mathcal{E}t/\hbar}. \tag{5}$$

It follows that $\psi^*(x, t)\psi(x, t) = \psi^*(x, 0)\psi(x, 0)$ so that the probability density is independent of time and is therefore stationary in space. For this reason, energy eigenfunctions are also called "stationary states". In (5) we see that the complex phase of the wave function varies with the angular frequency, $\omega = -\mathcal{E}/\hbar$; or, letting $\omega = -2\pi\nu$, we have

$$h\nu = \mathcal{E}. \tag{6}$$

This shows that the Planck relationship between energy and frequency holds for the frequencies in the eigenfunctions.

An especially important feature of the eigenfunctions is that they form a *complete set of functions* in terms of which any solution of equation (10) of Section 14.3 can be expressed. In this way, they are analogous to Fourier components in terms of which any function periodic in time can be expressed. We shall not discuss the question of completeness[1] but will demonstrate the *orthogonality* property of the eigenfunctions and derive the expansion formula for an arbitrary wave function in terms of the eigenfunctions.

Two wave functions, ψ_1 and ψ_2, are said to be *orthogonal*[2] if they satisfy the equation

$$\int \psi_1{}^*\psi_2 dx = 0. \tag{7}$$

If ψ_1 and ψ_2 are eigenfunctions of $\mathcal{H}$, corresponding to two distinct eigenvalues $\mathcal{E}_1$ and $\mathcal{E}_2$, then they are orthogonal as may be proved as follows: Let

$$\mathcal{H}\psi_1 = \mathcal{E}_1\psi_1, \quad \mathcal{H}\psi_2 = \mathcal{E}_2\psi_2; \tag{8}$$

then we have

$$\int \psi_1{}^*\mathcal{H}\psi_2 dx = \int \psi_1{}^*\mathcal{E}_2\psi_2 dx = \mathcal{E}_2\int \psi_1{}^*\psi_2 dx, \tag{9}$$

also

$$\int \psi_1{}^*\mathcal{H}\psi_2 dx = \int (\mathcal{H}\psi_1)^*\psi_2 dx = \mathcal{E}_1\int \psi_1{}^*\psi_2 dx; \tag{10}$$

hence by subtracting left side from left side and right from right

$$0 = (\mathcal{E}_2 - \mathcal{E}_1)\int \psi_1{}^*\psi_2 dx. \tag{11}$$

[1] The proof of completeness is a rather advanced topic. See E. C. Kemble, *Fundamental Principles of Quantum Mechanics*, McGraw-Hill Book Co., New York, 1937.

[2] The word "orthogonal" implies right angle so that in function space (7) resembles a scalar vector product which vanishes if the vectors are at right angles.

By hypothesis, $\mathcal{E}_2 - \mathcal{E}_1 \neq 0$, hence the integral vanishes, which establishes (7). It is evident that the same result could be proved for the eigenfunctions of any other operator.

If we have a complete set of normalized eigenfunctions $\psi_1, \psi_2, \cdots$, then any wave function ψ which satisfies the same boundary conditions as the eigenfunctions can be expanded in terms of them in the form

$$\psi = a_1\psi_1 + a_2\psi_2 + \cdots = \sum_s a_s\psi_s \tag{12}$$

where the coefficients $a_1, a_2, \cdots$, are readily evaluated as follows: Multiply by any eigenfunction ψ_r^* and integrate:

$$\int \psi_r^*\psi dx = \sum_s a_s \int \psi_r^*\psi_s dx. \tag{13}$$

The only non-zero term on the right side is $a_r \int \psi_r^*\psi_r dx = a_r$ since ψ_r is orthogonal to the other ψ's. Hence we obtain the result,

$$a_r = \int \psi_r^*\psi dx. \tag{14}$$

Our orthogonality proof depended upon the eigenvalue for ψ_r being different from that of ψ_s. In many cases two or more wave functions have the same eigenvalue which is then called a *degenerate eigenvalue.* If ψ_r and ψ_s are any two such ψ's, which are normalized, but not orthogonal, then they can be combined to form two new wave functions

$$\psi_a = a_r\psi_r + a_s\psi_s \quad \text{and} \quad \psi_b = b_r\psi_r + b_s\psi_s \tag{15}$$

and the coefficients chosen so that ψ_a and ψ_b form an orthogonalized, normalized pair.[3] This procedure can be extended if more than two ψ's are degenerate, and in this way any set of eigenfunctions can be brought to the *orthonormal* form, for which

$$\int \psi_r^*\psi_s dx = \begin{pmatrix} 1 & r = s \\ 0 & r \neq s \end{pmatrix} = \delta_{rs}. \tag{16}$$

The symbol $\delta_{rs}(= 1 \text{ for } r = s; \ = 0 \text{ for } r \neq s)$, called the *Kronecker delta symbol,* is frequently used to represent the right side of (16).

It is evident that, whenever the set of ψ_s are orthonormal, the expansion coefficients in (12) are given by (14).

[3] For example, ψ_r and $\psi_s - \left(\int \psi_r^*\psi_s dx\right)\psi_r$ are orthogonal and the latter is normalized by dividing by $\left(1 - \left|\int \psi_r^*\psi_s dx\right|^2\right)^{1/2}$.

14.6 SOLUTIONS FOR A SIMPLE "ATOM", TRANSMISSION LINE ANALOGUE

We shall illustrate the theory of energy levels for an electron in an atom and in a crystal by treating in this section a simple one-dimensional model of an atom and in the next section a periodic array of such atoms. The atom model to be discussed is represented by Figure 14.2(a) and consists

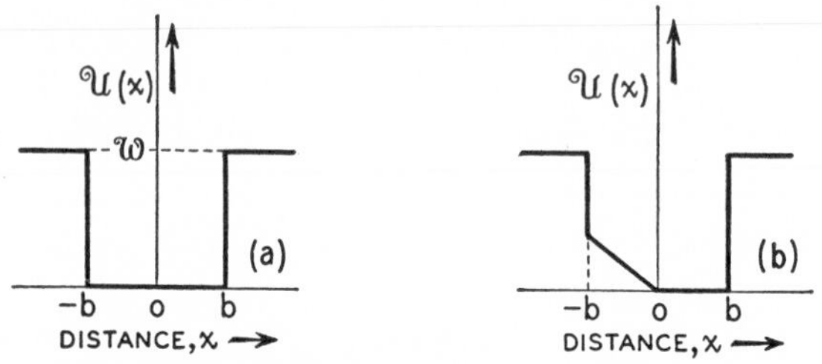

FIG. 14–2—Potential Wells Representing Simplified One-dimensional Atoms.
(a) The symmetrical well used in the text.
(b) An unsymmetrical well.

simply in a depression of the potential energy for the electron. This particular shape is frequently referred to as a *square potential well.* Our problem is to find solutions of Schroedinger's equation for an electron in this potential field,

$$\mathcal{H}\psi(x) = -\frac{\hbar^2}{2m}\frac{d^2\psi(x)}{dx^2} + \mathcal{U}(x)\psi(x) = \mathcal{E}\psi(x), \qquad (1a)$$

$$\mathcal{U}(x) = 0, \quad |x| < b, \quad \mathcal{U}(x) = \mathcal{W}, \quad |x| > b, \qquad (1b)$$

subject to the boundary conditions that $\psi(x)$ must vanish at $x = +\infty$ and $-\infty$, corresponding to the limiting case of boundary condition (1) for Section 14.4. We shall find that for any pair of fixed values of the depth $\mathcal{W}$ and width $2b$ of the well, there is a definite set of solutions which give wave functions in which the electron is bound to the well.

We shall first show that for a symmetrical well, the wave function ψ may be chosen to be either an "odd" function of x or an "even" function. This is done as follows: Suppose $\psi_1(x)$ is a solution of the equation with energy $\mathcal{E}_1$, then

$$\psi_2(x) = \psi_1(-x) \qquad (2)$$

is an equally good solution, since the well is symmetrical about $x = 0$, and has the same energy $\mathcal{E}_1$. Now consider

$$\mathbf{g}(x) = \psi_1(x) + \psi_2(x) \qquad (3a)$$

$$\mathbf{u}(x) = \psi_1(x) - \psi_2(x). \qquad (3b)$$

If ψ_1 is expanded in a power series in x, all of the odd terms will cancel out of **g** and all of the even terms will cancel out of **u**. Evidently

$$\mathbf{g}(x) = \mathbf{g}(-x) \quad \text{and} \quad \mathbf{u}(x) = -\mathbf{u}(-x). \tag{3c}$$

The symbols **g** and **u** correspond to the German *gerade* or even and *ungerade* or odd. Unlike the symbols χ and ϕ introduced below, $\mathbf{g}(x)$ and $\mathbf{u}(x)$ are considered to be defined for all values of x. We shall next obtain explicit expressions for $\mathbf{g}(x)$ and $\mathbf{u}(x)$ by first finding solutions to equation (1) inside the well and outside the well individually and then by joining these together at $x = +b$ and $x = -b$. We shall find that *eigenfunctions* satisfying the boundary conditions can occur only for certain values of the energy.

In the well, the wave equation is

$$\frac{d^2\psi}{dx^2} = -\frac{2m}{\hbar^2}\,\mathcal{E}\psi \tag{4}$$

for which the even and odd solutions are

$$\chi_{\mathbf{g}}(x) \equiv \cos\,(2m\mathcal{E}/\hbar^2)^{\frac{1}{2}}x \quad -b < x < b \tag{5a}$$

$$\chi_{\mathbf{u}}(x) \equiv \sin\,(2m\mathcal{E}/\hbar^2)^{\frac{1}{2}}x \quad -b < x < b \tag{5b}$$

where the symbols $\chi_{\mathbf{g}}$ and $\chi_{\mathbf{u}}$ are simply abbreviations and are defined only for $|x| \leq b$.

Outside of the well, the wave equation is

$$\frac{d^2\psi}{dx^2} = \frac{2m(\mathcal{W} - \mathcal{E})}{\hbar^2}\,\psi. \tag{6}$$

This equation has a sinusoidal solution for $\mathcal{E} > \mathcal{W}$ and does not vanish at infinity. Such solutions can be combined into wave-packets and used to represent free particles moving in space outside of the potential well. They do not, however, represent the bound states which we wish to consider. If $\mathcal{E}$ is less than $\mathcal{W}$, the particle does not have sufficient energy to escape from the well, even though its wave function, as we shall see, does not vanish outside of the well. For this case the possible solutions of the equation are

$$\phi_- = \exp\,[2m(\mathcal{W} - \mathcal{E})/\hbar^2]^{\frac{1}{2}}x \tag{7a}$$

$$\phi_+ = \exp\,-[2m(\mathcal{W} - \mathcal{E})/\hbar^2]^{\frac{1}{2}}x \tag{7b}$$

where the positive square root is to be understood by $[2m(W-E)^{\frac{1}{2}}/\hbar^2]^{\frac{1}{2}}$, the ϕ symbols, introduced for convenience, are defined only for $|x| > b$. The solution ϕ_+ increases without limit as $x \rightarrow -\infty$ and cannot be used as a solution for negative values of x; it can be used for positive values of x since it vanishes at $x = +\infty$; for this reason its subscript is a *plus sign.* Because of the symmetry properties of **g** and **u**, we need consider only the solution for $x > 0$ since the solution for $x < 0$ can readily be obtained from it by replacing x by $-x$; we shall illustrate this below.

We shall first look for solutions of the even type which are valid for $x>0$. These must evidently be of the form

$$\mathbf{g} = A\chi_{\mathbf{g}}(x) = A \cos (2m\mathcal{E}/\hbar^2)^{1/2}x, \qquad -b < x < b \tag{8}$$

$$\mathbf{g} = B\phi_{+}(x) = B \exp -[2m(\mathcal{W} - \mathcal{E})/\hbar^2]^{1/2}x, \qquad b < x. \tag{9}$$

How should these be joined at $x = b$? The requirement is that both **g** and $d\mathbf{g}/dx$ must be continuous functions of x at b. This follows directly from the differential equation which states that in any infinitesimal interval δx the change in $d\psi/dx$ is given by

$$\delta\left[\frac{d\psi}{dx}\right] = \int_{b-(\delta x/2)}^{b+(\delta x/2)} \frac{2m}{\hbar^2}[\mathcal{U}(x) - \mathcal{E}]\psi(x)dx. \tag{10}$$

Since ψ and $[\mathcal{U}(x) - \mathcal{E}]$ are both finite, although $\mathcal{U}(x)$ is discontinuous at $x = b$, the integral vanishes as $\delta x \to 0$ so that $d\psi/dx$ is continuous at b. It is even more obvious that ψ must be continuous at $x = b$ because a discontinuity in ψ would require an infinite value for $d\psi/dx$.

Thus if it is possible to choose A and B so that **g** and $d\mathbf{g}/dx$ are continuous at $x = b$, a wave function satisfying equation (1) and the boundary conditions at $x = +\infty$ and $-\infty$ will be obtained. Such a wave function, which is by definition an eigenfunction of equation (1), can exist only for certain *eigenvalues* of the energy which are obtained by requiring that (8) and (9) join properly. Continuity of **g** requires that

$$A\chi_{\mathbf{g}}(b) = B\phi_{+}(b) \tag{11a}$$

and continuity of $\mathbf{g}' = d\mathbf{g}/dx$ requires that

$$A\chi_{\mathbf{g}}'(b) = B\phi_{+}'(b) \tag{11b}$$

so that dividing the second equation by the first we obtain

$$\frac{1}{\chi_{\mathbf{g}}}\frac{d\chi_{\mathbf{g}}}{dx} = \frac{1}{\phi_{+}}\frac{d\phi_{+}}{dx} \quad \text{at } x = b. \tag{11c}$$

In order to deal with this condition analytically, we introduce the symbols

$$\alpha = \sqrt{2m\mathcal{W}}\, b/\hbar \tag{12a}$$

$$\beta = \sqrt{2m(\mathcal{W} - \mathcal{E})}\, b/\hbar = \sqrt{\alpha^2 - \theta^2} \tag{12b}$$

$$\theta = \sqrt{2m\mathcal{E}}\, b/\hbar. \tag{12c}$$

In terms of these we have

$$\frac{d\chi_g}{dx} \div \chi_g = -(\theta/b)\tan\theta \tag{13}$$

$$\frac{d\phi_+}{dx} \div \phi_+ = -\beta/b \tag{14}$$

so that the eigenfunction condition reduces to

$$-\theta\tan\theta = -\beta = -\sqrt{\alpha^2 - \theta^2}. \tag{15}$$

(We retain the minus signs so as to be in accord with the sign convention for $\boldsymbol{\gamma}$ and $\boldsymbol{\mu}$ in the following section.)

A similar treatment for $\mathbf{u}$ leads to

$$\theta\cot\theta = -\beta = -\sqrt{\alpha^2 - \theta^2}. \tag{16}$$

Both θ and β depend on $\mathcal{E}$, and the eigenvalue problem for $\mathcal{E}$ is thus to find the values of $\mathcal{E}$ for which the equation is satisfied. Figure 14.3 shows how this may be done graphically by plotting both sides as a function of θ, which is equivalent to $\mathcal{E}$ for this purpose, with negative values of $\mathcal{E}$ cor-

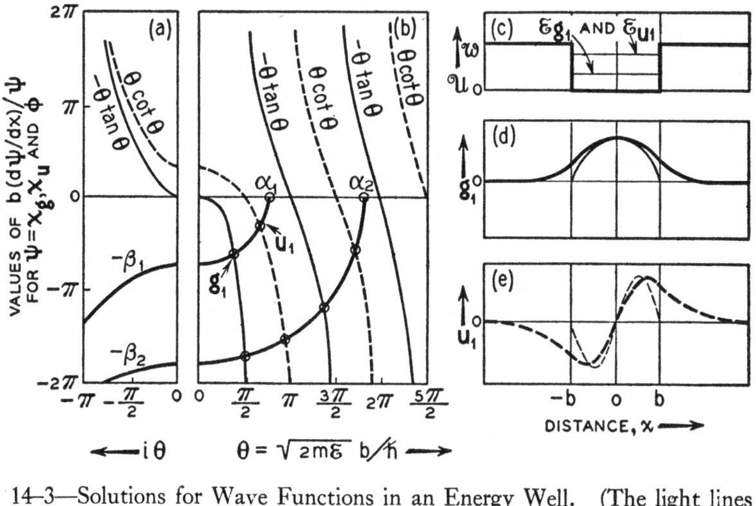

FIG. 14-3—Solutions for Wave Functions in an Energy Well. (The light lines for wave functions $\mathbf{g}_1$ and $\mathbf{u}_1$ correspond to $\mathcal{W} = \infty$. The figure is not quantitative but shows correctly the qualitative features.)

responding to imaginary values for θ. On this plot, $\beta = \sqrt{\alpha^2 - \theta^2}$ is the equation of a circle. The solid curves show $-\theta\tan\theta$ and the dashed curves $\theta\cot\theta$. There are no solutions for θ imaginary corresponding to negative values of $\mathcal{E}$. (Verify this by trying to draw $\mathbf{g}$ and $\mathbf{u}$ for negative $\mathcal{E}$.)

For the value $\alpha_1 \cong 2$ shown in the figure, there are only two solutions and their wave functions are plotted on the right. If α is increased, there will be more solutions and the character of $\mathbf{g}_1$ and $\mathbf{u}_1$ will change. For $\alpha \to \infty$, the values of θ are $\pi/2$ and π and the wave functions vanish at the edges of the well and are zero outside. This is the "vanishing" boundary condition (I) discussed in Section 14.4.

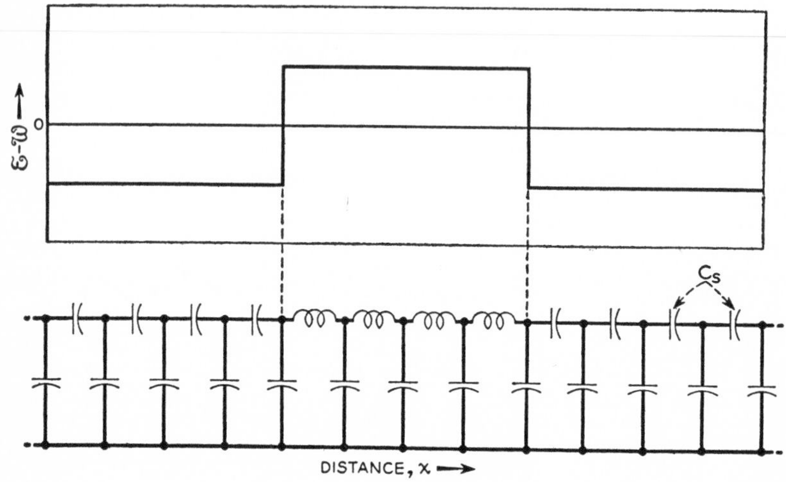

FIG. 14-4—Transmission Line Analogue Corresponding to a Bound State for an Electron in a Potential Well.

As discussed in Section 14.2, the wave equation is analogous to an electrical transmission line in which the circuit constants depend on the energy parameter in Schroedinger's equation. From Table 14.1, line 5, we see that, if C the capacitance to ground per unit length is kept constant, then current in the line is analogous to the wave function ψ in our case. We may take the solutions to have the form

$$\psi(x, t) = \psi(x)e^{i\omega t} \tag{17}$$

and

$$I(x, t) = I(x) \cos \omega t. \tag{18}$$

These lead to the equations of Table 14.1:

$$\frac{d^2I(x)}{dx^2} = -LC\omega^2 I \tag{19}$$

$$\frac{d^2\psi}{dx^2} = -\frac{2m}{\hbar^2}[\mathcal{E} - \mathcal{U}]\psi. \tag{20}$$

For a solution like $\mathbf{g}_1$, this leads to a positive value of $LC\omega^2$ for $|x| < b$

and negative for $|x| > b$. The negative value corresponds to taking $L\omega$ as negative which can be accomplished by replacing the series inductance by series capacitances C_s as shown in Figure 14.4, where the distributed constants are represented as lumped. For such a circuit, energy stored in the inductances cannot be lost at the ends since the capacitance attenuator networks cannot transmit power. This is analogous to the region of negative kinetic energy, in which probability density is present but into which the electron can penetrate only a short distance. Letting $\mathcal{W} \to \infty$ corresponds to letting $L\omega \to -\infty$ or letting the series capacitors C_s approach zero capacitance. For this case the inductive part of the line would be open-circuited at the ends so that I would vanish at $x = \pm b$.

14.7 SOLUTIONS FOR A SIMPLE "CRYSTAL"

In this section we shall discuss the wave functions in a one-dimensional crystal, such as might arise from repeating the "atoms" discussed in the last section periodically with a lattice constant a as shown in Figure 14.5. When this is done, it is found that there are bands of energy in which the wave function has the Bloch behavior and other bands in which the amplitude increases exponentially in one direction or the other of the crystal.

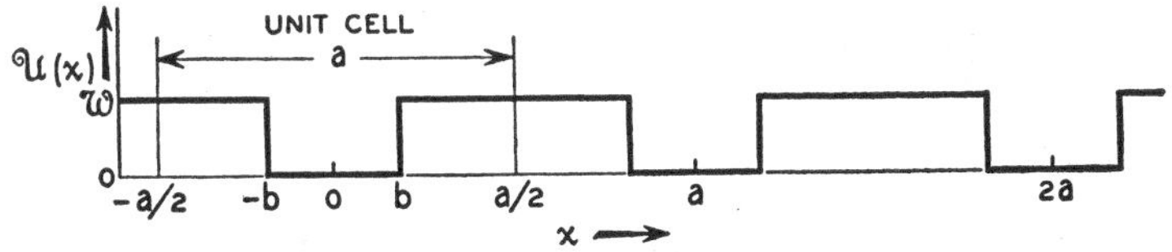

FIG. 14-5—Periodic Potential Composed of Symmetrical Unit Cells.

While these latter cases may be of interest in connection with surface states, they cannot lead to wave functions in the interior of a periodic lattice nor be used to satisfy periodic boundary conditions and they, therefore, correspond to energies in *forbidden* energy bands.

Several possible procedures may be followed in showing how the energy bands arise. One may start simply with an arbitrary finite periodic potential and prove by very general means that allowed energy bands must be present.[1] On the other hand, one may deal with a specific model and derive formulae and calculate the bands in detail.[2] We shall take an intermediate course and assume that the potential in each unit cell is symmetrical about the midpoint, the midpoints being located at $x = 0$,

[1] H. A. Kramers, *Physica* **2**, 483–490 (1935), summarized in F. Seitz, *The Modern Theory of Solids*. See Problem 13 of this chapter.

[2] Seitz, *loc. cit.;* V. Rojansky, *Introductory Quantum Mechanics*, Prentice-Hall; A. H. Wilson, *The Theory of Metals*, Cambridge.

$\pm a$, $\pm 2a \cdots$, so that we may deal with wave functions of the $\mathbf{g}(x)$ and $\mathbf{u}(x)$ form in the cell at $x = 0$. We shall then show that the problem can be defined in terms of two ratios of slope to value,

$$\left.\begin{aligned} \gamma &= \frac{(d\mathbf{g}/dx)}{\mathbf{g}} \\ \mu &= \frac{(d\mathbf{u}/dx)}{\mathbf{u}} \end{aligned}\right\} \text{evaluated at } x = a/2. \tag{1}$$

We shall then consider the general behavior of γ and μ as functions of the energy and discuss how the energy bands are related to the atomic levels of infinitely separated atoms.[3]

We shall first prove that any solution of the wave equation for energy $\mathcal{E}$,

$$\frac{d^2\psi}{dx^2} = \frac{2m}{\hbar^2}[\mathcal{U}(x) - \mathcal{E}]\psi, \tag{2}$$

can be expressed as the sum of two solutions ψ_1 and ψ_2 having the property

$$\psi_1(x + a) = e^{iPa/\hbar}\psi_1(x) \tag{3}$$

$$\psi_2(x + a) = e^{-iPa/\hbar}\psi_2(x) \tag{4}$$

where P is real in the allowed bands and imaginary in the energy gaps. We shall give a proof based on the mathematics of γ and μ, leaving the general poof of this important result, which is known as Floquet's theorem,[4] as a problem.

We shall introduce the quantity β (unrelated to β in any other section) by the equation

$$2\beta = Pa/\hbar \quad \text{so that} \quad e^{iPa/\hbar} = e^{2i\beta} \tag{5}$$

and will shortly obtain an equation for β in terms of γ and μ. It will be found that there are three possible ranges for β:

> β is real and $-\pi/2 < \beta < +\pi/2$, **allowed band;**
> β is imaginary, **forbidden band;**
> $\beta = (\pi/2) +$ an imaginary number, **forbidden band.**

For the allowed band, the complex phase changes by less than 180° from one cell to the next and the absolute value is the same in every cell. For the other two cases the wave function decays exponentially in one direction or the other, like ϕ_+ and ϕ_- of (7a) and (7b) of Section 14.6. The edges

[3] The problem of determining the "Surface States Associated with a Periodic Potential", namely, wave functions which decay away in both directions from the surface of a finite crystal, is conveniently treated in terms of γ and μ; W. Shockley, *Phys. Rev.* **56**, 317–323 (1939).

[4] See Whittaker and Watson, *Modern Analysis*, Cambridge, 1927, Section 19.4.

of the allowed bands come at energies for which real values of β are possible. The equation for β is obtained as follows:

In the cell at $x = 0$, we may write any wave function satisfying equation (2) in the form

$$\psi_1 = A\mathbf{g}(x) + B\mathbf{u}(x). \tag{6}$$

Furthermore, we require that the value of ψ_1 and $d\psi_1/dx$ both be multiplied by $\exp(iPa/\hbar) = \exp(2i\beta)$ as x varies from $-a/2$ to $+a/2$. If this is true, the wave function will start at the left edge of the cell "1", centered at $x = a$, with $\exp(2i\beta)$ times its slope and value at the left edge of cell "0". Since the wave equation is second order, the solution is determined (for a fixed value of $\mathcal{E}$) if slope and value are given; consequently, the value of the wave function in cell "1" is $\exp(2i\beta)$ times its value in cell "0". The same relationship will then hold for cell "2" with respect to "1" and so on; therefore, equation (3) will be true. Since $x = 0$ is the center of symmetry of the cell, we have from the symmetry properties [(3) of Section 14.6] of $\mathbf{g}(x)$ and $\mathbf{u}(x)$,

$$\psi_1(-a/2) = A\mathbf{g}(-a/2) + B\mathbf{u}(-a/2) = A\mathbf{g}(a/2) - B\mathbf{u}(a/2) \tag{7a}$$

$$\begin{aligned}\psi_1'(-a/2) &= A\mathbf{g}'(-a/2) + B\mathbf{u}'(-a/2)\\ &= -A\mathbf{g}'(a/2) + B\mathbf{u}'(a/2).\end{aligned} \tag{7b}$$

We shall hereafter omit the argument $a/2$, except where confusion would result, it being understood that $\mathbf{g}$, $\mathbf{g}'$, $\mathbf{u}$, and $\mathbf{u}'$ are all evaluated at $+a/2$. We next write

$$\psi_1(a/2) = A\mathbf{g} + B\mathbf{u} = \mathbf{e}^{2i\beta}\psi_1(-a/2) = \mathbf{e}^{2i\beta}(A\mathbf{g} - B\mathbf{u}) \tag{8a}$$

$$\psi_1'(a/2) = A\mathbf{g}' + B\mathbf{u}' = \mathbf{e}^{2i\beta}\psi_1'(-a/2) = \mathbf{e}^{2i\beta}(-A\mathbf{g}' + B\mathbf{u}'). \tag{8b}$$

These can be rewritten as

$$(1 - \mathbf{e}^{2i\beta})\mathbf{g}A + (1 + \mathbf{e}^{2i\beta})\mathbf{u}B = 0 \tag{9a}$$

$$(1 + \mathbf{e}^{2i\beta})\mathbf{g}'A + (1 - \mathbf{e}^{2i\beta})\mathbf{u}'B = 0. \tag{9b}$$

These equations can be solved for A and B if and only if the determinant of the coefficients vanishes. This gives

$$\begin{aligned}&(1 - \mathbf{e}^{2i\beta})^2\mathbf{g}\mathbf{u}' - (1 + \mathbf{e}^{2i\beta})^2\mathbf{g}'\mathbf{u}\\ &\qquad = 4\mathbf{e}^{2i\beta}[-\sin^2\beta\mathbf{g}\mathbf{u}' - \cos^2\beta\mathbf{g}'\mathbf{u}] = 0.\end{aligned} \tag{10}$$

The square bracket must vanish, which requires that

$$\tan^2\beta = -\mathbf{g}'\mathbf{u}/\mathbf{g}\mathbf{u}' = -\boldsymbol{\gamma}/\boldsymbol{\mu}. \tag{11}$$

This equation can be solved by a real value of β if and only if $\boldsymbol{\gamma}/\boldsymbol{\mu}$ is negative. Furthermore, if β is a solution so is $-\beta$ and so is $\beta + n\pi$ where n

is an integer. Thus if $P = 2\beta\hbar/a$ is one solution, so are

$$P + 2\pi\hbar/a = P + h/a, \quad P + 2h/a, \quad P + 3h/a, \text{ etc.,} \tag{12}$$

corresponding to the periodicity discussed at the end of Section 5.5. The reason is now seen to be that P enters the theory through equations (8a) and (8b) in the form of the factor exp $(iPa/\hbar)$ which is unaltered by adding integral multiple of h/a to P.

The fact that the solutions occur in P and $-P$ pairs shows for this case that *the quantum states corresponding to* P *and* $-P$ *in the Brillouin zone have equal energies*. This result was used in a three-dimensional form in Sections 6.4 and 8.4 to show that the current is zero for a full Brillouin zone and a partially filled Brillouin zone under conditions of thermal equilibrium. The three-dimensional result is readily obtained and is stated as a problem at the end of this chapter.

No matter what the value of γ/μ is, however, there will always be two solutions which give rise to two functions of the form given in equations (3) and (4). Since the wave equation can have only two linearly independent solutions for a given energy, any solution can be expressed in terms of their sum. A special case occurs if $P = 0$ or $h/2a$ is the solution, for in this case equations (3) and (4) become equivalent. For this special case, one solution is of type (3) and has either the same value in every cell or varies from cell to cell simply by being multiplied by -1. This solution represents the limiting case obtained by adding ψ_{+P} and ψ_{-P} so as to get a cosine-like function in the energy band or a cosh-like function in the forbidden band. The other solution is not of type (3) or (4) and represents an intermediate case between a sine-like and a sinh-like function. Its value increases not by a factor but by an additive constant in going from one cell to the next so that it satisfies the condition

$$\psi(x + na) = \psi(x) + nC \tag{13}$$

where C is a constant. The actual construction of these solutions is simple and is presented in a problem.

For positive values of γ/μ we must distinguish between two cases: If $\gamma/\mu < 1$, then the solution is $\beta = \pm i\chi$ which leads to[5]

$$-\tanh^2 \chi = -\gamma/\mu \tag{14}$$

which can always be solved for a real value of χ. For this case the factor from one cell to the next is

$$e^{2i\beta} = e^{\pm 2\chi} \tag{15}$$

so that one possible solution builds up in the $+x$-direction and the other decays. For $\gamma/\mu > 1$, the solution is $\beta = \pm i\chi + \pi/2$; this yields the

[5] χ is unrelated to χ of other sections.

equation

$$-\coth^2 \chi = -\gamma/\mu \tag{16}$$

and the decay factor is

$$e^{2i\beta} = e^{\pm 2\chi + \pi i} = -e^{\pm 2\chi} \tag{17}$$

so that the solution decays as for equation (15) but alternates in sign from cell to cell.

We shall next describe how the energy bands arise from the atomic energy levels of the model of the last section as the atoms are brought together from infinite separation (that is, as a varies from large to small

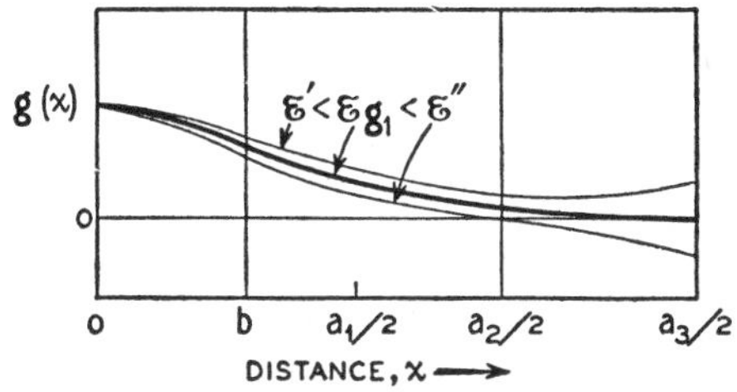

FIG. 14–6—Dependence of $\mathbf{g}(x)$ upon Changes in the Energy Parameter in Schroedinger's Equation. (Not exact.)

values), the method of attack being graphical in terms of the γ and μ curves. In Figure 14.6 we show the $\mathbf{g}(x)$ function for its "atomic" eigenvalue $\mathcal{E}_{g1}$, and for two slightly differing energies. For $\mathcal{E}_{g1}$, the value of γ is independent of a and has the value, deduced from ϕ_+ of equation (7b) of Section 14.6,

$$\gamma = \frac{d\phi_+/dx}{\phi_+} = -[2m(\mathcal{W} - \mathcal{E}_{g1})/\hbar^2]^{1/2}. \tag{18}$$

For $\mathcal{E}' < \mathcal{E}_{g1}$, the wave function is not an eigenfunction because the cosine function in the potential well does not join properly to the exponentially decaying function ϕ_+ but requires some of the "building-up" function ϕ_-. For large values of x, the ϕ_- part dominates and the value of γ is positive with a value of

$$\gamma = +[2m(\mathcal{W} - \mathcal{E}')/\hbar^2]^{1/2}. \tag{19}$$

A similar situation holds true for $\mathcal{E}'' > \mathcal{E}_{g1}$, the only difference being that ϕ_- has a coefficient of the opposite sign, a change which does not affect (19). The resultant behavior of γ as a function of $\mathcal{E}$ is shown in Figure 14.7. For a large spacing such as a_3, the value of γ is given by equation (19) except for very narrow ranges of energies centered about $\mathcal{E}_{g1}$ and about $\mathcal{E}_{g2}$, it being assumed that the well has two even energy levels. If $\mathcal{E}$ is *decreased* slightly from $\mathcal{E}_{g1}$ towards $\mathcal{E}'$, **g** increases and $d\mathbf{g}/dx$ vanishes so that γ *increases;* for a slight *increase* in $\mathcal{E}$, $\mathbf{g}(a_3)$ drops to zero, leading to $\gamma = -\infty$; and a further decrease leads to large positive values followed by a quick drop to the line of equation (19). For a smaller spacing, a_2, the effect of energy changes is less abrupt and the range $\gamma = 0$ to $\gamma = \pm\infty$

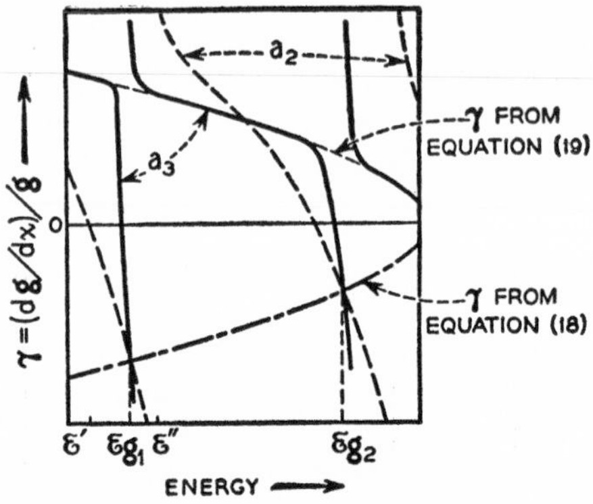

FIG. 14–7—Dependence upon Energy of $\gamma = (d\mathbf{g}/dx)/\mathbf{g}$ at the Edge of Cell, $x = a/2$. (Not exact.)

requires the range of energy $\mathcal{E}'$ to $\mathcal{E}''$. For all values of a larger than $2b$ of Figure 14.5, however, γ assumes the value of equation (18) when $\mathcal{E} = \mathcal{E}_{g1}$ and a similar statement is true for other energy levels.

Quite similar arguments apply to μ.

We must next consider the relationship between γ and μ and the resulting structure of the energy bands. In Figure 14.8 we show the structure for two values of a. On the left $a = a_3$, a large spacing. When $\mathcal{E} = \mathcal{E}_{g1}$, the **g** function decays exponentially outside the well but the **u** function does not. Consequently, μ will have the positive value given by equation (19) at this point. At $\mathcal{E} = \mathcal{E}_{u1}$, μ has a behavior similar to γ at $\mathcal{E}_{g1}$. The remaining behavior of the curves may be deduced by a set of general theorems. We shall prove the most essential theorem here and rely on qualitative arguments for the remainder of the demonstration, leaving the other theorems as problems.

THEOREM: *The* γ *and* μ *curves never have a point in common,* i.e. *they do not cross or touch.*

Proof: Suppose the two curves have a point in common. Then for the particular energy and value of a concerned,

$$\mathbf{g}'(a/2)/\mathbf{g}(a/2) = \mathbf{u}'(a/2)/\mathbf{u}(a/2). \qquad (20)$$

Next consider the function

$$\psi(x) = \frac{\mathbf{g}(a/2)}{\mathbf{u}(a/2)}\mathbf{u}(x). \tag{21}$$

Then at $x = a/2$, the two functions $\psi(x)$ and $\mathbf{g}(x)$ are equal and have equal derivatives. Since they satisfy the same second-order differential

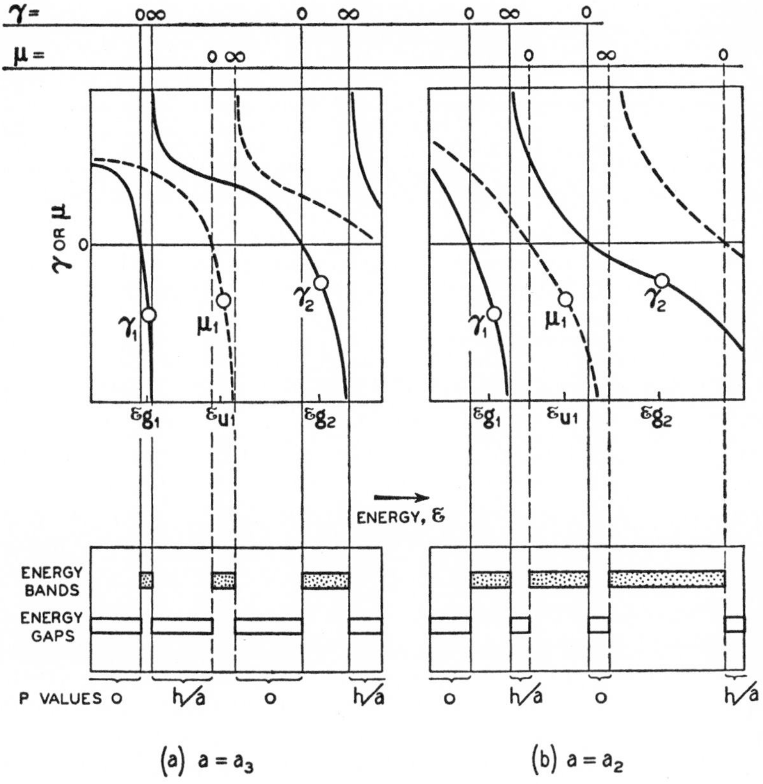

FIG. 14-8—The Structure of the Energy Bands for the Two Lattice Constants of Figures 14-6 and 14-7. (Not exact.)

equation, $\psi(x) = \mathbf{g}(x)$ for all x. But this cannot be true because at $x = 0$, $\mathbf{u} = 0$ but $\mathbf{g} \neq 0$. Therefore the assumption $\gamma = \mu$ leads to an absurdity. (The reader may generalize this by showing that $\gamma = \infty$ and $\mu = \infty$ cannot occur for the same value of $\mathcal{E}$.)

Another theorem, given in the problems, states that both $d\gamma/d\mathcal{E}$ and $d\mu/d\mathcal{E}$ are negative except when γ or $\mu = \pm\infty$.

By considering the situation near $\mathcal{E}_{g1}$ for $a = a_3$, we see that the allowed energy band arises from the region in which γ is negative (that is, where

γ/μ is negative). On the left edge of this band, $\gamma = 0$ and $P = 0$, corresponding to the center of the Brillouin zone. On the right edge, $\gamma = -\infty$ and $\beta = \pi/2$ and $P = h/2a$, corresponding to the edge of the Brillouin zone, and the wave function changes sign from cell to cell. For larger values of $\mathcal{E}$, γ is positive and larger than μ. Since the γ and μ curves cannot cross, γ remains greater than μ and γ/μ is > 1 until the situation $\mu = 0$ and $\gamma/\mu = \infty$ is reached. Between these two infinities in γ/μ, the value of β is $i\chi + \pi/2$ corresponding to the exponentially decaying or increasing type with alternating sign. In the next band, centered about $\mathcal{E}_{u1}$, P decreases from h/a to zero and, in the next gap, the exponential functions do not alternate in sign.

The behavior just discussed is in accord with diagrams like Figures 5.9 and 6.4 in which the energy curves have their minima alternately at $P = 0$ and $P = h/2a$.

It is also evident that the energy bands grow out of the atomic levels since $\mathcal{E}_{g1}$ occurs in the negative γ region, and, at least at large spacings, μ is positive there so that γ/μ is negative, corresponding to an allowed band.

From the theorem which shows that $d\gamma/d\mathcal{E}$ and $d\mu/d\mathcal{E}$ are negative and that they must go through cycles from $+\infty$ to $-\infty$, it follows that there must always be allowed bands of energy. This may be seen as follows: At any point where $\gamma = 0$, either μ is finite or infinite. If μ is finite, then γ/μ changes sign when $\gamma = 0$ so that this point corresponds to the boundary point between a band and a gap. If μ is ∞, then μ must change from $-\infty$ to $+\infty$ while γ changes from positive to negative. For this case then, energy bands exist to both sides of the point, since γ/μ is negative on both sides.

The case of the preceding sentence will actually occur for our model when a, the lattice constant, equals $2b$, the width of the well, so that the well bottom extends throughout the crystal. Under these conditions γ and μ become identical, except for a factor b, with the curves for $b\psi'(b)/\psi(b)$ of Figure 14.3. These curves have $\tan\theta$ and $\cos\theta$ as factors and one goes through zero as the other goes to infinity and vice versa. Thus for the condition $a = 2b$, there are no energy gaps and the energy spectrum is perfectly continuous, as it should be for an electron moving in free space. In this event, it is still possible, though somewhat artificial, to classify the wave functions as before, in terms of the Brillouin zone for lattice constant $2b$. If this is done, however, the energy curve becomes simply the free electron "parabola",

$$\mathcal{E} = \tfrac{1}{2}mv^2 = P^2/2m, \tag{22}$$

folded back on itself into the first Brillouin zone. (Figure 14.9 represents this case modified as discussed below.)

If a is slightly greater than $2b$, then very narrow walls will be left between the wells and the electrons will be almost free. However, when P is a multiple of $h/2a$, the Bragg reflection condition will be satisfied and the group velocity will be zero. This is produced in the γ, μ-diagram by a slight offset in either direction of the $\gamma = 0$ point from $\mu \pm \infty$ and vice versa. The effect on the free electron curves is shown in Figure 14.9.

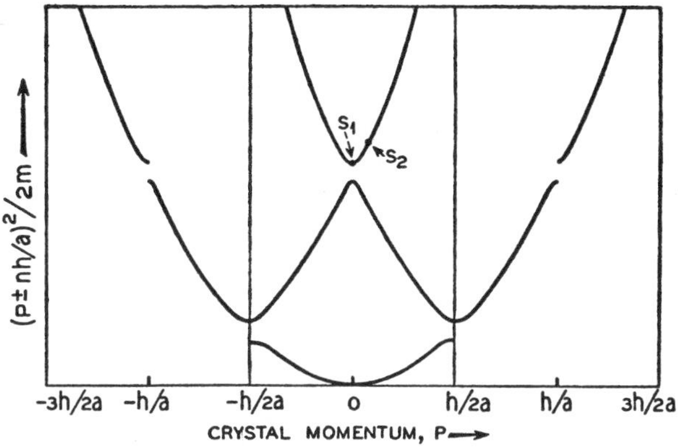

FIG. 14-9—Energy Bands for Almost Free Electrons. (Not exact.)

(See the Problems for details.) The curves are characterized by the sharp curvatures at the band edges. These may be interpreted[6] as corresponding to a very small effective mass, since near these points

$$v = d\mathcal{E}/dP \tag{23}$$

changes rapidly with small changes in P so that

$$\frac{dv}{dP} \gg 1/m \tag{24}$$

where m is the free electron mass and $1/m$ corresponds to the curvature of the free-electron parabola.

The smallness of the mass suggests the following problem: Suppose an electron is in a wave-packet state corresponding to point S_1, in Figure 14.9, the minimum of the third energy band. If a field is applied, the wave-packet will be rapidly accelerated and will move across the crystal in much less time than would a perfectly free electron initially at rest. How can this come about? Where does the electron in the third band get its extra energy? The answer is that it had the extra energy already. State S_1 can be shown to correspond to the superposition of two running waves corresponding to $P = +h/a$ and $P = -h/a$. Thus, although

[6] The basic theory for group velocity is given in Sections 15.2 and 15.4.

state S_1 has no net electron current, it represents a state of considerable energy. The effect of the applied field is to shift the electron to state S_2 by causing changes in the reflection conditions of the two running waves. As a result, all of the kinetic energy becomes associated with motion towards the right and the electron wave-packet is greatly accelerated. Once state S_2 is reached, however, the transfer of kinetic energy from one motion to another is complete, and the effective mass for additional acceleration has its normal value.

Ordinarily the masses for holes and electrons will be of the order of magnitude of the electron's mass. If the almost free electron approximation is used to represent the energy bands of the semiconductor, an erroneous impression is given of extreme smallness for the effective mass at the edges of the bands. The difference between the actual case and the almost free electron case comes about as follows: For the free electron case the zeros of $\boldsymbol{\gamma}$ are paired with the infinities of $\boldsymbol{\mu}$ and vice versa. The small disturbance in potential offsets these slightly and gives very narrow energy gaps with resultant very small effective masses at the band edges. (See Problems.) For true valence-bond wave functions and conduction-band wave functions the situation will be quite different and, in general, there will be no simple relationships between $\boldsymbol{\gamma}$ and $\boldsymbol{\mu}$, or rather their three-dimensional analogues. At atomic spacings the bands will spread by energies comparable to the binding energy, that is, to a few electron volts, and the energy gaps will be of the same general order. From dimensional arguments we thus expect the *effective mass* to satisfy approximately an equation of the form

$$\Delta\mathcal{E} = P_{\max}{}^2/2m_{\text{eff}} = h^2/8a^2m_{\text{eff}}.$$

Since $\Delta\mathcal{E} \doteq 0.1$ atomic unit and $h/a \doteq 2\pi\hbar/4a_0 = (\pi/2)\hbar/a_0 = (\pi/2)$ atomic units of momentum,[7] this leads to

$$m_{\text{eff}} \doteq (\pi/2)^2/8 \times 0.1 \doteq 3 \text{ electron masses.} \tag{25}$$

This argument is intended to show only that effective masses of the order of the electron's mass should be the rule rather than the exception in well-bonded semiconductors.

The discussion of the number of eigenfunctions in each energy band for periodic boundary conditions given in Chapter 5 is directly applicable to the wave functions of the one-dimensional periodic potential. The problem of counting for the case of an even number of unit cells involves the equivalence of the wave functions for $P = h/2a$ and $P = -h/2a$; for this condition there is only one solution which can satisfy periodic boundary conditions since the other solution varies as described in equation (13) and Problem 2.

[7] See Appendix A.

14.8 BLOCH FUNCTIONS FOR THREE DIMENSIONS

In this section we shall derive Bloch's theorem that the wave functions in a crystal are of the form

$$\psi(\boldsymbol{r}) = e^{i\boldsymbol{P}\cdot\boldsymbol{r}/\hbar}\mathbf{u}_{\boldsymbol{P}}(\boldsymbol{r}) \qquad (1)$$

where $\mathbf{u}_{\boldsymbol{P}}(\boldsymbol{r})$ has the periodicity of the lattice.[1] We shall restrict the proof to a crystal with periodic boundary conditions, for which case $\boldsymbol{P}$ is a real vector. We shall also verify in general that the volume of the Brillouin zone is h^3/V_a where V_a is the volume of the unit cell and that the density of allowed points is V_A/h^3 where V_A is the volume of the crystal, and hence that the number of states (neglecting the spin) in the Brillouin zone is $V_A/V_a = N$ where N is the number of unit cells in the crystal.

The basic property of the crystal is that it can be translated by certain vectors so that after translation each atom has been moved into the position occupied by an equivalent atom. If the potential energy of the electron is $\mathcal{U}(\boldsymbol{r})$, then for a permitted translation, for which we shall use the symbol $\boldsymbol{R}$ or $\boldsymbol{R}_\alpha$ in this section, the relationship

$$\mathcal{U}(\boldsymbol{r} + \boldsymbol{R}) = \mathcal{U}(\boldsymbol{r}) \qquad (2)$$

must hold.

The set of allowed translations $\boldsymbol{R}$ forms a *lattice,* which is defined as follows: Take any point in space as the origin and consider all points displaced from it by vectors of the form $\boldsymbol{R}$. The array so obtained is the lattice corresponding to the crystal structure. (In Figure 1.3, if a corner of the large cube shown is chosen as the origin of the lattice, then the atoms on the faces are at lattice points; the atoms on the interior, however, are not at lattice points.) The $\boldsymbol{R}$'s can always be described in terms of a set of *primitive vectors.*

$$\boldsymbol{a}_1, \boldsymbol{a}_2, \boldsymbol{a}_3 = \text{primitive vectors.} \qquad (3)$$

The primitive vectors themselves must be vectors of the lattice and, furthermore, it must be possible to express every vector $\boldsymbol{R}$ of the lattice in the form

$$\boldsymbol{R} = \alpha_1\boldsymbol{a}_1 + \alpha_2\boldsymbol{a}_2 + \alpha_3\boldsymbol{a}_3 \text{ (α's are integers)} \qquad (4)$$

where the α's are positive or negative integers. The primitive vectors define a parallelepiped, of which they are the edges, and this parallelepiped has a lattice point at every corner. Furthermore, it can have no lattice point except at corners because, if it did, in the set of α's for this point at least one α would lie between 0 and 1 in contradiction to (4). From this reasoning, it follows that there is exactly one lattice point for each parallele-

[1] This is the customary use of **u** in this book. **u** for *ungerade* is restricted to Sections 14.6 and 14.7 and the Problems.

piped having edges $\boldsymbol{a}_1$, $\boldsymbol{a}_2$, $\boldsymbol{a}_3$. Furthermore all space can obviously be filled by repeating these parallelepipeds by translation of the form (4). For this reason, the parallelepiped is referred to as a *unit cell.* The crystal can evidently be divided into such cells, each of which will have the same internal structure and it will be impossible to divide these cells further into equivalent parts which can be made to coincide by translation.

The volume of the unit cell is

$$V_a = \boldsymbol{a}_1 \cdot (\boldsymbol{a}_2 \times \boldsymbol{a}_3) = \boldsymbol{a}_2 \cdot (\boldsymbol{a}_3 \times \boldsymbol{a}_1) = \boldsymbol{a}_3 \cdot (\boldsymbol{a}_1 \times \boldsymbol{a}_2) \tag{5}$$

where it is assumed that $\boldsymbol{a}_1$, $\boldsymbol{a}_2$, $\boldsymbol{a}_3$ form a right-handed system. (If they do not, a new set with $\boldsymbol{a}_1' = -\boldsymbol{a}_1$ can be chosen.) From the previous discussion, it is evident that V_a is the volume per lattice point and must, therefore, have the same value for every set of primitive vectors. In

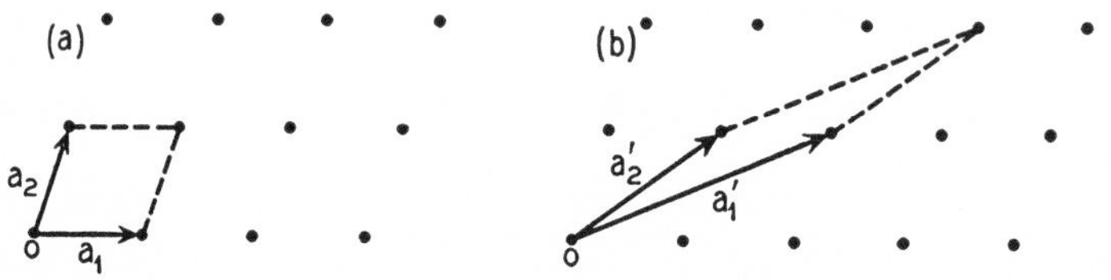

FIG. 14–10—Two Sets of Primitive Vectors in a Two-dimensional Lattice.

Figure 14.10 we show two ways of selecting primitive vectors in a two-dimensional lattice. The reader should verify that $\boldsymbol{a}_1$ and $\boldsymbol{a}_2$ can be expressed in form (4) in terms of $\boldsymbol{a}_1'$ and $\boldsymbol{a}_2'$ and vice versa.

We now suppose that the wave functions satisfy periodic boundary conditions, the periods being

$$\boldsymbol{A}_1 = N_1\boldsymbol{a}_1, \quad \boldsymbol{A}_2 = N_2\boldsymbol{a}_2, \quad \boldsymbol{A}_3 = N_3\boldsymbol{a}_3, \tag{6}$$

so that

$$\psi(\boldsymbol{r} + \boldsymbol{A}) = \psi(\boldsymbol{r}), \tag{7}$$

where $\boldsymbol{A}$ is any linear combination of the $\boldsymbol{A}_i$'s with positive or negative integers as coefficients. The volume of the crystal, denoted by V_A, is related to V_a as follows:

$$V_A = \boldsymbol{A}_1 \cdot (\boldsymbol{A}_2 \times \boldsymbol{A}_3) = N_1N_2N_3V_a = NV_a. \tag{8}$$

Suppose $\psi_0(\boldsymbol{r})$, satisfying the boundary conditions of equation (7), is a solution of Schroedinger's equation with eigenvalue $\mathcal{E}$. (The existence of such solutions can be proved without too much difficulty.) Then it is evident that

$$\psi_R(\boldsymbol{r}) = \psi_0(\boldsymbol{r} + \boldsymbol{R}) \tag{9}$$

is also a solution for the same value of $\mathcal{E}$ since it simply represents displacing ψ_0 by a vector $-\boldsymbol{R}$ and the values of $\mathcal{U}(\boldsymbol{r})$ and $\mathcal{U}(\boldsymbol{r} + \boldsymbol{R})$ are equal.

Our problem is *to prove that if $\psi_0(r)$ exists, it can always be expressed as a sum of one or more Bloch-type functions which satisfy the condition*

$$\phi_B(r + R) = e^{2\pi i B \cdot R}\,\phi_B(r) \tag{10}$$

$$B \equiv P/h \tag{11}$$

and are eigenfunctions with energy $\mathcal{E}$. This means that for the case of periodic boundary conditions it is unnecessary to consider functions other than those of the Bloch type, since all solutions of the problem can be expressed in terms of the latter.

In order to deal with this problem we introduce two additional sets of vectors g_i and G_i, defined as follows:

$$G_1 = a_2 \times a_3/V_a, \quad G_2 = a_3 \times a_1/V_a, \quad G_3 = a_1 \times a_2/V_a. \tag{12}$$

The G_i's are called the *reciprocal vectors* of the crystal lattice. The vectors

$$g_1 = A_2 \times A_3/V_A, \quad g_2 = A_3 \times A_1/V_A, \quad g_3 = A_1 \times A_2/V_A, \tag{13}$$

are reciprocal vectors of the periodic boundary periods and have no special name. It follows from these definitions that

$$a_i \cdot G_j = \delta_{ij} = A_i \cdot g_j \tag{14}$$

and also that the volumes in G-space, or reciprocal space, of the parallelepipeds with edges G_i and g_i are

$$V_G = G_1 \cdot G_2 \times G_3 \quad \text{and} \quad V_g = g_1 \cdot g_2 \times g_3. \tag{15}$$

It also follows that there are inverse relationships of the form

$$a_1 = G_2 \times G_3/V_G, \quad A_1 = g_2 \times g_3/V_g, \tag{16}$$

as can be proved by noting that a_1 satisfies $a_1 \cdot G_j = \delta_{1j}$ and that this uniquely determines a vector a_1. From the vector theorem

$$u \times (v \times w) = (u \cdot w)v - (u \cdot v)w \tag{17}$$

it follows that $a_1 \cdot a_2 \times a_3$ when expressed in terms of the G's reduces to

$$\begin{aligned} V_a &= (G_2 \times G_3) \cdot [(G_3 \times G_1) \times (G_1 \times G_2)]/V_G^3 \\ &= (G_2 \times G_3) \cdot \{[(G_3 \times G_1) \cdot G_2]G_1 - [(G_3 \times G_1) \cdot G_1]G_2\}/V_G^3 \\ &= V_G^2/V_G^3 = 1/V_G, \end{aligned} \tag{18}$$

since $(G_3 \times G_1) \cdot G_1$ vanishes. A similar relationship holds between V_A and V_g:

$$V_A V_g = 1. \tag{19}$$

There are N distinct translations subject to the periodic boundary conditions and these may be written as

$$R_\alpha = \alpha_1 a_1 + \alpha_2 a_2 + \alpha_3 a_3 \tag{20}$$

where the α's are integers running from 0 to $N_1 - 1$, $N_2 - 1$ and $N_3 - 1$ respectively. There is no need to consider $\boldsymbol{R}$'s with α's larger than these because such an $\boldsymbol{R}$, with $\alpha_1 = N_1 + 2$ for example, would produce the same effect as $\boldsymbol{R}' = \boldsymbol{R} - \boldsymbol{A}_1$ which has $\alpha_1' = 2$ lying in the range $0 \leq \alpha_1' \leq N_1 - 1$.

There is a similar set of $\boldsymbol{B}$'s, also N in number, defined by

$$\boldsymbol{B}_\gamma = \gamma_1 \boldsymbol{g}_1 + \gamma_2 \boldsymbol{g}_2 + \gamma_3 \boldsymbol{g}_3, \quad 0 \leq \gamma_i \leq N_i - 1. \tag{21}$$

We shall show that these $\boldsymbol{B}$'s are the appropriate $\boldsymbol{B}$'s for the Bloch functions.

In terms of the set (20) of $\boldsymbol{R}_\alpha$ and set (21) of $\boldsymbol{B}_\gamma$ we shall next define two sets of functions generated from $\psi_0(\boldsymbol{r})$ by translations and in terms of these new functions we shall prove that $\psi_0(\boldsymbol{r})$ can be expanded in terms of one or more functions of the Bloch type. The new functions, denoted by ψ_α and ϕ_γ, where α and γ are triplets of integers of the form (20) and (21), are defined as follows:

$$\psi_\alpha(\boldsymbol{r}) = \psi_0(\boldsymbol{r} - \boldsymbol{R}_\alpha) \tag{22}$$

and

$$\phi_\gamma = N^{-1/2} \sum_\alpha e^{2\pi i \boldsymbol{B}_\gamma \cdot \boldsymbol{R}_\alpha} \psi_\alpha \equiv \sum_\alpha S_{\gamma\alpha} \psi_\alpha \tag{23a}$$

where the $S_{\gamma\alpha}$ quantities are matrix elements for the transformation relating the ψ_α to the ϕ_γ and

$$S_{\gamma\alpha} \equiv N^{-1/2} e^{2\pi i \boldsymbol{B}_\gamma \cdot \boldsymbol{R}_\alpha}. \tag{23b}$$

It is evident that each of the set of ϕ's is an eigenfunction for the same energy $\mathcal{E}$ since each ψ_α is such an eigenfunction.

We shall next prove that

$$\phi_\gamma(\boldsymbol{r} + \boldsymbol{R}) = \exp(2\pi i \boldsymbol{B}_\gamma \cdot \boldsymbol{R}) \phi_\gamma(\boldsymbol{r}) \tag{24}$$

which shows that ϕ_γ *is of the Bloch type with* $\boldsymbol{P} = 2\pi\hbar \boldsymbol{B}_\gamma = h\boldsymbol{B}_\gamma$. *Finally, we shall prove that*

$$\psi_\alpha = N^{-1/2} \sum_\beta \exp(-2\pi i \boldsymbol{B}_\beta \cdot \boldsymbol{R}_\alpha) \phi_\beta \equiv \sum_\beta T_{\alpha\beta} \phi_\beta \tag{25}$$

which shows that any ψ_0, *or* ψ_α *obtained from it, can be expressed as a sum of the Bloch-type* ϕ*'s.*

To prove that ϕ_γ is of the Bloch type, we evaluate $\sqrt{N}\phi_\gamma(\boldsymbol{r} + \boldsymbol{R})$ as follows:

$$\begin{aligned} \sqrt{N}\phi_\gamma(\boldsymbol{r} + \boldsymbol{R}) &= \sum_\alpha \exp(2\pi i \boldsymbol{B}_\gamma \cdot \boldsymbol{R}_\alpha) \psi_0(\boldsymbol{r} + \boldsymbol{R} - \boldsymbol{R}_\alpha) \\ &= \sum_{\alpha'} \exp[2\pi i \boldsymbol{B}_\gamma \cdot (\boldsymbol{R}_{\alpha'} + \boldsymbol{R})] \psi_0(\boldsymbol{r} - \boldsymbol{R}_{\alpha'}) \\ &= \exp(2\pi i \boldsymbol{B}_\gamma \cdot \boldsymbol{R}) \sum_{\alpha'} \exp(2\pi i \boldsymbol{B}_\gamma \cdot \boldsymbol{R}_{\alpha'}) \psi_{\alpha'}(\boldsymbol{r}) \end{aligned} \tag{26}$$

where we have replaced $\boldsymbol{R}_\alpha - \boldsymbol{R}$ by $\boldsymbol{R}_{\alpha'}$ so that the sum over α' now extends outside the allowed set of α values. Each $\boldsymbol{R}_{\alpha'}$ outside is, of course, equivalent to an $\boldsymbol{R}_\alpha$ in the set which differs by an $\boldsymbol{A}$ vector. Replacing an outside $\boldsymbol{R}_{\alpha'} = \boldsymbol{R}_\alpha + \boldsymbol{A}_1$, for example, by the inside $\boldsymbol{R}_\alpha$ does not change the sum because

$$e^{2\pi i\boldsymbol{B}_\gamma\cdot\boldsymbol{A}_1} = \exp(2\pi i\gamma_1\boldsymbol{g}_1\cdot\boldsymbol{A}_1) = \exp(2\pi i\gamma_1) = 1 \tag{27}$$

so that the exponential is unchanged. Furthermore, $\psi_\alpha = \psi_{\alpha'}$, if $\boldsymbol{R}_{\alpha'} = \boldsymbol{R}_\alpha + \boldsymbol{A}_1$. Thus α' may be replaced by α in the sum of equation (26), so that (26) becomes

$$\sqrt{N}\phi_\gamma(\boldsymbol{r}+\boldsymbol{R}) = \exp(2\pi i\boldsymbol{B}_\gamma\cdot\boldsymbol{R})\sqrt{N}\phi_\gamma(\boldsymbol{r}), \tag{28}$$

which is the desired theorem of equation (24) except for the factor $N^{1/2}$.

Next we shall prove that any ψ_ϵ can be expanded in terms of the ϕ_β by proving that ψ_ϵ is equal to the sum

$$N^{-1/2}\sum_\beta \exp(-2\pi i\boldsymbol{B}_\beta\cdot\boldsymbol{R}_\epsilon)\phi_\beta = \frac{1}{N}\sum_\beta\sum_\gamma \exp[2\pi i\boldsymbol{B}_\beta\cdot(\boldsymbol{R}_\gamma - \boldsymbol{R}_\epsilon)]\psi_\gamma, \tag{29}$$

the latter equality resulting from the replacement of ϕ_β by equation (23). The term in the exponent may be written as

$$\begin{aligned} 2\pi i(\beta_1\boldsymbol{g}_1 + \beta_2\boldsymbol{g}_2 + \beta_3\boldsymbol{g}_3)\cdot(\alpha_1\boldsymbol{a}_1 + \alpha_2\boldsymbol{a}_2 + \alpha_3\boldsymbol{a}_3) \\ = 2\pi i[(\beta_1\alpha_1/N_1) + (\beta_2\alpha_2/N_2) + (\beta_3\alpha_3/N_3)], \end{aligned} \tag{30}$$

where the α's are integers (possibly negative) because of the form of $\boldsymbol{R}_\gamma$ and $\boldsymbol{R}_\epsilon$. For a fixed value of γ, the sum over β is equivalent to a triple sum over β_1, β_2, and β_3 and may be written as

$$\begin{aligned} \sum_{\beta_1}\sum_{\beta_2}\sum_{\beta_3} &\exp\{2\pi i[(\beta_1\alpha_1/N_1) + (\beta_2\alpha_2/N_2) + (\beta_3\alpha_3/N_3)]\} \\ &= \{1 + \exp[(2\pi i\alpha_1/N_1)\beta_1] + \cdots + \exp[(2\pi i\alpha_1/N_1)(N_1 - 1)]\} \\ &\times \{1 + \cdots + \exp[(2\pi i\alpha_2/N_2)(N_2 - 1)]\} \\ &\times \{1 + \cdots + \exp[(2\pi i\alpha_3/N_3)(N_3 - 1)]\}. \end{aligned} \tag{31}$$

The first term is N_1 if $\alpha_1 = 0$ or $\alpha_1 = \pm N_1$. However, the latter condition does not arise from $\boldsymbol{R}_\gamma - \boldsymbol{R}_\epsilon$. If α_1 is not zero, the first term vanishes as may be seen by writing it in the standard form for a geometric series;

$$\{1 - \exp[(2\pi i\alpha_1/N_1)\cdot N_1]\}/[1 - \exp(2\pi i\alpha_1/N_1)], \tag{32}$$

in which the numerator vanishes, while the denominator does not. Hence, the triple sum over β_1, β_2, β_3 vanishes unless $\alpha_1 = \alpha_2 = \alpha_3 = 0$, that is, unless $\boldsymbol{R}_\epsilon = \boldsymbol{R}_\gamma$. If $\boldsymbol{R}_\epsilon = \boldsymbol{R}_\gamma$, the sum becomes $N_1N_2N_3 = N$ so that we may write equation (29) in the form

$$N^{-1/2}\sum_\beta \exp(-2\pi i\boldsymbol{B}_\beta\cdot\boldsymbol{R}_\epsilon)\phi_\beta = \sum_\gamma \delta_{\epsilon\gamma}\psi_\gamma = \psi_\epsilon. \tag{33}$$

This proves that ψ_ϵ can be expanded in terms of the ϕ_β which was the basic aim set forth in equation (10).

Furthermore, for periodic boundary conditions characterized by $\boldsymbol{A}_1$, $\boldsymbol{A}_2$, and $\boldsymbol{A}_3$, Bloch-type functions ϕ_γ with non-integral values for the coefficients γ_i are not allowed because, if γ_1 were non-integral, then we should have

$$\phi_\gamma(\boldsymbol{r} + \boldsymbol{A}_1) = \exp\,(2\pi i\gamma_1)\phi_\gamma(\boldsymbol{r}), \tag{34}$$

and, if γ_1 were non-integral, the coefficient would not be unity and ϕ_γ would not satisfy the boundary conditions. Thus all *allowed* $\boldsymbol{B}$ vectors, that is, *allowed* by the boundary conditions, correspond to integer values for the γ_i's.

Equation (33) also shows that the matrices $S_{\beta\alpha}$ of equation (23) and $T_{\gamma\beta}$ of equation (25) are reciprocal since

$$\sum_\beta T_{\gamma\beta} S_{\beta\alpha} = \delta_{\gamma\alpha}. \tag{35}$$

In addition

$$T_{\gamma\beta} = S_{\beta\gamma}{}^* \tag{36}$$

so that the transformation is *unitary*, a feature which leads to simplifying results in many cases but which we do not use further here. (See Problem 3 of Section 17.3.)

Next we shall see how the Brillouin zone may be defined in terms of the $\boldsymbol{G}_i$ vectors and examine the equivalence of different $\boldsymbol{B}$ vectors. So far as their translational properties are concerned we shall show that $\phi_{\boldsymbol{B}}$ and $\phi_{\boldsymbol{B}'}$ are equivalent if $\boldsymbol{B}$ and $\boldsymbol{B}'$ differ by integral multiples of the $\boldsymbol{G}_i$'s, that is, by a *vector of the reciprocal lattice of the* $\boldsymbol{G}_i$'s, the $\boldsymbol{G}_i$'s being the *reciprocal vectors* of the crystal lattice as defined in equation (12). The result follows at once because in

$$\phi_{\boldsymbol{B}}(\boldsymbol{r} + \boldsymbol{R}) = \exp\,(2\pi i\boldsymbol{B}\cdot\boldsymbol{R})\phi_{\boldsymbol{B}}(\boldsymbol{r}) \tag{37}$$

the exponential is unaltered by adding $\boldsymbol{G}_1$, for example, to $\boldsymbol{B}$ since

$$\boldsymbol{G}_1\cdot\boldsymbol{R} = \boldsymbol{G}_1\cdot(\alpha_1\boldsymbol{a}_1 + \alpha_2\boldsymbol{a}_2 + \alpha_3\boldsymbol{a}_3) = \alpha_1 \tag{38}$$

so that the added exponent is a multiple of $2\pi i$. Hence we say that two $\boldsymbol{B}$ vectors which differ by multiples of the $\boldsymbol{G}_i$'s are *equivalent.* Consequently, it suffices to consider only those $\boldsymbol{B}$'s lying in the unit cell having edges $\boldsymbol{G}_1$, $\boldsymbol{G}_2$, and $\boldsymbol{G}_3$ and a volume

$$\boldsymbol{G}_1\cdot(\boldsymbol{G}_2\times\boldsymbol{G}_3) = V_G = 1/V_a. \tag{39}$$

$\boldsymbol{B}$'s lying outside of the cell are equivalent to $\boldsymbol{B}$'s inside it. This cell contains just the set of N vectors $\boldsymbol{B}$ of equation (21), since $\boldsymbol{B}$'s with $\gamma_1 = N_1$, etc., are equivalent to those with $\gamma_1 = 0$, etc. The density of these allowed points in $\boldsymbol{G}$-space is thus

$$N/V_G = NV_a = V_A \tag{40}$$

which is identical, of course, with the reciprocal of the volume of a unit cell of the $\boldsymbol{g}_i$ lattice upon which the allowed $\boldsymbol{B}$'s fall.

Since the $\boldsymbol{P}$ vectors are h times the $\boldsymbol{B}$ vectors,

$$\left.\begin{array}{c}\text{the density of allowed}\\ \text{points in } \boldsymbol{P}\text{-space}\end{array}\right\} = V_A/h^3, \tag{41}$$

or $2V_A/h^3$ if the two possibilities for the spin are taken into account. This formula is the same as that derived in Section 5.5 for the simple cubic lattice.

It is possible to translate the unit cell of $\boldsymbol{G}$- or $\boldsymbol{P}$-space so as to center it about the point $\boldsymbol{P} = 0$; it will then still contain N distinct allowed $\boldsymbol{B}$ vectors. However, the unit cell is somewhat arbitrary because the basic $\boldsymbol{a}_i$ and $\boldsymbol{G}_i$ vectors can be chosen in several ways. The first *Brillouin zone*, however, does not have this degree of arbitrariness and is defined in the following way: Make a lattice in $\boldsymbol{G}$-space by taking all points of the form $n_1\boldsymbol{G}_1 + n_2\boldsymbol{G}_2 + n_3\boldsymbol{G}_3$ for positive and negative integer values of the n's. This array of points is called the *reciprocal lattice*. Assign to each lattice point that part of $\boldsymbol{G}$-space nearer to it than to any other point. Then the space around the point at the origin is the first Brillouin zone. Its special importance is that for $\boldsymbol{B}$ vectors ending on its surface the conditions for Bragg reflection are satisfied. For the case of almost free electrons, the surface becomes an energy discontinuity, or energy gap, and the component of group velocity in a direction perpendicular to the surface vanishes when $\boldsymbol{B}$ ends on the surface.

It may be shown (see Problem 15) that no two points in the first Brillouin zone are equivalent. Furthermore, the volume of the zone is that of one unit cell in the reciprocal lattice. Hence, it must contain a complete set of N distinct $\boldsymbol{B}$'s.

The construction of Brillouin zones is treated in considerable detail in a number of standard texts.[2]

Problems

1. Discuss the relationship between the following manifestations of a common cause: For a large value of a, such as a_3 in Figures 14.6 and 14.7, (1) the energy bands are narrow, (2) the effective mass in the bottom band is large, and (3) the transmission of an electron wave-packet is small.

2. Show that at the edges and center of the Brillouin zone, the solutions are of the form given in equation (13) of Section (14.7). For example,

[2] A. H. Wilson, *The Theory of Metals*, Cambridge at the University Press, 1936. N. F. Mott and H. Jones, *The Theory of the Properties of Metals and Alloys*, Oxford at the Clarendon Press, 1936. F. Seitz, *Modern Theory of Solids*, McGraw-Hill Book Co., New York, 1940. L. Brillouin, *Wave Propagation in Periodic Structures*, McGraw-Hill Book Co., New York, 1946.

consider the lowest state of the first band corresponding to $\gamma = 0$, or $\mathbf{g}'(a/2) \equiv \mathbf{g}' = 0$. Show that one solution is

$$\psi_1(x) = A\mathbf{g}(x - na) \text{ for } x \text{ in cell centered at } x = na,$$

which corresponds to $P = 0$, and that another is, similarly,

$$\psi_2(x) = \mathbf{u}(x - na) + 2n[\mathbf{u}(a/2)/\mathbf{g}(a/2)]\mathbf{g}(x - na).$$

What happens if a multiple of ψ_1 is added to ψ_2? Find the corresponding solutions for other zone edges and show that they are transition types between cases for real P and cases for complex or imaginary P.

3. Prove that the extreme energies in each band come at $P = \pm h/2a$ and $P = 0$ and that the energy varies monotonically between these limits. Do this by using the relationship between velocity defined by $v = d\mathcal{E}(P)/dP$ and current discussed in connection with equation (4) of Section 14.3, which may be written as

$$\frac{\hbar}{im}\left[\psi^* \frac{d}{dx}\psi - \psi\frac{d}{dx}\psi^*\right].$$

If $v = 0$, corresponding to a flat region in the $\mathcal{E}$ versus P curve, then show that

$$\frac{d \ln \psi}{dx} = \frac{d \ln \psi^*}{dx}.$$

From this show that P is either zero or a multiple of $h/2a$. (It is thought by most students that this theorem has no three-dimensional analogue and that energy extremes may occur inside the Brillouin zone.)

4. Show from Problems 2 and 3 that at the edge of the energy band the average current is zero and hence by the group-velocity formula that $d\mathcal{E}/dP = 0$ at the edge of the band.

5. Consider the γ, μ-curves for an "almost free" electron. For this purpose, recenter the unit cell in the midpoint of the narrow section which is $\mathcal{W}$ high and $a - 2b = 2c$ wide. The potential is then

$$\mathcal{U}(x) = \mathcal{W} \qquad |x| < c$$

$$\mathcal{U}(x) = 0 \qquad c < |x| < a/2.$$

To simplify the problem, let $c \to 0$ and $\mathcal{W} \to \infty$ so that $c\mathcal{W} = V$ remains constant. Prove that

$$\mathbf{u}(x) = \sin\,(2m\mathcal{E}/\hbar^2)^{1/2}x \qquad |x| < a$$

$$\mathbf{g}(x) = \cos\,[(2m\mathcal{E}/\hbar^2)^{1/2}|x| - \alpha] \quad |x| < a$$

where

$$\alpha = \sqrt{2m/\mathcal{E}}\; V/\hbar.$$

From these show that the lowest energy state ($P = 0$) has energy $\mathcal{E} = 2V/a$, equal to the average potential energy throughout unit cell. Show that the energy gaps are $4V/a$ wide. Verify that the curves will have the shape of Figure 14.9. From this conclude that the effective mass is of the order of $mV/a\mathcal{E}$ where $\mathcal{E}$ is the free electron energy, corresponding to $V = 0$. Derive an exact formula for the effective mass at the edge of the zone. [The equality of the energy gaps is a consequence of letting $c \rightarrow 0$; for a more realistic potential the gaps become narrower at high energies.]

6. In Figures 14.7 and 14.8, cases in which $\mathcal{E}$ was less than $\mathcal{W}$ were stressed. How will the γ and μ curves behave for $\mathcal{E} > \mathcal{W}$? It may be helpful to consider Problems 9 to 13 first.

7. Prove that $\mathcal{E}(P) = \mathcal{E}(-P)$ for the one-dimensional case for a periodic potential which is not necessarily symmetrical about the center of the unit cell. Do this by writing ψ_P in the form $\psi_P = \exp\,(iPx/\hbar)\mathbf{u}_P(x)$ where $\mathbf{u}_P(x + a) = \mathbf{u}_P(x)$. Extend the result to three dimensions.

8. Show that, for a symmetrical potential, the function $\mathbf{u}_P(x)$ of Problem 7 can always be written as a real even function of x plus an imaginary odd function of x.

Remark: The following four problems give the steps necessary to prove that the γ and μ-curves used extensively in Section 14.8 decrease monotonically with energy from $+\infty$ to $-\infty$. The theorem that they do not cross was proved in Section 14.7. These features are all that is necessary to show that the energy band structure discussed in Figure 14.8 is a general phenomenon of a one-dimensional symmetrical periodic potential. Problem 13 shows how the results can be extended to a general periodic potential.

9. Given a periodic symmetrical potential $\mathcal{U}(x)$ where $\mathcal{U}(x) = \mathcal{U}(-x)$ and $\mathcal{U}(x + a) = \mathcal{U}(x)$ which is finite so that $0 \leq \mathcal{U}(x) \leq \mathcal{W}$, prove that, for $\mathcal{E} < 0$, γ and μ (defined in equation (1), Section 14.7) are both > 0 and that $\mathbf{g}(x)$ and $\mathbf{u}(x)$ have no nodes (zeros) in the unit cell. Prove that for $\mathcal{E}$ large enough, order of $\hbar^2/a^2m$, $\mathbf{g}$ and $\mathbf{u}$ must have nodes in the unit cell and large numbers of nodes for $\mathcal{E}$ very large.

10. Suppose that a wave function $\psi_1(x)$ which satisfies Schroedinger's equation for energy $\mathcal{E}$ in a bounded potential energy field, like that of Problem 9, has two consecutive nodes, at $x = a_1$ and $x = b_1 > a_1$. Consider ψ_2, also having energy $\mathcal{E}$, with a node at a_2 slightly less than a_1; prove that it has a node at $b_2 < b_1$. For this purpose, consider $\eta_1(x) = [d\psi_1(x)/dx]/\psi_1(x)$. Prove that η_1 varies from $+\infty$ for $x = a_1 + 0$ to $-\infty$ for $x = b_1 - 0$ and is continuous except for the end points where it varies as $\eta_1 = 1/(x - a_1)$ and $1/(x - b_1)$. Prove that $\eta_2(x)$, corresponding to ψ_2, and $\eta_1(x)$ are never equal. Hence prove that, since $\eta_2(a_1 + 0) < \eta_1(a_1 + 0)$, η_2 must go to $-\infty$ for $x < b_1$.

11. Consider the func[illegible]roblem 10, which corresponds to energy $\mathcal{E}_1$. Let ψ_2 also have a node at a_1 and be a solution for the wave equation for energy $\mathcal{E}_2$ where $\mathcal{E}_2$ is slightly greater than $\mathcal{E}_1$. Prove that the node of ψ_2 occurs at $b_2 < b_1$.

Hint: Make a diagram showing $\eta_1(x)$ varying from $+\infty$ at a_1 to $-\infty$ at b_1. Prove that $\eta_2 = \eta_1 = 1/(x - a_1)$ as $x \rightarrow a_1$ but that $\eta_2(x) < \eta_1(x)$ for $x - a_1$ finite. Show that if $b_2 \geq b_1$, then $\eta_2(x) > \eta_1(x)$ for $b_1 - x$ finite. Hence b_2 must be less than b_1. In other words, increasing the energy decreases the distance between nodes (like decreasing wave length). It may be helpful to derive the equation

$$\frac{d\eta_1}{dx} = \frac{2m}{\hbar^2}[\mathcal{U}(x) - \mathcal{E}_1] - \eta_1^2$$

for use in this problem.

12. From Problems 9, 10, and 11 show that γ and μ curves, like those of Figures 14.7 and 14.8, must decrease monotonically with increasing energy. (For example, consider $\mathbf{u}(x)$ which has a node at $x = 0$, and suppose that the energy is so low that it has no node inside the unit cell. Then the method of Problem 9 shows at once that increasing $\mathcal{E}$ causes μ to decrease at the edge of the cell. The cases with several nodes may be treated by considering the effect of increasing $\mathcal{E}$ upon the node spacing and upon the η-curve between the nodes.)

13. Prove for a general (possibly unsymmetrical) periodic potential $\mathcal{U}(x + a) = \mathcal{U}(x)$ that the solutions are of the form

$$\psi_1(x + a) = \lambda\psi_1(x)$$

$$\psi_2(x + a) = \lambda^{-1}\psi_2(x).$$

Do this by letting $\mathbf{s}(x)$ ($\mathbf{s}$ for slope) and $\mathbf{v}(x)$ ($\mathbf{v}$ for value) be the solutions which satisfy Schroedinger's equation and which take the values

$$\mathbf{s} = 0 \quad \text{and} \quad d\mathbf{s}/dx = 1 \quad \text{at } x = 0$$

$$\mathbf{v} = 1 \quad \text{and} \quad d\mathbf{v}/dx = 0 \quad \text{at } x = 0.$$

Denote the values and derivatives at $x = a$ by s, s', v, and v'. Prove that for any two solutions ϕ_1 and ϕ_2 of Schroedinger's equation,

$$\frac{d}{dx}\left[\phi_2 \frac{d\phi_1}{dx} - \phi_1 \frac{d\phi_2}{dx}\right] = 0$$

and hence that

$$s'v - sv' = 1.$$

Show that the periodicity requirement on $\psi_1 = A\mathbf{s}(x) + B\mathbf{v}(x)$ leads to

$$(\lambda - v)(\lambda - s') - sv' = \lambda^2 - \lambda(v + s') + 1 = 0.$$

From this conclude that if the two roots are λ_1 and λ_2, $\lambda_1 = 1/\lambda_2$ so that λ_1 and λ_1^{-1} are the transformation factors. Show that λ is real for

$|v + s'| > 2$ or else $|\lambda|^2 = 1$ for $|v + s'| < 2$. This proves there are no damped complex wave functions. Using Problems 9, 10, and 11 draw curves for $\mathbf{s}(x)$ and $\mathbf{v}(x)$ versus x and show that their nodes are alternating and that, consequently, $|v + s'| < 2$ is surely satisfied for certain energy ranges for any lattice constant $\boldsymbol{a}$.

14. Prove that if $\boldsymbol{R}_1$, $\boldsymbol{R}_2$, and $\boldsymbol{R}_3$ are three allowed translations of the lattice and if $\boldsymbol{R}_1 \cdot \boldsymbol{R}_2 \times \boldsymbol{R}_3 = V_a$, then these $\boldsymbol{R}$'s may be used as set of *primitive vectors*, equation (4) of Section 14.8.

15. Prove that the surfaces of the first Brillouin zone (see the end of Section 14.8) are planes which are perpendicular bisectors of vectors from the origin to other points of the reciprocal lattice. Prove that no two points in the interior of the zone are equivalent.

16. Consider a lattice in space with primitive vectors $\boldsymbol{a}_1$, $\boldsymbol{a}_2$, $\boldsymbol{a}_3$ and reciprocal vectors $\boldsymbol{G}_1$, $\boldsymbol{G}_2$, $\boldsymbol{G}_3$. Consider the family of planes defined by

$$\boldsymbol{r} \cdot \boldsymbol{H} = n, \quad \boldsymbol{H} = h_1\boldsymbol{G}_1 + h_2\boldsymbol{G}_2 + h_3\boldsymbol{G}_3,$$

where $\boldsymbol{r}$ is a vector from the origin of the crystal lattice, and h_i and n are integers. Show that any vector

$$\boldsymbol{R} = \alpha_1\boldsymbol{a}_1 + \alpha_2\boldsymbol{a}_2 + \alpha_3\boldsymbol{a}_3 \quad (\alpha\text{'s are integers})$$

satisfies the equation for a suitably chosen n, so that all lattice points lie on planes for various values of n. Show also that, if the h_i are relatively prime (have no common factor), then for every value of n there are points on the plane. Show for this case that the planes of lattice points are spaced $1/|\boldsymbol{H}|$ apart. This problem shows how the reciprocal lattice vectors can be used to determine systematically the various sets of planes upon which the atoms of a crystal lie.

CHAPTER 15

THEORY OF ELECTRON AND HOLE VELOCITIES, CURRENTS AND ACCELERATIONS

15.1 GROUP VELOCITY FOR WAVE-PACKETS

In Figure 15.1 two sine waves are shown. They are in phase at crest A of the upper wave and B of the lower wave. At other points they are out of phase and tend to cancel. If many more waves of about the same wave length are combined to make a wave-packet, there will be a high degree of

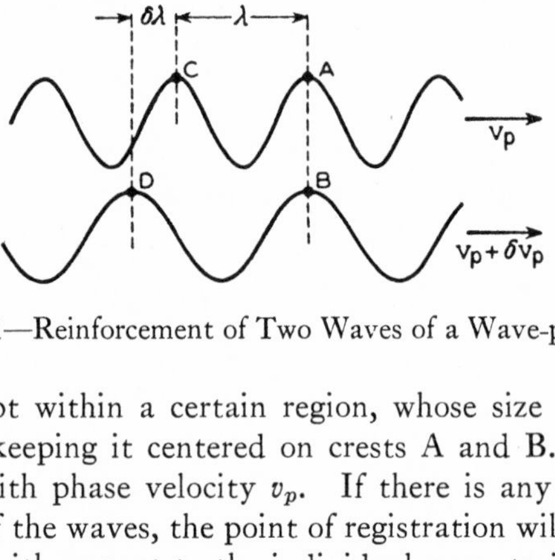

FIG. 15–1—Reinforcement of Two Waves of a Wave-packet.

cancellation except within a certain region, whose size may be adjusted mathematically, keeping it centered on crests A and B. The individual waves progress with phase velocity v_p. If there is any difference in the phase velocities of the waves, the point of registration will shift so that the group will move with respect to the individual wave trains. Thus, if the lower wave moves faster by an amount δv_p, then crest D will catch up with crest C in a time $t = \delta\lambda/\delta v_p$. If the wave-length difference is small, this expression may be expressed by the derivative $d\lambda/dv_p$. In this time, crest A has moved a distance $v_p t$, but the position of the reinforcement or center of the wave-packet has moved back an amount λ. Hence the velocity of the group is

$$v_g = (v_p t - \lambda)/t = v_p - \lambda dv_p/d\lambda. \tag{1}$$

This shows that if the phase velocity is independent of the wave length, the group and phase velocities are the same, a result valid for electromagnetic waves in free space, or for acoustical waves of long wave length in air.

The expression for v_g may be put into the desired form by suitably manipulating the derivatives. We write

$$v_p = \lambda\nu, \quad dv_p/d\lambda = \nu + \lambda d\nu/d\lambda \tag{2}$$

which reduces v_g to

$$v_g = v_p - \lambda\nu - \lambda^2 d\nu/d\lambda = -\lambda^2 d\nu/d\lambda. \tag{3}$$

By using the relationship $k = 1/\lambda$ we obtain

$$dk = -d\lambda/\lambda^2, \quad v_g = d\nu/dk \tag{4}$$

which is the form used in Chapter 6, with $\nu = \mathcal{E}/h$ and $k = P/h$.

In order to treat the three-dimensional case, we consider waves of the form $\exp 2\pi i[\mathbf{k} \cdot \mathbf{r} - \nu(\mathbf{k})t]$ which can be combined to form a packet[1]

$$\psi(r, t) = \int \{\exp 2\pi i[\mathbf{k} \cdot \mathbf{r} - \nu(\mathbf{k})t]\}\mathrm{w}(\mathbf{k})d\mathbf{k} \tag{5}$$

where $\int d\mathbf{k}$ represents integration over $\mathbf{k}$-space and $\mathrm{w}(\mathbf{k})$ is supposed to be real, positive, smoothly-varying function of $\mathbf{k}$ with values different from zero only near some particular value $\mathbf{k}_0$. The largest value of ψ for any t will thus occur for the value of $\mathbf{r}$ where the exponential is most nearly independent of $\mathbf{k}$, so that the integral over $\mathbf{k}$ adds together terms all of the same complex phase. This condition will be best met by satisfying

$$0 = \nabla_k[\mathbf{k} \cdot \mathbf{r} - \nu(\mathbf{k})t] = \mathbf{r} - \nabla_k\nu(\mathbf{k})t \tag{6}$$

for $\mathbf{k} = \mathbf{k}_0$. This leads to

$$\mathbf{r} = \nabla_k\nu(\mathbf{k})t, \tag{7}$$

which is equivalent to the expression for the components of group velocity given in Chapter 6.

15.2 APPLICATION TO ELECTRON WAVES

Two differences must be considered in applying the results just obtained to electron waves. In the first place, we must show that the results are independent of the choice of the zero point or origin of energy. Thus, for example, if a potential were applied to the semiconductor as a whole, all of the electronic energy levels should be thought of as shifting together. However, the internal behavior of the electrons should be unaffected. In the second place, the electron waves are not plane waves, but are modulated combinations of atomic wave functions and this difference must also be considered.

The unimportance of the choice of the origin of energy is readily shown from the Planck relationship $\nu = \mathcal{E}/h$. Changing the origin of energy is

[1] In Chapter 17, $\mathbf{k}$ corresponds to $2\pi/\lambda$ rather than to $1/\lambda$.

equivalent to adding a constant frequency ν_0 to all values of ν. Since only the derivatives of ν appear in ν_g, this added ν_0 has no effect.

The electron waves of interest are the Bloch waves

$$\psi_P = \mathbf{u}_P(\boldsymbol{r}) \exp \{2\pi i/h[\boldsymbol{P} \cdot \boldsymbol{r} - \mathcal{E}(\boldsymbol{P})t]\}. \tag{1}$$

Here $(1/h)\boldsymbol{P}$ plays the role of the wave vector $\boldsymbol{k}$ in the preceding section. Now if the wave-packet is made up of waves of almost the same value for $\boldsymbol{P}$, $\mathbf{u}_P(\boldsymbol{r})$ will be approximately the same function of $\boldsymbol{r}$ for all the waves and may be factored out of the integral over $\boldsymbol{P}$-space, thus giving for the wave-packet

$$\psi = \mathbf{u}_{P_0}(\boldsymbol{r}) \int \exp \{2\pi i/h[\boldsymbol{P} \cdot \boldsymbol{r} - \mathcal{E}(\boldsymbol{P})t]\} \mathbf{w}(\boldsymbol{P})d\boldsymbol{P}. \tag{2}$$

The same reasoning applies as before and, if $\mathbf{w}(\boldsymbol{P})$ is large only near $\boldsymbol{P}_0$, ψ will be large only near

$$\boldsymbol{r} = [\nabla_P \mathcal{E}(\boldsymbol{P})]_{\text{at } P_0} t \tag{3}$$

giving a group velocity

$$\dot{\boldsymbol{r}} = \frac{d}{dt}\boldsymbol{r} = [\nabla_P \mathcal{E}(\boldsymbol{P})]_{\text{at } P_0}. \tag{4}$$

15.3 AN ANALOGY FROM CIRCUIT THEORY

To elucidate the relationship between group velocity, density, and flow we shall treat the case of an artificial line with lumped constant inductances and capacitors. The relationship we shall establish is

power transmitted = energy per unit length × group velocity.

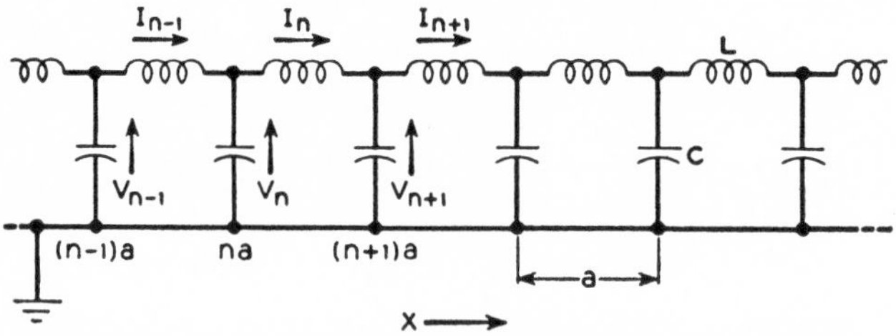

Fig. 15-2—Electrical Line Used to Show the Relationship between Power Transmission, Energy Density, and Group Velocity.

As has been explained in connection with Figure 6.1, this case furnishes one example of a general result: that rates of flow associated with wave equations may be calculated either by a density flow method or from the group velocity.

Figure 15.2 shows the line and the notation employed. We attempt to find a solution of the form

$$V_n = V_0 e^{2\pi i(na/\lambda - \nu t)} = V_0 e^{2\pi i(nka - \nu t)} \tag{1}$$

where $k = 1/\lambda$ and na is the position of the nth capacitor along the line. (These equations have been written to conform to the quantum-mechanical convention, with terms of the form exp $(-i\omega t)$; this convention introduces a minus sign into $i\omega L$ and $i\omega C$.) Accordingly, we obtain the following equations for the currents:

$$I_n = (V_n - V_{n+1})/(-i\omega L) = V_n(1 - e^{2\pi ika})/(-i\omega L) \tag{2a}$$

$$I_{n-1} = (V_{n-1} - V_n)/(-i\omega L) = -V_n(1 - e^{-2\pi ika})/(-i\omega L). \tag{2b}$$

Equating the difference in these currents to that furnished to the capacitor gives

$$\begin{aligned} -i\omega C V_n &= I_{n-1} - I_n = (V_n/-i\omega L)(e^{2\pi ika} - 2 + e^{-2\pi ika}) \\ &= (V_n/-i\omega L)(e^{\pi ika} - e^{-\pi ika})^2 \\ &= (4V_n/i\omega L)\sin^2 \pi ka \end{aligned} \tag{3}$$

This can be solved for sin πka giving

$$\sin \pi ka = \omega\sqrt{LC/4} = \pi\nu\sqrt{LC} \tag{4}$$

from which we conclude that the line can transmit up to $\pi\nu\sqrt{LC} = 1$ or $\nu = 1/\pi\sqrt{LC}$ but not above, since higher values of ν would require complex values of k. At the cutoff, $\sin \pi ka = 1$ and $\pi ka = \pi/2$ or 90°; thus the phase shift is 180° between unit cells as discussed in connection with Brillouin zones.

In the pass band, the group velocity is

$$v_g = \frac{d\nu}{dk} = \frac{d}{dk}\frac{\sin \pi ka}{\pi\sqrt{LC}} = \frac{a \cos \pi ka}{\sqrt{LC}}, \tag{5}$$

showing that the group velocity varies from $a/\sqrt{LC}$ at low frequencies (that is, for $ka \ll 1$) to zero at the cutoff frequency when $\pi ka = 90°$ and $\cos \pi ka = 0$.

Next we shall calculate the power flow down the line and the energy density along the line. The nth capacitor has a voltage

$$\text{Real } V_n = V_0 \cos (2\pi nka - \omega t), \tag{6}$$

leading to an energy of

$$(CV_0^2/2)\cos^2 (2\pi nka - \omega t) \tag{7}$$

which has peak and average values of $CV_0^2/2$ and $CV_0^2/4$ respectivelv The energy of the current I_n is similarly obtained from the amplitude of

I_n. The complex form of I_n may be rewritten as follows:

$$\begin{aligned} I_n &= V_n(1 - e^{2\pi ika})/-i\omega L \\ &= V_n e^{\pi ika}(-2i) \sin \pi ka/-i\omega L \\ &= (2V_n/\omega L)e^{\pi ika} \sin \pi ka \\ &= (2V_0/\omega L) \sin \pi ka \cdot e^{2\pi i(nka-\nu t) + \pi ika} \end{aligned} \tag{8}$$

which is evidently a current of peak value

$$(2V_0/\omega L) \sin \pi ka, \tag{9}$$

with a phase lag πka behind V_n. The energy stored in the inductance thus has a peak value of

$$\begin{aligned} \tfrac{1}{2}L(4V_0^2/\omega^2L^2) \sin^2 \pi ka &= \tfrac{1}{2}L(4V_0^2/\omega^2L^2)(\omega^2LC/4) \\ &= \tfrac{1}{2}CV_0^2, \end{aligned} \tag{10}$$

which is the same as that stored in the capacitor. Thus the energy of the line is $\frac{1}{4}CV_0^2$ on the average for each capacitor and each inductance or

$$CV_0^2/2a \text{ per unit length.}$$

If this energy flows with the group velocity, the power transmitted will be

$$\frac{CV_0^2}{2a} \cdot a \frac{\cos \pi ka}{\sqrt{LC}} = \frac{V_0^2(C/L)^{1/2} \cos \pi ka}{2}. \tag{11}$$

A direct calculation of power flow looking into the inductance which carries I_n is obtained from the amplitudes and phase angle of I_n and V_n. This leads to an average power flow of

$$\begin{aligned} \tfrac{1}{2}V \text{ peak} \cdot I \text{ peak} \cdot \cos \text{(phase difference)} &= \tfrac{1}{2}V_0 \cdot (2V_0/\omega L) \sin \pi ka \cdot \cos \pi ka \\ &= \frac{V_0^2}{\omega L} \cdot \frac{\omega\sqrt{LC}}{2} \cdot \cos \pi ka \\ &= \frac{V_0^2(C/L)^{1/2} \cos \pi ka}{2}, \end{aligned} \tag{12}$$

in agreement with the other form.

Our conclusion, in this particular example, is that consistent results are obtained when power flow is calculated (1) directly from the conventional expressions for power input at one point on the line and (2) from assuming that the energy stored in the line flows along the line with the group velocity.

15.4 THE PROBABILITY CURRENT AND BLOCH FUNCTION CURRENT

In this section we shall present the theory of the current operator and evaluate the current flow for electrons in the Brillouin zone.

The operator relationships $p_x = (h/2\pi i)\partial/\partial x$ and $\mathcal{H} = (ih/2\pi)\partial/\partial t$ are the starting point. Consider a volume V in space bounded by a surface S in which the one-electron wave function $\psi(x, y, z, t) \equiv \psi(r, t)$ satisfies

$$\mathcal{H}\psi = (ih/2\pi)\dot{\psi}. \tag{1}$$

Then the total probability current out of V is given by the rate of decrease of the probability that the electron is in V:

$$\begin{aligned} -\frac{d}{dt}\int_V \psi^*\psi\,dxdydz &= -\frac{2\pi}{ih}\int_V (\psi^*\mathcal{H}\psi - \psi\mathcal{H}\psi^*)dxdydz \\ &= \frac{2\pi i}{h}\left(\frac{-h^2}{8\pi^2 m}\right)\int_V (\psi^*\nabla^2\psi - \psi\nabla^2\psi^*)dxdydz \end{aligned} \tag{2}$$

since

$$\mathcal{H} = -\frac{h^2}{8\pi^2 m}\nabla^2 + \mathcal{U}(x, y, z) \tag{3}$$

and the integrals with $\mathcal{U}$ cancel. Noting that $\nabla \cdot (\mathbf{A}\nabla\mathbf{B} - \mathbf{B}\nabla\mathbf{A}) = \mathbf{A}\nabla^2\mathbf{B} - \mathbf{B}\nabla^2\mathbf{A}$ and letting $d\mathbf{S}$ be directed along the outward normal of S, we can reduce the integral to

$$\frac{-ih}{4\pi m}\int_S \nabla \cdot (\psi^*\nabla\psi - \psi\nabla\psi^*)dxdydz = \frac{1}{2m}\frac{h}{2\pi i}\int_S (\psi^*\nabla\psi - \psi\nabla\psi^*)\cdot d\mathbf{S}. \tag{4}$$

Thus the probability decreases as if there were a probability current of

$$I\ (\text{prob}) = \frac{1}{2m}[\psi^* p\psi + \psi(p\psi)^*] = (1/m)\text{ real part } \psi^* p\psi \tag{5}$$

flowing across S. [For a plane wave of momentum p_0, this gives $(1/m)p_0|\psi|^2$ just as if the probability density $|\psi|^2$ were flowing with velocity p_0/m.]

In order to gain insight into the nature of the current expression, it may be helpful to consider the following results: If $\psi(r, t)$ is energy eigenfunction, it takes the form $\psi(r)\exp(2\pi\mathcal{E}t/ih)$ so that $|\psi|^2$ is independent of time. This is consistent with the result that for this case $\nabla \cdot I$ (prob) contains $(\psi^*\nabla^2\psi - \psi\nabla^2\psi^*)$ as a factor which vanishes when ψ is an energy eigenfunction.[1] For the one-dimensional wave function of Figure 5.7(c), the node in $|\psi|$ shows that at that point the current vanishes; and hence, since $\nabla \cdot I = 0$, I vanishes at all points.

We shall next evaluate the current expression over one face of a unit cell and show that it reduces to the group velocity expression for average velocity.[2] Since I (prob) involves products of ψ and ψ^*, it is evident that a wave function of the Bloch type gives the same current for each unit

[1] Why? See Section 14.5.

[2] H. Jones and C. Zener, *Proc. Roy. Soc.* London, **144A**, 101 (1934).

cell since the changes in the exponential terms cancel. Suppose the unit cell has vector edges $\boldsymbol{a}$, $\boldsymbol{b}$, $\boldsymbol{c}$. We choose the x-axis perpendicular to the $(\boldsymbol{ab})$ face and evaluate the current across this face. This will obviously give us the x-component of current. For this case, the Bloch function is a function of $\boldsymbol{P}$ and $\boldsymbol{r}$, and $\mathcal{E}$ depends on $\boldsymbol{P}$. Differentiating the Schroedinger equation with respect to P_x gives

$$\frac{\partial}{\partial P_x}\left[\frac{p^2}{2m} + \mathcal{U}(r) - \mathcal{E}\right]\psi = 0 \tag{6}$$

or

$$\frac{\partial \mathcal{E}}{\partial P_x}\psi = \left[\frac{p^2}{2m} + \mathcal{U}(r) - \mathcal{E}\right]\frac{\partial \psi}{\partial P_x}. \tag{7}$$

In the following equations the right side of this equation will be manipulated so that it is expressed in terms of the probability current. Multiplying by ψ^* and integrating over the unit cell C gives

$$\frac{\partial \mathcal{E}}{\partial P_x}\int_C \psi^*\psi\, dV = \int_C \psi^*\left[\frac{p^2}{2m} + \mathcal{U}(r) - \mathcal{E}\right]\frac{\partial \psi}{\partial P_x}\, dV$$

$$= \int_C \frac{\partial \psi}{\partial P_x}\left[\frac{p^2}{2m} + \mathcal{U}(r) - \mathcal{E}\right]\psi^* dV + \int_C\left(\psi^*\frac{p^2}{2m}\frac{\partial \psi}{\partial P_x} - \frac{\partial \psi}{\partial P_x}\frac{p^2}{2m}\psi^*\right)dV$$

(and since if $\mathcal{H}\psi = \mathcal{E}\psi$, then $\mathcal{H}\psi^* = \mathcal{E}\psi$)

$$= 0 - \frac{h^2}{8\pi^2 m}\int_C\left(\psi^*\nabla^2\frac{\partial \psi}{\partial P_x} - \frac{\partial \psi}{\partial P_x}\nabla^2\psi^*\right)dV$$

$$= -\frac{h^2}{8\pi^2 m}\int_C \nabla\cdot\left(\psi^*\nabla\frac{\partial \psi}{\partial P_x} - \frac{\partial \psi}{\partial P_x}\nabla\psi^*\right)dV$$

$$= -\frac{h^2}{8\pi^2 m}\int_S\left(\psi^*\nabla\frac{\partial \psi}{\partial P_x} - \frac{\partial \psi}{\partial P_x}\nabla\psi^*\right)\cdot dS. \tag{8}$$

The last integral is taken over the six plane surfaces of the unit cell. As we shall see, the contributions on the faces not perpendicular to x cancel out, since the integrand is periodic, and even for the two $(\boldsymbol{ab})$ faces, there is considerable cancellation. To evaluate the integral, we write

$$\psi = \exp\,[2\pi i(\boldsymbol{P}\cdot\boldsymbol{r})/h]\mathfrak{u}_{\boldsymbol{P}}(\boldsymbol{r}) \tag{9}$$

$$\frac{\partial \psi}{\partial P_x} = \frac{2\pi i x}{h}\psi + \exp\left[2\pi i(\boldsymbol{P}\cdot\boldsymbol{r})/h\right]\frac{\partial \mathfrak{u}_{\boldsymbol{P}}(x)}{\partial P_x} \tag{10}$$

$$\frac{\partial}{\partial x}\frac{\partial \psi}{\partial P_x} = \frac{2\pi i x}{h}\frac{\partial \psi}{\partial x} + \frac{2\pi i}{h}\psi + \frac{\partial}{\partial x}\exp\left[2\pi i(\boldsymbol{P}\cdot\boldsymbol{r})/h\right]\frac{\partial \mathfrak{u}_{\boldsymbol{P}}(x)}{\partial P_x}. \tag{11}$$

From the form of $\partial\psi/\partial P_x$ it is evident that the integrand is periodic for cell edges $\boldsymbol{a}$ and $\boldsymbol{b}$ (which have no x-components) so that only the $(\boldsymbol{ab})$ face makes a contribution and for it only the x-component of ∇ is needed. For the $(\boldsymbol{ab})$ face, all terms are periodic except those with the factor x. Since x is larger by an amount c_x for the face of larger x, the net contribution is given by:

$$\frac{\partial \mathcal{E}}{\partial P_x}\int_C \psi^*\psi dv = \frac{-h^2}{8\pi^2 m}\int_{ab}\left(\psi^*\frac{2\pi i c_x}{h}\frac{\partial\psi}{\partial x} - \frac{2\pi i c_x}{h}\psi\frac{\partial\psi^*}{\partial x}\right)dxdy$$

$$= \frac{c_x h}{4\pi i m}\int_{ab}\left(\psi^*\frac{\partial\psi}{\partial x} - \psi\frac{\partial\psi^*}{\partial x}\right)dxdy$$

$$= c_x\int_{ab} I_x\,(\text{prob})\,dxdy$$

$$= c_x \boldsymbol{a}\cdot\boldsymbol{b}[\text{Average } I_x \text{ (prob) over } (\boldsymbol{ab}) \text{ face}]. \tag{12}$$

Hence the average probability current over the $(\boldsymbol{ab})$ face is given by

$$\frac{\partial \mathcal{E}}{\partial P_x}\int_C \psi^*\psi dV / c_x(\boldsymbol{a}\cdot\boldsymbol{b}). \tag{13}$$

This can be interpreted by saying that the average probability density

$$\left(\int_C \psi^*\psi dV\right) \Big/ \text{volume of cell}$$

flows across the x = constant face with an average velocity $v_x = \partial\mathcal{E}/\partial P_x$. Similar results for the other edges lead to the conclusion that $\boldsymbol{v} = \nabla_P \mathcal{E}(\boldsymbol{P})$.

15.5 SOME GENERAL PROPERTIES OF QUANTUM-MECHANICAL WAVE-PACKETS

As we have discussed in connection with Chapters 7 and 8, the wave-packet concept of the behavior of holes and electrons is essential to understanding the basic physics of transistor action and related topics in semiconductors. This appendix has been written as an introduction to a number of properties of wave-packets, which, although well known, are not readily available in textbooks. We shall describe first a minimum-uncertainty wave-packet. We do not give its derivation, however, since this is available in references quoted later. After that, we shall develop some general methods of modifying wave-packets and shall finally apply these to a study of the wave-packet shown in Figure 6.2.

One of the most publicized features of quantum mechanics is the Heisenberg uncertainty relation. We shall discuss the meaning of this relation in terms of wave-packets. A wave-packet, like that shown in Figure 6.2, for example, has a certain extension in space. Accordingly, it has a certain expected mean square deviation for measurements of its position. Formulae for calculating these are given in equations (11) and (14) of Section 14.3. Suppose that by the application of these we evaluate $\overline{x^2} - \bar{x}^2 =$

$\overline{(x - \bar{x})^2} = (\Delta x)^2$ so that we can call Δx the uncertainty in position of the electron for the wave-packet. There will also be a spread in expected momentum values from the mean value $\bar{p}$. If we for simplicity suppose $\bar{p} = 0$, then the mean value of $(p - \bar{p})^2$, called $(\Delta p)^2$, is simply $\overline{p^2} = 2m\ \overline{K.E.}$; we have shown in equation (9) of Section 14.4 that this quantity is positive unless ψ is constant—an impossible condition for a wave-packet. Straightforward calculations using the relationship $xp - px = i\hbar$, equation (20) of Section 14.4, lead to Heisenberg's uncertainty relation,

$$\Delta p \Delta x \geq \hbar/2 \tag{1}$$

for any wave function. Thus, if the wave-packet is made more compact, so as to decrease Δx, then Δp must increase. Mathematically, this follows from the fact that, if the wave function is made more compact, by simply compressing it along the x-axis for example, its derivative with respect to x will be greater and so will $\overline{p^2}$.

It is possible to construct a wave-packet for an electron in field-free space which has the minimum value of $\Delta p \Delta x$ for any particular time, say $t = 0$; however, it is found that thereafter the packet tends to spread. The equation for a normalized wave-packet, centered at $x = 0$, having $\bar{p} = 0$, and having a minimum value of $\Delta p \Delta x$, is:

$$\psi(x, t) = \frac{\exp\left[-\dfrac{x^2}{4a(a + iut)}\right]}{(2\pi)^{1/4}(a + iut)^{1/2}} \tag{2}$$

where

$$a = \Delta x \quad \text{for} \quad t = 0 \tag{3}$$

$$u = \frac{\hbar}{2m\Delta x} = \frac{\Delta p}{m}. \tag{4}$$

The quantity u has the dimensions of velocity. This equation can be derived by analogy with the diffusion equation, the Schroedinger time-dependent equation for the free particle being manipulated as follows:

$$i\hbar \frac{\partial \psi}{\partial t} = \mathcal{H}\,\psi = \frac{1}{2m}\left(\frac{\hbar}{c}\right)^2 \frac{\partial^2 \psi}{\partial t^2}$$

$$\frac{\partial \psi}{\partial t} = -\frac{\hbar}{2mi}\frac{\partial^2 \psi}{\partial x^2} = D\frac{\partial^2 \psi}{\partial x^2} \tag{5}$$

where $D = (\hbar/2mi)$ has the dimensions of a diffusion constant.[1] The

[1] This method is discussed by S. Dushman, *Elements of Quantum Mechanics*, John Wiley & Sons, New York, 1938, p. 405.

more conventional method is to resolve the $\psi(x, 0)$ minimum packet into momentum eigenfunctions, whose time dependence is simply exp $(tp^2/2m\hbar i)$, and obtain the time dependence of $\psi(x, t)$ as a consequence.[2]

We shall next derive some general theorems pertaining to all wave-packets for free electrons, after which we shall remove the limitation $\bar{p} = 0$, which holds for packet (2), so as to set it in motion and obtain the packet of Figure 6.2.

First of all, we shall prove the time-reversal feature of Schroedinger's equation which shows that for any $\psi(x, t)$ another ψ, called $\psi_-(x, t)$, can be found for which $|\psi_-(x, t)|^2 = |\psi(x, -t)|^2$. Thus ψ_- gives the same behavior for the probability density as does ψ except that time is reversed. The proof is very simple: We denote the function obtained by differentiating $\psi(x, t)$ in respect to its second argument by $\chi(x, t)$. We then have

$$i\hbar \frac{\partial \psi}{\partial t} = i\hbar\chi(x, t) = \mathcal{H}\psi(x, t). \tag{6}$$

Let

$$\psi_-(x, t) = \psi^*(x, -t).$$

Then

$$i\hbar \frac{\partial \psi_-(x, t)}{\partial t} = i\hbar \frac{\partial \psi^*(x, -t)}{\partial t} = -i\hbar\chi^*(x, -t). \tag{7}$$

However, from the complex conjugate of equation (6), we have

$$-i\hbar\chi^*(x, -t) = \mathcal{H}\psi^*(x, -t) = \mathcal{H}\psi_-(x, t) \tag{8}$$

where we have used the relationship $(\mathcal{H}\psi)^* = \mathcal{H}(\psi^*)$ which follows from the fact that $\mathcal{H}$ has no imaginary terms. From this equation and (7), we obtain

$$i\hbar \frac{\partial \psi_-}{\partial t} = \mathcal{H}\psi_- \tag{9}$$

which was to be proved. We shall show next that a wave-packet increases its value of Δx from its minimum value symmetrically for plus and minus time, a result which would appear unreasonable if Schroedinger's equation did not possess the time reversal feature. In particular, we shall prove that, for any wave-packet, Δx varies according to the equation

$$(\Delta x)^2 = \frac{\overline{p^2} - \bar{p}^2}{2m}(t - t_0)^2 + (\Delta x)_{\min}{}^2 \tag{10}$$

where t_0 is the instant at which Δx is a minimum. This equation shows that the packet spreads in space symmetrically to both past and future

[2] A very readable discussion of this procedure together with a derivation of $\Delta p \Delta x \geq \hbar/2$ is given in L. I. Schiff, *Quantum Mechanics*, McGraw-Hill Book Co., New York, 1949.

times measured from t_0. This result is proved from the expression for $(\Delta x)^2$, equation (13) of Section 14.3, and the formula for the time derivative of an operator, equation (11) of Section 14.4. We have

$$\frac{dx}{dt} = \frac{p}{m} \tag{11}$$

$$\frac{d}{dt}\left(\frac{d}{dt}x\right) = \frac{1}{m}\frac{d}{dt}p = \frac{1}{m}\frac{1}{i\hbar}(p\mathcal{H} - \mathcal{H}p) = 0 \tag{12}$$

since $\mathcal{H} = p^2/2m$ and $\mathcal{H}p = p\mathcal{H} = p^3/2m$. Next we evaluate dx^2/dt as follows

$$\begin{aligned} \frac{d}{dt}x^2 &\underset{(1)}{=} \frac{1}{2i\hbar m}(x^2p^2 - p^2x^2) \\ &\underset{(2)}{=} \frac{1}{2ihm}\left\{x^2p^2 - \left[x^2p^2 + 4\left(\frac{h}{i}\right)xp + 2\left(\frac{\hbar}{i}\right)^2\right]\right\} \\ &\underset{(3)}{=} \frac{-2(\hbar/i)}{2i\hbar m}\left(2xp + \frac{\hbar}{i}\right) \underset{(4)}{=} \frac{1}{m}(xp + px) \end{aligned} \tag{13}$$

where equality (**1**) is obtained directly from formula (11) of Section 14.4; equality (**3**) by simple algebra; equality (**4**) from the equation (20) of Section 14.4 $px = xp + (\hbar/i)$; equality (**2**) from repeated applications of (20) such as

$$\begin{aligned} p^n x &= p^{n-1}px = p^{n-1}\left(xp + \frac{\hbar}{i}\right) = p^{n-1}xp + \frac{\hbar}{i}p^{n-1} \\ &= p^{n-2}xp^2 + 2\frac{\hbar}{i}p^{n-1} = \text{etc.} = xp^n + n\frac{\hbar}{i}p^{n-1}. \end{aligned} \tag{14}$$

Proceeding similarly for d^2x^2/dt^2, reducing all terms containing $p^n x$ to xp^n terms by (14), we obtain

$$\begin{aligned} \frac{d^2x^2}{dt^2} &= \frac{d}{dt}\left(\frac{px + xp}{m}\right) = \frac{1}{i\hbar 2m^2}[(px + xp)p^2 - p^2(px + xp)] \\ &= \frac{1}{i\hbar 2m^2}\left(xp^3 + \frac{\hbar}{i}p^2 + xp^3 - xp^3 - 3\frac{\hbar}{i}p^2 - 2\frac{\hbar}{i}p^2\right) \\ &= \frac{1}{i\hbar 2m^2}\left(-\frac{4\hbar}{i}\right)p^2 = \frac{2p^2}{m^2}. \end{aligned} \tag{15}$$

Since $dp^2/dt = 0$, it is evident that $d^3x^2/dt^3 = 0$.

If the foregoing expressions are averaged in integrals of the form $\bar{x} = \int \psi^* x \psi\, dx$ (see RC. 3 in Section 14.3) we obtain:

$$\frac{d}{dt}\bar{x} = \frac{p}{m} = \bar{v} \quad \text{(constant in time)} \tag{16}$$

$$\frac{d}{dt}\overline{x^2} = \frac{1}{m}\overline{(px + xp)} = \overline{xv} \tag{17}$$

$$\frac{d^2}{dt^2}\overline{x^2} = \frac{2\overline{p^2}}{m^2} = 2\overline{v^2} \quad \text{(constant in time).} \tag{18}$$

The higher time derivatives of $\bar{x}$ and $\overline{x^2}$ vanish as do all time derivatives of p and $\overline{p^2}$. Since $(\Delta x)^2 = \overline{x^2} - \bar{x}^2$, we can write

$$\frac{d}{dt}(\Delta x)^2 = \frac{d\overline{x^2}}{dt} - 2\bar{x}\frac{d\bar{x}}{dt} = \overline{xv} - 2\bar{x}\bar{v} \tag{19}$$

$$\frac{d^2}{dt^2}(\Delta x)^2 = 2\overline{v^2} - 2\bar{v}^2 = 2\frac{\overline{p^2} - \bar{p}^2}{m^2} = \frac{2(\Delta p)^2}{m^2}. \tag{20}$$

These equations show that $(\Delta x)^2$ is a quadratic form in t, the coefficients of which may be expressed in terms of values at $t = 0$ (like $\overline{x_0^2}$) as follows:

$$(\Delta x)^2 = \frac{(\Delta p)^2}{m^2}t^2 + (\overline{xv} - 2\bar{x}\bar{v})_0 t + \overline{x_0^2} - \bar{x}_0^2. \tag{21}$$

$(\Delta x)^2$ will have its minimum at the time, call it t_m, for which

$$0 = \frac{d(\Delta x)^2}{dt} = 2\frac{(\Delta p)^2}{m^2}t_m + (\overline{xv} - 2\bar{x}\bar{v})_0. \tag{22}$$

This equation may be solved for t_m and the minimum value of $(\Delta x)^2$ obtained in the form

$$(\Delta x)_{\min}^2 = \frac{(\Delta p)^2}{m^2}t_m^2 + (\overline{xv} - 2\bar{x}\bar{v})_0 t_m + \overline{x_0^2} - \bar{x}_0^2. \tag{23}$$

In terms of t_m and $(\Delta x)_{\min}^2$, the expression for $(\Delta x)^2$ reduces to

$$(\Delta x)^2 = \frac{(\Delta p)^2}{m^2}(t - t_m)^2 + (\Delta x)_{\min}^2. \tag{24}$$

The equations just derived show that the center of gravity of the packet moves with constant velocity $\bar{v} = \bar{p}/m$ and that it spreads in space as expected for a spread in velocities $(\Delta p/m)$. However, the Δx of the packet never decreases below a minimum value. This may be compared with the wave-packet of (2) of which the exponent may be rewritten to give:

$$\psi(x, t) = \frac{\exp\left[+\dfrac{ix^2ut}{4a(a^2 + u^2t^2)} - \dfrac{x^2}{4(a^2 + u^2t^2)}\right]}{(2\pi)^{1/4}(a + iut)^{1/2}} \tag{25}$$

so that $\psi^*\psi$ is

$$|\psi(x, t)|^2 = \frac{\exp\left[-\dfrac{x^2}{2(a^2 + u^2t^2)}\right]}{\sqrt{2\pi}\,(a^2 + u^2t^2)^{1/2}} \tag{26}$$

which represents a normalized Gauss error function probability distribution with $(\Delta x)^2$ given by

$$(\Delta x)^2 = (a^2 + u^2t^2) = (\Delta x)_0{}^2 + \left(\frac{\Delta p}{m}\right)^2 t^2 \tag{27}$$

where the values of a and u from (3) and (4) have been used.

We shall next show that a wave-packet in free space may be given a Galilean transformation; that is, the whole solution may be given a velocity of translational motion and remain a solution of Schroedinger's time-dependent equation. This result holds in general for any solution of the one-dimensional equation,

$$\frac{1}{2m} p^2\psi(x, t) = i\hbar \frac{\partial}{\partial t} \psi(x, t). \tag{28}$$

The necessary transformation is

$$\phi(x, t) = \exp\left[\frac{i}{\hbar}\left(p_1 x - \frac{p_1{}^2}{2m} t\right)\right] \psi\left(x - \frac{p_1}{m} t, t\right) \tag{29}$$

where the added velocity of motion is p_1/m as may be seen from the fact that the new probability density is

$$|\phi|^2 = \left|\psi\left(x - \frac{p_1}{m} t, t\right)\right|^2, \tag{30}$$

which is simply the old probability density being translated with velocity p_1/m. It may readily be verified that ϕ satisfies the time-dependent equation. Writing $\phi = \exp(iA)\psi$ for brevity, we find, since $p \exp(iA) = [\exp(iA)](p_1 + p)$, that

$$\frac{1}{2m} p^2\phi = \frac{1}{2m} p[e^{iA}p_1 + e^{iA}p]\psi = \frac{e^{iA}}{2m} [p_1{}^2 + 2p_1p + p^2]\psi. \tag{31}$$

In evaluating $i\hbar\partial\phi/\partial t$, we must note that t occurs three times in ϕ and that $i\hbar$ times the derivative with respect to t of the first argument in ψ gives simply $(p_1/m)p\psi$. Thus we have

$$i\hbar \frac{\partial\phi}{\partial t} = e^{iA}\left[\frac{p_1{}^2}{2m} + \frac{p_1 p}{m} + \frac{p^2}{2m}\right]\psi, \tag{32}$$

the derivative of the last argument giving $(p^2/2m)\psi$, since $\psi(x, t)$ satisfies Schroedinger's equation. Thus we see that $(p^2/2m)\phi = i\hbar\partial\phi/\partial t$ so that the packet can undergo a Galilean transformation, Q.E.D.

We shall next compare the values of $\overline{\mathcal{H}}$ for ϕ and ψ. We have

$$\overline{\mathcal{H}} = \int \phi^* \mathcal{H} \phi dx = \int e^{-iA} \psi \frac{e^{iA}}{2m} [p_1^2 + 2p_1 p + p^2] \psi dx$$
$$= \frac{1}{2m} \int \psi^* [p_1^2 + 2p_1 p + p^2] \psi dx$$
$$= \frac{1}{2m} [p_1^2 + 2p_1 \bar{p} + \overline{p^2}] = \frac{1}{2m} [\overline{(p_1 + p)^2}], \qquad (33)$$

where the averages are taken over ψ. This represents simply the energy to be expected if the momentum distribution represented by ψ were simply increased by p_1. If $\bar{p} = 0$, the added energy is simply that of the motion of the packet of mass m with velocity $v = p_1/m$. From this it is evident that setting the packet in motion does not change $(\Delta p)^2$:

$$(\Delta p)^2 = \text{av. over } \phi \ (\overline{p^2} - \bar{p}^2)$$
$$= \text{av. over } \psi \ [\overline{(p_1 + p)^2} - (p_1 + \bar{p})^2]$$
$$= \text{av. over } \psi \ [p_1^2 + 2p_1 \bar{p} + \overline{p^2} - p_1^2 - 2p_1 \bar{p} - \bar{p}]$$
$$= \text{av. over } \psi \ [\overline{p^2} - \bar{p}^2]. \qquad (34)$$

Since the probability distribution $|\phi^2|$ is simply that of $|\psi|^2$ set in motion, (30), it is evident that ϕ has the same $(\Delta x)^2$ and $(\Delta p)^2$ values as does ψ.

We can now apply this transformation to the minimum-uncertainty packet in order to obtain the wave-packet used for illustration in Figure 6.2. Applying (29) to (25) and introducing the notation

$$p_1/\hbar = k, \quad p_1/m = v, \quad x = y + vt \qquad (35)$$

so that y measures distance from the center of the wave-packet, we obtain: $\phi(x, t)$

$$= \frac{\exp\left[-\dfrac{y^2}{4(a^2 + u^2 t^2)}\right]}{(2\pi)^{1/4}(a + iut)^{1/2}} \exp i \left[k\left(y + \frac{vt}{2}\right) + \frac{y^2 ut}{4a(a^2 + u^2 t^2)}\right]. \qquad (36)$$

From this we can see analytically several of the interesting results shown in Figure 6.2. Near $y = 0$, we see that $p\psi = \dfrac{\hbar}{i} k\psi = p_1 \psi$. However, near the leading edge or the back edge of the packet, corresponding to $y = \pm(a^2 + u^2 t^2)^{1/2}$ respectively, we have

$$p\psi = \frac{\hbar}{i}\left[k \pm \frac{2ut}{4a(a^2 + u^2 t^2)^{1/2}}\right] \psi \qquad (37)$$

so that for $ut \gg a$, the momentum operator gives

$$p\psi = \left(p_1 \pm \frac{\hbar}{2a}\right)\psi = (p_1 \pm \Delta p)\psi. \tag{38}$$

This variation of momentum between the front and back of the packet can readily be seen in Figure 6.2.

15.6 THE ACCELERATION OF AN ELECTRON WAVE-PACKET BY ELECTRIC AND MAGNETIC FIELDS[1]

In the text we stress the possibility of treating holes and electrons as particles which may be accelerated by electric and magnetic fields and scattered by irregularities in the crystal. The basis for this picture is that wave-packets for holes and electrons can be prescribed which move, to a good approximation, in accord with the classical picture. The relationship between velocity and momentum, however, is that discussed in Chapter 6 and between force and rate of change of momentum that discussed in Chapter 7. We shall establish these results for an electron wave-packet in this section and consider the modification for holes in the following sections.

The theorem which we shall prove is as follows: If we construct a wave-packet which represents a localized electron with a wave function corresponding to momentum $\boldsymbol{P}$ in an energy band, then this wave-packet will move with velocity $\boldsymbol{v} = \nabla_P \mathcal{E}(\boldsymbol{P})$ (a result previously established in Section 15.2) and will change its momentum $\boldsymbol{P}$ according to the formula

$$\dot{\boldsymbol{P}} = (-e)\left[\boldsymbol{E} + \frac{1}{c}\boldsymbol{v} \times \boldsymbol{H}\right]. \tag{1}$$

This result is in accord with the discussion of Chapter 7.

This theorem proves results which are partially derived in various texts in other ways. One of the simplest ways, although not rigorous, of reaching this conclusion is to make use of the group velocity formula and the principle of conservation of energy.[2] If the force on the electron is $\boldsymbol{F}$, one then equates the power input to the electron to its increase in energy. This gives

$$\boldsymbol{v} \cdot \boldsymbol{F} = \frac{d}{dt}\mathcal{E}(\boldsymbol{P}) = \dot{\boldsymbol{P}} \cdot \nabla_P \mathcal{E}(\boldsymbol{P}) = \dot{\boldsymbol{P}} \cdot \boldsymbol{v}$$

which is satisfied by

$$\dot{\boldsymbol{P}} = \boldsymbol{F}.$$

[1] The earliest references to proofs of the theorem derived in this appendix are Sommerfeld and Bethe, *Handbüch der Physik*, Vol. XXIV/2; H. Jones and C. Zener, *Proc. Roy. Soc.* London **144**, 101–117 (1934); H. Fröhlich, *Theorie der Metalle*, Julius Springer, Berlin, 1937.

[2] F. Seitz and N. F. Mott and H. Jones follow this procedure in their books.

However, it is equally well satisfied by $\dot{P} = F + (1/|v|)F \times v$ or any other vector whose component parallel to v is also F. Furthermore, the later use of $F = (-e)\left[E + \frac{1}{c} v \times H\right]$ represents an additional assumption since the forces due to H do no work.

A derivation of the acceleration law for electric fields using momentum eigenfunctions has been given by J. C. Slater.[3] A somewhat similar treatment including the effect of magnetic fields is given by A. H. Wilson.[4] However, neither of these references treat wave-packet behavior explicitly and it is easier for our purposes to start initially with the wave-packet.

The method we shall use is based upon obtaining an approximate solution for Schroedinger's equation for an electron in electric and magnetic fields and showing that this solution does behave in the predicted way. The treatment which we shall give is similar to one due to Houston,[5] except that we include magnetic fields and deal with a wave-packet. As an introduction we shall describe briefly the essential feature of Houston's method. We consider an electron in a wave function of the Bloch form

$$\psi_P = \exp\,[i(P \cdot r)/\hbar]\mathbf{u}_P(r)$$

where $\mathbf{u}_P(r)$ has the periodicity of the lattice and is normalized per unit volume.[6] This wave function satisfies the equation

$$\left[\frac{1}{2m}p^2 + \mathcal{U}(r)\right]\psi_P \equiv \mathcal{H}_0\psi_P = \mathcal{E}(P)\psi_P. \tag{2}$$

If an electric field E, producing a force $-eE$ and a potential energy $eE \cdot r$, is applied for time dt, the change in ψ is

$$d\psi_P = \frac{1}{i\hbar}\,dt\,\mathcal{H}\psi_P = \frac{dt}{i\hbar}\,[\mathcal{E}(P) + eE \cdot r]\psi_P \tag{3}$$

so that at $t + dt$, ψ has become

$$\begin{aligned}\{1 - i[\mathcal{E}(P) + e\mathcal{E} \cdot r]dt/\hbar\}\psi_P &= \exp\,\{-i[\mathcal{E}(P) + eE \cdot r]dt/\hbar\}\psi_P \\ &= \{\exp\,[-i\mathcal{E}(P)dt/\hbar]\}\{\exp\,[i(P - eEdt) \cdot r/\hbar]\}\mathbf{u}_P(r).\end{aligned}$$

This shows that in time dt, the effective P for ψ has changed by $(-e)Edt$ just as if $\dot{P} = (-e)E$. Houston finds that $\mathbf{u}_P$ also changes to $\mathbf{u}_{P-eEdt}$ so that the wave function as a whole varies according to $\dot{P} = (-e)E$. Except for fields so high that in one lattice constant the electron gains an appre-

[3] J. C. Slater, "The Electronic Structure of Metals", *Rev. Mod. Phys.* **6**, 209–280 (1934).

[4] A. H. Wilson, *The Theory of Metals*, Cambridge at the University Press, 1935.

[5] W. V. Houston, *Phys. Rev.* **57**, 184–186 (1940).

[6] The ψ_P used in Chapter 17 is normalized for the volume inside the periodic boundaries and is $1/\sqrt{V}$ times the ψ_P of this section.

ciable fraction of the energy required to jump the energy gap between bands, Houston shows that ψ_{P-eEt} is a good solution. Our method will also consist of finding the change in time dt (which we shall express as a derivative) but we shall deal with a wave-packet and also include the effect of a magnetic field.

In the presence of electric and magnetic fields, Schroedinger's equation[7] for an electron with charge $-e$ contains the scalar and vector potentials φ and A and is

$$\left[\frac{1}{2m}(p - p_A)^2 + \mathcal{U}(r) + (-e)\varphi\right]\psi = \mathcal{H}\psi = i\hbar\frac{\partial}{\partial t}\psi \tag{4}$$

where

$$p_A = \frac{(-e)}{c}A. \tag{5}$$

The scalar and vector potentials give electric and magnetic fields E and H:

$$E = -\nabla\varphi - \frac{1}{c}\frac{\partial}{\partial t}A; \quad H = \nabla \times A; \tag{6a}$$

the scalar and vector potentials are not unique and one may transform them by adding to φ a term $(-1/c)\partial\chi/\partial t$ and to A a term $\nabla\chi$ without changing E and H. If we suppose that φ arises from a stationary system of charges and that A arises from changing currents and charges outside of the region considered, then the *gauge transformation* function χ may be chosen so that $\nabla \cdot A = 0$. We shall suppose that χ has been so chosen and shall use the resulting equations

$$\nabla \cdot A = 0 \quad \text{and} \quad p \cdot p_A = p_A \cdot p \tag{6b}$$

in the subsequent analysis.

In the presence of a magnetic field it is impossible to have wave functions which are eigenfunctions for the velocity operator $\dot{r}$. In fact we find, by using $p_x x - x p_x = \hbar/i$, that in the presence of a magnetic field the operator corresponding to $\dot{r}$ is no longer p/m; and the new operator for $\dot{r}$ has such a form that no wave function can be simultaneously an eigenfunction for the three components of $\dot{r}$. Using the commutation relation $p_x x - x p_x = \hbar/i$ and the expression (11) of Section 14.4 for the time derivative of an operator, we obtain

$$\dot{r} = (1/i\hbar)(r\mathcal{H} - \mathcal{H}r) = (p - p_A)/m. \tag{7}$$

For a uniform field with $H = i_z H_z$, a possible choice for A is $i_y H_z x$ and for this it is evident that

$$\dot{x}\dot{y} = \left[\frac{\hbar}{i}\frac{\partial}{\partial x}\right]\left[\frac{\hbar}{i}\frac{\partial}{\partial y} + \frac{e}{c}H_z x\right] = \dot{y}\dot{x} + \frac{\hbar e}{ic}H_z \tag{8}$$

[7] See, for example, L. I. Schiff, *Quantum Mechanics*, McGraw-Hill Book Co., New York, 1949.

so that $\dot{x}$ and $\dot{y}$ do not commute and, therefore, no wave function can be an eigenfunction for both $\dot{x}$ and $\dot{y}$. This fact prevents dealing with indefinitely extended wave functions, such as Houston used, in the presence of a magnetic field; however, it does not interfere seriously with using wave-packets.

We must next consider the problem of constructing a wave function which represents an electron moving in a magnetic field and having the attributes associated with ψ_P, a wave function in the energy band for the case of no magnetic field. A clue to the solution is furnished by the *gauge transformation* of ψ which accompanies that of $\boldsymbol{A}$ and φ. Thus if ψ satisfies the equation

$$\left[\frac{1}{2m}(\boldsymbol{p} - \boldsymbol{p}_A)^2 + \mathcal{U} + (-e)\varphi\right]\psi = i\hbar\frac{\partial}{\partial t}\psi,$$

then the wave function

$$\psi' = e^{i(-e/c)\chi/\hbar}\,\psi \tag{9}$$

satisfies the equation with

$$\boldsymbol{A}' = \boldsymbol{A} + \nabla\chi, \quad \varphi' = \varphi - \frac{1}{c}\frac{\partial}{\partial t}\chi, \tag{10}$$

as the reader may verify by substitution. If we wish to center the wave-packet at $\boldsymbol{r}_0$, then we may regard $\boldsymbol{p}_A(\boldsymbol{r}_0, t)$ as arising in part from a gauge transformation with $\chi = \boldsymbol{A}(\boldsymbol{r}_0, t)\cdot\boldsymbol{r} = (c/-e)\boldsymbol{p}_A(\boldsymbol{r}_0, t)\cdot\boldsymbol{r}$. This reasoning leads us to investigate the properties of

$$\psi = e^{i\boldsymbol{p}_{A0}\cdot\boldsymbol{r}/\hbar}\psi_P(\boldsymbol{r}), \tag{11}$$

where $\boldsymbol{p}_{A0} \equiv \boldsymbol{p}_{A0}(t) = \boldsymbol{p}_A[\boldsymbol{r}_0(t), t]$. It is evident (A) that ψ gives the same change density $|\psi_P|^2$ as does ψ_P. We shall also verify (B) that, near $\boldsymbol{r}_0$, the wave function (11) is nearly an eigenfunction of $\mathcal{H}$ with the energy appropriate to state $\boldsymbol{P}$ in the absence of a magnetic field, and, furthermore, (C) that the average value of the velocity operator $\dot{\boldsymbol{r}} = (\boldsymbol{p} - \boldsymbol{p}_A)/m$, averaged over the unit cell at $\boldsymbol{r}_0$, is simply $\boldsymbol{v}(\boldsymbol{P}) = \nabla_P\mathcal{E}(\boldsymbol{P})$ appropriate to an electron in state $\boldsymbol{P}$. Thus near $\boldsymbol{r}_0$ the wave function (11) represents properly the state of motion described by $\boldsymbol{P}$ and $\boldsymbol{v}(\boldsymbol{P})$. After verifying assertions (B) and (C), we shall build a wave-packet based on (11).

To prove (B), we evaluate $\mathcal{H}\psi$, by carrying out the operations only on the first factor in equation (11):

$$\mathcal{H}\psi = e^{i\boldsymbol{p}_{A0}\cdot\boldsymbol{r}/\hbar}\left[\mathcal{H}_0 + (-e)\varphi + \frac{1}{m}(\boldsymbol{p}_{A0} - \boldsymbol{p}_A)\cdot\boldsymbol{p} + \frac{1}{2m}(\boldsymbol{p}_{A0} - \boldsymbol{p}_A)^2\right]\psi_P \tag{12}$$

where we have used the fact that both $\nabla\cdot\boldsymbol{p}_A$ and $\nabla\cdot\boldsymbol{p}_{A0}$ are zero in arranging the third term. At $\boldsymbol{r} = \boldsymbol{r}_0$, $\boldsymbol{p}_{A0} - \boldsymbol{p}_A$ is zero so that (12) gives

$$\begin{aligned}\mathcal{H}\psi &= e^{i\boldsymbol{p}_{A0}\cdot\boldsymbol{r}/\hbar}[\mathcal{H}_0 + (-e)\varphi]\psi_P = e^{i\boldsymbol{p}_{A0}\cdot\boldsymbol{r}/\hbar}[\mathcal{E}(\boldsymbol{P}) + (-e)\varphi]\psi_P \\ &= [\mathcal{E}(\boldsymbol{P}) + (-e)\varphi]\psi\end{aligned}$$

so that, at $\boldsymbol{r}_0$, ψ is an eigenfunction with energy $\mathcal{E}(\boldsymbol{P})$ due to its state in the band plus $(-e)\varphi$ due to its interaction with the added electrostatic potential. When $\boldsymbol{r}$ differs from $\boldsymbol{r}_0$, there are changes in φ and, as we shall see, these affect $i\hbar\dot{\psi}$ just as $-\boldsymbol{E}\cdot\boldsymbol{r}$ did in Houston's method. Also there are changes in $\boldsymbol{p}_{A0} - \boldsymbol{p}_A$ which produce the magnetic effects. The term in $(\boldsymbol{p}_{A0} - \boldsymbol{p}_A)^2$ is negligible compared to the term in $(\boldsymbol{p}_{A0} - \boldsymbol{p}_A)$, provided that the size of the wave-packet is not too large. We shall show this by a rough estimate of the two terms, using for the operator $\boldsymbol{p}$ the average value for a thermal electron and writing the radius of the packet as n times the lattice constant a and taking $\boldsymbol{A} = \boldsymbol{i}_y H_z x$. The quadratic term is then

$$\frac{1}{2m}\left[\frac{eH_z na}{c}\right]^2 = \frac{m}{2}(\omega na)^2, \quad \text{where} \quad \omega = \frac{eH_z}{mc}. \tag{13}$$

The linear term is obtained by writing $|\boldsymbol{p}| = mv$ where v is the thermal velocity, thus obtaining

$$\frac{1}{m}\frac{eH_z na}{c} mv = m(\omega na)v. \tag{14}$$

The ratio of (14) to (13) is $v/\omega na = (v/\omega)/na$, the latter form being (v/ω = radius of circular orbit in field H_z) divided by (radius of wave-packet). It is physically evident that, if this ratio is small, the motion of the electron will consist of a circular oscillation which is smaller than the size of the wave-packet, and in this case the internal motion of the packet may be more important than the bodily motion. For a thermal electron, however, $v = 10^7$ cm/sec and for a field of 10^3 gauss $\omega \doteq 10^{10}$ sec^{-1} so that the radius of the orbit is 10^{-3} cm and the two radii are equal for $na = 10^{-3}$ cm or $n \doteq 10^5$. We shall deal with packets much smaller than this and shall neglect the quadratic term.

The value of $\dot{\boldsymbol{r}}$, averaged over the cell of $\boldsymbol{r}_0$, is

$$\langle\dot{\boldsymbol{r}}\rangle_{\text{cell at } r_0} \overset{(a)}{=} \frac{\int_{\text{cell 0}} \psi^* \frac{1}{m}(\boldsymbol{p} - \boldsymbol{p}_A)\psi dV}{\int_{\text{cell 0}} |\psi|^2 dV}$$

$$\overset{(b)}{=} \frac{\int_{\text{cell 0}} \psi_P^* \frac{1}{m}(\boldsymbol{p} + \boldsymbol{p}_{A0} - \boldsymbol{p}_A)\psi_P dV}{\int_{\text{cell 0}} |\psi_P|^2 dV} \overset{(c)}{=} \boldsymbol{v}(\boldsymbol{P}) \tag{15}$$

where equality (a) follows from (7), (b) from the operation of $\boldsymbol{p}$ on the exponential factor of (11), and (c) from neglecting $\boldsymbol{p}_{A0} - \boldsymbol{p}_A$ compared to $\boldsymbol{p}$ and using the results of 15.4. Hence assertion (C) is correct, and the wave function ψ represents correctly the average velocity $\boldsymbol{v}$.

Wave-packet. We next construct a wave-packet by multiplying ψ_P by a weighting factor $\mathbf{w}(r, t)$ and a time factor $\exp[-if(t)/\hbar]$. We shall suppose that $\mathbf{w}$ and f may both be chosen to be real functions so that f contributes only to the complex phase and $\mathbf{w}$ simply gives the shape of the packet. We shall show later in equations (31) and (32) that such a choice gives a solution of Schroedinger's equation. Thus we write

$$\psi = e^{i[p_{A0}\cdot r - f]/\hbar}\mathbf{w}(r, t)\psi_P(r) \equiv e^{i\theta}\mathbf{w}\psi_P, \tag{16}$$

where the last form introduces abbreviations used in the subsequent analysis. The normalization condition on ψ is

$$1 = \int |\psi|^2 dV = \int |\psi_P|^2|\mathbf{w}|^2 dV = \int |\mathbf{w}|^2 dV$$

since we shall suppose that ψ_P is normalized per unit volume and that $\mathbf{w}$ is slowly varying so that we may replace $|\psi_P|^2$ by its average value unity in the integrand. We shall suppose that $\mathbf{w}$ is a bell-shaped function, such as $\exp[-(r - r_0)^2/(na)^2]$, so that $\nabla\mathbf{w} = 0$ at r_0 and the center of the packet is

$$r_0 = \int r|\mathbf{w}|^2 dV. \tag{17}$$

An alternative procedure would be to use, in place of $\mathbf{w}$, a sum over ψ_P; the two results would be equivalent, however, and the $\mathbf{w}$ method seems more convenient.

We next operate on ψ by $\mathcal{H}$. Proceeding as for (12) and neglecting $(p_{A0} - p_A)^2$, we obtain:

$$\begin{aligned}
\mathcal{H}\psi &= e^{i\theta}\left[\frac{1}{2m}[p^2 + 2(p_{A0} - p_A)\cdot p] + \mathfrak{U} + (-e)\varphi\right]\mathbf{w}\psi_P \\
&\qquad\qquad (a) \qquad\qquad\qquad\qquad\qquad (b) \qquad\qquad (c) \\
&= e^{i\theta}\left[\mathbf{w}[\mathcal{H}_0 + (-e)\varphi]\psi_P + \frac{1}{m}(p_{A0} - p_A)\cdot(\mathbf{w}p\psi_P + \psi_P p\mathbf{w})\right. \\
&\qquad\qquad\qquad (d) \qquad\qquad\qquad (e) \\
&\qquad\left. + \frac{1}{m}p\mathbf{w}\cdot p\psi_P + \frac{1}{2m}\psi_P p^2\mathbf{w}\right].
\end{aligned} \tag{18}$$

(The letters over the terms are for later reference.)

At $r = r_0$, we again obtain $\mathcal{H}\psi = [\mathcal{E}(P) + (-e)\varphi]\psi$ except for term (e), which we shall show is negligible to a first approximation. Similarly the average value of $\dot{r}$ is again nearly v over the cell at r_0; we shall show later that the whole packet itself also moves with the same velocity.

Except for terms (b) and (d), $\mathcal{H}\psi$ consists of slowly varying functions of r times ψ_P. Terms (b) and (d), however, which contain $p\psi_P$, have behavior in each unit cell quite different from ψ_P. They are, however, still functions of the form $\exp[i(\boldsymbol{P}\cdot\boldsymbol{r})/\hbar]$ times a periodic function of $\boldsymbol{r}$. They can, therefore, be expanded in a series of wave functions $\psi_{iP}(\boldsymbol{r})$ coming from other energy bands than does $\psi_P(\boldsymbol{r})$. If we suppose $\psi_P(\boldsymbol{r})$ comes from band "zero", and omit the subscript zero for simplicity so that $\psi_{0P}(\boldsymbol{r}) = \psi_P(\boldsymbol{r})$, then we can write

$$\boldsymbol{p}\psi_P = \sum_i (i|\boldsymbol{p}|0)\psi_{iP} \tag{19}$$

where

$$(i|\boldsymbol{p}|0) = \int \psi_{iP}^*\boldsymbol{p}\psi_P dV \tag{20}$$

integrated over unit volume. The term for $i = 0$ is

$$(0|\boldsymbol{p}|0) = m\boldsymbol{v} = m\nabla_P\mathcal{E}(\boldsymbol{P}) \tag{21}$$

in accordance with Section 15.4. We shall show that the contributions of ψ_{iP} are negligible for electric and magnetic fields of ordinary magnitudes. We shall postpone the proof of this until the end of the section and in the intermediate steps will omit effects depending on $\psi_{iP}(\boldsymbol{r})$.

Next we shall show that (c) and (e) are negligible compared to (b) and (d) respectively. Considering only the $(0|\boldsymbol{p}|0)$ term in (b) we have, for order of magnitude,

$$(b) : (c) = \mathrm{w}\psi_P m\boldsymbol{v} : \mathrm{w}\psi_P \frac{\hbar}{na} = mv : \hbar/na; \tag{22}$$

mv for a thermal electron corresponds to a wave length of about 50 angstroms and hence to $mv = h/50$ angstroms. Thus if $na > 50$ angstroms, meaning that the wave-packet is many wave lengths in size, (c) $\ll$ (b). Precisely similar estimates apply to (d) and (e) so that (e) may be regarded as negligible. As in Section 15.5, the term (e) contributes to the spreading of the wave-packet. However, we shall neglect the spreading effects in this treatment. Thus we may write for the approximate value of $\mathcal{H}\psi$:

$$\mathcal{H}\psi = e^{i\theta}\Big[\underset{(a)}{\mathrm{w}[\mathcal{H}_0 + (-e)\varphi]\psi_P} + \underset{(b)}{\frac{\mathrm{w}}{m}(\boldsymbol{p}_{A0} - \boldsymbol{p}_A)\cdot\boldsymbol{p}\psi_P} + \underset{(d)}{\frac{1}{m}\boldsymbol{p}\mathrm{w}\cdot\boldsymbol{p}\psi_P}\Big]. \tag{23}$$

We have left $\mathcal{H}_0$ as an operator so as to be able to use (23) later to estimate the contribution of the ψ_{iP} terms.

We shall next evaluate $i\hbar\partial\psi/\partial t$. In this we need to consider two terms specially:

$$\frac{\partial}{\partial t}\boldsymbol{p}_{A0}(t)\cdot\boldsymbol{r} = \frac{\partial}{\partial t}\boldsymbol{p}_A[\boldsymbol{r}_0(t), t]\cdot\boldsymbol{r} = (\dot{\boldsymbol{r}}_0\cdot\nabla)\boldsymbol{p}_A\cdot\boldsymbol{r} + \frac{\partial}{\partial t}\boldsymbol{p}_A\cdot\boldsymbol{r}$$

$$= \left[(\dot{\boldsymbol{r}}_0\cdot\nabla)\boldsymbol{p}_A + \frac{\partial}{\partial t}\boldsymbol{p}_A\right]_0\cdot\boldsymbol{r} \tag{24}$$

where the derivatives in the last line are obtained by writing p_A as $p_A(r, t)$ and evaluating the result at $r = r_0$ as suggested by the subscript. The other term is

$$i\hbar \frac{\partial}{\partial t} \psi_P = i\hbar \dot{P} \cdot \nabla_P \psi_P = i\hbar \dot{P} \cdot \nabla_P e^{i(P \cdot r)/\hbar} u_P$$
$$= -\dot{P} \cdot r\psi_P + i\hbar e^{i(P \cdot r)/\hbar} \dot{P} \cdot \nabla_P u_P(r). \tag{25}$$

The last term has the periodicity of ψ_P and can, like $p\psi_P$, be expanded in ψ_{iP}:

$$e^{i(P \cdot r)/\hbar} \nabla_P u_P(r) = \sum_i (i|\nabla_P|0)\psi_{iP} \tag{26}$$

where, by a convention different from (20) above, we let

$$(i|\nabla_P|0) = \int u_{iP}^* \nabla_P u_P dV. \tag{27}$$

These terms can be re-expressed[8] in terms of $(i|p|0)$; this is unnecessary for our purpose since rough qualitative features suffice. An upper limit for $(i|\nabla_P|0)$ may be derived as follows: As P varies from the center of the Brillouin to a point on the side, the wave function u_P will acquire approximately one additional wave length (see Figure 5.7, for example). As a result u_P will change by approximately 100 per cent while P varies by $h/2a$, the half-width of the Brillouin zone. Hence $\nabla_P u_P$ will be of the same order as $(2a/h)u_P$. This leads to the conclusion that $(i|\nabla_P|0)$ will be no larger than about $2a/h$. Actually $\nabla_P u_P$ will not vary in the same way as either u_P or u_{iP} so that the integral of the product may be much smaller than $(2a/h)$. This estimate is very crude but will suffice for our purposes. We can also conclude that $(0|\nabla_P|0)$ is a pure imaginary by showing that it is the negative of its complex conjugate. This conclusion follows directly from the fact that the wave functions are normalized for all values of P so that

$$0 = \nabla_P \int u_P^* u_P dV = (0|\nabla_P|0)^* + (0|\nabla_P|0). \tag{28}$$

We may now obtain $i\hbar(\partial\psi/\partial t)$ in terms of (24), (25) and (26):

$$i\hbar \frac{\partial}{\partial t} \psi = e^{i\theta} \Bigg\{ \dot{f} w - \underset{\text{(g)}}{\left[(\dot{r}_0 \cdot \nabla) p_A + \frac{\partial}{\partial t} p_A + \dot{P} \right]_0} \cdot rw\psi_P + \underset{\text{(h)}}{i\hbar \frac{\partial w}{\partial t} \psi_P}$$
$$+ \underset{\text{(k)}}{i\hbar \dot{P} \cdot \sum_i (i|\nabla_P|0)\psi_{iP}} \Bigg\}. \tag{29}$$

[8] A. H. Wilson, *The Theory of Metals*, Cambridge at the University Press, 1936, paragraph 2.81.

Again terms from other bands appear but with small coefficients. We shall neglect at first these terms and estimate them later.

In order to find the changes in the parameters f, $\mathbf{w}$, $\boldsymbol{P}$, and $\boldsymbol{r}_0$ (which depends on $\mathbf{w}$) of the wave-packet, we multiply $\mathcal{H}\psi = i\hbar\partial\psi/\partial t$ by $\exp(-i\theta)\psi_P^*$ and integrate over a unit cell at $\boldsymbol{r}$, considering all slowly varying functions to be constant and to have the values corresponding to $\boldsymbol{r}$. The integrals of ψ_P with ψ_{iP} are zero and with ψ_P give simply the volume of the unit cell, which can be factored out. In this way we obtain from $\mathcal{H}\psi$ and from $i\hbar\partial\psi/\partial t$, respectively, the left and right sides of the following equation:

$$\underset{(a_1)}{\mathbf{w}[\mathcal{E}(\boldsymbol{P}) + (-e)\varphi]_0} + \underset{(a_2)}{(\boldsymbol{r}-\boldsymbol{r}_0)\cdot[(-e)\nabla\varphi} - \underset{(b)}{\nabla(\boldsymbol{p}_A\cdot\boldsymbol{v})]_0\mathbf{w}} + \underset{(d)}{\frac{\hbar}{i}\nabla\mathbf{w}\cdot\boldsymbol{v}}$$

$$= \dot{f}\mathbf{w} - \underset{(g_0)}{\left[(\dot{\boldsymbol{r}}_0\cdot\nabla)\boldsymbol{p}_A + \frac{\partial}{\partial t}\boldsymbol{p}_A + \dot{\boldsymbol{P}}\right]_0} \cdot \underset{(g_1)}{[\boldsymbol{r}_0\mathbf{w}} + \underset{(g_2)}{(\boldsymbol{r}-\boldsymbol{r}_0)\mathbf{w}]} + \underset{(h)}{i\hbar\frac{\partial}{\partial t}\mathbf{w}}$$

$$+ \underset{(k)}{i\hbar\dot{\boldsymbol{P}}\cdot(0|\nabla_P|0)\mathbf{w}}. \tag{30}$$

The $\boldsymbol{v}$ terms arise from the $p\psi_P$ terms of (23) in accordance with (21). The $(\boldsymbol{r}-\boldsymbol{r}_0)$ terms arise from expanding the slowly varying functions φ, $\boldsymbol{p}_A$, and $\boldsymbol{r}$ in Taylor's series about $\boldsymbol{r} = \boldsymbol{r}_0$; all partial derivatives of these functions are thus obtained by expressing φ and $\boldsymbol{p}_A$ as functions of $\boldsymbol{r}$ and t and the values correspond to $\boldsymbol{r} = \boldsymbol{r}_0$, hence subscript zero. [See equation (24) for an example.] The antecedents in previous equations are indicated by the letters.

All the symbols in equation (30), except i itself, stand for real quantities. This equation thus contains **two sets of real terms:** (I) those which simply multiply $\mathbf{w}$ by a constant—these are $\dot{f}$, (a_1), $(g_0)\cdot(g_1)$ and (k); (II) real terms proportional to $\boldsymbol{r}-\boldsymbol{r}_0$; and (III) **two imaginary terms** involving derivatives of $\mathbf{w}$. Equating set (I) to zero gives, symbolically,

$$\dot{f} = (a_1) + (g_0)\cdot(g_1) - (k). \tag{31}$$

[The contribution of (k) can be seen to be very small, corresponding to an energy of say 10^{-4} election volt, by arguments given at the end of this section. In any event (k) does not affect the charge density and hence the location or motion of the wave-packet since it enters ψ only through θ.] The meaning of equation (31) can be seen most simply from $\mathcal{H}\psi = i\hbar(\partial\psi/\partial t)$; $\mathcal{H}\psi$ was evaluated in (18) and it was shown that at $\boldsymbol{r} = \boldsymbol{r}_0$ its value was $\mathcal{E}(\boldsymbol{P}) + (-e)\varphi(\boldsymbol{r}_0)$. The terms $\dot{f}$, $(g_0)\cdot(g_1)$ and (k) from the right side of (30) combine to give essentially $i\hbar(\partial\psi/\partial t)$. Thus choosing f in accordance with equation (31) corresponds to choosing the correct rate of change of complex phase angle in ψ so that the time derivative gives $\mathcal{E}(\boldsymbol{P}) + (-e)\varphi$. Because $(g_0) \cdot (g_1)$ depends on the origin of coordinates, there is no important physical interpretation for f itself.

From set (III) we find

$$\frac{\partial}{\partial t}\mathrm{w} = -\boldsymbol{v} \cdot \nabla \mathrm{w} \quad \text{which means} \quad \mathrm{w} = \mathrm{w}\left(\boldsymbol{r} - \int \boldsymbol{v}\,dt\right) \tag{32}$$

and indicates that the packet moves undistorted with velocity $\boldsymbol{v}$. From this it is evident that

$$\dot{\boldsymbol{r}}_0 = \boldsymbol{v}. \tag{33}$$

Actually the packet will spread owing to term (e) and to other small effects neglected in our approximations.

Set II is satisfied for arbitrary values of $\boldsymbol{r} - \boldsymbol{r}_0$ only if the vectors multiplying $\boldsymbol{r} - \boldsymbol{r}_0$ have a sum of zero. The resulting equation can be solved for $\dot{\boldsymbol{P}}$:

$$\overset{}{\dot{\boldsymbol{P}}} = -(-e)\nabla\varphi + \underset{}{\overset{(a_2)}{}}\ \nabla(\boldsymbol{p}_A \cdot \boldsymbol{v}) - \left[(\dot{\boldsymbol{r}}_0 \cdot \nabla)\boldsymbol{p}_A + \frac{\partial}{\partial t}\boldsymbol{p}_A\right].$$

$$\qquad\qquad (a_2) \qquad\qquad (b) \qquad\qquad (g_0)\cdot(g_2)$$

Using to simplify (b) the relationships

$$\nabla(\boldsymbol{p}_A \cdot \boldsymbol{v}) = (\boldsymbol{v} \cdot \nabla)\boldsymbol{p}_A + \boldsymbol{v} \times (\nabla \times \boldsymbol{p}_A)$$

$$\nabla \times \boldsymbol{p}_A = \frac{(-e)}{c}\nabla \times \boldsymbol{A} = \frac{-e}{c}\boldsymbol{H},$$

and for $(g_0) \cdot (g_2)$ the equation

$$\dot{\boldsymbol{r}}_0 = \boldsymbol{v}$$

and the expressions (6a) and (5)

$$(-e)\boldsymbol{E} = (-e)\left[-\nabla\varphi - \frac{1}{c}\frac{\partial \boldsymbol{A}}{\partial t}\right] = -(-e)\nabla\varphi - \frac{\partial}{\partial t}\boldsymbol{p}_A,$$

we obtain the desired expression for $\dot{\boldsymbol{P}}$:

$$\dot{\boldsymbol{P}} = (-e)\left[\boldsymbol{E} + \frac{1}{c}\boldsymbol{v} \times \boldsymbol{H}\right]. \tag{34}$$

Still reserving for a moment consideration of the effect of the other energy bands, let us review what has been done. We have prepared a wave function ψ which has the same charge density pattern, namely, $|\psi_P|^2$, near $\boldsymbol{r}_0$ as a wave function corresponding to state $\boldsymbol{P}$ of the Brillouin zone. At $\boldsymbol{r}_0$ this wave function satisfies, quite accurately, $\mathcal{H}\psi = [\mathcal{E}(\boldsymbol{P}) + (-e)\varphi]\psi$ corresponding to the energy of state $\boldsymbol{P}$ plus the added energy due to the potential φ; also it gives in the unit cell at $\boldsymbol{r}_0$ the same average velocity $\boldsymbol{v}$

normally associated with $\boldsymbol{P}$ by $\boldsymbol{v} = \nabla_P \mathcal{E}(\boldsymbol{P})$. This wave-packet then properly represents a localized wave function which corresponds to the extended wave function ψ_P in the absence of the fields. Under the influence of $\mathcal{H}\psi = i\hbar\partial\psi/\partial t$ this wave function ψ changes. For convenience we have described this change in terms of the symbols $\boldsymbol{P}, f$, and $\mathbf{w}$ which describe the wave-packet. The procedure of satisfying sets I, II, and III has been somewhat arbitrary. However, we have obtained a solution in this way and since the time dependence due to $\mathcal{H}\psi = i\hbar\partial\psi/\partial t$ is unique, our solution for the rate of change of ψ must be the only possible one. Because of the convenient form in which our solution is written, however, we can at once interpret it in terms of bodily motion and change in effective, $\boldsymbol{P}$ value. The results are in agreement with the interpretation of Chapter 7.

We now return to the terms from the other energy bands. The wave-packet in terms of ψ_P does not constitute a solution of $\mathcal{H}\psi = i\hbar\partial\psi/\partial t$ since terms from other bands enter through $p\psi_P$ in (b) and (d) and through ∇_P in (k). We must, therefore, modify the wave form by adding a series of the form $a_i\psi_{iP}$ to ψ_P. We can evaluate the coefficients a_i (which may depend on $\boldsymbol{r}$) in this series by multiplying $\mathcal{H}\psi = i\hbar\partial\psi/\partial t$ by $\exp(-i\theta)\psi_{iP}^*$ and integrating over the unit cell at $\boldsymbol{r}$. The terms of set I do not cancel now because $\mathcal{H}_0\psi_{iP} = \mathcal{E}_i(\boldsymbol{P})\psi_{iP}$ for band i. This leads to a term of the form

$$[\mathcal{E}_i(\boldsymbol{P}) - \mathcal{E}(\boldsymbol{P})]a_i$$

in the resulting equation. The other terms having a_i as a coefficient all involve the size of the wave-packet or the fields and are small. The terms coming from the approximate solution ψ_P through (b), (d), and (k) are of the following orders of magnitude:

$$\text{(b)} \qquad \frac{1}{m}(p_{A0} - p_A) \cdot (i|p|0) \approx \frac{1}{m}\frac{eH_z na}{ca} \cdot \frac{\hbar}{a} = n\hbar\omega;$$

$$\text{(d)} \qquad \frac{1}{m}(i|p|0) \cdot \frac{pw}{w} \approx \frac{1}{m}\frac{\hbar}{a} \cdot \frac{\hbar}{na} = \frac{1}{mn}\left[\frac{\hbar}{a}\right]^2;$$

$$\text{(e)} \qquad \hbar\dot{\boldsymbol{P}} \cdot (i|\nabla_P|0) \sim \hbar e\boldsymbol{E} \cdot \frac{2a}{h} = eEa/\pi.$$

The reader may verify, with the aid of Figure 10.2, that all of these energies are much smaller than $\mathcal{E}_i(\boldsymbol{P}) - \mathcal{E}(\boldsymbol{P})$, which is a few electron volts, so that $|a_i|$ is $\ll 1$ and the contribution of other bands to the wave-packet is small. (Questions of convergence in the sum over i may be investigated with the aid of the techniques of the "f-sum" rule. See Seitz and Wilson, *loc. cit.*) The effect of term (e) has been discussed by Houston, and earlier by Zener[9] in a different form. It causes appreciable transition

[9] C. Zener, *Proc. Roy. Soc.* London **145A**, 523–529 (1934).

probability from one band to another when the voltage drop eE across one unit cell a is comparable to the energy gap $\mathcal{E}_i(\mathbf{P}) - \mathcal{E}(\mathbf{P})$ where the bands are closest together. This leads to fields of about one volt per unit cell or 10^7 to 10^8 volts/cm. For smaller fields, the effect of other bands are shown by Houston to be negligible. We shall not repeat his rigorous treatment here but will be satisfied by the qualitative results obtained above.

15.7 ON THE PAULI PRINCIPLE AND ANTISYMMETRIC WAVE FUNCTIONS

We shall next consider in general terms the total wave function for M electrons occupying a certain set of quantum states having wave functions $\chi_1 \cdots \chi_M$. These wave functions will depend on the coordinates $r = x, y, z$ of the electron and upon its spin quantum number s, which may take on either of the values $+\frac{1}{2}$ or $-\frac{1}{2}$. It is convenient to deal with wave functions which are eigenfunctions of s and to represent the spin wave functions as $\alpha(s)$ and $\beta(s)$ where

$$\alpha(\tfrac{1}{2}) = \beta(-\tfrac{1}{2}) = 1, \quad \alpha(-\tfrac{1}{2}) = \beta(+\tfrac{1}{2}) = 0.$$

The wave functions χ are thus written symbolically in the form

$$\chi_n(q_i) = \phi_n(r_i)\alpha(s_i) \tag{1}$$

where the symbol q_i represents both coordinates and spin, and χ_n both the spatial and the spin wave function. In the next paragraph we shall consider certain general properties of the χ's.

If electron i occupies wave function χ_n, then the probability of finding it in volume $dV_i = dx_i dy_i dz_i$ with a specified spin s_i, a situation we refer to as dq_i, is represented as

$$|\chi_n|^2 dq_i = |\phi_n(r_i)|^2 dV_i \alpha^2(s_i). \tag{2}$$

The total probability regardless of spin value of finding electron i in dV_i is denoted by $\rho_n(r_i)dV_i$. It is obtained by summing (2) over the two values of s_i, thus obtaining

$$\rho_n(r_i)dV_i = |\phi_n(r_i)|^2 dV_i[\alpha^2(\tfrac{1}{2}) + \alpha^2(-\tfrac{1}{2})] = |\phi_n(r_i)|^2 dV_i. \tag{3}$$

The same result will be obtained for $\phi_n(r_i)\beta(s_i)$ so that the particle density ρ_n associated with χ_n is the same regardless of the spin function involved. In calculating average values, it is necessary to sum over the allowed values of s_i as well as integrating over dV_i; this is represented symbolically by $\int dq_i$. It is evident that, if the ϕ_n are an orthonormal set, then so are the χ_n and, furthermore, that wave functions of opposite spin are orthogonal since $\alpha(s)\beta(s)$ is zero when summed over s. In the following section we shall consider *a particular set of wave functions* ϕ *appropriate to dealing with a valence band with one electron missing.* In this section, however, we shall consider only general attributes of the problem of putting M electrons into M wave functions $\chi_1 \cdots \chi_M$. We shall, however, assume that the ϕ_n and χ_n are orthonormal sets, as is the case for the wave functions of an energy band.

A wave function which represents one electron in each of the M states[1] is

$$\chi_0 = \chi_1(q_1)\chi_2(q_2)\chi_3(q_3) \cdots \chi_M(q_M). \tag{4}$$

Other equally good wave functions would have electron (2) in χ_1 and (1) in χ_2 or other similar permutations, the most general of which can be represented as

$$\mathcal{Q}\chi_0 = \chi_1(\mathcal{Q}q_1)\chi_2(\mathcal{Q}q_2) \cdots \chi_M(\mathcal{Q}q_M) \tag{5}$$

where $\mathcal{Q}$ is a "permutation operator" which rearranges the q_i. For example $\mathcal{Q}q_1 = q_2$, $\mathcal{Q}q_2 = q_1$ and $\mathcal{Q}q_i = q_i$ for $i > 2$ defines an operator $\mathcal{Q}$ which interchanges electrons (1) and (2). There are $M!$ different ways of rearranging the M different q_i's and there are consequently $M!$ distinct possible $\mathcal{Q}$'s (including $\mathcal{Q} = 1$, the identical permutation which has no effect) and thus $M!$ wave functions of the form $\mathcal{Q}\chi_0$.

We next consider two closely related questions: (1) which $\mathcal{Q}\chi_0$ or what linear combination of $\mathcal{Q}\chi_0$ is correct to represent the M electrons occupying the M wave functions? (2) How should the Pauli exclusion principle be expressed so as to rule out wave functions such as

$$\chi_2(q_1)\chi_2(q_2)\chi_3(q_3) \cdots \chi_M(q_M)$$

for which two electrons use the same χ_2? Both of these questions are answered by the requirement that wave function for the M electrons must be such a function of the q's, denoted by $\mathbf{A}(q_1, q_2, \cdots, q_n)$, that, if any two electrons are interchanged, the result is equivalent to multiplying $\mathbf{A}$ by -1; thus

$$\mathbf{A}(q_2, q_1, q_3, q_4, \cdots, q_M) = -\mathbf{A}(q_1, q_2, q_3, q_4, \cdots, q_M),$$

or in general

$$\mathbf{A}(\mathcal{Q}q) = \epsilon(\mathcal{Q})\mathbf{A}(q) \tag{6}$$

where $\epsilon(\mathcal{Q}) = \pm 1$ depending on whether the permutation represented by $\mathcal{Q}$ can be produced by an even or an odd number of individual interchanges; as is familiar in the theory of determinants, to which we shall shortly refer more specifically, a given permutation $\mathcal{Q}$ can be produced in many ways by the cumulative effect of a set of interchanges of indi-

[1] Such a wave function will be good eigenfunction of $\mathcal{H}$ only under special circumstances; see 15.8. However, this question does not concern us here since we are interested in the interpretation of such wave functions independently of whether they are eigenfunctions or not.

vidual pairs of the q_i's. Furthermore, for all sets of interchanges in pairs which produce $\mathcal{Q}$, the number of interchanges is always even, or else always odd, depending upon $\mathcal{Q}$. Hence the $\mathcal{Q}$'s can be classified as "even" or "odd" and $\epsilon(\mathcal{Q})$ defined as above. A wave function of the form just considered is said to be *antisymmetrical*, hence the symbol **A**.

In terms of these definitions it can be shown (the proof is left as an exercise to the reader) that the only wave function of form **A** which can be produced by a linear combination of χ_0 and the permuted $\mathcal{Q}\chi_0$ functions is

$$\mathbf{A} = \frac{1}{\sqrt{M!}} \sum \epsilon(\mathcal{Q})\mathcal{Q}\chi_0; \tag{7}$$

other equivalent wave functions can be obtained from this **A** by multiplying it by a constant, but no linearly independent **A** can be produced. The factor $1/\sqrt{M!}$ serves to normalize the wave function, as we shall show below. [See text following equation (13).] This form of **A** is identical with the determinantal form

$$\mathbf{A} = \frac{1}{\sqrt{M!}} \begin{vmatrix} \chi_1(q_1) & \chi_2(q_1) & \cdots & \chi_M(q_1) \\ \chi_1(q_2) & \chi_2(q_2) & \cdots & \chi_M(q_2) \\ \cdots & \cdots & \cdots & \cdots \\ \chi_1(q_M) & \chi_2(q_M) & \cdots & \chi_M(q_M) \end{vmatrix} \tag{8}$$

as may be verified by expanding the determinant in terms of its permutations. From the determinantal form it is seen that, if two of the wave functions used in χ_0 are identical, two columns of the determinant will be equal and the determinant will vanish. This shows that the antisymmetric requirement on **A** excludes the possibility of the same wave function being used twice; this is the analytical equivalent of the simpler statement that no two electrons can occupy the same quantum state.

We shall next calculate the density of electrons produced by **A**. This is done as follows. The probability that electron 1 has its coordinates in the range

$$\begin{aligned} dq_1 &= dx_1dy_1dz_1 \quad \text{and} \quad s_1 = -\tfrac{1}{2} \quad \text{or} \quad s_1 = +\tfrac{1}{2} \\ &= dV_1 \qquad\qquad\; \text{and} \quad s_1 = -\tfrac{1}{2} \quad \text{or} \quad s_1 = +\tfrac{1}{2} \end{aligned}$$

and electron 2 has its coordinates in the range dq_2, etc., is, according to the basic quantum-mechanical assumptions,

$$|\mathbf{A}(q_1, q_2, \cdots, q_M)|^2 dq_1dq_2 \cdots dq_M. \tag{9}$$

If the wave function is properly normalized, the integral over all the dq_i should lead to unity. The total probability that electron (1) be in the range dq_1, regardless of the positions and spins of electrons $2, \cdots, M$, is obtained by integrating over dq_2, etc. (that is, integrating over dV_2 and summing over s_2), thus obtaining

$$\rho_1(\boldsymbol{q}_1)d\boldsymbol{q}_1 = d\boldsymbol{q}_1 \int_{\boldsymbol{q}_2} \cdots \int_{\boldsymbol{q}_M} |\mathbf{A}|^2 d\boldsymbol{q}_2 \cdots d\boldsymbol{q}_M. \tag{10}$$

The value of $\rho_1(\boldsymbol{q}_1)$ is the probability of finding electron 1 in the range $d\boldsymbol{q}_1$. If multiplied by the charge on the electron and summed over s_1, it gives the average charge density due to this electron. We can evaluate ρ_1 by using form (7) for **A**. We have

$$\rho_1(\boldsymbol{q}_1) = \frac{1}{M!}\int_{\boldsymbol{q}_2} \cdots \int_{\boldsymbol{q}_M} \Sigma\epsilon(\mathfrak{Q})\mathfrak{Q}\chi_0{}^*\Sigma\epsilon(\mathfrak{R})\mathfrak{R}\chi_0 d\boldsymbol{q}_2 \cdots d\boldsymbol{q}_M, \tag{11}$$

where the first sum extends over the $M!$ set of $\mathfrak{Q}$'s and the second independently over the set of $\mathfrak{R}$'s. Now since all the functions in the product

$$\chi_0 = \chi_1(\boldsymbol{q}_1)\chi_2(\boldsymbol{q}_2) \cdots \chi_M(\boldsymbol{q}_M)$$

are different and since they are assumed to be orthonormal, the only terms which integrate to a finite value are those for which $\mathfrak{Q} = \mathfrak{R}$; for all others the permutation of $\boldsymbol{q}$'s in $\chi_0{}^*$ will be different from those in χ_0 and consequently at least one integration of the form $\chi_m{}^*(\boldsymbol{q}_i)\chi_n(\boldsymbol{q}_i)d\boldsymbol{q}_i = 0$ will occur. Thus (11) reduces to a single sum over $\mathfrak{Q}$:

$$\rho_1(\boldsymbol{q}_1) = \frac{1}{M!}\Sigma \int_{\boldsymbol{q}_2} \cdots \int_{\boldsymbol{q}_M} \mathfrak{Q}\chi_0{}^*\mathfrak{Q}\chi d\boldsymbol{q}_2 \cdots d\boldsymbol{q}_M. \tag{12}$$

In the sum there will be $(M-1)!$ terms in which $\chi_1{}^*(\boldsymbol{q}_1)\chi_1(\boldsymbol{q}_1)$ occurs while $\boldsymbol{q}_2 \cdots \boldsymbol{q}_M$ are permuted through $\chi_2 \cdots \chi_M$. Each of the $M-1$ integrations for any one of these terms gives a factor $\int \chi_n{}^*(\boldsymbol{q}_i)\chi_n(\boldsymbol{q}_i)d\boldsymbol{q}_i = 1$, so that in the summation the term $\chi_1{}^*(\boldsymbol{q}_1)\chi_1(\boldsymbol{q}_1)$ is obtained $(M-1)!$ times over. The same result will be obtained for every $\chi_n{}^*(\boldsymbol{q}_1)\chi_n(\boldsymbol{q}_1)$, so that we obtain finally

$$\rho_1(\boldsymbol{q}_1) = \frac{(M-1)!}{M!}\sum_n \chi_n{}^*(\boldsymbol{q}_1)\chi_n(\boldsymbol{q}_1) = \frac{1}{M}\sum_n |\chi_n(\boldsymbol{q}_1)|^2. \tag{13}$$

If we now integrate $\rho_1(\boldsymbol{q}_1)d\boldsymbol{q}_1$, the value is unity since the χ_n are supposed to be normalized. (*This shows that* **A** *is properly normalized.*) We can then interpret ρ_1 as the probability density for electron 1, its value being simply the average for all the individual charge densities in the set of M wave functions χ_i. From the symmetry of **A** and of the result for $\rho_1(\boldsymbol{q}_1)$, it is evident that the same density will be obtained from every electron leading to a total average charge density $\rho_T(\boldsymbol{q}_1)$:

$$\rho_T(\boldsymbol{q}_1)d\boldsymbol{q}_1 = \sum_n |\chi_n(\boldsymbol{q}_1)|^2 d\boldsymbol{q}_1, \tag{14}$$

where $\rho_T(q_1)dq_1$ is the average number of electrons found in dq_1; that is, it is the probability of finding electron 1 in dq_1 plus similar probabilities for all the other electrons. The net charge density is thus just the same as if χ_0 itself, rather than **A**, had been used so that the Pauli principle does not affect the additivity of the charge density.

A precisely similar treatment can be carried out for the current operator corresponding to each electron. The same orthogonality conditions will eliminate terms with mixed $\chi_n{}^*$ and χ_m with $n \neq m$ and the end result is that the average value of the current due to the M electrons is simply the sum of M currents corresponding to the M wave functions χ_1 to χ_M.

We must next show that if the time dependence of each individual χ_i is obtained by requiring it to satisfy the equation

$$\mathcal{H}_1\chi_i(q_1, t) = i\hbar\frac{\partial}{\partial t}\chi_i(q_1, t) \tag{15}$$

where $\mathcal{H}_1$ is the Hamiltonian operator for the coordinates and momenta of electron 1 in the presence of the applied field, then the time dependence of **A** is properly given by expressing **A** in terms of the time-dependent χ's. The equation which **A** must satisfy is

$$\mathcal{H}_T\mathbf{A} = (\mathcal{H}_1 + \mathcal{H}_2 + \cdots + \mathcal{H}_M)\mathbf{A} = i\hbar\frac{\partial}{\partial t}\mathbf{A}. \tag{16}$$

We shall show that, if **A** is expressed in form (7), each individual term satisfies this equation and hence that **A** does. It suffices thus to investigate χ_0. We have

$$\begin{aligned}\mathcal{H}_T\chi_0 &= [\mathcal{H}_1\chi_1(q_1t)]\chi_2(q_2, t)\cdots\chi_M(q_M, t)\\ &\quad + \chi_1(q_1, t)[\mathcal{H}_2\chi_2(q_2, t)]\cdots\chi_M(q_M, t)\\ &\quad \cdots\\ &\quad + \chi_1(q_1, t)\chi_2(q_2, t)\cdots[\mathcal{H}_M\chi_M(q_M, t)]\\ &= i\hbar\dot{\chi}_1(q_1, t)\chi_2(q_2, t)\cdots\chi_M(q_M, t)\\ &\quad + i\hbar\chi_1(q_1, t)\dot{\chi}_2(q_2, t)\cdots\chi_M(q_M, t)\\ &\quad \cdots\\ &\quad + i\hbar\chi_1(q_1, t)\chi_2(q_2, t)\cdots\dot{\chi}_M(q_M, t)\\ &= i\hbar\frac{\partial}{\partial t}\chi_0\end{aligned} \tag{17}$$

which proves the desired result.

The Fock Equations. The determinantal wave functions are only approximate solutions for wave equations in which electronic interactions are included. The problem of finding the best set of χ's for a given Hamiltonian has been considered by Fock[2] and Slater[3] and the resulting equations which the individual one-electron χ's must satisfy are called the *Fock equations*. In these equations each electron is acted upon by the average potential of the other electrons, but in taking the average the influence of the Pauli exclusion principle is included. The Fock equations are treated in texts by Frenkel[4] and by Seitz;[5] the latter author makes special application to the problems treated in this appendix and uses *Koopmans' theorem*[6] to show that changes in the energy of the system are given by changes in $\mathcal{E}(\boldsymbol{P})$ for the individual wave functions.

15.8 THE CONSTRUCTION AND BEHAVIOR OF A HOLE WAVE-PACKET

In this section we shall deal with a valence band which is missing one electron so that in effect we shall study the properties of a hole. We shall, for simplicity, suppose that the missing wave function comes entirely from the states with one value of the spin so that the states with the other value of the spin are fully occupied. If the band of energy levels is completely filled, then all of the allowed quantum states will be occupied. Consequently, the application of electric or magnetic fields cannot change the way in which the electrons occupy the quantum states since it is required by the Pauli exclusion principle to use all the states to accommodate all the electrons. As a consequence, the electrons in a filled band are not strongly influenced by electric or magnetic fields but instead contribute only to the dielectric constant and the diamagnetic susceptibility. However, if there is a hole in the valence band, currents will be produced by electric fields which will be modified by magnetic fields in just the way expected for a positive charge. This is the theorem which we shall prove on the basis of equation (1) of Section 15.6 and the Pauli principle.

In this treatment we shall suppose that the spatial (not spin) parts of the wave functions of the electrons in the valence band all satisfy the same Schroedinger equation

$$\mathcal{H}_0\psi_\alpha = \left[\frac{1}{2m}p^2 + \mathcal{U}(r)\right]\psi_\alpha(r) = \mathcal{E}_\alpha\psi_\alpha(r). \tag{1}$$

If periodic boundary conditions are employed, a suitable set of ψ_α are the N wave functions ψ_P where N is the number of unit cells in the crystal. For other boundary conditions also, there will be N wave functions covering

[2] V. Fock, *Z. Physik* **61**, 126–148 (1930).
[3] J. C. Slater, *Phys. Rev.* **35**, 210–211 (1930).
[4] *Wave Mechanics, Advanced General Theory*, Oxford, 1934.
[5] *Modern Theory of Solids*, McGraw-Hill Book Co., New York, 1940.
[6] T. Koopmans, *Physica* **1**, 104–113 (1933).

the same band of energies. Since there are severe conceptual difficulties, however, in combining periodic boundary conditions and magnetic fields,[1] we shall consider a rectangular block of crystal outside of which the wave functions must vanish. We shall refer to this set of wave functions as ψ_α.

We first form a wave-packet function ϕ_1 by superposition of the wave functions ψ_α. From the completeness of the set of wave functions ψ_α, it follows that they can be used to produce as well defined a wave-packet for an electron in the valence band as the wave functions ψ_P used in Section 15.6. However, when the wave-packet formed with the ψ_α reaches the boundary of the crystal, it will be reflected since the ψ_α functions vanish outside the crystal; whereas if it were formed from the ψ_P with periodic boundary conditions, as it impinged on one face it would re-enter the opposite face. We shall suppose that we deal with wave-packets corresponding to a hole near the center of the crystal so that these matters need not concern us.

We consider, therefore, a wave-packet $\phi_1(r, t)$ which is composed of a sum of the ψ_α:

$$\phi_1(r, t) \equiv \sum_{\alpha=1}^{N} a_{1\alpha}(t)\psi_\alpha(r). \tag{2}$$

We suppose that ϕ_1 is normalized. Under the influence of electric and magnetic fields, the wave function ϕ_1 will transform so that the wave-packet is accelerated in accordance with equations (1) and (4) of Section 15.6. In the following equations, we shall use the symbol ϕ_1 to represent a time-dependent function which satisfies the wave equation

$$\mathcal{H}\phi_1 = i\hbar\frac{\partial}{\partial t}\phi_1 \tag{3}$$

where $\mathcal{H}$ includes terms due to the applied fields.

We next construct $N - 1$ other wave functions, $\phi_n(r, t)$ which are also solutions of (3) and together with ϕ_1 form an orthonormal set. It is possible to do this because at any particular time, say $t = 0$, we can always construct an orthonormal set from ϕ_1 and any $N - 1$ functions ψ_α which together with ϕ_1 form a linearly independent set of N functions.[2] These other ϕ's may be written as

$$\phi_n = \sum_{\alpha=1}^{N} a_{n\alpha}(t)\psi_\alpha(r) \tag{4}$$

where the dependence of $a_{n\alpha}$ on t makes them solutions of (3). Further-

[1] Probably closely related to the question of the existence of magnetic poles.

[2] See, for example, E. Madelung, *Die Mathematischen Hilfsmittel des Physikers*, p. 24, Dover Publications, New York, 1943, and Courant-Hilbert, *Methoden der Mathematische Physik*, Kap. II, par. 1, Dover Publications, New York, 1947, and the discussion in connection with formula (16) of Section 14.5.

more, if the set of other functions ϕ_n and ϕ_1 are orthonormal at $t = 0$ and all satisfy the same equation (3), then they are orthonormal at all later times. This follows from the Hermitian property of $\mathcal{H}$ as follows:

$$\frac{d}{dt}\int \phi_n{}^*\phi_m dV = \int \left[\left(\frac{1}{i\hbar}\mathcal{H}\phi_n\right)^* \phi_m + \phi_n{}^* \frac{1}{i\hbar}\mathcal{H}\phi_m\right] dV$$

$$= \frac{1}{i\hbar}\int [-(\mathcal{H}\phi_n)^*\phi_m + \phi_n{}^*\mathcal{H}\phi_m] dV = 0, \qquad (5)$$

which shows that the integral is constant so that the orthonormal relationship is preserved.

Charge Density and Current Density for a Hole Wave-packet. We shall next use the $2N$ wave functions, consisting of $\phi_1(r)\alpha(s), \cdots, \phi_N(r)\alpha(s)$ for one spin and $\phi_1(r)\beta(s), \cdots, \phi_N(r)\beta(s)$ for the opposite spin in determinantal wave functions in accordance with the principles of Section 15.7. In this analysis, ϕ_1, which represents the missing electron and corresponds to the hole, is imagined to be made from wave functions near the top of the valence band for which the energy decreases with increasing $\boldsymbol{P}$ as discussed in connection with equation (4) of Section 7.6 in the text. This feature is essential in determining the behavior of the hole.

We consider first the wave function for the filled band. The antisymmetrical wave function for this case is a determinant using all the ϕ's. We shall denote it by $\mathbf{A}_f(\phi)$, the f, for "full", meaning that all the ϕ's are used so that the band is full. However, since the ϕ's are simply linear combinations of the ψ_α's, this determinant has columns which are linear combinations of the columns of $\mathbf{A}_f(\psi)$ which is made by using all the ψ's. Since a determinant is unaltered by adding to one column linear combinations of other columns, $\mathbf{A}_f(\phi)$ and $\mathbf{A}_f(\psi)$ represent the same wave function except for a multiplying factor with absolute value unity.

Now under the influence of electric fields, $\mathbf{A}_f(\psi)$ will make contributions to the dielectric and diamagnetic susceptibility but will not contribute any conduction current. Since $\mathbf{A}_f(\phi)$ is the same wave function, it will, of course, vary in time in the same way and produce the same effects as $\mathbf{A}_f(\psi)$.

Next we consider $\mathbf{A}_p(\phi)$ which is made up of all the wave functions except $\phi_1\alpha$. It represents a charge density ρ_p. According to equation (14) of 15.7, ρ_p is equal to the sum of the charge densities for all the wave functions it contains. It is, therefore, related to the electron (not charge) density ρ_f of $\mathbf{A}_f(\phi)$ by the equation

$$\rho_p(r) + \rho_1(r) = \rho_f(r), \quad \rho_1(r) \equiv |\phi_1(r)|^2 \qquad (6)$$

since the sum in $\rho_p(r)$ differs from $\rho_f(r)$ only in having $|\phi_1(r)|^2 \equiv \rho_1(r)$ missing. Similarly the particle (not electric) currents due to $\mathbf{A}_p(\phi)$ will differ from those of $\mathbf{A}_f(\phi)$ only by the current of the wave-packet ϕ_1. If we let J represent the particle current (electrons/cm^2 sec), then the electric charge and current densities will be

$$-e\rho_p - e\rho_1 = -e\rho_f \quad \text{or} \quad -e\rho_p = -e\rho_f + e\rho_1 \tag{7}$$

$$-eJ_p - eJ_1 = -eJ_f \quad \text{or} \quad -eJ_p = -eJ_f + eJ_1. \tag{8}$$

Thus the added hole will add the charge and current corresponding to a charge $+e$ moving with the wave-packet motion of ϕ_1.

In order to prove that this behavior corresponds to a plus charge and a plus mass, we must show that the acceleration of the particle is given by the appropriate formula. Since the wave-packet is made from wave functions near the maximum energy in the band, the value of $\mathcal{E}(\boldsymbol{P})$ will be given by equation (4) of Section 7.6 and will lead to equation (5) of Section 7.6 or

$$\boldsymbol{v} = -\boldsymbol{P}/m, \tag{9}$$

where m is a positive quantity. Inserting this in the equation for $\dot{\boldsymbol{P}}$ derived as (34) of Section 14.6 leads to

$$m\dot{\boldsymbol{v}} = -\dot{\boldsymbol{P}} = +e[\mathcal{E} + \boldsymbol{v} \times \boldsymbol{H}/c] \tag{10}$$

which is appropriate to a positive charge.

Thus if the wave-packet function for the electron missing from the band comes from the top of the band, so that the minus sign occurs in the relationship [equation (9) above] between $\boldsymbol{v}$ and $\boldsymbol{P}$, then the motion of the remaining electrons is such that the currents and charge densities they produce are the same as if a wave-packet of plus charge and plus mass were moving under the influence of the fields.

These results can be given a more general expression in terms of the momentum and energy of all the electrons in the energy band. We represent by $\boldsymbol{P}_n$ and $\boldsymbol{P}_p$, $\mathcal{E}_c(\boldsymbol{P}_n)$ and $\mathcal{E}_v(\boldsymbol{P}_p)$, and $\boldsymbol{v}_n = \nabla_{\boldsymbol{P}_n}\mathcal{E}_c(\boldsymbol{P}_n)$ and $\boldsymbol{v}_p = \nabla_{\boldsymbol{P}_p}\mathcal{E}_v(\boldsymbol{P}_p)$ the values respectively for an excess electron and a hole of the momentum, the energy and the group velocity for an appropriate wave-packet wave function. The energy $\mathcal{E}_B$ and momentum $\boldsymbol{P}_B$ for the entire energy band are $\mathcal{E}_B = \mathcal{E}_0 + \mathcal{E}_c(\boldsymbol{P}_n)$ and $\boldsymbol{P}_B = \boldsymbol{P}_n$ for an electron and $\mathcal{E}_B = \mathcal{E}_0 - \mathcal{E}_v(\boldsymbol{P}_p)$ and $\boldsymbol{P}_B = -\boldsymbol{P}_p$ for a hole. In terms of these definitions, the equations for packet velocity and acceleration may be expressed in a symmetrical way in terms of $\mathcal{E}_B$ and $\boldsymbol{P}_B$ as follows:

Excess Electron	Hole	
$\boldsymbol{P}_B = \boldsymbol{P}_n$	$\boldsymbol{P}_B = -\boldsymbol{P}_p$	(a)
$\mathcal{E}_B = \mathcal{E}_0 + \mathcal{E}_c(\boldsymbol{P}_B)$	$\mathcal{E}_B = \mathcal{E}_0 - \mathcal{E}_v(-\boldsymbol{P}_B) = \mathcal{E}_0 - \mathcal{E}_v(\boldsymbol{P}_B)$	(b)
$\boldsymbol{v}_n = \nabla_{\boldsymbol{P}_n}\mathcal{E}_c(\boldsymbol{P}_n) = \nabla_{\boldsymbol{P}_B}\mathcal{E}_B(\boldsymbol{P}_B)$	$\boldsymbol{v}_p = \nabla_{\boldsymbol{P}_p}\mathcal{E}_v(\boldsymbol{P}_p) = \nabla_{\boldsymbol{P}_B}\mathcal{E}_B(\boldsymbol{P}_B)$	(11)
$\dot{\boldsymbol{P}}_B = \dot{\boldsymbol{P}}_n = -e\left[\mathcal{E} + \frac{1}{c}\boldsymbol{v}_n \times \boldsymbol{H}\right]$	$\dot{\boldsymbol{P}}_B = -\dot{\boldsymbol{P}}_p = +e\left[\mathcal{E} + \frac{1}{c}\boldsymbol{v}_p \times \boldsymbol{H}\right]$	(12)

where equality (a) holds because $\mathcal{E}_v(\boldsymbol{P}) = \mathcal{E}_v(-\boldsymbol{P})$ for any energy band and (b) follows because the same fact leads to $\boldsymbol{v}(\boldsymbol{P}) = -\boldsymbol{v}(-\boldsymbol{P})$. The last two rows show that, in terms of the energy and momentum of all the electrons, the group velocity and $\dot{\boldsymbol{P}}_B$ are given by formally identical equations except for the sign of the charge. Both velocities correspond to a positive mass since $\mathcal{E}_B(\boldsymbol{P}_B)$ increases in both cases as $\boldsymbol{P}_B$ varies from the point of minimum energy for $\mathcal{E}_B$, this point corresponding to the particle at rest.

Hole Wave-packets in Terms of Energy Eigenfunctions. The wave function $\mathbf{A}_p(\phi)$ for the hole wave-packet is not an eigenfunction for the Hamiltonian for the $2N - 1$ electrons in the valence band. However, it is possible to construct $\mathbf{A}_p(\phi)$ by superposition of a set of eigenfunctions for the $2N - 1$ electrons; we shall show how this is done, the procedure and the result both being instructive in regard to a number of the properties of the holes.

The wave functions ψ_α, which are superimposed to form ϕ_1 according to equation (2), which we repeat for reference here,

$$\phi_1(r, t) = \sum_{\alpha=1}^{N} a_{1\alpha}(t)\psi_\alpha(r),$$

are eigenfunctions of the one electron Hamiltonian operator $\mathcal{H}_0$ with eigenvalues $\mathcal{E}_\alpha$ [see equation (1)]. From this it follows that an antisymmetric wave function made of the ψ_α is an eigenfunction for $2N - 1$ electrons with an eigenvalue $\mathcal{E} = \sum \mathcal{E}_\alpha$ summed over the ψ_α used in the determinant; the proof of this statement is readily obtained by the procedures used in equation (17) of Section 15.7 in which the time dependence of a determinantal wave function was studied. Such determinantal wave functions can be conveniently designated by the quantum number β of the missing ψ_β, and we shall let $\mathbf{M}_\beta(\psi)$ represent the wave function of the $2N - 1$ electrons; we shall later in equation (13) give a rule which specifies the complex phase of $\mathbf{M}_\beta$. Compared to the energy of a full valence band, the energy of $\mathbf{M}_\beta(\psi)$ is evidently $-\mathcal{E}_\beta$.

Our problem is now to find an expression for $\mathbf{A}_p(\phi)$, which has ϕ_1 missing, in terms of $\mathbf{M}_\beta(\psi)$. The solution, as we shall verify below, is

$$\mathbf{A}_p(\phi) = \sum_\alpha a_{1\alpha}^* \mathbf{M}_\alpha(\psi) \tag{13}$$

where the $a_{1\alpha}^*$ are complex conjugates of the coefficients in (2).

This result is obtained by expanding the wave function for the full band in two ways and equating the resulting expressions. If the phases of $\mathbf{A}_f(\phi)$ and $\mathbf{A}_f(\psi)$ are properly chosen, we may write

$$\mathbf{A}_f(\psi) = \mathbf{A}_f(\phi). \tag{14}$$

Now both of these wave functions stand for a determinantal wave function of $2N$ electrons with the coordinates of electron 1 occurring across the top row [see equation (8) of 15.7]. They can then be expanded in products of the first row functions and their minors. The minors of $\mathbf{A}_f(\psi)$ are evidently just the functions $\mathbf{M}_\alpha(\psi)$ expressed as functions of the coordinates and spins of electrons 2, 3, $\cdots$ $2N$. Thus we may write

$$\mathbf{A}_f(\psi) = \sum_\alpha \psi_\alpha(q_1)\mathbf{M}_\alpha(\psi; q_2, \cdots, q_{2N}). \tag{15}$$

This serves to specify the phases of the $\mathbf{M}_\alpha$ functions. Similarly

$$\mathbf{A}_f(\phi) = \sum_s \phi_s(q_1)\mathbf{M}_s(\phi; q_2 \cdots q_{2N}) \tag{16}$$

where $\mathbf{M}_1(\phi, q_2 \cdots q_{2N})$ is evidently, except for a phase factor, $\mathbf{A}_p(\phi)$. In order to evaluate $\mathbf{M}_1(\phi)$ in terms of $\mathbf{M}_\alpha(\psi)$, we equate (16) to (15), multiply by $\phi_1{}^*(q_1)$ and integrate over q_1 (which includes summing over s_1). Since the ϕ's are an orthonormal set, we obtain

$$\mathbf{M}_1(\phi; q_2 \cdots q_{2N}) = \sum_\alpha \int \phi_1{}^*(q_1)\psi_\alpha(q_1)dq_1\mathbf{M}_\alpha(\psi; q_2 \cdots q_{2N}). \tag{17}$$

The coefficient of $\mathbf{M}_\alpha$ is readily found from (2) by multiplying by $\psi_\alpha{}^*$, integrating, and taking the complex conjugate. This leads to the value $a_{1\alpha}{}^*$ for the coefficients in (17), a result recognized as equivalent to equation (13) which was to be proved.

It is evident that, if the $\mathcal{E}_\alpha$ for the expansion functions in (13) are not equal, then the $\mathbf{M}_\alpha$'s correspond to different energies, so that, as stated earlier, $\mathbf{A}_p(\phi)$ will not be an energy eigenfunction.

The method of expanding $\mathbf{A}_p(\phi)$ in terms of the $\mathbf{M}_\alpha$ affords a way of dealing with holes without the one electron approximation introduced in equation (1) of this section. For a band with one missing electron there will be $N - 1$ eigenfunction solutions which are actually exact solutions and not approximations like the determinantal functions dealt with in this section and the previous one. When these exact solutions are combined according to (13), an exact solution representing a localized deficit of negative charge will be obtained. Dealing with these exact solutions will then lead to results like (11) and (12) in which $\boldsymbol{v}$ will be defined as $\nabla_{P_S}\mathcal{E}_S(\boldsymbol{P}_S)$, where $\boldsymbol{P}_S$ and $\mathcal{E}_S(\boldsymbol{P}_S)$ describe attributes of the exact solutions for the *S*ystem. From considerations of this sort, the behavior of a hole can be obtained purely from the characteristics of a band missing one electron without making use of one electron wave-function approximations. In other words, although the treatment here has been based on approximations, the fundamental attributes of a hole are more basic and would be obtained in an exact treatment as well.

15.9 PERIODIC BOUNDARY CONDITIONS AND ELECTRIC FIELDS

To a considerable degree, this section duplicates the results of 15.6. There are, however, important conceptual differences. In Chapter 6 we discussed the currents due to the Bloch wave functions, which represent uniform charge density in the crystal. The calculation of conductivity is often carried out on the basis of such a picture; however, the procedure used usually appears to be internally inconsistent. The inconsistency arises from the fact that the wave functions which initially satisfied the periodic boundary conditions change their $\boldsymbol{P}$ values uniformly under the influence of an electric field so that they no longer satisfy periodic boundary conditions. This apparent inconsistency is eliminated by a detailed consideration of the way in which the wave functions vary under the influence of a field when a model having truly periodic boundary conditions is employed.

The difference between this section and 15.6 is thus that we here deal with extended wave functions each corresponding to a single value of $\boldsymbol{P}$, with periodic boundary conditions and with electric fields only. To extend the method to include magnetic fields appears to offer considerable difficulties and we do not attempt it. We shall show that if the conditions considered in 15.6 are made to correspond to those considered here, the behavior of the wave functions become the same.

We shall consider in this section a model, similar to that of Figure 5.6 or 5.9, which physically has periodic boundary conditions in one direction and "vanishing" boundary conditions in the other two dimensions. An electric field in the direction of the periodic boundary conditions will be applied and the resulting behavior of the wave functions will be shown to be consistent with the description of Section 7.2 of shifting values of P_y, the component of crystal momentum parallel to the field.

We imagine that the crystal is in the form of a thin rectangular sheet which is bent to form a hollow cylinder. The periodic boundary condition applies in the ϕ-direction of a cylindrical coordinate system. The electric field in this direction is applied by making the cylinder surround an elongated concentric transformer core, so that the cylinder forms a single short circuited turn. If the thickness of the sheet is made small enough compared to the radius of the cylinder, the bend can take place elastically so that the atomic arrangement is perfect except for a curvature of the lattice. For a sufficiently large radius r_0 compared to the thickness δr_0 of the crystalline sheet, the wave equation can be made to differ by a negligible amount from that for a flat sheet with rectangular coordinates

$$x = r - r_0, \quad y = r_0\phi, \quad z = z. \tag{1}$$

We shall consider below the limiting case as $\delta r_0/r_0 \to 0$ so that set (1) become effectively rectangular coordinates.

We shall consider an electric field which varies slowly so that retardation effects can be neglected. If the flux in the core is represented by

$$\int B_z dA_z = -c \int V(t)dt, \tag{2}$$

where $V(t)$ is the potential difference which would appear across an open-circuited secondary, then the field outside the core may be represented by the vector potential alone and the vector potential and fields are

$$\boldsymbol{A} = \boldsymbol{i}_\phi \frac{c\int V(t)dt}{2\pi r}, \quad \boldsymbol{E} = -\frac{1}{c}\dot{\boldsymbol{A}} = \boldsymbol{i}_\phi \frac{-V(t)}{2\pi r}, \quad \boldsymbol{H} = \nabla \times \boldsymbol{A} = 0, \tag{3}$$

where $\boldsymbol{i}_\varphi$ is a unit vector in the φ direction.

The kinetic energy operator in the presence of the vector potential is

$$\frac{1}{2m}(\boldsymbol{p}^2 - \boldsymbol{p}\cdot\boldsymbol{p}_A - \boldsymbol{p}_A\cdot\boldsymbol{p} + \boldsymbol{p}_A)^2 = \frac{1}{2m}(\boldsymbol{p}^2 - 2\boldsymbol{p}_A\cdot\boldsymbol{p} + \boldsymbol{p}_A)^2 \tag{4}$$

where $\boldsymbol{p}_A = (-e/c)\boldsymbol{A}$ is the proper term to introduce for a particle with charge $-e$. Since $\nabla\cdot\boldsymbol{A} = 0$, it follows that $\boldsymbol{p}\cdot\boldsymbol{p}_A = \boldsymbol{p}_A\cdot\boldsymbol{p}$. In cylindrical coordinates (4) becomes

$$\frac{1}{2m}\left[(-\hbar^2)\left(\frac{\partial^2}{\partial r^2} + \frac{1}{r}\frac{\partial}{\partial r} + \frac{1}{r^2}\frac{\partial^2}{\partial\phi^2} + \frac{\partial}{\partial z^2}\right) - 2\frac{(-e)\int V(t)dt}{2\pi r}\frac{\hbar}{ir}\frac{\partial}{\partial\phi} + \boldsymbol{p}_A{}^2(r,t)\right], \tag{5}$$

where $\boldsymbol{p}$ has been expressed as $(\hbar/i)\nabla$. If we substitute the variables x, y, z of (1) into this expression, it becomes

$$\frac{1}{2m}\left[-\hbar^2\left(\frac{\partial^2}{\partial x^2} + \frac{\partial^2}{\partial y^2} + \frac{\partial^2}{\partial z^2}\right) - 2\frac{(-e)\int V(t)dt}{2\pi r_0}\frac{\hbar\partial}{i\partial y} + \boldsymbol{p}_A{}^2(r_0,t)\right]$$
$$+\frac{1}{2m}\left[-\frac{\hbar^2}{r}\frac{\partial}{\partial x} - \frac{2(-e)\int V(t)dt}{2\pi r_0}\left(\frac{r_0{}^2}{r^2} - 1\right)\frac{\hbar}{i}\frac{\partial}{\partial y} + \left(\frac{r_0{}^2}{r^2} - 1\right)\boldsymbol{p}_A{}^2(r_0,t)\right]. \tag{6}$$

The second line represents the correction term due to the finite radius of curvature. In it the term $(\hbar^2/r)\partial/\partial x$ stands approximately in the ratio a/r_0 to $\hbar^2\partial^2/\partial x^2$ of the first line where a is the lattice constant. The last two terms can be made negligible by making the sheet thin so that in it

$r - r_0 \ll r_0$. Thus the kinetic energy operator is well represented by

$$\frac{1}{2m}[p^2 - 2p_{Ay}p_y + p_{Ay}^2], \tag{7}$$

where p is now $(\hbar/i)\nabla$ in respect to the quasi cartesian coordinates xyz and

$$p_{Ay} = \frac{(-e)\int V(t)dt}{2\pi r_0} = (-e)\int -E_y(t)dt \text{ where } E_y(t) = -V(t)/2\pi r_0. \tag{8}$$

The quantity p_{Ay} is seen from (3) to be one component of p_A of 15.6 equation (5). For the elastically bent lattice we may take $\mathcal{U}$ as a periodic function $\mathcal{U}(x, y, z) = \mathcal{U}(x + n_x a, y + n_y a, z + n_z a)$ of the coordinates x, y, z where a is the lattice constant and the n's are integers. With this potential, the wave equation for $\psi(x, y, z, t)$ becomes

$$\mathcal{H}\psi = \left[\frac{1}{2m}(p^2 - 2p_{Ay}p_y + p_{Ay}^2) + \mathcal{U}\right]\psi = i\hbar\dot{\psi}. \tag{9}$$

A good solution to this equation can be obtained from the solutions for $p_{Ay} = 0$. These solutions correspond to a periodic boundary condition in the y-direction and vanishing boundary conditions in the x and z directions. For any fixed values of the x and z quantum numbers, which correspond to $|P_x|$ and $|P_z|$ in this case, we shall have solutions for values of P_y satisfying the periodic boundary conditions. These solutions can be written in the form

$$\psi_{P_y} = e^{iP_y y/\hbar} u_{P_y}(x, y, z) \tag{10}$$

where we have omitted $|P_x||P_z|$ which remain constant in the subsequent equations. Since ψ_{P_y} is an eigenfunction of $\mathcal{H}_0$, the Hamiltonian operator for $p_{Ay} = 0$, we have

$$\begin{aligned}\mathcal{H}_0\psi_{P_y} &= e^{iP_y y/\hbar}\left[\frac{1}{2m}(p^2 + 2P_y p_y + P_y^2) + \mathcal{U}\right]u_{P_y} \\ &= \mathcal{E}(P_y)\psi_{P_y} = e^{iP_y y/\hbar}\mathcal{E}(P_y)u_{P_y}.\end{aligned} \tag{11}$$

The approximate solution to (9) which reduces to ψ_{P_y} when $p_{Ay} = 0$, is

$$\begin{aligned}\psi &= e^{i(P_y y/\hbar)} u_{P_y - p_{Ay}} \exp \int \mathcal{E}(P_y - p_{Ay})dt/i\hbar \\ &\equiv e^{i(P_y y/\hbar) + i\theta} u_{P_y - p_{Ay}}\end{aligned} \tag{12}$$

where the integral in the exponential is represented by θ and P_y is a constant independent of t. We shall show that (12) is an approximate solution of $\mathcal{H}\psi = i\hbar\partial\psi/\partial t$ by evaluating the two sides of the equation. $\mathcal{H}\psi$ is

$$\mathcal{H}\psi = e^{i(P_y y/\hbar)+i\theta} \times\left[\frac{1}{2m}(p^2+2P_y p_y+P_y^2-2p_{Ay}p_y-2p_{Ay}P_y+p_{Ay}^2)+\mathfrak{U}\right]u_{P_y-p_{Ay}}$$

$$= e^{i(P_y y/\hbar)+i\theta}\left[\frac{1}{2m}[p^2+2(P_y-p_{Ay})p_y+(P_y-p_{Ay})^2]+\mathfrak{U}\right]u_{P_y-p_{Ay}}$$

$$= e^{i(P_y y/\hbar)+i\theta}\,\mathcal{E}(P_y - p_{Ay})u_{P_y-p_{Ay}} \tag{13}$$

and $i\hbar\partial\psi/\partial t$ is

$$i\hbar\frac{\partial}{\partial t}\psi = e^{i(P_y y/\hbar)+i\theta}\left[-i\hbar\dot{p}_{Ay}\frac{\partial}{\partial P_y}u_{P_y-p_{Ay}} + \mathcal{E}(P_y - p_{Ay})u_{P_y-p_{Ay}}\right]. \tag{14}$$

Thus the approximate solution satisfies the wave equation except for the term in $\partial u/\partial P_y$. We have discussed a term of this sort in Section 15.6. As was pointed out there, this term introduces negligible corrections for electric fields of reasonable magnitudes. We may, therefore, regard (12) as a good solution and consider its meaning.

Although the solution maintains the same value of P_y in the exponential term, so that its complex phase runs through the same number of cycles in the length of the crystal and thus continuously satisfies the periodic boundary conditions, the behavior of the energy and velocity associated with the wave function varies just as if P_y were increased at a rate $-eE_y$ as discussed in Section 7.2. This is seen for the energy directly from equation (13) which shows that

$$\int \psi^*\mathcal{H}\psi dx = \mathcal{E}(P_y - p_{Ay}) = \mathcal{E}(P_y - e\int E_y dt). \tag{15}$$

By applying the group velocity formula for v_y to this energy we obtain

$$v_y = \frac{\partial}{\partial P_y}\mathcal{E}(P_y - p_{Ay}) = v_y(P_y - p_{Ay}) \tag{16}$$

where $v_y(P_y) = \partial\mathcal{E}(P_y)/\partial P_y$ is the velocity in the absence of p_A. [The same result may be obtained by using the operator $(p_y - p_{Ay})/m$ for velocity [see Section 15.6, equation (7)] when a vector potential is present.] Thus although the change of phase of the wave function from one cell to the next is not altered by the field, the change in energy and velocity is just the same as if P_y varied at a rate $-\dot{p}_{Ay} = -eE_y$ so that the description given in 7.2, although not analytically correct, leads to the correct results.

Another way of viewing the effect of the field is to say that the quantum number P_y remains constant but that the energy and velocity are different functions of P_y at different times, the variation in these functions corresponding to shifting the energy surface and velocity diagram at a rate $+eE_y$ with respect to the P_y values.

We may now compare the wave function (12) with the wave-packet of 15.6. If the electric field is uniform and is produced through the vector potential only, the equation for $\boldsymbol{P}$ of 15.7 can be readily integrated. We have

$$\dot{\boldsymbol{P}} = -e\boldsymbol{E} = (-e)\left(-\frac{1}{c}\frac{\partial}{\partial t}\boldsymbol{A}\right) \tag{17}$$

or

$$\boldsymbol{P} = (-e)\left(-\frac{1}{c}\boldsymbol{A}\right) + \boldsymbol{P}_0 = \boldsymbol{P}_0 - \boldsymbol{p}_A \tag{18}$$

where $\boldsymbol{P}_0$ is an integration constant. The spatial part of the wave-packet then becomes [in accordance with equation (11) for ψ and (34) for $\boldsymbol{P}$ in Section 15.6]

$$\psi = \mathbf{w}\,[\exp\,(i\boldsymbol{p}_A \cdot \boldsymbol{r}/\hbar)]\,\psi_{\boldsymbol{P}_0-\boldsymbol{p}_A} \tag{19}$$

(since we assume $\boldsymbol{E}$ is uniform, $\boldsymbol{p}_A$ is independent of position and $\boldsymbol{p}_{A0} = \boldsymbol{p}_A$) and this may be rewritten as

$$\begin{aligned}\psi &= \mathbf{w}\,[\exp\,(i\boldsymbol{p}_A \cdot \boldsymbol{r}/\hbar)]\{\exp\,[i(\boldsymbol{P}_0 - \boldsymbol{p}_A)\cdot \boldsymbol{r}/\hbar]\}\mathbf{u}_{\boldsymbol{P}_0-\boldsymbol{p}_A} \\ &= \mathbf{w}\,[\exp\,(i\boldsymbol{P}_0 \cdot \boldsymbol{r}/\hbar)]\,\mathbf{u}_{\boldsymbol{P}_0-\boldsymbol{p}_A},\end{aligned} \tag{20}$$

a form equivalent to (12) of this section. Thus although $\boldsymbol{P}$ varies in 15.6, if the field arises from $\boldsymbol{A}$ only, the complex phase factor in front of the periodic $\mathbf{u}$ does not change. If, however, electric fields due to ϕ are present, the complex phase factor will change. For the periodic boundary conditions used in this section fields due to ϕ are excluded since ϕ must be single-valued. For the treatment of Section 15.6, the change from constant $\boldsymbol{P}_0$ to varying $\boldsymbol{P}_0$ may be brought about by changing ϕ, $\boldsymbol{A}$ and ψ by a gauge transformation; and the gauge transformation leaves the physical condition described by the wave function unaltered. To sum up, for a wave-packet in a uniform electric field the behavior of the spatial part of the wave function is the same as that of the extended wave function of this section, except for the factor $\mathbf{w}$.

Finally it may be mentioned that one can think of a model of the sort considered here for which periodicity in z as well as y is achieved. This can be accomplished by continuing the structure in the z-direction for a distance z_0 much greater than r_0. It then becomes a coaxial structure with a central transformer core and a cylindrical sheet of conductor.

This structure can then be bent into a toroidal structure so that the metal becomes a toroidal surface. If $z_0 \gg r_0$, the local effect of curvature can again be neglected. The core inside the metal surface will produce an electric field in the y-direction as before. A new field in the z-direction can be produced by changing the flux passing through the hole in the toroid itself.

CHAPTER 16

STATISTICAL MECHANICS FOR SEMICONDUCTORS

16.1 DERIVATION OF THE DISTRIBUTION LAWS FOR SIMPLE QUANTIZED SYSTEMS

16.1a. Basic Concepts. In this section we shall present a brief review of the application of statistical mechanics to quantized systems. The formulae will be derived for the Planck energy distribution for harmonic oscillators and for the Fermi-Dirac distribution. The relationship of statistical mechanics to thermodynamics will be touched on briefly. (A treatment of the relationship is presented in the Problems.)

In the statistical mechanics of a quantized system the concepts of the *system*, its *quantum states*, *distributions*, and *a priori probability* play a major role. We shall consider these in turn below:

By a system we mean a certain number of electrons and nuclei subject to certain boundary conditions. For example, we might have 4.52×10^{22} germanium atoms contained in a volume of 1 cm^3 with rigid, impenetrable walls. Usually it is a good approximation to consider the behavior of only part of some real system and to treat this part as if it were itself an entire system. One example of this sort is the 20 electrons and 36 quantum states considered in Section 10.2. As another example, the system may consist of a number Q of harmonic oscillators. The system of normal modes of atomic vibration for a crystal discussed in Section 17.3 is an example of this sort. A semiconductor contains parts of both types combined into a common system.

The all-important attribute of the system for statistical purposes is the energy scheme of its quantum states — in other words, the complete list of its quantum states and their energies. This list may also be referred to as the energy level scheme of the system, provided that proper allowance is made for the fact that several quantum states may belong to the same energy level. It is seldom possible to deal with an exact energy level scheme and it is usual to consider the energy of the system as made up of the energies of various parts, each of which is in its own quantum state. Approximations of this sort are frequently made and, as discussed in Chapter 13, furnish the best insight into the problems involved. The separate statistical treatment of electrons and lattice vibrations in Chapter 17 permits the neglected interactions to be interpreted as causing transitions. In fact, it is difficult to see how to formulate the theory of resistance in terms of the exact quantum states of the entire system and it is almost certain that the interpretation would be much more laborious. (Such an exact procedure may be necessary to explain superconductivity, however.[1]) Other examples of breaking the system into separate parts which are approximately independent are discussed in Section 5.2 and at the end of Section 15.7.

The approximation made in taking the quantum states of the entire system as being composed of the quantum states of the subsystems introduces a negligible error if the neglected interactions of the subsystems are not large. This result, which we shall accept as an additional assumption here, can be established with the aid of quantum-mechanical theorems relating to the invariance of the *trace* of a *Hermitian matrix* under a unitary transformation of the wave functions.

Once the system is considered to be broken into a set of subsystems, the concept of a *distribution* has meaning. We shall illustrate this concept first for the case of electrons governed by the Pauli principle (Sections 5.2 and 15.7 and Chapters 9 and 10). The energy of the system may be described as the sum of the energies of the quantum states occupied by the electrons. We classify the quantum states according to energy and divide them into groups of approximately equal energy. In terms of this classification, a distribution is specified by giving the number of occupied states in each group. To a given distribution there will usually correspond many quantum states for the system. For example, if one of the groups of states contains two states and the distribution specifies that only one is occupied, there will be two ways of choosing the occupied state. In general there will be more choices than this for each group so that there will be a very large number of ways of achieving the distribution, each way corresponding to a particular quantum state for the system.

The central problem of statistical mechanics consists of finding the distribution which can be achieved in the largest number of ways, for a given system with a given energy. The importance of this distribution is that it represents the properties of the system under equilibrium conditions. Several steps are necessary to show that the distribution having the most ways represents thermal equilibrium. The first step involves the introduction of the assumption of equal *a priori* probabilities for each quantum state of the system. According to this assumption, if a system is set up in the laboratory so that it has a certain amount of energy and is next isolated so that it remains in a particular quantum state, then it is equally likely to be in any one of the possible quantum states of this energy; in other words all quantum states for the system have equal *a priori* probability, provided that they have the right energy.

The assumption of equal *a priori* probability seems reasonable on its

[1] The recent theories of superconductivity of H. Frohlich [*Phys. Rev.*, late (1950)] and J. Bardeen [*Phys. Rev.*, 79, 167 (1950)] are concerned with showing that the lowest states of the combined electronic and atomic system can be superconducting under suitable conditions.

face value. It can, however, be made to seem still more plausible on the basis of the quantum-mechanical H-theorem. The proof of the H-theorem and other methods of examination of the foundations of the statistical mechanics of quantized systems represent advanced and difficult topics and will not be pursued further here.

The possible quantum states for a given energy can be grouped, of course, into various distributions; and no quantum state will occur in more than one distribution. Among all the possible distributions there will be one which has more quantum states than any other. There will be other distributions which differ slightly from this one and can be achieved in approximately the same number of ways. For systems with large numbers of particles it can be shown that only the most probable distribution need be considered. The reason is that distributions which differ appreciably from the most probable distribution contain very many less quantum states. In fact, if all these differing distributions are combined, the total number of quantum states in them is negligible compared to the number in the most probable distribution. Hence there is a negligible probability of finding the system in any state except one which belongs to the most probable distribution, or to distributions very similar to the most probable distribution.

A helpful means of thinking of the way in which the most probable distribution arises is furnished by the concept of the *microcanonical ensemble.* For this purpose we visualize a very large number of identical systems, one in each of the quantum states having the prescribed total energy. In order to avoid difficulties arising from the discrete nature of the energy levels, the prescribed energy is usually represented as a small range from $\mathcal{E}_1$ to $\mathcal{E}_2$ where the interval $\mathcal{E}_2 - \mathcal{E}_1$ is a very small fraction of $\mathcal{E}_1$. This large number or *ensemble* of systems is the *microcanonical ensemble* and of this large number all but a negligible fraction are in quantum states belonging to the most probable distribution. The observed value for any property of an actual system is, on the average, that obtained by averaging the property of interest over all the systems of the microcanonical ensemble, this conclusion being a direct consequence of the assumption of equal *a priori* probabilities. Because of the overwhelming importance of the most probable distribution, the average over the microcanonical ensemble of any quantity of practical interest is the same as that obtained for the most probable distribution. Deviations from the value for the most probable distribution do, however, become significant for a small system, such as an individual colloidal particle, and can be calculated with the aid of the theory of fluctuations, one example being given in the problems.

The procedure we shall follow is thus to set up a definition for the various possible distributions and then find the most probable distribution. The behavior of the system under equilibrium conditions will then correspond to this distribution. This procedure is arbitrary to the degree that defining the distributions is arbitrary. We shall find in the examples dealt with, however, that the arbitrary conventions used in defining the distribution drop out of the final answers and thus the choice of the convention need not concern us.

We shall consider below a system consisting of a number of independent harmonic oscillators, all having the same frequency. For this case we shall work out the most probable distribution. The number of ways in which this distribution can be achieved will be denoted by $W(\mathfrak{E})$, where $\mathfrak{E}$ is the energy of the system.

In order to define temperature on a statistical basis and to show the relationship of statistical mechanics to thermodynamics, we shall consider two systems in equilibrium. If the systems have energies $\mathfrak{E}_1$ and $\mathfrak{E}_2$ with $\mathfrak{E}_1 + \mathfrak{E}_2 = \mathfrak{E}_T$ where $\mathfrak{E}_T$ is the total energy, then the number of ways of achieving the division into $\mathfrak{E}_1$ and $\mathfrak{E}_2$ will be $W_1(\mathfrak{E}_1)W_2(\mathfrak{E}_2)$. We shall outline the arguments that show that temperature should be defined by

$$\frac{d \ln W_1(\mathfrak{E})}{d\mathfrak{E}} = \frac{1}{kT_1} \tag{1}$$

where k is Boltzmann's constant, and entropy is defined as

$$\mathfrak{S}_1 = k \ln W_1(\mathfrak{E}_1). \tag{2}$$

Using these results, we shall derive the Fermi-Dirac distribution.

16.1b. The Planck Harmonic Oscillator Distribution. We shall consider here a system composed of a large number Q of harmonic oscillators, each having the same frequency ν and energy levels $(n + \frac{1}{2})h\nu$, where $n = 0, 1, 2, \cdots$ is the quantum number and h is Planck's constant. We shall refer to the quantum states for the individual oscillators as *O-states* to distinguish them from states of the entire system, referred to as *S-states*. We shall define the distribution in terms of a set of numbers $Q_0, Q_1, Q_2, \cdots, Q_n, \cdots$ giving the number of oscillators in O-states having quantum numbers 0, 1, etc. Evidently we must have

$$Q = \sum_n Q_n. \tag{3}$$

If the total energy of the system is $\mathfrak{E}$, we must also have

$$\mathfrak{E} = \sum_n (n + \tfrac{1}{2})h\nu Q_n = Q\frac{h\nu}{2} + h\nu \sum_n nQ_n. \tag{4}$$

A quantum state for the system, or S-state, is described by telling which oscillators are in O-states with quantum numbers 0, 1, 2, etc. A particular S-state may thus be written as follows:

$$(0/abcde)(1/fgh) \cdots (s/ijk \cdots) \cdots \tag{5}$$

which implies that oscillators a, b, c, d, and e are in O-state $n = 0$ and f, g, and h are in O-state $n = 1$, etc. We shall refer to the expression above as a *writing*. The same S-state may, of course, be specified by other writings such as

$$(0/baedc)(1/ghf)\cdots(s/kji\cdots)\cdots. \tag{6}$$

Our problem is to find how many S-states correspond to a distribution such as $Q_0 = 5$, $Q_1 = 3, \cdots$, being careful not to count different writings of the same S-state as different S-states. We shall first show that the number of writings of a given S-state is

$$Q_0!\,Q_1!\cdots \equiv \prod_n Q_n! \tag{7}$$

as follows: Consider any given writing; the $n = 0$ parenthesis can be written in $Q_0!$ ways (that is, there are 5 choices for the first letter after the / and 4 for the second, etc., in the example above). Similarly each parenthesis can be written in $Q_n!$ ways so that each S-state can be written in $\prod_n Q_n!$ ways. Next we ask how many different writings are there for a given distribution? In any given writing, the symbols designating the oscillators are arranged in sequence and there are evidently just $Q!$ ways of permuting them in this sequence. Hence, there are $Q!$ writings in all. Since each S-state occurs written $\prod_n Q_n!$ times in this set of writings, the distribution can be achieved in

$$W(Q_0, Q_1, \cdots) = Q!/\prod_n Q_n! \tag{8}$$

ways; i.e. there are $Q!/\Pi Q_n!$ different quantum states corresponding to the distribution.

Which of the possible distributions is the most probable, that is, has the largest W, subject to a fixed total number of oscillators Q and total energy $\mathfrak{E}$? We shall find this distribution by requiring that, for any small change in the Q_n's, the value of W remains constant. This does not by itself prove that W is maximum; W might be a minimum equally well. However, the distribution so obtained can be shown by a separate calculation to maximize W. (This is left as an exercise for the reader.) Let us consider a perturbation, referred to as (Per) of the distribution specified as follows:

$$\text{(Per)}\begin{cases} Q_r \to Q_r + \delta Q_r & \text{(9a)} \\ Q_s \to Q_s + \delta Q_s, \quad \delta Q_s = -\delta Q_r & \text{(9b)} \\ \mathfrak{E} \to \mathfrak{E} + \delta\mathfrak{E}, \quad \delta\mathfrak{E} = h\nu(r - s)\delta Q_r. & \text{(9c)} \end{cases}$$

This corresponds to changing δQ_r oscillators from O-state r to O-state s. We shall deal, for convenience, not with W but with $\ln W$ and shall denote the change produced in ln W by δ ln W. This perturbation does not conserve the total energy $\mathfrak{E}$ so that we must supplement it with another which restores the energy. Suppose that the second perturbation, (Per)′, is designated by a prime and transfers δQ_v oscillators from O-state u to O-state v. Then, as we shall show below, we must have

$$\frac{\delta \ln W}{\delta \mathfrak{E}} = \frac{(\delta \ln W)'}{\delta \mathfrak{E}'}, \tag{10}$$

in other words, the change in the logarithm of W per unit energy change for the two perturbations must be equal and, since u and v for (Per)′ can be selected arbitrarily, the same value for $(\delta \ln W)'/\delta\mathfrak{E}'$ must be obtained for all perturbations. From this result we shall readily derive the Planck distribution and later the Fermi-Dirac distribution.

The proof of equation (10) is very simple. If W is to be a maximum, so is ln W. Now consider changes represented by (Per) and (Per)′:

$$\text{(Per)}: \quad \delta Q_r = -\delta Q_s = \delta\mathfrak{E}/h\nu(r - s) \tag{11a}$$

$$\text{(Per)}': \quad \delta Q_u = -\delta Q_v = \delta\mathfrak{E}'/h\nu(u - v). \tag{11b}$$

These will produce changes in ln W of δ ln W and $(\delta \ln W)'$. If $\delta\mathfrak{E} = -\delta\mathfrak{E}'$, these changes will not change $\mathfrak{E}$ and, if W is a maximum, they will not change ln W. The change in ln W will be proportional to $\delta\mathfrak{E}$ and $\delta\mathfrak{E}'$, hence

$$\begin{aligned} \delta \ln W &= \delta \ln W + (\delta \ln W)' \\ &= \frac{\delta \ln W}{\delta\mathfrak{E}}\,\delta\mathfrak{E} + \frac{(\delta \ln W)'}{\delta\mathfrak{E}'}\,\delta\mathfrak{E}'. \end{aligned} \tag{12}$$

For this to vanish when $\delta\mathfrak{E} = -\delta\mathfrak{E}'$, the two coefficients must be equal, so that we may introduce a symbol β defined as follows:

$$\beta = \frac{\delta \ln W}{\delta\mathfrak{E}} = \frac{(\delta \ln W)'}{\delta\mathfrak{E}'}\,. \tag{13}$$

We shall later show that $\beta = 1/kT$ where k is Boltzmann's constant and T is the absolute temperature.

The expression for $\delta \ln W/\delta\mathfrak{E}$ in terms of the parameters Q_n of the distribution may be readily obtained from the formula for ln W and the derivative of a factorial. The latter may be taken as the value for the minimum change $\delta x = 1$ in $\delta \ln x!/\delta x$; this leads to

$$\frac{d \ln x!}{dx} \doteq \frac{\ln (x + \delta x)! - \ln x!}{\delta x} = \ln (x + 1) \doteq \ln x, \tag{14a}$$

$$\delta \ln x! \doteq \delta x \ln x. \tag{14b}$$

From this we readily obtain [using also (11a) and 8]

$$\delta \ln W = -\delta \ln Q_r!\,Q_s! = -\delta Q_r \ln Q_r - \delta Q_s \ln Q_s$$
$$= \frac{\delta \mathcal{E} \ln (Q_s/Q_r)}{h\nu(r - s)}. \tag{15}$$

If we now imagine that u and v are held fixed and r and s are allowed to vary over all possible quantum numbers, we must have from (13)

$$\beta = \frac{\delta \ln W}{\delta \mathcal{E}} = \frac{\ln (Q_s/Q_r)}{h\nu(r - s)}. \tag{16}$$

From this we find that

$$\beta h\nu r + \ln Q_r = \beta h\nu s + \ln Q_s. \tag{17}$$

If we let $s = 0$, this shows that

$$Q_r = Q_0 e^{-\beta h\nu r} = Q_0 e^{-\gamma r} \tag{18}$$

$$\gamma \equiv \beta h\nu. \tag{19}$$

The quantities β and Q_0 are determined by the conditions on Q and $\mathcal{E}$ by the following steps:

$$Q = \sum Q_n = Q_0(1 + e^{-\gamma} + e^{-2\gamma} + \cdots) = \frac{Q_0}{1 - e^{-\gamma}} \tag{20}$$

$$\begin{aligned}
\mathcal{E} &= (h\nu Q/2) + h\nu \textstyle\sum nQ_n \\
&= (h\nu Q/2) + h\nu Q_0(0 \cdot 1 + 1e^{-\gamma} + 2e^{-2\gamma} + 3e^{-3\gamma} + \cdots) \\
&= (h\nu Q/2) + h\nu Q_0\left(-\frac{d}{d\gamma}\right)(1 + e^{-\gamma} + e^{-2\gamma} + \cdots) \\
&= (h\nu Q/2) - h\nu Q_0 \frac{d}{d\gamma}\frac{1}{1 - e^{-\gamma}} \\
&= (h\nu Q/2) + h\nu Q_0 \frac{e^{-\gamma}}{(1 - e^{-\gamma})^2} = h\nu Q\left[\frac{1}{2} + \frac{1}{e^{\gamma} - 1}\right].
\end{aligned} \tag{21}$$

These equations determine Q and $\mathcal{E}$ in terms of Q_0 and γ and can be solved, in principle, for γ and Q_0 in terms of Q and $\mathcal{E}$. We shall shortly show how $\gamma = \beta h\nu$ can be identified with $h\nu/kT$; this leads to the following expression for the average energy of an oscillator:

$$\langle \mathcal{E} \rangle_{\text{Av.}} = h\nu\left[\frac{1}{2} + \frac{1}{e^{h\nu/kT} - 1}\right], \tag{22}$$

and to an average value of n of

$$\langle n \rangle_{\text{Av.}} = \frac{1}{e^{h\nu/kT} - 1}. \tag{23}$$

The probability that any oscillator is in the state n is Q_n/Q and is given by

$$\frac{Q_n}{Q} = e^{-nh\nu/kT}(1 - e^{-h\nu/kT}). \tag{24}$$

The foregoing expressions give the Planck distribution for a system of harmonic oscillators.

By considering second-order changes produced by the perturbations δQ_r, δQ_s, etc., it is possible to show that ln W is actually a maximum and that it decreases very rapidly as the distribution deviates from the maximum probability distribution. The value of ln W for the most probable distribution is in general very large. This leads to a surprising and important result: The value of ln W and the value of ln (the total number of quantum states with energy $\mathfrak{E}$) differ by a negligible amount. This is surprising because the number of quantum states in the most probable distribution is, of course, only a small fraction of all the quantum states of energy $\mathfrak{E}$. A numerical example will help to show how the result comes about The value of ln W for one gram atom of oscillators with $\gamma \approx 1$ will be about 10^{23}. [This result may be derived from the expression for ln W by using Stirling's approximation $x! = (x/e)^x \sqrt{2\pi x}$ for x, with proper precautions, and more readily from the formulae in the problems for this chapter.] Suppose that only one state in 10^{20} is included in the most probable distribution. Then the total number of states will be 10^{20} exp (10^{23}) and the logarithm of this number will be $10^{23} + 46$. Evidently the term 46 is wholly negligible in ln W. The largeness of the terms in ln W frequently enables calculations to be made very inexactly so that W may be incorrectly estimated by large factors without significant errors being introduced into ln W.

16.1c. Relationship to Thermodynamics. We have shown that all perturbations of the system of harmonic oscillators must lead to the same value, which we call β, for

$$\frac{\delta \ln W}{\delta \mathfrak{E}} = \beta. \tag{25}$$

From this we shall show that if $W(\mathfrak{E})$ represents the number of ways in which the most probable distribution can be achieved, then

$$\frac{d \ln W(\mathfrak{E})}{d\mathfrak{E}} = \beta. \tag{26}$$

This is proved as follows: Consider the most probable distributions for energy $\mathfrak{E}$ and for energy $\mathfrak{E} + \Delta\mathfrak{E}$. We can evidently transform from one distribution to the other in series of steps of the form $(Q_r, Q_s) \rightarrow (Q_r + \delta Q_r, Q_s + \delta Q_s)$ considered before. These steps will introduce energy changes $\delta\mathfrak{E}$, $\delta\mathfrak{E}'$, $\delta\mathfrak{E}''$, etc., and for them the changes in ln W

will be $\beta\delta\mathfrak{E}$, $\beta\delta\mathfrak{E}'$, $\beta\delta\mathfrak{E}''$, etc. The total change in ln W will thus be $\beta(\delta\mathfrak{E} + \delta\mathfrak{E}' + \delta\mathfrak{E}'' + \cdots) = \beta\Delta\mathfrak{E}$ where $\Delta\mathfrak{E}$ is the total change in energy. Hence $\Delta \ln W/\Delta\mathfrak{E} = \beta$ for the change from the maximum distribution at energy $\mathfrak{E}$ to that at $\mathfrak{E} + \Delta\mathfrak{E}$; since the maximum probability distribution is a function of $\mathfrak{E}$, this ratio is the desired derivative.

Now suppose we have two systems, 1 and 2, which can share energy. We can then consider them to be a combined system with a total energy $\mathfrak{E} = \mathfrak{E}_1 + \mathfrak{E}_2$. The possibility of sharing energy corresponds to putting two bodies in thermal contact so that they can exchange heat and come to thermal equilibrium with equal temperatures and maximum total entropy.

As a result of their interaction neither system 1 nor system 2 has the same set of energy levels that it had when isolated. In fact the only energy levels are those of the system as a whole. However, if the interaction is very small each energy level for the combined system differs by a negligible energy from the sum of the energy levels of the two isolated systems. Thus it appears reasonable to treat the combined system as two separate systems while at the same time imagining that systems 1 and 2 may interact so as to transfer heat. This argument is far from rigorous, however, and it is difficult to prove rigorously that the behavior of two weakly interacting systems does properly simulate the thermodynamic behavior of two bodies in contact. For detailed examinations of these problems, the reader is referred to the literature.[1]

For any given division $\mathfrak{E}_1$ and $\mathfrak{E}_2$ of the energy, system 1 can be set up in $W_1(\mathfrak{E}_1)$ ways and system 2 in $W_2(\mathfrak{E}_2)$ ways, for the distributions of maximum probability. In fact, as discussed at the conclusion of the derivation of the Planck distribution, so far as $\ln W_1(\mathfrak{E}_1)$ is concerned, we can equally well include all quantum states of energy $\mathfrak{E}_1$. For similar reasons we can consider, if it is desirable to do so, the total number of states lying in a small energy interval $\mathfrak{E}_1$ to $\mathfrak{E}_1 + \delta\mathfrak{E}_1$ of system 1. For all of these choices, the large number aspect of W will lead to the same value ln W. The number of ways of dividing $\mathfrak{E}$ into $\mathfrak{E}_1$ and $\mathfrak{E}_2$ may thus be described, so far as its logarithm is concerned, by $\ln W_1(\mathfrak{E}_1)W_2(\mathfrak{E}_2)$; the most probable division is that for which this quantity is a maximum. (Because of the independence of ln W of the exact definition of W, we need not be concerned even with having $\mathfrak{E}_1 + \mathfrak{E}_2 = \mathfrak{E}$; it is sufficient to have $\mathfrak{E}_1 + \mathfrak{E}_2$ fall within a range $\delta\mathfrak{E}$ about $\mathfrak{E}$.) The maximum of $\ln W_1(\mathfrak{E}_1)W_2(\mathfrak{E}_2)$ corresponds to

$$\frac{d}{d\mathfrak{E}_1} \ln W(\mathfrak{E}_1, \mathfrak{E}_2) = 0 = \frac{d \ln W_1(\mathfrak{E}_1)}{d\mathfrak{E}_1} - \frac{d \ln W_2(\mathfrak{E}_2)}{d\mathfrak{E}_2}, \qquad (27)$$

the last equation following from $\mathfrak{E}_2 = \mathfrak{E} - \mathfrak{E}_1$. Hence under equilibrium

[1] R. H. Fowler, *Statistical Mechanics*, Cambridge, 1936. R. C. Tolman, *The Principles of Statistical Mechanics*, Oxford, 1938.

conditions

$$\beta_1 = \frac{d \ln W_1}{d\mathcal{E}_1} \quad \text{and} \quad \beta_2 = \frac{d \ln W_2}{d\mathcal{E}_2} \tag{28}$$

must be equal while $\ln W_1 + \ln W_2$ is a maximum. It can be shown that these conditions are equivalent to setting the entropy $\mathfrak{S}$ equal to $k \ln W$ and $\beta = 1/kT$. (See Problems.) The equality of β_1 and β_2 thus corresponds to uniform temperature and the maximizing of W to the maximum entropy condition.

We shall indicate the method of establishing the thermodynamic relationships, in terms of a simple example. Suppose that system 2 consists of harmonic oscillators of arbitrarily low frequency. Then $\gamma = \beta h\nu$ is $\ll 1$ and

$$\mathcal{E} = h\nu Q \left[\frac{1}{2} + \frac{1}{e^{\gamma} - 1}\right] \doteq h\nu Q \frac{1}{\beta h\nu} = \frac{1}{\beta} Q. \tag{29}$$

From the classical theory of statistical mechanics, we know that a system of oscillators has average energy of kTQ. Hence we identify kT with $1/\beta_2$ for system 2 and since $\beta_1 = \beta_2$ we identify β with $1/kT$ generally. This identification is not wholly satisfying, however, because the theorem $\mathcal{E} = kTQ$ itself involves a considerable amount of classical statistical theory. A better procedure would be to let system 2 represent an ideal monatomic gas. We should then find that

$$pV = N/\beta_2 \tag{30}$$

where p is the average pressure, V the volume, and N the number of atoms. Since the thermodynamic temperature scale can be defined for a perfect gas by the relationship

$$pV = NkT, \tag{31}$$

the identification of β_2 with $1/kT$ would be satisfactory by this means. To carry out this procedure for a quantum gas is relatively straightforward and would introduce few features not already discussed in Chapter 5 and in the next section of this chapter; the purpose of this section is to introduce ideas and formulae rather than to establish them rigorously and we shall not give the derivation. A still more general procedure is to introduce the dependence of the energy of the quantum states upon the volume V or some other mechanical parameter. In terms of this, the equation of state in terms of $\mathcal{E}$ and V can be given and a general thermodynamic equivalence established. This procedure is presented in the problems.

16.1d. Derivation of the Fermi-Dirac Distribution. We shall now obtain the formula for the Fermi-Dirac distribution using essentially the same method as that employed in obtaining the Planck distribution.

For this purpose we consider a system containing N electrons. The quantum states of the system are divided into groups, all the states of any group having approximately the same energy. Consider two such groups with energies $\mathcal{E}_r$ and $\mathcal{E}_s$; suppose that there are S_r states with energy $\mathcal{E}_r$ and that N_r of these are occupied by electrons and $P_r\ (= S_r - N_r)$ by holes. Our first problem is to find in how many ways the N_r electrons can be distributed among the S_r states. According to the Pauli principle (see Section 15.7), electrons are to be regarded as indistinguishable so that the quantum state of the system is specified by giving a list of the occupied one-electron states. Our problem is thus reduced to finding in how many ways the N_r occupied states can be selected from S_r states. To do this we list all of the S_r states giving each a symbol a, b, c, etc. Then one arrangement of electrons may be written as

$$(abef/dcgh \cdots) \tag{32}$$

where the states to the left of the / are occupied and those to the right are unoccupied. It is at once evident that each distinct selection of occupied states may be written in $N_r!\,P_r!$ ways; and also that there are $S_r!$ ways of permuting the S_r symbols in the entire expression while keeping N_r symbols to the left and P_r to the right of the /. Hence there are

$$S_r!/N_r!\,P_r! \tag{33}$$

distinct ways of choosing the N_r occupied states. The total number of ways of setting up a distribution $N_1, N_2, \cdots$, etc., is, therefore,

$$W = \prod_r (S_r!/N_r!\,P_r!). \tag{34}$$

We now consider a perturbation of the distribution which involves shifting δN_r electrons from group S_s to group S_r:

$$N_r \to N_r + \delta N_r \tag{35a}$$

$$N_s \to N_s + \delta N_s \quad \delta N_s = -\delta N_r \tag{35b}$$

$$\delta\mathcal{E} = (\mathcal{E}_r - \mathcal{E}_s)\delta N_r. \tag{35c}$$

All of the previous arguments relating to the equality of β for such changes hold for this case so that we may at once write that

$$\frac{1}{kT} = \beta = \frac{\delta \ln W}{\delta\mathcal{E}}. \tag{36}$$

The value of $\delta \ln W$ is readily found in terms of

$$\delta \ln N_r! = \delta N_r \ln N_r \tag{37}$$

$$\delta \ln P_r! = \delta P_r \ln P_r = -\delta N_r \ln P_r \tag{38}$$

and similar relationships. We thus obtain

$$\beta = \frac{\delta \ln W}{\delta \mathcal{E}} = \frac{1}{\mathcal{E}_r - \mathcal{E}_s}[-\ln N_r + \ln P_r + \ln N_s - \ln P_s]. \tag{39}$$

From this equation we obtain

$$\beta \mathcal{E}_r + \ln (N_r/P_r) = \beta \mathcal{E}_s + \ln N_s/P_s. \tag{40}$$

Since this equation must hold for all choices of s and r independently, both sides must be independent of r and s and can be written as $\beta \mathcal{E}_F$. Introducing the Fermi-Dirac distribution function $\mathbf{f}(\mathcal{E}_r) \equiv \mathbf{f}_r$ by the equations

$$N_r = \mathbf{f}_r S_r \quad \text{and} \quad P_r = (1 - \mathbf{f}_r) S_r, \tag{41}$$

we obtain

$$\ln \frac{\mathbf{f}_r}{1 - \mathbf{f}_r} = \beta(\mathcal{E}_F - \mathcal{E}_r) \tag{42}$$

which leads to

$$\mathbf{f}_r = \frac{1}{1 + \exp \beta(\mathcal{E}_r - \mathcal{E}_s)}. \tag{43}$$

Since $\beta = 1/kT$ this is the desired expression for the Fermi-Dirac distribution. It is to be noted that the arbitrary division into groups containing S_r states disappears from the final answer and that $\mathbf{f}$ is a function of the energy of the state, the Fermi level $\mathcal{E}_F$, and kT. The determination of $\mathcal{E}_F$ for semiconductors is discussed in Section 16.3.

The interpretation of the Fermi level as a *chemical potential* is readily obtained from equation (40). For this purpose we suppose that δN electrons are added to the system. If these electrons are added to states in group r, the increase in the *free energy*, $\mathfrak{F} \equiv \mathcal{E} - T\mathcal{S} = \mathcal{E} - kT \ln W$ is

$$\begin{aligned} \delta \mathfrak{F} &= \delta \mathcal{E} - kT\delta \ln W = \mathcal{E}_r \delta N + kT\delta N \ln (N_r/P_r) \\ &= \mathcal{E}_F \delta N \end{aligned} \tag{44}$$

by virtue of (34), (40) and the definition of $\mathcal{E}_F$. Thus the change $\delta \mathfrak{F}$ is independent of the group of states into which the electrons are put. $\mathcal{E}_F$ is similar to a chemical potential, which may be defined as the change in free energy of a system per added atom of a given chemical element.

This result may be used to prove the conclusion stated in Section 12.4 that the difference in Fermi levels φ is the voltage read on a voltmeter. For this purpose we suppose that δN electrons are transferred from conductor A at Fermi level $\mathcal{E}_{FA}$ to conductor B at Fermi level $\mathcal{E}_{FB}$, the transfer taking place reversibly through a motor so that

$$\text{the energy liberated} = -e\delta N(V_A - V_B) \tag{45}$$

where V_A and V_B are the voltages. Our problem is to show that $-e(V_A - V_B) = \mathcal{E}_{FA} - \mathcal{E}_{FB}$ from which it follows that $V_A - V_B =$

$\varphi_A - \varphi_B$ according to the definition of the Fermi level φ of Section 12.4. We shall prove this relationship by applying the thermodynamical theorem that $\delta\mathfrak{S} = 0$ for any reversible change in which no heat flows. This result is obtained in Problem 12, for the case in which work is done by pressure, in the form that $d\mathfrak{S} = k\beta d\mathfrak{Q}$ where $d\mathfrak{Q} = d$ (energy of system) + d (work done by the system). We shall suppose that the voltage difference is maintained by a battery connecting conductors A and B. If the system is thermally insulated, so that no heat can flow from outside, then the reversible shift of the electrons will leave $d\mathfrak{S} = 0$. Hence the work $-e\delta N(V_a - V_b)$ done by the system must equal $-\delta\mathfrak{E}$ and we may write

$$-e\delta N(V_A - V_B) = -\delta\mathfrak{E} = -(\delta\mathfrak{E} - T\delta\mathfrak{S}) = -\delta\mathfrak{F}. \tag{46}$$

However, $\delta\mathfrak{F}$ can be written as the sum of $\delta\mathfrak{F}_A = -\mathfrak{E}_{FA}\delta N$ due to removing δN electrons from A plus $\delta\mathfrak{F}_B = \mathfrak{E}_{FB}\delta N$ due to adding them at B. Inserting $\delta\mathfrak{F} = \delta\mathfrak{F}_A + \delta\mathfrak{F}_B$ in (46) leads to

$$-e(V_A - V_B) = \mathfrak{E}_{FA} - \mathfrak{E}_{FB}, \tag{47}$$

the desired equation.

16.2 CALCULATION OF THE EFFECTIVE NUMBER OF QUANTUM STATES IN AN ENERGY BAND

In order to calculate the density of electrons in the conduction band, we may evaluate the integral of equation (9) of Section 10.3 using for $N(\mathfrak{E})$ the appropriate value given in equation (9), Section 9.1. This leads to

$$n = \exp(\mathfrak{E}_F/kT)\int_{\mathfrak{E}_c}^{\mathfrak{E}_2} [4\pi(2m)^{3/2}/h^3](\mathfrak{E} - \mathfrak{E}_c)^{1/2} \exp(-\mathfrak{E}/kT)d\mathfrak{E}$$

$$= \exp[(\mathfrak{E}_F - \mathfrak{E}_c)/kT][4\pi(2m)^{3/2}/h^3]\int_0^{\mathfrak{E}_2-\mathfrak{E}_c} \exp(-\mathfrak{E}_1/kT)\mathfrak{E}_1^{1/2}d\mathfrak{E}_1 \tag{1}$$

where the substitution $\mathfrak{E}_1 = \mathfrak{E} - \mathfrak{E}_c$ has been used. Since the integral converges rapidly, the upper limit may be extended to infinity. The integral then becomes a standard form given in Pierce, "A Short Table of Integrals", Third Revised Edition, formula 496, the value being $(\pi kT)^{1/2}kT/2$. Inserting this in (1) gives

$$n = 2(2\pi mkT/h^2)^{3/2} \exp[-(\mathfrak{E}_c - \mathfrak{E}_F)/kT] \tag{2}$$

which is formula (11) of Section 10.3 quoted in the text. The effective number of states in the conduction band is thus

$$\begin{aligned} N_c &= 2(2\pi mkT/h^2)^{3/2} \\ &= 4.82 \cdot 10^{15}T^{3/2} \text{ cm}^{-3} \\ &= 2.41 \cdot 10^{19} \text{ cm}^{-3} \text{ at } 20°\text{C} \end{aligned} \tag{3}$$

$$N_c^2 = 2.32 \times 10^{31}T^3 \text{ cm}^{-6}. \tag{4}$$

[This same result can be obtained by integrating over p-space directly without making use of equation (9) of Section 9.1. For this method instead of $N(\mathcal{E})d\mathcal{E}$, we use $(2/h^3)dP_xdP_ydP_z$ with $\mathcal{E} = (P_x^2 + P_y^2 + P_z^2)/2m$. The integral in (1) then becomes the product of three integrals each of which contributes a factor $(2\pi mkT/h^2)^{1/2}$ to N_c.]

16.3 TEMPERATURE DEPENDENCE OF THE FERMI LEVEL

In this section we shall show in more analytical terms than employed in the text the way in which $\mathcal{E}_F$ and the electron and hole densities depend upon the temperature. We shall deal with the same distribution of

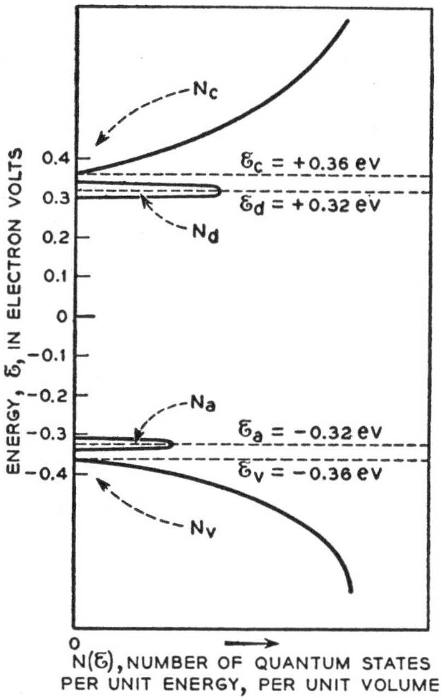

FIG. 16–1—Distribution of Quantum States in Energy.

quantum states that was discussed in Chapter 9. This is reproduced in Figure 16.1, with the effective number of quantum states in the filled and empty bands indicated. As explained in Chapter 10, the requirement for electrical neutrality is that the sum of the positive charges:

p = density of holes in the filled band

p_a = density of holes bound to acceptors

N_d = density of donors

must equal the sum of the negative charges:

n = density of electrons in conduction band
n_d = density of electrons bound to donors
N_a = density of acceptors.

This leads to the equation

$$p + p_a + N_d = n + n_d + N_a. \tag{1}$$

In Chapter 10, formulae for these densities are given in equations (8), (11), (13), and (14) of Section 10.3. These depend upon the nature of the semiconductor through the energy values of $\mathcal{E}_v$, $\mathcal{E}_a$, $\mathcal{E}_d$ and $\mathcal{E}_c$, the effective masses in N_v and N_c, and the impurity concentrations N_a and N_d. Given these quantities, our problem is to solve for the n's and p's as a function of T by solving equation (1) for $\mathcal{E}_F$. This is done, as shown below, by plotting the left and right sides of equation (1) each as a function of $\mathcal{E}_F$ and finding the solution graphically. The example chosen corresponds approximately to n-type germanium, which at room temperature would have a resistivity of 2.4 ohm-cm—the same as for the example discussed in Section 10.4.

The values used for this example are as follows:

$$\mathcal{E}_v = -0.36 \text{ electron volt}, \quad \mathcal{E}_a = -0.32 \text{ electron volt},$$

$$\mathcal{E}_d = +0.32 \text{ electron volt}, \quad \mathcal{E}_c = 0.36 \text{ electron volt}. \tag{2}$$

(This corresponds to arbitrarily setting the zero of energy at the midpoint between the bands.) The values of N_a and N_d are

$$N_d = 10^{15} \text{ per cm}^3, \quad N_a = 10^{14} \text{ per cm}^3 \tag{3}$$

and the electron and hole masses are taken to be the same as for a free electron so that equation (3) of Section 16.2 applies.

The densities obtained upon the basis of these assumptions are plotted in Figure 16.2 on a logarithmic scale as a function of $\mathcal{E}_F$ for $T = 300°\text{K}$. Positive charges are drawn as solid lines and their sum, which is the left side of equation (1), is shown as a heavy line. The densities of negative charges are similarly indicated by dashed lines. Where the heavy lines cross, equation (1) is satisfied and the condition for electrical neutrality is realized.

The behaviors of the individual curves for the negative charges can be understood as follows:

N_a. This is simply the fixed negative charge due to each acceptor. (This charge is not neutralized by its share of the valence bonds.)

n. This is the charge of the electrons in the conduction band. According to Section 16.2 it is about $2.41 \times 10^{19} \exp[(-\mathcal{E}_c + \mathcal{E}_F)/kT]$. Expressed in electron volts this is $2.41 \times 10^{19} \exp(-39\Delta V)$ where

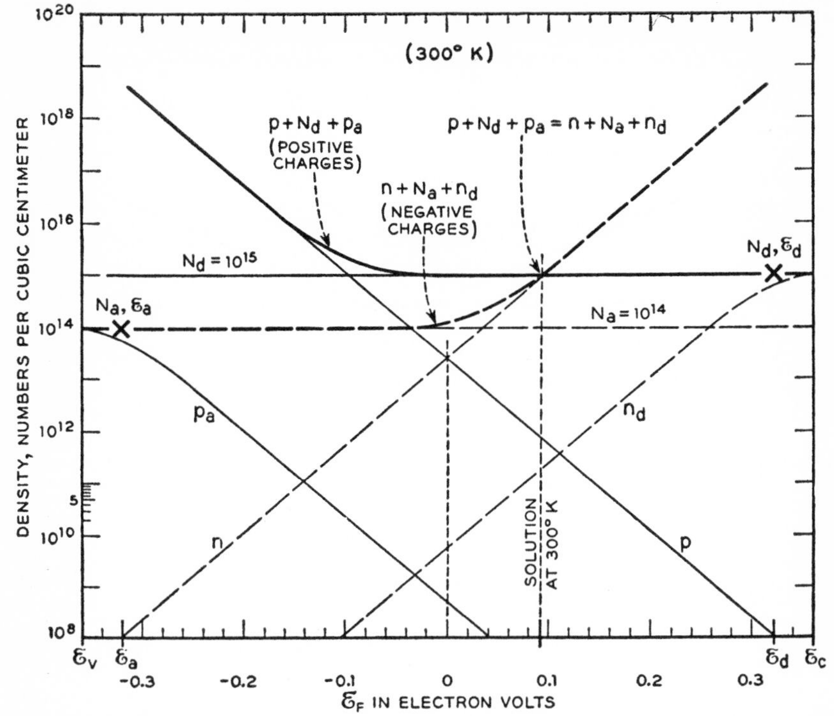

FIG. 16–2—Solution for $\mathcal{E}_F$ with Density of Acceptors, N_a, = 10^{14} cm^{-2}, Density of Donors, N_d, = 10^{15} cm^{-3} and $T \cong 300$°K.

ΔV is the voltage of $\mathcal{E}_F$ below $\mathcal{E}_c$ or

$$n = 2.41 \times 10^{19} \cdot 10^{-17\Delta V}. \tag{4}$$

This gives the straight line shown in Figure 16.2. (When $\mathcal{E}_F$ approaches within kT of $\mathcal{E}_c$, the approximation is not valid and n should deviate from the straight line as shown in Figure 10.5.)

n_d. This is the number of electrons in the donor states. For $\mathcal{E}_F$ much larger than $\mathcal{E}_d$ these states are all occupied and $n_d = N_d$. For $\mathcal{E}_F = \mathcal{E}_d$, they are half occupied and the curve lies below the point (N_d, $\mathcal{E}_d$), shown as a cross, by a factor of 2. For lower values of $\mathcal{E}_F$, the approximation $\mathbf{f} = \exp(\mathcal{E}_F - \mathcal{E}_d)/kT$ may be used which gives a straight line for n_d, with the same slope as for n. If extended, the straight line would pass through (N_d, $\mathcal{E}_d$).

Precisely similar conditions apply to the positive charges N_d, p, and p_a, with the distribution for holes used in place of that for electrons.

For the conditions of Figure 16.2, n is much larger than n_d for all values of $\mathcal{E}_F$. This situation arises from the fact that kT is comparable to the binding energy of an electron and N_c is much larger than N_d. As a result, the greater number of states available to electrons in the conduction band far more than overweighs the difference in the statistical factor f which favors the lower energy states. Under equilibrium conditions (where the heavy

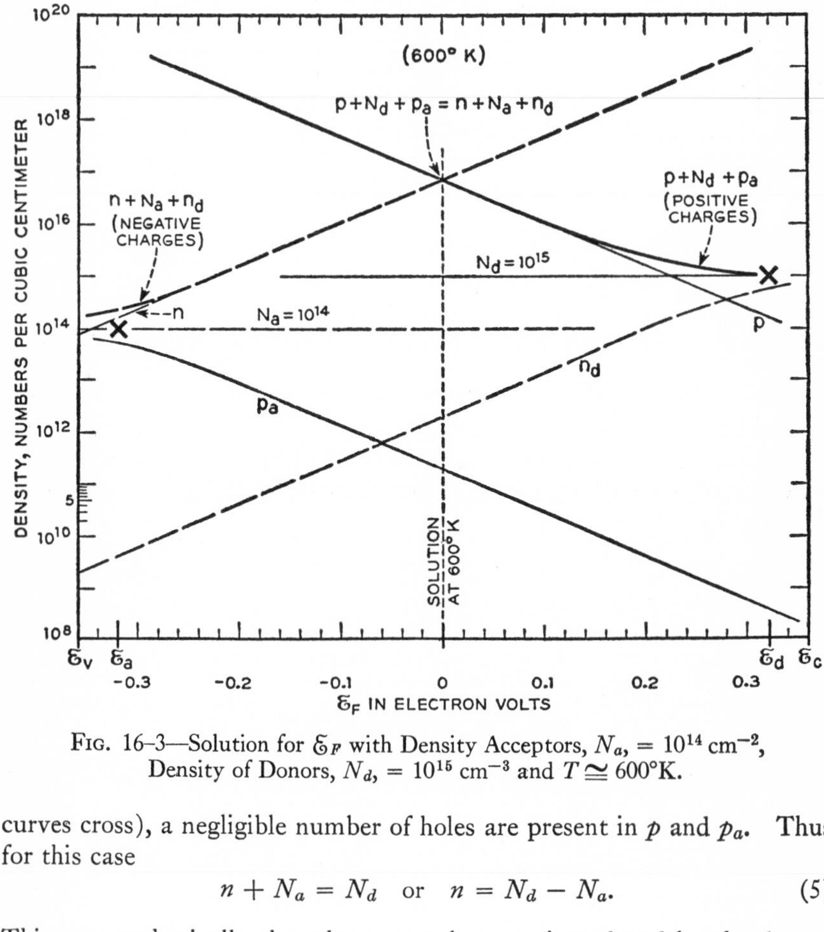

FIG. 16-3—Solution for $\mathcal{E}_F$ with Density Acceptors, N_a, = 10^{14} cm^{-2}, Density of Donors, N_d, = 10^{15} cm^{-3} and $T \cong 600$°K.

curves cross), a negligible number of holes are present in p and p_a. Thus for this case

$$n + N_a = N_d \quad \text{or} \quad n = N_d - N_a. \tag{5}$$

This means physically that the excess electrons introduced by the donors minus the acceptors have all been freed by thermal agitation and are in the conduction band. However, since p is negligible, the temperature is not high enough to break any appreciable number of bonds.

For $T = 600$°K, the situation is as shown in Figure 16.3. Under these conditions the dominant contributions come from n and p and the values of N_a and N_d do not influence the situation appreciably. This corresponds to a temperature for which the intrinsic behavior is reached. It is evident that neutrality requires that $\mathcal{E}_F \doteq (\mathcal{E}_v + \mathcal{E}_c)/2$ so that $p \doteq n$.

For Figure 16.4, $T = 150°K$, and n is more than 10 times larger than n_d for all values of $\mathcal{E}_F$ and contributes most of the variable negative charge. Charge balance occurs when

$$n + n_d = N_d - N_a. \tag{6}$$

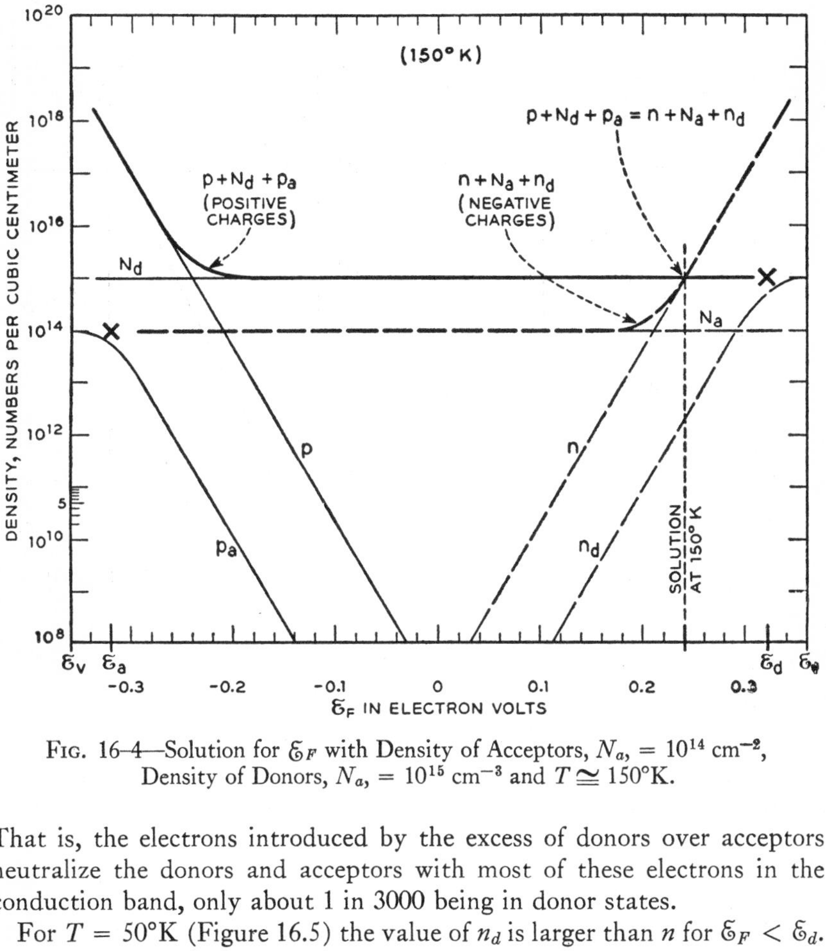

FIG. 16–4—Solution for $\mathcal{E}_F$ with Density of Acceptors, N_a, $= 10^{14}$ cm^{-3}, Density of Donors, N_a, $= 10^{15}$ cm^{-3} and $T \cong 150°K$.

That is, the electrons introduced by the excess of donors over acceptors neutralize the donors and acceptors with most of these electrons in the conduction band, only about 1 in 3000 being in donor states.

For $T = 50°K$ (Figure 16.5) the value of n_d is larger than n for $\mathcal{E}_F < \mathcal{E}_d$. For $\mathcal{E}_F > \mathcal{E}_d$ the donors are substantially filled; that is, most of the electrons have not been excited from the donor levels. The values of p and p_a are negligible and are not shown in the figure, which shows the energies near $\mathcal{E}_d$ on an expanded scale. Again equilibrium corresponds to

$$n + N_a + n_d = N_d. \tag{7}$$

However, in this case where the donor levels are almost completely filled, it is more convenient to consider the holes in the donor levels rather than

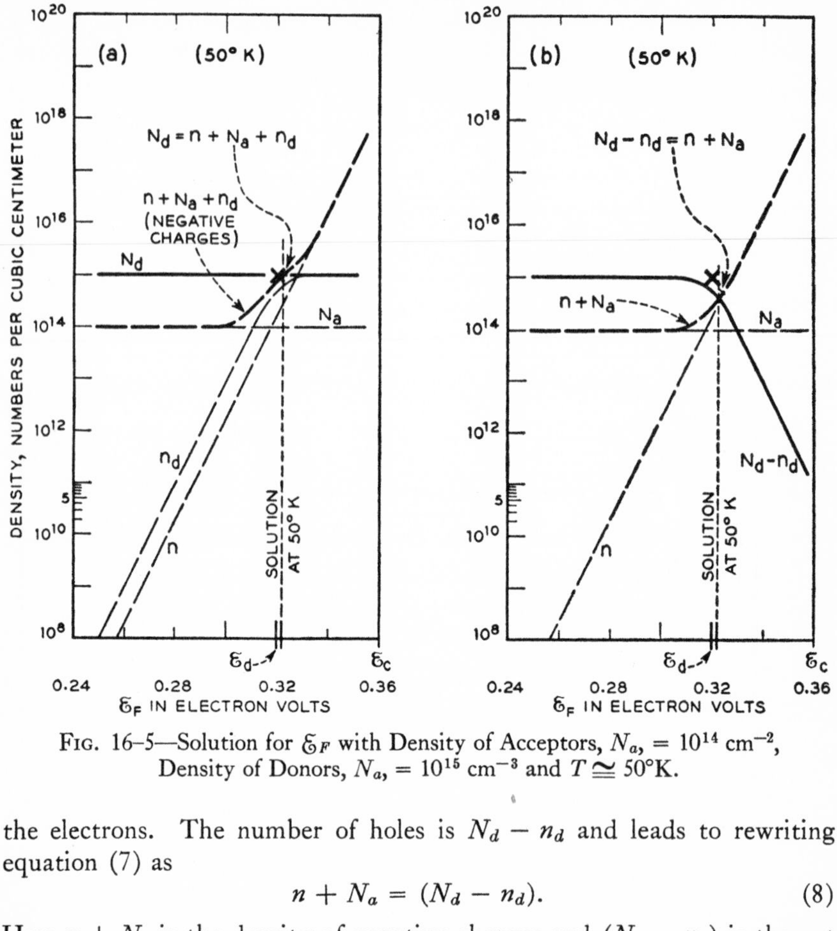

FIG. 16–5—Solution for $\mathcal{E}_F$ with Density of Acceptors, N_a, $= 10^{14}$ cm^{-2}, Density of Donors, N_a, $= 10^{15}$ cm^{-3} and $T \cong 50$°K.

the electrons. The number of holes is $N_d - n_d$ and leads to rewriting equation (7) as

$$n + N_a = (N_d - n_d). \tag{8}$$

Here $n + N_a$ is the density of negative charges and $(N_d - n_d)$ is the net density of positive charges, that is, the donor levels occupied by holes. The variation of $(N_d - n_d)$ with $\mathcal{E}_F$ is similar to that of p_a, and $(N_d - n_d)$ has its half value of $N_d/2$ when $\mathcal{E}_F = \mathcal{E}_d$. Figure 16.5(b) shows $(N_d - n_d)$ plotted versus $\mathcal{E}_F$ as a solid line and $n + N_a$ as a dashed line.

This method is still more appropriate at the lower temperatures of $37\frac{1}{2}$°K and 20°K [(Figure 16.6(a) and (b)]. Here we have considered an additional value of 10^{12} for N_a. For these temperatures, $\mathcal{E}_F$ approaches $\mathcal{E}_d$ so as to leave the donor levels partially filled as $T = 0$ is approached. However, if N_a were equal to zero, it is evident that $\mathcal{E}_F$ would lie halfway between $\mathcal{E}_c$ and $\mathcal{E}_d$ as T approaches zero, just as $\mathcal{E}_F$ lies halfway between the filled and conduction bands in the intrinsic range.

In the next section we shall analyze further the behavior of $\mathcal{E}_F$ and will discuss the relationship of the slope of ln n vs $1/T$ plots to energies.

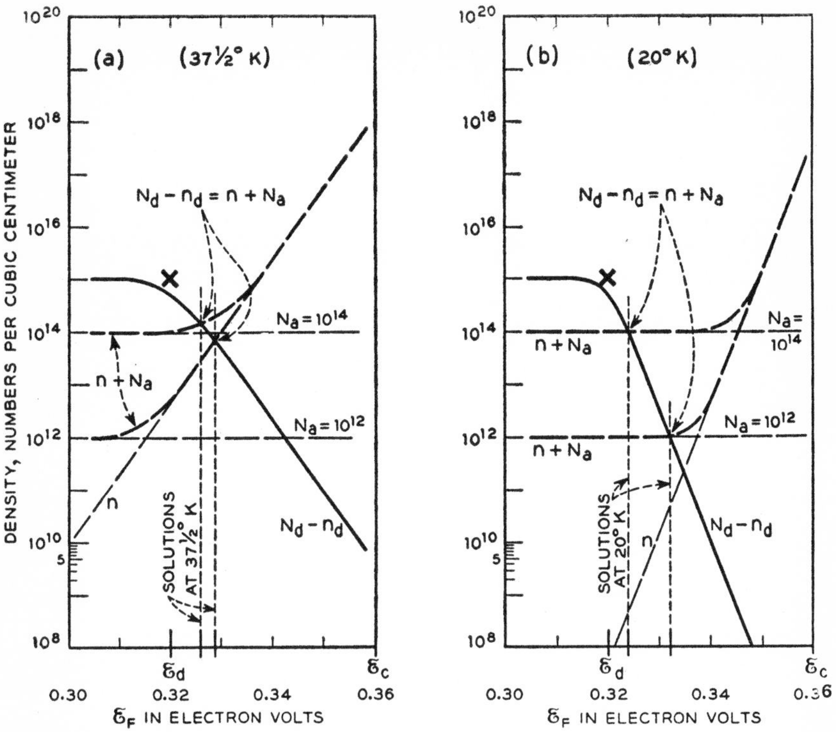

FIG. 16–6—Solutions for $\mathcal{E}_F$ with Densities of Acceptors, N_a, = 10^{14} cm^{-3} and 10^{12} cm^{-3}, Density of Donors, N_d, = 10^{15} cm^{-3}, for $T \cong 37.5$°K and 20°K.

16.4 ACTIVATION ENERGIES

The results obtained just above can be put in analytical form with instructive results. As T approaches zero, n_d approaches $N_d - N_a$ and n approaches zero. We wish to analyze the behavior of ln n as a function of T and to interpret this in terms of an activation energy. For low T's such that n is negligible compared to n_d, the neutrality condition is

$$n_d = N_d - N_a \tag{1}$$

and since

$$n_d = \mathbf{f}N_d \tag{2}$$

we find

$$\mathbf{f} = (N_d - N_a)/N_d. \tag{3}$$

Since by assumption $N_d > N_a$, this gives $\mathbf{f}$ a value between zero and unity which can be solved for $\mathcal{E}_F$ as follows: from

$$\mathbf{f} = \frac{1}{1 + \exp[(\mathcal{E} - \mathcal{E}_F)/kT]} \tag{4}$$

we obtain

$$\mathcal{E} - \mathcal{E}_F = kT \ln (1 - \mathbf{f})/\mathbf{f}, \tag{5}$$

and substituting $\mathcal{E}_d$ for $\mathcal{E}$ and $(N_d - N_a)/N_d$ for $\mathbf{f}$ we obtain

$$\mathcal{E}_F = \mathcal{E}_d + kT \ln [(N_d - N_a)/N_a]. \tag{6}$$

This shows that for temperatures so low that n is negligibly small compared to n_d (so that n_d may be taken to be $N_d - N_a$), $\mathcal{E}_F$ is linear in T. For $N_d = 2N_a$, the N_d states are just half filled at $T = 0$ and $\mathcal{E}_F$ does not depend on T since ln 1 is zero. For $N_d > 2N_a$, the slope is positive and $\mathcal{E}_F$ increases with increasing T; the physical interpretation is that for this case the donor states are more than half full and $\mathcal{E}_F$ must, therefore, lie above $\mathcal{E}_d$. For $N_d < 2N_a$, the states are less than half full and $\mathcal{E}_F$ always lies below $\mathcal{E}_d$.

When $\mathcal{E}_F$ is varying, what activation energy will be obtained from a $\ln n$ vs $1/T$ plot? The interesting result is that $\mathcal{E}_c - \mathcal{E}_d$ will be obtained as may be seen as follows: We have

$$\begin{aligned} n &= N_c \exp [(\mathcal{E}_F - \mathcal{E}_c)/kT] \\ &= 2(2\pi mkT/h^2)^{3/2} \exp [\mathcal{E}_d - \mathcal{E}_c + kT \ln (N_d - N_a)/N_a]/kT \\ &= 2(2\pi mk/h^2)^{3/2} T^{3/2} [(N_d - N_a)/N_a] \exp (\mathcal{E}_d - \mathcal{E}_c)/kT \end{aligned} \tag{7}$$

from equation (2) of Section 16.2. From this equation it is seen that the exponential dependence only involves $\mathcal{E}_d - \mathcal{E}_c$ and that the slope of $\mathcal{E}_F$ versus T affects the coefficient. Thus at sufficiently low temperatures, a plot of $\ln (nT^{-3/2})$ versus $1/T$ should give the correct value for $\mathcal{E}_c - \mathcal{E}_d$.

This result can be generalized in terms of a plot of $\mathcal{E}_F$ vs T, Figure 16.7. This plot summarizes the results obtained in Section 16.3 and just above in this section. When $N_a = 5 \times 10^{14}$, or half of N_d, $\mathcal{E}_F$ starts from $T = 0$ with zero slope. For N_a greater than this, it starts downward. For any particular curve of $\mathcal{E}_F$ versus T, the plot of $\ln nT^{-3/2}$ will vary as $(\mathcal{E}_F - \mathcal{E}_c)/kT$. It will, therefore, have as slope on the $1/T$ plot of

$$\begin{aligned} \frac{d}{d(1/T)}\left(\frac{\mathcal{E}_F - \mathcal{E}_c}{kT}\right) &= \frac{\mathcal{E}_F - \mathcal{E}_c}{k} + \frac{1}{kT} \cdot \frac{d\mathcal{E}_F}{d(1/T)} = \frac{\mathcal{E}_F - \mathcal{E}_c}{k} - \frac{1}{kT} \cdot \frac{d\mathcal{E}_F}{dT/T^2} \\ &= \frac{1}{k}\left[\mathcal{E}_F - T\frac{d\mathcal{E}_F}{dT} - \mathcal{E}_c\right] = \frac{1}{k}[\mathcal{E}_F{}^* - \mathcal{E}_c] \end{aligned} \tag{8}$$

where

$$\mathcal{E}_F{}^* \equiv \mathcal{E}_F - T\frac{d\mathcal{E}_F}{dT}. \tag{9}$$

This is just the slope associated with extrapolating a point on the $\mathcal{E}_F$ vs T curve tangentially back to $T = 0$ as is shown in Figure 16.8. The value $\mathcal{E}_F{}^*$ so obtained is the apparent Fermi level so far as activation energy

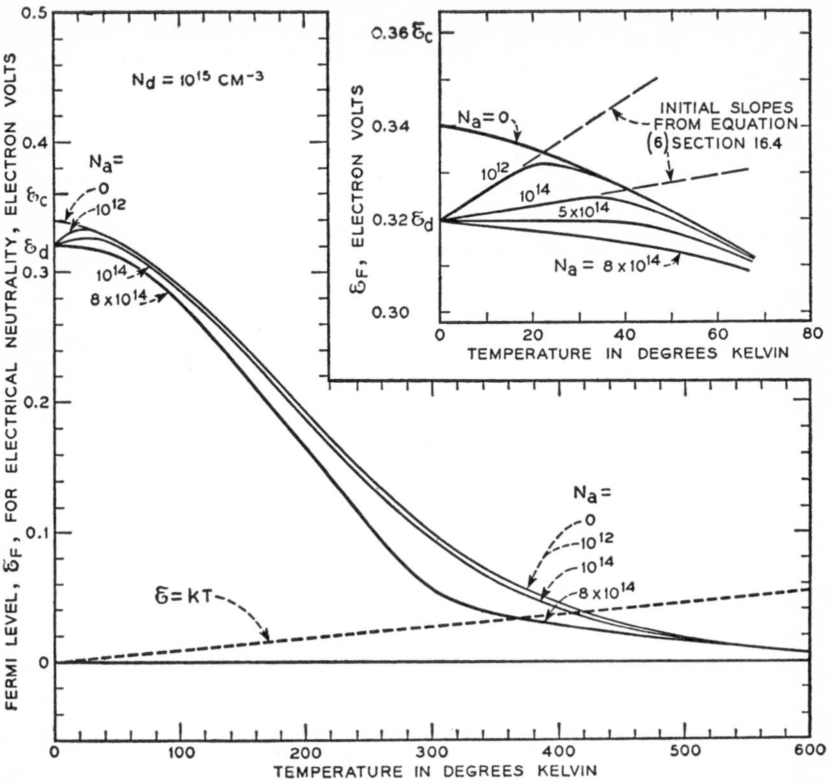

FIG. 16-7—The Fermi Level for Electrical Neutrality Plotted against Temperature for Several Densities of Acceptors.

slopes are concerned. For the example chosen, it comes above $\mathcal{E}_c$ so that the apparent activation energy is negative. This anomalous result is just what would be expected in the saturation range. In this range of temperature, the value of n is independent of T while N_c increases. Thus if N_c were treated as a lumped set of states all at energy $\mathcal{E}_c$, then increases in T correspond to decreases in **f** for these levels just as if the Fermi level were above $\mathcal{E}_c$. This shows why the apparent activation energy turns out negative as shown in Figure 16.8.

From these considerations it is evident that in general no simple interpretation of the activation energies is possible from the $1/T$ plots. Instead it is necessary to make detailed comparisons between theoretical curves based on exact solutions of the models discussed in Figures 16.1 to 16.7. Such comparisons are presented for a number of cases for silicon alloys in the paper of Pearson and Bardeen.[1]

[1] G. L. Pearson and J. Bardeen, *Phys. Rev.* **75**, 865–883 (1949).

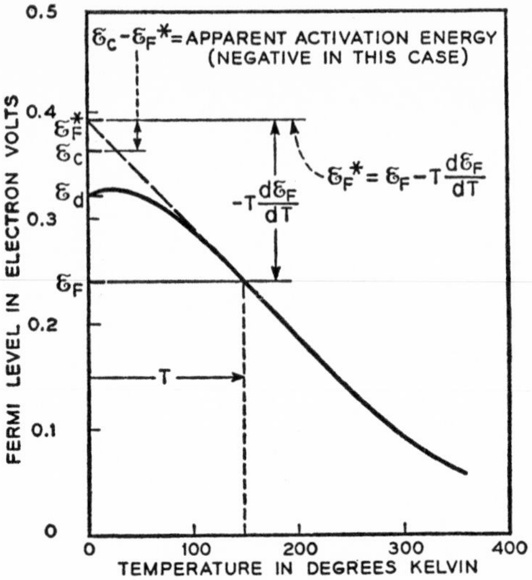

FIG. 16–8—Apparent Activation Energy for $N_a = 10^{14}$.

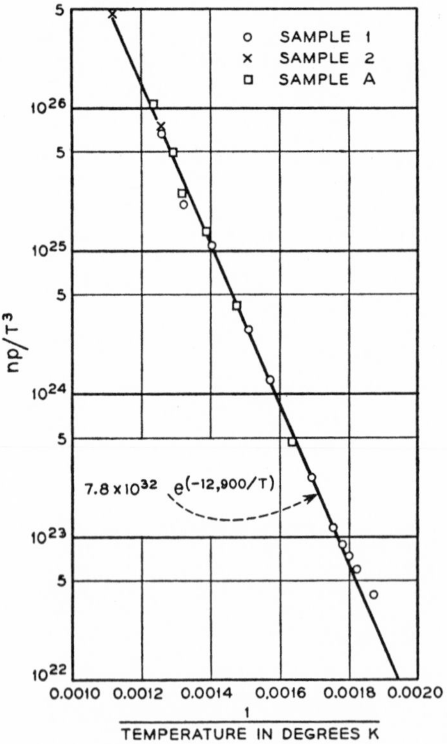

FIG. 16–9—Plot of Experimental Data for Several Silicon Samples for Comparison with Equations (10) and (11) of Section 16–4.

The simplest of these comparisons is made in the range in which both n and p are appreciable. Using the approximations of equations (11) and (14) of Section 10.3 for these gives

$$npT^{-3} = 4(2\pi mk/h^2)^3 \exp{(\mathcal{E}_v - \mathcal{E}_c)/kT}$$
$$= 2.32 \times 10^{31} \exp{(\mathcal{E}_v - \mathcal{E}_c)/kT} \text{ cm}^{-6} \, ^\circ\text{K}^{-3}. \quad (10)$$

The data obtained on the basis of this formula falls on a straight line as shown in Figure 16.9. However, the value of the coefficient obtained from the experimental data is found to be about 34 times larger than predicted from formula (10). This result can be explained by supposing that $\mathcal{E}_c - \mathcal{E}_v$, the energy gap between the bands, is temperature dependent and decreases, probably because of thermal expansion, as $\mathcal{E}_G - \beta T$. Using this value in equation (10) modifies the exponential as follows

$$\exp{(\mathcal{E}_v - \mathcal{E}_c)/kT} = e^{\beta/k} \exp{(-\mathcal{E}_G/kT)}. \quad (11)$$

The value of β has been estimated independently from measurements of photovoltaic effects in silicon $n - p$ junctions by F. S. Goucher and H. B. Briggs and appears to account for most of the factor of 34. This conclusion is in agreement with the other methods of estimating β discussed in Section 12.8. The remainder may be due to the electrons and holes having effective masses different from the free electron mass.

Problems

(See also Chapter 10)

1. Derive the Fermi-Dirac distribution law by considering two perturbations $\delta\mathcal{E}$ and $\delta\mathcal{E}'$ as was done for the Planck distribution.

2. Consider the distribution for electrons trapped on donor levels taking into account the effect of spin which allows two possibilities for each trapped electron. (Electrostatic effects will prevent more than one electron from being trapped on one donor.) Show that there will be

$$2^{n_d} N_d!/n_d!(N_d - n_d)!$$

ways of putting n_d electrons on N_d levels and hence that

$$n_d = N_d/[1 + (\tfrac{1}{2}) \exp{(\mathcal{E}_d - \mathcal{E}_F)/kT}].$$

This result is used in Problem 1 of Chapter 10.

3. Show that $\ln W$ for the Planck distribution is really a maximum by considering the value of $\delta \ln W$ for a large perturbation with $\delta\mathcal{E}' = -\delta\mathcal{E}$.

Remark

The following problems furnish an outline of one of the methods of showing how thermodynamics follows from the statistical mechanics of quantized systems.[1] They also serve to introduce the partition functions

[1] This treatment is similar to that of E. Schroedinger, *Statistical Thermodynamics*, Cambridge, 1946, but differs by laying emphasis on a more general heat bath.

Z and z which are very useful for calculating average properties and their fluctuations. The basic assumption is the assumption of equal *a priori* probabilities for all quantum states of the system. Problem 4 introduces the canonical ensemble and Z as artificial aids and shows that averages for the canonical ensemble are the same as averages for one particular microcanonical ensemble contained in the canonical ensemble. Thus the important terms in Z all arise from one particular microcanonical ensemble. In Problem 5, the number of states in the microcanonical ensemble is evaluated in terms of Z. In Problem 6, it is shown that the important terms in the canonical ensemble arise chiefly from one particular distribution and that the larger the system becomes, the less important are other distributions. This distribution must, of course, be the most probable distribution. These problems thus show that the canonical ensemble may be used in place of the microcanonical ensemble for calculating the average properties of a system, the great advantage of this result being the possibility of using the partition function Z. The canonical ensemble is next shown in Problems 9 and 10 to have a physical meaning; in fact it represents the distribution, not of an isolated system, but of a system in contact with a heat reservoir. In Problems 11 to 14, the average pressure of system in equilibrium with a heat reservoir is expressed in terms of Z, and the identification of statistical and thermodynamical properties is given.

4. Show that appreciable deviations from the equilibrium distribution are negligible by considering the average over the *canonical ensemble*. The canonical ensemble is introduced at this point as a mathematical aid; it is defined as follows: Give each quantum state for the system with energy $\mathcal{E}_a$ a weight $\exp(-\beta\mathcal{E}_a)$. Let

$$Z = \sum_{\text{all } a} e^{-\beta\mathcal{E}_a}.$$

Z is called the *states sum* (*Zustandssumme*) or *partition function*. The weighted average energy over the canonical ensemble is thus

$$\langle\mathcal{E}\rangle = \sum \mathcal{E}_a e^{-\beta\mathcal{E}_a}/Z = -\frac{d}{d\beta}\ln Z,$$

where $\langle\ \rangle$ is introduced to signify the weighted average for the canonical ensemble. The average squared energy is

$$\langle\mathcal{E}^2\rangle = \sum \mathcal{E}_a{}^2 e^{-\beta\mathcal{E}_a}/Z = \frac{1}{Z}\frac{d^2}{d\beta^2}Z.$$

Suppose that the system can be considered to be made up of a large number of similar and substantially independent subsystems, 1, 2, $\cdots$, Q. Then if the energy levels of one of those subsystems are ε_α, ε_β, ε_γ, etc., then each energy of the whole system may be written as a sum of these energies.

Show that

$$Z = z^Q$$

where z equals the states sum for one of the subsystems. Show that

$$\langle \mathcal{E}^2 \rangle - \langle \mathcal{E} \rangle^2 = Q[\langle \varepsilon^2 \rangle_z - \langle \varepsilon \rangle_z{}^2]$$

where the $\langle\ \rangle_z$ expressions represent averages over the terms in z. From this conclude that for a large system, for which $Q > 10^{10}$, for example, that a negligible contribution to Z is made by states which have $|\mathcal{E} - \langle \mathcal{E} \rangle|^2 > 10^{-6}\langle \mathcal{E} \rangle^2$, provided that $\langle \varepsilon^2 \rangle_z$ is not much greater than $\langle \varepsilon \rangle_z{}^2$. How does this prove that averages over the canonical ensemble of a property of the system equals the average over a microcanonical ensemble with $\mathcal{E} = \langle \mathcal{E} \rangle$?

Show also that the formula

$$\langle \mathcal{E} \rangle = -Q \frac{d}{d\beta} \ln z$$

(which is used in the preceding proof) leads to the harmonic oscillator energy, equation (22) of Section 16.1, when z is calculated for the energy levels $(n + \frac{1}{2})h\nu$.

5. From the result of Problem 4 show that

$$\ln W_m(\langle \mathcal{E} \rangle) = \beta \langle \mathcal{E} \rangle + \ln Z,$$

where $W_m(\langle \mathcal{E} \rangle)$ is the number of states in a microcanonical ensemble of energy in a small range near $\langle \mathcal{E} \rangle$. The quantities Z and $\langle \mathcal{E} \rangle$ are functions of β, the as yet unidentified parameter in Z. Do this by writing Z in the form

$$Z = \sum_{\mathcal{E}} W_m(\mathcal{E}) e^{-\beta \mathcal{E}}$$

where the states are lumped into groups of $W_m(\mathcal{E})$ states of approximately the same energy $\mathcal{E}$. Show that, for any reasonable choice of the energy intervals used for the groups, the value of $\ln W_m(\mathcal{E})$ is independent of the choice, provided that Q is large. [For example, show that if the energy interval is chosen as proportional to Q^1, Q^0, or Q^{-1}, the value of $\ln W_m(\mathcal{E})$ is not affected.]

6. Show that the major contribution in Z comes from states in substantially the same distribution. This is done by showing that there is an average distribution represented in Z and that deviations from it are negligible. Accordingly we find the average number $\langle Q_\alpha \rangle$ of subsystems with energy ε_α and also the average squared number $\langle Q_\alpha{}^2 \rangle$. The mean squared deviation of Q_α is then shown to be proportional to Q so that the fractional deviation is proportional to $1/Q$ and is negligible for large Q. The analysis is carried out as follows: Each term in Z is of the form $\exp -\beta[Q_\alpha \varepsilon_\alpha + Q_\beta \varepsilon_\beta + Q_\gamma \varepsilon_\gamma + \cdots]$ where the Q_α's represent the number of subsystems in each state. Evidently

$$\langle Q_\alpha \rangle = \frac{-1}{\beta Z} \frac{\partial}{\partial \varepsilon_\alpha} Z$$

and

$$\langle {Q_\alpha}^2 \rangle = \frac{1}{\beta^2 Z} \frac{\partial^2}{\partial {\varepsilon_\alpha}^2} Z.$$

Inserting z^Q for Z, obtain

$$\langle Q_\alpha \rangle = Q \frac{e^{-\beta \varepsilon_\alpha}}{z}$$

$$\langle {Q_\alpha}^2 \rangle = \langle Q_\alpha \rangle^2 + Q \frac{e^{-\beta \varepsilon_\alpha}}{z} \left[1 - \frac{e^{-\beta \varepsilon_\alpha}}{z} \right].$$

Show that for large values of Q, the fractional mean square deviation of Q_α is negligible so that the important contributions to Z all come from substantially the same distribution.

7. From Problems 5 and 6 prove that for large systems (and finite values of β) that

$$\ln W(\langle \mathfrak{E} \rangle) = \beta \langle \mathfrak{E} \rangle + \ln Z$$

where W applies to the most probable distribution.

8. Show that Problem 6 reduces directly to the Planck distribution [equation (24) of Section 16.1] if each oscillator is taken as a subsystem. Subsystems in general need not be so simple. (The same procedure cannot, of course, be applied to the Fermi-Dirac distribution since the Pauli principle precludes treating individual electrons as independent subsystems.)

9. Show that a heat reservoir, which may be taken to be a system in which Q is made arbitrarily large, satisfies the condition that

$$\delta \ln W(\mathfrak{E}) = \beta \delta \mathfrak{E}$$

for large changes $\delta \mathfrak{E}$. *Hint:* express Problem 7 in terms of Q and $z(\beta)$ and let $Q \to \infty$.

10. Show that a system in equilibrium with a heat reservoir with parameter β_R will have a relative probability $\exp(-\beta_R \mathfrak{E}_a)$ of being in a state of energy $\mathfrak{E}_a$. Do this by using the assumption of equal *a priori* probabilities, the method of combined systems used in Section 16.1 and the result of Problem 9. This shows that a system in equilibrium with a heat reservoir is distributed in a canonical ensemble with properties derived in Problems 4 to 6.

11. Suppose that the energy of each state $\mathfrak{E}_a$ of the system is a function of a mechanical variable, such as the volume V. Then the pressure exerted by the system in state $\mathfrak{E}_a$ will be

$$p_a = -\frac{d\mathfrak{E}_a}{dV}.$$

The state sum will then be a function $Z(V,\beta)$. Show that the average pressure is

$$\langle p \rangle = \frac{1}{\beta}\frac{\partial}{\partial V}\ln Z.$$

12. Next consider the state of the system as specified by V and β. Let the state of the system change by $\delta\beta$, δV. Show that $\delta \ln W$, as defined in Problems 5 or 7, can be written in the form

$$\delta \ln W = \beta\left[-\frac{\partial^2 \ln Z}{\partial\beta^2}\delta\beta - \frac{\partial^2 \ln Z}{\partial V\partial\beta}\delta V + \frac{1}{\beta}\frac{\partial \ln Z}{\partial V}\delta V\right]$$
$$= \beta[\delta\langle\mathfrak{E}\rangle + \langle p\rangle\delta V] = \beta\delta\mathfrak{Q},$$

where $\delta\mathfrak{Q}$ must be the heat furnished by reservoirs in order to satisfy conservation of energy. This shows that $\beta\delta\mathfrak{Q}$ is an exact differential, since it is equal to $\delta \ln W$ and $\ln W$ is a function of β and V. From this show that, in a Carnot cycle operating between reservoirs β_1 and β_2, the thermodynamic efficiency is

$$\frac{d\mathfrak{Q}_1 - d\mathfrak{Q}_2}{d\mathfrak{Q}_1} = 1 - \frac{\beta_1}{\beta_2}.$$

This shows that $\beta_1 : \beta_2 = T_2 : T_1$, according to the definition of the Kelvin or absolute temperature scale. We may, therefore, define a constant k as the ratio T_1/β_1. (The proof that k is Boltzmann's constant requires the treatment of a particular model such as an ideal gas or a harmonic oscillator, like that discussed in Section 16.1.)

Show that the reasoning of Problems 11 and 12 can be carried out for an electrical system, such as a storage battery, by replacing the volume V by the charge passed through the battery and the pressure p by the voltage.

13. Show that except for an additive constant the entropy of a system, defined as $\int d\mathfrak{Q}/T = k\int \beta d\mathfrak{Q}$, is

$$\mathfrak{S} = k \ln W = (1/T)\langle\mathfrak{E}\rangle + k \ln Z = -k\beta^2\frac{\partial}{\partial\beta}\left(\frac{\ln Z}{\beta}\right).$$

Show, using the relationship between $\beta d\mathfrak{Q}$ and the derivatives of Z in Problem 12, that the term on the extreme right of the last equation is a valid expression for the entropy even if T is so low and β so large that the approximation of Problem 5 is invalid. Hence show that the zero of entropy may conveniently be chosen so that at zero temperature the entropy is

$$\mathcal{S}_0 = k \ln W(\mathcal{E}_{\min}),$$

where $W(\mathcal{E}_{\min})$ is the number of quantum states of minimum energy for the system. (Is the entropy equal to zero at $T = 0$ for a hydrogen atom?)

14. Using the above definition of $\mathcal{S}$ in terms of Z, show that

$$-(1/\beta) \ln Z = \langle \mathcal{E} \rangle - T\mathcal{S} \equiv \mathcal{F},$$

where $\mathcal{F}$ is the quantity frequently called the *free energy*. Show the relationship of the definition of entropy of Problem 13 to the equation

$$\left(\frac{\partial \mathcal{F}}{\partial T}\right)_V = -\mathcal{S}.$$

15. Show that the mean square deviation of energy of a system in equilibrium with a heat reservoir is $\langle(\mathcal{E} - \langle \mathcal{E} \rangle)^2\rangle = kTc_VT$ where c_V is the specific heat of the system at constant volume. *Hint:* use

$$c_V = \frac{\partial \langle \mathcal{E} \rangle}{\partial T} = -k\beta^2 \frac{\partial \langle \mathcal{E} \rangle}{\partial \beta}$$

and express all quantities in terms of Z.

CHAPTER 17

THE THEORY OF TRANSITION PROBABILITIES FOR HOLES AND ELECTRONS

17.1 INTRODUCTION

This chapter covers the most involved material presented in this book, as may be judged from its length and the fact that all of its sections, save this introduction, are interdependent. In it Schroedinger's equation for a system of atoms and electrons is developed into a form suitable for interpretation in the language of transition probabilities. The argument falls into three main parts: (1) finding a set of wave functions suitable for describing a system consisting of an electron moving through a crystal whose atoms are vibrating, (2) finding how this wave function changes with time, and (3) interpreting the results in terms of transition probabilities.

It is not practical to follow the argument in this sequence, however, without making it seem unnecessarily abstract. It is pedagogically far preferable to treat first (3) and part of (2) and thus to show in general terms how solutions of Schroedinger's equation lead directly to transition probabilities and how the matrix elements obtained from the starting wave functions fit into this scheme. Accordingly transition probabilities are treated first in Section 17.2. The treatment makes use of the Weisskopf-Wigner method of dealing with transition probabilities, which has not to the author's knowledge been applied previously to the conductivity problem. It has been employed here, in spite of its complexity compared with simpler methods, in order to justify the details of the simple picture of transition probabilities used in Chapters 8 and 11.

On the basis of Section 17.2, it is seen that the transition probabilities depend upon the matrix elements for an electron-atomic vibration collision. These matrix elements are evaluated by carrying out part (1) in Sections 17.3 to 17.5. In terms of the wave functions so obtained, the matrix element is derived in Section 17.6.

Figure 17.1 has been prepared to show the logical connection between various parts of the treatment. As in the case of other figures of the same sort, it will probably be of most value in reviewing the argument rather than in foreseeing it.

Section 17.7 is concerned with showing the equivalence of the problem of one hole to that of one electron and then deals with the case in which many holes and electrons are present—one of the problems discussed in Section 11.2.

The transitions produced by impurity scattering are treated in Chapter 11 and are not enlarged upon in this chapter. It should be noted, however, that the procedures of Section 17.2 would apply equally well to the matrix elements for impurity scattering. It is also specifically pointed out in

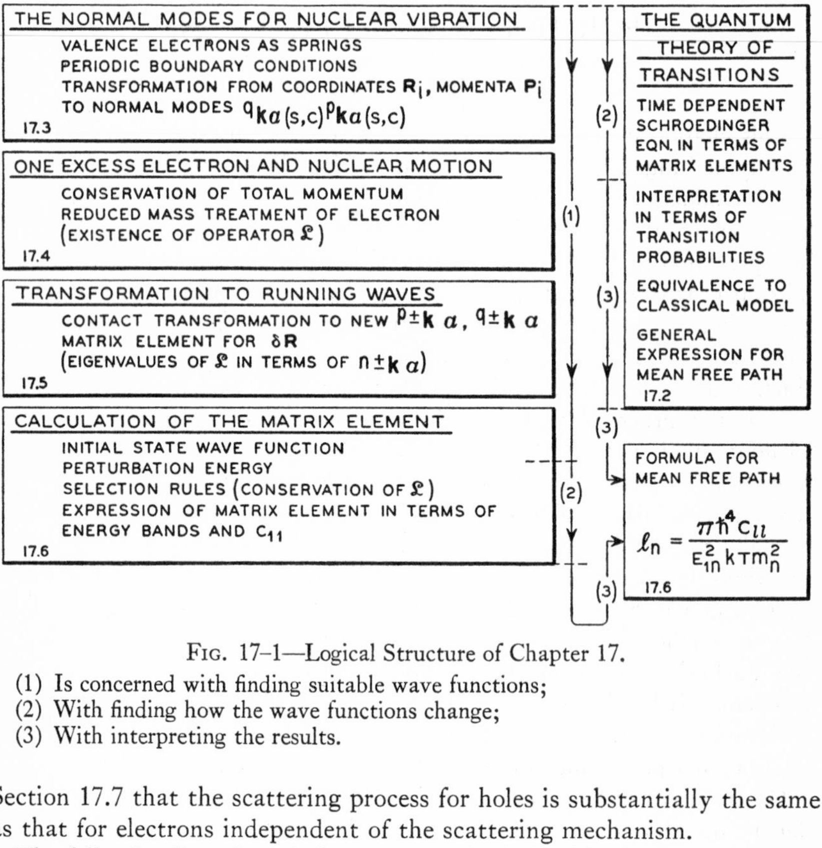

FIG. 17–1—Logical Structure of Chapter 17.
(1) Is concerned with finding suitable wave functions;
(2) With finding how the wave functions change;
(3) With interpreting the results.

Section 17.7 that the scattering process for holes is substantially the same as that for electrons independent of the scattering mechanism.

The following list of symbols used extensively in this chapter is repeated here for convenience. Equation numbers are in parenthesis and section numbers are not.

r_T = position of the center of gravity of the crystal plus the excess electron. 17.4.

r = position of the excess electron in respect to the center of the crystal. 17.4.

m_e = mass of the electron (used to distinguish this from m below). 17.4.

m = reduced mass of electron plus crystal (used in this meaning only in Chapter 17). 17.4.

M_c = mass of crystal. (1) 17.3.

M_T = mass of crystal plus electron. 17.4.

$M=\frac{1}{2}M_c$ = mass associated with a normal mode. (11) 17.3.

N_s = number of atoms in the crystal. 17.3.

Φ = wave function for entire system of crystal plus excess electron. (5) 17.5.

χ = factor of Φ corresponding to $\boldsymbol{r}_T$. 17.4.

ψ = factor of Φ corresponding to $\boldsymbol{r}$. 17.4.

$\phi(n\text{'s})$ = factor of Φ corresponding normal modes excluding motion of center of gravity. (22) 17.5 and (5) 17.6.

$\varphi_i = \varphi(n_{k\alpha}, n_{-k\alpha}; q_{k\alpha c}, q_{k\alpha s})$ = factor of $\phi(n\text{'s})$. 17.5 and (4) 17.6.

$\boldsymbol{G}_{k\alpha}, \boldsymbol{k}, \alpha$ = polarization unit vector, wave vector, polarization index. 17.3.

Some additional symbols such as $\boldsymbol{q}$, A and $\chi(a)$ are defined and used in Section 17.7.

17.2 THE THEORY OF TRANSITION PROBABILITIES; MEAN FREE PATH IN TERMS OF MATRIX ELEMENTS

In this section we shall derive the formula for the mean free path in terms of the general transition theory for a quantum-mechanical system. In this treatment the formula for the matrix elements, which determine the transition probabilities, is introduced as an added assumption, the derivation of the expression being given in Section 17.6 for atomic vibrations.

The system considered consists of an array of atoms, which are held together by valence electrons, plus one excess electron in the conduction band. Except for the excess electron, the electrons are regarded simply as furnishing a medium which binds the nuclei elastically to their equilibrium positions; the nuclei can vibrate according to quantum-mechanical laws about those positions (discussed in Sections 17.3 and 17.5) and their vibrations interact with the excess electron so as to cause it to make transitions from one quantum state to another. From any given state of the system, in which the electron has momentum $\boldsymbol{P}_0$, transitions may occur to states in which it has momenta $\boldsymbol{P}_j$; each change of this sort may occur in two ways, depending on whether the electron gains or loses energy to the lattice vibrations. It is found to be possible to treat these two possible transitions as a single transition by a simple means [Section 17.6, equation (33)]. For this reason we may consider the end states, to which a transition is made from a state with momentum $\boldsymbol{P}_0$, as uniquely specified by the final momentum $\boldsymbol{P}_j$. We may, therefore, describe the initial state of the system by $\boldsymbol{\Phi}_0$ and the final states by $\boldsymbol{\Phi}_j$, where the $\boldsymbol{\Phi}$ symbols are functions of the coordinates of the nuclei and of the position and spin of the excess electron.

We shall assume that the wave functions $\boldsymbol{\Phi}_j$ are eigenfunctions of a Hamiltonian $\mathcal{H}_0$ which differs by a small perturbation term, we shall call it $\mathcal{U}_1$, from the exact Hamiltonian $\mathcal{H}$. This assumption, which is a general requirement of perturbation theory, will be justified by the developments in later sections. Accordingly, we may write

$$\mathcal{H} = \mathcal{H}_0 + \mathcal{U}_1 \tag{1}$$

and

$$\mathcal{H}_0\boldsymbol{\Phi}_j = \mathcal{E}_j\boldsymbol{\Phi}_j. \tag{2}$$

If the system is initially in the state $\boldsymbol{\Phi}_0$ at $t = 0$, the presence of $\mathcal{U}_1$ will cause transitions to other states $\boldsymbol{\Phi}_j$, and our problem is to find the probability that the system is in any particular group of states as a function of time. This is done by obtaining a solution to the time-dependent Schroedinger equation for $\mathcal{H}$ by expressing $\boldsymbol{\Phi}$ as a series,

$$\boldsymbol{\Phi} = \sum a_j\boldsymbol{\Phi}_j$$

where the a's are functions of time and the initial state at $t = 0$ corresponds to

$$a_j = \delta_{j0}, \quad \text{that is,} \quad a_0 = 1, \quad a_{j\neq 0} = 0.$$

The probability of finding the system in any group of states at a later time is

$$\sum |a_j|^2, \quad \text{summed over states of interest.}$$

Before considering the mathematics of the perturbation theory, we shall show the relationship of this problem to the treatments of Chapters 8 and 11.

On the Basic Equivalence of the Quantum and the Classical Models. During a collision an electron gains or loses energy to the lattice vibrations and it is this process which converts the electrical work I^2R done on the electrons into heat vibrations. In any individual collision, however, the loss or gain of energy by the electron is negligible compared with its total energy, and we may neglect such changes while studying the scattering of the electrons. Thus, we suppose, as shown in Figure 17.2, that an electron initially in state $\boldsymbol{P}_0$ may make transitions to other states of the same energy, such as $\boldsymbol{P}_j$; and from these it can make transitions to other states, and so on.

We shall see later, under the heading *Quantum Theory of Transition Probabilities*, that the perturbation interaction introduces uncertainty in value of the energy after the transition. As a consequence of this, the end states do not lie exactly on the sphere of constant energy but instead are distributed chiefly in a thin shell, represented in Figure 17.3, with a spread in energy given by $\Delta\mathcal{E}\tau_0 \cong \hbar$ where τ_0 is lifetime of the initial state, discussed below. This uncertainty in energy is so small, however, that we may neglect its effects for most purposes and consider strict conservation of electronic energy to apply.

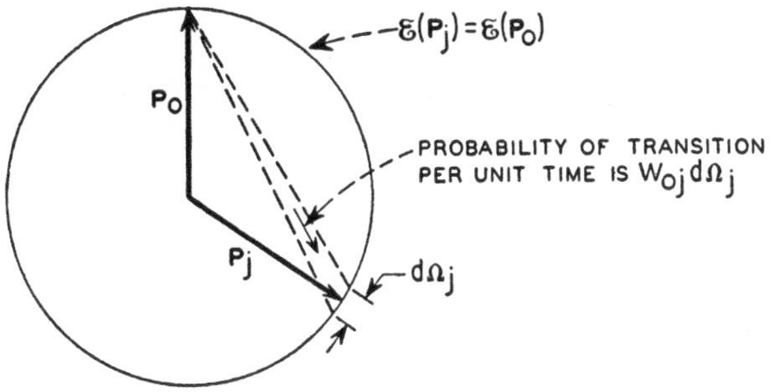

FIG. 17–2—An Electron Initial in State P_0 of the Brillouin Zone Makes Transitions to Other States of the Same Energy According to the Classical Model.

In accordance with implication (a) of assumption 1 of Chapter 8, the probability of transition from any one state to any other state should be independent of time. We shall now give this assumption mathematical expression in the form of equations, whose consequences (although not the equations themselves) we shall later show are equivalent to those of quantum-mechanical perturbation theory. The equivalence of these two treat-

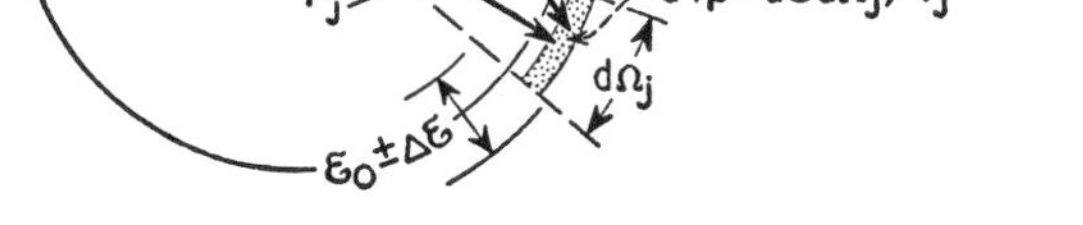

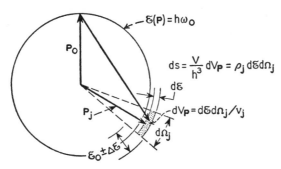

FIG. 17–3—The Relationships Between Element of Volume, Number of End States, and Velocity. (Discussed chiefly on page 489.)

ments justifies the use of the classical picture of collisions and mean free times in visualizing processes and making calculations.

In keeping with the procedure of Chapter 11, we consider in Figure 17.2 a thin shell in momentum space bounded by two closely neighboring surfaces of constant energy. Since changes in energy are neglected, we may treat this shell as a surface and speak of the number of states per unit area or number of electrons per unit area. Thus we consider transitions from the initial state $\boldsymbol{P}_0$ to the group of states near $\boldsymbol{P}_j$ lying in the element of surface area $d\Omega_j$ where $d\Omega_j$ has the dimensions of (momentum)^2. The probability of transition per unit time will be proportional to the area and may be written

$$W_{0j}d\Omega_j. \tag{3}$$

The total probability per unit time of making a transition from state $\boldsymbol{P}_0$ is then

$$W_0 = \int_{\Omega(\mathfrak{E})} W_{0j}d\Omega_j, \quad \Omega(\mathfrak{E}) = \text{energy surface.} \tag{4}$$

We similarly denote the probabilities of making transitions from state $\boldsymbol{P}_j$ by

$$W_{ji} \quad \text{and} \quad W_j = \int_{\Omega(\mathfrak{E})} W_{ji}d\Omega_i. \tag{5}$$

We shall next use these probabilities to describe the statistics of the scattering process.

It is helpful to think of the probabilities in terms of numbers of electrons; this can be done by supposing that a large number N of identical systems start with an electron in $\boldsymbol{P}_0$ at $t = 0$. Accordingly we shall use the following definitions

α_0 = number of electrons still in $\boldsymbol{P}_0$ at time t,

β_j = number of electrons per unit area (of the equal energy surface of Figure 17.2) near $\boldsymbol{P}_j$ which have been scattered once,

γ_i = number of electrons per unit area near $\boldsymbol{P}_i$ which have been scattered twice.

etc.

The number of electrons which have been scattered once and are in a range $d\Omega_j$ and the total number β which have been scattered once are

$$\beta_j d\Omega_j \quad \text{and} \quad \int_{\Omega(\mathfrak{E})} \beta_j d\Omega_j.$$

The classification of electrons in terms of the number of collisions they have had simplifies the comparison with the quantum-mechanical theory. From these definitions and the assumption that the W_{ji} are independent of time, it is evident that the rate of loss of electrons from α_0 is $W_0\alpha_0$, giving

$$\frac{d}{dt}\alpha_0 = -W_0\alpha_0, \quad \alpha_0 = \mathbf{e}^{-W_0t}, \quad W_0 = 1/\tau_0 \tag{6}$$

where τ_0 is the mean free time between collisions of Chapter 11. The rate of change of the singly scattered electrons in $d\Omega_j$ is

$$\frac{d}{dt}\beta_j d\Omega_j = \alpha_0 W_{0j} d\Omega_j - W_j \beta_j d\Omega_j$$

$$\dot{\beta}_j = W_{0j}\alpha_0 - W_j\beta_j; \tag{7}$$

and similarly

$$\dot{\gamma}_i = \int W_{ji}\beta_i d\Omega_j - W_j\gamma_i.$$

These equations express mathematically the idea that an electron in a given state $\boldsymbol{P}_j$ has definite probabilities of being scattered per unit time into various other states $\boldsymbol{P}_i$, these probabilities being independent of time. This is in agreement with the picture used in Chapter 11 to calculate the mobility. Our main objectives in this chapter are (1) to show that quantum mechanics leads to the behavior predicted by these equations, thus showing that the simple physical picture used in Chapters 8 and 11 is correct, and (2) to evaluate the probabilities W. We shall treat (1) in this section, verifying in addition that the form of the quantities W_{ij} is consistent with the principle of detailed balancing discussed in Chapter 11. The expressions for the W's in terms of the constants of the material will be derived in Section 17.6.

The Quantum Theory of Transition Probabilities. We shall next present a quantum-mechanical derivation of the collision formulae discussed above. The method employed is based on the Weisskopf and Wigner[1] treatment of radiation broadening. In their method an approximate solution is obtained for the decay of an initial excited state of an atom into the ground state with the radiation of a quantum of light. In our problem, we consider an electron in an initial state which may make a transition to another momentum state with radiation of a phonon. In the radiation case, the final state is specified by giving the characteristics of the emitted photon. In our case we can use instead the final momentum since the phonon emitted or absorbed is uniquely described in terms of the change in momentum. (The slight oversimplification of this statement is reconsidered in Section 17.6 following Table 17.2.) The electron-scattering problem is more complicated than the radiation problem, because, immediately after one transition occurs, others are possible. It is possible to show for the cases of principal interest, however, that these other transitions may be treated according to the scheme of equations (6) and (7).

[1] V. Weisskopf and E. Wigner, *Zeits. für Phys.* **63**, 54 (1930); **65**, 18 (1930). See also W. Heitler, *The Quantum Theory of Radiation*, Oxford University Press, New York, 1936.

We shall give a demonstration of some of these relationships in this section, leaving certain questions of rigor as problems at the end of Section 17.6.

The Weisskopf-Wigner method uses considerably more advanced techniques than those usually employed in deriving transition probabilities. The more conventional treatments lead only to an initial rate of decay which is linear in time and thus represent the first two terms, namely, $1 - (t/\tau)$, of equation (6). The Weisskopf-Wigner method leads to the proper exponential dependence. Both methods lead to the same formulae for τ, and the reader who is unfamiliar with the more elementary method may find it advantageous to study some of the standard texts.[2]

The Weisskopf-Wigner method starts by expanding the wave function in a series of the form[3]

$$\Phi = a_0 e^{-i\omega_0 t}\Phi_0 + \sum{}' b_j e^{-i\omega_j t}\Phi_j \tag{8}$$

where the coefficients are unknown functions of time satisfying the conditions

$$a_0 = 1, \quad b_j = 0 \quad \text{at} \quad t = 0 \tag{9}$$

and

$$\hbar\omega_0 = \mathcal{E}_0, \quad \hbar\omega_j = \mathcal{E}_j. \tag{10}$$

The Φ_j functions differ from Φ_0 by having one more or less phonon than Φ_0 in our case and in the radiation emission case by having one more photon. Thus Φ_0 is not one of the Φ's in the sum of Φ_j. We shall neglect in the first treatment the possibility of transitions from one Φ_j to another or to other functions in which more changes in the phonon field have occurred. With the aid of these assumptions the differential equation for the coefficient a_0 is obtained from the time-dependent equation,

$$\mathcal{H}\Phi = (\mathcal{H}_0 + \mathcal{U}_1)\Phi = (i\hbar)\dot{\Phi}, \tag{11}$$

by multiplying by Φ_0^* and integrating over all coordinates and by using $\mathcal{H}_0\Phi_0 = \mathcal{E}_0\Phi_0$:

$$i\hbar \dot{a}_0 = \sum{}' \mathcal{U}_{0j} e^{i(\omega_0 - \omega_j)t} b_j \tag{12}$$

where $\mathcal{U}_{0j}$ is the matrix element for transition from 0 to j:

$$\mathcal{U}_{0j} = \int \Phi_0^* \mathcal{U}_1 \Phi_j d\,(\text{all coordinates}). \tag{13}$$

The matrix elements of $\mathcal{U}_1$, as follows from the treatment of later sections, are Hermitian and in addition $\mathcal{U}_{00}$, which is omitted in (12), is shown to vanish in Section 17.6. [As the text just preceding equation (28) of this section indicates, the perturbation due to $\mathcal{U}_1$ is small compared to thermal

[2] Such as L. I. Schiff, *Quantum Mechanics*, McGraw-Hill Book Co., New York, 1949, Section 29. His equation (29.12) is equivalent to our equation (23) for W_{0j}.

[3] $\sum{}'$ represents a sum over all states omitting $j = 0$.

energies for electrons.] Similar manipulations on (11) with $\Phi_i{}^*$ give

$$i\hbar\dot{b}_j = \mathfrak{U}_{j0}\mathrm{e}^{i(\omega_j - \omega_0)t}a_0. \tag{14}$$

We shall next verify that an approximate solution of these equations is

$$a_0 = \mathrm{e}^{-\eta t} \tag{15}$$

$$b_j = \frac{\mathfrak{U}_{j0}}{i\hbar}\,\frac{\mathrm{e}^{[i(\omega_j - \omega_0) - \eta]t} - 1}{i(\omega_j - \omega_0) - \eta} \tag{16}$$

where η is a constant which we shall shortly evaluate.

In terms of η the probability that the electron is in state zero is $|a_0|^2 = \exp(-2\eta t)$ from which it is evident that the transition probability and mean free time for state Φ_0 are

$$W_0 = 2\eta \quad \text{and} \quad \tau_0 = 1/2\eta.$$

Equation (16) above for b_j follows at once from that for a_0 and the differential equation (14). In order to satisfy the equation for $\dot{a}_0$ we must have

$$\eta = \sum{}' \frac{\mathfrak{U}_{0j}\mathfrak{U}_{j0}}{\hbar^2}\,\frac{1 - \mathrm{e}^{[i(\omega_0 - \omega_j) + \eta]t}}{i(\omega_j - \omega_0) - \eta} \tag{17}$$

as may be found by direct substitution. (As we shall show below, this equation can be solved for η giving a constant value independent of t.) The summation is next replaced by an integration, it being assumed (justified in Section 17.6) that $\mathfrak{U}_{0j}\mathfrak{U}_{j0} = |\mathfrak{U}_{0j}|^2$ is a slowly varying function of the end state j. The density of end states is (V/h^3) since changes in spin do not occur. For integration coordinates we use elements of area $d\Omega_j$ of the constant energy surfaces and the energy itself. The element of volume in $\boldsymbol{P}$ space is thus

$$dV_{\boldsymbol{P}} = d\Omega_j g(\boldsymbol{P}_j)d\mathcal{E}$$

where

$$g(\boldsymbol{P}_j) = 1/|\nabla_{\boldsymbol{P}}\mathcal{E}(\boldsymbol{P})| = 1/v_j, \quad v_j = |\boldsymbol{v}_j|$$

is the separation of the energy surfaces per unit energy difference.[4] (As discussed above, we neglect the energy of the phonon.) For brevity we shall introduce ρ_j (a function of $\boldsymbol{P}_j$) which gives the number of end states ds per unit energy per unit surface; by definition we then have

$$\rho_j d\Omega_j d\mathcal{E} = \frac{V}{h^3 v_j}\,d\Omega_j d\mathcal{E}, \quad \rho_j = \frac{V}{h^3 v_j}. \tag{18}$$

This relationship, together with certain other features of the integration over $\boldsymbol{P}$-space, is shown in Figure 17.3 on page 485.

The integration may be carried out approximately making use of known

[4] See Section 11.2 for a further discussion of $dV_{\boldsymbol{P}}$.

integrals. It is convenient to replace $d\mathcal{E}$ by $\hbar d\omega_j$, which then brings equation (17) for η into the form

$$\eta = \int \frac{|\mathcal{U}_{0j}|^2}{\hbar} \frac{1 - e^{[i(\omega_0 - \omega_j) + \eta]t}}{i(\omega_j - \omega_0) - \eta} \rho_j d\omega_j d\Omega_j. \tag{19}$$

The factor involving $\omega_0 - \omega_j$ is largest for transitions which conserve energy, and for these its value is $(1 - e^{\eta t})/\eta$; for large differences between ω_j and ω_0 it is usually regarded as negligible. If we consider the important contributions to come from so small a range of ω_j near ω_0 that the other factors in (19) are constant, we can make use of the equations

$$\lim_{L\to\infty} \int_{-L}^{+L} \frac{1}{i(\omega_0 - \omega_j) - \eta} d\omega_j = \begin{cases} -\pi, & \eta > 0 \\ +\pi, & \eta < 0 \end{cases}$$

$$\lim_{L\to\infty} \int_{-L}^{+L} \frac{e^{[i(\omega_0 - \omega_j) + \eta]t}}{i(\omega_j - \omega_0) - \eta} d\omega_j = \begin{cases} -2\pi, & \eta > 0 \\ 0, & \eta < 0. \end{cases}$$

(See Problems for a discussion of the integrals and convergence questions, and with their aid carry out the integration over ω_j.) This procedure reduces equation (19) to

$$\eta = \int_{\Omega(\mathcal{E}_0)} \frac{|\mathcal{U}_{0j}|^2}{\hbar} \pi\rho_j d\Omega_j \tag{20}$$

where the integration is over the surface $\mathcal{E}_j = \mathcal{E}_0$. Thus for all values of t equation (17) is satisfied by the same value of η so that (15) and (16) constitute a good solution for the time-dependent Schroedinger equation (11).

We shall next show that the differential contributions to η can be interpreted as transition probabilities per unit area of the energy surface. To do this we consider the probability that the electron be found in the element of area $d\Omega_j$ with some unspecified energy $\hbar\omega_j$. This probability is the sum of the appropriate $|b_j|^2$ and is

$$d\Omega_j \int |b_j|^2 \rho_j \hbar d\omega_j = d\Omega_j \int \frac{|\mathcal{U}_{j0}|^2}{\hbar} \frac{i + e^{-2\eta t} - 2e^{-\eta t}\cos(\omega_j - \omega_0)t}{(\omega_j - \omega_0)^2 + \eta^2} d\omega_j \rho_j$$

$$= \frac{1}{2\eta}(1 - e^{-2\eta t}) \frac{|\mathcal{U}_{j0}|^2}{\hbar} 2\pi\rho_j d\Omega_j \tag{21}$$

where we have once more taken the other variables as constant over the important range of integration for ω_j and have allowed the limits to extend to $\pm\infty$; the necessary integrals are standard forms.[5] As is shown in Figure 17.3, the important range of integration is that for which $\Delta\mathcal{E} = |\mathcal{E}_j - \mathcal{E}_0| = \hbar\eta$ corresponding to $|\omega_j - \omega_0| = \eta$. This result is consistent

[5] B. O. Pierce, *A Short Table of Intervals*, Third Edition, formulae (480) and (490), Ginn and Co., New York, 1929.

with the uncertainty principle in the form $\Delta\mathcal{E}\Delta t = \hbar$ and with the interpretation $\Delta t = 1/\eta = 2\tau_0$ where τ_0, as we shall show below, is the lifetime of the state Φ_0. For cases of practical interest, $\Delta\mathcal{E}$ is much less than $\mathcal{E}_0$ so that the spread in energy is relatively unimportant; this point is also discussed shortly.

If in the classical equations we suppress second collisions, so as to have a situation corresponding to the quantum-mechanical case just treated, we find the fraction of the electrons which have been singly scattered by solving the following simplified form of (7)

$$\dot{\beta}_j = W_{0j}\alpha_0 = W_{0j}e^{-W_0 t},$$

thus obtaining

$$\beta_j = \frac{1}{W_0}(1 - e^{-W_0 t})W_{0j} \tag{22}$$

which is equivalent to (21) if we make the identifications

$$W_{0j} = \frac{|\mathfrak{U}_{j0}|^2}{\hbar} 2\pi\rho_j \tag{23}$$

$$1/\tau_0 = W_0 = 2\eta = \int_{\Omega(\mathcal{E}_0)} W_{0j} d\Omega_j. \tag{24}$$

Thus the behavior in time for the Weisskopf-Wigner approximate solution leads to the classical behavior with the interpretation that transition probability per unit area of the constant energy surface is given by (23) and the mean free time τ_0 is given by (24).

The Principle of Detailed Balancing. The form of equation (23) allows us to verify that the principle of detailed balancing holds for the collisions considered here. Thus if we consider a thin energy shell of the sort discussed in Chapter 11, Figure 11.1, we can show that under equilibrium conditions the number of electrons making transitions from an element of volume $d\mathcal{E}d\Omega_i$ to $d\mathcal{E}d\Omega_j$ is just equal to the number making transitions from $d\mathcal{E}d\Omega_j$ back to $d\mathcal{E}d\Omega_i$. The reasoning given in Section 11.2 is based on the equality of $\rho_i W_{ij}$ and $\rho_j W_{ji}$. This equality follows at once from equation (23) since we can write

$$\rho_i W_{ij} = \rho_i |\mathfrak{U}_{ji}|^2 2\pi\rho_j/\hbar = \rho_i\rho_j \mathfrak{U}_{ji}\mathfrak{U}_{ij} 2\pi/\hbar = \rho_j W_{ji}. \tag{25}$$

The third expression, which is symmetrical in i and j, follows from the Hermitian property of $\mathfrak{U}_1$ which leads to $\mathfrak{U}_{ij} = \mathfrak{U}_{ji}^*$.

The modifications required by the Pauli exclusion principle are discussed in Section 17.7 and also in Chapter 11 in which the result (25) is used.

Critique of the Quantum-mechanical Treatment. The derivation of (23) and (24) suffers from the flaw that second collisions were arbitrarily prevented. It is possible to extend the theory so as to include these but at the expense of additional questions of convergence. We shall outline the method here and give some additional details in a problem, leaving the actual manipulations as an exercise to the reader.

We have found that any given state, described by a_0, makes a transition to states b_j so that a_0 varies as exp $(-\eta t)$. Thus the effect of interaction with b states causes a_0 to vary according to the equation

$$i\hbar\dot{a}_0 = i\hbar\eta a_0.$$

This suggests that, in dealing with the decay of any one state, we can replace the effect of the states which it excites and to which it decays by writing $-i\hbar\eta a_0$ in place of the sum of $\mathcal{U}_{0j}b_j \exp [i(\omega_0 - \omega_j)t]$, which occurs in the equation for $i\hbar\dot{a}_0$ [that is, (12)]. If this simple procedure is legitimate, we may represent the effect of the states to which the $b_j\Phi_j$ series may decay by writing

$$i\hbar\dot{b}_j = \mathcal{U}_{j0}e^{i(\omega_j-\omega_0)t}a_0 - i\hbar\eta_j b_j, \tag{26}$$

the η_j term being obtained in just the same way as η except that the initial state is Φ_j instead of Φ_0. If this approximation is made, another solution of the Weisskopf-Wigner type[6] can be obtained for the coefficients b_j

$$b_j = \frac{\mathcal{U}_{j0}}{i\hbar}\cdot\frac{e^{[i(\omega_j-\omega_0)-\eta_j]t} - e^{-\eta_j t}}{i(\omega_j - \omega_0) - \eta + \eta_j} \tag{27}$$

and it can be shown that these b_j once more lead to the classical probabilities when a new set of transition probabilities W_{jk} are defined for the transitions from Φ_j to other states.

This process may be pushed farther on so that the states arising from decay of Φ_i are in turn considered, etc. In this way the analogy between the quantum theory and the classical collision theory may be made as complete as may be desired.

There are, however, some annoying questions about the validity of this procedure which we endeavor to dispose of in Problem 4 at the end of Section 17.6.

One point which we may discuss here is the validity of replacing $|\mathcal{U}_{0j}|^2\rho_j$ by a constant during the integration over ω_j. The important range of integration is that for which $|\omega_j - \omega_0| \doteq \eta = 1/2\tau_0$. This corresponds to a change in electronic energy of $\delta\mathcal{E} = \hbar/2\tau_0$. For an electron in germanium at room temperature, $\tau_0 \doteq 10^{-12}$ sec so that $\delta\mathcal{E} \doteq 10^{-15}$ erg or about $kT/50$. Hence the variation of energy over the important range is only a few per cent of the energy of the electron and, consequently, the changes in $|\mathcal{U}_{0j}|^2\rho_j$ may be neglected. (See, however, Problem 2, for other questions of integration.) The criterion that the important range of $\hbar\omega_j$ be small

[6] In fact very similar to the one they used in considering two successive radiation processes.

compared to kT has been discussed recently by Seitz[7] who expresses the requirement in the form $\mu > 30\ \text{cm}^2/\text{volt sec}$.

The Calculation of the Mean Free Time and Mean Free Path. As we shall show in Section 17.6, the value of the matrix element can be written[8]

$$|\mathfrak{U}_{j0}|^2 = AT/V \tag{28}$$

where T is the absolute temperature, V the volume of the crystal, and A is a constant depending on the elastic properties of the crystal, the mass of the atoms, and certain attributes of the wave functions but not upon $\boldsymbol{P}_0$ or $\boldsymbol{P}_j$. The formula states that $|\mathfrak{U}_{j0}|^2$ is independent of the change in direction of the electron's motion, so that, after being scattered, the electron is equally likely to have momentum in any direction. Since the values of W_{0j} do not depend on the direction of $\boldsymbol{P}_j$, the integration over $d\Omega_j$ in equation (24) in this case thus reduces simply to multiplication by $4\pi \boldsymbol{P}_0^2$. Using the formula $v_0 = |\boldsymbol{P}_0|/m_n$, which relates electron velocity v_0 to momentum and effective mass m_n, we may evaluate (24) as follows:

$$\frac{1}{\tau_0} = \int_{(\Omega\mathcal{E}_0)} W_{0j}d\Omega_j = \frac{2\pi}{\hbar}\rho_j|\mathfrak{U}_{j0}|^2 4\pi \boldsymbol{P}_0^2 = \frac{2\pi}{\hbar}\left[\frac{V}{h^3}\frac{1}{v_0}\right]\frac{AT}{V}4\pi \boldsymbol{P}_0^2$$

$$= \frac{ATm_n^2}{\pi\hbar^4}v_0 = v_0/l(T). \tag{29}$$

This expression shows that the transition probability is proportional to v_0, corresponding to a mean free path $l(T)$ independent of velocity and given by the formula

$$l(T) = \frac{\pi\hbar^4}{ATm_n^2}. \tag{30}$$

It is perhaps worth while to point out that the constant mean free path independent of *energy* of the electron results directly from the fact that the matrix element is independent of the electron's initial and final momenta. This independence of mean free path upon energy will apply even if the bands are degenerate, provided only that the energy in any direction of $\boldsymbol{P}$-space varies as $\boldsymbol{P}^2$. If this last is true, then the area of the constant energy surfaces will also vary as $\boldsymbol{P}^2$ or as the energy $\mathcal{E}$ and the spacing between surfaces will vary as $\mathcal{E}^{-1/2}$. Consequently, the transition probability will vary as $\mathcal{E} \times \mathcal{E}^{-1/2}$ or $\mathcal{E}^{1/2}$ and this is proportional to velocity. Hence even if the energy surfaces are quite complex, the mean free path for a given direction of motion will be independent of velocity.

As shown in Chapter 11, the expression for constant mean free path leads

[7] F. Seitz, *Phys. Rev.* **73**, 549–564 (1948).

[8] This effective value includes the contributions both of processes involving gain and of processes involving loss of energy by the electron.

to a mobility

$$\mu = 4l/3\sqrt{2\pi kT/m_n}. \tag{31}$$

The treatments of atomic vibrations and their interactions with the electrons in the following parts of this chapter establish the expression for $|\mathfrak{U}_{ij}|^2$, show that it has the form AT/V, and evaluate the constant A.

The equivalence of the scattering of holes to the scattering of electrons and related questions are discussed in Section 17.7.

Problems

1. Evaluate the integrals used in equation (19). This may be done by contour integration in the complex plane. The first integral is elementary and the second is obtainable from the contour from $-R$ to $+R$ on the real axis plus the semi-circle below the axis. (See also Whittaker and Watson, *Modern Analysis*, Fourth Edition, p. 123, problem 15. Cambridge at the University Press, 1927.)

2. Both integrals considered in connection with (19) are improper integrals and their values depend upon how the limits $\pm\infty$ for ω_j are approached. The exponential term will converge unless $|\mathfrak{U}_{0j}|^2$ increases more rapidly than $|\omega_0 - \omega_j|$. However, the other term may give an imaginary component of the form

$$\lim_{\epsilon\to 0}\int_{-\infty}^{\omega_0-\varepsilon} + \int_{\omega_0+\varepsilon}^{\infty} \frac{|\mathfrak{U}_{0j}|^2}{(\omega_j - \omega_0)}\, d\omega_j.$$

Compare this with the second-order perturbation energy [see, for example, Schiff, *loc. cit.*, equation (25.12)] and show that its effect is to correct the frequency of the state Φ_0 to account for the average energy of the system as a whole.

3. Substitute b_j of equation (27) into (12); show that (12) is satisfied and η given by (20). Calculate the value of β_j for these b_j's and show that the expression which replaces (21) is the correct solution for equation (7) of the collision theory provided the interpretation $W_i = 2\eta_i$ is used.

4. If $\Phi(t)$ is a solution of Schroedinger's equation (when magnetic fields are absent), then $\Phi^*(-t)$, which runs backwards, is also a solution. Yet equation (12), $a_0 = \mathbf{e}^{-\eta t}$, obviously cannot be run backwards indefinitely. Show that, at $t = 0$ when all $b_i = 0$, $\dot{a}_0 = 0$ and not $-\eta$ and hence that the Weisskopf-Wigner approximation must break down at $t = 0$. To what solution for negative t will the W–W approximation actually correspond? Why is it that the W–W solution can be reversed in time while those of the collision equations (6) and (7) cannot?

17.3 THE NORMAL MODES FOR ATOMIC VIBRATION

As outlined in Figure 17.1, several steps are required in calculating the matrix element for a transition involving changes of electronic and vibrational state of the crystal. In this section we initiate this treatment with an analysis of the motions of the nuclei, regarding the valence electrons simply as a medium for exerting internuclear forces. The justification for this procedure is based on the high ratio of nuclear mass to electron mass. As a consequence of this difference, the bonding electrons move so much more rapidly than the nuclei that their wave functions are the same as if the nuclei were momentarily stationary. The energies of the wave functions of the bonding electrons depend on the nuclear coordinates, however, and as the nuclei move these energies change. The work done by nuclei on the electrons as the nuclei move represents an effective potential energy for the nuclei and can be introduced as such in the Schroedinger equation for the nuclear coordinates. In this case the electrons can be thought of as moving so rapidly that the nuclei are affected only by the average force exerted by the electrons and this is given correctly by the use of the electronic energy as an effective potential energy.[1]

Accordingly we consider a crystal containing N_s atoms. We assume that the crystal structure has one atom per unit cell so that all atoms have the same mass and are otherwise equivalent. We shall assume that periodic boundary conditions apply so that we may think of atoms located at $x = A_x - a$, where A_x is the *boundary-period* in the x-direction and a, the lattice constant, as being neighbors of atoms at $x = 0$. If the crystal as a whole undergoes translational motion, atoms crossing the $x = A_x$ plane may be thought of as simultaneously re-entering at $x = 0$, no momentum being involved in the transfer back. Alternatively, we may imagine that the crystal as a whole moves outside of the block from $x = 0$ to $x = Ax$ by considering that the potential energy between any two nuclei may be represented as a periodic function of their coordinates $x_1y_1z_1$ and $x_2y_2z_2$ which satisfies the equation

$$\mathcal{V}(x_1 - x_2, y_1 - y_2, z_1 - z_2)$$
$$= \mathcal{V}(x_1 - x_2 + n_xA_x, y_1 - y_2 + n_yA_y, z_1 - z_2 + n_zA_z) \qquad (1)$$

where A_x, A_y, A_z are the periods and the n's are integers. These two possibilities are indicated in Figure 17.4(b) in which the atoms of the crystal and their periodic "images" are shown. The atoms a and b outside the boundary are equivalent to their periodically related images a' and b'' inside the boundary.[2]

A system of this sort can exert torques on itself so that the angular

[1] Detailed treatments of the separation of nuclear and electronic coordinates are given in many texts such as F. Seitz, *The Modern Theory of Solids*, Chap. XIV, McGraw-Hill Book Co., New York, 1940; H. Eyring, J. Walter, and G. E. Kimball, *Quantum Chemistry*, Chap. XI, John Wiley & Sons, New York, 1944. See also R. P. Feynman, *Phys. Rev.* **56**, 340–343 (1939).

[2] For a discussion of the effect of various boundary conditions see M. Born, *Proc. Phys. Soc. Lond.* **54**, 362–376 (1942).

momenta are not constants of motion. Thus if the crystal as a whole is rotated slightly as shown in Figure 17.4(c), the "images" produced outside the boundary by the periodic boundary conditions are seen to exert forces on the atoms inside and these forces tend to rotate the crystal back to parallelism with the periodic boundaries. This feature of the model is somewhat artificial; one can imagine, however, a two-dimensional crystal in which these boundary conditions would apply by following the line of reasoning given in Section 15.9. Furthermore, the general nature of the

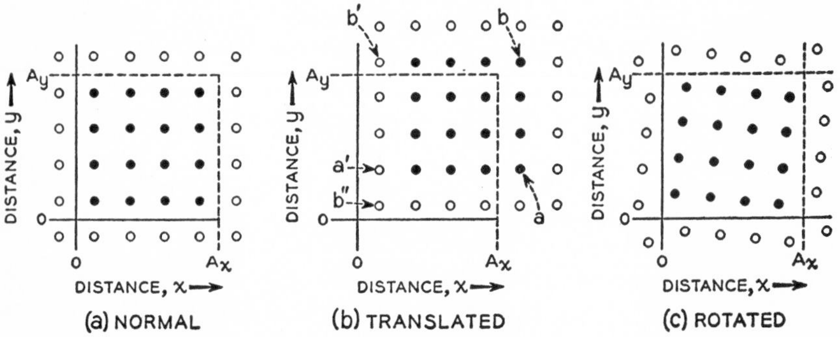

FIG. 17-4—The Effect of Periodic Boundary Conditions.
(a) Normal arrangement.
(b) Translated arrangement.
(c) Rotated arrangement showing how the crystal can exert a torque on itself.

results obtained with the periodic model would almost certainly be reproduced, although less conveniently, by using other more physical conditions.

The Normal Modes. Our problem in this section is to obtain a solution of the Schroedinger equation for the motion of the nuclei. In this treatment we shall use three sets of vectors, $\boldsymbol{R}_i$, $\boldsymbol{R}_i'$ and $\delta\boldsymbol{R}_i$, to describe the positions of the nuclei, the choice being dictated by convenience for the later development. These $\boldsymbol{R}$'s are defined as follows: A certain reference position in which the nuclei are perfectly arranged on a periodic lattice is selected for convenience. In this arrangement one of the atoms, called zero, is selected as the origin. The vector positions of the other atoms constitute the set $\boldsymbol{R}_i$ so that $\boldsymbol{R}_i$ is a translation of the lattice. The vector $\boldsymbol{i}_xA_x$, being an integral number of lattice constants, lies on an extended point of the $\boldsymbol{R}_i$ lattice. In the disturbed arrangement of the lattice, atom i moves by an amount $\delta\boldsymbol{R}_i$ so that

$$\delta\boldsymbol{R}_i = \boldsymbol{R}_i' - \boldsymbol{R}_i \tag{2}$$

or, in words,

displacement from reference position

= actual position minus reference position.

In terms of the position vectors $\boldsymbol{R}_i'$ and the conjugate momenta $\boldsymbol{P}_i$, Schroedinger's equation for the nuclear motion is

$$\left[\frac{1}{2M_a}\sum_i \boldsymbol{P}_i^2 + \mathcal{V}(\boldsymbol{R}_i', \cdots \boldsymbol{R}_{N_s}')\right]\Phi = \mathcal{E}\Phi \tag{3}$$

where $\boldsymbol{P}_i = (\hbar/i)\nabla_{\boldsymbol{R}i}'$ and the subscript indicates the coordinates of the nucleus or atom are involved. For small amplitudes of vibration, $\mathcal{V}$ may be expanded as a quadratic form and a set of normal modes obtained having coordinates and momenta q_i and p_i. For periodic boundary conditions these modes can be expressed in terms of running waves like those of Figure 5.6; for a three-dimensional crystal, however, the modes can propagate in space and, furthermore, each atom may vibrate in three dimensions. In order to arrive at these running wave solutions, we shall take two preliminary steps.

We consider first the classical problem of finding normal modes. For this case it is advantageous to use the set of quantities $\boldsymbol{\delta R}_i$ which represent the deviation of $\boldsymbol{R}_i'$ from the reference equilibrium position of the crystal for which the atomic positions are $\boldsymbol{R}_i$. The classical equation of motion, derived from Hamilton's equations, or from $\boldsymbol{F} = m\boldsymbol{a}$, are

$$M_a(d^2/dt^2)\boldsymbol{\delta R}_i = -\nabla_i\mathcal{V}(\boldsymbol{\delta R}\text{'s}) \tag{4}$$

where ∇_i implies taking the gradient in respect to $\boldsymbol{\delta R}_i$ and where $\mathcal{V}$ is now approximated as a quadratic form in the $\boldsymbol{\delta R}$'*s*. The normal modes for such structures, taking into account the effect of the periodic structure of the crystal upon the form $\mathcal{V}(\boldsymbol{\delta R}\text{'s})$ are treated in detail in many texts, the general methods being analogous to those employed in other parts of the chapter.[3] These normal modes are required to have the property that all atoms undergo simple harmonic oscillations with the same frequency and phase [except for reversals of direction corresponding to phase differences of 180°, as shown in Figure 5.3(j) to (n)]. We shall denote the amplitudes of the normal modes by the normal coordinates q. In terms of the normal modes the displacements of the atoms are found to be

$$\boldsymbol{\delta R}_i = \sum_{\boldsymbol{k},\alpha} G_{\boldsymbol{k},\alpha}(q_{\boldsymbol{k}\alpha c}\cos\boldsymbol{k}\cdot\boldsymbol{R}_i + q_{\boldsymbol{k}\alpha s}\sin\boldsymbol{k}\cdot\boldsymbol{R}_i) \tag{5}$$

where each normal mode coordinate undergoes simple harmonic motion independent of the others. The wave vector $\boldsymbol{k}$ is required to satisfy the periodic boundary conditions so that

$$k_xA_x = 2\pi n_x, \quad k_yA_y = 2\pi n_y, \quad k_zA_z = 2\pi n_z, \tag{6}$$

[3] Sections 14.7 and 15.3 for analogous problems; for textbooks see, for example, F. Seitz, *The Modern Theory of Solids*, Section 22, McGraw-Hill Book Co., New York, 1940; L. Brillouin, *Wave Propagation in Periodic Structures*, McGraw-Hill Book Co., New York, 1946; and A. H. Wilson, *The Theory of Metals*, Cambridge at the University Press, 1936, Chap. VI.

where the n's are integers. For each wave vector $\boldsymbol{k}$, there are three directions of polarization, corresponding to values 1, 2, 3 for subscript α; these three directions of polarization are specified by the unit polarization vectors $\boldsymbol{G}_{k\alpha}$. In an elastically isotropic crystal, one of the three $\boldsymbol{G}$'s for a given $\boldsymbol{k}$ is parallel to $\boldsymbol{k}$ and represents a longitudinal wave while the other two $\boldsymbol{G}$'s are perpendicular to $\boldsymbol{k}$ and to each other. For a cubic crystal, as is shown in Problem 1, the three $\boldsymbol{G}$'s are perpendicular although the property of being purely transverse and purely longitudinal is not preserved. For each $\boldsymbol{k}$ and α, there are two patterns of vibration one varying as $\cos \boldsymbol{k} \cdot \boldsymbol{r}$ with normal coordinate $q_{k\alpha c}$ and one as $\sin \boldsymbol{k} \cdot \boldsymbol{r}$ with normal coordinate $q_{k\alpha s}$.

The lattice of $\boldsymbol{k}$-values described by these equations is simply related to the basic lattice of $\boldsymbol{P}$ vectors in the Brillouin zone [see equations (9) to (11) section 5.5] by the equation

$$\hbar \boldsymbol{k} = \boldsymbol{P}. \tag{7}$$

The distinct values of $\boldsymbol{k}$ are also limited to the Brillouin zone and we shall refer to the limited set of $\boldsymbol{k}$'s *as the basic set of* $\boldsymbol{k}$'s. The relationship between the $\boldsymbol{k}$ and $\boldsymbol{P}$ values plays an essential role later in connection with electron transitions for which changes in $\boldsymbol{P}$ excite lattice waves whose $\boldsymbol{k}$-values are given by the changes in $\boldsymbol{P}/\hbar$.

So far as the standing waves described by $\cos \boldsymbol{k} \cdot \boldsymbol{R}_i$ and $\sin \boldsymbol{k} \cdot \boldsymbol{R}_i$ are concerned, it is evident that $\boldsymbol{k}$ and $-\boldsymbol{k}$ give rise to exactly the same displacements; hence, in the sum over the possible values of $\boldsymbol{k}$ we shall require that opposite values are excluded. (This can be done by excluding all negative n_x; for $n_x = 0$ excluding negative n_y; and for $n_x = n_y = 0$ excluding negative n_z. See Problem 1 for questions of counting.) It is also evident that equivalence relationships exist for $\boldsymbol{k}$-values on the surface of the Brillouin zone in the same way as they do for $\boldsymbol{P}$-values [see Figure 5.11(a)]. For a simple cubic crystal, the edge of the Brillouin zone for $\boldsymbol{k}$ corresponds to the value π/a where a is the lattice constant for $|k_x|$, $|k_y|$ and $|k_z|$. For each value of $\boldsymbol{k}$, there are three possible directions for the unit vector $\boldsymbol{G}_{k\alpha}$ distinguished by values 1, 2, and 3 for α. The frequencies of vibration of the normal modes are determined by $\boldsymbol{k}$ and by the direction of polarization and are denoted by $\omega_{k\alpha}$, the same frequency being obtained for both the sine and cosine modes. We shall not derive these results nor the other consequences, which we shall quote, of the analysis of the normal modes, but will instead outline the method of derivation in a set of problems at the end of this section.

Three normal modes are of special interest; they correspond to $\boldsymbol{k} = 0$, and to $\omega_{0,\alpha} = 0$. For one of them we may take $\boldsymbol{G}_{01}$ as $\boldsymbol{i}_x$, the unit vector along the x-axis. The sine term vanishes for $\boldsymbol{k} = 0$, and the cosine term gives a displacement $\delta \boldsymbol{R}_i = \boldsymbol{i}_x q_{0,1}$ corresponding simply to translation of the crystal. These three modes represent motion of the center of gravity of the crystal and may be expressed in the form

$$\boldsymbol{R}_c = \boldsymbol{i}_x q_{01} + \boldsymbol{i}_y q_{02} + \boldsymbol{i}_z q_{03} \tag{8}$$

where $\boldsymbol{R}_c = \sum \delta\boldsymbol{R}_i/N_s$ summed over all nuclei since the displacements due to each mode with $\boldsymbol{k} \neq 0$ satisfy the equation $\sum \delta\boldsymbol{R}_i = 0$.

The entire normal mode system corresponds to a transformation of coordinates of the form

$$x_i = \sum a_{ij} q_j$$

where x stands for a nuclear coordinate, such as the y-component of $\delta\boldsymbol{R}_i$ and q_j for one of the normal modes. From the general nature of the normal mode problem, it follows that the Hamiltonian function in terms of the normal modes is expressed in the form

$$\mathcal{H} = \frac{1}{2M_c}\boldsymbol{P}_c^2 + \sum \frac{1}{2M}(p_i^2 + M^2\omega_i^2 q_i^2) \tag{9}$$

where

$$M_c = N_s M_a = \text{the mass of the crystal}; \tag{10}$$

$$M = M_c/2; \tag{11}$$

$$\boldsymbol{P}_c = \sum_i \boldsymbol{P}_i. \tag{12}$$

The mass of each nucleus should have added to it the mass of its core and valence electrons which follow it in its motion; thus M_c will really be the mass of the entire crystal. The vector $\boldsymbol{P}_c$ is the momentum of the crystal as a whole and is conjugate to $\boldsymbol{R}_c$. The sum of the other terms extends over all $q_{k\alpha c}$ and $q_{k\alpha s}$ for which $k \neq 0$.

For future use we shall write the Hamiltonian in the abbreviated form

$$\mathcal{H} = \frac{1}{2M_c}\boldsymbol{P}_c^2 + \sum_i \mathcal{H}_i \tag{13}$$

where i stands for any of the $\boldsymbol{k}\alpha$ pairs for the standing waves so that

$$\mathcal{H}_i = \frac{1}{2M}[p_{ic}^2 + p_{is}^2 + M^2\omega_i^2(q_{ic}^2 + q_{is}^2)]. \tag{14}$$

We shall henceforth refer to the q's as the normal modes for the standing waves, or q_{std} for short, and to $\boldsymbol{R}_c$ as the center of gravity of the crystal.

The transformation to the q's and $\boldsymbol{R}_c$ is a contact transformation in both the classical and the quantum-mechanical sense (see Problems 2 and 3) so that the eigenvalue and wave function problem is equivalent to that of a free particle corresponding to $\boldsymbol{R}_c$ and to $3(N_s - 1)$ independent harmonic oscillators.

The following three problems amplify the reasoning behind these results.

Problem 1. This problem constitutes a proof that the waves of equation (5) are normal modes for a simple cubic lattice in which each atom is at a center of symmetry for the undeformed state. This simplified case will serve to illustrate the general features of the normal mode problem. (For more general crystals see references in footnote 3 of this section.)

Suppose that for the equilibrium reference position, one atom is at the origin and the other atoms are located at positions specified by lattice vectors $\boldsymbol{R}_i$. The displacement of atom i from its reference position $\boldsymbol{R}_i$ is

$$(x_i, y_i, z_i) = \boldsymbol{\delta R}_i.$$

The atom whose reference position is $\boldsymbol{R}_i + \boldsymbol{R}_s$ has a displacement which we may write as

$$(x_{i+s}, y_{i+s}, z_{i+s}) = \boldsymbol{\delta R}_{i+s}.$$

For small displacements of the atoms the force $\boldsymbol{F}_i$ on atom i will be a linear function of the displacements. Show that its x-component F_{ix} has the form:

$$\begin{aligned} -F_{ix} &= \sum_s \left[x_{i+s} \frac{\partial^2 V}{\partial x_i \partial x_{i+s}} + y_{i+s} \frac{\partial^2 V}{\partial x_i \partial y_{i+s}} + z_{i+s} \frac{\partial^2 V}{\partial x_i \partial z_{i+s}} \right] \\ &= \sum_s \left[(x_{i+s} - x_i) \frac{\partial^2 V}{\partial x_i \partial x_{i+s}} + (y_{i+s} - y_i) \frac{\partial^2 V}{\partial x_i \partial y_{i+s}} \right. \\ &\qquad \left. + (z_{i+s} - z_i) \frac{\partial^2 V}{\partial x_i \partial z_{i+s}} \right], \end{aligned}$$

the second equality following from

$$x_i \sum_s \frac{\partial^2 V}{\partial x_i \partial x_{i+s}} + y_i \sum_s \frac{\partial^2 V}{\partial x_i \partial y_{i+s}} + z_i \sum_s \frac{\partial^2 V}{\partial x_i \partial z_{i+s}} = 0;$$

an equation which states that translation of the crystal as a whole produces no restoring forces.

Show that the wave motion

$$\boldsymbol{\delta R}_i = \boldsymbol{G} \mathrm{e}^{i\boldsymbol{k}\cdot\boldsymbol{R}_i + i\omega t} \equiv \boldsymbol{G} \mathrm{e}^{i\theta_i}$$

satisfies the equations of motion, provided that

$$\begin{aligned} -M_a \ddot{x}_i &= \omega^2 M_a G_x \mathrm{e}^{i\theta_i} = -F_{ix} \\ &= \mathrm{e}^{i\theta_i} G_x \sum_s (\mathrm{e}^{i\boldsymbol{k}\cdot\boldsymbol{R}_s} - 1) \frac{\partial^2 V}{\partial x_i \partial x_{i+s}} + \mathrm{e}^{i\theta_i} G_y \sum_s (\mathrm{e}^{i\boldsymbol{k}\cdot\boldsymbol{R}_s} - 1) \frac{\partial^2 V}{\partial x_i \partial y_{i+s}} \\ &\qquad + \mathrm{e}^{i\theta_i} G_z \sum_s (\mathrm{e}^{i\boldsymbol{k}\cdot\boldsymbol{R}_s} - 1) \frac{\partial^2 V}{\partial x_i \partial z_{i+s}} \\ &= \mathrm{e}^{i\theta_i} [M_{xx} G_x + M_{xy} G_y + M_{xz} G_z]; \end{aligned}$$

and similar equations for G_y and G_z are satisfied. From the periodicity of the crystal lattice and the assumption that each atom is at a center of symmetry, show that

$$\frac{\partial^2 V}{\partial x_i \partial y_{i+s}} = \frac{\partial^2 V}{\partial x_i \partial y_{i-s}} = \frac{\partial^2 V}{\partial y_i \partial x_{i+s}}, \quad \text{etc.,}$$

and hence that the quantities

$$M_{xy} = M_{yx}, \quad M_{xz} = M_{zx}, \quad M_{yz} = M_{zy}$$

are real functions of $\boldsymbol{k}$. Thus the equations of motion reduce to

$$\omega^2 M_a G_x = M_{xx}G_x + M_{xy}G_y + M_{xz}G_z$$
$$\omega^2 M_a G_y = M_{yx}G_x + M_{yy}G_y + M_{yz}G_z$$
$$\omega^2 M_a G_z = M_{zx}G_x + M_{zy}G_y + M_{zz}G_z.$$

These equations are equivalent to the symbolic equation

$$\omega^2 M_a \boldsymbol{G} = M\boldsymbol{G}$$

where M is the matrix having the M's as components. This matrix is Hermitian since $M_{xy} = M_{yx}$ and $M_{yx} = M_{yx}{}^*$ because the M's are real. The eigenvalues $\omega^2 M_a$ of this Hermitian form are all positive, as may be shown by evaluating the potential energy. Consider the displacement

$$\boldsymbol{\delta R}_i = \text{Real part } q\boldsymbol{G}\mathbf{e}^{i\boldsymbol{k}\cdot\boldsymbol{R}_i}. \tag{A}$$

This produces a force on atom i for which the x-component, for example, is

$$F_{ix} = -\text{ Real part of } \mathbf{e}^{i\boldsymbol{k}\cdot\boldsymbol{R}_i}\,[M_{xx}G_x + M_{xy}G_y + M_{xz}G_z]$$
$$= -q\cos \boldsymbol{k}\cdot\boldsymbol{R}_i\,[M_{xx}G_x + M_{xy}G_y + M_{xz}G_z].$$

The energy required to increase q by dq is

$$-\sum \boldsymbol{F}_i \cdot d(\boldsymbol{\delta R}_i) = (\sum \cos^2 \boldsymbol{k}\cdot\boldsymbol{R}_i)q\,dq \times [M_{xx}G_x{}^2 + M_{xy}G_xG_y + M_{xz}G_xG_z$$
$$+ M_{yx}G_yG_x + M_{yy}G_y{}^2 + M_{yz}G_yG_z + M_{zx}G_zG_x + M_{zy}G_zG_y + M_{zz}G_z{}^2]$$

so that the stored energy due to (A) is [except for $\boldsymbol{k} = 0$ and $\boldsymbol{k} = (\pi/a, \pi/a, \pi/a)$]

$$q^2 \tfrac{1}{4} N_s (\boldsymbol{G}|M|\boldsymbol{G})$$

where $(\boldsymbol{G}|M|\boldsymbol{G})$ is the quadratic form of the $\boldsymbol{G}$'s. In proving this result obtain also for later use the relationships

$$\sum_i \cos(\boldsymbol{k}\cdot\boldsymbol{R}_i)\cos(\boldsymbol{k}'\cdot\boldsymbol{R}_i) = \sum_i \sin(\boldsymbol{k}\cdot\boldsymbol{R}_i)\sin(\boldsymbol{k}'\cdot\boldsymbol{R}_i)$$
$$= (N_s/2)\delta_{\boldsymbol{k},\boldsymbol{k}'}$$

unless $\boldsymbol{k} = 0$ or $\boldsymbol{k} = (\pi/a, \pi/a, \pi/a)$, for which cases the value is twice as large for the cosine sum, while the sine sum vanishes. Show also that terms in which mixed sines and cosines occur all vanish.

Since the equilibrium position (all $q = 0$) is assumed to be minimum energy, it is thus evident that $(\boldsymbol{G}|M|\boldsymbol{G})$ is a positive definite quadratic form so that the eigenvalues ω^2 are positive. Since the coefficients in M are real, the eigenvectors (G_x, G_y, G_z) can be chosen to be real.

Prove from the properties of the Hermitian form that the three eigenvectors $\boldsymbol{G}$, denoted by $\boldsymbol{G}_\alpha$, or $\boldsymbol{G}_1$, $\boldsymbol{G}_2$, $\boldsymbol{G}_3$, are mutually perpendicular. From the same properties prove that for the displacement

$$\boldsymbol{\delta R}_i = \boldsymbol{G}_1(q_{1c}\cos \boldsymbol{k}\cdot\boldsymbol{R}_i + q_{1s}\sin \boldsymbol{k}\cdot\boldsymbol{R}_i) + \boldsymbol{G}_2(q_{2c}\cos \boldsymbol{k}\cdot\boldsymbol{R}_i + q_{2s}\sin \boldsymbol{k}\cdot\boldsymbol{R}_1) + \boldsymbol{G}_3(q_{3c}\cos \boldsymbol{k}\cdot\boldsymbol{R}_i + q_{3s}\sin \boldsymbol{k}\cdot\boldsymbol{R}_i)$$

the stored energy is

$$\tfrac{1}{4}N_sM_a\,[\omega_1{}^2(q_{1c}{}^2 + q_{1s}{}^2) + \omega_2{}^2(q_{2c}{}^2 + q_{2s}{}^2) + \omega_3{}^2(q_{3c}{}^2 + q_{3s}{}^2)]$$

and that the kinetic energy is

$$\tfrac{1}{4}N_sM_a\,[\dot{q}_{1c}{}^2 + \dot{q}_{1s}{}^2 + \dot{q}_{2c}{}^2 + \dot{q}_{2s}{}^2 + \dot{q}_{3c}{}^2 + \dot{q}_{3s}{}^2].$$

Prove that no interaction terms arise between displacements coming from different values of $\boldsymbol{k}$ of the basic set, equation (6).

Show that for $\boldsymbol{k} = 0$ the sine terms should be omitted and that $\omega = 0$ so that appropriate coordinates are q_{01}, q_{02}, q_{03} giving

$$\boldsymbol{\delta R}_i = \boldsymbol{i}_x q_{01} + \boldsymbol{i}_y q_{02} + \boldsymbol{i}_z q_{03}$$

where the $\boldsymbol{i}$'s are unit vectors along the cartesian axes.

Count the number of normal modes produced by this process and show that including the motion of the center of gravity of the crystal there are just $3N_s$ normal modes corresponding to the three degrees of freedom per atom. [Consider especially the differences produced by odd and even numbers of atoms along the edges of the crystal and note the equivalence of points on the surface of the Brillouin zone; see Figures 5.11(a) and 5.6(d).]

This shows that the $3N_s$ coordinates of the nuclei can be expressed in terms of the $3N_s$ normal modes with the kinetic and potential energy terms becoming

$$K.E. = \frac{M_c}{2}(\dot{q}_{01}{}^2 + \dot{q}_{02}{}^2 + \dot{q}_{03}{}^2) + \frac{M}{2}\sum_{k\neq 0}\text{all } \dot{q}^2$$

$$P.E. = \frac{M}{2}\sum_{k\neq 0}\sum_{\alpha}\omega_{k\alpha}{}^2(q_{k\alpha s}{}^2 + q_{k\alpha c}{}^2)$$

where $M_c = N_sM_a$ is the mass of the crystal and $M = M_c/2$.

Problem 2. Show that Schroedinger's equation in the normal mode coordinates is correctly obtained by replacing the $\dot{q}$ terms by p/M and using $(\hbar/i)\partial/\partial q$ for p. (An outline of the proof is as follows:)

It is, of course, a basic assumption that for the nuclear coordinates themselves $\boldsymbol{P}_i = (\hbar/i)\boldsymbol{\nabla}_{\boldsymbol{R}_i}$. It must be proved that the transformation from the $\boldsymbol{R}_i$'s to the q's leads to the appropriate form for the kinetic energy operator for the q's. This can be proved by showing that

$$\frac{1}{M_a}\sum_i \nabla_{R_i}{}^2 = \frac{1}{M_c}\left[\frac{\partial^2}{\partial q_{01}{}^2} + \frac{\partial^2}{\partial q_{02}{}^2} + \frac{\partial^2}{\partial q_{03}{}^2}\right] + \frac{1}{M}\sum_{k\neq 0}\frac{\partial^2}{\partial q^2}.$$

Prove this in more general terms by showing that if a real linear transformation A from one set of coordinates q_i to another set Q_j has the form

$$Q_i = \sum_j a_{ij}q_j$$

and preserves the diagonal form of the kinetic energy

$$\sum_i M_i\dot{Q}_i{}^2 = \sum_j m_j\dot{q}_j{}^2,$$

then the transformation B given by

$$b_{ij} = \sqrt{M_i/m_j}\, a_{ij}$$

has the property

$$\sum_l b_{li}b_{lj} = \delta_{ij}$$

so that, since the b_{ij} are real, B is unitary. (See Problem 3.) From the unitary property it follows that

$$\sum_l b_{il}b_{jl} = \delta_{ij}.$$

From this prove that

$$\sum_i \frac{1}{M_i}\frac{\partial^2}{\partial Q_i{}^2} = \sum_j \frac{1}{m_j}\frac{\partial^2}{\partial q_j{}^2}$$

which shows that Schroedinger's equation takes its proper form in the normal mode coordinates.

Prove that the classical momenta $P_i = M_i\dot{Q}_i$ and $p_j = m_j\dot{q}_j$ transform according to

$$P_i = \sum_j c_{ij}p_j \quad \text{where} \quad c_{ij} = (M_i/m_j)a_{ij} = \sqrt{M_i/m_j}\, b_{ij}$$

and that this transformation is a *contact transformation* since

$$\sum_i P_i dQ_i = \sum_j p_j dq_j.$$

Show that the operators $(\hbar/i)\partial/\partial Q_i$ and $(\hbar/i)\partial/\partial q_j$ also transform the same way, that is,

$$\frac{\hbar}{i}\frac{\partial}{\partial Q_i} = \sum_j c_{ij}\frac{\hbar}{i}\frac{\partial}{\partial q_i}.$$

Show that, if a set of operators p_i, q_i which satisfy the commutation relations

$$(p_iq_j - q_jp_i) = (i/\hbar)\delta_{ij}$$

are transformed according to

$$P_i = \sum_j c_{ij} p_j \quad \text{and} \quad Q_i = \sum_j a_{ij} q_j,$$

then the operators P_i and Q_i satisfy similar commutation relations.

Problem 3. Establish the properties of a unitary transformation of n variables x_j to n variables y_i. Let the transformation A be

$$y_i = \sum_i a_{ij} x_j, \quad \text{or symbolically,} \quad y = Ax \tag{A}$$

where all quantities may be complex. If the transformation satisfies the requirement

$$\sum_i y_i^* y_i = \sum_j x_j^* x_j,$$

it is said to be unitary. From this show that

$$\sum_i a_{ij}^* a_{ik} = \delta_{jk}$$

so that the reciprocal transformation A^{-1} has the matrix

$$(A^{-1})_{ji} = a_{ij}^* \quad \text{where} \quad A^{-1} y = x.$$

Show that, if A is a unitary transformation, so is A^{-1} and hence that the a_{ij}'s also satisfy

$$\sum_i a_{ji}^* a_{ki} = \delta_{jk}.$$

17.4 THE INTERACTION BETWEEN ATOMIC MOTION AND ONE EXCESS ELECTRON

In this section we shall consider the effect of adding one extra electron to the crystal over and above those necessary to complete the valence bonds. We require that its wave function satisfy periodic boundary conditions with the same boundary-periods A_x, A_y, A_z that apply to the nuclei. The electron interacts with the nuclei and with the valence electrons. The state of the latter, we shall suppose in keeping with the last section, is determined by the nuclear positions. Thus the interaction energy may be taken to be $\mathcal{U}(\boldsymbol{r}_e, \boldsymbol{\delta R}\text{'s})$ where $\boldsymbol{r}_e$ is the vector position of the electron and the $\boldsymbol{\delta R}$'s are the nuclear coordinates, equation (2) of Section 17.3. Since our model is periodic, $\mathcal{U}$ must satisfy the equation

$$\mathcal{U}(\boldsymbol{r}_e + \boldsymbol{A}, \boldsymbol{\delta R}\text{'s}) = \mathcal{U}(\boldsymbol{r}_e, \boldsymbol{\delta R}\text{'s}) \tag{1}$$

where $\boldsymbol{A}$ is any vector whose components are integral multiples of the boundary-periods A_x, A_y, A_z. In terms of the normal coordinate q's and the position of the center of gravity of the crystal, $\mathcal{U}$ may be written as

$$\mathcal{U} = \mathcal{U}(\boldsymbol{r}_e - \boldsymbol{R}_c, q\text{'s}) = \mathcal{U}_0(\boldsymbol{r}_e - \boldsymbol{R}_c) + \mathcal{U}_1(\boldsymbol{r}_e - \boldsymbol{R}_c, q\text{'s}) \tag{2}$$

as may be seen as follows. If the q's have certain values, corresponding to a deformation of the lattice which leaves the center of gravity of the nuclei fixed so that $\boldsymbol{R}_c = 0$, then the potential energy will be $\mathcal{U}(\boldsymbol{r}_e, q\text{'s})$. If $\boldsymbol{R}_c \neq 0$, the effect is to shift the entire potential field by $\boldsymbol{R}_c$; this is taken account of by replacing $\boldsymbol{r}_e$ by $\boldsymbol{r}_e - \boldsymbol{R}_c$. The $\mathcal{U}_0$ term corresponds to the perfect lattice and does not depend on the q's. The $\mathcal{U}_1$ term vanishes when the q's are all zero and represents the effect of deformations; it is the term which causes transitions. In making calculations we shall consider only the terms in $\mathcal{U}_1$ which are linear in the q's. (Since the electron at position $\boldsymbol{r}_e$ is probably not at a minimum potential energy, linear terms will be present.)

This form for the interaction of the electron with the lattice leads to the Hamiltonian

$$\mathcal{H} = \frac{1}{2M_c}\boldsymbol{P}_c^2 + \sum_i \mathcal{H}_i + \frac{1}{2m_e}\boldsymbol{p}_e^2 + \mathcal{U}(\boldsymbol{r}_e - \boldsymbol{R}_c, q\text{'s}) \tag{3}$$

where the subscript e refers to the actual, not effective, mass of the electron. Since $\boldsymbol{R}_c$ now occurs in $\mathcal{H}$, $\boldsymbol{P}_c$ no longer commutes with $\mathcal{H}$ and is not a constant of motion. However, since $\boldsymbol{r}_e$ and $\boldsymbol{R}_c$ occur only in the form of the difference vector $\boldsymbol{r} = \boldsymbol{r}_e - \boldsymbol{R}_c$, it is possible to replace $\boldsymbol{r}_e$ and $\boldsymbol{R}_c$ by two new coordinates and $\boldsymbol{P}_c$ and $\boldsymbol{p}_e$ by two new momenta according to the following transformations:

$$\boldsymbol{p}_T = \boldsymbol{p}_e + \boldsymbol{P}_c, \quad \boldsymbol{p} = \frac{M_c}{M_T}\boldsymbol{p}_e - \frac{m_e}{M_T}\boldsymbol{P}_c \tag{4}$$

$$\boldsymbol{r}_T = \frac{m_e}{M_T}\boldsymbol{r}_e + \frac{M_c}{M_T}\boldsymbol{R}_c, \quad \boldsymbol{r} = \boldsymbol{r}_e - \boldsymbol{R}_c \tag{5}$$

$$M_T = M_c + m_e. \tag{6}$$

In these equations $\boldsymbol{p}_T$ and $\boldsymbol{r}_T$ are evidently the momentum and the position of the center of gravity of the (T for) total system of electron plus crystal. As the reader may verify, this reduced mass transformation brings $\mathcal{H}$ to the form

$$\mathcal{H} = (1/2M_T)\boldsymbol{p}_T^2 + \sum_i \mathcal{H}_i + (1/2m)\boldsymbol{p}^2 + \mathcal{U}(\boldsymbol{r}, q\text{'s}) \tag{7}$$

where m is the reduced mass:

$$m = m_eM_c/(m_e + M_c) = m_eM_c/M_T. \tag{8}$$

The inverse transformation is

$$\boldsymbol{p}_e = (m_e/M_T)\boldsymbol{p}_T + \boldsymbol{p} \qquad \boldsymbol{P}_c = (M_c/M_T)\boldsymbol{p}_T - \boldsymbol{p} \tag{9}$$

$$\boldsymbol{r}_e = \boldsymbol{r}_T + (M_c/M_T)\boldsymbol{r} \qquad \boldsymbol{R}_c = \boldsymbol{r}_T - (m_e/M_T)\boldsymbol{r}. \tag{10}$$

The reader should verify for the x-components, for example, of this transformation that the conditions of Problem 2 of Section 17.3 are satisfied so that the operators $(\hbar/i)\nabla_r$ and $(\hbar/i)\nabla_{r_T}$ may be substituted for p and p_T.

The expression for $\mathcal{H}$ may advantageously be written as the sum of four terms

$$\mathcal{H} = \mathcal{H}_{r_T} + \mathcal{H}_r + \sum_i \mathcal{H}_i + \mathcal{U}_1(r, q\text{'s}) \tag{11}$$

where the $\mathcal{H}_i$ terms correspond to the normal modes, equation (14) of Section 17.3, and the new Hamiltonian expressions are

$$\mathcal{H}_{r_T} \equiv (1/2M_T)p_T^2 \tag{12}$$

$$\mathcal{H}_r \equiv (1/2m)p^2 + \mathcal{U}_0(r). \tag{13}$$

Except for the term $\mathcal{U}_1$, each term involves only one set of conjugate coordinates and momenta. For such cases the coordinates are said to be separated and the solutions for the equation $(\mathcal{H} - \mathcal{U}_1)\Phi = \mathcal{E}\Phi$ can be obtained in terms of products of eigenfunctions for $\mathcal{H}_{r_T}$, $\mathcal{H}_r$ and the various $\mathcal{H}_i$, a procedure which forms the basis of the calculation of the matrix elements for transitions given in Section 17.6. We shall deal in this section with the wave functions for r_T and r, leaving the q's for Section 17.5 and the entire wave function Φ for Section 17.6.

We shall thus consider a wave function of the form $\chi(r_T)\psi(r)$ which satisfies

$$(\mathcal{H}_{r_T} + \mathcal{H}_r)\,\chi(r_T)\psi(r) = \mathcal{E}\chi(r_T)\psi(r)$$

and the proper boundary conditions.

In accordance with the customary method of separating variables, we divide this equation by $\chi\psi$; noting that $\mathcal{H}_{r_T}$ does not operate on r in ψ nor does $\mathcal{H}_r$ operate on r_T in χ, we obtain

$$(\mathcal{H}_{r_T}\chi/\chi) + (\mathcal{H}_r\psi/\psi) = \mathcal{E}.$$

From which we conclude that both terms on the left are constants (independent of r and r_T) so that we have

$$\mathcal{H}_{r_T}\chi = \mathcal{E}'\chi; \quad \mathcal{H}_r\psi = \mathcal{E}''\psi$$

and

$$\mathcal{E}' + \mathcal{E}'' = \mathcal{E}.$$

It is evident that p_T commutes with $\mathcal{H}_{r_T}$ so that p_T is a constant of motion. This leads to

$$\chi = V^{-1/2} \exp\,(iP_T \cdot r_T/\hbar), \quad \mathcal{E}' = P_T^2/2M_T \tag{14}$$

where $V = A_xA_yA_z$ is a normalization factor so that the integral of $\chi^*\chi dV_{r_T}$ over the volume is unity.

The equation for ψ is identical with that of an electron moving in a fixed potential field $\mathcal{U}_0(r)$ except that m occurs in place of m_e. Since

$$m = m_eM_c/(M_c + m_e) = m_e/[1 + (m_e/M_c)],$$

the difference between m and m_e is negligible. The solutions of this equation may then be taken to be the customary Bloch functions $\psi_P(r)$ where

$$\mathcal{H}_r \psi_P(r) = \mathcal{E}(P)\psi_P(r) \tag{15}$$

$$\psi_P(r) = V^{-1/2} \exp(iP \cdot r/\hbar) u_P(r) \tag{16}$$

where $u_P(r)$ is normalized to unit volume and has the periodicity of the undistorted lattice.

The boundary conditions for r_T and r are deduced from those of the electron and the nuclei. If any of these particles is advanced by one of the periods A_x, A_y, A_z, then the wave function must be unaltered since the new configuration is equivalent to the old.

The boundary condition for $\chi(r_T)$ is found as follows: If all of the particles are simultaneously advanced by one of the periods A_x, A_y, A_z, the same change takes place in r_T, but no changes occur in r or in the q's. It is evident, therefore, that $\chi(r_T)$ must satisfy the periodic boundary conditions and consequently P_T must be a vector of the form

$$P_T = \hbar[i_x(2\pi n_x/A_x) + i_y(2\pi n_y/A_y) + i_z(2\pi n_z/A_z)] \tag{17}$$

where the n's are integers. These vectors lead to periodic behavior for χ and include the basic lattice of the Brillouin zone. However, their values are not restricted to the interior of the zone, and the energy $P_T{}^2/2M_T$ increases indefinitely with increasing $|P_T|$.

We shall have other occasions to deal with vectors of the form (17) which are not restricted by the Brillouin zone boundary. We shall refer to them as *vectors of the extended* ***P****-lattice.* We shall similarly refer to vectors $1/\hbar$ times as large as *vectors of the extended* ***k****-lattice.*

To find the factor associated with increasing r by one of the periods A_x, A_y, A_z, of the boundary conditions, we consider the effect of advancing the electron alone by i_xA_x. This must have no effect on the wave function. However, the change of i_xA_x in r_e and zero in R_c is equivalent [equation (5)] to changing

$$r_T \text{ by } i_xA_x(m_e/M_T) \text{ and } r \text{ by } i_xA_x. \tag{17a}$$

These change $\chi\psi$ by the factor

$$\exp\left[i\left(P_T \cdot i_xA_x\frac{m_e}{M_T} + P \cdot i_xA_x\right)\Big/\hbar\right] = \exp\left[i\left(\frac{m_e}{M_T}P_T + P\right)\cdot i_xA_x/\hbar\right].$$

From this and the similar equations for A_y and A_z, we conclude that the vector in parenthesis must be a vector of the extended P-lattice, that is, of form (17). Thus we may write

$$P = P_e - (m_e/M_T)P_T \tag{18}$$

where P_e is a vector of the extended lattice. It will be important in later calculations that the difference in $\boldsymbol{P}$ for constant values of $\boldsymbol{P}_T$ are vectors of the extended lattice, even though the $\boldsymbol{P}$'s themselves are not. (The $\boldsymbol{P}_T$ term corresponds to correcting $\boldsymbol{P}_e$ for the motion of the center of gravity to obtain $\boldsymbol{P}$ for the relative motion; see Problem 1.)

In Section 8.6 we shall prove that a selection rule holds for transitions due to $\mathfrak{U}_1$. This rule resembles closely the requirement of conservation of momentum, the change in $\boldsymbol{P}$ for the electron being compensated by a change in a similar quantity for the lattice vibrations. Actually conservation of momentum is not concerned at all since the transitions do not involve $\boldsymbol{r}_T$ or $\boldsymbol{P}_T$, which represent the total momentum of the system. The conservation law is fundamentally associated with the existence of a set of operators $\mathfrak{L}(\boldsymbol{R}_j)$, which we shall treat in the following paragraphs. $\mathfrak{L}(\boldsymbol{R}_j)$ commutes with $\mathcal{H}$ and, consequently, its eigenvalue is a constant of motion. It is really to the eigenvalue scheme of the $\mathfrak{L}(\boldsymbol{R}_j)$ operators that the conservation law applies.

The Lattice Displacement Operator.[1] We shall, therefore, consider a group of operators which produce displacements not by the periods A_x, A_y, A_z of the boundary conditions but instead by the vectors $\boldsymbol{R}_j$ of the crystal lattice. The operator $\mathfrak{L}(\boldsymbol{R}_j)$ has the effect of advancing the entire wave function by $\boldsymbol{R}_j$ in the following sense: the wave function $\mathfrak{L}(\boldsymbol{R}_j)\Phi$ has the same value as $\Phi(\boldsymbol{R}_c, \boldsymbol{r}_e, q\text{'s})$ when the electron coordinate in $\mathfrak{L}(\boldsymbol{R}_j)\Phi$ is $\boldsymbol{r}_e + \boldsymbol{R}_j$ and the nuclear coordinates in $\mathfrak{L}(\boldsymbol{R}_j)\Phi$ have values which produce the same deformation as those for Φ except that the deformation pattern (but not the nuclei) is advanced by $\boldsymbol{R}_j$. It is evident from the periodic properties of $\mathcal{V}$ and $\mathfrak{U}$ that if Φ is an eigenfunction of $\mathcal{H}$, so is $\mathfrak{L}(\boldsymbol{R}_j)\Phi$.

It may be proven generally that $\mathfrak{L}(\boldsymbol{R}_j)$ commutes with $\mathcal{H}$ and the eigenvalues of $\mathfrak{L}$ may also be determined. For this purpose we express Φ in terms of a set of "deformation coordinates" $\boldsymbol{S}_i$

$$\boldsymbol{S}_i = \delta\boldsymbol{R}_i - \boldsymbol{R}_c = \boldsymbol{R}_i' - \boldsymbol{R}_i - \boldsymbol{R}_c.$$

Thus the $\boldsymbol{S}$'s represent the deformation which takes place in addition to simple translation of the crystal. The wave function may then be written as $\Phi(\boldsymbol{R}_c, \boldsymbol{S}_i, \boldsymbol{r}_e)$ where the $N_s + 1$ variables $\boldsymbol{R}_c$ and $\boldsymbol{S}_i$ are linear functions of the N_s nuclear vectors $\boldsymbol{R}_i'$. The effect of $\mathfrak{L}(\boldsymbol{R}_j)$ upon Φ is evidently to replace the variable $\boldsymbol{S}_i$ in Φ by the variable $\boldsymbol{S}_l$ where

$$\boldsymbol{R}_l = \boldsymbol{R}_i + \boldsymbol{R}_j$$

and to replace $\boldsymbol{r}_e$ by $\boldsymbol{r}_e - \boldsymbol{R}_j$. If the deformation pattern and the electron are advanced by $\boldsymbol{R}_j$, then $\boldsymbol{S}_l$ will take on the values formerly associated with $\boldsymbol{S}_i$; and $\boldsymbol{r}_e - \boldsymbol{R}_j$ will take on its former value $\boldsymbol{r}_e$ so that

$$[\mathfrak{L}(\boldsymbol{R}_j)\Phi](\text{shifted pattern}) = \Phi\ (\text{unshifted pattern}).$$

[1] The treatment of this operator, which was suggested to the author by C. Herring, puts the selection rules dealt with in Section 17.6 on a more fundamental basis. It is not, however, an essential step in obtaining the formulae for mean free path.

This equation means that the wave function denoted by $\mathfrak{L}(\boldsymbol{R}_j)\boldsymbol{\Phi}$ takes on the same values as $\boldsymbol{\Phi}$ when the pattern of nuclei and the electron are advanced by $\boldsymbol{R}_j$.

The kinetic energy operator commutes with $\mathfrak{L}(\boldsymbol{R}_j)$ since it has the effect of taking ∇^2 in respect to all the coordinate positions in $\boldsymbol{\Phi}$ and it makes no difference whether this is done before or after the coordinates are changed by $\mathfrak{L}(\boldsymbol{R}_j)$. Since the values of the potentials $\mathcal{V}$ and $\mathcal{U}$ are unchanged by changing the coordinates by $\mathfrak{L}(\boldsymbol{R}_j)$, $\mathfrak{L}(\boldsymbol{R}_j)$ commutes with them and hence with $\mathcal{H}$ for the entire system.

The operators $\mathfrak{L}(\boldsymbol{R}_i)$ obviously satisfy the equations

$$\mathfrak{L}(\boldsymbol{R}_i)\mathfrak{L}(\boldsymbol{R}_j) = \mathfrak{L}(\boldsymbol{R}_j)\mathfrak{L}(\boldsymbol{R}_i) = \mathfrak{L}(\boldsymbol{R}_i + \boldsymbol{R}_j)$$

and

$$\mathfrak{L}(\boldsymbol{i}_x A_x) = \mathfrak{L}(\boldsymbol{i}_y A_y) = \mathfrak{L}(\boldsymbol{i}_z A_z) = 1$$

since the last operations merely move the electron by one of the boundary periods. Consequently, if the eigenvalues of $\mathfrak{L}(\boldsymbol{i}_x a)$, $\mathfrak{L}(\boldsymbol{i}_y a)$ and $\mathfrak{L}(\boldsymbol{i}_z a)$, where a is the lattice constant of a cubic structure, are written in the form[2]

$$\exp(-iP_{\mathfrak{L}x}a/\hbar), \quad \exp(-iP_{\mathfrak{L}y}a/\hbar), \quad \exp(-iP_{\mathfrak{L}y}a/\hbar),$$

then the eigenvalue of $\mathfrak{L}(\boldsymbol{R}_i)$ must be

$$\exp(-i\boldsymbol{P}_{\mathfrak{L}} \cdot \boldsymbol{R}_i/\hbar)$$

and, consequently, $\boldsymbol{P}_{\mathfrak{L}}$ must be a vector of the extended lattice, equation (17). In this case, however, since only translations by lattice vectors are involved, the distinct values of $\boldsymbol{P}_{\mathfrak{L}}$ are contained in the Brillouin zone.

In the next section we shall see how the running waves for the nuclear vibrations fit into the $\boldsymbol{P}_{\mathfrak{L}}$ scheme and in Section 17.6 how the conservation of $\boldsymbol{P}_{\mathfrak{L}}$ leads to the selection rules for transitions.

Problems

1. Show that when the $\mathcal{U}_1$ term is neglected, the wave function may be written in the form

$$\exp[i(\boldsymbol{P}_{c0} \cdot \boldsymbol{R}_c + \boldsymbol{P}_e \cdot \boldsymbol{r}_e)/\hbar]\boldsymbol{\Phi}(\boldsymbol{r}, q\text{'s})$$

where $\boldsymbol{P}_{c0}$ and $\boldsymbol{P}_e$ are vectors of the set $\hbar\boldsymbol{k}$ and $\boldsymbol{\Phi}(\boldsymbol{r}, q\text{'s})$ has the periodicity of the lattice in $\boldsymbol{r}$. What is the relationship of $\boldsymbol{P}_{c0}$ and $\boldsymbol{P}_e$ to the vectors $\boldsymbol{P}_T$ and $\boldsymbol{P}$ discussed in the text?

2. Show that if one of the nuclei is a heavy isotope, the operator $\mathfrak{L}(\boldsymbol{R}_i)$ does not commute with the nuclear kinetic energy. (The fact that $\mathfrak{L}$ does not commute with $\mathcal{H}$ does not indicate a serious failure of the selection rule of Section 17.6 since the latter depends on a linear approximation for

[2] The minus signs are introduced as a convention to simplify an interpretation made at the end of Section 17.5.

$\mathfrak{U}_1$, which should be valid at low temperatures independent of the isotopic content of the crystal.)

3. Show that an operator $\mathfrak{L}(\boldsymbol{R}_i)$ exists for a crystal like NaCl which has two different atoms per unit cell.

4. Consider the operators $\mathfrak{L}(\boldsymbol{R}_i)$ from the group viewpoint. Why are their irreducible representations one-dimensional?

17.5 TRANSFORMATION TO RUNNING WAVES

The coordinates q introduced in Section 17.3 correspond to normal modes or standing waves. By a contact transformation a new set of coordinates can be obtained which correspond to running waves. The required transformation has the property of mixing the p's and q's belonging to one of the sine and cosine pairs for particular values of the wave vector $\boldsymbol{k}$ and direction of polarization α. The Hamiltonian function $\mathcal{H}_i$ for these two degrees of freedom, which we denote simply by q_c and q_s, omitting $\boldsymbol{k}$ and α, is

$$\mathcal{H}_i = [p_c^2 + p_s^2 + M^2\omega^2(q_c^2 + q_s^2)]/2M. \tag{1}$$

The transformation from the old q's and p's, referred to as the standing wave set or q_{std}, to the running set, referred to as the q_{run}, is given by

$$q_c = (2M\omega)^{-1/2}(q_- + p_+) \qquad q_s = (2M\omega)^{-1/2}(q_+ + p_-)$$
$$p_c = (M\omega/2)^{1/2}(p_- - q_+) \qquad p_s = (M\omega/2)^{1/2}(p_+ - q_-); \tag{2}$$

which transforms the Hamiltonian into the form

$$\mathcal{H}_i = \tfrac{1}{2}\omega(p_+^2 + q_+^2 + p_-^2 + q_-^2). \tag{3}$$

From the form of the transformation, or its inverse,

$$q_+ = (M\omega/2)^{1/2}q_s - (2M\omega)^{-1/2}p_c \quad q_- = (M\omega/2)^{1/2}q_c - (2M\omega)^{-1/2}p_s$$
$$p_+ = (M\omega/2)^{1/2}q_c + (2M\omega)^{-1/2}p_s \quad p_- = (M\omega/2)^{1/2}q_s + (2M\omega)^{-1/2}p_c, \tag{4}$$

it can be shown that, regarded as operators, the new p, q_{run} set must satisfy the quantum-mechanical commutation rules such as

$$p_+q_+ - q_+p_+ = \hbar/i, \quad p_+q_- - q_-p_+ = 0, \tag{5}$$

provided that the old p, q_{std} set does. (The reader should verify this as an exercise.) Furthermore, the p, q_{std} set does satisfy the commutation rules in accordance with the results of Problem 2 of Section 17.3. The transformation is also a contact transformation in the classical sense since

$$p_+dq_+ + p_-dq_- = p_cdq_c + p_sdq_s + dS \tag{6}$$

where dS is an exact differential. (As an exercise, evaluate the function S.)

The classical motion produced by p_+q_+ will now be shown to be a running wave moving in the $+\boldsymbol{k}$-direction while p_-q_- is a wave running in the $-\boldsymbol{k}$-direction. On the basis of this result it is natural to describe the new variables, obtained by transforming any pair $q_{k\alpha c}$ and $q_{k\alpha s}$, not as $q_{k\alpha+}$ and $q_{k\alpha-}$ but instead as $q_{k\alpha}$ and $q_{-k\alpha}$ simply by including the negative values of the $\boldsymbol{k}$ contained in the sums over the q_{std} set. (It will be recalled that of the possible $\boldsymbol{k}$-values in Brillouin zone which satisfy the periodic boundary conditions, only half were used with the q_{std} set since $-\boldsymbol{k}$ gave the same standing waves as $+\boldsymbol{k}$.) This will extend the sums in the q_{run} system over all $\boldsymbol{k}$-values in the Brillouin zone and to each $\boldsymbol{k}\alpha$ pair there will correspond only one coordinate.

The classical motion corresponding to p_+ and q_+ is obtained by solving the Hamiltonian equations

$$\dot{q}_+ = \partial\mathcal{H}/\partial p_+ = \omega p_+ \quad \text{and} \quad \dot{p}_+ = -\partial\mathcal{H}/\partial q_+ = -\omega q_+ \tag{7}$$

which lead to

$$q_+ = A_+ \cos(\omega t + \theta_+) \tag{8a}$$

$$p_+ = -A_+ \sin(\omega t + \theta_+). \tag{8b}$$

When this is introduced into the expression for $\boldsymbol{\delta R}_i$, omitting $\boldsymbol{k}$ and α subscripts, we obtain

$$\boldsymbol{\delta R}_i = \boldsymbol{G}[\cos(\boldsymbol{k}\cdot\boldsymbol{R}_i)q_c + \sin(\boldsymbol{k}\cdot\boldsymbol{R}_i)q_s] \tag{9a}$$

$$= A_+(2M\omega)^{1/2}\boldsymbol{G}\sin[\boldsymbol{k}\cdot\boldsymbol{R}_i - \omega t - \theta_+], \tag{9b}$$

which is a wave moving in the $+\boldsymbol{k}$-direction with phase velocity $\omega/|\boldsymbol{k}|$.

It may similarly be shown that the p_-q_- solution represents a wave running in the $-\boldsymbol{k}$-direction. The proof is left as an exercise for the reader.

We shall next show how q_+ and q_- partake of the aspects of running waves in the quantum-mechanical sense. For that purpose we shall consider in a preliminary way how the interaction of the electron with the nuclear vibrations is calculated. As discussed in Section 17.4, the energy of the electron with the deformed lattice depends on $\boldsymbol{r}$, the position of the electron in respect to the center of gravity of the nuclei, and on the deformation expressed by the variables q_{std}. We expand this interaction in a power series in the q's and assume that the deformation is small enough so that we may retain only the linear terms. This leads to the equations

$$\mathcal{U}(\boldsymbol{r}, q_{\text{std}}) = \mathcal{U}_0(\boldsymbol{r}) + \mathcal{U}_1(\boldsymbol{r}, q_{\text{std}}) \tag{10}$$

where the perturbation term $\mathcal{U}_1$ is

$$\mathcal{U}_1(\boldsymbol{r}, q_{\text{std}}) = \sum_{\boldsymbol{k}\alpha}[q_{\boldsymbol{k}\alpha c}(\partial\mathcal{U}/\partial q_{\boldsymbol{k}\alpha c}) + q_{\boldsymbol{k}\alpha s}(\partial\mathcal{U}/\partial q_{\boldsymbol{k}\alpha s})], \tag{11}$$

the partial derivatives being evaluated at zero deformation. For a given $\boldsymbol{k}\alpha$ pair, the values of q_c and q_s specify a sinusoidal distortion with $\boldsymbol{\delta R}_i$ given by (9a). In the next section we shall consider how the interaction of the electron with such a sinusoidal disturbance may be evaluated. In this section we shall consider certain general aspects of the matrix elements for the p's and q's and for $\delta \boldsymbol{R}_i$.

Although we wish to think of the deformation in terms of the q_{std}, it is advantageous to use the quantum numbers for the running waves to describe the nuclear vibrations. Specifically we shall write the part of the wave function involving the q_s and q_c (corresponding to a particular $\boldsymbol{k}\alpha$) in the form $\phi(n_+, n_-; q_c, q_s)$ and require it to be an eigenfunction individually for $\mathcal{H}_+$ and $\mathcal{H}_-$ [with eigenvalues $\hbar\omega(n_+ + \frac{1}{2})$ and $\hbar\omega(n_- + \frac{1}{2})$ respectively] as well as for the total energy $\mathcal{H}_i = \mathcal{H}_+ + \mathcal{H}_-$ of two modes. By expressing q_+ and p_+ in terms of q_{std}, we obtain

$$\mathcal{H}_+ = \tfrac{1}{2}\omega(p_+^2 + q_+^2) = \tfrac{1}{2}\mathcal{H}_i + \tfrac{1}{2}\omega(p_s q_c - p_c q_s) \tag{12}$$

and

$$\mathcal{H}_- = \tfrac{1}{2}\omega(p_-^2 + q_-^2) = \tfrac{1}{2}\mathcal{H}_i - \tfrac{1}{2}\omega(p_s q_c - p_c q_s). \tag{13}$$

Since $\mathcal{H}_+$ and $\mathcal{H}_-$ are the Hamiltonian operators for harmonic oscillators, we must have

$$\begin{aligned} \mathcal{H}_+\phi(n_+, n_-; q_c, q_s) &= \hbar\omega(n_+ + \tfrac{1}{2})\phi(n_+, n_-; q_c, q_s) \\ \mathcal{H}_-\phi(n_+, n_-; q_c, q_s) &= \hbar\omega(n_- + \tfrac{1}{2})\phi(n_+, n_-; q_c, q_s), \end{aligned} \tag{14}$$

so that the eigenvalue of $\mathcal{H}_i$ is

$$\mathcal{E}_i = \hbar\omega(n_+ + n_- + 1) = \hbar\omega(n_i + 1). \tag{15}$$

For a given value of $\mathcal{E}_i$, there are $n_i + 1$ ways of dividing the energy between n_+ and n_-.

The matrix element of $\delta \boldsymbol{R}_i$ for transitions in n_+ and n_- is linear in q_c and q_s and hence in p_+, q_+, p_-, and q_-. The matrix element for p_+ and q_+ vanish if n_- changes, since p_+ and q_+ commute with $\mathcal{H}_-$, and if n_- does not change, they vanish unless n_+ changes by ± 1. For such changes they have values given by the customary formulae:[1]

$$(n_+, n_-|p_+ \text{ or } q_+|n_+', n_-)$$

$$= \iint \phi^*(n_+, n_-; q_c, q_s)[p_+ \text{ or } q_+]\phi(n_+', n_-; q_c, q_s)\,dq_c\,dq_s \tag{16}$$

$$(n_+, n_-|p_+|n_+ \pm 1, n_-) = \begin{cases} \sqrt{\hbar(n_+ + 1)/2} \\ \sqrt{\hbar n_+/2} \end{cases} \tag{17}$$

$$(n_+, n_-|q_+|n_+ \pm 1, n_-) = \begin{cases} i\sqrt{\hbar(n_+ + 1)/2} \\ -i\sqrt{\hbar n_+/2}. \end{cases} \tag{18}$$

[1] See P. A. M. Dirac, *Quantum Mechanics*, Oxford at the Clarendon Press, 1930 or 1935, for a treatment of the harmonic oscillator. The phase factors exp $(i\gamma')$ in equation (34), 88, of the first edition enter the same way in both the p_+ and q_+ matrix elements and do not affect the results derived later; they may also be eliminated by proper choices of the phases of the representation.

The formulae for p_- and q_- are the same and can be obtained by interchanging the subscripts (leaving the differences due to ± 1 as they are).

The matrix element for $\delta \boldsymbol{R}_i$ in terms of these matrix elements is found by expressing the matrix elements for q_c and q_s in terms of (17) and (18) and inserting the results in (9a); the value is

$$(n_+, n_-|\delta \boldsymbol{R}_i|n_+ \pm 1, n_-) = G\sqrt{\frac{\hbar}{4M\omega}}\begin{cases}\sqrt{n_+ + 1}\exp(i\boldsymbol{k}\cdot\boldsymbol{R}_i)\\ \sqrt{n_+}\exp(-i\boldsymbol{k}\cdot\boldsymbol{R}_i).\end{cases} \tag{19}$$

The quantity M, defined in Section 17.3, equation (11), and at the end of Problem 1, is one half the mass of the crystal. The similar expression of a transition in n_- is found to be

$$(n_+, n_-|\delta \boldsymbol{R}_i|n_+, n_- \pm 1) = G\sqrt{\frac{h}{4M\omega}}\begin{cases}i\sqrt{n_- + 1}\exp(-i\boldsymbol{k}\cdot\boldsymbol{R}_i)\\ -i\sqrt{n_-}\exp(i\boldsymbol{k}\cdot\boldsymbol{R}_i).\end{cases} \tag{20}$$

In both of these expressions the real and imaginary parts correspond to the q_c and q_s terms individually.

Since these matrix elements depend only on the quantum number which changes, we may express them as functions of this number and describe it simply as $n_{k\alpha}$ corresponding to $q_{k\alpha}$. Both elements then take the form

$$(n_{k\alpha}|\delta \boldsymbol{R}_i|n_{k\alpha} + \delta n_{k\alpha}) = G_{k\alpha}[h(2n_{k\alpha} + 1 + \delta n_{k\alpha})/8M\omega_{k\alpha}]^{1/2}\exp(i\delta n_{k\alpha}\boldsymbol{k}\cdot\boldsymbol{R}_i) \tag{21}$$

except for a factor $\pm i$ which may be disregarded since only the absolute value of (21) is needed for calculating the transition probabilities. Unless $|\delta n_{k\alpha}| = 1$, the matrix element vanishes.

The general wave function which describes the nuclear vibrations and is an eigenfunction for all the $\mathcal{H}_{\text{run}}$ with quantum numbers $n_{k\alpha}$ can then be written as a product

$$\phi(n_{k\alpha}\text{'s}) = \prod_{k_i\alpha_j}\phi(n_{k_i\alpha_j}, n_{-k_i\alpha_j}; q_{k_i\alpha_jc}, q_{k_i\alpha_js}) = \prod_i \phi_i, \tag{22}$$

there being as many terms in the product as there are distinct $\boldsymbol{k}_i\alpha_j$ pairs of values in the set of q_{std}. The last expression is an abbreviated form used in the next section. There will be matrix elements of the q_{std}, and consequently of $\delta \boldsymbol{R}_i$, only for transitions in which one and only one n_k changes by ± 1.

The Effect[2] *of the Operators* $\mathcal{L}(\boldsymbol{R}_s)$ *on* $\boldsymbol{\Phi}$. The fundamental reason why the eigenvalues $n_{k\alpha}$ for the running waves are suitable is that in terms of them and the functions ψ_P and χ_{PT} the entire wave function $\boldsymbol{\Phi}$ becomes an eigenfunction of $\mathcal{L}(\boldsymbol{R}_s)$. The meaning of this statement is that the wave function $\boldsymbol{\Phi}'$, obtained by the operation $\mathcal{L}(\boldsymbol{R}_s)\boldsymbol{\Phi}$, is simply $L_s\boldsymbol{\Phi}$ where L_s is

[2] As for the section on the operators $\mathcal{L}(\boldsymbol{R}_s)$ in Section 17.4, it is not necessary to use this method to calculate the mean free path.

the eigenvalue (shown below to be a number with absolute value unity). The wave function $\mathcal{L}(\mathbf{R}_s)\Phi \equiv \Phi'$ was defined in Section 17.4 as follows: Consider the value of Φ for some set of values, denoted symbolically by Q, of the coordinates of the entire system; then shift the deformation pattern of the nuclei and the position of the electron by $\mathbf{R}_s$ and call the new set of values of the coordinates Q'. Evidently Q' is a function of Q and $\mathbf{R}_s$ and Q has the same functional dependence on Q' and $-\mathbf{R}_s$; these relationships may be written as

$$Q' = \mathcal{Q}(Q, \mathbf{R}_s) \quad \text{and} \quad Q = \mathcal{Q}(Q', -\mathbf{R}_s). \tag{23}$$

(We shall later add subscripts to $\mathcal{Q}$ when dealing with particular coordinates such as $\mathbf{r}$ and $\mathbf{r}_T$.) The definition of $\mathcal{L}(\mathbf{R}_s)\Phi$ is then

$$[\mathcal{L}(\mathbf{R}_s)\Phi](Q') \equiv \Phi'(Q') = \Phi(Q) \tag{24a}$$

so that

$$\Phi'(Q') = \Phi[\mathcal{Q}(Q', -\mathbf{R}_s)]. \tag{24b}$$

Showing that Φ is an eigenfunction of $\mathcal{L}(\mathbf{R}_s)$ reduces to showing that

$$\Phi'(Q') = \Phi[\mathcal{Q}(Q', -\mathbf{R}_s)] = L_s\Phi(Q') \tag{24c}$$

i.e. that Φ' of any set of values of the coordinates, Q' being such an arbitrary set, is simply L_s times Φ of the same set of values.

In order to establish the above result, we must determine the functional relationship $\mathcal{Q}$ for the coordinates $\mathbf{r}_T$, $\mathbf{r}$ and the q's. The relationships for $\mathbf{r}$ and $\mathbf{r}_T$ are simply

$$\begin{aligned} \mathbf{r}' &= \mathcal{Q}_r(\mathbf{r}, \mathbf{R}_s) = \mathbf{r} + \mathbf{R}_s; \\ &\quad \mathcal{Q}_r(\mathbf{r}', -\mathbf{R}_s) = \mathbf{r}' - \mathbf{R}_s \end{aligned} \tag{25a}$$

$$\begin{aligned} \mathbf{r}_T' &= \mathcal{Q}_{rT}(\mathbf{r}_T, \mathbf{R}_s) = \mathbf{r}_T + (m_e/M_T)\mathbf{R}_s; \\ &\quad \mathcal{Q}_{rT}(\mathbf{r}_T', -\mathbf{R}_s) = \mathbf{r}_T' - (m_e/M_T)\mathbf{R}_s \end{aligned} \tag{25b}$$

as may be seen from equation (5) of Section 17.4 and the fact that the shift of the pattern of nuclear distortion does not alter $\sum \delta\mathbf{R}$ so that $\mathbf{R}_c$ is unchanged.

In order to deal with the effect of the shift of $\mathbf{R}_s$ upon the q's we consider one $\mathbf{k}\alpha$ pair, denoted by q_c and q_s the subscripts $\mathbf{k}\alpha$ being omitted for brevity. We introduce a new pair of coordinates:

$$\rho = (q_c^2 + q_s^2)^{1/2}, \qquad \theta = \tan^{-1}(q_s/q_c) \tag{26a}$$

$$q_c = \rho\cos\theta \qquad\qquad q_s = \rho\sin\theta. \tag{26b}$$

In terms of these coordinates, the shift $\mathbf{R}_s$ of the deformation pattern has a very simple effect which we shall now derive, the result being given in (32a). The coordinates themselves amount to expressing the two harmonic oscillators q_c and q_s as a two-dimensional isotropic oscillator. For such an oscillator, the angular momentum operator $(\hbar/i)\partial/\partial\theta$ commutes with $\mathcal{H}$ and has eigenvalues varying from $\hbar n$ to $-n\hbar$ in steps of $2\hbar$ where n prescribes the total energy in the form $\hbar\omega(n + 1)$. Without referring to this interpretation of θ, however, we see that purely formally we have

$$\frac{\hbar}{i}\cdot\frac{\partial}{\partial\theta} = \frac{\hbar}{i}\left[\frac{\partial q_c}{\partial\theta}\frac{\partial}{\partial q_c} + \frac{\partial q_s}{\partial\theta}\frac{\partial}{\partial q_s}\right] = \frac{\hbar}{i}\left[-q_s\frac{\partial}{\partial q_c} + q_c\frac{\partial}{\partial q_s}\right]$$
$$= q_c p_s - q_s p_c. \tag{27}$$

From this we see that equation (12) becomes

$$\mathcal{H}_+ = \tfrac{1}{2}\mathcal{H}_i + (\hbar\omega/2i)(\partial/\partial\theta) \tag{28}$$

and consequently that the wave function $\phi(n_+, n_-; q_c q_s)$ is an eigenfunction of $\partial/\partial\theta$ with eigenvalue $i(n_+ - n_-)$. We may, therefore, write

$$\frac{\partial}{\partial\theta}\phi(n_+, n_-; q_c, q_s) = i(n_+ - n_-)\phi(n_+, n_-; q_c, q_s) \tag{29}$$

so that

$$\phi(n_+, n_-; q_c, q_s) = \exp i(n_+ - n_-)\theta \times \text{function of } \rho. \tag{30}$$

Expressed in terms of ρ and θ, the contribution of $k\alpha$ to $\boldsymbol{\delta R}_i$ may be rewritten as follows:

$$\boldsymbol{\delta R}_i = \boldsymbol{G}_{k\alpha}[\cos \boldsymbol{k}\cdot\boldsymbol{R}_i\,\rho\cos\theta + \sin\boldsymbol{k}\cdot\boldsymbol{R}_i\,\rho\sin\theta]$$
$$= \boldsymbol{G}_{k\alpha}\,\rho\cos[\boldsymbol{k}\cdot\boldsymbol{R}_i - \theta]. \tag{31}$$

Thus the effect of the "shift" $\boldsymbol{R}_s$ upon the pattern is to change θ to $\theta' = \theta + \boldsymbol{k}\cdot\boldsymbol{R}_s$ so that the value of $\boldsymbol{\delta R}$ formerly obtained at $\boldsymbol{R}_i$ is now obtained at $\boldsymbol{R}_i + \boldsymbol{R}_s$. Accordingly we may write that

$$\theta' = \mathfrak{Q}_\theta(\theta, \boldsymbol{R}_s) = \theta + \boldsymbol{k}\cdot\boldsymbol{R}_s; \quad \mathfrak{Q}_\theta(\theta', -\boldsymbol{R}_s) = \theta' - \boldsymbol{k}\cdot\boldsymbol{R}_s \tag{32a}$$

$$\rho' = \mathfrak{Q}_\rho(\rho, \boldsymbol{R}_s) = \rho; \quad \mathfrak{Q}_\rho(\rho', -\boldsymbol{R}_s) = \rho'. \tag{32b}$$

Similar relationships will hold for all values of $\boldsymbol{k}$ and α.

In terms of the properties of $\chi_{PT}(r_T)$, $\psi_P(r)$ defined by equations (14) and (16) of Section 17.4 and those of the ϕ's established in (30), the value of

$$\boldsymbol{\Phi} = \chi_{PT}\psi_P\phi(M_{k\alpha}\text{'s}) \tag{33}$$

is readily evaluated for the set of coordinate values $\mathfrak{Q}(Q', -\boldsymbol{R}_s)$ as required in equation (24c). In the evaluation each factor of $\boldsymbol{\Phi}$ gives a term of the form $\exp(-i\boldsymbol{P}\cdot\boldsymbol{R}_s/\hbar)$ or $\exp[-i(n_+ - n_-)\boldsymbol{k}\cdot\boldsymbol{R}_s]$. The terms $(n_+ - n_-)\boldsymbol{k}$ may be rewritten as $n_k\boldsymbol{k} + n_{-k}(-\boldsymbol{k})$, so that the eigenvalue L_s, defined in Section 17.3 as $\exp(-i\boldsymbol{P}_{\mathfrak{L}}\cdot\boldsymbol{R}_s/\hbar)$, becomes

$$L_s = \exp(-i\boldsymbol{P}_{\mathfrak{L}}\cdot\boldsymbol{R}_s/\hbar)$$
$$= \exp -i[\boldsymbol{P} + (m_e/M_T)\boldsymbol{P}_T + \hbar\textstyle\sum n_{k\alpha}\boldsymbol{k}]\cdot\boldsymbol{R}_s/\hbar \tag{34}$$

where the sum extends over all $\boldsymbol{k}\alpha'$ pairs, including $\boldsymbol{k}$ and $-\boldsymbol{k}$ as discussed early in this section.

The vector $\boldsymbol{P} + (m_e/M_T)\boldsymbol{P}_T$ is a vector of the extended $\boldsymbol{P}$-lattice, as discussed in connection with equation (18) of Section 17.4, and so is the sum of the $\hbar\boldsymbol{k}$ vectors according to equation (7) of Section 17.3. Consequently, $\boldsymbol{P}_{\mathfrak{L}}$ is a vector of the extended $\boldsymbol{P}$-lattice, in agreement with the conclusion reached at the end of Section 17.4.

The minus sign in L_s arises from the fact that we have dealt with this problem from the viewpoint of advancing the wave function by $\boldsymbol{R}_s$ rather than from the viewpoint of changing the set of values of the coordinates from Q to $\mathfrak{Q}(Q, \boldsymbol{R}_s)$ in one and the same wave function. The effect of the latter change is readily shown to be

$$\Phi[\mathfrak{Q}(Q, \boldsymbol{R}_s)] = \exp\,[+i\boldsymbol{P}_{\mathfrak{L}} \cdot \boldsymbol{R}_s/\hbar]\Phi(Q). \tag{35}$$

Thus $\boldsymbol{P}_{\mathfrak{L}}$ as defined in this section and the last plays a role similar to $\boldsymbol{P}$, $\boldsymbol{P}_T$ and $\hbar n_{k\alpha}\boldsymbol{k}$ so far as its sign is concerned.

Since both $\boldsymbol{P}_{\mathfrak{L}}$ and $\boldsymbol{P}_T$ are constants of motion for the complete Hamiltonian $\mathcal{H}$, *the vector* $\boldsymbol{P} + \hbar\sum n_{k\alpha}\boldsymbol{k}$ *must be conserved during transitions produced by* $\mathcal{U}_1$, even if the effect of nonlinear terms in the q's are included. *It is this conservation law for* $\boldsymbol{P}_{\mathfrak{L}}$ *that underlies the selection rules derived in the next section.*

17.6 CALCULATION OF THE MATRIX ELEMENT

As discussed in Section 17.2, the transition probabilities arise from matrix elements of $\mathcal{H}$ between different states of the system. The states are approximately eigenfunctions of $\mathcal{H}$ and are exact eigenfunctions for an $\mathcal{H}_0$ which differs only slightly from $\mathcal{H}$. In the case which we shall treat, the wave functions considered are eigenfunctions of

$$\mathcal{H}_0 = \mathcal{H} - \mathcal{U}_1 \tag{1}$$

where $\mathcal{U}_1$ is the term discussed in Section 17.4, equation (2), which gives the added interaction of the excess electron due to the deformation of the lattice described by the q's.

The eigenfunctions Φ for $\mathcal{H}_0$ may be formed from products of wave functions, already described, for $\boldsymbol{r}_T$, $\boldsymbol{r}$ and the q_c, q_s pairs for various values of $\boldsymbol{k}\alpha$. These individual wave functions for $\boldsymbol{r}_T$ and $\boldsymbol{r}$, derived in (14) and (16) of Section 17.4, are

$$\chi_{P_T}(\boldsymbol{r}_T) = V^{-\frac{1}{2}} \exp\,(i\boldsymbol{P}_T \cdot \boldsymbol{r}_T/\hbar) \tag{2}$$

$$\psi_P(\boldsymbol{r}) = V^{-\frac{1}{2}} \exp\,(i\boldsymbol{P} \cdot \boldsymbol{r}/\hbar)\mathbf{u}_P(\boldsymbol{r}) \tag{3}$$

where the factor involving $V = A_xA_yA_z$ normalizes the wave function over the region considered, it being supposed that $\mathbf{u}_P(\boldsymbol{r})$ is normalized over unit volume. The normalized wave functions for the q's are written as

$$\phi(n_{k\alpha}, n_{-k\alpha}; q_{k\alpha c}, q_{k\alpha s}) \equiv \phi_j \tag{4}$$

where ϕ_j is an abbreviation for use later, the index j running over all $\boldsymbol{k}\alpha$ pairs of the standing wave scheme. The wave function for the system may then be written

$$\boldsymbol{\Phi}(\boldsymbol{P}_T, \boldsymbol{P}, n\text{'s}) = \chi_{\boldsymbol{P}_T}\psi_{\boldsymbol{P}}\Pi\phi_i, \tag{5}$$

where the variables $\boldsymbol{r}_T$, $\boldsymbol{r}$ and the q's have been omitted for brevity. The Hamiltonian $\mathcal{H}_0$ is

$$\begin{aligned}\mathcal{H}_0 &= (1/2M_T)\boldsymbol{p}_T^2 + (1/2m)\boldsymbol{p}^2 + \mathcal{U}_0(\boldsymbol{r}) + \sum_i \mathcal{H}_i \\ &= \mathcal{H}_{\boldsymbol{r}_T} + \mathcal{H}_{\boldsymbol{r}} + \sum_i \mathcal{H}_i\end{aligned} \tag{6}$$

in the notation of Section 17.4, equations (11) to (13). Each term in $\mathcal{H}$ operates on only one of the factors of $\boldsymbol{\Phi}$ so that the value of the energy $\mathcal{E}$ is

$$\mathcal{E} = (1/2M_T)\boldsymbol{P}_T^2 + \mathcal{E}(\boldsymbol{P}) + \sum_i \hbar\omega_i[n_{+i} + n_{-i} + 1], \tag{7}$$

the individual terms being those discussed in 17.4 (14) and (15) for $\mathcal{H}_{\boldsymbol{r}_T}$ and $\mathcal{H}_{\boldsymbol{r}}$ and 17.5 (15) for $\mathcal{H}_i$.

The $\boldsymbol{\Phi}$'s for the entire system form an orthonormal set, the only term requiring consideration being $\psi_{\boldsymbol{P}}$. The reader may verify that the orthogonality relationship between the $\psi_{\boldsymbol{P}}$'s is not disturbed by the displacement due to $\boldsymbol{P}_T$ [see Section 17.4, equation (18)] of the allowed $\boldsymbol{P}$-values from the basic $\boldsymbol{P}$-lattice of the Brillouin zone.

$\mathcal{H}$ has, of course, no matrix elements between its orthogonal eigenfunctions $\boldsymbol{\Phi}$ so that all transitions are a result of $\mathcal{U}_1(\boldsymbol{r}, q\text{'s})$. We shall suppose that the lattice vibrations are so small in amplitude that $\mathcal{U}_1$ can be expanded in a series in the first powers of the q's as follows:

$$\begin{aligned}\mathcal{U}_1(\boldsymbol{r}, q\text{'s}) &= \sum_i q_{ic}(\partial\mathcal{U}/\partial q_{ic}) + q_{is}(\partial\mathcal{U}/\partial q_{is}) \\ &= \sum_i q_{ic}\mathcal{U}_{ic}(\boldsymbol{r}) + q_{is}\mathcal{U}_{is}(\boldsymbol{r})\end{aligned} \tag{8}$$

where the last form expresses the fact that the partial derivatives are evaluated for all q's $= 0$ so that they are functions of $\boldsymbol{r}$ only.

The two terms in this series corresponding to a particular $\boldsymbol{k}\alpha$ pair, denoted by i, will have a matrix element between two $\boldsymbol{\Phi}$'s only if:

(a) all $n_{+j}n_{-j}$ for $j \neq i$ do not change;
(b) only one n_{+i} or n_{-i} changes by ± 1.

If (a) is violated, the term vanishes because of the orthogonality of the two ϕ_j's in the integral. For the ϕ_i terms themselves, the matrix element of q_{ic}

and q_{is} are nonvanishing only for condition (b), the values then being those given in equations (17) and (18) of Section 17.5.

Since $\boldsymbol{r}_T$ does not occur in $\mathcal{U}_1$, $\boldsymbol{P}_T$ is constant of motion, as discussed in Section 17.4, and there are no matrix elements for which $\boldsymbol{P}_T$ changes.

We thus need to consider only cases in which $\boldsymbol{P}$ and only one of the n's change. We shall denote the change in n symbolically by n_i and n_i' and will later consider specific $\boldsymbol{k}\alpha$ pairs in place of $\boldsymbol{i}$. For this case the matrix element is

$$(\boldsymbol{P}n_i|\mathcal{U}_1|\boldsymbol{P}'n_i') = \int \Phi^*\mathcal{U}_1\Phi' d \text{ (all coord.)}$$

$$= \frac{1}{V}\int \psi_{\boldsymbol{P}}^*(\boldsymbol{r})\phi^*(n_{i+}, n_{i-}; q_{ic}, q_{is})[q_{ic}\mathcal{U}_{ic} + q_{is}\mathcal{U}_{is}]$$

$$\times\ \psi_{\boldsymbol{P}'}(\boldsymbol{r})\phi([n_{i+}, n_{i-}]'; q_{ic}, q_{is})dV_r dq_{ic} dq_{is} \quad (9)$$

where the symbol $[n_{i+}, n_{i-}]'$ means that only one of the two n's has changed by ± 1 and dV_r is the element of volume for $\boldsymbol{r}$-space.

Later in this section we shall deal directly with the integral of $\mathcal{U}_{ic}$ and $\mathcal{U}_{is}$. The techniques used are somewhat specialized, however, and there are certain advantages in treating (9) initially by a more direct method in which we consider the effect of one of the q waves as the sum of the individual effects of the displacements $\boldsymbol{\delta R}_i$ of single nuclei. We define a "single nucleus deformation" as follows: For a given value of $\boldsymbol{r}$, for the deformed crystal, we imagine the electron and all the nuclei save $\boldsymbol{i}$ to be held fixed while nucleus $\boldsymbol{i}$ undergoes a displacement $\boldsymbol{\delta R}_i$. The change $\delta\mathcal{U}_i$ in the potential energy $\mathcal{U}_1$ for this "single nucleus deformation" will be a function of $\boldsymbol{r}$, the position of the electron in respect to the center of gravity of the deformed crystal, and of $\boldsymbol{\delta R}_i$, being for small deformations a linear function of the three components of the latter. Accordingly we may write

$$\delta\mathcal{U}_i = \boldsymbol{\delta R}_i \cdot \boldsymbol{W}_{\boldsymbol{R}_i}(\boldsymbol{r})$$

to express the change in energy due, we repeat, to displacing nucleus $\boldsymbol{i}$ by $\boldsymbol{\delta R}_i$, holding the position of the electron and all the other nuclei fixed. The net effect of all the "single nucleus deformations" is identical with the deformation of the q_c wave.[1] Thus the value of $\mathcal{U}_1$ for the deformation of the q_c wave can be expressed in two equivalent ways:

$$\sum_i \delta\mathcal{U}_i = \sum_i \boldsymbol{\delta R}_i \cdot \boldsymbol{W}_{\boldsymbol{R}_i}'(\boldsymbol{r}) = q_c\mathcal{U}_c(\boldsymbol{r}). \quad (10)$$

[1] It may be noted that the sum of all the $\boldsymbol{\delta R}_i$ due to the q_c wave does not move the center of gravity of the crystal [see 17.3 equation (8) and text], so that although each "single nucleus deformation" does produce a small change in $\boldsymbol{R}_c$, and hence in $\boldsymbol{r}$ for a fixed position of the electron, the cumulative effect of all the "single nucleus deformations" does not change $\boldsymbol{r}$.

From the periodicity of the undeformed lattice it is evident that

$$W_{R_i}(r) = W_0(r - R_i), \tag{11}$$

which states that the effect on the potential of moving any nucleus is the same as that of moving nucleus zero except for a translation by the lattice vector from 0 to R_i. Integrals of $W_0(r - R_j)$ occur in the matrix element of $\mathfrak{U}_1$ and, by making use the formula for the ψ_P functions and a change in variable of integration, we reduce these to simpler forms as follows:

$$\begin{aligned} I_j &\equiv \int \psi_P{}^*(r) W_0(r - R_j) \psi_P{}'(r) dV_r \\ &= \int \psi_P{}^*(r' + R_j) W_0(r') \psi_P{}'(r' + R_j) dV_{r'} \\ &= e^{i(P'-P)\cdot R_j/\hbar} \int \psi_P{}^*(r') W_0(r') \psi_{P'}(r') dV_{r'} \\ &= e^{i(P'-P)\cdot R_j/\hbar} I_0 \end{aligned} \tag{12}$$

since the integral over r' is equivalent to that over r.[2]

In terms of I_0 and the matrix element for δR_i as given in (21) of Section 17.5, for a transition in which $n_{k\alpha} \to n_{k\alpha} + \delta n_{k\alpha}$ we obtain

$$\begin{aligned} (P, n_{k\alpha}|\mathfrak{U}_1|P', n_{k\alpha} + \delta n_{k\alpha}) &= \sum_j \int \Phi^* \delta R_j \cdot W_{R_j} \Phi' d \text{ (all coord.)} \\ &= \sum_j (n_{k\alpha}|\delta R_j|n_{k\alpha} + \delta n_{k\alpha}) I_j \end{aligned}$$
$$= G_{k\alpha} \cdot I_0[\hbar(2n_{k\alpha}+1+\delta n_{k\alpha})/8M\omega_{k\alpha}]^{1/2} \sum_j \exp[i(\hbar\delta n_{k\alpha}k + P' - P)\cdot R_j/\hbar]. \tag{13}$$

The sum of the exponential terms vanishes unless[3]

$$P' + \hbar\delta n_{k\alpha} k = P \tag{14}$$

so that there are matrix elements only between states having the same value for

$$P + \hbar \sum n_{k\alpha} k \tag{15}$$

just as if a phonon momentum $\hbar\sum n_{k\alpha}k$ were combined with an electron momentum P in a conservation law. We have seen, however, that P_T represents the momentum of the entire system so that this conservation

[2] This conclusion depends on the fact that, although in general neither ψ_P nor $\psi_{P'}$ has the periodicity of the boundary conditions, $\psi_P{}^*\psi_{P'}$ does. The reader should verify this as an exercise.

[3] For the theory of conduction in metals cases occur for which $P' \pm \hbar k - P$ is a vector whose components are multiples of the edges of the Brillouin zone so that this vector gives periodic behavior, that is, it corresponds to 2π times a vector of the reciprocal lattice. Such transitions have been called "Umklappprozesse" by R. Peierls, *Ann. d. Physik* **12**, 154–168 (1932).

law has another meaning, the one which is given at the ends of Sections 17.4 and 17.5.

To summarize: There are matrix elements only for changes in Φ which

(a) do not change $\boldsymbol{P}_T$;
(b) change only one of the n_k and that to $n_{k\alpha} \pm 1 \equiv n_{k\alpha} + \delta n_{k\alpha}$;
(c) change $\boldsymbol{P}$ to $\boldsymbol{P}'$ so that $\boldsymbol{P} = \boldsymbol{P}' + \hbar\delta n_{k\alpha}\boldsymbol{k}$.

For any values of $\boldsymbol{P}$ and $\boldsymbol{k}$, there will always be an allowed value of $\boldsymbol{P}'$ which will satisfy (c) because of the fact that, like $\hbar\boldsymbol{k}$, differences between $\boldsymbol{P}$ and $\boldsymbol{P}'$ are vectors of the extended lattice of the Brillouin zone.

If these conditions are satisfied, all the terms in the sum over $\boldsymbol{R}_j$ are equal and add to N_s so that, except for a phase factor of absolute value unity, we obtain

$$(\boldsymbol{P}n_{k\alpha}|\mathfrak{U}_1|\boldsymbol{P}'n_{k\alpha} + \delta n_{k\alpha})$$
$$= \boldsymbol{G}_{k\alpha} \cdot \boldsymbol{I}_0(\boldsymbol{P}, \boldsymbol{P}')N_s[\hbar(2n_{k\alpha} + 1 + \delta n_{k\alpha})/8M\omega_{k\alpha}]^{1/2} \qquad (16)$$

where it will be recalled N_s is the number of cells in the crystal and $M = N_s M_a/2 = M_c/2$ is where M_a is the mass of the atom in each cell. The dependence of $\boldsymbol{I}_0$ upon $\boldsymbol{P}$ and $\boldsymbol{P}'$ is indicated.

Several methods have been employed to evaluate the matrix element $\boldsymbol{G}_{k\alpha} \cdot \boldsymbol{I}_0$ based on various approximate treatments of the way the potential energy $\mathfrak{U}$ varies with the displacement of the nuclei. We shall discuss only the most recent of these proposed by Bardeen and Shockley,[4] which evaluates the matrix element in terms of certain general properties of the energy band picture in a way which permits correlation with other experimental data.

17.6a. The Method of Deformation Potentials. The essence of the method described here is to think of the deformation of the phonon wave as producing local variations in the size and shape of the unit cell. Figure 17.5 shows qualitatively how the band structure may be expected to vary with lattice constant for a crystal like silicon or germanium. If the deformation waves simply changed the lattice constant, without altering the crystal symmetry, then the energy for an electron at the bottom of the conduction band would vary as curve $\mathcal{E}_c$. If the electron were in a state near the bottom of the band its energy, shown by curve $\mathcal{E}_c(\boldsymbol{P})$, would also rise by almost the same amount. We shall shortly consider the effects of more general deformations than the pure compression or dilation of Figure 17.5.

There is, unfortunately, a difficulty in applying the slopes of the curves of

[4] W. Shockley and J. Bardeen, *Phys. Rev.*, **77**, 407–408 (1950) and a subsequent article planned for 1950. In the latter article it is shown that the method of the "effective mass" may be applied to the deformation potentials. Also Sections 11.3b and 12.8.

Figure 17.5 to the changes in energy of an electron in different parts of a deformed crystal. These have to do with the lack of a precise meaning for the zero reference value of the potential $\mathcal{U}_0(r)$ in a homogeneously deformed crystal. When curves like those of Figure 17.5 are computed numerically, some arbitrary procedure must be adopted. For example, in calculating the diamond curves of Figure 5.5, Kimball used a potential

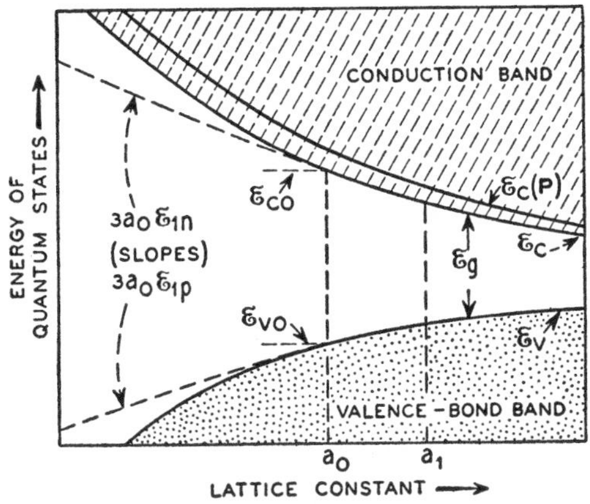

FIG. 17-5—Variation of Energy with Lattice Constant.

which took on the same numerical value near the interior of the atom for all values of the lattice constant and, as he points out, this choice is purely arbitrary.

An absolute meaning can, however, be attached to the curves of Figure 17.5 in the following way. Start with an undeformed crystal of lattice constant a_0. Imagine a deformation which leaves one portion of the crystal unaltered but, as a result of a continuously varying deformation, produces another large region in which the crystal is uniformly dilated to a value a_1. Under these conditions, one might think that the values of the energy at the bottom of the conduction band, namely, $\mathcal{E}_c(a_0)$ and $\mathcal{E}_c(a_1)$ would have definite and unique values. This will not be the case in general, however, because accompanying the deformation there will usually be electric fields and, although for any particular deformation the value of $\mathcal{E}_c(a_1)$ may be uniquely related to $\mathcal{E}_c(a_0)$, another distortion also leading to the same value for a_1, may produce different electric fields, different average potentials in the dilated region and, consequently, different values for $\mathcal{E}_c(a_1)$. These electric fields are hard to estimate but are not excluded even for non-polar crystals like diamond. However, the experimental data on mobility versus temperature in silicon and germanium suggests that they are not important in the ordinary range of observation.

In a problem at the end of this section we show how the deformations of the longitudinal phonon waves produce potentials which vary as Δ/k^2 where Δ is the dilatation of the q-wave, but that for covalent crystals this effect is small compared to the effect of shifting the bands. (See problem 1 of this section.)

Unlike the individual energies $\mathcal{E}_c$ and $\mathcal{E}_v$, the energy gap $\mathcal{E}_g$ is an absolute function of the deformation of the lattice since any variation in the zero of potential has simply the effect of shifting all the energies equally. We have discussed in Chapter 12 the ways in which the variation in the energy gap with lattice constant may be determined experimentally.

According to the assumption that the long-range electric field effects, which make the potential depend on the strain pattern, are negligible, we shall consider $\mathcal{E}_c$ and $\mathcal{E}_v$ each to be a unique function of the local strain. For a cubic crystal up to linear terms in the strain, this leads to the conclusion that

$$\mathcal{E}_c = \mathcal{E}_{c0} + \mathcal{E}_{1n}\Delta \tag{17a}$$

[abbreviated as $\mathcal{E} = \mathcal{E}_0 + \mathcal{E}_1\Delta$ for equations up to (35)] (17a′)

$$\mathcal{E}_v = \mathcal{E}_{v0} + \mathcal{E}_{1p}\Delta \tag{17b}$$

$$\mathcal{E}_{G'} = \mathcal{E}_{G0} + \mathcal{E}_{1G}\Delta \tag{17c}$$

$$\mathcal{E}_{1G} = \mathcal{E}_{1n} - \mathcal{E}_{1p} \tag{17d}$$

where Δ is the dilatation

$$\Delta = \varepsilon_{xx} + \varepsilon_{yy} + \varepsilon_{zz} = \nabla_{R_i} \cdot \delta R_i$$

and the $\mathcal{E}_1$-quantities are constants, the subscripts p and n being used for consistency with later expressions which involve other quantities associated with holes and electrons. The other strain components vanish by symmetry in the linear approximation. This may be illustrated by two pure shear deformations having $\varepsilon_{xy}' > 0$ for one equal to $-\varepsilon_{xy}''$ for the other. Viewed in the $z = 0$ plane, these have the effect of changing the angle between the x and y axes by $\pm\delta\theta$ and, by symmetry for a cubic crystal, these two changes should produce equal changes in $\mathcal{E}_c$. Hence $\mathcal{E}_c$ must be an even function of ε_{xy} and the linear term in ε_{xy} must vanish.

In the remainder of this section we shall deal with the transitions of an electron and will use the abbreviations of (17a′), the modifications for a hole being discussed in Section 17.7. We shall also neglect terms involving changes in the effective mass with strain. These terms can be seen to be of order $\boldsymbol{P}^2$ times a function of the strains from curve $\mathcal{E}_c(\boldsymbol{P})$ of Figure 17.5 as follows: The minimum curve for $\mathcal{E}_c$ for the conduction band corresponds to $\boldsymbol{P} = 0$ (or to a corner of the Brillouin zone instead). Curve $\mathcal{E}_c(\boldsymbol{P})$ lies higher than this by an amount $m_n\boldsymbol{P}^2/2$. If the curvature of $\mathcal{E}$ vs $\boldsymbol{P}^2$ changes with strain, the effect is equivalent of a change in m_n and this change should be linear in the strains for small strains. For such a state we may write:

$$\mathcal{E}(\boldsymbol{P}, \varepsilon) = \mathcal{E}_0(\boldsymbol{P}) + \mathcal{E}_1\Delta + \text{terms in } \boldsymbol{P}^2 \times \text{strains}. \tag{18}$$

We shall refer to $\mathcal{E}_1\Delta$ as the *deformation potential.*

(In the diamond structure there are also "optical modes" in which the two atoms in the unit cell move in opposite directions. If the energy band considered is non-degenerate so that there is only one $\mathcal{E}$ function of the strain, one can conclude that these optical modes will have no first-order terms. This follows from the fact that for the long wave lengths of interest in semiconductors, the optical waves are produced by displacing one face-centered cubic sublattice in respect to the other, and symmetry permits no single-valued function which is linear in the displacement components. If the band is degenerate, however, there may be linear terms for the individual components[4a]; however, the average interaction will still be independent of the optical waves. The pursuit of these problems may best be made by the representation theory of the symmetry groups involved[5] and lies beyond the scope of this book.)

We shall next prove the basic theorem of the Bardeen-Shockley method: *The matrix element for* $\mathcal{U}_1$ *is equal to the matrix element of the* DEFORMATION POTENTIAL, *provided that we deal with electron transitions involving small values of* $\boldsymbol{P}$ *and* $\boldsymbol{P}'$. The integral which we must evaluate to prove the theorem is obtained from (9) by integrating over the q's:

$$(\boldsymbol{P}', n|\mathcal{U}_1|\boldsymbol{P}, n \pm 1) = (n|q_c|n \pm 1)\int \psi_{P'}{}^*(r)\mathcal{U}_c(r)\psi_P(r)dV_r$$
$$+ (n|q_s|n \pm 1)\int \psi_{P'}{}^*(r)\mathcal{U}_s(r)\psi_P(r)dV_r \tag{19}$$

where we have omitted the subscript i for $\boldsymbol{k}\alpha$, and **have reversed the order of $\boldsymbol{P}$ and $\boldsymbol{P}'$**, both for the purpose of simplifying the following manipulations in which the ψ on the right side is chiefly involved.

The two integrals of (19) are very similar and we shall consider the first. In it $\mathcal{U}_c$ corresponds to a perturbation in potential due to a deformation of the form $\delta\boldsymbol{R}_i = \boldsymbol{G}\eta \cos \boldsymbol{k}\cdot\boldsymbol{R}_i$ where η is a parameter which we shall let approach zero so as to obtain the effect of $\partial/\partial q_c$. We shall thus consider η to be a small quantity and neglect its square in subsequent expressions. The value of $\mathcal{U}_c$ is evidently $\partial\mathcal{U}_1/\partial\eta$ as $\eta \to 0$. The deformation itself is

[4a] J. Bardeen has proposed that the "optical modes" may play an important role in the absorption of infrared photons by conduction electrons or holes, *Phys. Rev.*, **79**, 216 (1950). E. J. Ryder and W. Shockley have also concluded that the "optical modes" are important in accounting for the energy losses of electrons in high electric fields, *Phys. Rev.*, probably late 1950.

[5] See, for example, F. Seitz, *Phys. Rev.* **73**, 549–564 (1948).

defined only in terms of the nuclear motions but, for the relatively small k-values and long wave lengths of interest in the excess electron case, the deformation varies slowly from one unit cell to the next and we may define a continuously varying dilatation

$$\Delta(r) = \nabla \cdot G\eta \cos k \cdot r = -\eta G \cdot k \sin k \cdot r.$$

We evaluate the integral by considering a crystal in which such a deformation is actually present with η having such a small value that non-linear terms can be neglected. We shall then prove by an approximate method that

$$\int \psi_{P'}{}^*(r)\eta \mathcal{U}_c(r)\psi_P(r)dV_r = \frac{\eta}{V}\int \exp\,[i(P' - P')\cdot r/\hbar]\mathcal{E}_1\Delta(r)dV_r$$
$$+ \text{ terms of order } \eta(P^2 \quad \text{or} \quad P'^2)/m. \qquad (20)$$

Hence *for small values of* P *and* P', *the matrix element may be evaluated simply in terms* $\mathcal{E}_1$ *just as if the perturbation in potential were* $\delta\mathcal{U}(r) = \mathcal{E}_1\Delta(r)$. This proof involves considering the exact wave functions in *homogeneously strained* crystals.

At any point in the crystal deformed by the η-wave, there is a state of strain called $\eta\varepsilon(r)$ where ε is symbolic for $\varepsilon_{xx}, \varepsilon_{yy}, \varepsilon_{zz}$, etc., together with any bodily translation and rotation of the material. The unit cell at r_a is then deformed by $\eta\varepsilon(r_a)$. We shall define a wave function appropriate to this cell by imagining the deformed cell to be periodically repeated in all directions to make a lattice of homogeneous strain in which the cell at r_a is one cell. The proper wave function in such a lattice would be

$$\psi_P[r, \eta\varepsilon(r_a)] = e^{iP\cdot r/\hbar}u_P[r, \eta\varepsilon(r_a)] \qquad (21)$$

where for small deformations $u_P[r, \eta\varepsilon(r_a)]$ will differ in cell r_a only very slightly from $u_P(r - \delta R_a)$. In the cell at r_a the potential is $\mathcal{U}_0(r) + \eta\mathcal{U}_c(r)$ and within this cell $\psi[r, \eta\varepsilon(r_a)]$ satisfies the wave equation

$$[(1/2m)p^2 + \mathcal{U}_0(r) + \eta\mathcal{U}_c(r)]\psi_P[r, \eta\varepsilon(r_a)] \equiv [\mathcal{H}_r + \eta\mathcal{U}_c(r)]\psi_P[r, \eta\varepsilon(r_a)]$$
$$= [\mathcal{E}_0(P) + \mathcal{E}_1\eta\Delta(r_a)]\psi_P[r, \eta\varepsilon(r_a)] \qquad (22)$$

where we have neglected terms, like those of equation (18), of the order of $\eta P^2/m_n$ which arise from changes in the effective mass m_n with strain. It is to be noted that in the first line of (22) p^2 operates only on r and not on r_a.

This last equation is one of salient points towards which the attack of the last section and this has been directed. It is to be noted that $\mathcal{U}_c(r)$ is a real function of position and that in the physical reasoning associated with the form $\mathcal{E}_0 + \mathcal{E}_1\Delta$ we have dealt only with such real functions. We shall later combine the $\mathcal{U}_c$ and $\mathcal{U}_s$ terms and obtain an expansion involving the complex factor exp $(ik \cdot r)$; this procedure will be used, however, only after the integrals of $\mathcal{U}_c$ and $\mathcal{U}_s$ are individually considered.

We next define a function which takes account of the variation in $\varepsilon(r)$. This function is defined as

$$\begin{aligned}\psi_P[r, \eta\varepsilon(r)] &= e^{iP\cdot r/\hbar}\mathcal{U}_P[r, \eta\varepsilon(r)] \\ &= \psi_P(r) + \delta\psi_P(r). \qquad (23)\end{aligned}$$

This function changes its form in every unit cell to conform to a periodic structure made by repeating that unit cell; in fact it changes its form slightly even in the unit cell itself because $\varepsilon(r)$ varies slightly throughout any unit cell. The term $\delta\psi$ is evidently linear in η for small η and we may, therefore, neglect its product with other terms linear in η.

We prove the desired theorem of equation (20) by considering the integral

$$I = \int \psi_{P'}^*(r)[\mathcal{H}_r + \eta\mathcal{U}_c(r)]\psi_P[r, \eta\varepsilon(r)]dV_r. \qquad (24a)$$

The result of the operation on the right-hand ψ gives

$$I = \int \psi_{P'}^*[\mathcal{E}_0(P) + \mathcal{E}_1\eta\Delta(r)](\psi_P + \delta\psi_P)dV_r + \text{terms involving } \frac{\partial u_P[r, \eta\varepsilon(r)]}{\partial\varepsilon}\eta\nabla\varepsilon(r). \qquad (24b)$$

These last terms can be shown to be of the order $\eta P^2/m$, some of the steps in the proof being indicated in problem 2. Since the ψ's are orthogonal, and $\eta\delta\psi_P$ (being proportional to η^2) is negligible we obtain:

$$I \doteq \eta\mathcal{E}_1\int \psi_{P'}^*\Delta\psi_P dV_r + \mathcal{E}_0(P)\int \psi_{P'}^*\delta\psi_P dV_r. \qquad (25)$$

We now evaluate I again by expanding the last two factors and using the fact that ψ_P and $\psi_{P'}$ are orthogonal eigenfunctions of $\mathcal{H}_r$. This gives, up to terms in η^2,

$$\begin{aligned}I &= \int \psi_{P'}^*\eta\mathcal{U}_c\psi_P dV_r + \int \psi_{P'}^*\mathcal{H}_r\delta\psi_P dV_r \\ &= \eta\int \psi_{P'}^*\mathcal{U}_c\psi_P dV_r + \mathcal{E}_0(P')\int \psi_{P'}^*\delta\psi_P dV_r; \qquad (26)\end{aligned}$$

the second step involves using the Hermitian property of $\mathcal{H}_r$. (See problem 3.) Solving (25) and (26) for the desired integral gives

$$\eta\int \psi_{P'}^*\mathcal{U}_c\psi_P dV_r = \eta\mathcal{E}_1\int \psi_{P'}^*\psi_P\Delta V_r + [\mathcal{E}_0(P) - \mathcal{E}_0(P')]\int \psi_{P'}^*\delta\psi_P dV_r. \qquad (27a)$$

The last term is of the order $\eta P^2/2m_n$ and can be neglected. The desired expression for the integral of $\mathfrak{U}_c$ is then given by dividing (27a) by η:

$$\int \psi_{P'}^* \mathfrak{U}_c \psi_P dV_r = \mathfrak{E}_1 \int \psi_{P'}^* \psi_P \Delta dV_r. \tag{27b}$$

This result shows that the interaction integral is just what would be obtained by introducing a deformation potential $\mathfrak{E}_1\Delta$, as was done in Chapter 11.

We shall next combine the $\mathfrak{U}_c$ and $\mathfrak{U}_s$ terms of (19) in a form similar to that for $\delta \boldsymbol{R}_i$ in equation (2) of Section 17.5. **We also interchange the positions of P and P'** in the equations to be consistent with earlier notation. Equation (19) may then be rewritten as follows:

$$(\boldsymbol{P}, n_{k\alpha}|\mathfrak{U}_1|\boldsymbol{P}', n_{k\alpha} + \delta n_{k\alpha})$$

$$= \mathfrak{E}_1 \int \psi_P^* \psi_{P'} (n_{k\alpha}|\nabla \cdot \boldsymbol{G}_{k\alpha}(q_c \cos \boldsymbol{k}\cdot\boldsymbol{r} + q_s \sin \boldsymbol{k}\cdot\boldsymbol{r})|n_{k\alpha} + \delta n_{k\alpha}) dV_r$$

$$= \mathfrak{E}_1 \boldsymbol{G}_{k\alpha}\cdot\boldsymbol{k} \int \psi_P^* \psi_{P'} \exp(i\delta n_{k\alpha}\boldsymbol{k}\cdot\boldsymbol{r}) dV_r [\hbar(2n_{k\alpha}+1+\delta n_{k\alpha})/8M\omega_{k\alpha}]^{1/2} \tag{28}$$

where the first equality comes from writing the dilatations for the q_c and q_s waves as divergences and the second equality from the form of the matrix element for $q_c \cos \boldsymbol{k}\cdot\boldsymbol{r} + q_s \sin \boldsymbol{k}\cdot\boldsymbol{r}$ which is obtained from (20) of Section 17.5 by replacing $\boldsymbol{R}_i$ by $\boldsymbol{r}$. Some factors with absolute value unity such as i and i^2 have been omitted; these omissions are unimportant, however, since only $|(28)|^2$ occurs in the transition probability.

The integral in (28) vanishes unless

$$\boldsymbol{P} = \boldsymbol{P}' + \hbar\boldsymbol{k}\delta n_{k\alpha}$$

in agreement with equations (14) and (15). When the integral does not vanish, its value is

$$\frac{1}{V}\int \mathbf{u}_P^*(\boldsymbol{r})\mathbf{u}_{P'}(\boldsymbol{r}) dV_r \tag{29}$$

where, it will be recalled, $\mathbf{u}_P$ is normalized per unit volume. For a non-degenerate band and small values of $\boldsymbol{P}$, $\mathbf{u}_P(\boldsymbol{r})$ may be written in the form

$$\mathbf{u}_P(\boldsymbol{r}) = \mathbf{u}_0(\boldsymbol{r}) + P_x\mathbf{u}_x(\boldsymbol{r}) + P_y\mathbf{u}_y(\boldsymbol{r}) + P_2\mathbf{u}_2(\boldsymbol{r})$$

where the $\mathbf{u}$'s may be chosen to be orthogonal, and from this the reader may prove that the integral (29) is unity up to terms of the order of $\boldsymbol{P}^2$. Thus we obtain finally

$$(\boldsymbol{P}, n_{k\alpha}|\mathfrak{U}_1|\boldsymbol{P}', n_{k\alpha} + \delta n_{k\alpha}) = \mathfrak{E}_1 \boldsymbol{G}_{k\alpha}\cdot\boldsymbol{k}[\hbar(2n_{k\alpha} + 1 + \delta n_{k\alpha})/8M\omega_{k\alpha}]^{1/2} \tag{30}$$

for the allowed transitions. The allowed values of $\delta n_{k\alpha}$ are ± 1 [see equations (16) to (21) of Section 17.5] and M is one half the mass of the crystal [see equation (11) of Section 17.3]. This formula is equivalent to that appearing in the literature except for the interpretation of $\mathfrak{E}_1$.[6]

17.6b. Final Expression in Terms of Elastic Constants. For the waves of interest in semiconductors, the $\hbar \boldsymbol{k}$ values are small compared to $h/2a$, since the changes in $\boldsymbol{P}$ are small, and consequently the wave lengths of the $\boldsymbol{k}$-waves are large compared to $\boldsymbol{a}$. For these waves the solid may be treated as a continuum so that macroscopic elastic theory may be used.

As has been evident from the treatment in deriving the matrix element, the important waves are those which produce dilatation and have a large component of their unit vector $\boldsymbol{G}_{k\alpha}$ parallel to $\boldsymbol{k}$. In an isotropic solid, the three possible polarizations for long wave-length waves consist of a purely longitudinal mode and two purely transverse modes; and in a cubic crystal the same situation is true for waves propagating in the [100], [110], and [111] directions, which lie along symmetry axes. For other directions there is a slight mixing of longitudinal and transverse polarizations. For germanium the elastic constants and sound velocities are given in Tables 17.1 and 17.2. [For data on Si see *Phys. Rev.* **83**, 1080 (1950).]

TABLE 17.1 ELASTIC CONSTANTS OF GERMANIUM
(dynes/cm²)

$c_{11} = 1.292 \times 10^{12}$ $\quad c_{12} = 0.479 \times 10^{12}$ $\quad c_{44} = 0.670 \times 10^{12}$

TABLE 17.2 VELOCITIES OF ELASTIC WAVES IN GERMANIUM
(cm/sec)

Polarization Direction		Direction of Propagation (Parallel to $\boldsymbol{k}$) [100]	[110]	[111]
Longitudinal		4.92×10^5 (a)	5.39×10^5 (b)	5.54×10^5 (c)
Transverse	[001]	3.54×10^5 (d)	3.54×10^5 (e)	—
	[110]	—	2.75×10^5 (f)	3.03×10^5 (g)

The formulae for ρc^2 are:
(a) c_{11}, (b) $(c_{11} + c_{12} + 2c_{44})/2$, (c) $(c_{11} + 2c_{12} + 4c_{44})/3$, (d) c_{44},
(e) c_{44}, (f) $(c_{11} - c_{12})/2$ (g) $(c_{11} - c_{12} + c_{44})/3$

$\rho = 5.35$ gm/cm³ as deduced from the lattice constant.

Data from W. L. Bond, W. P. Mason, H. J. McSkimin, K. M. Olsen, and G. K. Teal, *Phys. Rev.*, **78**, 176 (1950).

[6] $\mathfrak{E}_1$ is equal to C as used by A. H. Wilson, *The Theory of Metals*, Cambridge University Press, 1936, and $\mathfrak{E}_1 = \frac{2}{3}C$ for C as used by A. Sommerfeld and H. Bethe, *Handbuch der Physik*, Vol. XXIV$_1$, and as used by F. Seitz, *Phys. Rev.* **73**, 549–564 (1948).

For other directions of propagation, the longitudinal velocity will lie between the extremes at [100] and [111] and, although the waves will not be perfectly longitudinal, the value of $(\boldsymbol{G}_{k\alpha} \cdot \boldsymbol{k})^2$, which enters the transition probability, will be nearly equal to $\boldsymbol{k}^2$ and since

$$(\boldsymbol{G}_{k1} \cdot \boldsymbol{k})^2 + (\boldsymbol{G}_{k2} \cdot \boldsymbol{k})^2 + (\boldsymbol{G}_{k3} \cdot \boldsymbol{k})^2 = \boldsymbol{k}^2,$$

only very small contributions will be made by the chiefly transverse waves to the $(\boldsymbol{G}_{k\alpha} \cdot \boldsymbol{k})^2$ terms. We see, therefore, that the customary approximation of treating a cubic crystal as isotropic and using only the longitudinal waves for scattering will introduce only relatively small errors.

Since the longitudinal velocities do not differ greatly in different directions, we shall introduce an average longitudinal elastic constant c_{ll}. For c_{ll} we take arbitrarily the value for a longitudinal wave in the [110] direction, this value being intermediate between the extremes of [100] and [111]. We also introduce an average velocity $c = (c_{ll}/\rho)^{1/2}$.

If we neglect the transverse waves, we see that, for each transition $\boldsymbol{P}$ to $\boldsymbol{P}'$, or $\boldsymbol{P}_0$ to $\boldsymbol{P}_j$ in the notation of Section 17.2, there are only two possible changes in the phonon field given by

$$\hbar\boldsymbol{k} = \boldsymbol{P}' - \boldsymbol{P} \quad \text{for} \quad \delta n_k = +1$$

$$\hbar\boldsymbol{k} = \boldsymbol{P} - \boldsymbol{P}' \quad \text{for} \quad \delta n_k = -1.$$

Thus, except for this ambiguity, the end state is uniquely determined by the value of $\boldsymbol{P}'$. We shall shortly discuss how the two matrix elements can be treated as the single matrix element $|\mathfrak{U}_{0j}|^2$ of Section 17.2.

In terms of c_{ll}, the matrix element (30) can be put in a form which is more easily remembered. For this purpose we shall introduce the average energy $\langle \mathcal{E}_{k\alpha} \rangle$ of the normal mode before and after collision, where $\langle \mathcal{E}_{k\alpha} \rangle = \hbar\omega_{k\alpha}(2n_{k\alpha} + 1 + \delta n_{k\alpha})/2 = \hbar\omega_{k\alpha}(n_{k\alpha} + \frac{1}{2} + n_{k\alpha}' + \frac{1}{2})/2$. In addition we note that (I) $M = M_c/2 = \frac{1}{2}\rho V$ where ρ is the density of the crystal and V its volume and (II) $\omega_{k\alpha}/k = c_{k\alpha} \doteq \sqrt{c_{ll}/\rho}$ where $c_{k\alpha}$ is the speed of wave $\boldsymbol{k}\alpha$. In terms of these relationships, equation (30) leads to

$$|(\boldsymbol{P}n_{k\alpha}|\mathfrak{U}_i|\boldsymbol{P}'n_{k\alpha} + \delta n_{k\alpha})|^2 = \mathcal{E}_1{}^2\hbar\omega_{k\alpha}(2n_{k\alpha} + 1 + \delta n_{k\alpha})/4Vc_{ll}$$

$$= \mathcal{E}_1{}^2\frac{\langle \mathcal{E}_{k\alpha} \rangle}{2Vc_{ll}} = \mathcal{E}_1{}^2\frac{\langle \Delta^2 \rangle}{2}. \qquad (31)$$

In this expression $\langle \Delta^2 \rangle$ is the average squared dilatation associated with energy $\langle \mathcal{E}_{k\alpha} \rangle$ and is defined as follows: For a mechanical wave, the energy is, on the average, equally divided between potential and kinetic energies. The potential energy density is $\frac{1}{2}c_{ll}(\varepsilon_{kk})^2$ where ε_{kk} is the longitudinal strain and is equal to Δ the dilatation for a longitudinal wave. Equating the average potential energy to half the total energy gives

$$\tfrac{1}{2}Vc_{11}\langle\Delta^2\rangle = \tfrac{1}{2}\langle\mathcal{E}_{k\alpha}\rangle$$

which leads to the last equation above.

The result may be put in words as follows: *The matrix element squared is obtained by computing the mean squared dilatation associated with the energy in the running wave averaged before and after the collision. This mean squared dilatation* $\langle\Delta^2\rangle$ *corresponds to a mean squared deformation potential* $\mathcal{E}_1{}^2\langle\Delta^2\rangle$ *and the matrix element is half this latter value.* Stated in this form, the result will apply to waves whose polarization is only partly longitudinal, as the reader may verify.

In Section 17.2 we used the approximation that the change in energy of the transition $\boldsymbol{P} \rightarrow \boldsymbol{P}'$ could be attributed entirely to the changing energy of the electron. This can be shown to be a good approximation for the conditions of interest by considering first the requirement that energy be conserved in the transition so that the change in $\mathcal{E}(\boldsymbol{P})$ equals $\pm\hbar\omega_{k\alpha}$. For the spherical approximation with $\mathcal{E}(\boldsymbol{P}) = \boldsymbol{P}^2/2m$, we may write the energy in terms of the change $\delta|\boldsymbol{P}|$ in the form

$$\delta\mathcal{E}(\boldsymbol{P}) = |\boldsymbol{P}|\delta|\boldsymbol{P}|/m = |\boldsymbol{v}|\delta|\boldsymbol{P}|.$$

For the lattice wave we have a change in energy of

$$\hbar\omega = \hbar 2\pi c/\lambda = \hbar c|\boldsymbol{k}| = c|\boldsymbol{P}' - \boldsymbol{P}|.$$

Equating these two energies gives

$$\delta|\boldsymbol{P}| = \frac{c}{|\boldsymbol{v}|}|\boldsymbol{P}' - \boldsymbol{P}|\cdot \tag{32}$$

At room temperature, $|\boldsymbol{v}|$ is about 10^7 cm/sec whereas c is about 20 times smaller. Thus the changes in $|\boldsymbol{P}|$ will be only a few per cent of $|\boldsymbol{P}|$. Figure 17.6 shows the end states $\boldsymbol{P}'$ resulting from $\boldsymbol{P}$. For large values of $|\boldsymbol{P} - \boldsymbol{P}'|$, $\boldsymbol{k}$ is large and the two surfaces are separated so that $\delta\mathcal{E}(\boldsymbol{P}) = 2\hbar\omega_{\boldsymbol{k}}$. If the value of $\boldsymbol{P}$ deviates from one of the surfaces, the change in total energy will vary owing to $\mathcal{E}(\boldsymbol{P})$ as $\boldsymbol{v}\cdot\delta\boldsymbol{P}$ and owing to $\hbar\omega_{k\alpha}$ as $c \times$ (component of $\delta\boldsymbol{P}$ parallel to $\boldsymbol{k}$). From this we see that the changes in energy arise largely from changes in $|\boldsymbol{P}|$. Since the separation of the two surfaces is only a few per cent of $|\boldsymbol{P}|$, we may take the two surfaces as being a common sphere. For very low energy electrons, however, $|\boldsymbol{v}|$ will be comparable to c. Under these conditions the two surfaces take the form shown in (b). For still lower energies, the surface takes the form (c) and no transitions in which the electron loses energy are possible. These interesting cases are of importance only at very low temperatures and lie outside of the range of phenomena with which we are concerned.[7]

[7] See F. Seitz, *Phys. Rev.* **73**, 549–564 (1948) for a further discussion of these points and a list of references.

Accordingly we approximate the end states for $\boldsymbol{P}'$ by a sphere. Furthermore, two independent processes, corresponding to $\delta n = +1$ and $\delta n = -1$, contribute to the transitions from $\boldsymbol{P}$ to $\boldsymbol{P}'$. These two probabilities should be calculated separately using the absolute values of the squares of the

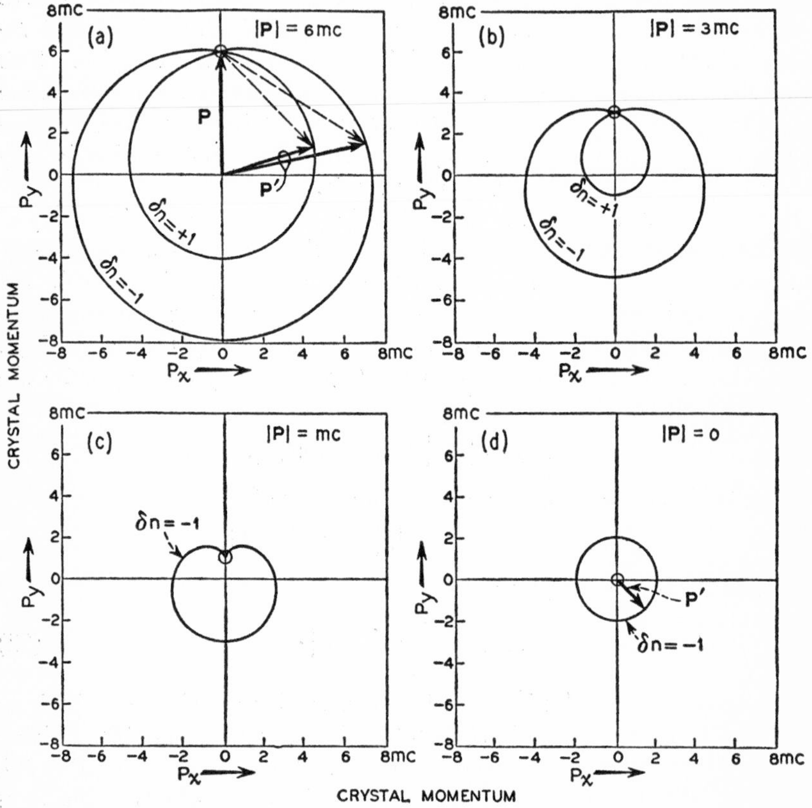

FIG. 17–6—Initial and Final Values of $\boldsymbol{P}$ for Transitions which Conserve Energy.

(a) The two nearly spherical sheets corresponding to emission and absorption of a phonon with $|\boldsymbol{P}_0| = 6mc$.

(b) The two sheets for an electron with $|\boldsymbol{P}_0| = 3mc$.

(c) For an electron with $|\boldsymbol{P}_0| = mc$, energy loss is impossible.

(d) Case of $\boldsymbol{P}_0 = 0$.

matrix elements and then added. The result will be the same, however, as that obtained by simply adding the squares of the absolute values of the two matrix elements. The effective value of $|\mathfrak{U}_{j0}|^2$ used in Section 17.2, in which $\boldsymbol{P}$ and $\boldsymbol{P}'$ are replaced by $\boldsymbol{P}_0$ and $\boldsymbol{P}_j$, is thus

$$|\mathfrak{U}_{j0}|^2 = |(\boldsymbol{P}_0 n_{k\alpha}|\mathfrak{U}_1|\boldsymbol{P}_j n_{k\alpha} + 1)|^2 + |(\boldsymbol{P}_0 n_{-k\alpha}|\mathfrak{U}_1|\boldsymbol{P}_j n_{-k\alpha} - 1)|^2$$

$$= \mathcal{E}_1{}^2 \frac{\hbar\omega_{k\alpha}(n_{k\alpha} + 1) + \hbar\omega_{-k\alpha} n_{-k\alpha}}{2Vc_{11}}. \tag{33}$$

Since the waves $\boldsymbol{k}\alpha$ and $-\boldsymbol{k}\alpha$ have the same frequency, the two n's will be equal. Furthermore, the energy of the phonons for the important transitions is of the order of $\hbar\omega = \hbar c 2\pi/\lambda = \hbar c|\boldsymbol{k}| \doteq c|\boldsymbol{P}|$ and, since this is much less than $|\boldsymbol{v}|\,|\boldsymbol{P}| \doteq kT$, the lattice waves of interest are well above their characteristic temperature and consequently

$$\hbar\omega_{\boldsymbol{k}\alpha}(n_{\boldsymbol{k}\alpha} + \tfrac{1}{2}) = kT.$$

Using this equation to eliminate $(2n_{\boldsymbol{k}\alpha} + 1)$, we obtain the expression used in 17.2 except for the change in notation, with $\boldsymbol{P}, \boldsymbol{P}'$ in place of $\boldsymbol{P}_0, \boldsymbol{P}_j$ of 17.2:

$$|\mathfrak{U}_{j0}|^2 = \mathcal{E}_1{}^2 \frac{kT}{Vc_{ll}} = \mathcal{E}_1{}^2 \langle\Delta^2\rangle \equiv \frac{AT}{V}\,; \quad A = \frac{\mathcal{E}_1{}^2 k}{Vc_{ll}} \tag{34}$$

where A is independent of the temperature. The value of $\langle\Delta^2\rangle$ corresponds to an elastic energy of $kT/2$ in form of $c_{ll}\varepsilon_{kk}{}^2/2$ where the use of c_{ll} for all directions of propagation follows from the isotropic approximation. Equation (34) is twice as large as (32) as a result of combining phonon absorption and emission transitions.

Inserting this in the expression for the mean free path derived in 17.2 gives

$$\frac{1}{l(T)} = \frac{ATm_n{}^2}{\pi\hbar^4} = \frac{\mathcal{E}_{1n}{}^2 kTm_n{}^2}{\pi\hbar^4 c_{ll}} \tag{35}$$

where we have replaced $\mathcal{E}_1$ by $\mathcal{E}_{1n}$ to remind the reader that this formula was derived for an excess electron. In the next section we shall show that the formula for a hole differs only by replacing $\mathcal{E}_{1n}$ by $\mathcal{E}_{1p}$ and m_n by m_p.

Problems

1. Calculate the electrostatic potential variations in the crystal due to a $q_{\rm std}$ wave of long wave length.

Consider the potential $\boldsymbol{W}_0(\boldsymbol{r}) \cdot \boldsymbol{\delta R}_0$. This is the change in potential in the crystal due to giving nucleus zero a vector displacement $\boldsymbol{\delta R}_0$. Of course, $\boldsymbol{W}_0$ will be largest in cell zero where the displacement occurs. At large distances from the atom, the nuclear motion alone would be equivalent to producing an electric dipole of magnitude $+Ze\boldsymbol{\delta R}_0$. This dipole is largely neutralized by compensating shifts of the core electrons, $Z - 4$ in number, and the valence electrons which are 4 in number. The core electrons will move by almost exactly the same amount as the nucleus and will reduce the moment to $4e\boldsymbol{\delta R}_0$. If there were no interaction between one and another of the four covalent bonds around the nucleus, the valence electrons would exactly cancel the remaining $4e\boldsymbol{\delta R}_0$; this follows from the fact that the two bonding electrons between atom zero and its neighbor, atom one, have their center of gravity midway between the two atoms and will thus move half as far as atom zero and produce a dipole moment of $2(-e)\frac{1}{2}\boldsymbol{\delta R}_0 = -e\boldsymbol{\delta R}_0$; consequently, four covalent bonds would just cancel the $4e\delta\boldsymbol{R}_0$ due to moving the core. There will, however, be interactions among the bonds so that if atom zero (say the central dark atom of Figure 1.3) moves directly towards one of its neighbors (say the atom in Figure 1.3 nearest the reader), it will tend to decrease the angles between the bonds to the three atoms it moves away from and this squeezing together of the bonds will cause, through the exchange effect, the electron density of the three bonds to shift by less than $\frac{1}{2}\boldsymbol{\delta R}_0$. Consequently, a very small net dipole moment, denoted by $e'\boldsymbol{\delta R}_0$ where $e' \ll e$, will be left over because of the nuclear motion.

Suppose, therefore, that at large distances from $\boldsymbol{R}_0$, $\boldsymbol{\delta R}_0 \cdot \boldsymbol{W}_0(\boldsymbol{r})$ is the field due to a dipole $e'\boldsymbol{\delta R}_0$, embedded, of course, in medium of dielectric constant κ. Then the wave $\boldsymbol{\delta R}_i = \boldsymbol{G}q_c \cos \boldsymbol{k} \cdot \boldsymbol{R}_i$ gives rise to a wave of polarization. Show that this polarization produces a potential

$$\begin{aligned}\varphi(\boldsymbol{r}) &= (4\pi e' N_s/\kappa V k^2)(\boldsymbol{k} \cdot \boldsymbol{G})q_c \sin \boldsymbol{k} \cdot \boldsymbol{r} \\ &= -(4\pi e' N_s/\kappa V k^2)\Delta(\boldsymbol{r})\end{aligned}$$

where N_s/V is the density of nuclei per unit volume and $\Delta(\boldsymbol{r})$ is the dilatation of the wave. Show that for the wave whose $\boldsymbol{k}$ will reverse an electron of energy kT at room temperature in germanium, φ is equal to:

$$\begin{aligned}\varphi &= 5.1 \times 10^{-2}(e'/e)\Delta \text{ absolute e.s.u. volts} \\ &= 15(e'/e) \text{ practical volts.}\end{aligned}$$

Since e' is probably only a few per cent of e, this term is negligible compared to the terms considered for the slope of the bands. However, $\boldsymbol{k}^2$ varies directly as T, the absolute temperature, so that this term might be important at temperatures of liquid hydrogen or below.

2. Show that the terms in $\partial \mathfrak{u}_{\boldsymbol{P}}(\boldsymbol{r}, \varepsilon)/\partial\varepsilon$ are of order $\eta \boldsymbol{P}^2/m$. (Equation (24b) of Section 17.6.)

First note that the terms involving $(\partial/\partial\varepsilon)$ are multiplied by $\nabla\varepsilon$ and that $\nabla\varepsilon \propto \boldsymbol{k}$ and that $\hbar\boldsymbol{k}$ and $\boldsymbol{P}$ are of the same order. Hence note that the only terms which do not lead directly to order $\eta\boldsymbol{P}^2$ have factors like

$$\frac{\partial^2}{\partial x \partial\varepsilon} \mathfrak{u}_{\boldsymbol{P}}(x, y, z, \varepsilon)$$

and that these lead to integrals similar to those for $\bar{\boldsymbol{p}}$ in the expression for $\boldsymbol{v} = \nabla_{\boldsymbol{P}}\mathfrak{S}(\boldsymbol{P})$ and hence to terms no larger than $\eta\boldsymbol{P}^2/m_n$.

3. Prove that

$$\int \psi_{\boldsymbol{P}}{}^* \mathcal{H}_r \delta\psi_{\boldsymbol{P}'} dV_r = \int (\mathcal{H}_r \psi_{\boldsymbol{P}})^* \delta\psi_{\boldsymbol{P}'} dV_r$$

by considering the periodicity of $\delta\psi_P$ and the effect of this upon integrating

$$\int \psi_P{}^* \frac{\partial^2}{\partial x^2} \delta\psi_{P'} dV_r$$

by parts over the volume of the crystal.

4. In equation (26) of Section 17.2 the effect of the states into which a state b_j might decay was replaced by a term $-i\hbar\eta_j b_j$ which leads to an exponential decay of the form $\exp(-\eta_j t)$ for b_j. Although this procedure is correct, as shown by equations (11) to (17) of Section 17.2, when a single initial state decays to a set of final states, it requires scrutiny when a group of initial states decay to the same set of final states. We shall use the coefficients b for states to which a_0 can make transitions and c for those to which the b-states can make transitions. Then it is evident that there are four classes of c-states: (i) those having two more phonons than a_0 (corresponding to two transitions with phonon emission); (ii) those having one more phonon of one type and one less of another type; and (iii) those having two less phonons. Consider a state of type (i) with coefficient denoted by $c(\boldsymbol{k}, \boldsymbol{k}')$ which has one more phonon each of type $\boldsymbol{k}$ and type $\boldsymbol{k}'$ than a_0. This may arise in either of two ways as may be seen from the following scheme:

a_0	$b(\boldsymbol{k})$	$b(\boldsymbol{k}')$	$c(\boldsymbol{k}, \boldsymbol{k}')$
$\boldsymbol{P}$	$\boldsymbol{P} - \hbar\boldsymbol{k}$	$\boldsymbol{P} - \hbar\boldsymbol{k}'$	$\boldsymbol{P} - \hbar(\boldsymbol{k} + \boldsymbol{k}')$
$n_{\boldsymbol{k}}$	$n_{\boldsymbol{k}} + 1$	$n_{\boldsymbol{k}}$	$n_{\boldsymbol{k}} + 1$
$n_{\boldsymbol{k}'}$	$n_{\boldsymbol{k}'}$	$n_{\boldsymbol{k}'} + 1$	$n_{\boldsymbol{k}'} + 1$

Each column above describes a state and the state $c(\boldsymbol{k}, \boldsymbol{k}')$ may arise either by

$$a_0 \rightarrow b(\boldsymbol{k}) \rightarrow c(\boldsymbol{k}\boldsymbol{k}')$$

or by

$$a_0 \rightarrow b(\boldsymbol{k}') \rightarrow c(\boldsymbol{k}\boldsymbol{k}').$$

This shows that the decay of any state $b(\boldsymbol{k})$ into a set of states $c(\boldsymbol{k}\boldsymbol{k}')$ may be influenced by the effect of other $b(\boldsymbol{k}')$ which may also decay into the same set of states. If this effect is sufficiently pronounced, it will mean that errors may be introduced by assuming that each b decays with its own characteristic η as expressed in equation (26) of Section 17.2.

The problem left to the reader is as follows: The important b's and c's are those for which the energy of the system is the same as for state a_0, all other b's and c's being relatively small. Each $b(\boldsymbol{k})$ may be said to excite its own set of $c(\boldsymbol{k}, \boldsymbol{k}')$ which cause it to decay. Show that any important $c(\boldsymbol{k}, \boldsymbol{k}')$ arises in general chiefly from only one $b(\boldsymbol{k})$ and that the contribution from the one other possible $b(\boldsymbol{k}')$ is small. Hence show that the c's which make any one $b(\boldsymbol{k})$ decay arise chiefly from that $b(\boldsymbol{k})$, so that the effect of the other $b(\boldsymbol{k}')$ upon the decay of $b(\boldsymbol{k})$ is negligible. Show that similar conditions apply to (ii) and (iii).

This reasoning indicates that the decay of any b may be represented by a term like $i\hbar\eta b$ in (26) of Section 17.2 and hence that the equivalence of the classical and quantum collision theories can be extended arbitrarily far.

5. Consider the allowed transitions as represented in Figure 17.6 for cases in which $|\boldsymbol{P}|$ is comparable to mc. Show that, for $\boldsymbol{P}_0$ initially in the y-direction and for collisions with $\boldsymbol{P}'$ in the P_xP_y plane, the curves of Figure 17.6 are the solutions of the equation

$$\pm c|\boldsymbol{P}' - \boldsymbol{P}_0| = \frac{1}{2m}(\boldsymbol{P}'^2 - \boldsymbol{P}_0^2)$$

where $\delta n = \mp 1$ and that these are intersections of a parabola of revolution $\mathcal{E} = (P_x^2 + P_y^2)/2m$ with a circular cone having its vertex on the parabola at $P_y = |\boldsymbol{P}_0|$, $P_x = 0$, $\mathcal{E} = \boldsymbol{P}_0^2/2m$. For values of $\boldsymbol{P}'$ along the P_y axis show that there are two solutions, one of which always corresponds to phonon absorption with $\delta n = -1$, and that the other changes from phonon emission to phonon absorption when $|\boldsymbol{P}_0| = mc$, corresponding to tangency between a line of the cone and the surface of the parabola of revolution. Verify from this that the curves of Figure 17.6 are qualitatively correct.

17.7 THE EFFECT OF THE PAULI PRINCIPLE UPON THE MATRIX ELEMENTS

In this section, we shall investigate the changes in the scattering theory which are required when many electrons are present. We shall treat first the simplest case, that of a single hole, and show that it reduces to a form substantially equivalent to that for an electron. On the basis of the manipulations used in establishing the transition probability for a hole, we shall in Section 17.7(b) make a generalization to the case of a partially filled band and obtain results which form the basis for the treatment given in Section 11.2.

17.7a. The Case of a Single Hole. For the case of a single hole, we consider the set of $2N$ wave functions constituting the valence band and represent them by the following expressions (which are slightly modified from previous sections in order to avoid three levels of subscripts):

$$\chi(a;\; \boldsymbol{q}_c) = \psi(\boldsymbol{P}_a;\; \boldsymbol{r}_c)\gamma_a(s_c), \tag{1}$$

where $\boldsymbol{q}_c = (\boldsymbol{r}_c,\; s_c)$ represents the position vector $\boldsymbol{r}_c$ of electron c with respect to the center of gravity of the crystal (not including the mass of the electrons under consideration) and s_c the spin. The index a runs over the basic lattice of $\boldsymbol{P}$ values twice, once for plus spin with $\gamma_a = \alpha$

and once with minus spin with $\gamma_a = \beta$, α and β being the spin functions defined in Section 15.7. When only the functional form of χ need be considered, χ is written as $\chi(a)$. (We shall use indices a, b, c for electrons, reserving i and j for atomic vibrations.) The definition of $\psi(P_a;\ r) \equiv \psi_{P_a}(r)$ given in equation (3) of Section 17.6 normalizes $\chi(a;\ q)$. As for Section 15.7, we mean by $\int dq$ integration over volume and sum over spin.

The electronic wave function for the electrons must be an antisymmetrical function; and since only one hole is present, it can be characterized by the missing wave function, as was done in Section 17.8 in connection with equation (11). We shall write this wave function as follows:

$$\mathbf{A}_a(q_e\text{'s}) = \frac{1}{\sqrt{(2N-1)!}}\sum\{\mathfrak{Q}\}\varepsilon(\mathfrak{Q})\prod_{c=1}^{2N-1}\chi(e_c;\ \mathfrak{Q}q_c) \tag{2}$$

where e_c includes all χ's except $\chi(a)$, and $\sum\{\mathfrak{Q}\}$ means the sum over the permutations $\mathfrak{Q}$ of the $2N - 1$ sets of coordinates in the determinantal wave function discussed in equations (7) and (8) of Section 15.7.

The wave function for the normal modes of vibration may be taken to be that used in the last section [equations (4) and (5) of Section 17.6 or (22) of Section 17.5]:

$$\begin{aligned}\phi(n\text{'s}) &= \prod_{i,j}\phi(n_{k_i\alpha j}, n_{-k_i\alpha j};\ q_{k_i\alpha jc}, q_{k_i\alpha js}) \\ &= \prod_i \phi_i.\end{aligned} \tag{3}$$

As in the last section, each elementary process will be seen to involve the change of only one of the n's. (We shall use ('s) to indicate complete sets of n-numbers and the complete sets of coordinates.)

The wave function, except for the motion of the center of gravity of the entire system, is thus

$$\Phi(P_a, n\text{'s}) = \mathbf{A}_a\prod_i\phi_i = \mathbf{A}_a\phi(n\text{'s}). \tag{4}$$

This is an eigenfunction for the Hamiltonian

$$\mathcal{H}_0 = \sum_c \mathcal{H}_{r_c} + \sum_i \mathcal{H}_i \tag{5}$$

where the $\mathcal{H}_i$ terms are the harmonic oscillator Hamiltonians and

$$\mathcal{H}_{r_c} = (1/2m)p_c^2 + \mathcal{U}_0(r_c). \tag{6}$$

For the case of an almost filled band, each $\mathcal{U}_0$ term must include the average effect of all the other electrons in the band as well as the potential due to the nuclei, core electrons, and valence electrons in other bands, if any. Thus one term omitted from $\mathcal{H}_0$ arises from variations in $\mathcal{U}_0(r_c)$ from its average value due to fluctuations in the positions of the other electrons. We shall refer to this term as $\mathfrak{U}_2$. The exact potential energy for all the electrons can then be written as

$$\begin{aligned}\mathfrak{U}(r_c\text{'s}, q_i\text{'s}) &= \sum_c \mathfrak{U}_0(r_c) + \sum_c \mathfrak{U}_1(r_c, q_i\text{'s}) + \mathfrak{U}_2(r_c\text{'s}) \\ &= \sum_c \mathfrak{U}_0(r_c) + \mathfrak{U}_{1T}(r_c\text{'s}, q_i\text{'s}) \qquad (7)\end{aligned}$$

where $\mathfrak{U}_{1T}$ represents the total perturbation energy. The pair of terms $\mathfrak{U}_0(r_c) + \mathfrak{U}_1(r_c, q_i\text{'s})$ represents the potential energy of electron c due to interaction with the nuclei and the *average* positions of all other electrons. This gives to $\mathfrak{U}_0$ and $\mathfrak{U}_1$ the same meaning as that introduced in Section 17.4, except that we shall now consider that some of the electrons whose average behavior is included in these terms may also undergo transition. $\mathfrak{U}_2$ represents the additional effect of fluctuations in the instantaneous positions of the electrons in the valence band. Contributions to $\mathfrak{U}_1$ produced by displacements of the nuclei will include shifts in the average charge density of the valence band electrons, which are bound to the nuclei and whose wave functions deform as the lattice vibrates. (See problem 1 of Section 17.6 for a discussion of such deformations.) The $\mathfrak{U}_1$ terms are the important terms and we shall show that for the case of one hole, no transitions are produced by the perturbation $\mathfrak{U}_2$.

(Another term which has been omitted from $\mathcal{H}_0$ arises from the inertial interaction between the electrons in the reduced mass treatment. This is negligibly small and is disposed of in problem 1.)

We shall now calculate the matrix element connecting two states

$$\mathbf{\Phi} = \mathbf{A}_a \prod_i \phi_i \quad \text{and} \quad \mathbf{\Phi}' = \mathbf{A}_b \prod_i \phi_i{}'. \qquad (8)$$

The perturbing energy is

$$\mathfrak{U}_{1T} = \sum_c \mathfrak{U}_1(r_c, q_i\text{'s}) + \mathfrak{U}_2(r_c\text{'s}) \qquad (9)$$

and the matrix element is

$$(\boldsymbol{P}_a\, n\text{'s}|\mathfrak{U}_{1T}|\boldsymbol{P}_b\, n''\text{'s}) = \int \mathbf{\Phi}^* \mathfrak{U}_{1T} \mathbf{\Phi}' d\,(\text{all } \boldsymbol{q}\text{'s and } q\text{'s}) \qquad (10a)$$

$$= \int \mathbf{A}_a{}^* \boldsymbol{\phi}^*(n\text{'s}) \mathfrak{U}_{1T} \mathbf{A}_b \boldsymbol{\phi}(n''\text{'s}) d\,(\text{all } \boldsymbol{q}\text{'s and } q\text{'s}) \qquad (10b)$$

$$\begin{aligned} &= \frac{1}{(2N-1)!} \int \Big[\sum\{\mathcal{Q}\}\varepsilon(\mathcal{Q}) \prod_c \chi(e_c{}^*;\, \mathcal{Q}\boldsymbol{q}_c)\Big] \prod_i \phi_i{}^* \\ &\quad \times \Big[\sum\{\mathcal{S}\}\varepsilon(\mathcal{S}) \prod_d \chi(f_d;\, \mathcal{S}\boldsymbol{q}_d)\Big] \prod_j \phi_j{}^1 \\ &\quad \times \Big[\sum_e \mathfrak{U}_1(r_e, q_i\text{'s}) + \mathfrak{U}_2(r_c\text{'s})\Big] d(\text{all } \boldsymbol{q}\text{'s and } q\text{'s}) \end{aligned} \qquad (10c)$$

where a is omitted from the e_c series of χ's in the first bracket and b from f_d series in the third bracket. In this formidible expression, almost everything integrates either to zero or unity.

First we shall eliminate $\mathfrak{U}_2$; evidently $\mathfrak{U}_2$ will give zero if any n changes. Furthermore, it will be zero unless $\sum \boldsymbol{P}$ is the same when summed over all states in $\mathbf{A}_a$ and $\mathbf{A}_b$; if the sums are different, a change in coordinates made by displacing every electron by a period $\boldsymbol{R}_i$ of the crystal lattice will multiply each wave function $\mathbf{A}$ by exp $[i(\sum \boldsymbol{P}) \cdot \boldsymbol{R}_i/\hbar]$, but will not affect $\mathfrak{U}_2$. Because of the periodic boundary conditions, this displacement must not change the integral; hence if $\mathbf{A}_a{}^*\mathbf{A}_b$ is changed by the $\boldsymbol{R}_i$ displacement, the integral must be zero. For wave functions corresponding to one hole, however, $\sum \boldsymbol{P}$ is simply $-\boldsymbol{P}_a$ for $\mathbf{A}_a$ and $-\boldsymbol{P}_b$ for $\mathbf{A}_b$ so that unless $\boldsymbol{P}_a = \boldsymbol{P}_b$ the $\mathfrak{U}_2$ term vanishes. If $\boldsymbol{P}_a = \boldsymbol{P}_b$, there is, however, no transition since the n's cannot change. Thus all that $\mathfrak{U}_2$ does is to affect the average energy of the state and this effect, presumably, has already been included in $\mathfrak{U}_0(\boldsymbol{r}_c)$. Hence we can forget $\mathfrak{U}_2$ for the case of one hole.

Next we shall show that each of the $\mathfrak{U}_1(\boldsymbol{r}_c, q_i\text{'s})$ terms makes the same contribution. This result appears obvious from symmetry considerations, since all electrons are treated alike in the wave functions $\mathbf{A}_a$ and $\mathbf{A}_b$. It may be established rigorously by considering the contribution of any representative term in (10b), for example, $\mathfrak{U}_1(\boldsymbol{r}_f, q_i\text{'s})$. We shall show that the contribution will be the same as that of $\mathfrak{U}_1(\boldsymbol{r}_g, q_i\text{'s})$. To do so we first simply relabel $\boldsymbol{q}_f = (\boldsymbol{r}_f, s_f)$ and $\boldsymbol{q}_g = (\boldsymbol{r}_g, s_g)$ by interchanging the subscripts. This does not affect the value of the integral since in both cases $\boldsymbol{r}_f$ and $\boldsymbol{r}_g$ are integrated over the interior of the periodic crystal with volume $V = A_xA_yA_z$ and the spins are summed over $\pm\frac{1}{2}$. The integral obtained from $\mathfrak{U}_1(\boldsymbol{r}_f, q_i\text{'s})$ by the interchange differs from that due to $\mathfrak{U}_1(\boldsymbol{r}_g, q_i\text{'s})$, however, in having $\boldsymbol{q}_f$ and $\boldsymbol{q}_g$ interchanged in both $\mathbf{A}_a{}^*$ and $\mathbf{A}_b$; since both of these interchanges multiply the integral by (-1), the integral from $\mathfrak{U}_1(\boldsymbol{r}_f, q_i\text{'s})$ may be made identical with that from $\mathfrak{U}_1(\boldsymbol{r}_g, q_i\text{'s})$ by interchanging $\boldsymbol{q}_f$ and $\boldsymbol{q}_g$ in $\mathbf{A}_a{}^*$ and $\mathbf{A}_b$. The same reasoning shows that all $\mathfrak{U}_1$ terms give the same value.

We may, therefore, consider only one term in the $\mathfrak{U}_1(\boldsymbol{r}_e, q_i\text{'s})$ series, say $\mathfrak{U}_1(\boldsymbol{r}_1, q_i\text{'s})$ and then multiply the result by $2N - 1$. There are $(2N - 1)!$ products in the sum in the first bracket of (10c) and $(2N - 1)!$ products in the second bracket, so that $\mathfrak{U}_1(\boldsymbol{r}_1, q_i\text{'s})$ is involved in $(2N - 1)!^2$ terms altogether. Since the χ's are an orthonormal set, the integral over $\boldsymbol{q}_2 \cdots \boldsymbol{q}_{2N-1}$ will vanish for each of these terms unless the two sets of χ^*'s and χ's (each $2N - 2$ in number) in which these $\boldsymbol{q}$'s occur in (10c) are identical and arranged in the same order. When they are in the same order, the integral over these $\boldsymbol{q}$'s is unity. The only set of χ's common to $\mathbf{A}_a$ and $\mathbf{A}_b$ is that obtained by eliminating $\chi(a)$ and $\chi(b)$ from the total set of $2N$ χ's. Hence (10c) vanishes unless $\chi^*(b; \boldsymbol{q}_1)$

occurs in the first bracket of (10c) and $\chi(a;\ q_1)$ in the second. It still vanishes unless the permutations $\mathfrak{Q}$ and $\mathfrak{S}$ are such as to pair properly the $2N - 2$ remaining q's in the common χ's sets. There will be, of course, $(2N - 2)!$ ways of choosing simultaneously $\mathfrak{Q}$ and $\mathfrak{S}$ so as to do this; and for each of these $\varepsilon(\mathfrak{Q})\varepsilon(\mathfrak{S})$ will have the same value (a result which the reader may verify); this value will be either $+1$ or -1 depending upon the number of interchange necessary to bring q_1 into $\chi^*(b)$, and upon the number then required to bring q_1 into $\chi(a)$ and in addition to arrange the other q's in the χ's of (10c) to match the q's in the χ^*'s. Since the sign of the matrix element is not important in the transition probabilities, we need not establish rules for determining it.

We shall now rewrite the matrix element (10) by dropping $\mathfrak{U}_2$, evaluating $\mathfrak{U}_1(q_1, q_i\text{'s})$ only, multiplying by $(2N - 1)$ to include the effect of the omitted $\mathfrak{U}_1$'s, and multiplying by $(2N - 2)!$ to account for the different ways of pairing q_2 to q_{2N-1}. These factors give $(2N - 1)(2N - 2)! = (2N - 1)!$ which just cancels the coefficient in (10c). The matrix element thus becomes

$$(P_a, n\text{'s}|\mathfrak{U}_{1T}|P_b, n'\text{'s})$$
$$= \pm\int \chi^*(b; r_1, s_1)\prod_i \phi_i^* \mathfrak{U}_1(r_1, q_i\text{'s})\chi(a;\ r_1, s_1)\prod_j \phi_j' dq_1 d\,(\text{all } q\text{'s}). \quad (11)$$

The $\pm$ sign, arising from $\varepsilon(\mathfrak{Q})\varepsilon(\mathfrak{S})$ as discussed above, does not affect the transition probability. It may be noted that, in the integration, $\mathbf{A}_a$ reduces to $\chi^*(b)$ and $\mathbf{A}_b$ to $\chi(a)$. The interpretation of this interchange of a and b is that the transition probability from a to b for the hole is identical with that of an electron from b to a. Further inspection of (11) shows that it is of essentially the same form as the matrix element treated in Section 17.6; if $\mathfrak{U}_1$ is considered as linear in the q_i's, then the same selection rules will apply and (11) will reduce to (9) of Section 17.6 in which only one n of the running wave scheme changes.

There is also a selection rule for the spin. Since $\mathfrak{U}_1$ does not involve the spin, the integral vanishes unless $\chi(a)$ and $\chi(b)$ both contain α or both contain β. If spin is conserved, the sum over the spin coordinate eliminates the spin functions and replaces the χ's by the corresponding $\psi(P) \equiv \psi_P$ functions.

It is thus evident that the matrix elements for the transitions of a hole from P_a to P_b are identical with those of an electron from P_b to P_a. This latter may be calculated by precisely the same methods as those used in Section 17.6, the only difference being that wave functions near the top of the valence band are used instead of wave functions near the bottom of the conduction band.

Because of the equivalent form of the matrix element, it is evident that the methods of Section 17.2 can be applied to the transitions of a hole.

This will lead to an expression like (35) of Section 17.6 for the mean free path of a hole, except that the rate of change of $\mathcal{E}_v$ with dilation will replace the rate of change of $\mathcal{E}_c$ with dilation in Section 17.6, equations (17) and (27). The two expressions for the mean free path will thus be

$$\frac{1}{l_n(T)} = \frac{\mathcal{E}_{1n}{}^2 kTm_n{}^2}{\pi\hbar^4 c_{ll}} \tag{12}$$

$$\frac{1}{l_p(T)} = \frac{\mathcal{E}_{1p}{}^2 kTm_p{}^2}{\pi\hbar^4 c_{ll}}. \tag{13}$$

One important difference between the hole transition and the electron transition should be mentioned, however. For the electron transition the quantity

$$\boldsymbol{P} + \hbar\sum n_{ka}\boldsymbol{k} \tag{14}$$

is conserved; owing to the interchange of subscripts a and b on the right and left sides of (11), the conservation law for holes applies instead to

$$-\boldsymbol{P} + \hbar\sum n_{ka}\boldsymbol{k}. \tag{15}$$

As pointed out in Section 15.8, however, the natural momentum to associate with the hole is $\boldsymbol{P}_B = -\boldsymbol{P}$ where $\boldsymbol{P}$ is the momentum of the omitted wave function and $\boldsymbol{P}_B$ the momentum of the remaining electrons in the energy band. This value of $\boldsymbol{P}_B$ is consistent with the picture of positive charge and positive mass for the hole. We see now that it is also consistent with the conservation of the quantity

$$\boldsymbol{P}_B + \hbar\sum n_{ka}\boldsymbol{k}. \tag{16}$$

This shows that the formal equivalence of the behavior of holes and electrons can be made complete, so far as scattering and acceleration are concerned, by making use of $\boldsymbol{P}_B$. Considerations of this formal equivalence do not appear to afford any greater insight into the mechanism, however, and we shall not develop them further here.

Before proceeding to the case of many electrons or many holes, it should be pointed out that the reduction of the many-electron matrix element (10) to the one-electron form (11) did not involve any reliance on the specific form of $\mathcal{U}_1$, save that the individual terms depended on the coordinates of a single electron. The fact that $\mathcal{U}_1$ was linear in the normal modes entered only in showing how (11), or its counterparts in Section 17.6, lead to selection rules for $\boldsymbol{P}$ and the n_i's. Consequently, (10) will reduce to (11) for scattering due to impurity ions, lattice imperfections, such as dislocations, or any other causes of deviation from the perfectly periodic potential. From this we conclude that the scattering of a hole by these imperfections will be the same as the scattering of an electron, except that wave functions of the valence band rather than of the conduction band will be involved. This fact is used in Chapter 11 in a simpler form in discussing the impurity scattering of holes.

17.7b. The Intermediate Case—Many Electrons and Holes. If the band is partially filled so that an appreciable fraction of the quantum states are occupied, the effect of the Pauli principle in preventing transitions of electrons to occupied states becomes important. We shall show here that the effect of the Pauli principle on the matrix elements is consistent with the description given in Chapter 11, in which the effect of having the states partially occupied was accounted for by reducing the rate of electron transitions by a factor $1 - \mathbf{f}$, corresponding to the fraction of end states available. The treatment of this section does not go as deeply into its problem as did the treatment of Section 17.2, in which the equivalence of the results of quantum theory and the classical model was shown with considerable generality by deriving and comparing the time dependence of the distribution of end states for the two cases. For the case of many holes and electrons, the detail involved in carrying out a similar comparison appears prohibitive compared to its value for expositional purposes and we shall proceed only to the calculation of the matrix elements. From these matrix elements we shall infer the desired transition probabilities by applying the formulae which give the transition probabilities for one electron in terms of the matrix elements for that case.

Accordingly, we shall deal with M electrons which may occupy a fraction of a set of $2N$ quantum states which arise from an energy band, or possibly from several overlapping energy bands. The antisymmetric wave functions for the electrons will now be

$$\mathbf{A} = \mathbf{A}(a_1, \cdots a_M;\, q_1, \cdots, q_M) \tag{17}$$

where the indices a describe the M wave functions of form $\chi(a)$, see equation (1). The order of the indices a_1 to a_M in $\mathbf{A}$ is immaterial since the only effect of altering it is to introduce a factor ± 1 in the determinant which $\mathbf{A}$ represents [see Section 15.8, equation (8)].

The wave functions for the normal modes of atomic vibration will once more be expressed in terms of the quantum numbers for the running waves, equation (3) of this section [or (4) and (5) of Section 17.6 or (22) of Section 17.5]:

$$\begin{aligned} \phi(n\text{'s}) &= \prod_{ij} \phi(n_{k_i\alpha_j}, n_{-k_i\alpha_j};\, q_{k_i\alpha_jc}, q_{k_i\alpha_js}) \\ &= \prod_i \phi_i. \end{aligned} \tag{18}$$

The wave function representing the initial state is then

$$\Phi = \mathbf{A}(a_1, \cdots, a_M;\, q_1, \cdots, q_M)\phi(n\text{'s}) = \mathbf{A}\phi. \tag{19}$$

The perturbation energy will be the same as before and can be written as

$$\mathfrak{U}_{1T} = \sum_{c=1}^{M} \mathfrak{U}_1(r_c, q_i\text{'s}) + \mathfrak{U}_2(r_1, \cdots, r_M), \tag{20}$$

the terms having the same interpretation as before for equation (9).

The matrix elements of interest will be those connecting two Φ's in which one or more a's change and one n changes by ± 1; this last selection rule evidently follows at once from the linear dependence of $\mathfrak{U}_1(r_c, q_i\text{'s})$ upon the q's. For the case of one hole, it was possible to disregard $\mathfrak{U}_2$. This may still be done for the purposes of conductivity theory if the band is nearly full or nearly empty and the spherical energy surface approximation is made; for transitions induced by $\mathfrak{U}_2$, there will be, as discussed above in connection with (10b), conservation of $\sum_i \boldsymbol{P}_{a_i}$. For the spherical approximation for the energy surfaces, there is a linear relationship between $\boldsymbol{P}_a$ and $\boldsymbol{v}_a$, so that

$$\sum \boldsymbol{v}_a = (1/m_n)\sum \boldsymbol{P}_a, \quad \text{or} \quad \sum \boldsymbol{v}_a = (-1/m)_p \sum \boldsymbol{P}_a. \tag{21}$$

Hence electron-electron (or hole-hole) collisions, which are produced by $\mathfrak{U}_2$, do not alter the net motion of the electrons or the current and, therefore, do not contribute to the relaxation of current. [The reader may verify that this significance of (21) is not altered by recentering the Brillouin Zone at a corner.] If, however, the energy surfaces are not spherical, or if the bands are degenerate so that several values of $\mathcal{E}(\boldsymbol{P})$ and $\boldsymbol{v}(\boldsymbol{P})$ must be considered for each value of $\boldsymbol{P}$, $\mathfrak{U}_2$ might have an important influence. The dependence of mobility upon temperature indicates, however, that such effects are not significant in germanium and silicon nor in ordinary metals and we shall disregard them.[1]

If we can neglect $\mathfrak{U}_2$, considerations precisely the same as those used in going from equation (10) to equation (11) may be applied. This leads to the conclusion that the matrix element vanishes if more than one of the a's changes, because if two or more change, then there will be no common set of $M - 1$ χ's in $\mathbf{A}^*$ and $\mathbf{A}'$ in the matrix element. Hence we conclude that matrix elements occur only for transitions for which one of the a's and one of the n's change. For these the matrix element will evidently reduce to form (11) which, because of the selection rule for n, is equivalent to (9) of Section 17.6 and involves the initial and final wave functions for the electron and the harmonic oscillator wave functions for the only mode which changes. The allowed matrix elements are thus identical with the set obtained by considering one-electron transitions from the

[1] Scattering in metals in which several bands are involved are discussed for example, in N. F. Mott and H. Jones, *The Properties of Metals and Alloys*, Oxford at the Clarendon Press, 1936, p. 265. See also J. Bardeen, "Conductivity of Monovalent Metals," *Phys. Rev.* **52**, 688–697 (1937) for a discussion of electron-electron interactions.

initial occupied states, represented by the set $a_1, \cdots, a_r, \cdots, a_M$ to a new set $a_1, \cdots, a_r', \cdots, a_M$ in which only one a_r changes, and that to an unoccupied state of the same spin.

As was seen in equation (23) of Section 17.2, the probability of transition from an initial state to a set of final states is proportional to the density of final states and the square of the absolute value of the matrix element. Applying this result to the case of many electrons, we conclude that the density of available final states for any one electron transition $(a_r \rightarrow a_r')$ will be reduced by $(1 - \mathbf{f})$ since a fraction $\mathbf{f}$ of the states which might be selected for a_r' are already occupied. Thus the transition probability will be the same as it would be for one single electron in state a_r, except for the factor $(1 - \mathbf{f})$. This is physically equivalent to the argument used in Section 11.2 which stated that the distribution of electrons made transitions just as if each electron acted independently, except for the restriction that transitions to occupied states were forbidden.

We have thus justified to a certain extent the use of the classical model of Section 11.2. To review and criticize the argument we point out that what we have done on the analytic side is to show that the matrix element connecting two states of the entire system differing by having one electronic state $\chi(a_r)$ replaced by another (a_r') is identical (except for a factor ± 1) with the matrix element for a system with one electron only in which the one electron makes a transition from $\chi(a_r)$ to $\chi(a_r')$. (It is assumed, very artificially, that the potential energy for one electron is the same for both cases.) It also follows that because there are no possible wave functions in which two electrons are in the same end state, there are no matrix elements for transitions to already occupied states. It was pointed out that the transition-probability formula involves the density of end states so that a factor $(1 - \mathbf{f})$ is introduced to account for the reduction of available end states.

However, formula (23) of Section 17.2 for the transition probability was justified by showing the equivalent behavior of the classical and quantum-mechanical models. This equivalence was demonstrated for the case in which only one excess electron (or hole) was involved. In order to establish the same degree of equivalence for the case of many electrons, it would be necessary to extend the Weisskopf-Wigner treatment to many electrons and, as stated earlier, the increased rigor obtained by such a treatment does not appear to warrant the complications involved.

Problem 1. The reduced mass problem for many electrons interacting with the crystal leads to a Hamiltonian with non-diagonal terms in the electronic momenta. Let the coordinates of the electrons in respect to a fixed origin be $\boldsymbol{r}_{e1}$, $\boldsymbol{r}_{e2}$, etc., and that of the crystal be $\boldsymbol{R}_c$. Then $\boldsymbol{r}_a = \boldsymbol{r}_{ea} - \boldsymbol{R}_c$ is the proper coordinate to use for the position of an electron in respect to the crystal and is the $\boldsymbol{r}_a$ referred to in this section. Show that the equations which should be used in place of (4) to (7) of Section 17.4 are

$$\boldsymbol{p}_T = \boldsymbol{P}_c + \sum_a \boldsymbol{p}_{ea}$$

$$\boldsymbol{p}_a = \boldsymbol{p}_{ea} - (m_e/M_T)\boldsymbol{P}_T$$

$$\boldsymbol{r}_T = (M_c/M_T)\boldsymbol{R}_c + (m_e/M_T)\sum \boldsymbol{r}_{ea}$$

$$\boldsymbol{r}_a = \boldsymbol{r}_{ea} - \boldsymbol{R}_c$$

and that the Hamiltonian becomes

$$\mathcal{H} = (1/2M_T)\boldsymbol{p}_T^2 + (1/2m)\sum \boldsymbol{p}_a^2 + (1/2M_c)\sum_a \sum_{b \neq a} \boldsymbol{p}_a \cdot \boldsymbol{p}_b + \mathcal{U}(\boldsymbol{r}_a\text{'s}, q_i\text{'s})$$

where $m = m_e M_c/(m_e + M_c)$ as before. The effective mass in the $\boldsymbol{p}_a \cdot \boldsymbol{p}_b$ terms is so large that it represents a negligible correction.

APPENDIX A

UNITS

(a) The Basic Rule for Changing Units.[1] The problem of changing the measured value of some quantity from one system of units to another continually occurs. The work can be systematized with the aid of the following idea: Measurement consists of comparing quantitatively the thing studied with reference units of length, mass, time, temperature, etc. In order to discuss the relationship between measurements made in two systems $\mathcal{S}_1$ and $\mathcal{S}_2$ we shall introduce the following conventions: measured numerical magnitudes of scalar and vector quantities will appear in *italics* L whereas symbols representing the units themselves will appear in **boldfaced roman L.** Thus the length of a 12-in. ruler measured in the c.g.s. system is written as

$$L_1\mathbf{L}_1 \tag{1}$$

where $L_1 = 30.5$ and $\mathbf{L}_1 = 1$ cm. Conversion factors between units will also be written in boldfaced roman with double subscripts; for example if $\mathbf{L}_2 = 1$ in. we write

$$\mathbf{L}_1 = \mathbf{L}_{12}\mathbf{L}_2, \quad \mathbf{L}_{12} = (2.54)^{-1}. \tag{2}$$

Conversion factors between the measured magnitudes of the same quantity measured in two units will be written in lightfaced italic with two subscripts; for the 12-in. ruler we have

$$L_1 = L_{12}L_2, \quad L_1 = 30.5, \quad L_2 = 12, \quad L_{12} = 2.54. \tag{3}$$

If one and the same quantity is measured in the same way in two sets of units, relationships of the form

$$L_1\mathbf{L}_1 = L_2\mathbf{L}_2 \tag{4}$$

always hold. From this we deduce that

$$L_{12} = \mathbf{L}_{12}^{-1} = L_{21}^{-1} = \mathbf{L}_{21}. \tag{5}$$

The example given above applies to situations in which the units are changed while the measurements are carried out by the same operational procedure. Another possibility, however, is to keep the units the same while changing the operational procedure. For example, consider a circle: If it is an automobile tire, its size might be measured by calipering its

[1] For more detail see P. W. Bridgman, *Dimensional Analysis*, Yale University Press, 1922.

diameter and dividing by two thus obtaining

$$(\text{Size}) \text{ equals } L_1\mathbf{L}_1 = 16 \text{ in.} \tag{6a}$$

which measure its radius. Equally well its circumference might be measured giving

$$(\text{Size})' \text{ equals } L_1'\mathbf{L}_1 = 100.5 \text{ in.} \tag{6b}$$

It is evident that no simple rule of changing units relates L_1 and L_1'; the difference between the two arises from a difference between the two operational procedures defining size. In the next section we shall consider $\mathfrak{S}_2$, the M.K.S. system of electrical units, and compare it with $\mathfrak{S}_1$, the c.g.s. system. For that case the transformation of mechanical quantities (i.e. force, velocity, energy etc.) is a straightforward problem in changing units. For the electrical units, however, the change is of the form (Size) → (Size)′ discussed above since different definitions of the operations used for measuring charge, magnetic field etc. are used in the two systems. The fact that these two transformations are carried out simultaneously is probably the principle cause of confusion in seeing the relationships between the two systems.

We shall next derive the formulae for the transformation of measured quantities for cases in which the method of measurement remains the same. Accordingly we consider some physical entity $\mathfrak{E}$ such as a solid or region of space. We measure in system $\mathfrak{S}_1$ some particular property $\mathcal{P}$ of the entity and obtain a value X_1. This result is expressed by writing[2]

$$\text{entity } \mathfrak{E} \text{ has } X_1\mathbf{L}_1{}^a\mathbf{M}_1{}^b\mathbf{T}_1{}^c \text{ of } \mathcal{P}. \tag{7}$$

Example:

$$\text{water has } 1.0 \text{ cm}^{-3} \text{ gm}^{+1} \text{ of density.} \tag{7a}$$

If we measure $\mathcal{P}$ of entity $\mathfrak{E}$ in $\mathfrak{S}_2$, we write

$$\text{entity } \mathfrak{E} \text{ has } X_2\mathbf{L}_2{}^a\mathbf{M}_2{}^b\mathbf{T}_2{}^c \text{ of } \mathcal{P} \tag{8}$$

$$\text{water has } X_2 \text{ ft}^{-3} \text{ lbs}^{+1} \text{ of density.} \tag{8a}$$

The amount of property $\mathcal{P}$ should be independent of the method of measurement so that we may write

$$X_1\mathbf{L}_1{}^a\mathbf{M}_1{}^b\mathbf{T}_1{}^c = X_2\mathbf{L}_2{}^a\mathbf{M}_2{}^b\mathbf{T}_2{}^c \tag{9}$$

$$1.0 \text{ cm}^{-3} \text{ gm}^1 = X_2 \text{ ft}^{-3} \text{ lb}^1. \tag{9a}$$

It is frequently convenient in manipulation of expressions like (9a) in order to find X_2 to proceed by multiplying the left side by factors of the form

$$1 = (\mathbf{L}_{12}\mathbf{L}_2/\mathbf{L}_1), \quad 1 = (\mathbf{M}_{12}\mathbf{M}_2/\mathbf{M}_1) \tag{10}$$

where we may use any convenient form of writing unity which accomplishes

[2] We shall use T rather than t for time in this appendix in order to be consistent with L and M. The use of M is intended to reduce confusion with m for mass and m for meter.

the desired cancellation shown below. This cancellation process appears as follows:

$$X_1\mathbf{L}_1{}^a(\mathbf{L}_{12}\mathbf{L}_2/\mathbf{L}_1)^a\mathbf{M}_1{}^b(\mathbf{M}_{12}\mathbf{M}_2/\mathbf{M}_1)^b\cdots$$
$$= X_1\mathbf{L}_{12}{}^a\mathbf{M}_{12}{}^b\cdots\mathbf{L}_2{}^a\mathbf{M}_2{}^b\cdots = X_2\mathbf{L}_2{}^a\mathbf{M}_2{}^b\cdots \quad (11)$$

$$1.0\text{ cm}^{-3}\left(\frac{2.54\times 12\text{ cm}}{\text{ft}}\right)^3\text{gm}\left(\frac{2.2\text{ lb}}{1000\text{ gm}}\right)$$
$$= 1.0\,(30.5)^3\,\frac{2.2}{1000}\,\text{ft}^{-3}\text{ lb} = X_2\text{ ft}^{-3}\text{ lb}. \quad (11a)$$

Since the unit symbols are the same, we have

$$X_1\mathbf{L}_{12}{}^a\mathbf{M}_{12}{}^b\cdots = X_2 \quad (12)$$
$$1.0\times 2.85\times 10^4\times 2.2\times 10^{-3} = 62.7 = X_1. \quad (12a)$$

Sometimes it is convenient to proceed in cascade: The life of a radioactive element is 2 years, convert this to seconds.

$$2\text{ years}\cdot\frac{365\text{ days}}{1\text{ year}}\cdot\frac{24\text{ hrs}}{1\text{ day}}\cdot\frac{3600\text{ sec}}{1\text{ hr}}$$
$$= 2\cdot 365\cdot 24\cdot 3600\text{ sec} = 6.31\times 10^7\text{ sec}. \quad (13)$$

By virtue of the relationships (5), the transformation of measured magnitudes can also be written in terms of the L_{12} quantities. Thus if X is measured in units of $\mathbf{L}^a\mathbf{M}^b\mathbf{T}^c$, the relationship between X_1 and X_2 may be written

$$X_1 = L_{12}{}^aM_{12}{}^b\cdots X_2. \quad (14)$$

A number of the manipulations in the next section will be carried out in this form.

(b) Electromagnetic Units. The electromagnetic units used in the text are in

absolute electrostatic units for $\boldsymbol{E}, \boldsymbol{D}, \rho, q, \boldsymbol{I}$ (15)

absolute electromagnetic units for $\boldsymbol{H}$ and $\boldsymbol{B}$. (16)

In these units Maxwell's equations and the force equation are

Absolute c.g.s. (System 1)

$$\nabla\times\boldsymbol{H}_1 = (1/c_1)[\dot{\boldsymbol{D}}_1 + 4\pi\boldsymbol{I}_1] \quad (a) \qquad \nabla\times\boldsymbol{E}_1 = (-1/c_1)\dot{\boldsymbol{B}}_1 \quad (b)$$
$$\nabla\cdot\boldsymbol{B}_1 = 0 \quad (c) \qquad \nabla\cdot\boldsymbol{D}_1 = 4\pi\rho_1 \quad (d)$$
$$\boldsymbol{B}_1 = \kappa_m\boldsymbol{H}_1 \quad (e) \qquad \boldsymbol{D}_1 = \kappa_e\boldsymbol{E}_1 \quad (f)$$
$$\boldsymbol{I}_1 = \sigma_1\boldsymbol{E}_1 \quad (g) \qquad \boldsymbol{F}_1 = q_1[\boldsymbol{E}_1 + (1/c_1)\boldsymbol{v}_1\times\boldsymbol{B}_1] \quad (h)$$
$$c_1 = 2.998\times 10^{10}\text{ cm/sec} \quad (i). \qquad (17)$$

For $\kappa_m = 1$, $\boldsymbol{B}$ can be replaced by $\boldsymbol{H}$ in the force equation and the subscript e dropped from κ_e, the specific inductive capacity or dielectric constant,

In this system unit electric charges and magnetic poles produce unit field at one cm in free space. In terms of these requirements the units of $\boldsymbol{E}$, $\boldsymbol{H}$, $\boldsymbol{B}$ and $\boldsymbol{D}$ are the same and are $\mathbf{M}^{1/2}\mathbf{L}^{-1/2}\mathbf{T}^{-1}$ and while q_1 is $\mathbf{M}^{1/2}\mathbf{L}^{3/2}\mathbf{T}^{-1}$. Using these dimensional formulae, equations (17) can be changed to other units for mass, length and time in accordance with the basic rule of part (a).

The M.K.S. electromagnetic system cannot, however, be obtained from (14) simply by changing the units of length, mass and time. The reason is that Maxwell's equations in the M.K.S. system are not in absolute M.K.S. units, which could be obtained from (14) by the basic rule, but instead are modified so that none of the units in which the electrical quantities are measured are absolute M.K.S. units in the same sense that the units of equations (17) are absolute. For the purposes of this exposition we shall regard the choice of the M.K.S. system of electrical and magnetic units as being dictated by two requirements: First, electrical quantities shall be in practical units so that unit charge is the coulomb (coul) and unit electric field is the volt per meter; these choices lead to a unit of electrical energy equal to the watt sec or joule which is also the unit of mechanical energy in the M.K.S. system. Second, the units of $\boldsymbol{D}$, $\boldsymbol{B}$ and $\boldsymbol{H}$ shall be chosen so that the coefficients $(1/c)$ and 4π in parts (a), (b) and (d) of (17) disappear. This is accomplished by the following transformation scheme where subscript 2 implies the M.K.S. system:

TABLE A.1 CONVERSION BETWEEN (1) ABSOLUTE c.g.s. AND (2) M.K.S. SYSTEMS

Mechanical	
$100\ \text{cm} = 100\mathbf{L}_1 = \mathbf{L}_2 = \text{meter}$	
$L_1 = 100L_2$	$L_{12} = 10^2$
$10^3\ \text{gm} = 10^3\mathbf{M}_1 = \mathbf{M}_2 = \text{kg}$	
$M_1 = 10^3M_2$	$M_{12} = 10^3$
$T_1 = T_2$	$T_{12} = 1$
$v_1 = (L_{12}/T_{12})v_2$	$v_{12} = 10^2$
$F_1 = (M_{12}L_{12}/T_{12}^2)F_2$	$F_{12} = 10^5$
Coulomb-Volt	
$c_1 10^{-1}$ stat. coulombs $= c_1 10^{-1}$	$\mathbf{q}_1 = \mathbf{q}_2 = \text{coulomb}$
$q_1 = c_1 10^{-1} q_2$	$q_{12} = c_1/10$
$(10^8/c_1)$ stat. volts/100 cm $= (10^6/c_1)\mathbf{E}_1 = \mathbf{E}_2 = \text{volt/meter}$	
$E_1 = (10^6/c_1)E_2$	$E_{12} = 10^6/c_1$
Additional Conversion Factors	
$\rho_1 = q_1/V_1 = (q_{12}/L_{12})(q_2/V_2)$	$\rho_{12} = c_1/10^7$
$I_1 = q_1/L_1^2T_1 = (q_{12}/L_{12}^2)I_2$	$I_{12} = c_1/10^5$
$\sigma_1 = I_1/E_1 = (I_{12}/E_{12})\sigma_2$	$\sigma_{12} = c_1^2/10^{11}$
$D_1 = 4\pi c_1 10^{-5} D_2$	$D_{12} = 4\pi c_1 10^{-5}$
$B_1 = 10^4 B_2$	$B_{12} = 10^4$
$H_1 = 4\pi 10^{-3} H_2$	$H_{12} = 4\pi 10^{-3}$

In terms of these conversions equations (12) become:

M.K.S. System (System 2)

$$\nabla \times \boldsymbol{H}_2 = \dot{\boldsymbol{D}}_2 + \boldsymbol{I}_2 \quad \text{(a)} \qquad \nabla \times \boldsymbol{E}_2 = -\dot{\boldsymbol{B}}_2 \quad \text{(b)}$$

$$\nabla \cdot \boldsymbol{B}_2 = 0 \quad \text{(c)} \qquad \nabla \cdot \boldsymbol{D}_2 = \rho_2 \quad \text{(d)}$$

$$\boldsymbol{B}_2 = \kappa_m \mu_0 \boldsymbol{H}_2 \quad \text{(e)} \qquad \boldsymbol{D}_2 = \kappa_e \varepsilon_0 \boldsymbol{E}_2 \quad \text{(f)}$$

$$\boldsymbol{I}_2 = \sigma_2 \boldsymbol{E}_2 \quad \text{(g)} \qquad \boldsymbol{F}_2 = q_2[\boldsymbol{E}_2 + \boldsymbol{v}_2 \times \boldsymbol{H}_2] \quad \text{(h)}$$

$$\mu_0 = 4\pi 10^{-7} = 1.257 \times 10^{-6}, \quad 1/\mu_0 = 7.958 \times 10^5$$
$$1/4\pi\mu_0 = 6.332 \times 10^5 \qquad \text{(i)}$$

$$\varepsilon_0 = 10^{11}/4\pi c_1{}^2 = 8.854 \times 10^{-12}, \quad 1/\varepsilon_0 = 1.129 \times 10^{11}$$
$$1/4\pi\varepsilon_0 = 8.988 \times 10^9 = c_2{}^2 10^{-7} \qquad \text{(j)}$$

$$c_2 = (\varepsilon_0\mu_0)^{-1/2} = 2.998 \times 10^8 \text{m sec}^{-1} \qquad \text{(k)} \qquad (18)$$

where κ_m and κ_e have the same meaning and values in both systems. For purposes of illustration we shall verify that (17a) transforms to (18a):

$$(\nabla \times \boldsymbol{H})_1 = (H_{12}/L_{12})(\nabla \times \boldsymbol{H})_2 = 4\pi 10^{-5}(\nabla \times \boldsymbol{H})_2 \qquad (19a)$$

$$(1/c_1)[\dot{\boldsymbol{D}}_1 + 4\pi \boldsymbol{I}_1] = (1/c_1)[4\pi c_1 10^{-5}\dot{\boldsymbol{D}}_2 + 4\pi c_1 10^{-5}\boldsymbol{I}_2] \qquad (19b)$$

the left sides are equal by (17a) and the right sides are $4\pi 10^{-5}$ times the left and right sides of (18a) which proves that (17a) converts to (18a). We shall also verify that (17f) converts to (18f):

$$\boldsymbol{D}_1 = 4\pi c_1 10^{-5}\boldsymbol{D}_2 = \kappa_e \boldsymbol{E}_1 = \kappa_e(10^6/c_1)\boldsymbol{E}_2 \qquad (20a)$$

so that

$$\boldsymbol{D}_2 = \kappa_e(10^{11}/4\pi c_1{}^2)\boldsymbol{E}_2 = \kappa_e\varepsilon_0\boldsymbol{E}_2 \qquad (20b)$$

if ε_0 has the value given in (18j).

The quantities in the M.K.S. form of Maxwell's equations have simple units when expressed in terms of the volt, coulomb (coul), sec system:

TABLE A.2 UNITS IN M.K.S. SYSTEM

q_2 coulombs	$\boldsymbol{D}_2$ coul m^{-2}
ρ_2 coul m^{-3}	$\boldsymbol{B}_2$ webers m^{-2} = $\boldsymbol{B}_2$ volt sec m^{-2}
$\boldsymbol{I}_2$ coul m^{-2} sec^{-1}	$\boldsymbol{H}_2$ amp turns m^{-2}
σ_2 ohm^{-1} m^{-1}	μ_0 volt sec amp^{-1} m^{-1} = μ_0 henry m^{-1}
$\boldsymbol{E}_2$ volt m^{-1}	ε_0 farads m^{-1}
Unit of force:	newton = 10^5 dynes
Unit of energy:	joule = newton m = watt sec = 10^7 dyne cm = 10^7 ergs.

An alternative method of deriving the relationships of Table A.1 is to consider the repulsion between corresponding ends of two similar thin solenoids each carrying flux $m_2 = B_2 \times$ (area of solenoid) and thus correspond to magnetic poles of strength m_2. In free space the force is required to be

$$m_2^2/4\pi\mu_0 r_2^2 = 10^7 m_2^2/16\pi^2 r_2^2. \qquad (21)$$

This equation fixes m_2 in an absolute sense just as does the absolute c.g.s. system of requiring unit force. From this condition the ratio of B_2 to B_1 and to H_2 etc. may be determined leading to the definitions of Table A.1.

A quantity of considerable use is the M.K.S. value of e_M/m_M, the charge to mass ratio of the electron where we have introduced the subscript M in place of 2 to facilitate comparison with Section 8.8. This has the value

$$(e_M/m_M) = 1.60 \times 10^{-19}/9.03 \times 10^{-31} = 1.77 \times 10^{11} \text{ coul/Kg}. \qquad (22)$$

The acceleration of an electron in a field of E_M volts/meter is

$$a = (e_M/m_M)E_M = 1.77 \times 10^{11} E_M \text{m sec}^{-1}. \qquad (23)$$

The circular frequency in a field B_M is found by equating the acceleration to the force divided by m_M thus obtaining

$$a = \omega^2 r = (e_M/m_M)\omega r B_M \quad \text{or} \quad \omega = (e_M/m_M)B_M. \qquad (24)$$

In a field of $B_M = 1$ or 10,000 gauss, the frequency is

$$f = \omega/2\pi = (e_M/m_M)/2\pi = 2.82 \times 10^{10} \text{ sec}^{-1} \qquad (25)$$

which agrees with the value in Fig. 10.2.

(c) Atomic Units (System 3). Another system of units, due to Hartree,[3] and frequently employed in calculations of wave function is that in which *the values of electronic charge, electronic mass* and *Dirac's* $\hbar$ *are unity.* We shall relate this system, called $\mathfrak{S}_3$, to the c.g.s. system:

$$\text{cm} = \mathbf{L}_1 = \mathbf{L}_{13}\mathbf{L}_3 = \mathbf{L}_{13}a_0 = (0.528 \times 10^{-8})^{-1} \times \text{Bohr orbit} \qquad (26)$$

$$\text{gm} = \mathbf{M}_1 = \mathbf{M}_{13}\mathbf{M}_3 = (9.03 \times 10^{-28})^{-1} m_e \qquad (27)$$

$$\text{sec} = \mathbf{T}_1 = \mathbf{T}_{13}\mathbf{T}_3 = (2.419 \times 10^{-17})^{-1}\mathbf{T}_3 \qquad (28)$$

where to avoid confusion with the meter m_e is used for the mass of the electron and a_0 is the Bohr orbit. The formulae for a_0 and $\mathbf{T}_3$ are

$$a_0 = \hbar^2 m_e^{-1} e^{-2} = 0.528 \times 10^{-8} \text{ cm} \qquad (29a)$$

$$\mathbf{T}_3 = \hbar^3 m_e^{-1} e^{-4} = 2.419 \times 10^{-17} \text{ sec} \qquad (29b)$$

$$m_e = 9.03 \times 10^{-28} \text{ gm}. \qquad (29c)$$

In these units the speed of light has the value

$$c_1 \text{ cm/sec} = c_3 a_0/T_3 = 137.29 a_0/T_3 \qquad (30)$$

[3] D. R. Hartree, *Proc. Comb. Phil. Soc.*, **24**, 89 (1926). The values agree with those in Condon and Shortley, *The Theory of Atomic Spectra*, Cambridge University Press, 1935.

where $c_3 = 137.29 = c_1\hbar_1/e_1^2$ is the reciprocal of the fine structure constant. The unit of energy is

$$(\text{energy})_3 = m_e a_0^2 T_3^{-2} = \text{atomic unit} = 27.07 \text{ electron volts.} \quad (31)$$

The Bohr orbit calculation for a hydrogen atom becomes in these units

$$\begin{array}{lll} \mathfrak{S}_1 & \mathfrak{S}_3 & \\ \oint p\,dq = m_e r^2 \omega 2\pi = h = 2\pi\hbar & r^2\omega = 1 & (32a) \\ \text{Force} = e^2/r^2 = m_e r \omega^2 & r^3\omega^2 = 1 & (32b) \end{array}$$

so that $r = \omega = 1$ is the solution and the energy is

$$-e^2/r + m\omega^2 r^2/2 = -1 + \tfrac{1}{2} = -\tfrac{1}{2} \quad \text{or} \quad -13.53 \text{ electron volts.} \quad (33)$$

In atomic units the wave equation for an electron in the field of a nucleus of charge Z is

$$\tfrac{1}{2}\nabla^2\psi - (Z/r) = i\partial\psi/\partial t = \mathcal{E}\psi \quad (34)$$

for which some of the lowest energy solutions are

$$\psi_{1s} = (Z^{3/2}/\sqrt{\pi})\, e^{-Zr} \quad (35a)$$

$$\psi_{2s} = (Z^{3/2}/4\sqrt{2\pi})(2 - Zr)e^{-Zr/2} \quad (35b)$$

$$\psi_{2p_z} = (Z^{3/2}/4\sqrt{2\pi})Zre^{-Zr/2} \cos\theta \quad (35c)$$

$$\psi_{2p_x} = (Z^{3/2}/4\sqrt{2\pi})Zre^{-Zr/2} \sin\theta \cos\phi \quad (35d)$$

$$\psi_{2p_y} = (Z^{3/2}/4\sqrt{2\pi})Zre^{-Zr/2} \sin\theta \cos\phi, \quad (35e)$$

the energy for $1s$ being $-\frac{1}{2}Z^2$ and for the others $-\frac{1}{8}Z^2$.

NAME INDEX

Collected references will be found in the lists at the beginning of each part.

SUBJECT INDEX